Dryland Climatology

Dryland Climatology provides a comprehensive review of dryland climates and their relationship to the physical environment, vegetation, hydrology, and inhabitants. Chapters are divided into four major sections on background meteorology and climatology; the nature of dryland climates in relation to precipitation and hydrology; the climatology and climate dynamics of the major dryland regions on each continent; and life and change in the world's drylands. It includes chapters on key environmental and ecological topics such as vegetation, geomorphology, dust, drought, desertification, microhabitats, and adaptation to dryland environments.

This interdisciplinary volume draws on the author's experience over several decades to provide an extensive review of the primary literature (covering nearly 2000 references) and a guide to the conventional and satellite data sets that form key research tools for dryland climatology. Illustrated with over 300 author photographs that demonstrate physical processes and the evolution of dryland landscapes, this book presents a unique view of dryland climates for a broad spectrum of researchers, environmental professionals, and advanced students in climatology, meteorology, geography, environment science, earth system science, ecology, hydrology, and geomorphology.

SHARON E. NICHOLSON is a professor of meteorology at Florida State University, where she holds the rank of Distinguished Research Professor and also serves as the H. and K. Lettau Professor of Climatology. She has previously held positions at the University of Bonn (West Germany), Clark University (Massachusetts), and the National Center for Atmospheric Research (Colorado). She is acknowledged as an international expert on the climate of arid and semi-arid regions, having been active in arid lands research for 40 years, and is best known for her work on climatic variability in Africa. Professor Nicholson's work has been acknowledged by awards and medals from the American Meteorological Society and the Royal Meteorological Society of the UK. Her photographic skills have seen her placed as a finalist in the National Geographic Travel Photography Contest and the American Geophysical Union's Geophysical Images Competition.

Dryland Climatology

Sharon E. Nicholson
Florida State University

CAMBRIDGE
UNIVERSITY PRESS

CAMBRIDGE
UNIVERSITY PRESS

University Printing House, Cambridge CB2 8BS, United Kingdom

One Liberty Plaza, 20th Floor, New York, NY 10006, USA

477 Williamstown Road, Port Melbourne, VIC 3207, Australia

4843/24, 2nd Floor, Ansari Road, Daryaganj, Delhi - 110002, India

79 Anson Road, #06-04/06, Singapore 079906

Cambridge University Press is part of the University of Cambridge.

It furthers the University's mission by disseminating knowledge in the pursuit of
education, learning and research at the highest international levels of excellence.

www.cambridge.org
Information on this title: www.cambridge.org/9781108446549

First published 2011
First paperback edition 2017

A catalogue record for this publication is available from the British Library

Library of Congress Cataloging in Publication data
Nicholson, Sharon E.
    Dryland climatology / Sharon Nicholson.
        p.   cm.
    Includes bibliographical references and index.
    ISBN 978-0-521-51649-5
    1. Arid regions climate.   I. Title.
    QC993.7.N53 2011
    551.65–dc23
    2011026304

ISBN 978-0-521-51649-5 Hardback
ISBN 978-1-108-44654-9 Paperback

Additional resources for this publication at www.cambridge.org/nicholson

This book is dedicated to my son Ronnie, who has given much joy, support, and patience during the production of this book, and to Mary Seely, the former Director of the Namib Desert Ecological Research Unit (currently, Gobabeb Research and Training Centre). Dr. Seely's hospitality at Gobabeb and the fascinating research carried out at DERU were instrumental in creating this author's fascination with the dryland environment.

Contents

Preface

Climate was long defined as the mean weather conditions in a given location. A corollary to this definition is that climate is an inherent and invariant aspect of environment. More and more, climate is being recognized instead as an environmental variable, and an overwhelming body of evidence points to accelerated rates of climate change during the past century. At the same time, unmistakable environmental changes have occurred in dryland regions in response to human behavior. Some of the human factors include the sedentarization of nomads, the continual increase in urbanization, population increase, land-use changes, and technological means of exploiting the environment. The result is worldwide changes in the land surface, soils, and vegetation. One of the future challenges is to understand the interplay between these various aspects of the environment and managing the environmental resources in ways that provide for human well-being while conserving and protecting the resources.

There are pressing reasons for focusing on the drylands. First, because the essential resource – water – is discontinuously available in time and space, environmental processes are quite fragile. The semi-arid regions, in particular, are expected to be among those most sensitive to future climate change and increasing intensity of land use (IPCC 1996). Thus, the ability to predict changes in dryland landscapes is one of the top priorities for global change research (Breshears and Barnes 1999). Further, perhaps more than anywhere else, the environment of drylands is a consequence of closely tuned feedbacks among biological, geomorphological, hydrological, and human systems. Changes in any of these systems can readily upset the feedback loops, creating serious disturbances in the environment (Graetz 1991). Finally, an increasing body of evidence suggests that this sensitivity is such that, when critical thresholds of certain variables are surpassed, abrupt and irreversible changes in the ecosystem can occur (e.g., Rietkerk et al. 2004; Scheffer et al. 2001). Perhaps more importantly, misinformation concerning this sensitivity abounds, mainly as a result of an incomplete understanding of the role played by climate in the dryland environment. The issue of desertification, discussed in detail in this book, is a case in point.

Despite the relative paucity of vegetation in drylands, compared with humid areas, these regions may be more crucial in the context of understanding global change. The reason is that, unlike wetter, radiation-limited environments, vegetation in the drylands responds abruptly and intensely to changes in water availability; in turn, vegetation dramatically alters the distribution of water. Thus, one might argue that the feedbacks that couple vegetation and climate are much stronger and more apparent in the drylands.

Consistent with these interactions, this book takes the view that climate can be understood only in the context of the whole earth system. This system can be represented by five geophysical realms termed the lithosphere (i.e., solid earth), the atmosphere, the biosphere, the hydrosphere (liquid water), and the cryosphere (solid water) (Fig. 0.1). There are constant exchanges of energy, momentum, particulates, moisture, and solutes between these spheres. The atmosphere, and hence climate, is controlled by fluxes from below (the exchange of these various materials) and above (solar radiation).

The fluxes at the lower boundary are particularly important because the atmosphere is relatively transparent to solar radiation. Hence, solar heating of the atmosphere is indirect: the solar beam is absorbed at the surface and transformed to long-wave radiation, which is effectively absorbed by the greenhouse gases in the atmosphere. Moisture and greenhouse gases, such as carbon dioxide, also originate at the surface and vegetation plays a critical role in the exchange processes. This is clearly illustrated by the annual cycle of carbon dioxide (Fig. 0.2), which shows the phenology of vegetation growth and the contrast between the water and land hemispheres. Hence, climate drives vegetation, but vegetation also drives climate.

This interaction and sensitivity suggests a new definition of climate: "climate encompasses the mean weather conditions, their variability, causes, and interrelationships with the global earth system." This definition is at the heart of the

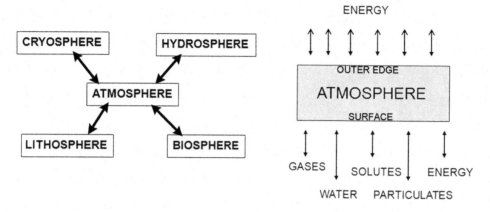

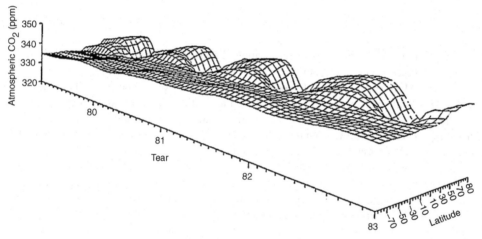

Fig. 0.1 Diagrams illustrating the relationship of climate to the whole earth system and the relevant exchanges between system components.

Fig. 0.2 Atmospheric carbon dioxide concentration (ppm) as a function of latitude and month.

interdisciplinary framework of this book. It also fits well into emerging interdisciplinary areas such as hydroclimatology and ecohydrology (Rodriguez-Iturbe *et al.* 1994; Rodriguez-Iturbe 2000; Breshears 2005; Newman *et al.* 2006).

Some pressing issues involve the interplay between climate and other geophysical spheres: atmospheric dust and its influence on climate, dwindling water resources, desertification and land degradation, and biomass burning. In writing this book the author hopes to provide a repository of information that can help researchers meet the challenges posed by these issues and aims to provide simple explanations that can facilitate the exchange of information among the relevant disciplines. Such interdisciplinary understanding is at the heart of designing experiments, creating models, and understanding and interpreting climatic variability and change.

For this reason, *Dryland Climatology* provides not only basic climate information on the earth's drylands, but also puts them into the context of the geophysical system "earth." This requires consideration not only of climate, but of the biogeophysical characteristics of the environment and how they interact with climate. The book therefore includes detailed, up-to-date treatments of new research areas not only in climate dynamics, but also in hydrology, ecology, and geomorphology. Some of the

book's recurrent themes are the complexity of the dynamics governing dryland climate; the spatial heterogeneity of characteristics and processes; the episodic nature of the water cycle; and the inextricable interactions of the vegetation, soils and water cycle.

Dryland Climatology was first conceived nearly 30 years ago. At the time, the author was a traditional climatologist who found herself assigned to teach a course entitled the Climatology and Geomorphology of Arid Lands. She had the good fortune not only to learn about the deserts from her then colleagues Larry Lewis and Len Berry, but also to meet and learn from pioneering researchers of an earlier generation, including Jean Dubief, Heinz Lettau, Heinrich Schiffers, Théodore Monod, Pierre Rognon, Mohammed Kassas, Nicole Petit-Maire, Raymond Bonnefille, Francoise Gasse, Dick Grove, Hermann Flohn, David Sharon, Hubert Lamb, Ken Hare, Horst Mensching, and many others. She gratefully acknowledges the inspiration she has gained from these scientists and hopes that, through these associations, she can provide a nearly century-long look at our evolving knowledge of the global drylands. The most noteworthy trends are the proliferation of information on deserts during the last one or two decades and the change in the tools we possess to study them. Hopefully, *Dryland Climatology* will serve as a guide to these new resources.

REFERENCES

Breshears, D. D., 2005: An ecologist's perspective of ecohydrology. *Bulletin of the Ecological Society of America*, **86**, 296–300.

Breshears, D. D., and F. J. Barnes, 1999: Interrelationships between plant functional types and soil moisture heterogeneity for semiarid landscapes within the grassland/forest continuum: a unified conceptual model. *Landscape Ecology*, **14**, 465–478.

Graetz, R. D., 1991: Desertification: a tale of two feedbacks. In *Ecosystem Experiments* (H. A. Mooney, E. Medina, D. W. Schindler, E.-D. Schulze, and B. H. Walker, eds.), Wiley, New York, pp. 59–87.

IPCC (Intergovernmental Panel on Climate Change), 1996: *Climate Change 1995: The Science of Climate Change* (J. T. Houghton, L. G. Meiro Filho, B. A. Callander, N. Harris, A. Kattenburg, K. Maskell, eds.), Cambridge University Press, Cambridge, UK.

Newman, B. D., and Coauthors, 2006: Ecohydrology of water-limited environments: a scientific vision. *Water Resources Research*, **42**, W06302, doi:10.1029/2005WR004141.

Rietkerk, M., S. C. Dekker, P. C. de Ruiter, and J. van de Koppel, 2004: Self-organized patchiness and catastrophic shifts in ecosystems. *Science*, **305**, 1926–1929.

Rodriguez-Iturbe, I., D. Entekhabi, and R. L. Bras, 1994: *MIT Colloquium on Hydroclimatology and Global Hydrology, Cambridge, Massachusetts, 7–8 April 1993*. Special Issue of *Advances in Water Resources*, **17**(1).

Rodriguez-Iturbe, I., 2000: Ecohydrology: a hydrologic perspective of climate–soil–vegetation dynamics. *Water Resources Research*, **36**, 3–9.

Scheffer, M., S. Carpenter, J. A. Foley, C. Folke, and B. Walker, 2001: Catastrophic shifts in ecosystems. *Nature*, **413**, 591–596.

Acknowledgments

The author is extremely grateful to her assistant, Douglas Klotter, who created the graphics for this book. His painstaking attention to detail and his facility with Photoshop have greatly enhanced the numerous photos in the book. She would also like to thank her brother, Ron Nicholson, for proofreading the entire manuscript and Simon Berkowicz and Mary Seely for reviewing several chapters. The author would also like to thank Dean Graetz, of CSIRO, for labeling of Australian landscape photos and the following individuals for contributing photographs: Doug Klotter, Dara Entekhabi, Carol Breed, Tracie Mitchell, Nicole Petit-Maire, and Simon Berkowicz.

Much of the book was written during various sabbaticals. The author would like to acknowledge her hosts: Pauline Dube and the University of Botswana, Steve Running and the University of Montana, Bob Dickinson and the University of Arizona, and M. V. K. Sivakumar of ICRISAT Niger. Support for these sabbaticals was provided by Fulbright and by NSF, NASA, NOAA, and Florida State University. Much material for the book was gathered during field trips and conferences. The author is grateful for the hospitality shown by Mary Seely, Mark Stafford-Smith, David Sharon, Simon Berkowicz, and Stefan Kröpelin and colleagues.

The author would like to acknowledge her former students and a number of colleagues. Peter Webster, Keil Soderberg, Stefan Hastenrath, and Steve Prince have shared their wisdom on numerous occasions. Former students and collaborators have been a constant source of inspiration, and their work is featured prominently in this book. These include Dara Entekhabi, Tammy Farrar, Tom Schonher, Mike Davenport, Ada Malo, Andrew Lare, Jeremy Grist, Jose Marengo, Mamoudou Ba, Jeeyoung Kim, Janet Selato, Dorcas Leposo, and Xungang Yin.

The production of the book was partially supported by a long list of grants from NOAA, NASA, and the National Science Foundation. Of particular importance were two special grants from NSF: FAW Grant ATM 9024340 and a POWRE Award ATM 0074961. The author is grateful for the long-term support of Jay Fein of NSF, who provided the NSF grants.

Permission to reproduce figures was generously granted by Anne Verhoef, Alan Strahler, John Ludwig, James MacMahon, Avi Shmida, M.J.A. Werger, Nick Lancaster, Massimo Meneti, Steve Running, Qiaozhen Mu, the Royal Meteorological Society, the American Meteorological Society, the Geological Society of America, US Department of Agriculture, Ecological Society of America, and the International Association of Hydrological Sciences. Figures 3.21, 10.54, 11.11 to 11.14, 11.30, 11.37, 11.39, 12.22, 19.14 and 19.22 have been reproduced with the kind permission of Springer Science and Business Media.

The author would also like to acknowledge the extent to which she has relied upon the published works of a handful of scientists in other disciplines. These have tremendously helped her to gain an understanding of dryland environments well beyond meteorology. Some of the sources upon which she has heavily relied include books by Goudie and Wilkinson, Cooke and Warren, Dunne and Leopold, Anne Verhoef, the excellent geography textbooks of Arthur and Alan Strahler, and numerous articles by Nick Lancaster, Ian Livingstone, Mary Seely, and David Breshears, to name but a few.

Special thanks are due to editor Susan Francis for her patience in seeing this book through to completion, past several deadlines, and to Laura Clark, Jeremy Toynbee, and Nancy Boston for their valuable editorial assistance.

Part I The dryland environment

1 Introduction to dryland environments

1.1 EXPLORATION AND AWARENESS OF DRYLANDS

Drylands, simply defined, are areas where potential evaporative water loss balances or exceeds the meager annual rainfall; a shortage of water inherently plagues these regions. The drylands comprise one-third of the global land surface; they support 14% of the world's inhabitants and a significant share of world agriculture. The climates of the drylands attracted little attention until the 1970s, when the combination of recurrent drought and intense land use demonstrated both the fragility and the importance of these ecosystems. The effects of climatic fluctuations and human-induced pressures have been especially adverse in semi-arid lands: agricultural regions of the American Midwest and Central Asia, India, the African Sahel, and much of southern Africa regularly have to contend with severe and often persistent drought. In the Sahel unusually dry conditions prevailed from the late 1960s until the turn of the twenty-first century. In some dryland regions climatic fluctuations and human mismanagement have interacted to initiate the process of desertification, a long-term degradation of the environment. Concern for the future is enhanced by the expansion of development into drylands as population growth continues, and by the prospect of global and regional climatic change. More unsettling are claims that human activities, including land use, may induce climatic change.

Initially, knowledge about the world's drylands was achieved in the context of exploration and exploitation, rather than scientific inquiry. German geographers such as Passarge (1904) and Jaeger (1921) wrote extensively about Namibia and South Africa. Missionaries such as Moffat and Livingstone (Fig. 1.1) kept detailed diaries that chronicled year-to-year environmental events at mission stations. French geographers and explorers, such as Capot-Rey (1953), Monod (1958), and Tilho (1911), vastly expanded our geographical knowledge of the Sahara. Huntington (1907) and Hedin (1904–5) journeyed into the Gobi Desert and other Asian regions. Powell (1875) and Gilbert (1975) surveyed the American West. Their journals included information on climate and weather, but the focus was on description, not physical processes. The field of meteorology was likewise slow to focus on dryland climates, understand their dynamics, or put them into a global context.

Fortunately, this picture has changed. A number of developments have brought deserts and semi-arid regions into the mainstream of modern meteorological and ecological research. Electronic data-recording systems have extended meteorological networks into more remote areas. Remote sensing techniques have vastly expanded global information gathering and have provided a means of monitoring the status of drylands. The catastrophic Sahel drought of the early 1970s and synchronous droughts elsewhere evoked worldwide interest in climatic fluctuations in dryland areas. Recurrent El Niño episodes have produced dramatic, newsworthy events, from floods in Peru and California to snow in Israel, as well as drought in many semi-arid areas. Scientists began expressing concern that human activities could influence global climate and numerous studies suggested that drylands would be the regions most severely affected. Finally, the concurrent development of theories of global climate and sophisticated numerical models to test these theories served both to increase our understanding of climate and its variability and to point out the necessity of including the world's drylands in this new global perspective on climate.

The result is an increased awareness of drylands. These regions are now recognized as a dynamic component of the global environmental system. They are also recognized as an endangered global resource with importance in areas as diverse as global economics, worldwide agriculture, and dynamic climatology. The need for protection and proper management, in order to maintain the drylands as renewable resources in the face of potential climatic changes and increasing land-use pressure, is clearly acknowledged. Some of the specific concerns include climatic changes induced by increased greenhouse gases; groundwater depletion by extensive irrigation; and long-term effects of desertification and land-use change. These concerns are supported by evidence that in some dryland regions

3

Fig. 1.1 (left) The Kuruman (South Africa) mission station of Robert Moffat, whose daughter married the missionary and explorer David Livingstone. (right) Plaque commemorating the Livingstone mission at Kolobeng, near Gaborone, Botswana.

the combination of climatic variability and imprudent land use can adversely alter the region's climate.

The demand for development in drylands will continue. At the same time, global climate change could reduce the potential resources of the drylands. In view of this, *Dryland Climatology* focuses on the climates of arid and semi-arid regions with the aim of underlining their environmental potential and limitations. Common dryland characteristics include low and highly variable rainfall, seasonal water availability, thermal extremes, and sensitivity to both climatic fluctuations and human intervention.

These characteristics represent the potentials of, and limitations to, developing and sustaining drylands. The differences associated with the diverse climatological characteristics (e.g., tropical and temperate) are of secondary importance. The common features determine the strategies suitable for coping with these less than optimal environments. An understanding of these environments, their characteristics and their dynamics, can lead to better management practices that will help preserve the dryland regions as a global "resource."

1.2 GEOGRAPHIC EXTENT OF DRYLANDS

Drylands, semi-arid and arid regions where evapotranspiration potentially balances or exceeds rainfall, comprise about one-third of the earth's land surface. Precise climatic limits have been prescribed by numerous authors in order to define an arid or a semi-arid environment (e.g., Wallén 1967; Koeppen 1931; Thornthwaite, 1931), but such a definition remains a complex problem compounded by the inconstancy of climatic parameters.

The key factor – the one that limits human activities, vegetation growth, land use, and surface hydrology – is moisture availability. This, in turn, is dependent on a number of climatic and environmental factors, the most important of which are rainfall, its concentration in time and space, and evaporation. Quantitative assessments of moisture availability by a variety of

schemes, as discussed in Chapter 9, provide the basis for classifying dryland regions. Unfortunately, these schemes produce quite divergent distributions of arid and semi-arid lands. Similarly, there is little consensus on the criteria establishing boundaries on the basis of vegetation, soil, or streamflow, although these features are more constant than climatic parameters.

Meigs' (1957) system, developed for UNESCO, is the best-known and most widely accepted dryland classification scheme. It distinguishes semi-arid, arid, and extremely arid regions; the last two classes being considered the true deserts. Meigs' criteria are based on the Thornthwaite moisture index (TMI, Thornthwaite 1948), which compares rainfall and potential evapotranspiration. According to this system, 4% of the world's land surface is extremely arid, 15% is arid, and 14.6% is semi-arid.

Despite diversity in the dryland boundaries established by various classification schemes, the principal deserts recognized by Meigs are universally accepted as arid lands, if not true deserts. These include the Kalahari-Namib, Somali-Chalbi and Sahara on the African continent; the Arabian, Iranian and Turkestan deserts and the Thar, Taklamakan and Gobi regions of Asia; the Monte-Patagonian and Atacama-Peruvian deserts of South America; the Australian desert; and the North American desert, including the Great Basin and the Sonoran, Mojave and Chihuahuan deserts (Fig. 1.2). Of these, only the Sonoran, Peruvian, Arabian, Taklamakan, Sahara and Namib deserts include lands classed as extremely arid. Vast tracts of semi-arid land, including the American Great Plains and the African Sahel, border each of the deserts. Extensive semi-arid regions exist also in Europe (on the Iberian Peninsula) and in Northeast Brazil. Thus, dryland areas are found on six continents (Table 1.1), although they are concentrated in Africa, Asia and Australia, where about half the land is arid or semi-arid.

Drylands, especially true deserts, tend to be situated in subtropical latitudes. This is partially a consequence of the prevailing subtropical high-pressure cells. However, other factors, such as topography or coastal effects, produce dryland climates.

Table 1.1. *Area (million km²), by continent, of arid lands (from Meigs 1957). Final column indicates the percentage of the continent that is arid or semi-arid.*

Continent	Extremely arid	Arid	Semi-arid	Total	Percentage of continent
Africa	4.56	7.30	6.08	17.94	60
Asia	1.05	7.91	7.52	16.48	37
Australia	–	3.86	2.52	6.38	84
North America	0.03	1.28	2.66	3.97	16
South America	0.17	1.22	1.63	3.02	17
Europe	–	0.20	0.80	1.00	10

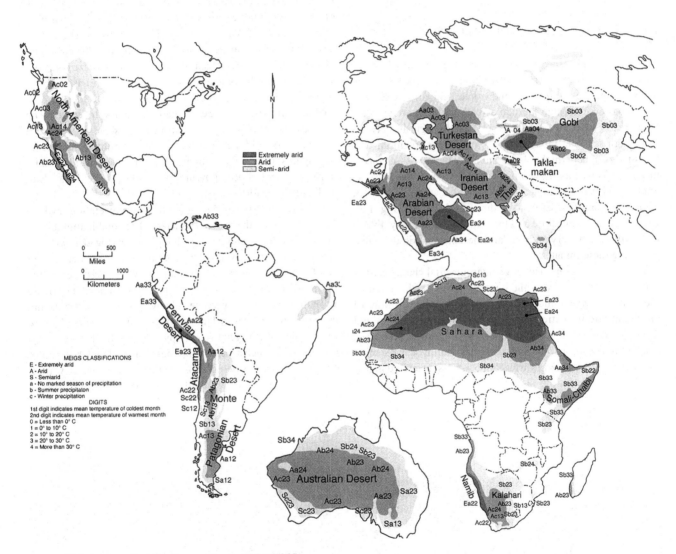

Fig. 1.2 Global extent of arid lands according to Meigs (1966).

As a result the latitudinal location and extent of the drylands vary greatly. In Africa, they lie approximately between 15° and 30° N and 6° and 33° S. Major Asian deserts are concentrated between 15° and 35° N in Arabia, 22°–48° N over southeastern and Middle Asia, and 36°–46° N in Central Asia (Petrov 1976).

Together, the African and Asian systems constitute a subtropical lower–mid-latitude dry zone, extending longitudinally over nearly half a hemisphere. The North American desert region extends from about 22° to 44° N; the Australian from 15° to 30° S. The South American deserts have the greatest latitudinal extent (1°–52° S) but the east–west extent is small because they are largely confined by coasts and mountains to narrow zones.

There are, of course, dry climates created by extreme altitude and near-polar location. These differ from those discussed so far in two basic respects: the limited moisture supply is primarily the result of cold conditions, and evapotranspiration is inherently low and of minor importance in determining water availability. In consequence, other aspects of their climatic environments differ greatly from those dryland areas discussed above. For this reason, generalizations in this text concerning dryland regions and common environmental concerns do not apply to polar or altitudinal deserts. Such regions are therefore omitted from this book.

1.3 PHYSICAL FEATURES OF DRYLANDS

The climatic and surface characteristics of the world's drylands are extremely diverse. Petrov (1976) points out that deserts fall within temperate, subtropical, and tropical climatic zones, even when cold deserts (high latitude or high altitude) are ignored. Similarly, Tricart and Cailleux (1969) distinguish between those with seasonal precipitation, sporadic precipitation, or a humid atmosphere (high relative humidity because of a coastal location), as well as those with seasonal or sporadic freezing. Shmida's (1985) classes are somewhat broader: warm deserts, cold deserts, and fog deserts (i.e., the coastal deserts) (Fig. 1.3). Clearly, both the moisture and thermal regions are very diverse within the earth's drylands, as are the atmospheric dynamics governing these climates.

There are, however, distinctive environmental characteristics common to dryland areas, and it is these traits to which human use of the ecosystem must adapt. The most general, and most important, are the extreme moisture and thermal regimes and their pronounced and rapid fluctuations. Water supply is limited and highly variable in time and space; the rhythm of moistening and desiccation is generally irregular, or at least brief and highly

seasonal. Rainfall is generally in the form of short, intense showers that can suddenly flood a desert landscape. The thermal climate is extreme because the factors that moderate the thermal environments in humid climates – vegetation, cloudiness, and soil moisture – are inherently sparse or almost absent in dryland regions. The sparse vegetation cover also leads to relatively windy conditions that, like extreme temperatures, favor high evaporation.

The intensity and rhythm of these climatic extremes vary greatly within the diverse dryland habitats. In the subtropical savanna, where summer rains usually prevail and temperatures are relatively constant, seasonal cycles of rainfall and aridity are dominant determinants of flora, fauna, and agriculture. In the low-latitude deserts, aridity and the irregular timing of the rare rainfall events are the most characteristic features. Torrential rains break long periods of drought. The harshness of the environment is accentuated by the diurnal cycles of temperature: extreme heat during the day and extreme cold at night. In mid-latitude deserts the seasonal cycle of temperature is equally extreme, but deserts in the low latitudes have no cold season, with night being the only analog to winter. The semi-arid steppes (temperate-latitude grasslands) experience hot summers, cold winters, and low annual rainfall that occurs mainly in summer. Thus, the harsh environment is dictated by marked annual cycles of rainfall and drought, heat and cold, and longer or shorter days.

Climate is the prime determinant of the vegetation, hydrology, and soils of the arid land surface. The combination of climatic, hydrologic, and geomorphic processes, together with the sparse vegetation cover, determines the surface materials and landforms in the very arid environments. Wind and water are the primary forces in landscape development. Unhindered by dense vegetation cover, they become effective geomorphic agents and create distinctive desert landforms and surfaces (Fig. 1.4). In the absence of a vegetative canopy to break the force of raindrops and root networks to bind materials, soil material cannot accumulate, so that desert soils are rare. The action of wind is likewise accentuated. In humid areas, vegetation cover reduces

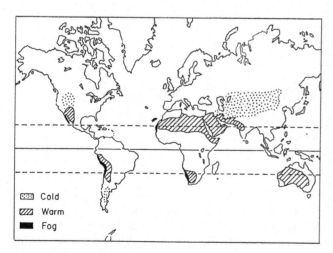

Fig. 1.3 Location of cold, warm and fog (coastal) deserts (after Shmida 1985).

Fig. 1.4 Erosional landform in the Libyan Desert of Egypt.

Fig. 1.5 Typical desert surface features: (a) bedrock in the Egyptian Sahara, (b) Namib dunes, (c) stone pavement in the Namib, (d) ephemeral stream channel of the Kuiseb River, Namibia, (e) inselberg in the savanna of northern Nigeria, (f) badlands in Bryce Canyon of the western USA.

wind velocity near the surface, absorbs the force of the wind, and prevents it from being directed against the land surface: it also protects surface material from the full erosional force of wind and water.

Contrary to the common perception of deserts as vast sand seas, sand dunes and ergs constitute a relatively small proportion of desert surfaces. Australia has the largest proportion of sand and that is only 30%. Other highly distinctive surface types result from the characteristic hydrologic and geomorphic processes operating in drylands (Fig. 1.5). These include bedrock (hamada desert), stone pavements (e.g., reg and serir), desert crusts, and depositional flats (such as playas) (Fig. 1.6). These are discussed more fully in Chapter 2. Sand fields are common in the Sahara and the Australian deserts, but the Syrian and Gobi deserts are largely hamada and reg. As the desert gives way to semi-arid landscape, soils become more common and widespread. The surface also contains hydrological networks: ephemeral stream channels and exotic streams (those originating in wetter, external regions) in deserts, and drainage systems in semi-arid regions. Other common landscape features include alluvial fans deposited by streams, pediments (eroded bedrock platforms at the foot of hillslopes), inselbergs (huge, isolated,

Fig. 1.6 Playa surface in southern Africa.

steep-sided rock outcrops), and erosional features such as mesas and badlands.

Landforms and surfaces represent the combined forces of weathering (which breaks down materials), wind and water erosion (which transports and removes finer materials), and depositional processes. Stone pavements – fields of coarse material and rock fragments set in or on finer materials – are erosional features formed when water and wind strip away finer, more transportable material. Sand surfaces are depositional, left behind when the forces of the water or wind transporting the sand ebbs. Deposition by water also produces alluvial fans and desert flats, the residue left after water has evaporated.

Playas are characterized by repeated cycles of inflow and evaporation, which leave behind crystalline materials such as chlorides, carbonates, and sulfates (e.g., rock salt, gypsum, lime, and sodium salts). Cycles of moistening and evaporation also produce the desert crusts, but these result from smaller-scale processes within desert soils. Vertical water movement, which is preferentially upward in dry regions, leaves crusts of hard materials such as gypsum, lime, laterites (iron-rich material), and silicates. Although these can accumulate in lower soil horizons, they are commonly at or near the surface.

1.4 ACQUISITION OF DATA AND INFORMATION

1.4.1 *IN SITU* CLIMATIC AND METEOROLOGICAL DATA

The availability of climatic and meteorological data is relatively limited in most dryland regions because observing stations are mainly situated in areas of dense population. This is illustrated by the station network in the Global Historical Climatology Network (GHCN) data set compiled by the National Climatic Data Center in the USA (Fig. 1.7). This data set is one of the most widely used in climatic research. The decrease in station density toward the dry interiors of Africa, Australia and Asia and along the coastal deserts is clearly apparent.

The trend in modern climatological research is to use gridded global data sets. This includes both analyses solely based on station data (e.g., Legates 1995; New *et al.* 2002) and "blended" analyses that combine station data with remotely sensed data

and modeling analysis. Examples of the latter are the NCEP/NCAR Reanalysis data (Kistler *et al.* 2001) and ERA-40 (Uppala *et al.* 2005). These are electronically available and easy to use. However, these data sets have large error bars where station data are sparse. This includes most arid and semi-arid regions. Figure 1.8 illustrates the large apparent differences in mean annual rainfall over the Namib coastal desert, as evaluated from three different data sets.

Unfortunately, in most regions the reporting networks have declined over the last few decades. There are many reasons for this, ranging from changing economic conditions in many developing countries and decreased international data exchange to increased reliance on satellite data for weather information. Whereas meteorological data were once freely exchanged and regularly published in monthly bulletins and annual reports, many countries in Africa and South America have started to charge exorbitant sums even for routine information on temperature and precipitation.

As a consequence, much of the best climatological information comes from sources compiled many years ago. These include such sources as the multi-volume series *Ecosystems of the World* (Goodall 1983–2000), *World Climatic Data* (Wernstedt 1972), the six volumes of meteorological tables published by the British Meteorological Office (1958–1967), works by Dubief (1959, 1963), Meigs (1966), Hastings and Humphrey (1969a, 1969b), Amiran and Wilson (1973), and the multi-volume series *World Survey of Climatology* (Landsberg 1971 onward) and *World Weather Records* (see NOAA website below). Useful data sets can also be found online from various websites of the US National Oceanographic and Atmospheric Administration (NOAA) (e.g., www.ncdc.noaa.gov/), and NASA (http://rain.atmos.colostate.edu/CRDC/) and in the monthly publication *Monthly Climatic Data for the World*. A few national or regional precipitation data sets have also been compiled, such as those for the Middle East (Yatagai *et al.* 2008), Australia (Lavery *et al.* 1997), Africa (Nicholson 1986), China (Hong *et al.* 2005), the Former Soviet Union (Groisman *et al.* 1991) and Siberia in particular (Yang and Ohata 2001), the USA (Di Luzio *et al.* 2008, Guirguis and Avissar 2008), Mexico (Graef *et al.* 2000), and South America (Liebmann and Allured 2005), Brazil in particular (Silva *et al.* 2007).

A wealth of climate information was also compiled over the years by various arid lands research institutes. Some of the major ones include the Repetek Desert Research Center in Turkmenistan; the Training and Research Centre Gobabeb, Namibia; the Arid Lands Office at the University of Arizona; the Lanzhou Institute of Desert Research in China; the Arid Lands Ecosystem Research Center at the Hebrew University in Jerusalem; the Blaustein Institute for Desert Research at Ben-Gurion University in Tel Aviv; and the Institute for Arid Zone Research at the University of the Negev in Beer Sheva. Our knowledge of arid lands was also rapidly advanced by UNESCO's Arid Lands Research Programme, which commenced in 1951.

Various field programs have also contributed to our understanding of arid lands. One of the first took place in the Pampa

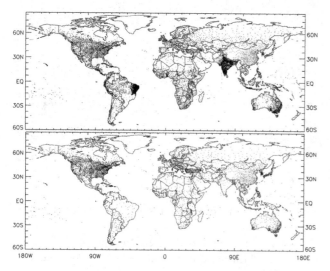

Fig. 1.7 Precipitation (top) and temperature (bottom) station network of the Global Historical Climatology Network (GHCN).

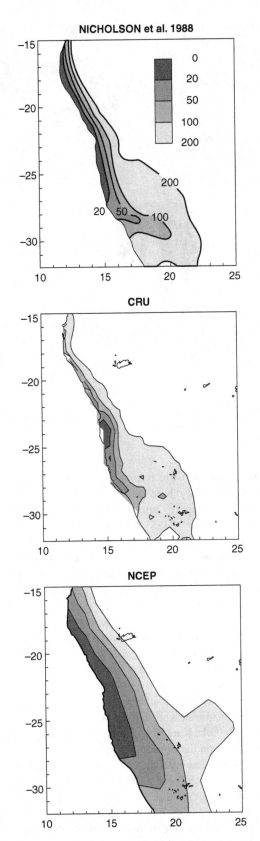

Fig. 1.8 Comparison of mean annual rainfall (mm) in the Namib coastal desert, as assessed with data from an expanded, local gauge data set (top; Nicholson *et al.* 1988), the East Anglia (CRU) gridded data set (center), and NCEP Reanalysis data (bottom).

de la Joya in Peru in the early 1960s (Lettau and Lettau 1978). Conducted by the University of Wisconsin, it provided unprecedented information on the causes of coastal aridity, desert energy balances, and dune dynamics. With the development of remote sensing, numerous field experiments in dryland regions emphasized the creation of algorithms for assessing the land surface from space and deriving ground-truth for observations from aircraft, radar, and satellites. The impetus for many of the field experiments was concern about desertification, as well as an emerging interest in land–atmosphere interactions.

A landmark international experiment, the FIFE campaign (First ISCLSCP Field Experiment) took place in the US Great Plains during 1987–1989 (Sellers and Hall 1992), setting the stage for a similar ground-truth/remote sensing campaign in the African Sahel in 1990–1992. Termed HAPEX-Sahel, this international effort emphasized hydrology in this semi-arid region (Goutorbe 1997). Elsewhere in Africa, the TRACE-A and SAFARI experiments in southern Africa (1992), and later SAFARI-2000 (Privette and Roy 2005; Swap *et al.* 2003), investigated the impact of savanna fires on atmospheric chemistry. Numerous field experiments relating land surface and climate took place in the American Southwest in the 1990s: Monsoon '90 (Kustas and Goodrich 1994), Walnut Gulch '92 (Moran *et al.* 1994), and JORNEX (Ritchie *et al.* 1998). Concerted cooperative field efforts in drylands have also taken place under the auspices of programs such as SALSA (Semi-Arid Land–Surface Atmosphere program; Goodrich and Chehbouni 1998), the Kalahari Transect (Scholes and Parsons 1997), and the Miombo Network (Desanker *et al.* 1997).

1.4.2 REMOTE SENSING

The development of remote sensing techniques during the last three decades has markedly enhanced our ability to investigate and monitor environmental parameters on a large scale, especially in regions where conventional data acquisition networks are not extensively employed. These techniques have been particularly important for drylands in developing countries, where the demise of the influence and activity of colonial powers, together with economic and practical factors, has led to a breakdown of conventional monitoring programs. The methods of remote sensing rely on large-scale data collection from airborne and space-borne platforms.

The central principles in remote sensing are that all natural substances vary in their relative capacities to absorb, reflect, emit, and transmit electromagnetic radiation and that these capacities vary with the wavelength of the incident or emitted radiation. These techniques involve the measurement of (1) *reflected* solar radiation, (2) radiation *emitted* from the earth's surface, and (3) *reflected* microwave radiation. Solar radiation spans the visible and adjacent ultraviolet and infrared portions of the electromagnetic spectrum. Earth, or terrestrial radiation, is in the thermal infrared or microwave wavelengths.

Natural surfaces have a unique "spectral signature": the reflectivity and emission of radiation vary systematically and

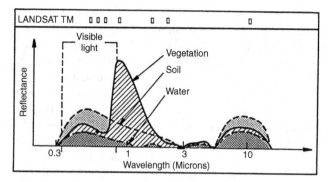

Fig. 1.9 Spectral reflectance signatures of various natural surfaces, along with the spectral detection bands of the Landsat thematic mapper (TM).

characteristically with wavelength (Fig. 1.9). This signature provides a means to identify the nature of the surface material being monitored, many of its physical properties, and many properties of the atmosphere. The brightness of a radar image, for example, depends on the intensity of microwave backscatter, which in turn depends on the physical and electrical properties of the targeted surface. The former include slope, roughness, and vegetation cover, while electrical properties (chiefly conductivity) are affected by such factors as soil porosity and water content. The reflection of visible and near-infrared solar radiation from the earth's surface is most strongly dependent on chemical composition. The emission and reflectivity of longer, thermal infrared wavelengths are functions of temperature and heat capacity. The intensity of emitted microwave radiation varies with chemical and physical properties. This is useful, for example, in distinguishing between ice and water. New techniques used to derive quantitative interpretations of radar and satellite imagery are rapidly expanding their applicability. Some of the characteristics that can be determined include surface cover (e.g., soil, vegetation, water, sand, even rock type), surface topography, soil moisture, and soil and air temperatures.

Satellites have become the most useful and sophisticated tool for remote sensing. These fall into two broad categories: environmental satellites and earth resources satellites. The former are generally geostationary, i.e., in an equatorial orbit, they remain over a selected geographic location. Examples are the American geostationary operational environmental satellites (GOES) and the European Meteosat. The resolution of environmental satellites is relatively low, however, ranging from about 1 to 11 km. The earth resources satellites, designed for monitoring a slowly varying land surface, provide high-resolution but relatively infrequent coverage at individual locations. An example is the American Landsat, launched in 1972. The earth resources satellites are useful tools for environmental mapping.

Most satellites rely on three kinds of sensor systems: cameras, radiometers, and radar. Cameras provide visual images usually derived from visible light or infrared radiation. Radiometers measure the amount of radiation reflected or emitted by the targeted area; individual sensors detect radiation within narrow bands of the spectrum. Radar emits microwaves or radiowaves and detects the amount of backscatter. Recently LiDAR (light

Fig. 1.10 Landsat image of Lake Kara Kul, a water-filled volcanic crater in Tajikistan. White areas are snow cover (from USGS, http://landsat.usgs.gov/).

detection and ranging), which emits light via laser pulses instead of radio waves, has also been used.

Monitoring techniques can be simply illustrated with the multispectral scanner (MSS) instrument carried aboard the first Landsat. This includes sensors that detect radiation in four bandwidths, two in the visible range (a red and a green channel), one at the visible/near-infrared transition, and one near-infrared channel. Four separate images, each corresponding to one of these bands, are transmitted from the satellite. The "green" band is best able to penetrate water and determine turbidity, to distinguish green vegetation, and to identify geologic structures, while the "red" band can distinguish various vegetation types, topographic features, and areas of human settlement. The red/infrared band ratio is useful in monitoring land use and vegetation cover. The near-infrared helps to delineate land–water and soil–crop boundaries. Figure 1.10 shows a composite image of the four bands: Lake Kara Kul in Tajikistan shows up clearly, as does snow cover on the surrounding mountains.

Advancement came with the development of Landsat's thematic mapper, which included the seven bands indicated in Fig. 1.9. The current satellite, Landsat-7, uses a more advanced instrument termed the ETM+ (enhanced thematic mapper plus). It measures radiation in eight bands, with spatial resolution varying from 15 to 60 m, depending on the band. Other earth resources satellites include the French SPOT (with up to 10 m resolution), the European ERS, and the Russian RESURS. The latter, with 160 km resolution, fills the gap between the geosynchronous satellites and the high-resolution earth resources satellites. A commercial satellite, IKONOS, provides resolution of 1 m or better.

In 1999 an instrument specifically designed for land surface monitoring was launched into space aboard the Terra satellite, a joint venture between the USA, Canada and Japan. Termed MODIS (moderate resolution imaging spectroradiometer), this instrument detects radiation in 36 spectral bands and has a spatial resolution of 250–1000 m, depending on the band. MODIS transmits information on surface reflectance, snow cover, surface temperature, land cover and dynamics, vegetation indices, fire, burned areas, leaf area index (LAI), photosynthetically active radiation, net photosynthesis, and vegetation cover change (Justice *et al.* 2002). It provides particularly useful information for savanna regions.

Radar systems have certain advantages over cameras and radiometers. Because of the long wavelength of the beam, radar can penetrate through clouds, and often to some depth below the land surface. This depth is proportional to the wavelength of the signal and is strongly dependent on soil moisture, so that a penetration of several meters is possible in very arid regions, compared with a few centimeters in humid regions. Also, because the intensity of the backscattered signal is highly sensitive to varying physical properties, microwave radar is equally suited for mapping landforms, drainage systems, and geologic features, especially in arid regions. Topography, for example, is best observed when the illumination is almost perpendicular to the direction of the topographic trend. Unlike solar illumination, the direction and angle of incidence of the radar beam can be varied to optimize the detection of individual features. The device is so sensitive to topography that a change of slope of a few degrees can change the radar backscatter by a factor of two or more. For similar reasons, microwave radar is highly sensitive to surface roughness, making it a useful tool in detecting such features as erosional processes (rough surfaces), "desert varnish" (smooth), or sheets of sand (highly reflective and smooth). Thus, it shows excellent promise for the environmental mapping of arid and semi-arid regions.

Early space-borne radar systems were flown on the Seasat spacecraft in 1978 and on the Columbia space shuttle flight of November 1981. These produced surprising observations of several arid regions. The salt pans in the Great Kavir region of the Iranian desert were clearly visible as wavelike patterns and ellipses. Images of sand dunes in the North American desert showed that the brightness in the microwave region of the spectrum varies with the direction of illumination and vegetation cover. The shuttle imaging radar (SIR-A) system aboard the Columbia space shuttle (Elachi *et al.* 1982) produced images of the Lake Chad Basin with clearly discernible dune ridges, interdunal flats, former drainage channels, lake beds, playas, and vegetation. In some cases, even desert wadis were visible. The most striking result was obtained in the hyper-arid eastern Sahara, where the lack of soil moisture allowed the microwaves to penetrate several meters beneath the vast sand fields. The radar detected buried stream channels (some nearly as wide as the Nile) and other subsurface drainage features, as well as possible Stone Age occupation sites (McCauley *et al.* 1982).

1.4.3 ENVIRONMENTAL APPLICATIONS OF REMOTE SENSING IN DRYLAND REGIONS

Both satellite and radar imagery can be applied to a broad range of environmental situations. In dryland regions, remote sensing is especially useful for deriving information on soils, vegetation, and water resources and for monitoring environmental changes, dust outbreaks, and fire. In true deserts, detailed pictures of dunes and other surface materials, and even surface features, can be derived. The accuracy and resolution of remote sensing techniques are still insufficient to provide a complete substitute for aerial photography and field observations, but they provide an important tool for surveying and monitoring remote regions where more conventional data are scarce. Similarly, they provide excellent weather information (rainfall, cloudiness, winds, surface and air temperatures) for remote dryland regions where there is no network of observing stations.

Remote sensing in dryland regions presents special challenges because of the sparseness of the vegetation cover and the spatial heterogeneity of surface features. Most satellite pixels include a combination of vegetation, soil, and other features and special techniques must be developed to unravel the various components. The size of the areas of spatially homogeneous surface is on the order of meters to tens of meters (Bhark and Small 2003). Useful reviews of the challenges and approaches are provided by Okin and Roberts (2004), Ustin (2004), Asner (2004) and Jafari *et al.* (2007).

GEOLOGY, SURFACE FEATURES, AND SOILS

Multispectral scanners and microwave radar have three primary geologic applications. They can discern topography and landforms, provide basic data for geologic mapping, and produce information useful in locating mineral deposits. New techniques used to derive quantitative information from radar images are rapidly expanding its geologic applicability, especially in delineating surface features. The sensitivity of visible and near-infrared reflectivity to chemical composition is sufficient to allow individual rock types to be discerned.

Remote sensing has numerous applications in the identification of soils, soil processes, and soil characteristics such as salinity (Huete 2004). Roughness is the main factor determining soil reflectance and this is highly variable among desert soils (Ustin 2004). The spectral signatures of soils in arid regions are also distinct, sufficiently so that the Landsat image mosaic can serve as a base map for soil types (Satterwhite and Ponder Henley 1984). Landsat has been used for this purpose in the Big Desert of Idaho and in the Indo-Gangetic plain of northern India. SPOT has been used in the Chihuahuan Desert of the USA (Franklin *et al.* 1993). Satellite imagery is also useful in monitoring soil erosion (e.g., Omuto and Vargas 2009), soil moisture, and irrigation. Techniques for assessing soil moisture are described in a later section. In dryland areas, where evaporation is generally high and fields are often irrigated, satellites can identify salt-affected areas and problematic saline soils. A good review is provided by Metternicht and Zinck (2003).

Techniques have also been developed to study dunes (Stephen and Long 2005; al-Dabi *et al*. 1997), sand transport (Ramsey *et al*. 1999), and other features of arid lands. Biogenic crusts, which can be early signals of environmental change (Ustin *et al*. 2005), can be detected from their spectral signature (Karnieli 1997). Space-borne radar facilitates the determination of dune forms and morphology and even the internal structure of dunes (e.g., Bristow *et al*. 2005).

VEGETATION

The remote sensing of vegetation is based on a simple physical principle: while most natural substances show a gradual increase in reflectivity with wavelength in the solar bands of the spectrum, green vegetation shows a dramatic increase between the red and near-infrared wavelengths (Fig. 1.9). This differential reflection in the two bands is described by a number of indices (e.g., Huete *et al*. 2002), the most common of which is the normalized difference vegetation index (NDVI), generally derived from data from the NOAA satellites. It is defined as:

$$(CH_2 - CH_1) / (CH_1 + CH_2) \qquad (1.1)$$

where CH_1 is the reflectance in channel 1 (visible/red, 0.55–0.68 microns) and CH_2 is the reflectance in channel 2 (near-infrared, 0.73–1.1 microns).

NDVI is ultimately a measure of the total absorption of photosynthetically active radiation (PAR), but in semi-arid regions it correlates well with such parameters as percentage surface cover, biomass, and leaf area index, as well as rainfall (Fig. 1.11). In wetter regions, however, the index tends to saturate, leveling off to some maximum value. In most of Africa this tends to happen in regions where annual rainfall exceeds about 1000–1200 mm, but in parts of southern Africa, the limit can be as low as about 500 mm/year (Fig. 1.12).

The apparent vegetation signal is altered by such factors as satellite calibrations, changes in orbit, and atmospheric effects such as clouds and dust. The use of a ratio minimizes these effects. Sophisticated models have been developed to deal with such problems and to resolve and interpret images with more than one major surface component (e.g., bare soil and vegetation or grass and tree cover).

In arid and semi-arid regions the effect of the underlying soil has a large impact on the vegetation index. When vegetation cover falls below 40%, the soil dominates the reflectance signature (Smith *et al*. 1990). The effect of soils is threefold. Soils vary greatly in overall brightness, or magnitude of reflectivity (Fig. 1.13). This problem is greatly reduced by the use of ratios, such as the NDVI. The secondary variation of soils is associated with "color" differences, i.e., the shape of the spectral

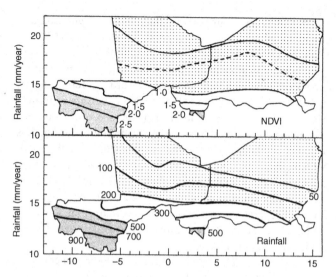

Fig. 1.11 Maps of rainfall (mm/year) (lower panel) and annually integrated NDVI (upper panel) over West Africa (Malo and Nicholson 1990).

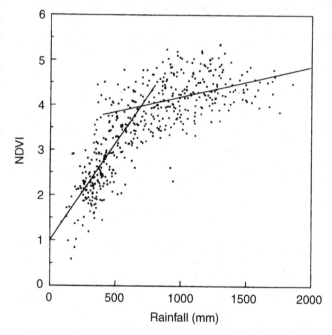

Fig. 1.12 Monthly rainfall (mm) versus monthly NDVI for dryland locations in western, eastern and southern Africa (Nicholson *et al*. 1996). This indicates the "saturation" of NDVI (point where the lines cross) as an index of either rainfall or vegetation growth.

signature; this is related to soil biochemistry and its effect on the absorption of solar radiation. The third problem relates to cases of mixed vegetation and bare soil and is due to multiple scattering. Much of the radiation incident on the bare ground has already been reflected by the vegetation layer. Since the vegetation layer absorbs relatively more of the red than near-infrared (NIR), the light that is scattered toward the ground is "enriched" in the NIR wavelengths, so that, when re-reflected, it gives anomalously high values of NDVI. The effect can be

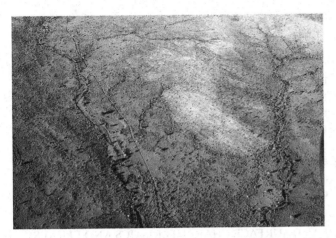

Fig. 1.13 Contrast in albedo of two Botswana soils, the darker being sandy and the lighter being a saline soil.

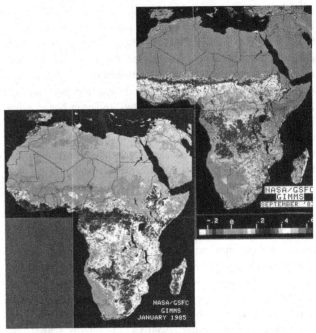

Fig. 1.14 The "green wave" and "brown wave" of vegetation growth and senescence over Africa, as assessed with NDVI. The area of maximum vegetation growth, the very bright area, shifts between the Northern and Southern Hemispheres. Gray areas are those with little or no vegetation growth.

so large that some bare soils have an NDVI as large as a wheat cover of 25% (Huete *et al.* 1984).

Although resource monitoring satellites were intended for such applications as vegetation monitoring, the NOAA satellites proved to be more useful in many cases because they provide much more frequent assessment. Each location is viewed once a day, compared with once every 18 days for earth resources satellites. The higher frequency of coverage has two important advantages. One is that it can more easily capture vegetation changes in areas of annual or ephemeral growth, which can green up or senesce in a few days. This is particularly useful in arid lands. Another advantage is that a shorter time span is needed to obtain a relatively cloud-free image. By choosing the highest value of the vegetation index within a certain period (e.g., 10 days), there is a high probability that the image corresponds to a relatively cloud-free day. Several long-term NDVI data sets based on the NOAA-AVHRR (advanced very high resolution radiometer) instrument have been developed (e.g., Tucker *et al.* 2005). These have the disadvantage of relatively poor spatial resolution, compared with earth resources satellites. The resolution is 1.1 km for LAC (local area coverage) data and 4 km for GAC (global area coverage) data (Masseli and Rembold 2002), compared with 15 m for the current Landsat-7 and 10 m for the French SPOT satellite.

In early studies, vegetation indices were used in semi-arid regions to detect the passage of the "green wave" and the "brown wave," associated with the onset and end of the rainy season (Fig. 1.14), thereby furnishing indirect information on drought and irregular seasonal cycles (Kogan 1995). The dry stage and early growth stage of ephemeral vegetation produce such distinct signatures that spectral imagery can be used to study the influence of rainfall events on the ephemeral arid zone vegetation cover.

The techniques have been greatly refined during the past decade. In dryland regions, the vegetation cover and type (Xie *et al.* 2008), leaf area index (LAI, Myneni *et al.* 2002), diversity (John *et al.* 2008), resilience (Washington-Allen *et al.* 2008), primary production, herbaceous production, foliage, woody biomass, diversity, and various other biophysical properties (Geerken *et al.* 2005; Hong *et al.* 2007) can all be determined with reasonable accuracy. One application of such work is rangeland management, which requires close monitoring of the feed-base for grazing animals. This is particularly important for the delicate pastoral systems along the desert fringe, where the impact of climatic fluctuations and human activity can dramatically alter the landscape. In Australia, Landsat has been used to produce a broad survey of arid rangelands (Pickup *et al.* 1994), an impossible task with conventional surveys. Remote sensing can also provide detailed data on crops, such as maturation, amount, and condition (including disease and drought stress).

In many arid regions, remote sensing can provide a far more comprehensive survey of the extent, condition, distribution, and development of both ephemeral and perennial vegetation than conventional ground and aerial observations. These methods have direct application in crop yield forecasting and in monitoring the response of natural vegetation to natural hazards, climatic fluctuations, and land-use pressure (such as overgrazing and overcultivation), land management practices, and rangeland management.

LAND USE, LAND QUALITY, AND LAND-SURFACE CHANGE

Remote sensing provides excellent surveys of land cover and land use. The AVHRR instrument on the NOAA satellites has been used to distinguish 11 land cover classes at 1° resolution, including cultivated crops, and can distinguish urban or built-up areas at 1 km and 8 km resolution (Hansen *et al.* 1998, 2000). Landsat accurately distinguishes between major categories of use, such as urban areas, fields, and crops, although its accuracy and resolution are still insufficient to provide a complete substitute for aerial photography. Studies in Niger, using Landsat in conjunction with intensive field studies to detect settlement and land use, show the feasibility of this approach (Reining 1979). In the Taklamakan Desert, satellite imagery has been used to assess 40 years of land cover change in the region's oases (Ishiyama *et al.* 2007).

Arid and semi-arid lands are highly sensitive to climatic fluctuations and human impact; consequently, environmental change is a major concern. Remote sensing is useful in monitoring the environment, especially in developing countries (e.g., Karnieli *et al.* 2008), where land-use change is rapid but the lack of economic resources prohibits the extensive use of aerial reconnaissance or ground surveys. In the case of human impact, the factors involve improper management of agriculture, vegetation, and water resources. The relative scarcity of these resources compounds the problem.

A number of studies have demonstrated the sensitivity of NDVI to climatic fluctuations, namely precipitation changes. Figure 1.15, for example, shows the average NDVI in the Sahel zone between 1980 and 1997 (Tucker and Nicholson 1999). This closely parallels changes of rainfall in the Sahel, along the southern margin of the Sahara. Close relationships between interannual fluctuations of NDVI and rainfall have been shown in numerous arid and semi-arid regions of Africa (e.g., Nicholson *et al.* 1990; Farrar *et al.* 1994). Consequently, the remote sensing of vegetation is useful in drought monitoring (Bajgiran 2008).

One critical environmental change is desertification, a general loss of the productivity of the land. Both its causes and impact can be detected remotely: slash-and-burn agriculture, land clearance, overgrazing, intense cultivation, erosion and erosional features such as arroyos, deterioration of vegetation and soils, reduction of ground cover, dune encroachment, and dust. Landsat has been used to study the desertification process in Africa and the Middle East by analyzing the carrying capacity, cultivation, soils, vegetation, and settlement patterns. Satellite estimates of airborne dust (Fig. 1.16) serve as indicators of both drought and desertification. Satellite imagery, combined with field study to provide ground-truth, has been used to quantify desertification in a number of semi-arid regions in the western USA, China (Luk 1983), the Kalahari (Dube 1994), the Sahel and sub-Saharan Africa (Wessels *et al.* 2008), southern Africa (Wessels *et al.* 2004), Australia (Bastin *et al.* 2004), the Mediterranean (Hill *et al.* 1995), India (Tripathy *et al.* 1996) and many other semi-arid locations. Some of these studies are discussed in more detail in Chapter 21.

LAND SURFACE CHARACTERISTICS AND FLUXES

A variety of techniques are also available to monitor various physical characteristics of the surface, including surface temperature, soil moisture, evapotranspiration, and surface albedo

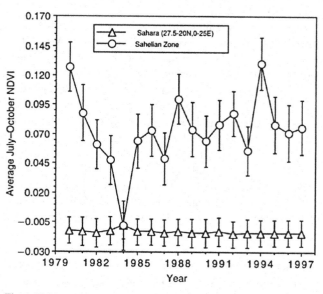

Fig. 1.15 Mean NDVI in the Sahelian zone of West Africa 1980 to 1997 (from Tucker and Nicholson 1999).

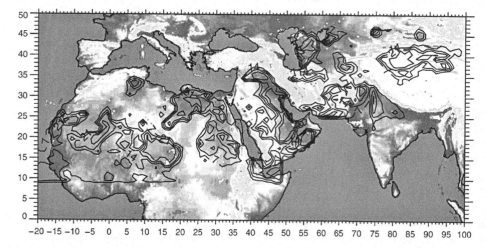

Fig. 1.16 Contours of TOMS (total ozone mapping spectrometer) aerosol index in the desert belt of Africa and Asia (from Prospero *et al.* 2002).

(reflectivity). Temperature is assessed by way of longwave emission, although corrections are necessary to remove the effects of the atmosphere overlying the surface (e.g., Nagol *et al.* 2009). Albedo is estimated from reflected solar radiation (e.g., Ba *et al.* 2001). Three very different techniques are used to assess soil moisture. The thermal approach relies on the thermal inertia of the surface (resistance to temperature change) or the relationship between surface brightness, temperature and NDVI (Wang *et al.* 2004). Soil surface is the primary control on both surface thermal inertia and the NDVI/temperature relationship. Soil moisture can also be assessed from active and passive microwave methods (Anderson *et al.* 2007; Pampaloni and Sarabandi 2004).

These parameters can be used to assess surface energy balance and evapotranspiration (Anderson and Kustas 2008; Courault *et al.* 2005; Diak *et al.* 2004), although independent methods have also been developed to monitor the latter (e.g., Eichinger *et al.* 2006; Wang *et al.* 2006). Several articles provide good reviews of remote sensing models used to assess evaporation (Norman *et al.* 1995; Caparrini *et al.* 2003; Su 2005). Several classes of models are described in these sources, but all of them rely on the use of surface radiative temperatures to assess surface energy balance. Zhao *et al.* (2005) concluded that these methods could not create the balance between accuracy and simplicity that is required for global land surface modeling.

Cleugh *et al.* (2007) utilized a new approach that eliminates the reliance on surface temperature measurements, using instead the Penman–Monteith equation (see Chapter 8) in conjunction with remotely sensed vegetation indices. The results of this approach not only agreed well with measurements, but the algorithm continued to perform well when meteorological data with coarser spatial and temporal resolution were used as input. This study produced continental evaporation maps for Australia and is likely to be a major step forward toward global monitoring.

HYDROLOGY AND WATER RESOURCES

Radiometric sensors and radar are capable of detecting surface and subsurface water (Engman and Gurney 1991). This has great potential for drylands. In many of them, excess water use is depleting subsurface storage; prudent management requires that this resource be continually monitored. Remote sensing also facilitates the study of the water availability, and thus the resource potential, of less developed areas and desert margins. Satellite information can be used to locate wadis, oases, and wells; to depict drainage systems, irrigation networks, and water courses; and to help in groundwater exploration. In arid and semi-arid watersheds it can be applied to deriving runoff estimates and predicting runoff and to surveying the occasional floods (e.g., Bastin *et al.* 2004). Spectral images even provide information from which water temperature, mixing, turbidity and sediment load, and water depth can be estimated. Hand-held sensors are being developed to assess soil moisture and dew (Mouazen 2004; Heusinkveld *et al.* 2006).

In evaluating water availability, the most important measurements of surface water include its spatial extent, depth

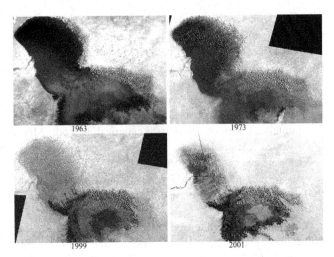

Fig. 1.17 The extent of Lake Chad during four years (1963, 1973, 1999, 2001), commencing when the lake was nearly full in 1963 and extending past the extended period of drought commencing in 1968 (images from USGS Eros Data Center website).

and depth changes, and the slope of water bodies (Cazenave *et al.* 2004; Alsdorf *et al.* 2007). Together these allow formulations of numerous hydrological quantities, including discharge, runoff, and storage loss. Water surface area (Fig. 1.17) can be derived from various visible sensors (Landsat, MODIS, SPOT) and from synthetic aperture radar (SAR) (Smith 1997). Both profiling systems – which derive vertical profiles at individual points, and imaging systems – which provide two-dimensional views, are useful in measuring surface elevation from space (Alsdorf and Lettenmaier 2003; Farr *et al.* 2007). This permits the estimation of the depth, depth changes, and slope of surface waters. Profilers include radar altimeters, such as that aboard the TOPEX/Poseidon satellite, and LiDAR.

So far radar altimeters produce the most accurate estimates of surface elevation. Generally, the larger the system over which elevation measurements are averaged, the greater the accuracy of measurement. The altimeter aboard the TOPEX/Poseidon satellite can measure the height of large lakes to within 3–4 cm (Birkett and Doorn 2004). For rivers, the accuracy is an order of magnitude lower (Alsdorf *et al.* 2007).

Other measurements are more experimental. Some success in assessing flow velocity and stream discharge has been achieved using ground-penetrating radar to measure channel cross-section, and Doppler radar to measure velocity (Costa *et al.* 2006). The GRACE satellite is now providing assessments of the total water mass in the earth system (atmosphere, surface, and subsurface) using gravimetric measurements (Han *et al.* 2005), but so far the technology is useful only for very large basins (larger than 200,000 km²).

RAINFALL MONITORING

Global rainfall is routinely monitored using satellite imagery. Early methods relied upon empirical relationships between rainfall and clouds or outgoing infrared radiation. In a very general sense, the brighter the clouds in the visible channel and the colder in the infrared channel, the more rainfall they produce. A number of projects have been carried out in the drylands of Africa and the Mediterranean (e.g., Dugdale and Milford 1986; Grist et al. 1997; Ba and Nicholson 1998; Feidas et al. 2009). These have established that seasonal and geographical differences exist in the relationships between cloud characteristics and rainfall, a shortcoming of such empirical methods. More sophisticated satellite methods have been developed that rely upon the microphysical properties of clouds. These generally rely on the measurement of microwave emission or reflection. Rudolf and Rubel (2005) present an excellent review of state-of-the art methods.

Commonly used satellite-based precipitation estimates are the GPI (Global Precipitation Index), based on infrared radiation, and GPCP (Global Precipitation Climatology Project) blended product, based on a combination of gauge data plus infrared and microwave satellite data, and the microwave-based SSM/I estimates. These are available on a monthly basis, GPCP and SSM/I since 1979 and GPI since 1986. A number of daily products are also available from satellites and numerical models. Limited products are available at temporal resolutions as high as 30 minutes and spatial resolutions as high as 12 km (Sapiano and Arkin 2009; Ebert et al. 2006). A 0.25° product has also been developed from passive microwave estimates (Joseph et al. 2009).

GPI and GPCP are excellent indicators for global-scale problems, but their use in regional monitoring is limited by the coarse spatial resolution of these products (2.5° latitude/ longitude). Also, some validation with gauge data is advisable for regional-scale studies. This is illustrated in Fig. 1.18, which compares seasonal rainfall over West Africa derived from a very dense gauge network with estimates based on GPI, GPCP and the microwave product SSM/I. GPI and SSM/I markedly overestimate seasonal rainfall. GPCP overestimates rainfall in the most humid region, but produces reasonably good estimates elsewhere. However, it systematically overestimates August rainfall (Nicholson et al. 2003a).

A tremendous advancement in the remote sensing of precipitation was made with the launch of the Tropical Rainfall Measuring Mission (TRMM) satellite in 1997. Its instrumentation includes precipitation microwave radar, a passive microwave radiometer, and visible and infrared sensors. Because of its limited temporal sampling, TRMM is commonly used as a "flying rain gauge" (Adler et al. 2000). Its estimate of an instantaneous rain rate is used to "calibrate" less accurate measurements from other satellites that have a better time and space resolution.

Examples of such blended products are the TRMM-adjusted GPI (or AGPI), a pure satellite product, and "TRMM-merged," which combines AGPI with gauge data. A TRMM multi-satellite precipitation data set with 3-hourly and 0.25° resolution was produced more recently (Huffman et al. 2007). Figure 1.19 shows a validation of the TRMM-merged product at 1° latitude/longitude resolution over Africa, and at 2.5° latitude/longitude resolution over Australia. For annual rainfall estimates, the root-mean-square error (RMSE) for West Africa averages 1.1 mm/day at the 1° scale and 0.4 mm/day at the 2.5° scale (Nicholson et al. 2003b). For Australia it is 0.7 mm/day and 0.6 mm/day on the 1° and 2.5° scales, respectively. For both continents, it is on the order of 1 mm/day for rainy season months at the 2.5° scale.

DUST, SMOKE AND FIRE

Instruments aboard several satellites are useful for dust monitoring (King et al. 1999). These vary with respect to the spatial resolution and temporal frequency of the images, the method by which dust is detected, and the nature of the aerosol information provided. The monitoring capabilities depend on whether the satellite is a polar orbiter (sun-synchronous) or geostationary. The latter affords the advantage of continuous coverage, but relatively low spatial resolution. The polar orbiters can provide higher spatial resolution than the geostationary satellites, but less frequent coverage. Ideally, a multi-sensor approach is used.

The methods are based on reflected solar radiation, thermal infrared emission, and atmospheric absorption. In the visible and infrared channels, dust is detected via brightness contrast or reflectance signature. The resultant products represent aerosol optical thickness. Dust assessment via thermal emission uses the atmospheric "window" channels around 10–12 microns, in which little atmospheric absorption takes place. The absorption by dust varies with time but that of the atmospheric background does not. By separating the variable and constant components the presence of dust can be determined and its frequency of occurrence calculated. Both methods rely upon separating the background signal from the dust signal. This makes them particularly useful over the oceans. The spatially and temporally varying albedo of the land surface limits their usefulness over continents, although appropriate approaches are being developed (e.g., Holben et al. 1992; Hagolle et al. 2008). Over land the most useful method involves the absorption of ultraviolet radiation by atmospheric dust and ozone. The ozone absorption

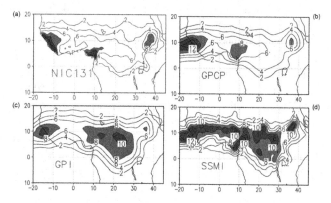

Fig. 1.18 Maps of June–August rainfall over the African continent from (a) the author's data set (NIC131), (b) the GPCP satellite-gauge blended data set, (c) the GPI (Global Precipitation Index) based on satellite infrared measurements, and (d) SSMI based on satellite microwave measurements (from Nicholson et al. 2003a).

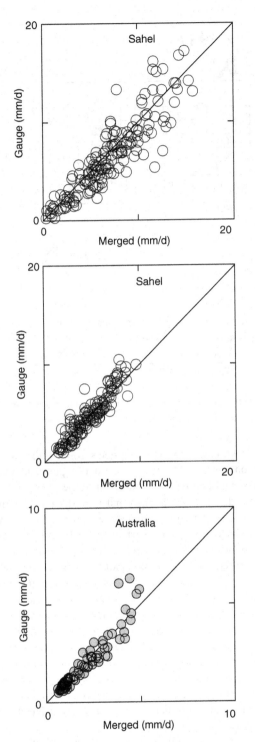

Fig. 1.19 TRMM validation: gauge-measured rainfall versus the TRMM-merged data set for (top) Sahel August, (center) Sahel annual, and (bottom) Australia annual (from Nicholson *et al.* 2003b).

is calculated and the dust signal is a residual (Fishman *et al.* 1992).

Instruments aboard polar orbiters that are useful for aerosol measurements include the NOAA-AVHRR, the Landsat thematic mapper (TM), and the Nimbus-7 total ozone mapping spectrometer (TOMS). The AVHRR measures reflected solar radiation and monitors visible and infrared channels. It provides 1-km resolution and daily coverage, but only over ocean regions (Husar *et al.* 1997). TOMS measures atmospheric absorption by aerosols; it provides 50-km resolution and daily images (Herman *et al.* 1997). The Landsat TM and enhanced-TM provide much higher spatial resolution (30 km and 15 km, respectively), but each location is covered only once in 16 days (Tanré *et al.* 1988).

The geostationary satellites that have been used for aerosol studies include GOES (geostationary operational environmental satellites) and the European Meteosat. In both cases aerosol retrievals can be derived using either reflection in the visible channel or thermal emission. GOES has a spatial resolution of 1.1 km and Meteosat provides 30-km resolution in the visible spectrum. The continuous tracking by these satellites affords the opportunity to examine many dynamic aspects of dust that cannot be readily seen from the polar-orbiting satellites. As an example, Dunion and Velden (2004) used GOES imagery to show the interaction of the Saharan air layer with tropical cyclones, demonstrating its power to suppress cyclone activity over the Atlantic.

Recently launched instruments aboard polar-orbiting satellites in the EOS (earth observing satellite) series also show much promise for detailed dust monitoring. The sea-viewing wide field-of-view sensor (SeaWiFS) instrument for detecting ocean color, aboard the Aura satellite, has produced remarkable images of dust outbreaks (see Fig. 13.21; Eckhardt and Kuring 2005). The MODIS instrument aboard the Terra spacecraft is providing dust assessments from both the solar and thermal channels. Both provide daily images with a spatial resolution on the order of 0.25–1 km. Other satellites carrying instruments specially suited for aerosol monitoring include the French microsatellite Parasol, which carries the Polder radiometer, and the joint French–American Calipso satellite, which carries a LiDAR instrument.

The products derived from the various sensors are not interchangeable. The various instruments provide diverse parameters, and several different algorithms may be used to convert the instrument's radiance information to a particular dust parameter. Some instruments permit quantitative assessments of dust optical thickness, while others, such as TOMS, provide only semi-quantitative indices. Also, because of the limited temporal coverage, the images from different satellites may cover different periods of time.

TOMS is most sensitive to aerosols in the middle and upper troposphere and in the stratosphere, where aerosols reside for long periods of time (Prospero *et al.* 2002). Aerosols below roughly 500–1000 m, which are generally short-lived, are unlikely to be detected by TOMS (Torres *et al.* 2002). TOMS is also limited in its ability to compare different seasons or regions with very different climates. Other satellites provide information for the whole atmospheric column. In contrast, visibility data preferentially detects surface dust, although this generally shows a good relationship with broader atmospheric concentrations.

Several studies have compared the global aerosol information produced by various approaches. TOMS is perhaps the most widely used satellite index. It compares reasonably well with optical depths derived from the AERONET surface network when the optical depth is high (Torres *et al.* 2002), with the Meteosat dust index (Chiapello and Moulin 2002), and with AVHRR-derived aerosol optical thickness on seasonal time scales (Cakmur *et al.* 2001). It does not compare well with AVHRR on interannual time scales, probably because of the limited temporal sampling of both. On the other hand, the interannual variability evident in TOMS compares well with the record of dustfall in Barbados (Chiapello *et al.* 2005). Washington *et al.* (2003) also compare TOMS with surface visibility records, noting that TOMS appears to miss some of the major source areas in the US Great Plains, the Middle East and Mongolia's Gobi Desert. The authors present several possible reasons for the disparity, such as the dominance of low-level dust that is not readily detected by TOMS. Despite the limitations of the individual sources, combined use of the various data sets has resulted in a global capability for long-term aerosol monitoring.

1.4.4 GEOGRAPHIC INFORMATION SYSTEMS

An important means of deriving and evaluating information on arid lands is the geographic information system (GIS). A GIS is essentially computer software designed to handle spatially organized (i.e., geographic) data. These systems integrate remote sensing, conventional databases, database management, computer cartography, and computer-assisted design into a spatial data handling system that supersedes the traditional manual methods of map analysis and processing. The first GIS systems were developed in the mid-1960s by various government agencies as a response to a new sense of awareness and urgency in dealing with problems of environment and natural resources. Systems were developed, for example, by the state of Minnesota and by Environment Canada. Clark University in the USA developed a system called IDRISI (Eastman 2006), which is used extensively in developing countries.

A GIS is, in essence, a database consisting of spatially distributed observations of earth-referenced variables, which can be defined in space as points, lines, and areas. Unlike traditional systems of information storage, a GIS utilizes a computer to store, retrieve, manipulate, and displace the spatial data. This allows for markedly quicker and more efficient information processing. In general, each observation in the GIS database is stored together with information to specify it in reference frames of both space and time. There are two basic models to the system. These differ in how a variable is defined in space: raster representation and vector representation (Fig. 1.20). The basis of the raster model is a finite number of geographical grid boxes to store information. The vector model represents geographical features as geometrical shapes: points, lines and polygons, such as used in a typical map. The two models differ in their ease of production and analysis, storage requirements, and quality of data representation. The raster form facilitates the production of map overlays, but often requires more storage and output may take on an unrealistic block-like appearance.

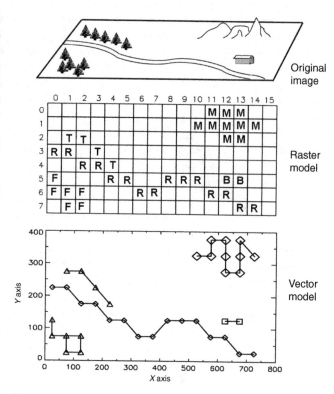

Fig. 1.20 Raster and vector models for geographic information systems (GIS).

All geographic information systems contain the following major components (Marble 1990): a data input system to collect and/or process data; a storage and retrieval system, which organizes data and facilitates rapid retrieval and updating; a data manipulation and analysis system; and a data reporting system, which displays all or part of the database. Software must include all of these in order to be considered a GIS. Much of the data input is derived from remote sensing and photogrammetry (i.e., analysis of aerial photography). For this reason, geographic information systems are particularly useful in arid regions, where conventional observations are generally scarce.

The steps in applying a GIS to a geographical problem or region include identification of spatial elements, determination of data location on a standard coordinate system, measurement of spatial attributes, and storage and portrayal on a map. For example, transparent overlays for each data set might be created, then registered so that the coordinates are aligned, producing a useful composite image.

The advantages of GIS systems are their physically compact form and the relatively low cost and rapidity of information maintenance and retrieval (Dangermond 1990). They also permit quick and efficient manipulation, analysis, and integration of data sets. Disadvantages are the high initial cost of the technology and the skill required to utilize such systems.

GIS systems have many applications in arid land management (e.g., Agatsiva and Oroda 2000). Some of the earliest examples are the use of a GIS to evaluate fuelwood availability in Botswana (van Heist and Kooiman 1991) and to evaluate agricultural suitability on a continental scale (van der Laan 1992). Other uses of GIS systems in arid lands include evaluating flood hazard (Al-Rawas *et al.* 2001), desertification (Tripathy *et al.* 1996), climate (Alijani *et al.* 2008), erosion (Chen *et al.* 2008), and geomorphic change (Koch *et al.* 1998).

REFERENCES

Adler, R. F., G. J. Huffman, D. T. Bolvin, S. Curtis, and E. J. Nelkin, 2000: Tropical rainfall distributions determined using TRMM combined with other satellite and rain gauge information. *Journal of Applied Meteorology*, **39**, 2007–2023.

Agatsiva, J., and A. Oroda, 2000: Remote sensing and GIS in the development of a decision support system for sustainable management of the drylands of eastern Africa: a case of the Kenyan drylands. *International Archives of the Photogrammetry, Remote Sensing and Spatial Information Sciences*, **34**(6), 42–49.

Al-Dabi, H., M. Koch, M. Al-Sarawi, and F. El-Baz, 1997: Evolution of sand-dune patterns in space and time in northern Kuwait using Landsat TM images. *Journal of Arid Environments*, **36**, 15–24.

Alijani, B., M. Ghohroudi, and N. Arabi, 2008: Developing a climate model for Iran using GIS. *Theoretical and Applied Climatology*, **92**, 103–112.

Al-Rawas, G., M. Koch, and F. El-Baz, 2001: Using GIS for flash flood hazard mapping in Oman. *Earth Observation Magazine*, **10**, 18–20.

Alsdorf, D. E., and D. P. Lettenmaier, 2003: Tracking fresh water from space. *Science*, **301**, 1485–1488.

Alsdorf, D. E., E. Rodriguez, and D. P. Lettenmaier, 2007: Measuring surface water from space. *Reviews of Geophysics*, **45**, RG2002.

Amiran, D. H. K., and A. W. Wilson (eds.), 1973: *Coastal Deserts: Their Natural and Human Environments*, University of Arizona Press, Tucson, AZ.

Anderson, M., and W. Kustas, 2008: Thermal remote sensing of drought and evapotranspiration. *EOS Transactions of the American Geophysical Union*, **89**(26), 1–2.

Anderson, M. C., J. M. Norman, J. R. Mecikalski, J. A. Otkin, and W. P. Kustas, 2007: A climatological study of evapotranspiration and soil moisture stress across the continental United States based on thermal remote sensing. 2. Surface moisture climatology. *Journal of Geophysical Research*, **112**, doi:10.1029/2006JD007507.

Asner, G. P., 2004: Biophysical remote sensing signature in arid and semi-arid regions. In *Remote Sensing for Natural Resource Management and Environmental Monitoring* (S. Ustin, ed.), Wiley, 53–109.

Ba, M. B., and S. E. Nicholson, 1998: Analysis of convective activity and its relationship to the rainfall over the Rift Valley lakes of East Africa during 1983–1990 using the METEOSAT infrared channel. *Journal of Climate and Applied Meteorology*, **10**, 1250–1264.

Ba, M. B., G. Dedieu, S. E. Nicholson, and R. Frouin, 2001: Temporal and spatial variability of surface radiation budget over the African continent as derived from METEOSAT. I. Derivation of global solar irradiance and surface albedo. *Journal of Climate*, **14**, 45–58.

Bajgiran, P. R., 2008: Using AVHRR-vegetation based vegetation indices for drought monitoring in the Northwest of Iran. *Journal of Arid Environments*, **72**, 1086–1096.

Bastin, G., V. Chewings, J. Ludwig, R. Eager, and A. Liedloff, 2004: Potential of multi-scale video imagery to indicate the leakiness of rangelands. *Range Management Newsletter*, **43**, 1–5.

Bhark, E. W., and E. E. Small, 2003: Association between plant canopies and the spatial patterns of infiltration in shrubland and grassland of the Chihuahuan desert, New Mexico. *Ecosystems*, **6**, 185–196.

Birkett, C. and B. Doorn, 2004: A remote sensing tool for water resources management. *Earth Observation Magazine*, **13**, 20–21.

Bristow, C. S., N. Lancaster, and G. A. T. Duller, 2005: Combining ground penetrating radar surveys and optical dating to determine dune migration in Namibia. *Journal of the Geological Society, London*, **161**, 315–361.

Cakmur, R. V., R. L. Miller, and I. Tegen, 2001: A comparison of seasonal and interannual variability of soil dust aerosols over the Atlantic Ocean as inferred by the TOMS AI and AVHRR AOT retrievals. *Journal of Geophysical Research*, **106**, 18287–18303.

Caparrini, F., F. Castelli, and D. Entekhabi, 2003: Mapping of land–atmosphere heat fluxes and surface parameters with remote sensing data. *Journal of Hydrometeorology*, **5**, 145–159.

Capot-Rey, R., 1953: *Le Sahara Français*. R.U.F., Paris.

Cazenave, A., P. C. D. Milly, H. Douville, J. Benveniste, P. Kosuth, and D. Lettenmaier, 2004: Space techniques used to measure change in terrestrial waters. *EOS Transactions of the American Geophysical Union*, **85**(6), 59.

Chen, H., A. E. Garouani, and L. A. Lewis, 2008: Modelling soil erosion and deposition within a Mediterranean mountainous environment utilizing remote sensing and GIS: Wadi Tlata, Morocco. *Geographica Helvetica*, **63**, 36–46.

Chiapello, I., and C. Moulin, 2002: TOMS and METEOSAT satellite records of the variability of Saharan dust transport over the Atlantic during the last two decades (1979–1997). *Geophysical Research Letters*, **29**, doi:10.1029/2001GL013767.

Chiapello, I., C. Moulin, and J. M. Prospero, 2005: Understanding the long-term variability of African dust transport across the Atlantic as recorded in both Barbados surface concentrations and large-scale total ozone mapping spectrometer (TOMS) optical thickness. *Journal of Geophysical Research*, **110**, doi:10.1029/2004JD005132.

Clayton, H. H., 1944–1947: *World Weather Records*. Smithsonian Institution, Washington, DC, 3 volumes (continued in additional volumes by US National Climatic Data Center).

Cleugh, H. A., R. Leuning, Q. Mu, and S. W. Running, 2007: Regional evaporation estimates from flux tower and MODIS satellite data. *Remote Sensing of Environment*, **106**, 285–304.

Costa, J. E., and Coauthors, 2006: Use of radars to monitor stream discharge by non-contact methods. *Water Resources Research*, **42**, W07422, doi:10.1029/2005WR004430.

Courault, D., B. Seguin, and A. Olioso, 2005: Review on estimation of evapotranspiration from remote sensing data: from empirical to numerical modleing approaches. *Irrigation and Drainage Systems*, **19**, 223–239.

Dangermond, D. F., 1990: A classification of software components commonly used in Geographic Information Systems. In *Introductory Readings in Geographic Information Systems* (D. J. Peuquet and D. F. Marble, eds.), Taylor and Francis, London, pp. 30–51.

Desanker, P. V., P. G. H. Frost, C. O. Frost, C. O. Justice, and R. J. Scholes (eds.), 1997: *The Miombo Network: Framework for a Terrestrial Transect Study of Land-Use and Land Cover Change in the Miombo Ecosystems of Central Africa*. IGBP Report 41, IGBP Secretariat, Stockholm, 109 pp.

Di Luzio, M., G. L. Johnson, C. Daly, J. K. Eischeid, and J. G. Arnold, 2008: Constructing retrospective gridded daily precipitation and temperature datasets for the conterminous United States. *Journal of Climate and Applied Meteorology*, **47**, 475–497.

Diak, G. R., and Coauthors, 2004: Estimating land surface energy budgets from space. *Bulletin of the American Meteorological Society*, **85**, 65–78.

Dube, O. P., 1994: Monitoring land degradation in semi-arid regions using high-resolution satellite data. In *National Vegetation as a Resource: A Regional Remote Sensing Workbook for East* and *Southern Africa* (J. M. O. Scurlock, M. J. Wooster, and G. d'Souza, eds.), National Research Institute, pp. 125–135.

Dubief, J., 1959: *Le Climat du Sahara*. Vol. 1. Mémoires de l'Institut de Recherche Saharienne, Algiers, 312 pp.

Dubief, J., 1963: *Le Climat du Sahara*. Vol. II. Institut de Recherches Sahariennes, Université d'Alger, Algiers, 275 pp.

Dugdale, G., and J. R. Milford, 1986: Rainfall estimates over the Sahel using METEOSAT thermal infrared data. Proceedings of the ISLSCP conference on parameterization of land surface characteristics. 1985 ESA SP 248.

Dunion, J. P., and C. S. Velden, 2004: The impact of the Saharan air layer on Atlantic tropical cyclone activity. *Bulletin of the American Meteorological Society*, **85**, 353–365.

Eastman, J. R., 2006: *IDRISI 15: The Andes Edition*. Clark University, Worcester, MA.

Ebert, E. E., J. Janowiak, and C. Kidd, 2006: Comparison of near real time precipitation estimates from satellite observations and numerical models. *Bulletin of the American Meteorological Society*, **87**, 47–64.

Eckhardt, F. D., and N. Kuring, 2005: SeaWIFS identifies dust sources in the Namib. *International Journal of Remote Sensing*, **26**, 4159–4167.

Eichinger, W. E., D. I. Cooper, L. E. Hipps, W. P. Kustas, C. M. U. Neale, and J. H. Prueger, 2006: Spatial and temporal variation in evapotranspiration using Raman Lidar. *Advances in Water Resources*, **29**, 369–381.

Elachi, C., and Coauthors, 1982: Shuttle Imaging Radar Experiment. *Science*, **218**, 996–1003.

Engman, E. T., and R. J. Gurney, 1991: *Remote Sensing in Hydrology*. Chapman and Hall, London, 225 pp.

Farr, T., and Coauthors, 2007: The Shuttle Radar Topography Mission. *Reviews of Geophysics*, **45**, RG2004, doi:10.1029/2005RG000183.

Farrar, T. J., S. E. Nicholson, and A. R. Lare, 1994: The influence of soil type on the relationships between NDVI, rainfall, and soil moisture in semiarid Botswana. II. NDVI response to soil moisture. *Remote Sensing of Environment*, **50**, 121–133.

Feidas, H., and Coauthors, 2009: Validation of an infrared-based satellite algorithm to estimate accumulated rainfall over the Mediterranean basin. *Theoretical and Applied Climatology*, **95**, 91–109, doi:10.1007/s00704–007–0360-y.

Fishman, J., V. G. Brackett, and K. Fakhruzzaman, 1992: Distribution of tropospheric ozone in the tropics from satellite and ozonesonde measurements. *Journal of Atmospheric and Terrestrial Physics*, **54**, 589–597.

Franklin, J., J. Duncan, and D. L. Tanner, 1993: Reflectance of vegetation and soil in Chihuahuan desert plant communities from ground radiometry using SPOT wavebands. *Remote Sensing of Environment*, **46**, 291–304.

Geerken, R., N. Batikha, D. Celis, and E. DePauw, 2005: Differentiation of rangeland vegetation and assessment of its status: field investigations and MODIS and SPOT vegetation data analyses. *International Journal of Remote Sensing*, **26**, 499–526.

Gilbert, G. K., 1875: Report on the geology of portions of Nevada, Utah, California and Arizona. Part 1 of *Geographical and Geological Explorations and Surveys West of the 100th Meridian*. Engineers Dept., U.S. Army, 3, 21–187.

Goodall, 1983–2000: *Ecosystems of the World*, Elsevier, Amsterdam, 30 volumes.

Goodrich, D., and A. Chehbouni, 1998: An overview of the 1997 activities of the semi-arid land-surface-atmosphere (SALSA) program. In *Special Session on Hydrology*, American Meteorological Society, Phoenix, AZ, pp. 1–7.

Goutorbe, J. P. (ed.), 1997: *HAPEX-Sahel*. Elsevier, 1088 pp.

Graef, F., E. G. Pavis, and J. Reyes, 2000: A new temperature and precipitation climatology of 1000 stations in Mexico. *EOS Transactions of the American Geophysical Union*, **81**, F744.

Grist, J., S. E. Nicholson, and A. Mpolokang, 1997: On the use of NDVI for estimating rainfall fields in the Kalahari of Botswana. *Journal of Arid Environments*, **35**, 195–214.

Groisman, P. Y., V. V. Koknaeva, T. A. Belokrylova, and T. R. Karl, 1991: Overcoming biases of precipitation measurement: a history of the USSR experience. *Bulletin of the American Meteorological Society*, **72**, 1725–1733.

Guirguis, K. J., and R. Avissar, 2008: A precipitation climatology and dataset intercomparison for the western United States. *Journal of Hydrometeorology*, **9**, 825–841.

Hagolle, O., G. Dedieu, B. Mougenot, V. Debaecker, B. Duchemin, and A. Meygret, 2008: Correction of aerosol effects on multi-temporal images acquired with constant viewing angles: application to Formosat-2 images. *Remote Sensing of Environment.*, **112**, 1689–1701.

Han, S. C., C. K. Shum, C. Jekeli, and D. Alsdorf, 2005: Improved estimation of terrestrial water storage changes from GRACE. *Geophysical Research Letters*, **32**, L07302, doi:10.1029/2005GL022382.

Hansen, M., R. DeFries, J. R. G. Townshend, and R. Sohlberg, 1998: *UMD Global Land Cover Classification, 1 Degree, 1.0, 1981–1994*, Department of Geography, University of Maryland, Baltimore, MD.

Hansen, M., R. DeFries, J. R. G. Townshend, and R. Sohlberg, 2000: Global land cover classification using a decision tree classified. *International Journal of Remote Sensing*, **21**, 1331–1365.

Hastings, J. R., and R. R. Humphrey, 1969a: *Climatological Data and Statistics for Baja California. Technical Reports on the Meteorology and Climatology of Arid Regions*, No. 18, University of Arizona, Tucson, AZ, 96 pp.

Hastings, J. R., and R. R. Humphrey, 1969b: *Climatological Data and Statistics for Sonora and Northern Sinaloa. Technical Reports on the Meteorology and Climatology of Arid Regions*, No. 19, University of Arizona, Tucson, AZ, 96 pp.

Hedin, S., 1904–5: *The Scientific Results of a Journey in Central Asia 1899–1902. Vol. 1, The Tarim Basin*. Lithographic Institute of the General Staff of the Swedish Army, Stockholm.

Herman, J. R., P. K. Barthia, O. Torres, C. Hsu, C. Seftor, and E. Celarier, 1997: Global distribution of UV absorbing aerosol from Nimbus7/TOMS data. *Journal of Geophysical Research*, **102**, 16,911–16,922.

Heusinkveld, B. G., S. M. Berkowicz, A. F. G. Jacobs, A. A. M. Holtslag, and W. C. A. M. Hillen, 2006: An automated microlysimeter to study dew formation and evaporation in arid and semi-arid regions. *Journal of Hydrometeorology*, **7**, 825–832.

Hill, J., J. Mégier, and W. Mehl, 1995: Land degradation, soil erosion and desertification monitoring in Mediterranean ecosystems. *Remote Sensing Reviews*, **12**, 107–130.

Holben, B. N., E. Vermote, Y. J. Kaufman, D. Tanre, and V. Kalb, 1992: Aerosol retrieval over land from AVHRR data-application for atmospheric correction. *IEEE Transactions on Geoscience and Remote Sensing*, **30**, 212–222.

Hong, S., V. Lakshmi, and E. E. Small, 2007: Relationship between vegetation biophysical properties and surface temperature using multisensor satellite data. *Journal of Climate*, **20**, 5593–5606.

Hong, Y., H. A. Nix, M. F. Hutchinson, and T. H. Booth, 2005: Spatial interpolation of monthly mean climate data for China. *International Journal of Climatology*, **25**, 1369–1679.

Huete, A., 2004: Remote sensing of soils and soil processes. Chapter 1 in *Manual of Remote Sensing* (S. Ustin, ed.), Wiley.

Huete, A., Post, R. D. F., and Jackson, R. D., 1984: Soil spectral effects on 4-space vegetation discrimination. *Remote Sensing of Environment*, **15**, 155–165.

Huete, A., and Coauthors, 2002: Overview of the radiometric and biophysical performance of the MODIS vegetation indices. *Remote Sensing of Environment*, **83**, 195–213.

Huffman, G. J., and Coauthors, 2007: The TRMM multi-satellite precipitation analysis: quasi-global, multi-year, combined-sensor precipitation estimates at fine scale. *Journal of Hydrometeorology*, **8**, 38–55.

Huntington, E., 1907: *The Pulse of Asia*. Houghton, Mifflin and Co., Boston, 415 pp.

Husar, R. B., J. M. Prospero, and L. L. Stowe, 1997: Characterization of tropospheric aerosols over the oceans with the NOAA advanced very high resolution radiometer optical thickness operational product. *Journal of Geophysical Research*, **102**, 16889–16909.

Ishiyama, T., N. Saito, S. Fujikawa, K. Ohkawa, and S. Tanaka, 2007: Ground surface conditions of oases around the Taklimakan Desert. *Advances in Space Research*, **39**, 46–51.

Jaeger, F., 1921: *Deutsch Südwestafrika*. Wissenschaftlische Gesellschaft, Breslau, 251 pp.

Jafari, R., M. M. Lewis, and B. Ostendorf, 2007: Evaluation of vegetation indices for assessing vegetation cover in southern arid lands in South Australia. *Rangeland Journal*, **29**, 39–49.

John, R., and Coauthors, 2008: Predicting plant diversity based on remote sensing products in the semi-arid region of Inner Mongolia. *Remote Sensing of Environment*, **112**, 2018–2032.

Joseph, R., T. M. Smith, M. R. P. Sapiano, and R. R. Ferraro, 2009: A new high-resolution satellite-derived precipitation dataset for climate studies. *Journal of Hydrometeorology*, **10**, 935–952.

Justice, C. O., J. R. G. Townshend, E. Vermote, R. Wolfe, N. El Saleous, and D. Roy, 2002: An overview of MODIS land data processing and product status. *Remote Sensing of Environment*, **83**, 3–15.

Karnieli, A., 1997: Development and implementation of spectral crust index over sand dunes. *International Journal of Remote Sensing*, **18**, 1207–1220.

Karnieli, A., U. Gilad, M. Ponzet, T. Svoray, R. Mirzadinov, and O. Fedorina, 2008: Assessing land cover change and degradation in the Central Asian deserts using satellite image processing and geostatistical methods. *Journal of Arid Environments*, **72**, 2093–2105.

King, M., Y. Kaufman, D. Tanre, and T. Nakajima, 1999: Remote sensing of tropospheric aerosols from space: past, present, and future. *Bulletin of the American Meteorological Society*, **11**, 2229–2259.

Kistler, R., and Coauthors, 2001. The NCEP-NCAR 50-year Reanalysis monthly means CD-ROM and documentation. *Bulletin of the American Meteorological Society*, **82**, 247–267.

Koch, M., F. El-Baz, and A. Abuelgasim, 1998: Application of a Geographic Information System (GIS) methodology to modeling geomorphological change in Kuwait's desert. In *Sustainable Development in Arid Zones: Assessment and Monitoring of Desert Ecosystems* (S. A. S. Omar, R. Misak, and D. Al-Ajmi, eds.), A. A. Balkema, Rotterdam, Netherlands, pp. 333–340.

Koeppen, W., 1931: Klimakarte der Erde. In *Grundriss der Klimakunde*. de Gruyter, Berlin.

Kogan, F. N., 1995: Droughts of the late 1980s in the United States as derived from NOAA polar-orbiting satellite data. *Bulletin of the American Meteorological Society*, **76**, 655–668.

Kustas, W. P., and D. C. Goodrich, 1994: Preface to special section on MONSOON '90. *Water Resources Research*, **30**, 1211–1225.

Landsberg, M., 1969–1995: *World Survey of Climatology*, 16 vols, Elsevier.

Lavery, B., G. Joung, and N. Nicholls, 1997: An extended high-quality historical rainfall data set for Australia. *Australian Meteorological Magazine*, **46**, 27–38.

Legates, D. R., 1995: Global and terrestrial precipitation: a comparative assessment of existing climatologies. *International Journal of Climatology*, **15**, 237–258.

Lettau, H. H., and K. Lettau, 1978: *Exploring the World's Driest Climate*. Institute for Environmental Studies, University of Wisconsin, Madison, WI, 264 pp.

Liebmann, B., and D. Allured, 2005: Daily precipitation grids for South America. *Bulletin of the American Meteorological Society*, **86**, 1567–1570.

Luk, S. H., 1983: Recent trends of desertification in the Maowusu Desert, China. *Environmental Conservation*, **10**, 213–224.

Malo, A. R., and S. E. Nicholson, 1990: A study of rainfall and vegetation dynamics in the African Sahel using normalized difference vegetation index. *Journal of Arid Environments*, **19**, 1–24.

Marble, D. F., 1990: Geographic Information Systems: An overview. In *Introductory Readings in Geographic Information Systems* (D. J. Peuquet and D. F. Marble, eds.), Taylor and Francis, London, pp. 8–17.

Masseli, F., and F. Rembold, 2002: Integration of LAC and GAC NDVI data to improve vegetation monitoring in semi-arid environments. *International Journal of Remote Sensing*, **23**, 2475–2488.

McCauley, J. F., and Coauthors, 1982: Subsurface valleys and geoarchaeology of the eastern Sahara revealed by shuttle radar. *Science*, **218**, 1004–1020.

Meigs, P., 1957: Arid and semiarid climate types of the world. In *Proceedings of the International Geographical Union, 17th Congress, 8th General Assembly*, Washington, DC, pp. 135–138.

Meigs, P., 1966: *Geography of Coastal Deserts*. UNESCO, Arid Zone Research Vol. 28, 140 pp.

Meteorological Office, 1958–1967: *Tables of Temperature, Relative Humidity and Precipitation for the World*. Her Majesty's Stationery Office, London, 6 volumes.

Metternicht, G. I., and J. A. Zinck, 2003: Remote sensing of soil salinity: potentials and constraints. *Remote Sensing of Environment*, **85**, 1–20.

Monod, T., 1958: Majabat Al-Koubrá. *Memories de l'institut Francies d'Afrique Norie*, **52**, 406 pp.

Moran, M. S., D. C. Goodrich, and W. P. Kustas, 1994: Integration of remote sensing and hydrologic modeling through multidisciplinary semi-arid campaigns: MONSOON '90, WALNUT GULCH '92, and SALSA-MEX. In *Proceedings of the Spectral Signatures Conference, Val d' l Sere, France, January 17–21*.

Mouazen, A. M., 2004: Towards development of on-line soil moisture content sensor using a fibre-type NIR spectrophotometer. *Soil and Tillage Research*, **80**, 171–183.

Myneni, R. B., and Coauthors, 2002: Global products of vegetation leaf area and fraction absorbed PAR from year one of MODIS data. *Remote Sensing of Environment*, **83**, 214–231.

Nagol, J. R., E. F. Vermote, and S. D. Prince, 2009: Effect of atmospheric variation on AVHRR NDVI data. *Remote Sensing of Environment*, **113**, 392–397.

New, M., D. Lister, M. Hulme, and I. Makin, 2002: A high-resolution data set of surface climate over global land areas. *Climate Research*, **21**, 1–25.

Nicholson, S. E., 1986: The spatial coherence of African rainfall anomalies: interhemispheric teleconnections. *Journal of Climate and Applied Meteorology*, **25**, 1365–1381.

Nicholson, S. E., M L. Davenport, and A. R. Malo, 1990: A comparison of the vegetation response to rainfall in the Sahel and East Africa, using normalized difference vegetation index from NOAA AVHRR. *Climatic Change*, **17**, 209–214.

Nicholson, S. E., A. R. Lare, J. A. Marengo, and P. Santos, 1996: A revised version of Lettau's evapoclimatonomy model. *Journal of Applied Meteorology*, **35**, 549–561.

Nicholson, S. E., and Coauthors, 2003a: Validation of TRMM and other rainfall estimates with a high-density gauge data set for West Africa. I. Validation of GPCC rainfall product and pre-TRMM satellite and blended products. *Journal of Applied Meteorology*, **42**, 1337–1354.

Nicholson, S. E., and Coauthors, 2003b: Validation of TRMM and other rainfall estimates with a high-density gauge data set for

West Africa. II. Validation of TRMM rainfall products. *Journal of Applied Meteorology*, **42**, 1355–1368.

Norman, J. M., W. B. Kustas, and K. S. Humes, 1995: Source approach for estimating soil and vegetation energy fluxes in observations of directional radiometric surface temperature. *Agricultural and Forest Meteorology*, **77**, 263–293.

Okin, G. S., and D. A. Roberts, 2004: Remote sensing in arid regions: challenges and opportunities. In *Remote Sensing for Natural Resource Management and Environmental Monitoring* (S. Ustin, ed.), Wiley, pp. 111–146.

Omuto, C. T., and R. R. Vargas, 2009: Combining pedometrics, remote sensing and field observations for assessing soil loss in challenging drylands: a case study of Northwestern Somalia. *Land Degradation and Development*, **20**, 101–115.

Pampaloni, P., and K. Sarabandi, 2004: Microwave remote sensing of land. *Radio Science Bulletin*, **308**, 30–48.

Passarge, S., 1904: *Die Kalahari*. Reimer, Berlin, 2 vols.

Petrov, M. P., 1976: *Deserts of the World*. Halsted (Wiley and Sons), New York, 447 pp.

Pickup, G., G. N. Bastin, and V. H. Chewings, 1994: Remote sensing based condition assessment for non-equilibrium rangelands under large-scale commercial grazing. *Ecological Applications*, **4**, 497–517.

Powell, J. W., 1875: *Exploration of the Colorado River of the West (1869–72)*, Washington, 291 pp.

Privette, J., and D. Roy, 2005: Southern Africa as a remote sensing test bed: the SAFARI 2000 Special Issue overview. *International Journal of Remote Sensing*, **26**, 4141–4158.

Prospero, J. M., P. Ginoux, O. Torres, S. E. Nicholson, and T. E. Gill, 2002: Environmental characterization of global sources of atmospheric soil dust identified with the Nimbus 7 Total Ozone Mapping Spectrometer (TOMS) absorbing aerosol product. *Reviews of Geophysics*, **40**(1), 1002, doi:10.1029/2000RG000095.

Ramsey, M. S., P. R. Christensen, N. Lancaster, and D. A. Howard, 1999: Identification of sand sources and transport pathways at the Keso Dunes, California using thermal infrared remote sensing. *Geological Society of America Bulletin*, **111**, 646–662.

Rencz, A. N. (ed.), 1998: *Remote Sensing for the Earth Sciences. Manual of Remote Sensing*, Vol. 3, John Wiley, 707 pp.

Reining, P., 1979: *Challenging Desertification in West Africa: Insights from LANDSAT into Carrying Capacity, Cultivation, and Settlement Sites in Upper Volta and Niger*. Center for International Studies, Ohio University, Athens, OH.

Ritchie, J. C., and Coauthors, 1998: JORNEX: a remote sensing campaign to study plant community response to hydrologic fluxes in desert grasslands. In *Rangeland Management and Water Resources* (D. F. Potts, ed.), American Water Resources Association, Middleburg, VA, pp. 65–74.

Rudolf, B., and F. Rubel, 2005: Global precipitation. In *Observed Global Climate* (M. Hantel, ed.), Landolt-Börnstein V/6, Springer, Berlin, pp. 11–22.

Sapiano, M. R. P., and P. A. Arkin, 2009: An intercomparison and validation of high-resolution satellite precipitation estimates with 3-hourly gauge data. *Journal of Hydrometeorology*, **10**, 149–166.

Satterwhite, M. B., and J. Ponder Henley, 1994: Spectral characteristics of selected soils and vegetation in Northern Nevada and their discrimination using band ratio technique. *Remote Sensing of Environment*, **23**, 155–175.

Scholes, R. J., and D. A. B. Parsons (eds.), 1997: *The Kalahari Transect: Research on Global Change and Sustainable Development in Southern Africa*. IGBP Report 42, IGBP Secretariat, Stockholm, 61 pp.

Sellers, P. J., and F. G. Hall, 1992: FIFE in 1992: results, scientific gains, and future research directions. *Journal of Geophysical Research*, **97**, 19001–19009.

Shmida, A., 1985: Biogeography of the desert flora. In *Hot Deserts and Arid Shrublands* (M. Evenari, I. Noy-Meir, and D. W. Goodall, eds.), *Ecosystems of the World*, Vol. 12A, Elsevier, Amsterdam, pp. 23–77.

Silva, V. B. S., V. E. Kousky, W. Shi, and R. W. Higgins, 2007: An improved gridded historical daily precipitation analysis for Brazil. *Journal of Hydrometeorology*, **8**, 847–861.

Smith, L. C., 1997: Satellite remote sensing of river inundation area, stage, and discharge: a review. *Hydrological Processes*, **11**, 1427–1439.

Smith, M. O., S. L. Ustin, J. B. Adams, and A. R. Gillespie, 1990: Vegetation in deserts: I. Environmental influences on regional abundance. *Remote Sensing of Environment*, **31**, 1–26.

Stephen, H., and D. G. Long, 2005: Microwave backscatter modeling of erg surfaces in the Sahara Desert. *IEEE Transactions on Geoscience and Remote Sensing*, **43**, 238–247.

Swap, R. J., and Coauthors, 2003: Africa burning: a thematic analysis of the Southern African Regional Science Initiative (SAFARI 2000). *Journal of Geophysical Research*, **108**, No. D13, 8465, doi:10.1029/2003JD003747.

Su, Z., 2005: Hydrological applications of remote sensing. Surface fluxes and other derived variables: surface energy balance. In *Encyclopedia of Hydrological Sciences* (M. Anderson, ed.), John Wiley and Sons.

Tanré, D., C. Devaus, M. Herman, and R. Saster, 1988: Radiative properties of desert aerosols by optical ground-based measurements at solar wavelengths. *Journal of Geophysical Research*, **93**, 14223–14231.

Thornthwaite, C. W., 1931: The climates of North America according to a new classification. *Geographical Review*, **21**, 633–655.

Thornthwaite, C. W., 1948: An approach toward a rational classification of climate. *Geographical Review*, **38**, 55–94.

Tilho, J., 1911: *Documents scientifique de la mission Tilho*, 3 vols, Ministère des Colonies, Imprimeure Nationale.

Torres, O., P. K. Bhartia, J. R. Herman, A. Sinyuk, P. Ginoux, and B. Holben, 2002: A long-term record of aerosol optical depth from TOMS observations and comparison to AERONET measurements. *Journal of the Atmospheric Sciences*, **59**, 398–413.

Tricart, J., and Cailleux, A., 1969; *Modelé des Régions Sèches*. S.E.D.E.S., Paris, 472 pp.

Tripathy, G. K., Ghosh, T. K., and Shah, S. D., 1996. Monitoring of desertification process in Karnataka state of India using multi-temporal remote sensing and ancillary information using GIS. *International Journal of Remote Sensing*, **17**, 2243–2257.

Tucker, C. J., and S. E. Nicholson, 1999: Variations in the size of the Sahara desert from 1980 to 1997. *Ambio*, **28**(2), 587–591.

Tucker, C. J., and Coauthors, 2005: An extended AVHRR 8-km NDVI data set compatible with MODIS and SPOT vegetation NDVI data. *International Journal of Remote Sensing*, **26**, 4485–4498.

Uppala, S. M., and Coauthors, 2005: The ERA-40 Re-analysis. *Quarterly Journal of the Royal Meteorological Society*, **131**, 2961–3012.

Ustin, S. L. (ed.), 2004: *Remote Sensing for Natural Resource Management and Environmental Monitoring*. Wiley and Sons, New York, 848 pp.

Ustin, S. L., S. Jacquemond, A. Palacios-Orueta, L. Li, and M. L. Whiting, 2005: Remote sensing based assessment of biophysical indicators for land degradation and desertification. *Proc. Internat. Conf. Remote Sensing and Geoinformation Processing in the Assessment and Monitoring of Land Degradation and Desertification*. State Of the Art and Operational Perspectives, Trier, Germany, 28 pp.

van der Laan, F., 1992: Raster GIS allows agricultural suitability modeling at a continental scale. *GIS World*, **5**, 42–50.

van Heist, M., and A. Kooiman, 1991: The fuelwood availability for settlements in south eastern Botswana. *Journal of the Forestry Association of Botswana*, **1991**, 21–35.

Wallén, C. C., 1967: Aridity definitions and their applicability. *Geografiska Annaler: Series A, Physical Geography*, **49**, 367–384.

Wang, C., J. Qi, S. Moran, and R. Marsett, 2004: Soil moisture estimation in a semiarid rangeland using ERS-2 and TM imagery. *Remote Sensing of Environment*, **90**, 178–189.

Wang, K., Z. Li, and M. Cribb, 2006: Estimation of evaporative fraction from a combination of day and night land surface temperatures and NDVI: a new method to determine the Priestley–Taylor parameter. *Remote Sensing of Environment*, **102**, 293–305.

Washington, R., M. Todd, N. Middleton, and A. Goudie, 2003: Dust-storm source areas determined by the total ozone monitoring spectrometer and surface observations. *Annals of the Association of American Geographers*, **93**, 297–313.

Washington-Allen, R. A., R. D. Ramsey, N. E. West, and B. E. Norton, 2008: Quantification of the ecological resilience of drylands using digital remote sensing. *Ecology and Society*, **13**, 33–53.

Wernstedt, F. L. 1972: *World Climatic Data*. Climatic Data Press, Lemont, PA.

Wessels, K. J., S. D. Prince, and I. Reshef, 2008: Mapping land degradation by comparison of vegetation production to spatially derived estimates of potential production. *Remote Sensing of Environment*, **91**, 47–67.

Wessels, K. J., S. D. Prince, P. E. Frost, and D. V. Zyl, 2004: Assessing the effects of human-induced land degradation in the former homelands of northern South Africa. *Journal of Arid Environments*, **72**, 1940–1949.

Xie, Y., Z. Sha, and M. Yu, 2008: Remote sensing imagery in vegetation mapping: a review. *Journal of Plant Ecology*, **1**, 9–23.

Yang, D. and T. Ohata 2001: A bias-corrected Siberian regional precipitation climatology. *Journal of Hydrometeorology*, **2**, 122–139.

Yatagai, A., P. Xie, and P. Alpert, 2008: Development of a daily gridded precipitation data set for the Middle East. *Advances in Geosciences*, **12**, 165–170.

Zhao, M., F. A. Heinsch, R. Nemani, and S. W. Running, 2005: Improvements of the MODIS terrestrial gross and net primary production global data set. *Remote Sensing of Environment*, **95**, 164–176.

2 The geomorphologic background

2.1 THE PHYSICAL SETTING

Much of the traditional literature on deserts focuses on categorizing and classifying them on the basis of such physical characteristics as morphology, process, and climate. Classification schemes abound, two of the most common being the largely topographic/structural distinction between shield-and-platform deserts and basin-and-range deserts and the distinction between aggradational and degradational deserts. The latter is based on the dominance of erosion or deposition but largely corresponds to the morphological classes as well. The shield-and-platform deserts (Fig. 2.1) are broad plains of low relief, covered with stone surfaces, sand seas, or finer materials. They tend to be areas of warm continental climates, like the Sahara or the Australian deserts. The basin-and-range deserts (Fig. 2.1), such as those in the cordillera of the Americas and in Asia, are areas of high relief superimposed on low plains and are often tectonically active. The associated climates are generally those of the mid-latitudes, with cold winters.

These distinctions are convenient, but somewhat inappropriate. They stemmed from a belief both in uniformity of process and character in the desert environment and in a clear-cut distinction between the processes and characteristics of desert and humid environments. This paradigm has been replaced by one that recognizes the tremendous diversity of desert environments and the commonalities that exist between dry and wet environments. The classical categorizations are further misleading because the desert environment is a continuum of surface types, processes, and climatic conditions, with considerable overlap between niches. Moreover, the environment is dynamic, continually changing both to attain an equilibrium with the forces shaping it and in response to changes in these forces, e.g., climatic change. This state of dynamic equilibrium reflects the integration of a multitude of forces and factors acting on a spectrum of time and space scales, including the long-term geologic past. The deserts can only be understood within this context.

Nonetheless some categorization is necessary for pedagogical purposes. The framework adapted here is largely the physiographic setting summarized by Mabbutt (1979) and Cooke *et al.* (1993), with a range of desert provinces from the uplands of the traditional range-and-basin desert to the lowland plains of the traditional shield-and-platform desert. This synthesis is chosen because it recognizes the continuum of forces and processes and provides a convenient paradigm for relating the physical characteristics of the desert to the underlying forces of climate. The discussion will focus on these interrelationships. This emphasis is not meant to diminish the importance of other factors, such as tectonics and regional geology; rather, the connection with climate is the focus of this text and hence its relationship to the geomorphic background of the desert environment is emphasized.

This physiographic framework is illustrated in Fig. 2.2. The desert uplands are areas of high relief and exposed bedrock where erosion is dominant and where the desert watershed originates. The remaining settings, in a downslope sequence, are the piedmont, the stony deserts, the lake basin, and the sand deserts. The piedmont is a gentler slope, marked by an abrupt change of gradient from the uplands. It is an area of both erosion and deposition. The flatter, lowland environments are largely depositional. This sequence is evident in all deserts, but is geographically compressed in the basin-and-range deserts. In contrast, the piedmont and upland settings are markedly diminished in the shield-and-platform deserts. Each environment is functionally related to the upslope and downslope, with desert rivers, drainage channels and floodplains linking the uplands with the piedmont and lowlands. The uplands control the runoff and the supply of detritus that reach the piedmont and the plains below.

Along this continuum of physiographic settings are associated sequences of aridity, drainage patterns, energy supply, and relative dominance of the forces of wind and water (Fig. 2.2). There is a general increase in aridity downslope, largely because of the orographic enhancement of rainfall in the uplands. Correspondingly, there is a change in the nature of the drainage from connected, branching systems with distinct channels

Fig. 2.1 (left) Shield-and-platform desert of Namibia. (right) Basin-and-range landscape in Australia.

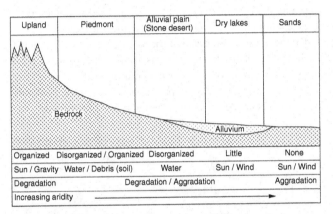

Fig. 2.2 Continuum of desert physiographic settings from the mountains to the plains and the characteristics and processes associated with these settings.

in the uplands, to disorganized and disintegrated networks with sparse, thin channels of flow in the lower regions.

The importance of fluvial activity wanes downslope, as aeolian activity increases; the desert lakes often mark the termination of desert drainage and the transition from the fluvial to the aeolian regime. In regimes of erosion and weathering in the uplands, the energy for geomorphological processes derives from insolation and from water flowing under the influence of gravity (Cooke *et al.* 1993). On the hillslopes, the water is of primary importance, but the energy of abrasion by loose soil plays a secondary role. In the drainage channels, energy is almost exclusively supplied by water. In the playas, the energy that sculpts the surface is supplied by the sun and wind. Thus, the various settings of the desert environment have a close relationship with atmospheric processes.

2.2 THE PHYSIOGRAPHIC CONTINUUM

2.2.1 DESERT UPLANDS AND SLOPES

The hillslope collectively refers to all land surfaces contributing runoff and sediment to stream channels, i.e., the segment of the

landscape between the drainage divide and the stream channel below. The landscape consists of steep and rugged exposed rocks and debris-covered slopes; degradation or erosion is the dominant process (Fig. 2.3). At the top is a relatively flat exposed surface (waxing slope), abutting a steep cliff or free-face; here movement is gravity-controlled. Below is the debris-mantled slope; in arid regions this is rarely continuous or uniform in slope. In the steeper parts, debris is coarse and moves downslope by mass movement (debris-controlled slopes), but on gentler slopes the debris is finer and moved by slope wash (wash-controlled or rain-washed slopes). The debris usually forms only a thin layer; these are slopes of transportation where there is an equilibrium between the supply of material from the cliff face above and its removal downslope by gravity and water. On the hillslope, flow is unconfined or overland; water moves through shallow and impermanent rills rather than through permanent channels.

The piedmont is a gentler slope, marked by an abrupt change of gradient from the debris-mantled slopes above it. The greater the aridity, the more pronounced is the junction of the piedmont and debris slope, and the lower is the gradient of the piedmont. The piedmont intervenes between the connected upland drainage patterns and the disintegrated drainage of the plains below. Few permanent drainage channels continue from the hills through the piedmont. The piedmont is the transition between degradation on the slopes and aggradation on the plains, so its landforms can be either erosional or depositional. The former are termed pediments, plains cut in bedrock where drainage takes the form of networks of tiny distributary rills, representing diminishing channels of runoff. Inselbergs are also common on the piedmont (Fig. 2.4).

The depositional forms are alluvial fans and bajadas. Alluvial fans (Fig. 2.5) are cone-shaped bodies of alluvial detritus formed at the outlet of a mountain valley or canyon. They may be a few hundred meters to tens of kilometers in extent and are formed when a wadi leaves the confines of a mountain channel and emerges on the plains below. The energy of the stream is dissipated by surface friction, but the unconfined water also spreads or diverges; both factors lead to deceleration of the flow. The

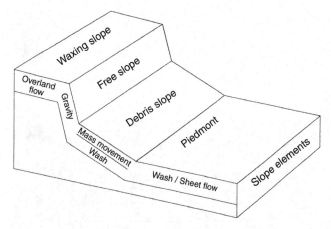

Fig. 2.3 Highly idealized slope model of desert uplands and the associated processes of water movement (based on Wood 1942, Mabbutt 1979, and others).

lower velocity cannot support the suspended load, and material is deposited on the surface. Where the fans are compound or where they coalesce to form unstructured alluvial plains, the term bajada is used.

2.2.2 DESERT LOWLANDS

Below the piedmont are the lowlands or plains; these consist of stony deserts, riverbeds and floodplains, desert lake basins, and ultimately sand deserts. The stone deserts have quite varied surfaces. The stone mantles are termed pavements when stones are closely packed on relatively flat surfaces. The term hamada is used for boulder-rich terrain and tablelands; the term reg (serir in the Sahara) is applied to pavements of finer materials.

Many of the rocks and stones in the desert lowlands reflect the powerful force of the wind. Ventifacts are stones shaped by sand blast and dust abrasion and oriented with respect to the wind regime. Yardangs (Fig. 2.6) are large, wind-sculpted ridges of exposed rock. The stony deserts also include regions of hard and impenetrable duricrusts, formed as a result of restricted leaching in deserts. Some, such as laterites (iron-rich crusts) and silcretes (crusts of siliceous materials like sands and quartz grains), tend to be relicts of past climates. Lime-rich calcrete crusts or caliche are currently forming in dryland regions.

Fig. 2.4 Ayers Rock, Australia, is an example of an inselberg.

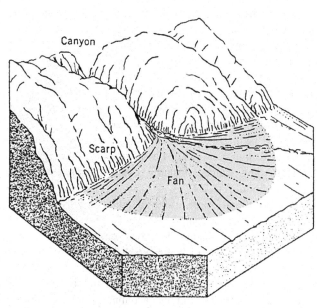

Fig. 2.5 Diagram of an alluvial fan (copyright © A. N. Strahler 1994).

Fig. 2.6 Yardang in the Western Desert of Egypt.

Table 2.1. *Names used commonly used for saline and non-saline playas (Menenti 1984).*

Country/Region	Playa	Non-saline playa	Saline playa
USA	playa, dry lake	dry playa, clay playa	salt flat/marsh, salina
Mexico	laguna, salina	laguna	salina
Chile	–	–	salina, salar (very salty)
Australia	playa, lake	clay pan	salt pan, salina
Russia	pliash	takir	tsidam
Mongolia	gobi, nor	takyr	tsaka, nor
Iran	daryacheh	daqq	kavir
South Africa	pan, vloer, mbuga	clay pan, kalpfannen	salt pan
North Africa	sebkha	qarat, garaet, khabra	sebkha, chott
Arabia	–	khabra	mamlahah, sabkhah
Jordan	ghor	qa	–
Iraq	hawr	faydat	sabkhat
India	rei	–	–
Pakistan	hamun	–	–

 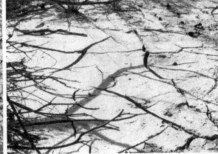

Fig. 2.7 Extensive cracking in desiccated clay soils.

The stone pavements may exhibit interesting surface colors and patterns. In some cases, the stones are covered with a lustrous dark-stained patina, called desert varnish. Deep orange, red, brown, and black, these are hydrous forms of iron and manganese oxides derived from the soil solution or atmospheric dust. Lichens, algae, and microorganisms are catalysts in their formation. The sorting of materials by differential expansion and contraction by salt and water may leave perceptible geometric patterns. There are diverse forms of patterned ground such as hummocks (gilgai in the Australian desert), desiccation cracks, or stone polygons (Fig. 2.7).

2.2.3 DESERT LAKE BASINS

Scattered over the desert floor are lake basins occupying the lowest areas of the desert drainage network or lying beyond its termination (Fig. 2.8). The most common term for these basins is playas, but they are also called sebkhas or chotts in northern Africa, dry lakes in North America, and pans in southern Africa. There are numerous local names as well; Menenti (1984) lists 35, with different names being given to saline and non-saline playas (Table 2.1). The nature of these basins is quite diverse, even within an individual desert. The

playa sediments are generally fine-grained, with the coarser sediments having been deposited when the intensity of runoff diminishes as it traverses the plains. The sediments also tend to be saline, a characteristic that demonstrates the role of evaporation in playa formation.

Playas vary in size and number, ranging from a few square meters to thousands of square kilometers. The largest is Lake Eyre in Australia, occupying an area of 9300 km². Their occurrence is frequent, but most are quite small, so that playas represent only about 1% of desert surfaces worldwide. There are over 300 playas in the western USA, over 9000 in the Kalahari of Botswana, and over 1000 in North Africa. Their frequent occurrence provides evidence of the disorganized drainage pattern of the desert (Mabbutt 1979).

These lake basins differ widely with respect to the amount of water they contain and the length of time they contain water. Motts (1970) has suggested applying the terms lakes, playa lakes, and playas in accordance with the length of the usual period of inundation. Most of these lake basins remain dry except for brief periods of seasonal or ephemeral flooding. Water supply comes from surface runoff, direct precipitation, or groundwater discharge. Some, such as those in the Siwa Oasis of Egypt, are produced by spring flow (Fig. 2.9).

Fig. 2.8 Playa surface of Sossusvlei in the Namib, covered with desiccation cracks of the clay and saline material of the playa crust. Remains of vegetation that grew during the last flooding are also apparent.

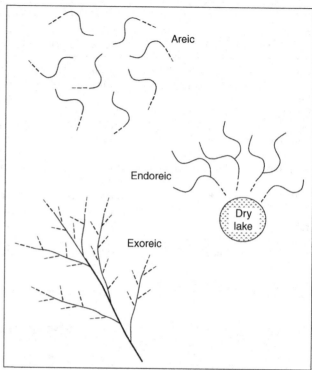

Fig. 2.10 Diagram showing exoreic, endoreic and areic drainage.

Fig. 2.9 One of three large, salt lakes in the Siwa Oasis of Egypt.

2.3 DRAINAGE NETWORKS AND STREAMS

The drainage networks in arid and semi-arid regions have distinctive characteristics that distinguish them from drainage in humid regions. The differences are in geomorphic shape and in the hydrological regime (runoff frequency, duration, and magnitude). In drylands, runoff tends to be episodic and of brief duration, but is often high volume. Also, the drainage networks are generally areic, with no integrated network of surface flow, or endoreic, draining into internal basins rather than reaching the ocean (Fig. 2.10). The streams can be perennial (permanently containing water) or ephemeral (episodically containing water). Streams in humid regions have exoreic drainage, eventually reaching the sea.

Endoreic drainage is advantageous in that it concentrates runoff in small areas, and thereby reduces water loss through evaporation. The final destination of the runoff can be tributary flow or closed basins. Drainages with perennial runoff usually become tributaries, sometimes of large rivers. Those with

ephemeral runoff frequently terminate in an alluvial fan or closed basin, typically associated with a desert playa or temporary lake. Figure 2.11, showing discharge frequencies for both types, indicates a clear distinction between the two.

The degree of endoreism is largely a function of aridity (Fig. 2.12 and Table 2.2), with areas of endoreic and areic drainage (Fig. 2.13) corresponding closely to the arid zones in Meig's maps (see Fig. 1.2 in Chapter 1). The latitudinal distribution of these drainage types is the inverse of rainfall, with a concentration in the subtropics. Another characteristic of dryland drainage is that the actual area of the drainage basin is often considerably larger than its effective area (that where runoff originates) because the basins were formed during pluvial periods of the past.

Other important characteristics of the drainage network are its degree of convergence or divergence, the proportion of total area occupied by floodplain or riparian (river) corridor, and overall drainage density. Convergent flow results in channel incision, but divergent flow is associated with the deposition of materials (Bull 1997). The amount of area occupied by floodplains and riparian corridors is important because these tend to be regions where surface–groundwater interactions take place. Such interactions are important to the ecosystem (see Chapter 12).

Drainage density, the ratio of the total length of stream channels to drainage area, is generally small in dryland regions. It tends to maximize in semi-arid regions, and to decrease or remain steady as rainfall further increases (Tucker and Bras 1998). In the southwestern USA drainage density increases from about

Table 2.2. *Endoreic and areic drainage, as a percentage of the continental area (adapted from Mabbutt 1979).*

Continent	Arid and semi-arid (%)	Endoreic drainage (%)	Areic drainage (%)
Australia	83	21	43
Africa	64	12	39
Asia	39	12	24
South America	17	6	8
North America	16	4	4

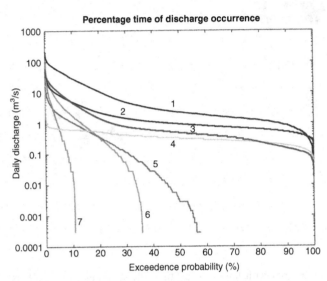

Fig. 2.11 Flow duration curves for various drainage basins in New Mexico, illustrating the contrast between perennial (1–4) and ephemeral (5, 6, 7) streams (from Newman *et al.* 2006).

0.5 to 0.6 km/km² where rainfall is between 250 and 300 mm/year to about 1 km/km² at higher levels of mean annual rainfall (Newman *et al.* 2006).

2.4 FLUVIAL AND AEOLIAN PROCESSES

The processes of erosion and weathering are strong in dryland regions, where the energy of the wind and surface water flow is high and vegetation is sparse. In the desert, terrain is sculpted primarily by the forces of water and wind, sometimes aided by other chemical and mechanical forces. Aeolian and fluvial activity forms the desert pavements and the intricate patterns of sand dunes in the sand deserts. The energy is proportional to velocity and mass; hence flowing water is more effective than wind and carries different types of materials.

2.4.1 WEATHERING

Weathering is the slow breakdown and disintegration of rock material. The process occurs continually in all climates, but particular types are favored in dry regions where the energy of insolation is intense and erratic streamflow is high (Sperling and Cooke 1985). Weathering in deserts is characterized by

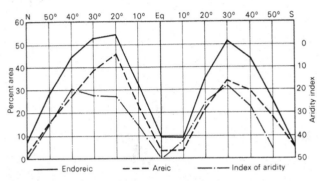

Fig. 2.12 Extent of endoreic (solid line) and areic (dashed line) drainage with relation to latitude and aridity (from de Martonne and Aufrère 1927).

superficiality and selectivity. Soil profiles and weathered mantles are shallow and weathering is localized, limited to areas with a conducive microclimate.

In arid regions, various weathering processes act to chip away larger rock materials. Disintegration commonly results from infinite cycles of expansion and contraction caused by heating and cooling, frost formation, or successive moistening and desiccation (Warke 2000; Viles 2005). Salt, common in the drylands, similarly acts as a weathering agent by successively expanding and contracting in response to heat, hydration, or crystal growth and dissolution. Wind, chemical transformations, lichens (Fig. 2.14), and internal pressure in rocks also have a weathering action on parent materials.

Striking examples of weathering can be seen in the Namib Desert (Fig. 2.15). Fog rills are formed in rocks when fog condenses high on the rocks and runs down their sides. Huge rocks are split as a result of extreme temperature fluctuations or gradients near their surface. Rock surface temperatures can be as high as 93°C in some deserts (Roof and Callagan 2003) but the extreme heating disappears within millimeters of the surface.

2.4.2 TRANSPORT OF MATERIAL BY WIND AND WATER

Size determines the mobility of materials in air and water. This serves as the basis for textural divisions. The most important distinction is between clays, silts and sands, all of which can be held in suspension, and large materials such as granulars, pebbles, cobbles and boulders, which cannot. In most classification

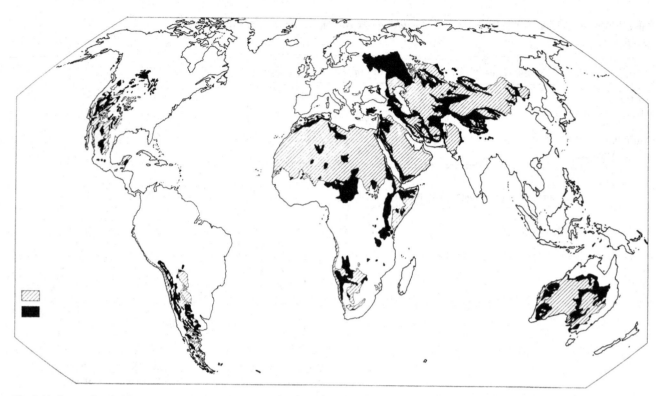

Fig. 2.13 Areas of areic (diagonal shading) and endoreic (black shading) drainage worldwide (from de Martonne and Aufrère 1927).

Fig. 2.14 Lichens on a rock in the Namib.

systems, sand particles have diameters between 0.05 and 2 mm; silt ranges from 0.05 to 0.002 mm; and clay particles are less than 0.002 mm in diameter. Silt and clay particles are small enough to be held in suspension by the wind. Sand can be held in suspension by water, but not by wind. The force of the wind is sufficient to bounce or drag sand particles, but they are only briefly airborne and rise just above the surface. The bouncing of sand grains in the wind is called saltation (Leenders *et al.* 2005). Movement by the force of surface drag is called surface creep or roll. It can be initiated by saltating particles or by gravity. The grains dislodged by saltating particles can also be bounced

along, but with lower energy than the saltating particles. This form of motion is called reptation.

The size of the particles that can be mobilized and held in suspension in either air or water depends on the velocity of flow. Particles will be mobilized when certain threshold velocities are exceeded. To understand this, it is useful to consider the forces acting on particles, starting with the example of a grain of sand under the force of wind. Three forces exert pressure on the grain (Fig. 2.16). The velocity pressure (P_{VE}) is related to impact on the windward side of the grain. The viscosity of air results in lower pressure on the lee side; this is termed viscosity pressure (P_{VI}). Third is the static pressure of air (P_S), which is reduced above the grain as the air flowing over it is compressed and therefore accelerates (the Bernoulli effect). All three increase with the speed of the wind. The lift on the grain is $P_{Sair} - P_{Sgrain}$; the drag is $P_{VE} - P_{VI}$.

The threshold velocities required to mobilize the grain (Fig. 2.17) represent the minimum velocity at which the aerodynamic forces producing lift or drag are strong enough to overcome the forces holding particles together. The fluid threshold – the velocity required to set the grain in motion by surface drag – decreases rapidly with grain size down to a limit of about 0.1 mm diameter. This represents the wind velocity required for surface creep. Below particle sizes of about 0.1 mm (very fine sands, silts, and clays), the fluid threshold increases rapidly with decreasing particle size, because such forces as ionic and chemical bonds and the surface tension associated with moisture increase the bonding between grains. At the same time, surface

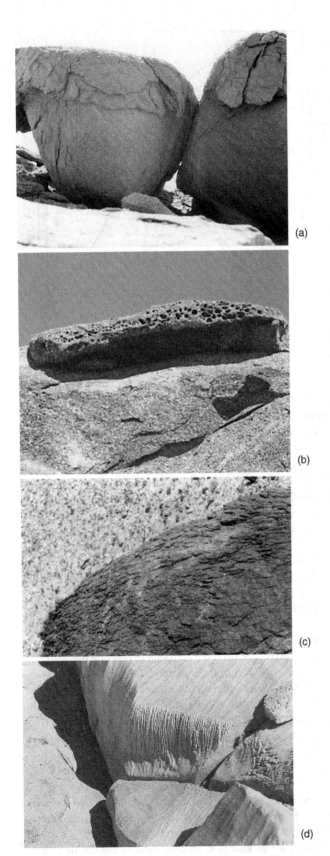

Fig. 2.15 Patterns of weathering in the Namib Desert: (a) rocks cracked by thermal weathering, (b) honeycomb pattern of chemical weathering, (c) exfoliation, fracturing in layers via weathering. (d) Rillenkarren (fog rills) are depressions weathered by fog-water that condenses on the rocks and flows downwards.

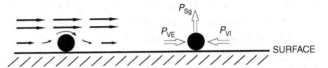

Fig. 2.16 Airflow over a sand particle and the resultant forces: VE = velocity force, VI = viscosity force, and S_g = static pressure over the grain.

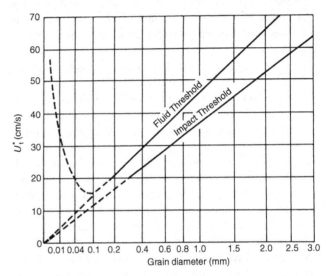

Fig. 2.17 Fluid and impact threshold velocities (U_t^*, cm/s) for sand movement in relation to grain diameter (mm) (from Bagnold 1941).

roughness is lower over smaller particles and the drag force of the wind is reduced. The impact threshold – that required for the processes of saltation and reptation – steadily increases with particle size, but it is consistently lower than the fluid threshold. The greater effectiveness of impact is due to increased velocity pressure in the case of a dense impacting grain, compared with the velocity pressure of air. Since impact can mobilize materials at lower wind velocities than fluid flow, saltation contributes more than creep to total sand transport and surface creep is enhanced by saltation.

A number of formulae have been developed to relate sediment flux to the third power of the shear velocity U^* (also called friction velocity). Shear velocity is associated with surface shear stress (the driving force of erosion) and is proportional to the gradient of the velocity profile plotted on a log scale (Livingstone *et al.* 2007) (see also Section 6.4). The most general formula for sediment transport Q is

$$Q \propto (U^*)^a (U^* - U^*_t)^b \tag{2.1}$$

where U^*_t is a threshold friction velocity and a and b are exponents that total 3 (Livingstone and Warren 1996). The friction velocity is a function of surface drag, wind speed, and roughness, being inversely proportional to the latter. Based on sand transport studies in the Peruvian-Atacama Desert, Lettau and Lettau (1978) suggested that the mean wind at 10 m could be a reasonable substitute for friction velocity. A variant of Lettau's formula is used in the widely utilized "Fryberger model" of sediment transport (Pearce and Walker 2005). However, recent research has shown that this relationship may be too simplified (Baas and Sherman 2005; Walker and Nickling 2003).

Once in motion, particles will be held airborne as long as their terminal velocity of fall does not exceed the upward velocity of air currents. Since terminal velocity is a function of size, dust (consisting of silts and clays) tends to remain airborne, while the larger sand grains fall back to the surface. Sand generally rises at most a few feet from the surface, while dust can be carried thousands of meters upward and thousands of kilometers horizontally.

The above discussion makes it clear that the mobilization of surface particles depends not only on wind and particle size, but also on other surface characteristics. One is bonding by soil moisture (Ravi *et al.* 2004); erodibility is roughly inversely proportional to soil moisture. Another is sorting of grains, since the presence of saltating sand helps to mobilize silt and clay. The presence of non-erodible roughness elements (such as rocks and pebbles) reduces erodibility by absorbing some of the force of the wind. Vegetation cover also plays a role, since it both bonds the soil surface and decreases the force of the surface wind. In addition to wind speed, meteorological factors that play a role include vertical velocities, thermal stability (which determines vertical velocities and turbulence or surface drag), temperature, and atmospheric humidity.

Many of the characteristics of the dryland environment are conducive to the mobilization and transport of surface particles. The sparse vegetation and the bare, often sandy, soil increase the erodibility of the surface. The low surface roughness of bare ground reduces surface drag, thus enhancing wind speed near the surface. The intense heating of the ground surface produces instability and turbulence, enhancing vertical velocities of air. The dryness of the air reduces particle bonding.

The transport of material by water is analogous to that by air, with the same forces involved. It is much more effective, however, because water is denser than air and the pressures involved are therefore greater. The three primary mechanisms are rainsplash, overland flow (sheet or slope wash), and gullying (the creation of channels when erodibility or erosion potential change). The effectiveness of rainsplash is a function of the kinetic energy of raindrops. This, in turn, is related to their terminal velocities (and hence size) and their number (a function of rainfall intensity), as well as wind. Generally, a threshold intensity of rainfall must be exceeded before it is erosive.

2.4.3 EROSION

Wind and water act collectively to erode the landscape. The rate of erosion, especially that by wind, is difficult to assess. However, the factors that promote erosion are well understood, and their relationship to regional climate has been established in a number of studies (e.g., Angel *et al.* 2005). Studies of sediment yield suggest that erosion rates vary widely in the drylands, from zero to over 300 m³/km² per year. These estimates ignore wind

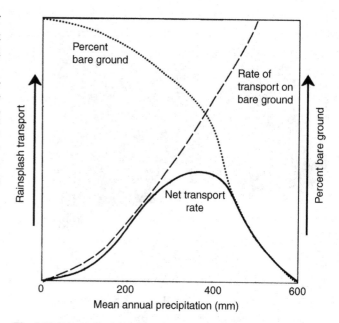

Fig. 2.18 Rates of erosion by rainsplash versus annual precipitation (modified from Kirkby 1969).

erosion and dissolved sediment load, which can both be considerable in arid and semi-arid regions (Cooke *et al.* 1993).

The classic work of Langbein and Schumm (1958) and others has established relationships between mean annual sediment yield and mean annual "effective rainfall" (the amount of rainfall required to produce a given amount of runoff under specified temperature conditions). Data from the western USA (Fig. 2.18) indicate that sediment yield is extremely low in arid regions and that maximum sediment yield occurs where annual rainfall is between about 200 and 500 mm, ranges that are generally associated with semi-arid climates. Although based only on data from the USA, the general form of the sediment yield–rainfall relationship shown in Fig. 2.18 has been confirmed by studies in other regions.

Several factors explain the high erosion rates in semi-arid regions. As rainfall increases, so does its direct impact on soil (rainsplash) (Fig. 2.18) and its capacity to generate runoff, which also erodes the surface). In contrast, vegetation cover, which generally increases with rainfall, reduces the vulnerability of the surface to erosion by shielding the surface from rainsplash, by retarding lateral runoff, and by binding the soil elements (Fig. 2.19). The seasonality of rainfall also contributes to the high sediment yield in semi-arid regions. Soil is highly susceptible to erosion when rainfall commences abruptly after the end of a long dry season. A third factor enhancing sediment yield in semi-arid regions is the intensity of rain events. The most intense storms in the world are not in the wettest regions, such as the rainforests, but occur largely in semi-arid regions (see Chapter 5).

A number of studies examining erosion worldwide and its relationship to climate have concluded that total erosion (wind plus water) in deserts is relatively small, owing to the presence of relatively coarse surface material, low rainfall, and often low

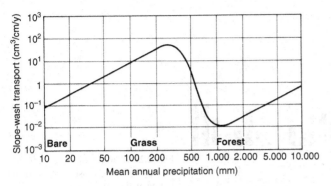

Fig. 2.19 Rates of sheet erosion (slope wash transport) versus annual precipitation (from Carson and Kirkby 1972).

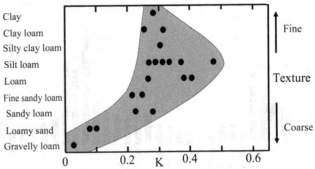

Fig. 2.20 Water erodibility K factor versus soil texture (modified from Dunne and Leopold 1978). Dots indicate measurements; shaded area indicates range of values.

wind speeds. According to Corbel (1964), for example, erosion from arid lands accounts for only about 4% of total world erosion. However, erosion can be intense in some parts of the arid landscape. The rate of erosion on the uplands is about 2–5 times greater than on the plains (Mabbutt 1979). It is also high in dry lake beds. Those in North Africa may supply as much as half of the global atmospheric loading of mineral dust (Prospero et al. 2002).

Erosion is notoriously difficult to measure, so that numerous models have been developed to produce quantitative estimates. The earliest were empirical, based merely upon the statistical relationships between environmental variables and erosion. Later on, process models were developed that assessed the physical, chemical, and biological processes involved in erosion (e.g., Bulygina et al. 2007). One of the earliest models of water erosion was the universal soil loss equation (USLE) (Wischmeier and Smith 1978). This has since been revised and upgraded (USDA 2003). The revised version (RUSLE) relates total soil erosion by rainfall (E_r) to six variables:

$$E_r = f(R,K,L,S,C,P). \tag{2.2}$$

These variables are rainfall erosivity (R), soil erodibility (K), length of slope (L), steepness of slope (S), and two factors (C and P) involving land management and erosion control. Soil type significantly affects water erodibility. The relationship with texture is complex but, in general, erodibility appears to be highest with loams and lowest in the case of coarse soils with high proportions of sand and gravel (Fig. 2.20). This is in marked contrast to wind erosion, which is high for coarser soils. It is important to note that the RUSLE is empirical and the functional relationships of the variables are not well established and differ with climate type. However, it is routinely and successfully used to estimate erosion, with revisions allowing an extension to cover sediment yield as well (Mutua et al. 2006).

An analogous equation for wind erosion (Woodruff and Siddoway (1965), termed the wind erosion equation (WEQ), takes the form

$$E_w = f(I,K,C,L,V) \tag{2.3}$$

where the relevant variables are soil erodibility by wind (I), surface roughness (K), local climate (C), a length (L) related to field length downwind, and vegetative cover (V). The climate term includes winds, precipitation, and evaporation (and hence, implicitly, soil moisture). As with the universal soil loss equation, the applicability of the equation is difficult because of the necessity of establishing the functional relationships of the variables. As with the USLE, the WEQ has been revised and there has also been a move to development of process-based models of wind erosion (Leys 1999).

The primary factors affecting wind erodibility are soil moisture, cementing agents (such as salts), surface roughness, and soil cohesiveness (related to packing density, texture, and aggregation) (Pye 1987; Gillette 1981). Even a moisture content of less than 1% can double the threshold velocity, compared with dry soil (Belly 1964). In general, soils with larger particles, such as sand, loamy sand, and sandy loam, are most readily eroded.

In the USA, water produces more erosion than wind, but since the 1930s several hundred thousand square kilometers have suffered serious wind erosion (Kimberlin et al. 1977). Wind erosion was particularly strong during the droughts of the 1930s (the "Dust Bowl" years) and the 1950s (Fig. 2.21). In 1934 in the Great Plains, high winds from one storm transported an estimated 272×10^6 tons of soil.

2.4.4 FORMATION OF DESERT PAVEMENTS

The stone pavements (Fig. 2.22) which cover much of the desert surface, are concentrations of coarse materials that have been sorted by the action of wind and water. Deflation, the removal of fine-grained material by wind, is the most common explanation (Figs. 2.23 and 2.24). Given a deposit of stream sediment of various sizes, the fine materials are progressively blown away, leaving the coarse residue. In deserts, of course, the sediment field must be a product of ancient streamflow and probably a more humid climate. A similar sorting and pavement formation occurs when running water flushes away the fine materials. Sometimes wind or water will be insufficient to set material in motion, because of inertial forces; the impact of raindrops can

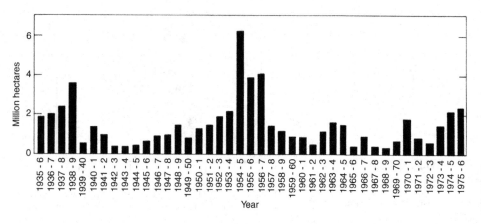

Fig. 2.21 Amount of land annually damaged by wind erosion in the Great Plains of the USA from 1935 to 1976 (from Pye 1987, copyright Elsevier).

be forceful enough to overcome the static force of friction and initiate the activity of wind and surface water.

A desert surface in western Australia illustrates the complexity of the processes forming desert pavements. The surface of the Pinnacle Desert is covered with large limestone pillars (Fig. 2.25). The pinnacles were formed from high sand dunes composed of lime-rich sand weathered from seashells. Rain leached the lime from the sand, cementing the lower levels of the dune into a soft limestone, part of which became a hard calcrete layer. Encroaching plant roots created deep cracks that allowed water to seep into the limestone, eroding most of it and leaving the pillars in areas not reached by the seeping water. The surface was further sculpted by the winds blowing around the pinnacles, forming deep hollows in the lee of each pinnacle (Fig. 2.26).

Fig. 2.22 Stone pavement in the Namib Desert.

2.5 SAND DUNES

2.5.1 DUNES AND OTHER AEOLIAN FEATURES

The most striking aspects of deserts are the large sand seas and the varied but regular patterns of dunes within them. The term sand sea, or erg, is usually limited to areas exceeding 30,000 km² in size (Cooke *et al.* 1993). The world's largest, the Rub' al Khäli in Saudi Arabia, covers 560,000 km² (Wilson 1973). Other large sand seas are the Erg Oriental (192,000 km²) and the Erg Chech-Adrar (319,000 km²) in the Sahara, and the Simpson Desert of Australia (300,000 km²).

Sand seas are generally confined to basins that are separated by topographic features such as plateaus, massifs, and mountain ridges (Fryberger and Ahlbrandt 1979). They are distinguished from sand sheets and streaks by the presence of overlying dunes of various forms and sizes. They occupy about one-quarter to one-third of the surface area of the world's deserts, mainly in Africa, Asia and Australia. Their extent ranges from less than 1% in North American deserts to about 28% in the Sahara and 40% in Australia.

The formation of sand seas is well understood (Lancaster 1999). In most cases, sand is transported from areas with high

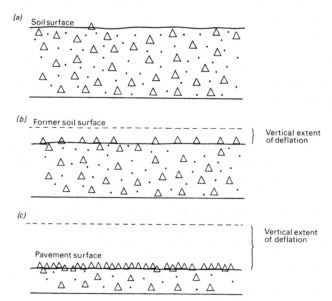

Fig. 2.23 Schematic of the process of deflation (from Goudie and Wilkinson 1980).

Fig. 2.24 Two examples of deflated surfaces in the Namib. The sand lies below in each case. (left) Coarse heterogeneous material overrides uniform sand. (right) Markedly uniform coarse grains lie above the sand surface.

Fig. 2.25 Limestone formations in the Pinnacle Desert of western Australia.

Fig. 2.26 Hollows sculpted by the wind in the lee of limestone pinnacles.

wind energy and high non-aeolian sand supply. Sand particles are bounced by the wind and pushed along, although never really suspended, and are left behind when the force of the wind decreases. Thus, sand tends to accumulate in areas of less wind energy and more variable direction. Such conditions exist in the vicinity of the subtropical highs linked to desert regions such as the Kalahari and the western Sahara. In some cases, the sand may be transported over large distances, with the limits of the sand seas imposed by climatic controls (e.g., a change in the wind regime) or by topography.

Dunes occupy roughly 60% of the vast sand seas (Fig. 2.27). A variety of dune forms exist, based on size and on dune morphology and its relationship to the sand-transporting winds (Fig. 2.28). Time scales for formation increase with the size of the dune. The largest features are generally on the scale of kilometers. The time scale for formation is on the order of millennia. In the Kalahari, Namib Desert, and parts of Asia they attain a height of 200–250 m or more.

Ripples are much smaller features, generally superimposed upon dunes (Fig. 2.29). These are on the order of 0.05–10 cm. In some cases (granular ripples) the crest contains material of larger size than the rest of the ripple, making these features particularly visible. Ripples are formed from chance irregularities in the surface. The irregularities promote deposition on the windward side and bombardment by saltating particles. This loosens particles, which are then transported mostly by reptation and deposited at some distance downwind, with additional ripples building up consecutively downstream (Livingstone and Warren 1996).

Megaripples (Yizhaq 2008) have wavelengths on the order of 30 cm to 20 m and a time scale of days, compared with minutes for normal ripples. One spectacular example comes from the Carachi Pampa of Argentina, where the wind reaches 400 km/hour. Here the ripples reach 1.5 m in height with a wavelength of 18 m.

2.5.2 DUNE FORMS

The control of dune formation is closely linked with both wind regime and sand supply (Fig. 2.30). Classically, three types of dunes are distinguished: longitudinal (or linear), transverse, and star (Livingstone and Warren 1996). Linear dunes form where

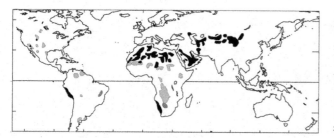

Fig. 2.27 Map of areas of major sand seas and dunes (based on Cooke and Warren 1975, Mabbutt 1979 and others).

Table 2.3. *Morphodynamic dune classification (Livingstone and Thomas 1993).*

Dune type	Wind regime	Mode of activity
Transverse	Unimodal	Migrating
Linear	Bimodal	Extending
Star	Complex	Sedentary

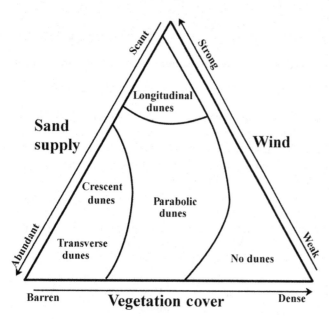

Fig. 2.28 Dune classification system proposed by Livingstone and Warren (1996).

Fig. 2.29 Aeolian ripples in the Namib near Gobabeb.

the wind regime is bimodal, with one primary direction, or where it is predominantly from one sector but with much directional variation within that sector. They are aligned more or less parallel to the prevailing wind. Transverse and barchan dunes are associated with high-energy, unidirectional wind, with barchans forming in areas where sand supply is limited. They run more or less perpendicular to the prevailing wind. Star dunes form in a multidirectional wind regime. They are therefore sedentary and are thus areas of sand accumulation. These dune types also differ with respect to their mode of activity: migrating, extending, or sedentary (Table 2.3).

Longitudinal dunes are arranged in long, linear, parallel ridges (Fig. 2.31), with interdune spacing on the order of 1–3 km, somewhat smaller in width, and up to 20 km or more in length (Lancaster 1982). They comprise about 50% of the total area of dunes worldwide (Fryberger and Goudie 1981). Simple, compound, and complex linear dunes are also distinguished on the basis of dune morphology (Lancaster 1983). Simple linear dunes consist of a single ridge with a narrow, single crest; compound dunes have multiple, narrow ridges along the crest. This class includes the classical seif dune (Fig. 2.32) with a

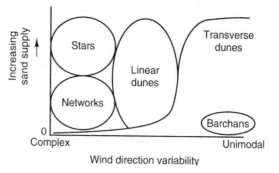

Fig. 2.30 Schematic of dune types with respect to wind regime and sand supply (Livingstone and Warren 1996).

Fig. 2.31 Simple linear dunes of the Namib. These start abruptly at the Kuiseb River in the north and extend some 20 km southward. The linear dunes lie inland, in the bimodal wind regime. The general location of the dunes is shown in Fig. 20.23.

Fig. 2.32 Classic seif dune with sinuous curving crests.

Fig. 2.33 Complex linear dunes in the Namib. These lie further inland from the linear dunes of Fig. 2.32 and are indicative of a more complex wind regime where topography and other factors play a role. The general location of the dunes is shown in Fig. 20.23.

sharp and sinuous crest (prevalent in the Sinai and the Sahara) and long, straight dune ridges (common in the Kalahari and in the Simpson Desert of Australia). Compound linear dunes include two superimposed sets of dunes of the same type, with

one type generally much smaller. Complex linear dunes (Fig. 2.33) have two dissimilar dune types superimposed. Frequently secondary dunes form at oblique angles to the orientation of the primary ridge. Linear dunes are the prevailing dune forms in the Namib, Australia, and parts of the Kalahari and the Sahara–Arabian desert belt, but they are rare in the sand seas of Asia and the Americas.

More complex dune forms include star dunes (Fig. 2.34) or rhourds and network dunes. These dunes form patterns with a confused set of slipfaces pointing in several different directions. They generally form where the wind regime includes several persistent wind directions or where there is a directional shift with the seasons. Without a prevailing direction of transport, these dune types are relatively stationary.

Transverse dunes are likewise parallel ridges and the ridges lie roughly perpendicular to the wind (Fig. 2.35); those with sinuous crests are termed aklé. The crescent shaped barchan dunes (Fig. 2.36) are also oriented normal to the sand-moving winds and result when sand supply is too limited for the development of transverse ridges. Small barchans are common but large ones (greater than 2 m) are rare. They occupy only about 0.01% of the area of dunes worldwide (Goudie and Wilkinson 1980). Barchans are situated with the horns of the crescent facing upwind; when the horns face downwind, the term parabolic or linguoid dune applies. These occur together in the sinuous ridges of the transverse aklé dunes (Fig. 2.36). In some cases, termed a reversing dune, a thin ridge exists along the crest. The reversal generally indicates a seasonal wind reversal. An example of a reversing dune is shown in Chapter 20 (Fig. 20.37), with the reversal due to the development of an easterly berg wind. The reversal is also seen on a dune in the background of Fig. 2.36.

2.5.3 CONTROLS ON DUNE MORPHOLOGY

Dunes are clearly the result of the prevailing wind regime (e.g., Livingstone 2003; Saqqa and Atallah 2004), but the particular details relating wind to such characteristics as dune form, size, and spacing are more difficult to account for (Lancaster 1982). Probably both geomorphic and aerodynamic factors play a role. The size of the dune must be in equilibrium with the sand supply/wind energy ratio (Lancaster 1981), with larger and more widely spaced dunes existing where the ratio is high.

The close link between the wind regime and sand movement is illustrated by studies of the movement of barchan dunes in the Peruvian desert. Barchans move in a downwind direction, with a speed (also termed celerity) on the order of 10–50 m/year. The motion commences when wind power (related to the cube of wind speed) exceeds a particular threshold. This clearly demonstrates the need for mechanisms of dune formation to be evaluated not on the basis of prevailing winds, but on winds capable of transporting sand (Fig. 2.37). The threshold tends to be exceeded when a strong lapse condition exists in the temperature gradient (i.e., temperature decreases sharply with height) and the friction velocity of wind is high. In Peru, the

Fig. 2.34 (left) Star dune in the Namib, indicative of a wind regime with multiple, persistent directions. (right) A star dune is evident in a field of compound dunes, south of the Kuiseb, near the Tsondab Flats. The general location of the dunes is shown in Fig. 20.23.

Fig. 2.35 Transverse dune field in the Namib, near Swakopmund. These dunes lie near the coast, in the unimodal wind regime dominated by land and sea breezes. The general location of the dunes is shown in Fig. 20.23.

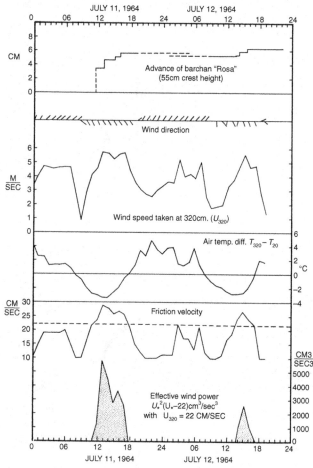

Fig. 2.36 A field of merging barchans. Both lingoid and barchanoid elements are apparent. A reversing dune is visible in the background. The classic form of an individual barchan is seen in Fig. 2.38.

Fig. 2.37 Movement of a barchan in the Peruvian desert as a function of various characteristics of the wind regime and the surface temperature gradient (temperature difference between 20 and 320 cm above ground (from Lettau and Lettau 1978). Movement occurs only when wind power is adequate, which in turn occurs only when relatively unstable conditions occur. Movement is halted during the nocturnal inversions.

threshold wind speed is about 22 cm/s (Lettau and Lettau, 1969, 1978). The celerity is thus a function of wind speed, but it is also closely and inversely related to dune height.

The simple classification illustrated in Table 2.3 persisted for a long time. Remote sensing imagery demonstrated the far greater complexity of dune forms, indicating that dune-forming processes are also much more complex. Within a given sand sea, generally several generations of dunes exist as a result of several periods of dune construction, stabilization, and reworking. The oldest are often remnants of a wind regime from thousands of years ago. Consequently, active and stabilized dunes may coexist under the same climatic conditions (Yiqhaq *et al.* 2007).

The various generations become apparent by way of abrupt transitions in morphology, size, alignment, and composition (Lancaster 1999). The Grand Desierto sand sea of northern Mexico contains at least five generations of dunes (Beveridge *et al.* 2006). In parts of the Sahara, the dunes are an amalgam of late Pleistocene, Holocene, and recent deposits, with the largest/smallest being the oldest/most recent. Regional variations in sand supply and wind regimes likewise create abrupt discontinuities in dunes.

2.5.4 DUNE FORMATION

The mechanism for barchan formation is probably the best understood. Barchans form from winds elongating a mass of sand, with accelerated flow around the sandy "obstacle" elongating the horns downwind (Fig. 2.38). This, together with a lee vortex transporting sand in the upwind direction between the ridges (Lettau and Lettau 1969), produces the characteristic crescent shape (Fig. 2.36). The shape rapidly becomes exaggerated, since the horns are narrower than the bulk of the dune and hence respond more readily to the winds.

Bagnold (1941) postulated the mechanisms for the formation of longitudinal and transverse dunes. His suggested mechanism for transverse dunes involves Kelvin–Helmholtz waves sculpting the surface at preferred intervals, corresponding to their wavelength (Fig. 2.39). Bagnold (1953) further theorized that linear dunes result from helical roll vortexes in the atmosphere blowing essentially along the ridges and piling up sand where they converge (Fig. 2.40). Such vortices are responsible for the formation and alignment of cloud streets, and the analogy is tempting. Bagnold's explanation for the seif form of linear dune is that it results from an elongation of crescent-shaped barchans, when these dunes advance into a bidirectional wind regime. Accordingly, the oblique winds would progressively elongate one horn. Observations in the Namib Desert (Lancaster 1980) support this mechanism.

Bagnold's ideas were widely accepted for decades. However, the formation of linear dunes is considerably more complex and not yet completely understood (Wang *et al.* 2004). Recent studies have suggested that much of Bagnold's hypothesis, in particular the roll vortex concept, is incorrect. In his comprehensive review of linear dunes, Lancaster (1982) evaluates this hypothesis and demonstrates that it is inconsistent with observations. However,

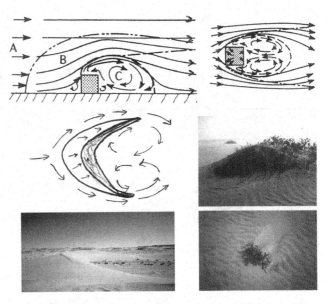

Fig. 2.38 Schematic of flow around an obstacle (partially adapted from Oke 1987), around a barchan dune, and compared with tracks of sand movement around a nebkha dune.

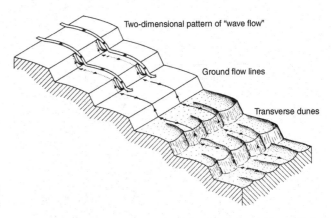

Fig. 2.39 Schematic showing the formation of transverse dunes (from Cooke and Warren 1975).

Tseo (1993) found field evidence to support the helical vortex hypothesis in his study of linear dunes in the Strzelecki Desert of Australia. Corbett (1993) also concluded that this mechanism operates in the southern Namib Desert.

Tsoar (1989) and Livingstone (1989) offered alternate explanations based on the interaction between the dunes and the wind. The details of the proposed mechanisms vary, but both hypothesized that linear dunes extend along a resultant of the winds from the two directions. When wind blows obliquely toward a dune, flow separation occurs as the air traverses the crest. The wind is deflected at the dune's surface and tends to blow parallel to it on the leeward side, transporting material in a parallel direction and elongating the dune (Fig. 2.41). The central issue is the balance of deposition and erosion by the wind; in a unidirectional wind, they take place at the same location and no elongation results. The wind patterns in the Namib

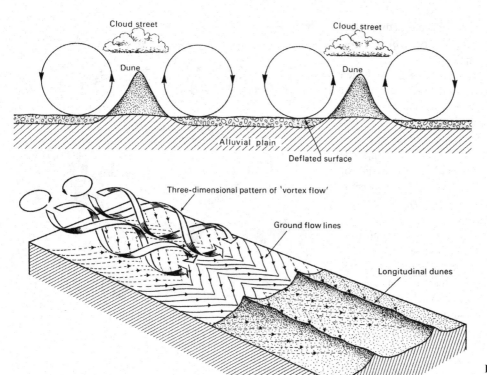

Fig. 2.40 Schematic of the helical vortex theory of dune formation (from Bagnold 1953).

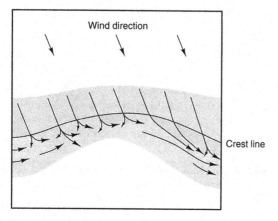

Fig. 2.41 Tsoar's concept of linear dune formation (from Livingstone 1990, based on Tsoar 1983).

linear dunes support this hypothesis, as do observations from a number of other regions (Lancaster 1982).

Differences in the two proposed mechanisms involve the importance of the flow separation vortex created in lee of the dune. Tsoar, based on extensive field work in the Sinai, says the flow separation is essential to dune dynamics. Livingstone says it is incidental, with the dune form resulting from variations in wind speed (i.e., shear stress) across the dune. He emphasizes the effect of increasing wind speed at the crest and decreasing wind speed in the lee, on the erosion and deposition of sand. He also suggests that the two theories are mutually compatible, with the mechanisms being complementary and perhaps operating in tandem.

2.5.5 DUNE FIELDS AND AIR FLOW AROUND DUNES

Recent research on dunes has shown that the relationship to wind is considerably more complex than implied by the dune form/wind relationships described in Section 2.5.3 (Tsoar *et al.* 2004; Lancaster 2007). One reason is that these relationships and other aspects of dune dynamics are based on the study of individual dunes, rather than the ensemble within a dune field. Dune-scale processes do not contribute to understanding processes at the scale of dune fields or sand seas (Livingstone and Nickling 2004). Also, turbulence rather than mean or instantaneous wind speed may be an important factor in dune dynamics (Schönfeldt and von Löwis 2005).

The complex patterns in a dune field result from many factors, including the non-linearity of the relevant physical processes and interactions among dunes and between dunes and their physical environment. There is feedback among the dunes and among the wind fields that each individual dune creates. Consequently, it has been argued that dunes are actually self-organized complex systems (Kocurek and Ewing 2005). This suggestion is particularly interesting in light of recent research suggesting that the vegetation patterns in arid regions are also self-organized (see Chapter 3).

Dunes do not just respond to the wind regime, but create their own secondary flow regimes (Fig. 2.42) that interact with the dune itself and play a role in its dynamics and development (Lancaster *et al.* 1996; Sharon *et al.* 2002). The impact of a dune on the wind field is best illustrated using the case of a transverse dune, with wind blowing perpendicular to the dune. Turbulence is very high at the windward edge of a dune (Fig. 2.43). As air

Fig. 2.42 A "smoking" dune. The sand transport across the crest is indicative of the high wind speeds of the crest and the flow separation as air blows across it. Another example of the dune's impact on airflow is seen in the curving pattern of sand flow at the base of the dune. The dune creating this pattern is not visible.

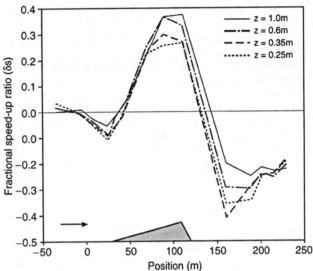

Fig. 2.44 Wind speeds along a transect perpendicular to a transverse dune (Wiggs *et al.* 1996).

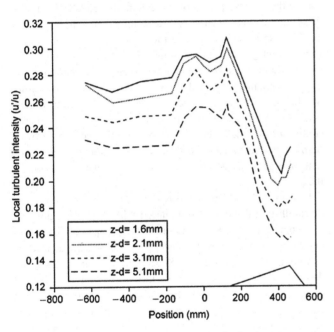

Fig. 2.43 Turbulence along a transect perpendicular to a transverse dune (Wiggs *et al.* 1996).

flows upslope toward the crest of a dune, the airstream is compressed, causing the flow to accelerate near the crest of the dune (Fig. 2.44). Furthermore, the ambient wind speed at a height equivalent to that of the dune is considerably higher than near the surface. As a consequence, wind speeds are very high at the crest. Flow separation occurs there and in the lee of the dune.

Feedbacks result between airflow, dune morphology, and sediment transport. Sand flux is greatest near the crest of the dune, where wind speeds are high. However, the changes in wind speed across the transverse dune are not commensurate with changes in sand flux (Wiggs *et al.* 1996). This may be indicative

of turbulent eddies created at the points of flow separation (Walker and Nickling 2003). Intense sand flux events might also be created by turbulent bursts (Baas and Sherman 2005).

2.6 DRYLAND SOILS

The soils of arid and semi-arid regions are quite diverse but have distinctive characteristics resulting from the common factors affecting the soil formation process. The nature of the soil is determined to a large extent by climate, but in many cases there are *relict* soils – those formed under past climatic conditions. These may be remnants of tropical soils, such as laterites (those with hard, impenetrable layers of iron or aluminum), but in many cases the topsoil has been eroded away and only the relatively infertile lower horizons are left. Soils also reflect their parent material, likewise a product of past environments. At the same time, vegetation and drainage affect the process of soil development.

Cooke *et al.* (1993) summarize several ways in which dryland soils differ physically from those in more humid regions (Table 2.4). They are often covered with large areas of bare rock or gravel and patinated surfaces such as desert varnish. The rates of weathering, erosion, and leaching are relatively low. Consequently, desert soils tend to be coarse-textured and shallow and they generally retain soluble substances. Weathering and erosion are especially low in ancient shield deserts, such as those in Australia and the Sahara, so that many very old soils exist in these environments.

Vegetation cover also affects soil processes. In arid regions, biomass and productivity are low, so little organic matter is produced. Contents as low as 0–2% are common. The species typical of dry regions promote the accumulation of potassium and calcium and low carbon/nitrogen ratios. Phreatophytic (deep-rooted) types also tend to keep the water table low.

Table 2.4. *Ways in which dryland soils differ from those of wetter environments (Cooke et al. 1993).*

Less weathering and leaching
Consequently, coarse textures, shallow depth, and retention
  of soluble substances
Slow soil formation
Large areas of bare rock
Great expanses of patinated surfaces (e.g., desert varnish)
Major inputs of aeolian material
Low rates of erosion
Abrupt soil boundaries

Dixon (1994) also points to the widespread occurrence of vesicular A horizons (i.e., near-surface horizons with high porosity), surficial crusts, and diagnostic subsurface horizons that contain soluble mineral constituents, such as gypsum, sodium, silicates, calcium carbonate, and various salts. Many of these become impermeable hardpan layers, generally composed of silica, iron, or calcium carbonate. These layers limit drainage and water flow in the soil. Very common among these is caliche, a calcium carbonate layer that is particularly prevalent in the desert soils of the USA (Schlesinger 1985). Its formation is controlled by soil hydrology: sufficient moisture to introduce dissolved calcium carbonate but insufficient moisture to leach it through the soil. Consequently, the depth of the calcium carbonate layer increases with decreasing aridity (Akin 1991).

Other characteristics of the dryland soils primarily reflect the influence of climatic factors, particularly low rainfall and relatively high temperatures. Birkeland (1984) presents a good review. Temperature controls the rates of organic matter decay, chemical reactions, and evaporation (and hence salt and mineral accumulation). It also affects the nature of the vegetation cover, which likewise influences soil processes. Moisture affects the transport of soil material and solutes and the rates of weathering. Low rainfall means that water and solutes penetrate only to a limited depth, where the salts and silicates tend to be deposited when drying occurs and water movement is upward. This process of deposition of soluble minerals from upper horizons in lower horizons is called illuviation. The result is salt deposits with characteristic profiles, based on the solubility (of the salts), low pH, high base saturation, and low cation exchange capacity (Cooke and Warren, 1975). The lack of water inhibits weathering, hence in arid regions there tends to be little clay (except in alluvial soils) and the soils tend to be thin, in stark contrast to more humid regions.

Parent material is a major factor in the overall texture of the soil. In semi-arid regions, the nature of drainage frequently results in surface deposition of clays, and many soils in semi-arid regions are thus clay-rich, with clay content roughly increasing with mean annual rainfall. These soils tend to retain water, and salts do not move freely within them. The soils dry out and crack during the long dry season (Fig. 2.7). In drier regions, soils often form on alluvial deposits on floodplains, such as deltas, desert flats, or lake basins. In the deltas and desert flats, the soil will contain much rocky and coarse material and infiltration will be high. In the lake basin the soils will be fine-textured and saline or alkaline; the fine texture generally results in poor drainage.

In his monumental text on arid soils, Dregne (1976) identifies five major soil orders common in desert regions: alfisols, aridisols, entisols, mollisols, and vertisols. These categories are distinguished on the basis of such characteristics as soil horizons, texture, base status, mineral accumulations, and organic matter content. Regosols, very rocky soils, are common, especially in the more arid regions. The vertisols and alfisols are more often found in more humid regions. Vertisols are characterized by high clay content and are subject to cracking upon desiccation. They also retain moisture, making them very productive in some regions (Farrar and Nicholson 1994).

Desert soils are most often the sandy soils: aridisols, with well-developed profiles, and entisols, with poorly developed profiles. Collectively these are frequently termed arenosols and comprise almost 80% of the soil surface in arid lands (Table 2.5) and about 24% of the global land surface. Most of the soils in the desert expanses of the Sahara, the Kalahari, and the Australian desert are entisols. The sand content can be as high as 95% in some regions, such as the Kalahari. Sandy soils are, in general, highly permeable, although water movement is not necessarily rapid. These soils tend to moderate the temperature and moisture regime of the subsurface, reducing the intense diurnal fluctuations that characterize dryland regions to near zero below 20 or 30 cm in depth (Cooke and Warren 1975). These conditions provide a better environment for vegetation growth than the heavier soils in the same region, despite their low content of organic matter, and growth can be relatively dense on the arenosols.

In semi-arid and subhumid regions, the most common soils are mollisols. These soils dominate such grassland regions as the Great Plains of North America, the pampas of Argentina and Uruguay, and many Asian steppes. Mollisols are defined by a combination of properties (Strahler and Strahler 1994), the defining one being a very dark surface horizon of considerable thickness (generally 15–30 cm) and loose structure. The dark rich colors of the surface horizon are due to the high accumulation of organic matter. Mollisols are also high in bases and nutrients. Hence they are very productive, especially during abnormally wet years in semi-arid regions.

The diagram in Fig. 2.45 shows a highly idealized progression of various soil characteristics as a function of climate. It must be noted, however, that these are just generalities for "typical" soils in each of the regions; by no means do all soils exhibit these distinctive characteristics. The contrast between humid and arid regions is quite clear. The former are characterized by high clay content, low pH, downward transport of water and materials through infiltration and percolation, a dominance of eluviation (downward transport of fines, both mineral and organic), and, often, hard crusts (like laterites). Organic matter content is low because the humid conditions favor decay and leaching of organics. Since the depth of accumulation of soluble salts is

Table 2.5. *Dominance of various soil orders: area (in millions of km²), percent of arid lands, and percent of world land areas (from Dregne 1976).*

Soil order	Area (million km²)	% of arid lands	% of world land area
Alfisols	3.1	6.6	2.1
Aridisols	16.6	35.9	11.3
Entisols	19.2	41.5	13.1
Mollisols	5.5	11.9	3.7
Vertisols	1.9	4.1	1.3

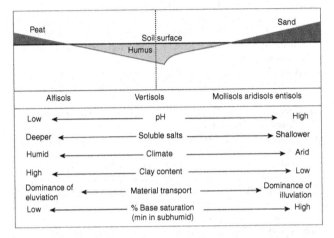

Fig. 2.45 Typical characteristics of principal soil types and relationship to prevailing climate (based on Dregne 1976, Heathcote 1983, and others).

related to the depth of moisture percolation and hence rainfall, the accumulation is quite deep. The base saturation is relatively low, having a minimum in subhumid regions. The arid soils tend to have high pH, upward movement of water, and hence a dominance of illuviation (i.e., deposition). During the brief rains, water penetrates downward but to a limited depth; most of the time it seeps upward by capillary action as the surface dries out, leaving behind soluble salts and bases. Consequently, base saturation is quite high. Like the humid soils, however, arid soils tend to have low organic matter content, since there is little vegetation cover to form humus. They also tend to be very thin and have large accumulations of salt, calcium, and silica close to the surface. Leaching is insufficiently strong to transport bases downward to the water table.

The semi-arid soils, such as the mollisols, represent a balance between the processes in arid and humid region soils: neutral pH, eluviation and illuviation, water movement up and down seasonally, and accumulation of salts and minerals at a moderate depth. They are also rich in organic matter because vegetation is dense enough to produce much organic material but the regions are too dry for rapid decay or leaching. They tend to be fragile, because the rate of erosion is greatest in semi-arid regions (see Figs. 2.18 and 2.19) and because the delicate dynamic equilibrium in soil chemistry and in moisture transport is highly sensitive to anthropogenic impact. The balance is readily upset by irrigation or removal of the natural vegetation, as in replacing deep-rooted savanna species with shallow-rooted crops.

In addition to these soil orders, various types of saline soils are also common in arid regions. Major categories are solonchak – a white alkali soil, and solonetz – a dark or black alkali soil. Solonchaks have a saline horizon of sodium chloride (NaCl), generally at the surface, giving it the characteristic white color. The solonetz contain Na_2CO_3 (sodium carbonate) in the upper layers. The dark color is a result of dissolved organic matter, which is soluble in alkaline solutions of sodium. The salts differ in their solubility, hence when they are transported in soil water or in water evaporated in desert drainage basins, the various types tend to be sequentially deposited, with the most soluble remaining in the solution. This results in characteristic salinity profiles and layering in soils and zonation of salt deposits in playas. The carbonates tend to be least soluble, chlorides most soluble.

In general, saline soils are not productive because they have detrimental effects on vegetation growth; only water availability is a more important factor in growth. The influence is threefold. Salt affects the physical structure of soil by deflocculating soil aggregates; this reduces the porosity and saline soils or horizons become impervious. The osmotic effect of the high salt concentrations opposes that of water in the root zone, increasing soil moisture stress. Finally, the soils affect the nutrient balance and can even be toxic.

REFERENCES

Akin, W. E., 1991: *Global Patterns: Climate, Vegetation and Soils.* University of Oklahoma Press, Norman, OK, 370 pp.

Angel, J. R., M. A. Palecki, and S. E. Hollinger, 2005: Storm precipitation in the United States. Part II: Soil erosion characteristics. *Journal of Applied Meteorology*, **44**, 947–959.

Baas, A. C. W., and D. J. Sherman, 2005: Formation and behavior of aeolian streams. *Journal of Geophysical Research – Earth Surface*, **110**(F3), F03011.

Bagnold, R. A., 1941: *The Physics of Wind-Blown Sand and Desert Dunes.* Methuen, London, 265 pp.

Bagnold, R. A., 1953: The surface movement of blown sand in relation to meteorology. In *Desert Research*, Research Council of Israel Special Publication 35, Research Council of Israel and UNESCO, pp. 23–32.

Belly, Y.-P., 1964: *Sand Movement by Wind.* Technical Memorandum 1. U. S. Army Corps of Engineers, Coastal Engineering Research Center, Vicksburg, MS, 38 pp.

Beveridge, C., and Coauthors, 2006: Development of spatially diverse and complex dune-field patterns: Gran Desierto Dune Field, Sonora, Mexico. *Sedimentology*, **53**, 1391–1409.

Birkeland, P. W., 1984: *Soils and Geomorphology.* Oxford University Press, New York.

Bull, W. B., 1997: Discontinuous ephemeral streams. *Geomorphology*, **19**, 227–276.

Bulygina, N. S., M. A. Nearing, J. J. Stone, and M. H. Nichols, 2007: DWEPP: a dynamic soil erosion model based on WEPP source terms. *Earth Surface Processes and Landforms*, **32**, 998–1012.

Carson, M. A., and M. J. Kirkby, 1972: *Hillslope Form and Process.* Cambridge Geographical Studies. Cambridge University Press, Cambridge, UK, 484 pp.

Cooke, R. U., and A. Warren, 1975: *Geomorphology in Deserts.* Batsford, London, 394 pp.

Cooke, R. U., A. Warren, and A. Goudie, 1993: *Desert Geomorphology.* UCL Press, London, 526 pp.

Corbel, J., 1964: L'érosion terrestre, étude quantitative. *Annales de Géographie,* **73,** 385–412.

Corbett, I., 1993: The modern and ancient pattern of sandflow through the southern Namib deflation basin. In *Aeolian Sediments: Ancient and Modern* (K. Pye and N. Lancaster, eds.), International Association of Sedimentologists Special Publication 16, Blackwell, Oxford, pp. 45–60.

de Martonne, E., and L. Aufrère, 1927: Regions of inter-basin drainage. *Geographical Review,* **17,** 397–414.

Dixon, J. C., 1994: Aridic soils, patterned ground and desert pavements. In (A. D. Abrahams and A. J. Parsons, eds.) *Geomorphology of Desert Environments.* Chapman and Hall, London.

Dregne, H., 1976: *Soils of Arid Regions.* Elsevier, Amsterdam.

Dunne, T., and L. B. Leopold, 1978: *Water in Environmental Planning.* W. H. Freeman, New York, 818 pp.

Farrar, T. J., and S. E. Nicholson, 1994: The influence of soil type on the relationships between NDVI, rainfall and soil moisture in semi-arid Botswana. Part I. Response to rainfall. *Remote Sensing of Environment,* **50,** 107–120.

Fryberger, S. G., and T. S. Ahlbrandt, 1979: Mechanisms for the formation of eolian sand seas. *Zeitschrift für Geomorphologie,* **23,** 440–460.

Fryberger, S. G., and A. S. Goudie, 1981: Arid geomorphology. *Progress in Physical Geography,* **5,** 420–428.

Gillette, D. A., 1981: Production of dust that may be carried great distances. *Geological Society of America,* **186,** 11–26.

Goudie, A., and J. Wilkinson, 1980: *The Warm Desert Environment.* Cambridge University Press, New York, 88 pp.

Heathcote, R. I., 1983: *Arid Lands: Their Use and Abuse.* Longman, London.

Jenny, H., 1941: *Factors of Soil Formation.* McGraw-Hill, New York.

Kimberlin, L. W., A. L. Hidelbaugh, and A. R. Grunewald, 1977: The potenteial wind erosion problem in the United States. *Transactions of the ASAE,* **20,** 873–879.

Kirkby, M. J., 1969: Erosion by water on hillslopes. In *Water, Earth and Man* (R. J. Chorley, ed.), Methuen, London, pp. 229–238.

Kocurek, G., and R. C. Ewing, 2005: Aeolian dune field self-organization: implications for the formation of simple versus complex dune-field patterns. *Geomorphology,* **72,** 94–105.

Lancaster, N., 1980: The formation of seif dunes from barchans: supporting evidence for Bagnold's model from the Namib Desert. *Zeitschrift für Geomorphologie NF,* **24,** 160–167.

Lancaster, N., 1981: Aspects of the morphology of linear dunes of the Namib desert. *South African Journal of Science,* **77,** 366–368.

Lancaster, N., 1982: Spatial variations in linear dune morphology and sediments in the Namib sand sea. *Palaeoecology of Africa,* **15,** 173–182.

Lancaster, N., 1983: Controls on dune morphology in the Namib sand sea. In *Eolian Sediments and Processes* (M. E. Brookfield and T. S. Ahlbrandt, eds.), Elsevier, Amsterdam, pp. 261–289.

Lancaster, N., 1999: Geomorphology of desert sand seas. In *Aeolian Environments, Sediments and Landforms* (A. S. Goudie, I. Livingstone, and S. Stokes, eds.), Wiley and Sons, Chichester, UK, pp. 49–69.

Lancaster, N., 2007: Low latitude dune fields. In *Encyclopedia of Quaternary Science* (S. A. Elias, ed.), Elsevier, Amsterdam, pp. 626–642.

Lancaster, N., W. G. Nickling, C. K. M. Neuman, and V. E. Wyatt, 1996: Sediment flux and airflow on the stoss slope of a barchan dune. *Geomorphology,* **17,** 55–62.

Langbein, W. B., and S. A. Schumm, 1958: Yield of sediment in relation to mean annual precipitation. *Transactions – American Geophysical Union,* **39,** 1076–1084.

Leenders, J. K., J. H. van Voxel, and G. Sterk, 2005: Wind forces and related saltation transport. *Geomorphology,* **71,** 357–372.

Lettau, K., and H. H. Lettau, 1969: Bulk transport of sand by the barchans of the Pampa de Joya in Southern Peru. *Zeitschrift für Geomorphologie NF,* **13,** 182–195.

Lettau, H. H., and K. Lettau, 1978: *Exploring the World's Driest Climate.* Institute for Environmental Studies, University of Wisconsin, Madison, WI, 264 pp.

Leys, J., 1999: Wind erosion on agricultural lands. In *Aeolian Environments, Sediments and Landforms* (A. S. Goudie, I. Livingstone, and S. Stokes, eds.), Wiley and Sons, Chichester, UK, pp. 143–166.

Livingstone, I., 1989: Monitoring surface change on a Namib linear dune. *Earth Surface Processes and Landforms,* **14,** 317–332.

Livingstone, I., 1990: Desert sand dune dynamics: review and prospect. In *Namib Ecology* (M. K. Seely, ed.). Pretoria, Transvaal Museum, pp. 47–53.

Livingstone, I., 2003: A twenty-one year record of surface change on a Namib linear dune. *Earth Surface Processes and Landforms,* **28,** 1025–1031.

Livingstone, I., and W. G. Nickling, 2004: Preface: Aeolian research. *Geomorphology,* **59,** 1–2.

Livingstone, I., and D. S. G. Thomas, 1993: Modes of linear dune activity and their palaeo-environmental significance: an evaluation with reference to southern African examples. In *The Dynamics and Environmental Context of Aeolian Sedimentary Systems* (K. Pye, ed.), Geological Society Special Paper 72, pp. 91–101.

Livingstone, I., and A. Warren, 1996: *Aeolian Geomorphlogy: An introduction.* Longman, Singapore.

Mabbutt, J. A., 1979: *Desert Landforms.* MIT Press, Cambridge, MA, 340 pp.

Menenti, M., 1984: *Physical Aspects and Determination of Evaporation in Deserts Applying Remote Sensing Techniques.* Institute for Land and Water Management, Wageningen, The Netherlands, 202 pp.

Motts, W. S. (ed.), 1970: *Geology and Hydrology of Selected Playas in Western United States.* University of Massachusetts, Amherst, MA.

Mutua, B. M., A. Klik, and W. Loiskandl, 2006: Modelling soil erosion and sediment yield at a catchment scale: the case of Masinga catchment, Kenya. *Land Degradation and Development,* **17,** 557–570.

Newman, B. D., E. R. Vivoni, and A. R. Groffman, 2006: Surface water–groundwater interactions in semiarid drainages of the American southwest. *Hydrological Processes,* **20,** 3371–3394.

Oke, T. R., 1987: *Boundary Layer Climates.* 2nd ed., Halsted, New York.

Pearce K. I., I. J. Walker, 2005: Frequency and magnitude biases in the 'Fryberger model', with the implications for characterizing geomorphically effective winds. *Geomorphology,* **68,** 39–55.

Prospero, J. M., P. Ginoux, O. Torres, and S. E. Nicholson, 2002: Global soil dust sources. I. Environmental characterization. *Journal of Geophysical Research,* **40(2).**

Pye, K., 1987: *Aeolian Dust and Dust Deposits.* Academic Press, London, 334 pp.

Ravi, S., P. d'Ororico, T. M. Over, and T. M. Zobeck, 2004: On the effect of air humidity on soil susceptibility to wind erosion: the case of air dry soils. *Geophysical Research Letters,* **31,** doi:10.1029/2004GL019485.

Roof, S., and C. Callagan, 2003: The climate of Death Valley, California. *Bulletin of the American Meteorological Society,* **84,** 1725–1739.

Saqqa, W., and M. Atallah, 2004: Characterization of the Aeolian terrain facies in Wadi Araba Desert, southwestern Jordan. *Geomorphology,* **62,** 63–87.

Schlesinger, W. H., 1985: The formation of caliche in soils of the Mojave Desert, California. *Geochimica et Cosmochimica Acta*, **49**, 57–66.

Schönfeldt, H.-J., and S. von Löwis, 2005: Turbulence-driven saltation in the atmospheric surface layer. *Meteorologische Zeitschrift*, **12**, 257–268.

Sharon, D., A. Margalit, and S. M. Berkowicz, 2002: Locally modified surface winds on linear dunes as derived from directional rain-gauges. *Earth Surface Processes and Landforms*, **27**, 867–889.

Sperling, C. H. B., and R. U. Cooke, 1985: Laboratory simulation of rock weathering by salt crystallization and hydration processes in hot arid environments. *Earth Surface Processes and Landforms*, **10**, 541–555.

Strahler, A. H., and A. N. Strahler, 1994: *Introducing Physical Geography*, Wiley and Sons, New York, 535 pp.

Tseo, G., 1993: Two types of longitudinal dune fields and possible mechanisms for their development. *Earth Surface Processes and Landforms*, **18**, 627–643.

Tsoar, H., 1983: Dynamic processes acting on a longitudinal (seif) dune. *Sedimentology*, **30**, 567–578.

Tsoar, H., 1989: Linear dunes: forms and formation. *Progress in Physical Geography*, **13**, 508–528.

Tsoar, H., D. G. Blumberg, and Y. Stoler, 2004: Elongation and migration of sand dunes. *Geomorphology*, **57**, 293–302.

Tucker, G. E., and R. L. Bras, 1998: Hillslope processes, drainage density, and landscape morphology. *Water Resources Research*, **3**(4), 2751–2764.

U.S. Department of Agriculture, 2003: Agricultural Research Service, National Sediment Laboratory (USDA-ARS-NSL). RUSLEI.06c and RUSLE2.

Viles, H. A., 2005: Microclimate and weathering in the central Namib Desert, Namibia. *Geomorphology*, **67**, 189–209.

Walker, I. J., and W. G. Nickling, 2003: Simulation and measurement of surface shear stress over isolated and closely spaced transverse dunes in a wind tunnel. *Earth Surface Processes and Landforms*, **28**, 1111–1124.

Wang, X. M., Z. B. Dong, L. C. Liu, and H. J. Qu, 2004: Sand sea activity and interactions with climatic parameters in the Taklimakan Sand Sea, China. *Journal of Arid Environments*, **57**, 225–238.

Warke, P. A., 2000: Micro-environmental conditions and rock weathering in hot arid regions. *Zeitschrift für Geomorphologie*, **120**, 83–95.

Wiggs, G. F. S., I. Livingstone, and A. Warren, 1996: The role of streamline curvature in sand dune dynamics: evidence from field and wind tunnel measurements. *Geomorphology*, **17**, 29–46.

Wilson, I. G., 1973: Ergs. *Sedimentology*, **10**, 77–106.

Wischmeier, W. H., and D. D. Smith, 1978: *Predicting Rainfall Erosion Losses. A Guide to Conservation Planning*. Agriculture Handbook No. 537. USDA-SEA, US Government Printing Office, Washington, DC, 58 pp.

Wood, A., 1942: The development of hillside slopes. *Proceedings of the Geologists' Association*, **53**, 128–138.

Woodruff, N. P., and Siddoway, F. H., 1965: A wind erosion equation. *Proceedings – Soil Science Society of America*, **29**, 602–608.

Yizhaq, H., 2008: Aeolian megaripples: mathematical model and numerical simulations. *Journal of Coastal Research*, **24**, 1369–1378.

Yizhaq, H., Y. Ashkenazy, and H. Tsoar, 2008: Why do active and stabilize dunes co-exist under the same climatic conditions? *Phys. Rev. Lett.*, **98**, 188001.

3 Vegetation of the dryland regions

3.1 OVERVIEW OF DRYLAND VEGETATION

3.1.1 VEGETATION TYPES/CLASSIFICATION

Vegetation and climate are so fundamentally linked that the first climate classification systems were actually based on vegetation. In one sense vegetation is a response to climate, but it is far from a passive end-product. Rather, there are complex interactions and feedbacks between vegetation and climate, the intricate processes of the global energy, mass and water balances. On the space scales associated with climate, the most meaningful categories of vegetation are based on such concepts as life forms and biomes. These concepts are closely linked, both relating to the physical structure of the vegetation.

The principal life forms of plants are *trees, shrubs, lianas,* and *herbs*. Both trees and shrubs are woody and erect; lianas and herbs are not. Trees have a single, upright main trunk with branching in the upper part to form a crown. The foliage is concentrated in the crown. A shrub consists of several stems branching near the ground, such that foliage is concentrated in a mass starting close to the ground. Lianas are woody vines that climb on trees. Herbs, consisting of grasses and forbs (broad-leaved herbs), are usually small and lack woody stems.

Vegetation is also classified into a series of biomes – assemblages of plants with characteristic life forms. A biome consists of all those plant communities that have similar ecological functions in terms of such processes as primary production and nutrient recycling (Odum 1971). Ecosystems of the same biome will tend to look similar but may have no similarity of species composition. Worldwide, 13 terrestrial biomes are generally recognized (Scholes 1990a) (Table 3.1). It must be emphasized that there are no fixed or precise limits to the biomes; instead one gradually grades into another and boundaries shift over time.

The major dryland biomes include *savanna, grassland*, and *desert*. Several others are important in arid regions, such as

the karoo and fynbos biomes of southern Africa (Ellery *et al.* 1991). Succulents (plants which can impound water in their leaves, roots, and/or stems) are important components of arid ecosystems. These plants can fall into nearly any of the major dryland biomes.

"Savanna" is a term widely used with a number of different connotations. In general, it is a landscape with a combination of trees, shrubs, and grasses in various proportions. The term is reserved for tropical ecosystems. The wetter savanna habitats, representing the transition to forest, are termed woodlands. A grassland is an assemblage of herbaceous species, but this biome can include trees in wetter parts of the habitat. Generally, tropical grasslands are included in the savanna classification. In the higher latitudes, two types of grassland are distinguished, the tall-grass prairie and the short-grass steppe.

The desert is an environment of thinly dispersed plants with a large proportion of bare ground. Growth is water-limited in most cases. Semi-desert environments have a few widely scattered trees, but the true desert consists of species especially adapted to dry conditions plus occasional shrubs.

A rough correspondence exists between biomes and climate. This is readily seen through a comparison of the global pattern of vegetation (Fig. 3.1) with any map of climate distribution. Shmida (1985) generalizes this relationship in terms of macro-gradients, describing a north–south progression in which the winter rains diminish and the summer rains increase, and showing associated sequences of vegetation types along this gradient. Moving southward, from the subtropics of the Northern Hemisphere, rainfall shifts from the winter season to negligible then to the summer season. Correspondingly, the vegetation shifts from sclerophyll forest to steppe to desert, then to scrub/grassland. As the summer rainy season progressively lengthens and gives way to year-round rainfall, the grassland gives way to savanna, woodland and, finally, forest.

Biomes can be delineated, even on a small scale, when either temperature or net radiation is used in conjunction with rainfall (Ellery *et al.* 1991; Budyko 1986). Whittaker (1975) delineates the major global biomes using only mean annual precipitation and mean annual temperature (Fig. 3.2). The grasslands and savannas exist where mean annual rainfall is on the order of a few hundred millimeters to about 1500 mm (Ripley 1992), with the range more restricted and the limits lower in higher latitudes. Conditions of temperature or radiation generally determine which type of savanna or grassland prevails.

Budyko (1986) emphasizes the importance of radiation, as opposed to temperature, and uses net radiation and a "dryness ratio" based on net radiation and rainfall to delineate the major biomes (Fig. 3.3). Based on the link with water balance, he also calculates typical runoff values for these ecosystems.

Table 3.1. *Thirteen major terrestrial biomes (from Olson* et al. *2001).*

Tropical and subtropical moist broadleaf forests
Tropical and subtropical dry broadleaf forests
Tropical and subtropical coniferous forests
Temperate broadleaf and mixed forests
Temperate conifer forests
Boreal forests/taiga
Tropical and subtropical grasslands, savannas and shrublands
Temperate grasslands, savannas and shrublands
Flooded grasslands and savannas
Montane grasslands and shrublands
Tundra
Mediterranean forests, woodlands, and scrub
Deserts and xeric shrublands

Recent satellite data show that Budyko's net radiation estimates are not very accurate (see Chapter 7), so that the limits he indicates are not valid, although his dryness concept is. Recalculations for African ecosystems show that they clearly fall along the dryness ratio/radiation axes (Fig. 3.4), but the dryness ratio that delineates biomes is strongly influenced by net radiation.

The associations described above are highly idealized. In reality, the earth's drylands contain widely varying proportions of trees, shrubs, succulents, herbs, and grasses and they differ greatly with respect to the number and variety of species. Tree species are more dominant in tropical dryland areas, while low, woody shrubs and succulents characterize temperate-latitude deserts, such as those in the Americas and Central Asia. Grasses typically cover the semi-arid regions, but the Australian desert is also particularly rich in grass species.

3.1.2 GENERAL ASPECTS OF DRYLAND VEGETATION

The dryland ecosystems exist along a continuum that ranges from forest cover to bare ground (Meron *et al.* 2007). The position along this continuum affects many ecosystem properties, including near-ground energy input, water balance, erosion rates, and nutrient cycling. For a given site, its position depends in large part on soil moisture availability which, along with rainfall, generally decreases along the continuum (Meron *et al.* 2004).

Dryland environments differ greatly in the amount of biomass, productivity, characteristic species, variety and richness of these species, percentage surface cover, and size and variety

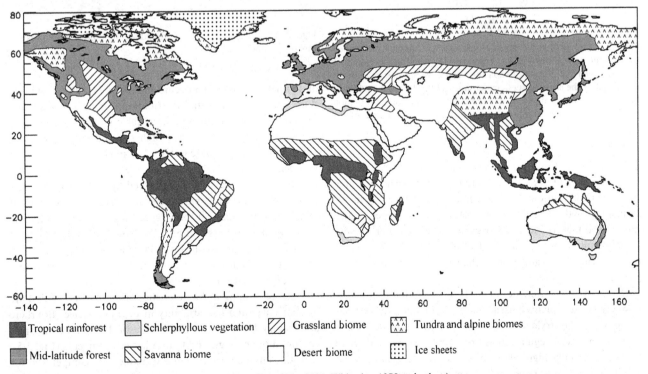

Fig. 3.1 Global vegetation patterns (based on Strahler and Strahler 1992, Whittaker 1975 and others).

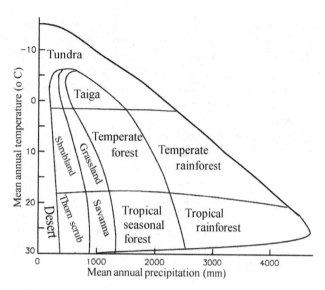

Fig. 3.2 Relation of world biome types to mean annual temperature and mean annual precipitation (modified from Whittaker 1975).

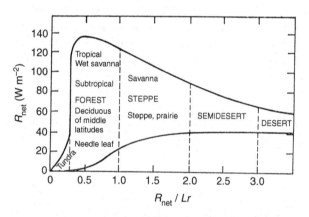

Fig. 3.3 Budyko's (1986) concept of geobotanical zonality: the principal ecosystems as a function of net radiation (R_{net}) and the dryness ratio (ratio between annual net radiation and annual average precipitation (r) multiplied by the latent heat of condensation (L)). The upper and lower bounds of the figure are thought to represent the limits of net radiation conditions found in natural environments. Runoff as a function of net radiation and the dryness ratio is also shown.

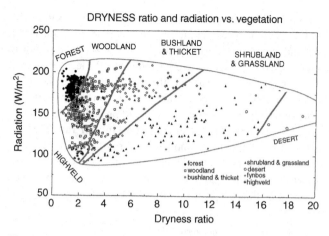

Fig. 3.4 Net radiation versus the dryness ratio for various African biomes, based on roughly 1300 stations.

Fig. 3.5 Large amounts of bare ground are an almost ubiquitous feature of drylands.

of life forms. This diversity reflects, in part, the diversity of dryland climates. But dryland vegetation is even more diverse than the climatic conditions because climate is not the only factor in plant growth. The distribution of vegetation depends also on physical factors, such as topography, surface materials, and local hydrology; on evolutionary factors such as climatic change or continental drift; and even on the vegetation in surrounding regions.

Distinctive characteristics of dryland vegetation do emerge, however, because plants must adapt to environmental conditions common to the drylands. These include moisture deficiency, fluctuating and irregular moisture supply, extreme thermal stress and, usually, high salinity of the water and ground. More specifically, the factors controlling vegetation growth are: low

and variable rainfall that occurs with irregular timing and random but restricted spatial distribution; highly seasonal rainfall and therefore a limited and variable growing season; high evaporation and the resulting high salt concentrations; large vertical movements of the water table, and extreme fluctuations between summer or daytime heat and winter or nocturnal cold, the temperature range being enhanced by high insolation, low ground moisture, and sparse vegetation. Thus, the rhythm of plant growth is adapted to cycles of heating and cooling and of moistening and desiccation of the soils, cycles that fluctuate markedly in timing and intensity. Plant forms must develop tolerance to moisture deficiency, thermoregulatory mechanisms, and protection against excess salt.

The most common characteristics of dryland vegetation are its sparseness and its variability in time and space (Bourlière and Hadley 1983; Goudie and Wilkinson 1980). Dryland areas contain more bare ground than other environments (Fig. 3.5), with this characteristic serving to reduce the competition for soil moisture. However, ground cover varies considerably

Table 3.2. *Biomass in various vegetation zones (Goudie and Wilkinson 1980).*

Zone	Biomass (kg ha^{-1})
Tropical rainforest	100+
Broad-leaved temperate forest	74–82
Northern taiga spruce	20
Savanna	14
Dry savanna	4–6
Steppe grassland	5
Dry steppes	4
Semi-shrub desert	.8 -.9
Subtropical desert	.2-.3

among the dryland regions because it depends not only on available moisture, but also on the dominant flora, being larger with grasses, for example, than with shrubs or succulents. The growth of most dryland species is seasonal or episodic, with plant life and life processes adjusting to the water supply. In true deserts, for example, most plants are annuals or ephemerals. The former respond to a regular annual cycle of rainfall, whereas ephemerals grow during irregular periods when moisture is available.

Both coverage and biomass fluctuate in response to varying moisture during the year or over longer periods. This is particularly true for ephemeral vegetation (see Section 3.3). A brief shower can turn barren ground into a field of verdure. Interannual changes in rainfall, even small ones, similarly alter the surface. In more arid regions, a series of dry years may leave a dry, sandy, and brittle soil that the rains of a humid year transform into a field of grass and flowers. During a wet year in the Namib, following a long sequence of dry years, desert biomass increased nearly tenfold (Seely and Louw 1980).

A number of generalizations are often made concerning biomass, productivity, root systems, life forms, and species diversity of dryland vegetation, but most are true only to a limited extent (Shmida 1985). For example, biomass (amount of living plant material above and below ground) and primary productivity (annual production of biomass) are frequently assumed to be low in deserts because both increase sharply with increasing moisture availability (Table 3.2). However, these characteristics vary considerably among the world's drylands and even within a desert province. Desert productivity and biomass are also highly variable in time, primarily because of rainfall variability. Thus, at certain times and places, desert productivity and biomass might be within the same range as grasslands, shrublands, and even woodlands (Noy-Meir 1985).

Desert plants tend to have unusually extensive root systems, so that the ratio of below- to above-ground biomass is high, particularly in cold deserts. For water-limited ecosystems in general, plants tend to have deep roots, and root depth tends to increase with decreasing moisture (Guswa 2008). Unlike the root systems in wetter environments, root depth is more sensitive to rainfall intensity than to rainfall frequency.

Species diversity also varies widely among the world's deserts. It is generally highest in semi-arid regions, decreasing with rainfall as one moves to the semi-desert then desert (Noy-Meir 1985). The North American deserts contain at least 48 grass species, as well as numerous types of herbaceous and succulent species and low-growing woody and semi-woody plants, including the giant cactus, the mesquite tree and yucca plant, and short-lived multicolored wildflowers. The Sonoran Desert alone, an area of 310,000 km^2, supports a flora of about 2500 species of seed plants (Shreve and Wiggins 1964).

Throughout the dryland regions there are a number of common families of flora. The most important is *Chenopodiaceae* (Shmida 1985). These plants are widespread in the Irano-Turanian region, the Sahara, and the deserts of North and South America and Australia. There are also numerous genera common to several different dryland regions, such as *Artemisia* (sagebrush) and the *Aristida* and *Stipa* grasses, but the species found in dryland habitats are extremely divergent. Mesquite, creosote shrubs, and giant cactus species inhabit both the Sonoran Desert and the Monte Desert of Argentina, regions with similar soils and climate. Overall, however, only 5% of the species found in the Monte overlap with the plant life of the Sonoran Desert. Other shared genera include *Acacia*, common throughout the Old World, and *Prosopis* (a tree and shrub group that includes mesquite), which replaces it in North and South America.

3.1.3 DESERT VEGETATION

Along the gradient of decreasing rainfall, the savanna and grassland environments gradually give way to semi-deserts and true deserts. The transition is gradual and vegetation of the transition zone may resemble the drier parts of the semi-arid environments. The dominant vegetation in most deserts includes dwarf and low-growing shrubs (such as *Artemisia* or sagebrush) and succulents (cacti and *Euphorbia* species).

Desert vegetation consists of two general classes, based on the primary method of resisting moisture stress: perennials, which *avoid* drought, and annuals or ephemerals, which *evade* it (Goudie and Wilkinson 1980; Petrov 1976). Perennials adapt to the average climatic conditions and their seasonal cycles; while annuals and ephemerals grow profusely during periods of favorable precipitation and produce seeds that lie dormant during drier periods, until wetter conditions return.

The perennials are commonly dwarf and woody or succulent (water-storing). They include phreatophytes, which develop long taproots penetrating downward to the water table, and xerophytes, which adapt to low water supply and high salinity. Typical phreatophytes are the date palm, tamarisk, and mesquite. Roots often extend 10–15 m or more below ground, and for the mesquite 50 m is not unusual. An ancient mesquite found in Argentina was 10 m high with an 80 m root system (Stone and Kalisz 1991). Phreatophytes often seek wetter habitats, near stream channels and wadis, springs, or lakeshores. Xerophytes arrange their life cycles and processes to suit drought conditions of varying lengths. One class resists drought either by reducing their water requirements or by impounding water in leaves, roots, and stems. The latter, termed succulents, are typified by the cacti of the American deserts, but include many other plants as well. The second class of xerophytes is drought-tolerant; these plants survive long periods of dehydration by spending them in the vegetative stage. These types, which are relatively uncommon, look brown and dead when devoid of water.

An additional category of dryland vegetation is the halophytes or salt-tolerant species. These have evolved in such a way that high salinity, which is detrimental to most plants, stimulates growth (Petrov 1976). The salt content of leaves may reach 45% of their dry weight, increasing osmotic pressure in the cells and retarding transpiration. Common halophytes include the salt-bushes and saltworts of the American, Asian and Australian deserts. Salt being vital, they grow on saline soils, in playas, and on the edge of saline lakes.

3.1.4 THE SAVANNAS

Savanna environments are characterized by the coexistence of woody species and grasses. Structurally, the vegetation (Fig. 3.6) is composed of an open stratum of trees and/or shrubs of variable height and density and a ground layer of grasses (Cole 1986). These environments frequently mark the transition between humid and dry climates. Extreme events such as droughts and floods are inherent constraints on their existence. The savanna ecosystem is of global importance because the type and pattern of woody cover affects streamflow and groundwater recharge, biophysical interaction between vegetation and the atmosphere, carbon source–sink relationships, and tropospheric chemistry. The world's savannas are thought to be the regions most susceptible to impending global climate change (Breshears and Barnes 1999), with the potential for irreversible, catastrophic change (Newman *et al.* 2006).

The classic factors in the savanna ecosystem are herbivory, fire, nutrients, and water (Meyer *et al.* 2007). The relative importance of these factors has long been debated. Geographers once

Fig. 3.6 Typical savanna in Niger, with three structural elements (grasses, shrubs, and trees).

thought that savannas were a man-made environment, primarily an end-product of fires. It is now thought that some savannas are, indeed, anthropogenic in origin, but that others are natural vegetation biomes. Scholes and Walker (1993) consider water and nutrients to be primary factors in the existence of savannas, with disturbances such as fire and herbivory playing a secondary role. Overall, the type of savanna, its vegetation associations, and the species within it are determined by complex interactions of factors and overall the system is in a state of dynamic equilibrium. The core regions of each category are relatively stable, but the peripheries are "tension zones" where the environment shifts markedly in response to drought or exceptional rains or other stresses.

Savannas (Fig. 3.7) cover roughly one-eighth of the global land surface (Scholes and Archer 1997), predominantly in regions of semi-arid climate. Cole (1986) estimates that savannas cover about 23 million km² and occupy about 20% of the global land surface. Savannas cover about 65% of Africa, 60% of Australia, 45% of South America, and about 10% of India and Southeast Asia.

No precise numbers can be given because there is no universally accepted definition of the term savanna and its common usage varies. The term is most often associated with grasslands in South America, with open woodlands and shrublands in Australia and Asia, and with parklands and low tree and shrub savanna in Africa. Many authors include tropical grasslands in the savanna category (Cole 1986). While the savannas are very diverse in species and form, they share structural and functional characteristics. They also exhibit a distinctive seasonal cycle of growth in response to the seasonality of moisture availability; a characteristic that allows the vegetation to tolerate the seasonal drought.

The seasonal rainfall regime, with moisture deficiency and plant stress during the cooler season, is the only commonality of the physical environments of savannas. The annual dry season is of varying length and intensity within the various savannas, ranging from about 3 or 4 months in the wetter regions to about 8 or 9 months in the drier regions. The dry season is of sufficient duration and intensity to cause woody plants to shed their leaves and grasses to dry out (Nix 1983).

In most savannas, conditions of relatively low cloudiness, moderate temperatures and high insolation (140–190 kcal/cm² per year) favor photosynthesis and plant growth; limiting factors are water deficiency and either inadequate or excessive drainage. Mean maximum temperatures of the warmest months are in the range of 30–35°C near the desert margin, 25–30°C at the forest boundary. Temperatures of the cooler months are on the order of 13–18°C and 8–13°C along these same boundaries. Potential evapotranspiration is generally on the order of 1000–1500 mm annually, but can exceed 2000 mm in some areas.

The savanna environment is subdivided on the basis of canopy cover; size, spacing, and arrangement of woody elements; and the degree of dominance of trees or grasses. Generally five or six classes are distinguished: *savanna woodland, savanna parkland, savanna grassland, low tree and shrub savanna*, and *thicket and scrub* (Table 3.3). Some classification

Table 3.3. *Classification of savanna biomes (Cole 1986).*

Savanna woodland	Deciduous and semi-deciduous woodland of tall trees (>8 m high) and tall mesophytic grasses (>80 cm high); spacing of the trees more than a diameter of the canopy
Savanna parkland	Tall mesophytic grassland (grasses 40–80 cm high) with scattered deciduous trees (<8 m high)
Savanna grassland	Tall tropical grasslands without trees or shrubs
Low tree and shrub savanna	Communities of widely spaced low-growing perennial grasses (<80 cm high) with abundant annuals and widely spaced, low-growing trees and shrubs
Thicket and scrub	Communities of trees and shrubs without stratification

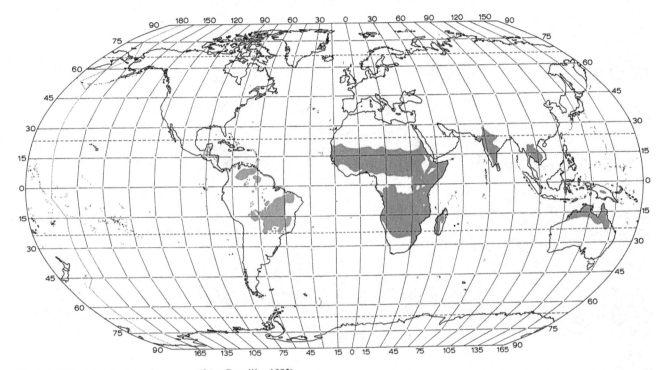

Fig. 3.7 Global distribution of savannas (from Bourlière 1983)

systems differentiate between tree savanna and shrub savanna. Savanna parkland is a tree–grass landscape in which circular clumps or groves of woody plants are dispersed throughout a grassy matrix (Scholes and Archer 1997). Thicket is a very dense growth of small trees and shrubs. Scrubland is predominantly tall and relatively dense shrubs. Figure 3.8 shows some examples of the various types of savanna.

Climatic and edaphic (soil-related) factors (Fig. 3.9), as well as environmental history, control the distribution of the five categories. The most important factors are water availability (largely a function of rainfall and soil texture) and nutrient availability, but other soil characteristics, such as pH and organic matter content, also play a role. Walter and Breckle (1999) thus distinguish climatic savannas and edaphic savannas, based on the factors controlling their distribution, and secondary savannas, controlled by factors such as fire, large mammals, and anthropogenic interference.

A more common distinction is the simple dichotomy between moist and dry savannas. Typically, the rainy season is 5–7 months in the latter, but 7–9 months in the former. The moist savanna is generally associated with nutrient-poor (i.e., dystrophic) soils, with nutrients being the limiting factor in growth. The dry savanna typically has nutrient-rich (eutrophic) soils and water is generally the limiting factor. Scholes (1990b) suggests that wet and dry savannas be distinguished instead on the basis of whether or not they are water-limited. The dividing criterion would be where the strong linear dependence of annual herbaceous production on annual rainfall begins to fall off, which generally occurs at about 700 mm on sandy soils and 900 mm on finer-textured soils. In general, the woodlands would be classified as moist/dystrophic, the parklands and low tree and shrub savanna as dry/eutrophic, and the savanna grasslands, thicket and shrub savannas as either.

The broadest contrast is between the savanna woodlands, on relatively infertile, highly leached soils at the moist end of the spectrum, and the low tree and shrub savanna on unleached, nutrient-rich soils at the more arid margins. Climate, notably

Fig. 3.8 Examples of the diversity of savanna environments, including woodland.

the lengthening dry season and greater extremes of temperature, determines the broad gradation between woodlands and low tree and shrub savanna. On a continental scale, the woodlands extend roughly from the forest, where mean annual rainfall is on the order of 1500–2000 mm and the winter dryness and coolness become negligible, to areas with annual rainfall of about 500 mm concentrated in the summer half-year. The low tree and shrub savanna commences there and extends to the desert margin, where rainfall is generally less than 250 mm and occurs erratically, often without preferred seasonality.

3.1.5 MID-LATITUDE GRASSLANDS

Like the savannas, mid-latitude grasslands (Fig. 3.10) mark the transition between arid and humid climates. The vegetation is herbaceous, primarily grasses and forbs, with grasses being dominant. Several categories of grassland are distinguished by various authors, using criteria such as climate, height of dominant elements, or species composition, but all are to some extent arbitrary. A common distinction is between regions of tall grasses, termed prairies, and those with shorter grass, commonly called steppes. The latter occupy drier and/or colder environments. Estimates of the extent of grassland are quite varied, ranging from about 6% to 25% of the global land surface.

The mid-latitude grasslands generally occupy areas where mean annual rainfall is between about 300 and 1000 mm and the dry season ranges from 0 to 8 months. In the drier steppes, potential evapotranspiration exceeds rainfall most of the year, but in the prairies, long periods of moisture surplus occur within the year. Some grasslands occupy more humid regions where soils or drainage dictate their occurrence.

Grasslands (Fig. 3.11) are well developed in North America, occupying most of the Great Plains, with prairie in the more humid eastern portions and steppe in the west (Coupland 1992). A long expanse of steppe extends from the Ukraine of

European Russia east into Siberia, spanning some 50° of longitude across Asia. There is also extensive steppe in Mongolia and northern China (Singh *et al.* 1983). South America also has extensive grasslands, which are termed *pampas* in drier regions and *campos* in regions of more humid climate (Soriano 1992). Those in South Africa lie in the summer rainfall region in the eastern parts of the country (Scholes 1991).

The mid-latitude grasslands differ from the low-latitude savanna grasslands in several fundamental ways. One is the relatively harsh thermal stress of the cold season and the large annual temperature range, summers being relatively hot. Another is the occurrence of a proportion of the annual rainfall during the winter season, making it more effective than the rainfall in savanna grasslands. Additionally the grasses of the temperate latitudes tend to follow the C4 photosynthetic pathway (see Section 3.2.2). Those of the savanna generally follow the C3 pathway, which favors the high light intensity and temperature of the low-latitude environment (see Section 3.2.2). Consequently, the optimum conditions for growth differ between the mid-latitude steppes and prairies and the savanna grasslands. The temperate species have optimum growth at about 20°C, but growth continues with temperatures as low as about 5–10°C. By comparison, for tropical grass species growth is optimum at about 30–35°C and it is limited by temperatures less than about 15°C (Singh *et al.* 1983).

3.1.6 MEDITERRANEAN FOREST

Like climates in the savanna environment, the climates on the poleward side of the subtropical deserts have a strong seasonal

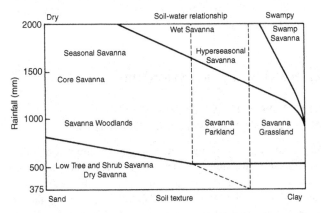

Fig. 3.9 Climatic and edaphic factors associated with various types of savannas (from Cole 1986, copyright Elsevier).

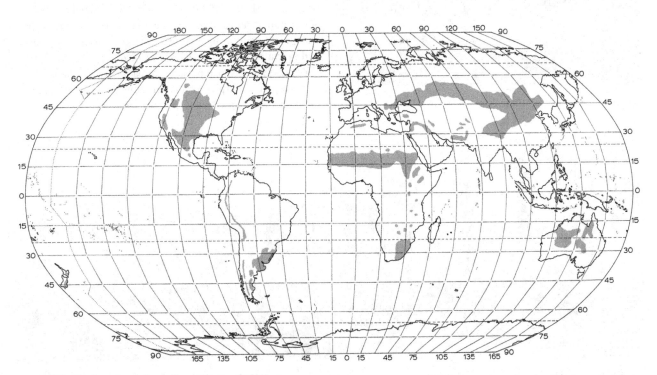

Fig. 3.10 Distribution of grasslands (from Coupland 1992).

cycle in moisture availability. In this case, however, precipitation occurs primarily during the cooler winter season. Moisture stress is less extreme in winter-rainfall regions than in the summer-rainfall savannas, because less water is lost to evaporation. These climates are best developed in the Mediterranean region, and hence the term "Mediterranean" is globally applied to the wet winter–dry summer climate and to the typical dominant vegetation of these regions. The Mediterranean forest also occurs in a number of west-coastal regions of continents, such as in central and southern California, coastal Chile, and the winter-rain Cape Province of South Africa, and this ecosystem is well developed throughout southern Australia.

The Mediterranean climate differs from that of the savannas in other ways. As a consequence of the dry summer, the moisture stress is coupled with the thermal stress of the hot season. On the other hand, most areas of Mediterranean climate are reasonably close to a coast, so that the climate is overall fairly moderate, with mild winters and cool summers. This is particularly true of west coasts, where cold currents tend to further moderate climate. Thus, thermal stress is not extreme.

In view of these differences in climate, the adaptations of the Mediterranean vegetation to climate are considerably different from those typical of savanna vegetation. The most characteristic adaptation is hard, thick, leathery leaves. This both reduces water loss and insulates plants from the heat of the dry season. Thus, such plants, termed sclerophylls, do not need to lose their leaves during the dry or cold season. This gives them a growth advantage over deciduous types when favorable conditions for growth resume. Typical sclerophyllous trees include the live oak and certain species of *Acacia* and *Eucalyptus*.

The term "Mediterranean forest" (or "sclerophyll forest") is used for a number of diverse vegetation formations, such as evergreen forest, woodland, scrub, and dwarf forest. An example of the last is the chaparral common in the Mediterranean climate of California (Fig. 3.12). These various types occupy areas with a wide range of moisture availability. In the driest regions, mean annual rainfall is on the order of 300–400 mm. It is as high as 1000 mm on the poleward margins where the mid-latitude forest begins.

3.2 PLANT–WATER RELATIONSHIPS

3.2.1 MOISTURE AVAILABILITY AND ITS RELATIONSHIP TO DRYLAND VEGETATION

The main factors limiting vegetation growth are solar radiation, moisture availability, and nutrients. The level of radiation intensity required for peak growth is usually attained in dryland regions, so that moisture availability is the limiting factor in growth in most dryland environments. Plant size, percent ground cover, and biomass all tend to increase with rainfall, both for ecosystems as a whole and for individual formations or species (Shmida 1985). The interplay between water availability and terrain and soil characteristics controls such ecosystem processes as transpiration, growth, species composition, and mortality (Gutiérrez-Jurado *et al.* 2006).

Because water is so important, the pattern of vegetation in drylands responds sharply to small variations in soil moisture and to other microclimatic effects. Generally, plant cover is concentrated

Fig. 3.11 A typical mid-latitude grassland in the Great Plains of the USA.

Fig. 3.12 Sclerophyll vegetation: (left) eucalyptus of Australia; (right) chaparral of California. The eucalypts are young trees regrowing after a burn.

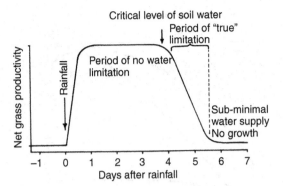

Fig. 3.13 Grass growth (net productivity) with relation to moisture availability (from Scholes 1990b).

in wetter niches in the drylands. Dense stands of trees may exist in wadis, while scrub covers nearby sandy or stony soils. Water supply tends to be greater at higher elevations (particularly over "island" situations like Tibesti or the Hoggar, where rainfall is enhanced by topography), at the foot of slopes and inselbergs, in depressions and oases, and near exotic streams.

Moisture availability represents the interplay of rainfall, evaporation and the underlying ground surface, with surface temperature, ground chemistry and soil moisture conditions all playing a role. The last factor is strongly dependent on soil type, slope and topography, the types and characteristics of surface materials, the nature of the drainage system, and the proximity to groundwater, the water table, and seasonal and exotic streams (Goudie and Wilkinson 1980).

Soil influences productivity through its effects on moisture availability and on nutrients. The interplay between rainfall and edaphic factors can be seen when the growth cycle is considered in detail, using grasslands as an example. The grasses respond only to rain events that exceed some threshold that is largely dependent on soil characteristics. Below the threshold, all water input is lost as runoff or evaporation (Fernández 2007).

When a rain event occurs that is sufficient for growth, the immediate growth is not water-limited but proceeds at a rate that is largely dependent on nutrient status. Growth continues at this rate (Fig. 3.13) until soil water drops below a critical threshold (a function of soil type). Then growth occurs at a continually decreasing rate, limited by moisture availability. In this scenario (Scholes 1990b), water supply (together with soil type) controls the duration of grass production, but nutrients determine the growth rate during productive periods. On a seasonal basis, once the soil becomes wet enough that growth is not water-limited, grass production will proceed at a nutrient-controlled level until, toward or after the end of the rainy season, the soil dries out and reaches the threshold for water-limited growth (Justice and Hiernaux, 1986).

Although some growth can occur within days of a rain event, the overall productivity within the season is strongly related to the bulk soil moisture content during the season. Within African savannas and grasslands, vegetation is therefore most

strongly correlated not with rainfall in the current month, but with accumulated rainfall for a two- or three-month period (e.g., Nicholson *et al.* 1990). A similar lag is observed in India (Chandrasekar *et al.* 2006). However, there is also a "memory" in the vegetation cover, such that growth in one season may be influenced by the rainfall during a previous season (Malo and Nicholson 1990; Prince *et al.* 1998; Martiny *et al.* 2005). This is particularly true when a given year is preceded by extreme drought. Martiny *et al.* refer to this as the "recovery effect."

The relationship between net primary production (P) and rainfall (R) can be approximately expressed as $P = k(R - R_t)$ where k is a constant and R_t is a threshold value of rainfall representing the minimum amount of rainfall that is effective for plant growth (Noy-Meir 1985). This threshold is not homogeneous within a landscape, even one dominated by a relatively uniform vegetation cover. The critical precipitation threshold R_t varies as a result of differences in soil, nutrients, topography, and runoff patterns (Scheffer *et al.* 2005). This introduces numerous complexities into the interaction between precipitation and vegetation cover.

3.2.2 THE EFFICIENCY OF WATER USE IN DRYLAND VEGETATION

Vegetation links the water, energy and carbon cycles through its basic growth process, photosynthesis. In this process, water (H_2O) and carbon dioxide (CO_2) are combined to produce carbohydrates ($C_6H_{12}O_6$) and molecular oxygen (O_2):

$$6CO_2 + 12H_2O \xrightarrow[\text{chlorophyll}]{\text{light}} C_6H_{12}O_6 + 6H_2O + 6O_2. \quad (3.1)$$

Sunlight is required to catalyze the reaction. The opposite occurs in respiration; carbohydrate is broken down and combined with oxygen to yield carbon dioxide and water:

$$C_6H_{12}O_6 + 6H_2O + 6O_2 \rightarrow 6CO_2 + 12H_2O + \text{chemical energy.} \quad (3.2)$$

Net photosynthesis is the gross photosynthesis minus respiration.

Photosynthesis is affected by a large number of environmental characteristics, the most important being climate. The most relevant climatic variables are insolation, wind, temperature, and precipitation. The portion of solar insolation that is useful for the process is termed photosynthetically active radiation (PAR) and generally consists of the wavelengths in the range 0.38–0.71 µm. Net photosynthesis increases log-linearly with the amount of PAR intercepted (IPAR) until a peak growth rate is achieved and maintained as PAR continues to increase (Fig. 3.14). The growth rate reaches a maximum because photosynthesis is proportional to the amount of PAR intercepted by plants, which in turn is proportional to the leaf area exposed to sunlight. The area ceases to increase once the canopy is so dense that leaves lower down in the canopy are completely shielded from solar radiation (Asrar *et al.* 1984).

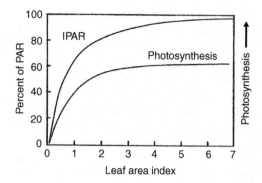

Fig. 3.14 Simplified diagram of the relationships between intercepted photosynthetically active radiation (IPAR), photosynthesis, and leaf area index (from Nicholson 1999, based on Asrar *et al.* 1984 and Sellers 1987). IPAR is given as a percentage of the incident PAR: units for photosynthesis are mg CO_2 dm^{-1} $hour^{-1}$.

The radiation intensity required for peak growth is usually attained in drylands, so that PAR is seldom a limiting variable. Water, a raw material for photosynthesis, is therefore the primary determinant of the rate of photosynthesis. Growth is also influenced by temperature and by wind, which promotes both the drying and cooling of vegetation.

The conversion of carbon dioxide and water to carbohydrate described above includes intermediate steps involving many different compounds. Three main sequences, or "photosynthetic pathways," have been identified. These differ with respect to the enzymes involved in the conversion (Odum 1993). Most species use the C3 or C4 pathways, named for the number of carbon atoms in the first photosynthetic product. A third pathway is termed CAM (crassulacean acid metabolism). The three pathways are favored by different environmental conditions and may represent adaptive strategies for plants. C3 and C4 plants respond differently to light and temperature (Fig. 3.15). The photosynthetic rate of C3 plants (per unit leaf surface) peaks at moderate light intensities and temperature, so that high light and temperature inhibit the growth of C3 species. C4 plants utilize water more efficiently under these more extreme conditions. CAM species absorb carbon dioxide at night and hold it in reserve until light is available the next day. This allows CAM plants to close their stomata during the day, thus conserving water. Some plant species can shift from one photosynthetic pathway to another in times of high moisture or salinity stress. A good example is *Welwitschia* in the Namib Desert (see Chapter 24).

The photosynthetic pathways of C4 and CAM plants and the stomatal behavior of CAM plants give them an ecological advantage over C3 plants under arid conditions. In a very general sense, the dominant species change from C3 to C4 to CAM under increasingly arid conditions. Thus, in forests and at high latitudes C3 plants prevail, while C4 are generally dominant in desert and grassland communities in warm temperate and tropical climates. CAM species are found in the most arid conditions; some evidence indicates that they require fog-water for growth and are hence common in coastal deserts. However, these are only generalizations and the distribution of the various species

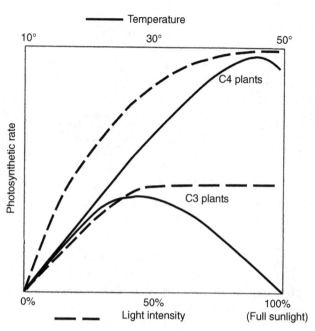

Fig. 3.15 Response of C3 and C4 plants to temperature and light (based on Odum 1993 and others).

is complex. The dominance of C3, C4, or CAM species depends on the proportions of annuals and perennials, rainfall seasonality, and environmental history (Fig. 3.16). A good example comes from the Mojave and Sonoran deserts of the southwestern USA. Winter season rainfall prevails in the Mojave and all species of winter annuals are C3. Just to the east in the Sonoran, with summer rainfall, C4 species dominate.

Transpiration and photosynthesis are intimately linked by their common regulatory mechanism, the opening and closing of stomata (Tucker and Sellers 1986). Therefore the ratio of stomatal resistances for carbon dioxide and water vapor is the inverse ratio of their diffusion coefficients. This suggests a constant relationship between the rates of photosynthesis dP and transpiration dT,

$$dP/dT = W \qquad (3.3)$$

where W is a species-specific constant termed "water-use efficiency." This relationship holds generally, but not strictly, because the behaviors of water vapor and carbon dioxide are not analogous in the passage through the mesophyll or cuticle. In general, W is higher by a factor of two for C4 than for C3 plants. Therefore, C4 plants have a more rapid and abundant response to more favorable conditions (e.g., a wet year). W is highest for CAM plants, since these can photosynthesize at night and close their stomata during the day, thus reducing water loss. Also some desert species have the capacity to control transpiration and photosynthesis independently, so that W varies with moisture stress (Evenari 1985).

The net production of plant material by photosynthesis is measured in terms of biomass, the dry weight of the organic

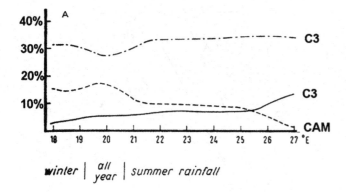

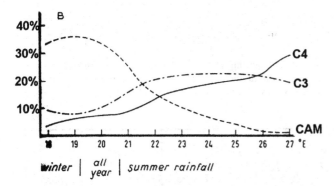

Fig. 3.16 Relationship between the prevalence of C3, C4 and CAM plants in the South African karoo versus seasonality of rainfall (modified from Werger 1986). Top: Prevalence is based on the number of species. Bottom: Prevalence is based on the percent cover.

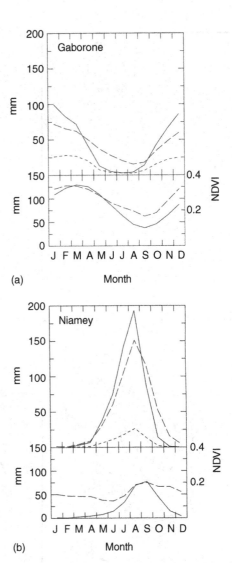

Fig. 3.17 (top) Monthly precipitation (solid lines), runoff (dotted lines), and evaporation (dashed lines) for Gaborone (a) and Niamey (b), in mm. Runoff is multiplied by a factor of 10. (bottom) normalized difference vegetation index NDVI (right hand axis, dashed line) versus soil moisture (left hand axis in mm, solid line). Note that background noise increases NDVI in the dry season in West Africa (Nicholson *et al.* 1997).

matter above and below ground. The net annual increase of biomass, or net primary production, is a measure of the productivity of an ecosystem. It is linearly related to rainfall in the drier regions. However, the rate of primary production decreases then levels off as rainfall increases and water ceases to be the limiting factor in growth. This happens around the transition to sub-humid or humid climates, on the order of 700–900 mm mean annual rainfall (Scholes 1990b). As with other plant–water relationships, the productivity/rainfall curve varies with the type of vegetation and soil and with moisture availability.

Noy-Meir (1985) suggests that the ratio of primary production to rainfall (*P/R* or "rain-use efficiency") is a better parameter than biomass or productivity for characterizing and comparing different arid regions. It is less variable in time and space; moreover it is a constant over a broad range that can be interpreted as the range where rainfall is the main limiting factor for plant growth. Rain-use efficiency varies between biomes and it decreases across biomes as mean annual precipitation increases. This parameter also appears to converge to a common maximum among arid biomes during the driest years (Huxman *et al.* 2004). This characteristic has implications for the response of ecosystems to climate change.

Rain-use efficiency depends on two main variables, the proportion of rainfall returned to the atmosphere by transpiration

(*T/R*) and the ratio of productivity to transpiration. The latter variable is essentially the water-use efficiency *W*. All three of these ratios are controlled by a number of physical and physiological factors and they vary with the type of plant. The ratio *P/R* is also influenced by the nature of the precipitation regime and how this controls the efficiency of soil moisture generation. This is illustrated using Gaborone, Botswana, in southern Africa, and Niamey, Niger, in the Sahel. At these stations mean annual rainfall is 531 mm and 559 mm, respectively, but at Gaborone it is spread over a much longer season and daily totals are less intense. Consequently, considerably more soil moisture is generated and vegetation growth is much higher, despite similar annual totals (Fig. 3.17).

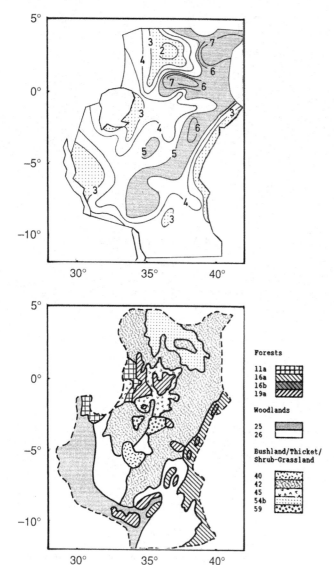

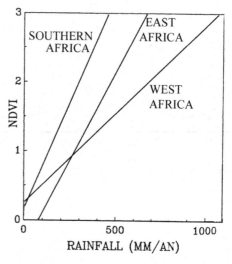

Fig. 3.19 Relationship between NDVI (normalized difference vegetation index) and rainfall for savanna regions of Botswana, East Africa, and the West African Sahel (from Farrar *et al.* 1984).

Fig. 3.18 Maps indicating for East Africa (top) the NDVI–rainfall ratio, an index of rain-use efficiency (from Davenport and Nicholson 1993), and (bottom the prevailing vegetation types after white.

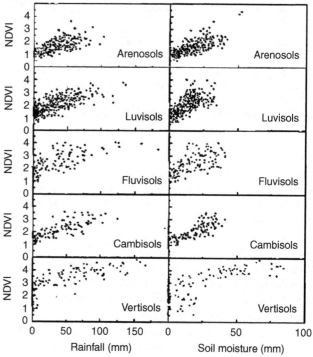

Fig. 3.20 Relationship between NDVI and annual rainfall (mm) (left) and NDVI and soil moisture (mm) (right) for five soil types in Botswana. Soil types are arranged from top to bottom in the order of increasing clay content/decreasing sand content (from Farrar *et al.* 1984).

Quantities analogous to rain-use efficiency and water-use efficiency have been evaluated on a biome basis for several dryland regions in Africa (Nicholson *et al.* 1990; Farrar *et al.* 1994). These ratios are geographically distinct and are soil- and vegetation-dependent. In East Africa (Fig. 3.18), for example, the NDVI–rainfall ratio clearly distinguishes vegetation types. A comparison of three regions shows that the efficiency of water use in Botswana and in East Africa is markedly higher than in the Sahel (Fig. 3.19), even when climatic contrasts are taken into account.

The rain/water-use efficiency also depends on soil type. In Botswana, the rate of growth per unit of rainfall or soil moisture is markedly higher in the vertisols, with very high clay content, than in the Kalahari sands (arenosols) (Fig. 3.20). Consequently, some areas of closed forest exist over the vertisols only 100–200

km from areas of much sparser savanna vegetation but with similar amounts of rainfall (Nicholson and Farrar 1994). On sandy soils, productivity increases much less rapidly with rainfall but is higher under conditions of low rainfall. This leads to the "inverse texture" hypothesis: that productivity is higher on sand than on vertisols in arid and semi-arid regions (Austin

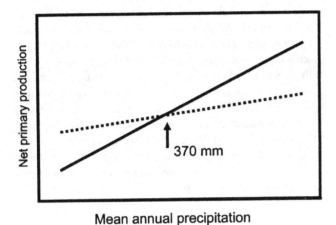

Fig. 3.21 The effect of mean annual precipitation on the rate of vegetation growth in fine-textured (solid line) and coarse-textured (dotted line) soils (from Austin *et al.* 2004).

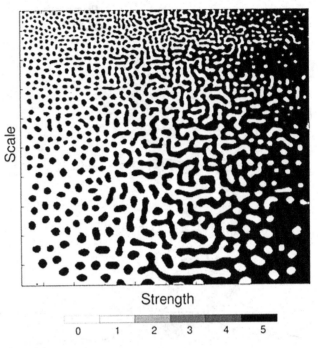

Fig. 3.22 Modeled representation of patchiness in dryland ecosystems. Black indicates maximum vegetation cover and white indicates minimum cover. Feedbacks govern shifts between varying degrees of patchiness, including shifts between spots, labyrinths, and gaps (from Rietkerk *et al.* 2004).

et al. 2004). The reason is that most water loss is from evaporation, which is lower over sand because soil moisture infiltrates more efficiency and resides in deeper layers than in a clay soil. Consequently, productivity is higher on (coarse-textured) sand under conditions of relatively low rainfall, but at higher levels of rainfall, productivity is greater on finer-textured soils (Fig. 3.21).

3.3 VEGETATION–CLIMATE INTERACTION AND THE DYNAMICS OF DRYLAND ECOSYSTEMS

A common characteristic of savannas and arid lands is the heterogeneity, or "patchiness," of the vegetation cover. The patches are localized concentrations of biomass induced mainly by water stress (Rietkerk *et al.* 2004; Gilad *et al.* 2007). Patches differ in spatial form and ecological function, depending on species, precipitation, soil, and other environmental factors (Meron *et al.* 2007). Savannas consist of trees, shrubs, and grasses in various proportions, the "patches" being individual or clustered woody elements spaced within a mostly continuous grass strata. The ratio of grasses to trees is a major determinant of ecosystem properties (Breshears and Barnes 1999). Arid ecosystems consist of vegetation patches scattered within areas of bare or nearly bare ground. Specific feedbacks between vegetation and moisture shape the development of regular, organized patterns of patchiness (Fig. 3.22). Irregular patterns also occur when these feedbacks amplify the influence of small, topographic irregularities (Klausmeier 1999).

The feedbacks that control the structure and pattern of dryland ecosystems involve the extensive impact of vegetation on microclimatic conditions. The various feedbacks are often interactive (Schymanski *et al.* 2009). Any vegetation cover reduces thermal amplitude, radiation exposure, wind desiccation, and soil erosion. It modifies moisture availability by extracting soil water, intercepting precipitation, and modulating percolation and runoff. In arid regions, vegetation prevents the formation

of soil crusts that reduce the infiltration capacity of the bare patches. The physical structure of the woody canopy modifies the environment beneath it. The subcanopy microclimate is therefore quite different from that of the intercanopy patches (Breshears and Barnes 1999). Since the degree of environmental modification by woody plants decreases with distance from the plant, a mixed tree–grass community consists of a spatial patchwork of different degrees of competition and facilitation (a positive feedback).

Vegetation influences both the surface energy balance, via shading, and the surface water balance, via the redistribution of precipitation. Leaves and woody structures intercept and retain a portion of the precipitation. The vegetation patches also capture runoff, and roots promote infiltration, thus increasing soil moisture; shading of soil by the plants reduces evaporation from its surface. Stem flow and leaf drip concentrate moisture beneath the tree. Intercepted water contributes to the evaporation, but also to soil moisture via stemflow and leaf drip (Breshears *et al.* 1998). Interception can be on the order of 5–50% of annual rainfall. Evapotranspiration is modulated by these processes and by plant transpiration.

The collective effect of these processes is that, in the drylands, water tends to flow from sparsely to densely vegetated patches or from intercanopy to canopy spaces, in part because of the formation of impenetrable crusts in the barren patches. Thus, vegetation patches collect not only water but also nutrients, essentially becoming "islands of fertility" (Ridolfi *et al.* 2008)

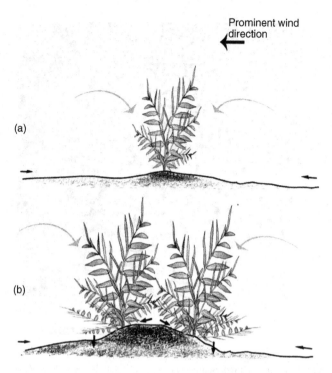

Prominent wind direction

(a)

(b)

Fig. 3.23 Conceptual diagram showing the interaction of hydrological and aeolian processes (straight black arrows indicate hydrological processes and curved gray arrows indicate aeolian processes). (a) Formation of islands of fertility around the shrub by the deposition of wind-borne fines. (b) Changes in hydrological processes (infiltration and runoff), which result in changes in the growth pattern of the shrubs (from Ravi *et al.* 2007).

Fig. 3.24 An "island of fertility" in North Africa.

(Figs. 3.23 and 3.24). The impact on nutrients may be greater in the drier savannas than in the wetter ones, where trees have a large impact on the spatial patterning of soil resources (Okin *et al.* 2008).

The characteristic patchiness of dryland ecosystems is believed to result from feedback mechanisms related to these microclimatic effects. Regular (as opposed to random) patchiness is

considered to be "self-organized" (i.e., resulting from dynamic processes within the ecosystem, rather than imposed externally). Two opposing feedbacks control the water balance in the patches (e.g., Gilad *et al.* 2007): infiltration concentrates water at the vegetation patch, accelerating (i.e., facilitating) vegetation growth. At the same time, water uptake by the vegetation acts to deplete water near the patch. A dominance of uptake favors intraspecies and interspecies competition for the available moisture (Meron *et al.* 2007). A dominance of infiltration favors facilitation. The dominant feedback is generally a function of scale. Facilitation occurs near the patches, but competition prevails at a distance from the patch.

3.4 HYPOTHESES FOR THE COEXISTENCE OF TREES AND GRASSES IN THE SAVANNA ECOSYSTEM

Many hypotheses have been proposed for the coexistence of grasses and trees. Most fall into the categories of competition-based mechanisms or "demographic bottlenecks" (Sankaran *et al.* 2004; Meyer *et al.* 2007). The latter focus on disturbances (e.g., fire or extreme climatic events) and how they affect vegetation at various stages in the life cycle. The competition-based hypotheses principally involve balanced competition or niche separation by rooting depth or phenology.

The concept of balanced competition relies on the assumption that the superior competitor becomes self-limiting at a biomass insufficient to eliminate the poorer competitor. Since mature trees generally dominate over grasses, the balanced competition model predicts that all savannas should trend toward a woodland with a sparse understory of grasses. Those not in that state are unstable. This model predicts two metastable states: one a dense woodland with little or no grass, and the other a dense grassland with no trees (e.g., Eagleson and Segarra 1985).

The most commonly accepted hypothesis for tree–grass coexistence has long been niche separation by rooting depth. Its origin lies in a two-soil-layer model developed by Walter (1971). Trees are assumed to be deep-rooted species drawing water from the lower layer, and grasses are assumed to have shallow roots within the upper layer. Trees are favored on soils of low water-holding capacity, such as sands, and under wetter conditions, because both of these situations facilitate water seepage into the deeper layer. The concept of niche separation by rooting depth is at the core of most land surface schemes incorporated into climate models (e.g., Pitman 2003).

Niche separation by phenology relates to the temporal contrast in the growth and water utilization patterns of trees and grasses. Trees usually achieve full leaf development within weeks of onset of the rainy season. In moist savannas, C3 species may even commence leaf expansion several weeks prior to the rainy season. Grasses respond to availability during the rainy season and then senesce long before most deciduous trees. Hence trees monopolize the available moisture early and late in the growing season, but allow for grass development during the peak season.

Demographic bottleneck models take life stages into account explicitly, but they exclude competition. The models focus on disturbances and climatic variability as limiting factors in tree recruitment and growth. Higgins *et al.* (2000) state that tree recruitment is pulsed in time following the stochastic rainfall patterns that typify arid and semi-arid regions. The longevity of trees enables them to persist over periods with rainfall that is sufficient for the growth of grass. Such a situation exists in north-central Chile and northwest Peru. Forest regeneration is limited to the high rainfall events associated with El Niño (Holmgren *et al.* 2006).

Other hypotheses for coexistence are based on spatial heterogeneities in soil or moisture (Jeltsch *et al.* 1998) or spatial competition for moisture (Rodriguez-Iturbe *et al.* 1999). Spatial heterogeneities in surface moisture are caused by the spottiness of rainfall in time and space, soil type, and topography. Species differ with respect to their ability to utilize these heterogeneities (Breshears *et al.* 1997).

The aforementioned models all assume that the environment reaches an equilibrium state. Other models of tree–grass coexistence do not assume a stable equilibrium. These predict continual changes of state due to frequent environmental disturbances (e.g., drought or fire) that alternatively give competitive advantage to either trees or grasses. The implicit question in judging the various models is whether the tree–grass mix in the savannas is stable or unstable. Scholes and Archer (1997) review the various hypotheses and argue that the answer to this question depends on the scale of observation. An environment may be stable at the landscape scale, with a given savanna persisting over millennia, while at the same time consisting of a shifting mosaic of many patches in various states of transition between grassy and woody dominance. These ideas have been incorporated into the relatively new concept of "patch dynamics" of ecosystems (see Section 3.5). The basis for this concept is the feedback between woody species and grasses (see Section 3.3).

Sankaran *et al.* (2005) proposed a comprehensive model that can incorporate many of the hypotheses on tree and grass coexistence. The basis for the model is the relationship between tree cover and mean annual precipitation shown in Fig. 3.25. Accordingly, they concluded that in the dry savannas precipitation provides an upper limit to the amount of tree cover, with disturbances and soil characteristics reducing the cover below this maximum amount. This occurs below a certain threshold (650 mm in this case), below which precipitation drives ecosystem dynamics and savannas are "climatically determined." Above the threshold, disturbances such as herbivory and fire become more dominant; the resulting savannas are "disturbance driven." Their conclusion is that, in the absence of such disturbances, a nearly complete woody canopy could be maintained.

3.5 THE EMERGING CONCEPT OF "PATCH DYNAMICS"

Patch dynamics is a new and potentially unifying mechanism to explain the tree–grass coexistence in savannas (Meyer *et al.*

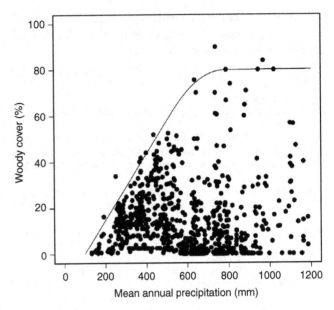

Fig. 3.25 Relationship between woody cover in savannas and mean annual rainfall, based on 854 African savanna sites (from Sankaran *et al.* (2005). The solid line indicates the maximum cover sustainable under the given rainfall conditions. Disturbances and soil characteristics result in a less dense cover.

2007). This mechanism can integrate all of the aforementioned hypotheses. The underlying concept of patch dynamics is that a savanna consists of patches in which a cyclical succession between woody and grassy dominance proceeds spatially asynchronously. The four classic savanna factors (herbivory, fire, nutrients, and water) are all candidates for the disturbance that drives the successional cycles. Pulsed resources (nutrients and water) might contribute to cyclical successions in arid ecosystems (Cheeson *et al.* 2004).

A given site is a two-phase vegetation mosaic: individual or aggregated woody plants and intercanopy space. These environments exist along a forest/grass continuum. The position along the continuum depends in large part on soil moisture availability, which changes in response to feedbacks and water stress (Rietkerk *et al.* 2004). Five basic vegetation states exist along the rainfall gradient: uniform vegetation, gaps, stripes, spots, and bare soil (Meron *et al.* 2007).

A number of studies have attempted to predict the occurrence and nature of changes along this continuum. Consistent with the much earlier model of Eagleson and Segarra (1985), one hypothesis is that two stable states may occur, separated by an unstable equilibrium. Such a state of "bistability" exists as a result of positive feedback resulting from local facilitation (Kefi *et al.* 2007). The implication is that discontinuous transitions ("catastrophic shifts") can occur, resulting from a small change in some parameter (Scheffer and Carpenter 2003). The heterogeneity of typical dryland vegetation cover could facilitate such shifts (van Nes and Scheffer 2005). Patchiness is considered to be an indicator of the potential for such transitions, with certain shapes predicting imminent discontinuous transition toward the bare state. Once this state exists, the feedbacks between trees and grasses that sustain the savanna environment are no longer possible, resulting in irreversible change.

REFERENCES

Asrar, G., Fuchs, M., E. T. Kanemasu, and J. L. Hatfield, 1984: Estimating absorbed photosynthetic radiation and leaf area index from spectral reflectance in wheat. *Agronomy Journal*, **76**, 300–306.

Austin, A. T., and Coauthors, 2004: Water pulses and biogeochemical cycles in arid and semiarid ecosystems. *Oecologia*, **141**, 221–235.

Bourlière, F. (ed.), 1983: *Tropical Savannas*. Ecosystems of the World 13, Elsevier Science and Technology, Amsterdam, 730 pp.

Bourlière, F., and M. Hadley, 1983: Present–day savannas: an overview. In *Tropical Savannas* (F. Bourlière, ed.), *Ecosystems of the world*, vol. 13, Elsevier, Amsterdam, pp. 1–17.

Breshears, D. D., and F. J. Barnes, 1999: Interrelationships between plant functional types and soil moisture heterogeneity for semiarid landscapes within the grassland/forest continuum: a unified conceptual model. *Landscape Ecology*, **14**, 465–478.

Breshears, D. D., O. B. Meyers, S. R. Johnson, C. W. Meyer, and S. N. Martens, 1997: Differential water use of heterogeneous soil moisture by two semiarid woody species: *Pinus edulis* and *Juniperus monosperma*. *Journal of Ecology*, **85**, 289–299.

Breshears, D. D., J. W. Nyhan, C. E. Heil, and B. P. Wilcox, 1998: Effects of woody plants on microclimate in a semiarid woodland: soil temperature and soil evaporation in canopy and intercanopy patches. *International Journal of Plant Sciences*, **159**, 1010–1017.

Budyko, M. I., 1986: *The Evolution of the Biosphere*. Reidel, Dordrecht, The Netherlands, 423 pp.

Chandrasekar, K., M. V. R. S. Sai, A. T. Jayaseelan, R. S. Dwivedi, and P. S. Roy, 2006: Vegetation response to rainfall as monitored by NOAA-AVHRR. *Current Science*, **91**, 1626–1633.

Cheeson, P., and Coauthors 2004: Resource pulses, species interactions, and diversity maintenance in arid and semi-arid environments. *Oecologia*, **141**, 236–253.

Cole, M. M., 1986: *The Savannas*. Academic Press, London, 438 pp.

Coupland, R. T., 1992: Overview of the grasslands of North America. In *Natural Grasslands* (R. T. Coupland, ed.), Ecosystems of the World 8A, Elsevier Science and Technology, Amsterdam, pp. 147–149.

Davenport, M. L., and S. E. Nicholson, 1993: On the relationship between rainfall and the normalized difference vegetation index for diverse vegetation types in East Africa. *International Journal of Remote Sensing*, **14**, 2369–2389.

Dekker, S. C., M. Rietkerk, and M. F. P. Bierkens, 2007: Coupling microscale vegetation-soil water and macroscale vegetation-precipitation feedbacks in semiarid ecosystems. *Global Change Biology*, **13**, 671–678.

Eagleson, P. S., and R. I. Segarra, 1985: Water-limited equilibrium of savanna vegetation systems. *Water Resources Research*, **21**, 1483–1493.

Ellery, W. N., R. J. Scholes, and M. T. Mennis, 1991: An initial approach to predicting the sensitivity of the South African grassland biome to climate change. *South African Journal of Science*, **87**, 499–503.

Evenari, M., 1985: Adaptations of plants and animals to the desert environment. In *Hot Deserts and Arid Shrublands* (M. Evenari, I. Noy-Meir, and D. W. Goodall, eds.), Ecosystems of the World 12A, Elsevier, Amsterdam, pp. 79–92.

Farrar, T. J., S.E. Nicholson, and A. R. Lare, 1994: The influence of soil type on the relationships between NDVI, rainfall and soil moisture in semi-arid Botswana. Part II. Response to soil moisture. *Remote Sensing of Environment*, **50**, 121–133.

Fernández, R. J., 2007: On the frequent lack of response of plants to rainfall events in arid areas. *Journal of Arid Environments*, **68**, 688–691.

Gilad, E., M. Shachak, and E. Meron, 2007: Dynamics and spatial organization of plant communities in water-limited systems. *Theoretical Population Biology*, **72**, 214–230.

Goudie, A., and J. Wilkinson, 1980: *The Warm Desert Environment*. Cambridge University Press, Cambridge, UK, 88 pp.

Guswa, A. J., 2008: The influence of climate on root depth: a carbon cost–benefit analysis. *Water Resources Research*, **44**, doi:10.1029/2007WR006384.

Gutiérrez-Jurado, H. A., E. R. Vivoni, J. B. J. Harrison, and H. Guan, 2006: Ecohydrology of root zone water fluxes and soil development in complex semiarid rangelands. *Hydrological Processes*, **20**, 3289–3316.

Higgins, S. I., W. J. Bond, and S. W. Trollope, 2000. Fire, resprouting, and variability: a recipe for grass–tree coexistence in savanna. *Journal of Ecology*, **88**, 213–229.

Holmgren, M., B. C. López, J. R. Gutiérrez and F. A. Squeo, 2006: Herbivory and plant growth rate determine the success of El Niño Southern Oscillation-driven tree establishment in semi-arid South America. *Global Change Biology*, **12**, 2263–2271.

Huxman, T. E., and Coauthors, 2004: Convergence across biomes to a common rain-use efficiency. *Nature*, **429**, 651–654.

Jeltsch, F., G. Weber, W. R. J. Dean, and S. J. Milton, 1998: Disturbances in savanna ecosystems: modelling the impact of a key determinant. In *Ecosystems and Sustainable Development* (J. L. Usó, C. A. Brebbia, and H. Power, eds.), Computational Mechanics Publications, Southampton, UK, pp. 233–242.

Justice, C. O., and P. H. Y. Hiernaux, 1986: Monitoring the grasslands of the Sahel using NOAA AVHRR data: Niger 1983. *International Journal of Remote Sensing*, **7**, 1475–1498.

Kefi, S., M. Rietkerk, M. van Baalen, and M. Loreau, 2007: Local facilitation, bistability and transitions in arid ecosystems. *Theoretical Population Biology*, **71**, 367–379.

Klausmeier, C. A., 1999: Regular and irregular patterns in semiarid vegetation. *Science*, **284**, 1826–1828.

Malo, A. R., and S. E. Nicholson, 1990: A study of rainfall and vegetation dynamics in the African Sahel using normalised difference vegetation index. *Journal of Arid Environment*, **19**, 1–24.

Martiny, N., Y. Richard, and P. Camberlin, 2005: Interannual persistence effects in vegetation dynamics of semi-arid Africa. *Geophysical Research Letters*, **32**, doi:10.1029/2005GL024634.

Meron, E., E. Gilad, J. von Hardenberg, M. Shachak, and Y. Zarmi, 2004: Vegetation patterns along a rainfall gradient. *Chaos, Solitons and Fractals*, **19**, 367–376.

Meron, E., H. Yizhaq, and E. Gilad, 2007: Localized structures in dryland vegetation: forms and functions. *Chaos*, **17**, 037109.

Meyer, K. M., K. Wiegand, D. Ward, and A. Moustakas, 2007: The rhythm of savanna patch dynamics. *Journal of Ecology*, **95**, 1306–1315.

Newman, B. D., and Coauthors, 2006: Ecohydrology of water-limited environments: a scientific vision. *Water Resources Research*, **42**, W06302, doi:10.1029/2005WR004141.

Nicholson, S. E., 1999: The physical–biotic interface in arid and semi-arid systems: a climatologist's viewpoint. In *Arid Lands Management: Towards Ecological Sustainability* (T. W. Hoekstra, and M. Shachak, eds.), University of Illinois Press, Urbana, IL, pp. 31–47.

Nicholson, S. E., and T. J. Farrar, 1994: The influence of soil type on the relationships between NDVI, rainfall and soil moisture in semi-arid Botswana. Part I. Relationship to rainfall. *Remote Sensing of Environment*, **50**, 107–120.

Nicholson, S. E., M. L. Davenport, and A. R. Mole, 1990: A comparison of the vegetation response to rainfall in the Sahel and East Africa, using Normalized Difference Vegetation Index from NOAA AVHRR. *Climatic Change*, **17**, 209–215.

Nicholson, S. E., J. Kim, M. B. Ba, and A. R. Lare, 1997: The mean surface water balance over Africa and its interannual variability. *Journal of Climate*, **10**, 2981–3002.

Nix, H. A., 1983: Climate of the tropical savannas. In *Tropical Savannas* (H. Bourlière, ed.), Ecosystems of the World 13, Elsevier, Amsterdam, pp. 37–62.

Noy-Meir, I., 1985: Desert ecosystem structure and function. In *Hot Deserts and Arid Shrublands* (M. Evenari, I. Noy-Meir, and D. W. Goodall, eds.), Ecosystems of the World 12A, Elsevier, Amsterdam, pp. 93–103.

Odum, E. P., 1971: *Fundamentals of Ecology*. Saunders, Philadelphia, PA, 574 pp.

Odum, E. P., 1993: *Ecology and Our Endangered Life-Support Systems*. Sinclair, Sunderland, MA, 301 pp.

Okin, G. S., N. Mladenov, L. Wang, D. Cassel, K. K. Caylor, S. Ringrose, and S. A. Macko, 2008: Spatial patterns of soil nutrients in two southern African savannas. *Journal of Geophysical Research*, **113**, doi:10.1029/2007JG000584.

Olson, D. M., E. Dinerstein, and E. D. Wikramanayake, 2001: Terrestrial ecoregions of the world: a new map of life on earth. *BioScience*, **51**, 933–938.

Petrov, M. P., 1976: *Deserts of the World*. Halsted (Wiley and Sons), New York, 447 pp.

Pitman, A. J., 2003: The evolution of, and revolution in, land surface schemes designed for climate models. *International Journal of Climatology*, **23**, 479–510.

Prince, S. D., E. Brown de Colstoun, and L. Kravitz, 1998: Evidence from rain use efficiencies does not support extensive Sahelian desertification. *Global Change Biology*, **4**, 359–373.

Ravi, S., P. D'Odorico, and G. S. Okin, 2007: Hydrologic and aeolian controls on vegetation patterns in arid landscapes. *Geophys. Res. Lett.*, **34**, doi: 10.1029/2007 GL031023.

Ridolfi, L., F. Laio, and P. D'Odorico, 2008: Fertility island formation and evolution in dryland ecosystems. *Ecology and Society*, **13(1)**, 5.

Rietkerk, M., S. C. Dekker, P. C. de Ruiter, and J. van de Koppel, 2004: Self-organized patchiness and catastrophic shifts in ecosystems. *Science*, **305**, 1926–1929.

Ripley, E. A., 1992: Grassland climate. In *Natural Grasslands* (R. T. Coupland, ed.), Ecosystems of the World 8A, Elsevier, Amsterdam, pp. 7–24.

Rodriguez-Iturbe, I., P. D'Odorico, A. Porporato, and L. Ridolfi, 1999: On the spatial and temporal links between vegetation, climate and soil moisture. *Water Resources Research*, **35**, 3709–3722.

Sankaran, M., J. Ratnam, and N. P. Hanan, 2004: Tree–grass coexistence in savannas revisited: insights from an examination of assumptions and mechanisms invoked in existing models. *Ecology Letters*, **7**, 480–490.

Sankaran, M., N. P. Hanan, R. J. Scholes, J. Ratnam, D. J. Augustine, B. S. Cade, and Coauthors, 2005: Determinants of woody cover in African savannas. *Nature*, **438**, 846–849.

Scheffer, M. M., and S. R. Carpenter, 2003: Catastrophic regime shifts in ecosystems: linking theory to observation. *Trends in Ecology and Evolution*, **18**, 648–656.

Scheffer, M., M. Holmgren, V. Brovkin, and M. Claussen, 2005: Synergy between small- and large-scale feedbacks of vegetation on the water cycle. *Global Change Biology*, **11**, 1003–1012.

Scholes, R. J., 1990a: Change in nature and the nature of change: interactions between terrestrial ecosystems and the atmosphere. *South African Journal of Science*, **86**, 350–354.

Scholes, R. J., 1990b: The influence of soil fertility on the ecology of southern African dry savannas. *Journal of Biogeography*, **17**, 415–419.

Scholes, R. J., 1991: An initial approach to predicting the sensitivity of the South African grassland biome to climate change. *South African Journal of Science*, **87**, 499–503.

Scholes, R. J., and S. R. Archer, 1997: Tree–grass interactions in savannas. *Annual Review of Ecology and Systematics*, **28**, 517–544.

Scholes, R. J., and B. H. Walker, 1993: *An African Savanna: Synthesis of the Nylsvley Study*. Cambridge University Press, Cambridge.

Schymanski, S. J., M. Sivapalan, M. L. Roderick, L. B. Hutley, and J. Beringer, 2009: An optimality-based model of the dynamic feedbacks between natural vegetation and the water balance. *Water Resources Research*, **45**, W01412.

Seely, M. K., and G. N. Louw, 1980: First approximation of the effects of rainfall on the ecology and energetics of a Namib Desert dune ecosystem. *Journal of Arid Environments*, **1**, 117–128.

Sellers, P. J., 1987: Canopy reflectance, photosynthesis and transpiration. II. The role of biophysics in the linearity of their dependence. *Remote Sensing of Environment*, **21**, 143–183.

Shmida, A., 1985: Biogeography of the desert flora. In *Hot Deserts and Arid Shrublands* (M. Evenari, I. Noy-Meir, and D. W. Goodall, eds.), *Ecosystems of the World* 12A, Elsevier, Amsterdam, pp. 23–77.

Shreve, F., and I. L. Wiggins, 1964: *Vegetation and Flora of the Sonoran Desert*. Stanford University Press, Palo Alto, CA.

Singh, J. S., W. K. Lauenroth, and D. G. Milchunas, 1983: Geography of grassland ecosystems. *Progress in Physical Geography*, **7**, 46–80.

Soriano, A., 1992: Río de Plata Grassland. In Natural Grasslands (R. T. Coupland, ed.), *Ecosystems of the World*, Vol. 8B, Elsevier, Amsterdam, pp. 367–408.

Stone, E. L., and P. J. Kalisz, 1991: On the maximum extent of tree roots. *Forest Ecology and Management*, **46**, 59–102.

Strahler, A. H. and A. N. Strahler, 1992: *Modern Physical Geography*, 4th edn. John Wiley and Sons, New York, 638 pp.

Tucker, C. J., and P. J. Sellers, 1986: Satellite remote sensing of primary production. *International Journal of Remote Sensing*, **7**, 1395–1416.

van Nes, E. H., and M. Scheffer, 2005: Implications of spatial heterogeneity for catastrophic regime shifts in ecosystems. *Ecology*, **86**, 1797–1807.

Walter, H., 1971: *Ecology of Tropical and Subtropical Vegetation*. Longman, London, 539 pp.

Walter, H., and S.-W. Breckle, 1999: *Vegetation und Klimazonen*. Verlag Eugen Ulmer, Stuttgart, 544 pp.

Whittaker, R. H., 1975: *Communities and Ecosystems*, 2nd edn. Macmillan, New York, 352 pp.

Part II The meteorological background

4   The general atmospheric circulation

4.1 RECEIPT OF SOLAR RADIATION

The ultimate source of the earth's energy is the sun. It provides energy in the form of electromagnetic radiation that is converted within the earth–atmosphere system to the mechanical motion of the winds and the thermal energy of the earth and atmosphere. The amount of solar radiation received at the top of the atmosphere at a given point on the earth depends on four factors: (1) the magnitude of solar output, (2) the distance from the sun, (3) the obliquity or tilt of the sun's rays, and (4) the length of day. Thus, the insolation at the top of the atmosphere varies with season, time of day, and latitude.

However, the amount received at the top of the atmosphere on a surface perpendicular to the solar beam is relatively constant. This amount, which is 2 cal/cm² per minute, is called the *solar constant*. The beam's effectiveness is reduced when it hits the atmosphere or earth's surface at an oblique angle, as is the case at higher latitudes. This is illustrated in Fig. 4.1. In both diagrams the beam is equally intense, but when it strikes obliquely, the radiation is distributed over a greater area. The obliquity affects the amount of radiation reaching the earth's surface in another way; the path length of oblique rays through the atmosphere is greater than those striking perpendicularly. Therefore, the attenuation of the beam by scattering and absorption in the atmosphere is also greater.

As the solar beam passes through the atmosphere, it is attenuated by processes of scattering, absorption, and reflection. In a cloud-free atmosphere, very little of the beam is depleted because the atmospheric gases are selective absorbers. That is, they absorb only certain wavelengths of radiation. The principal atmospheric gases, such as oxygen, nitrogen, water vapor, and carbon dioxide, are relatively transparent to the visible and ultraviolet wavelengths that constitute most of the solar beam. These molecules scatter solar radiation, but most of the scattered radiation eventually reaches the ground as diffuse radiation. When clouds are present, however, a large proportion of the beam is reflected back to space.

The radiation received in the summer hemisphere varies little with latitude, because the increasing length of day at higher latitudes compensates for the latitudinally decreasing intensity of the solar beam. In the winter hemisphere, a strong radiation gradient exists because both beam intensity and daylength decrease with latitude. The result is a strong latitudinal temperature gradient in the winter hemisphere, a weak gradient in the summer hemisphere and, in the annual mean, high temperature gradients in the mid- and high-latitudes, but low gradients in the tropics (Fig. 4.2). These gradients are one of the fundamental causes of the earth's global patterns of winds and pressure, termed the general atmospheric circulation.

4.2 GLOBAL PATTERNS OF WIND AND PRESSURE

The general atmospheric circulation is the planetary system of major wind and pressure features. It is composed of systems on space scales of 5×10^3 km and is associated with processes acting on time scales of months. It includes such features as the westerlies, the trade winds, the mid-latitude and subtropical westerly jet streams, the tropical easterly jet stream, and the major pressure systems such as the subtropical and polar highs or anticyclones and the mid-latitude and equatorial belts of low pressure. The planetary-scale circulation spawns the development of synoptic-scale features such as the cyclones, anticyclones, and tropical storms depicted on weather maps. It interacts with these and smaller mesoscale systems, such as sea breezes and mountain winds, to produce the essence of local climate.

The causes of the general atmospheric circulation are quite simple. The primary factors are the differential heating of the earth's surface as a function of latitude (i.e., the latitudinal or equator-to-pole temperature gradient illustrated in Fig. 4.2) and the earth's rotation. The former factor is manifested as the pressure gradient force, which pulls air parcels toward low pressure. The latter is manifested as the Coriolis force, which is directed perpendicular to the motion of an air parcel. The Coriolis force increases with latitude and is zero at the equator. Acting together, these two forces produce air motion that is parallel to the

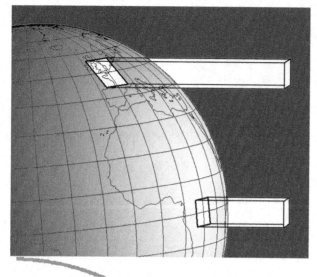

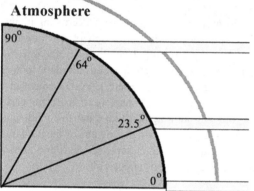

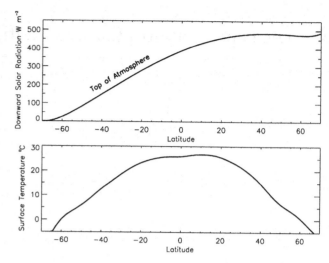

Fig. 4.2 (top) Latitudinal distribution of solar energy received at the top of the atmosphere on the June solstice. (bottom) Mean annual temperature as a function of latitude. Based on NCEP data.

Fig. 4.1 Effects of the obliquity of the solar beam on the receipt of solar radiation (modified from Trewartha and Horn 1980). Top diagram illustrates the larger surface area over which an oblique ray is distributed, compared with a vertical ray. The bottom diagram illustrates the longer passage of the oblique ray through the atmosphere.

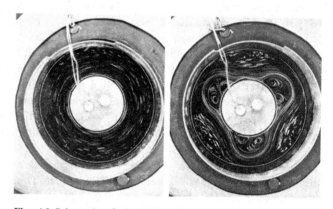

Fig. 4.3 Schematic of the "dishpan" experiment and the ensuing Rossby circulations (from Douglas *et al.* 1972). The circulation on the left is typically "zonal" (east–west oriented) flow. That on the right is a typical "meridional" pattern, with strong north–south components.

contours of atmospheric pressure, i.e., isobars. This is referred to as geostrophic motion and occurs at elevations high enough that the friction from the earth's surface is negligible. Note that in the lower latitudes, the Coriolis force is so weak that geostrophic flow, such as the mid-latitude westerlies, cannot occur.

These two primary factors are so fundamental that, even operating alone, they can prescribe the major features of the general atmospheric circulation. This fact was clearly demonstrated by the classical "dishpan" experiments conducted in the 1950s at the University of Chicago. The dishpan was a rotating annulus which could be heated along its edges and cooled in the center and which contained a dense fluid. In the absence of heating and cooling, rotation of the annulus resulted in horizontal flow moving west to east (Fig. 4.3). Termed the Rossby circulation, this is analogous to the westerly winds prevailing in mid-latitudes. Under some circumstances the east–west circulation developed waves that produce north–south variations in the flow (Fig. 4.3). The former is referred to as "zonal" flow and the latter is referred to as "meridional."

The result was quite different when the annulus did not rotate but was heated along the outside and cooled in the center, to simulate differential heating with latitude, In this case, a vertical circulation resulted, with rising fluid along the edges and sinking fluid in the center of the annulus. This overturning is called a Hadley circulation. When the annulus was subjected to both rotation and differential heating, the Rossby circulation prevailed in the center (i.e., at higher latitudes) and the vertical Hadley overturning was limited to outer portions (i.e., at lower latitudes). This is precisely the situation that prevails, to a first approximation, in the actual atmosphere.

The strength of the westerlies increases with the magnitude of the mid-latitude temperature gradient. The reason for this is the density contrast between warm and cold air. This contrast causes pressure to decrease more rapidly in the cold air, so that the pressure gradient increases with altitude when a horizontal thermal gradient exists. This relationship, termed the thermal

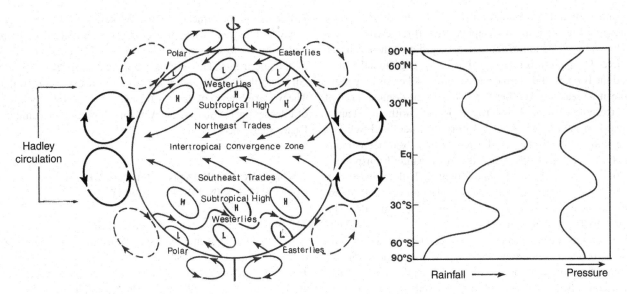

Fig. 4.4 Schematic representation of the major features of the atmospheric general circulation, compared with the mean latitudinal distributions of rainfall and pressure.

wind, prescribes an increase in westerlies/easterlies when the cold air is poleward/equatorward of the warm air. Thus easterly winds increase with height over Asia, where the Tibetan plateau (average elevation over 4500 m) provides an elevated heat source that lies adjacent to a relatively cool air mass over the Indian Ocean to the south.

The primary factors of rotation and differential heating create the essence of the atmospheric circulation, but they prescribe features that are solely a function of latitude. This is an unrealistic situation. Inhomogeneities within latitudinal belts, and regional detail, are produced by a set of secondary factors. The foremost of these are the distribution of oceans and continents and the thermal contrast associated with this distribution. Because pressure reflects the weight of air and cold air is denser, high pressure tends to override cooler areas and low pressure tends to override warmer areas. Therefore the seasonal change of the land–water thermal contrast (land warmer in summer, colder in winter) produces much of the seasonal change of the general atmospheric circulation. This contrast acts together with the distribution of large-scale topographic features, such as major mountain ranges, to create one additional aspect of the general atmospheric circulation, waves that transport air in a north–south direction.

The general atmospheric circulation is illustrated in Fig. 4.4. It includes the major circulation features observed in the dishpan experiment: vertical overturning in the low latitudes and an extensive belt of westerly winds in the mid-latitudes. Overall the general atmospheric circulation consists of three major wind systems, associated with particular latitudinal zones, and four major pressure systems that give rise to the wind systems. The dominant pressure systems are the subtropical highs and the mid-latitude lows. The former are relatively stable features, but the latter are the mean patterns associated with migrating cyclones in the mid-latitudes with lifetimes of generally a few

days. The polar highs and the general belt of low pressure in the equatorial regions complete the picture.

Since air tends to blow counterclockwise around lows and clockwise around highs in the Northern Hemisphere (the reverse in the Southern Hemisphere), the major wind systems are flanked by these pressure systems. Thus, westerlies prevail in mid-latitudes, easterlies in the subtropical and polar latitudes. These easterlies are called the trade winds and they are northeasterly in the Northern Hemisphere, southeasterly in the Southern. The trades converge near the equator in a region referred to as the equatorial trough or the Intertropical Convergence Zone (ITCZ). In the vicinity of the subtropical highs, winds are generally weak and variable.

This pattern of pressure also gives rise to a series of three vertical circulation cells. Figure 4.5 shows these for the March–May season. The annual average looks similar, but the cells vary greatly in magnitude and location during the winter and summer seasons. These vertical cells occur because air converges into the center of lows, forcing it to rise, but diverges in the vicinity of high-pressure cells, resulting in a sinking motion. These cells influence the general global pattern of precipitation because rising motion promotes precipitation while the sinking motion suppresses its formation. The largest and strongest of these vertical cells is the tropical Hadley cell, which is the dominant mean air movement in the low latitudes (Webster 2004). Its ascending branch is associated with the convergence of the trade winds in the ITCZ and equatorial heating; its descending branches are associated with the centers of the subtropical highs. In the mid-latitudes, the rising air in the prevailing cyclones is coupled with sinking motion in both subtropical and polar highs. This produces two other vertical cells in each hemisphere, but these are relatively weak, especially in comparison with the prevailing wind systems.

The picture of two Hadley cells on both sides of the equator shown in Fig. 4.5 is an oversimplification. It represents the circulation as a function of latitude, averaged over the earth as a whole. The pattern is interrupted in many longitudinal sectors and the two-celled pattern depicted in Fig. 4.4 exists only in the transition seasons. At other times there is one dominant cell of overturning, that of the winter hemisphere, with an extensive area of rising motion in the equatorial latitudes and well into the summer hemisphere and a region of subsidence (sinking motion) in the subtropical latitudes of the winter hemisphere.

In addition to the north–south oriented Hadley cells, a vertical circulation oriented east–west also plays a major role in producing the earth's patterns of climate in the low latitudes. This cell, termed the Walker circulation or equatorial zonal circulation, owes its existence to two factors. The first is that, for complex dynamic reasons, atmospheric stability is greater along the eastern flank of subtropical highs than along the western flank. Consequently, a circulation results with sinking motion in the east but a tendency to rising motion in the west. This is enhanced by orographic effects, with ascent over major mountain ranges, and by cold currents (see section 4.3.1).

The classic picture of this circulation is depicted in Fig. 4.6. Again, this is a simplification, with the patterns existing only in some seasons and years (Hastenrath 2001). However, the concept is useful in understanding the global distribution of climates and interannual variations in global climate. The Walker circulation is best developed over the Pacific Ocean, where it is evident year-round. Major vertical cells are also apparent over the Atlantic Ocean in the boreal winter and over the Indian Ocean in the boreal autumn (Hastenrath *et al.* 2002; Hastenrath 2007). Minor cells of east–west overturning are evident over the continents.

The major consequence of the Walker circulation is a tendency for the west coasts of continents in the subtropical and mid-latitudes to be considerably drier than the eastern parts of the continents. This is clearly apparent over Africa, South America, the United States and, to a lesser extent, Australia. The western branch of each of the major cells is linked to areas of intense convection over the Amazon, Central Africa, and the Pacific maritime continent.

4.3 MAJOR COMPONENTS OF THE GENERAL CIRCULATION IN THE LOW LATITUDES

The major components of the general atmospheric circulation in low latitudes include the subtropical highs, the ITCZ, the trade winds, the monsoons, and the equatorial westerlies. These are fundamentally related to the patterns of rainfall and aridity in the low latitudes and to the seasonality of rainfall in the semi-arid regions. These systems are critical in understanding the climates of most of the dryland regions, especially the causes of aridity. This is particularly true of the coastal deserts, which are intimately linked to the subtropical highs.

4.3.1 SUBTROPICAL HIGH

The subtropical highs are semi-permanent circulation features that shift seasonally some 5° of latitude. They are closed circulation cells, especially in summer when they are situated over the ocean areas. In the Southern Hemisphere winter, when high pressure builds up over the continents as well, the cells merge to form a latitudinal high-pressure belt in the subtropics. The subtropical highs are basically regions of subsiding (sinking) air and low horizontal wind velocity. These characteristics produce stable and dry air masses associated with fair weather and clear skies.

Figure 4.7 shows the patterns of surface airflow around regions of high pressure and low pressure in the extra-tropical latitudes. There is clockwise/counterclockwise motion around a high in the Northern/Southern Hemisphere and the reverse around a low. Near the surface, where friction is present, air flows at a small angle across the isobars, away from the center of the high. The net result is divergent airstreams. Because air is relatively incompressible, divergence near the surface must be counteracted by subsidence. In contrast, air converges into the center of a low-pressure cell, creating rising motion.

The stability of the subtropical highs results from adiabatic warming of air as it subsides. The warming occurs because

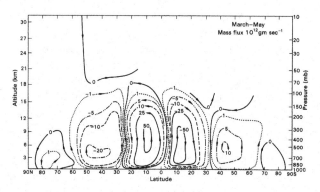

Fig. 4.5 The global mean vertical circulation for March-to-May (from Newell *et al.* 1969).

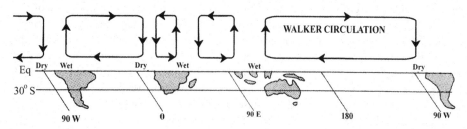

Fig. 4.6 Schematic view of the equatorial east–west Walker circulation around the globe (based on Webster 1983 and others).

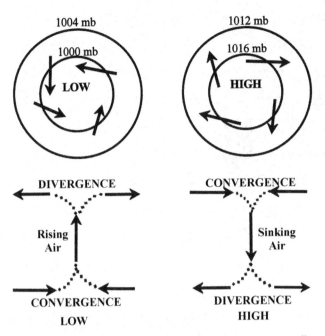

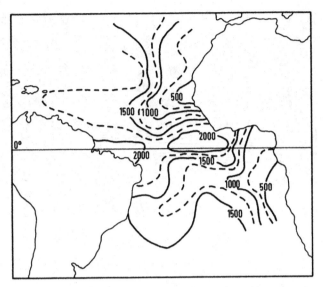

Fig. 4.10 Height of the base of the trade wind inversion over the Atlantic (after Riehl 1979, with permission of Elsevier).

Fig. 4.7 Schematic of the airflow around low- and high-pressure cells.

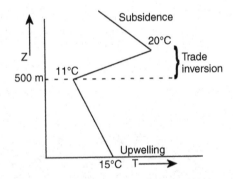

Fig. 4.8 A typical temperature profile through the trade inversion layer.

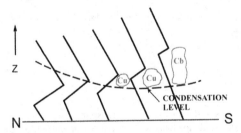

Fig. 4.9 North–south variation in the structure of the trade inversion, compared with cumulus development and lifting condensation level (the level at which rising air cools to the dew point, allowing clouds to form). This schematic represents the change from the centers of the subtropical highs to the ITCZ.

pressure steadily increases downward, so that a subsiding air mass is progressively compressed. This creates a temperature inversion (called the trade inversion) (Fig. 4.8). Inversions are associated with extreme thermal stability, which greatly limits cloud development and convective activity (Fig. 4.9). The subsiding air within the high-pressure cells does not penetrate to

the surface, but instead overlies a surface layer of relatively cool air. This enhances the inversion, particularly in the eastern part where the airflow over the ocean surface causes the upwelling of cold water from below. These inversions are well developed along the arid coasts of South America, Africa, and North America.

The height of the base of the inversion (Fig. 4.10) increases from the center of the high toward the equator. The height of the base also decreases toward its eastern edge. At the same time, the thickness and intensity of the inversion and the lifting condensation level (the level where clouds develop) steadily increase. The result is generally clear skies in the center of the subtropical high, but cloud development steadily increasing equatorward toward the ITCZ. Over the Atlantic, the base of the inversion varies from about 2000 m near the equator to 500 m on the eastern flank of the two high-pressure cells. The base is at about 450 m over southern California on the eastern flank of the North Pacific High, and 700–800 m along the Pacific coast of South America (Hastenrath 1988a).

Over the global oceans the drag of the wind produces a pattern of surface currents that strongly resemble the airflow around the highs (Fig. 4.11). The current continually changes direction with depth, for some distance below the surface, because each deeper layer is progressively further from the drag force of the surface wind (Fig. 4.12). This directional change is referred to as the Ekman spiral.

The net effect of the spiral is that the net transport of water in the subsurface layer is perpendicular to the surface wind (to the left in the Southern Hemisphere and to the right in the Northern Hemisphere). When wind blows parallel to a coastline, the net transport of water in the Ekman layer can be away from the coast (depending on the wind direction). The water transported seaward is replaced by colder, "upwelled" water from below. This results in cold water currents along the eastern sides of the subtropical highs (Fig. 4.11). The cooling is enhanced by

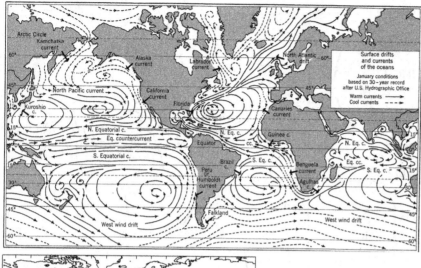

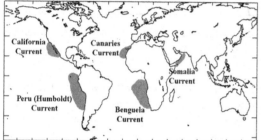

Fig. 4.11 (upper) Global pattern of surface ocean currents (from the US Navy Oceanography Office). Major cold and warm currents are indicated. (lower) Location of cold surface ocean currents resulting from coastal upwelling. Most lie on the eastern flank of the subtropical highs. Indicated temperatures are approximate anomalies from the temperatures further seaward.

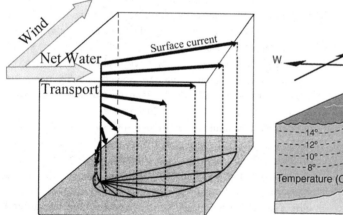

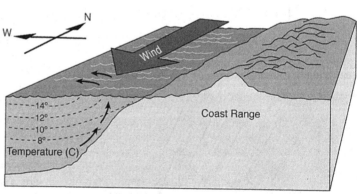

Fig. 4.12 The development of upwelling. (left) The Ekman spiral in the near-surface layers of the ocean. The net transport of water is perpendicular to the surface wind direction. (right) When wind blows parallel to the coast and from the south/north in the Southern/Northern Hemisphere, the net transport of water is away from the coast. In such cases, cold water from below rises to replace the water transported seaward (from Warner 2004).

advection, as the current flows from higher to lower latitudes. The cold water stabilizes the surface air and reinforces the effect of the trade inversion on the eastern flank of the subtropical highs, especially on its equatorward side.

4.3.2 THE TRADE WINDS

The trade winds are the easterly surface flow on the equatorial flanks of the subtropical highs (Fig. 4.13). They blow northeasterly in the Northern Hemisphere, southeasterly in the southern. The trade winds converge in the equatorial latitudes to produce the Intertropical Convergence Zone (ITCZ), an area of generally rising air that favors the production of rainfall. The trades do not extend into the convergence zone; there they are replaced by low, variable wind (doldrums) (Fig. 4.14) or by equatorial westerlies.

The trade winds are best developed over the eastern portions of the oceans. They cover approximately 20° of latitude in the summer hemisphere and 30° of latitude in the winter

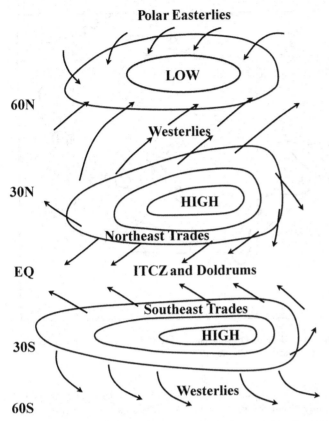

Fig. 4.13 Diagram illustrating the relationship between the subtropical high, the trade winds, and the Intertropical Convergence Zone.

Commonly, the region is thought of as a continuous zone of cloudiness and ascending air (Fig. 4.15). This picture is inaccurate. The ITCZ and the convective activity coupled with it are not continuous in either time or space; day-to-day variations are common. A quasi-continuous cloud band is sometimes apparent, but at other times no convection is evident at all. Much of the ITCZ consists of mesoscale cloud clusters (see Chapter 5), with subsidence in the clear air surrounding the clusters. In many cases, the cloud clusters are not even linked to the convergence of the trade winds, but instead to the propagating wave disturbances described in Section 5.4. This is the case over West Africa (Nicholson 2009). Sometimes, two or more convergence zones are evident, especially in parts of the western Pacific or western Atlantic.

Figure 4.16 shows the classical image of the ITCZ and its seasonal changes. The ITCZ has a mean position at about 5° north of the equator. Various theories to explain this fact have been proposed. One of the simplest suggests that the ITCZ is also the earth's thermal equator, dividing areas of the earth receiving equal amounts of heat. Because the Southern Hemisphere is colder than the northern, the ITCZ lies on average in the Northern Hemisphere. Accordingly, for similar reasons, the ITCZ always moves into the summer hemisphere, following the sun by a month or two. This theory would also explain why the seasonal changes are markedly greater over the continents than over the water, since the summer/winter temperature change is markedly greater over land.

It is now recognized that this picture is valid primarily over the oceans (Barry and Chorley 1999). But even there the rainfall maximum does not necessarily coincide with the ITCZ itself (Tomas and Webster 1997). Over the continents surface convergence zones may exist. However, they generally do not mark the convergence of trade winds (which are not well developed over continents), they are seldom continuous and are generally not co-located with the rainfall maximum. Over India, for example, the maximum rainfall is considerably north of the ITCZ over the Indian Ocean (Webster and Fasullo 2003). Over South America the term "equatorial trough" is generally used instead of ITCZ (Satyamurty *et al.* 1998). Nicholson (2009) likewise suggests using the term "tropical rainbelt" for the rainbelts within the equatorial trough over Africa. Over West Africa the rainbelt lies some 10 degrees of latitude south of the surface convergence marking the ITCZ. Many have also suggested distinguishing between the "marine ITCZ," which bears some similarity to Fig. 4.16, and the "land ITCZ," which essentially marks the loci of large numbers of convective disturbances (Holton *et al.* 1971).

hemisphere (Fig. 4.14). The trades are strongest in winter when the subtropical highs are most intense. Typical speeds are 2–3 m/s in summer and 4–5 m/s in winter. Their extent varies with both season and longitude because the latitude of the equatorial trough, which marks their equatorward boundary, exhibits large seasonal variations.

The trade winds are noted for their extreme constancy in both speed and direction, characteristics that reflect the permanence of the subtropical highs. The trade winds are only occasionally interrupted by atmospheric disturbances. Over the ocean, they blow within ±45° of ENE or ESE 95% of the time. In the region of the trades, there is little convective activity. Cumulus clouds will often form but with little vertical development. Convection is hindered both by the trade inversion and by weak subsidence within the trades.

4.3.3 EQUATORIAL TROUGH AND INTERTROPICAL CONVERGENCE ZONE

Near the equator is a region of low pressure and converging airflow. The terms equatorial trough and ITCZ are both used in referring to this region. However, the lowest pressure (the core of the trough) does not coincide with the region of maximum wind convergence. This region is cloudy, compared with the trade wind zone, and is the site of much convective activity.

4.3.4 EQUATORIAL WESTERLIES

The trade wind system is complicated by the changes in dynamic forces as air flows across the equator. The sign of the earth's rotational, or Coriolis force, is different in the two hemispheres. Consequently easterly flow veers and takes on a westerly direction as an air parcel crosses the equator. The result is extensive

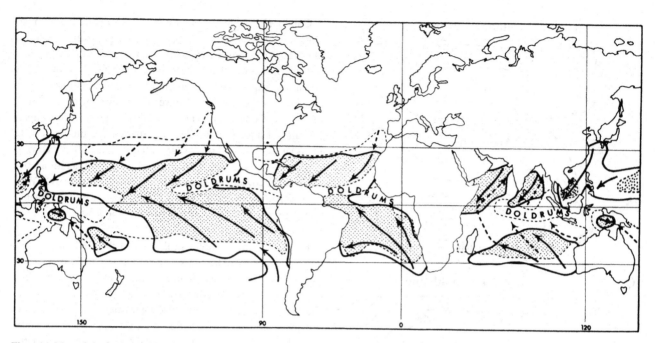

Fig. 4.14 Map of the trade wind belts and the doldrums. The limits of the trades are delimited by solid (January) and dashed (July) lines; the shaded area is affected by trade winds in both months; arrows indicate streamlines (from Barry and Chorley 1999).

Fig. 4.15 Satellite photo showing a well-developed Intertropical Convergence Zone (ITCZ) over the Atlantic.

regions of surface westerlies in the equatorial latitudes (Flohn 1960).

The equatorial westerlies are a shallow system overlain by the tropical easterlies. They are best developed over the continents and in the Northern Hemisphere during its summer season (Fig. 4.17). The equatorial westerlies include the southwest monsoon flow in areas such as India and Africa. In some cases, they may be transient features.

The importance of the equatorial westerlies is that, in contrast to the trade winds, they are often associated with cloudiness and rainfall. The reason for this relates to complex aspects of the operative dynamic forces. One study of the equatorial Atlantic found that westerlies were accompanied by precipitation about 25% of the time (Flohn 1960). In contrast, rainfall occurred on only about 8% of the occasions when the easterly trades prevailed. The equatorial westerlies influence the rainfall patterns in a number of semi-arid regions, particularly in Africa and Australia, and are a major component of the Asian monsoon system.

Over West Africa the equatorial westerlies occasionally become a very deep system, extending into the mid-troposphere. In such cases, a low-level westerly jet with a core around 850 mb develops. This jet is a very important factor in the interannual variability of rainfall in the Sahel region (Nicholson 2009; Nicholson and Webster 2007).

4.3.5 THE MONSOONS

A common definition of monsoonal climates requires that three criteria be met: a seasonal wind shift of at least 120° between January and July; a certain minimum intensity of wind; and directional persistence such that the wind must be in the prevailing direction at least 40% of the time (Ramage 1971). By this definition, a number of regions over the earth are defined as monsoonal, including West Africa, the southwestern USA, Asia, and northern Australia. These are semi-arid regions where the monsoon of one season brings rainfall, the other dryness. Many sources (e.g., Vera 2006) also refer to the "South American monsoon," but there is no universal agreement that a monsoon climate exists over equatorial South America (Zhou and Lau 1998). Here only the Asian monsoon is described because it is the largest and most intense and because it affects numerous dryland regions; the surface pressure pattern corresponding to it is illustrated in Chapter 20.

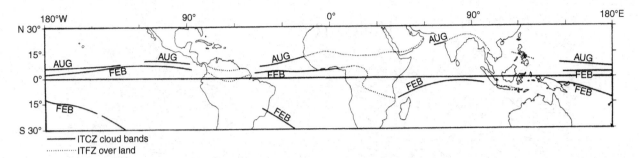

Fig. 4.16 The position of the Intertropical Convergence Zone (ITCZ) in February and August (after Barry and Chorley 1999). Note that it is drawn as a dashed line over land, where it becomes very indistinct.

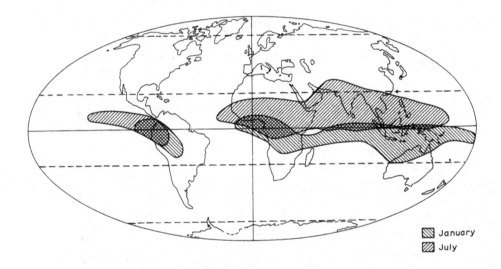

Fig. 4.17 Distribution of the equatorial westerlies in the layer below 3 km (after Flohn 1965).

A principal cause of the monsoons is differential heating of land and water (Webster 2006). However, other factors play a role, such that the monsoon represents a coupled, self-regulating system of steady interaction between the ocean and atmosphere (Webster and Fasullo 2003). In winter, cold anticyclones develop over the land while heat lows develop over the land in summer. This explains the great intensity of the Asian monsoon: the tremendous contrast between the intense summer heating of the Tibetan plateau at high altitudes and the intense cold that develops over Siberia in winter. The cold is extreme because the prevailing westerlies and the highlands of Asia to the south completely insulate Siberia from receiving any warmer flow from southerly latitudes.

The southeast summer monsoon is a warm, humid air mass more than 3000 m deep. It consists of three major airstreams: the Indian monsoon, the Australian monsoon, and the Pacific Southeast monsoon. During the monsoon season, frequent thunderstorms and heavy rainfall and occasional tropical disturbances, such as monsoon depressions and typhoons, occur. The rainfall distribution within the season is made up of sequences of rainy periods ("active" periods) interrupted by dry periods ("break" periods), each lasting 10–30 days (Webster and Hoyos 2004). There is no preferred timing of these intraseasonal variations, but the frequency of occurrence is on the order of every 25–80 days (Hoyos and Webster 2007).

The winter monsoon is coupled with a very shallow high-pressure cell over the Tibetan plateau and Siberia. Subsidence is associated with the high and with the prevailing westerlies. A continual stream of stable and relatively cold air, called the Northeast monsoon, prevails over Asia. It generally brings fair weather, but is occasionally interrupted by extra-tropical disturbances associated with the westerlies.

4.4 SEASONAL CHANGES OF THE GENERAL ATMOSPHERIC CIRCULATION

The general circulation changes seasonally in a regular fashion dictated by the changing planetary thermal conditions. In comparison with the summer hemisphere, the winter hemisphere has steeper latitudinal gradients of radiation and temperature and the land–water contrast is reversed (the land is colder than the water). Furthermore, the ITCZ advances into the summer hemisphere (see Section 4.3.3), so that the winter hemisphere is the larger of the two.

Wind and pressure patterns for January and July (Fig. 4.18) underscore the contrast between the summer and winter

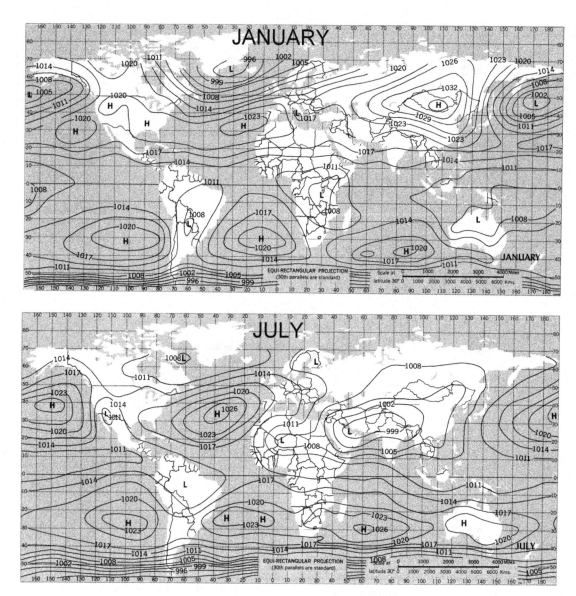

Fig. 4.18 Surface circulation patterns in January and July, with sea-level pressure in mb indicated (after Mintz and Dean 1952).

hemispheres. In the summer hemisphere, the mid-latitude continents are regions of low pressure and the subtropical highs lie over the oceans. These highs are the dominant winter pressure features: the Azores and North Pacific Highs in the Northern Hemisphere, the St. Helena and Mascarene Highs in the South Atlantic and Indian Ocean, respectively, and the South Pacific High. Between winter and summer the subtropical highs expand and move poleward. Other circulation features are also displaced poleward in summer, as the latitude of maximum heating moves into that hemisphere. This includes the vertical circulations (Fig. 4.5).

In the winter hemisphere, the mid-latitude lows lie over the high latitude oceans and they are the dominant pressure features in that hemisphere. High pressure prevails over the mid-latitude continents and the subtropical highs weaken and shift equatorward. In general, the surface circulation is considerably more intense in the winter hemisphere than in the summer hemisphere, a direct consequence of the strong mid-latitude winter temperature gradient.

4.5 UPPER AIR PATTERNS

Above the earth's surface, the character of the general circulation changes markedly as the thermal effect of the land–ocean contrast weakens and the effect of friction from the earth's surface wanes. As a result, the flow aloft (Fig. 4.19) is more zonal (more directly east–west with fewer undulations) and more geostrophic (i.e., the winds blow along surfaces of equal pressure rather than toward low pressure).

The vertical structure of the earth's major wind systems is shown in Fig. 4.20. The westerlies are dominant throughout the

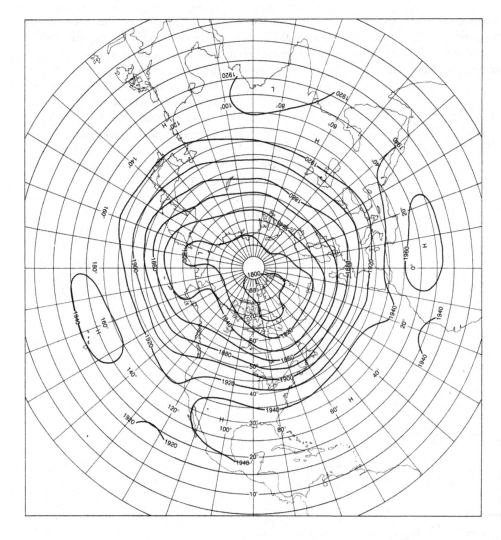

Fig. 4.19 Average height of the 500 mb surface in the Northern Hemisphere in January and July (after US Dept. of Commerce).

depth of the atmospheric column in winter, but are overlain by easterlies in the summer hemisphere. In the winter, when temperature gradients are strong, the westerlies are clearly more intense and they are displaced equatorward. Over the ocean the tropical easterlies prevail throughout the tropical atmosphere to the tropopause, but they generally override surface westerlies over land. The tropical easterlies are considerably weaker than the mid-latitude westerlies. They are more expansive in the Northern Hemisphere summer, when the extreme heating of the Tibetan plateau and the Sahara produces a strong north–south temperature gradient between the land and the Indian Ocean to the south. The polar easterlies are a very shallow system and are stronger in the winter hemisphere.

4.6 JET STREAMS

Important features of the upper-level circulation are the major jet streams, evident in Fig. 4.20. Dominant are westerly jet streams of the extra-tropical latitudes, with maxima in the upper troposphere. In the tropics, an easterly jet (see below) is apparent in the upper troposphere. The annual mean of the zonal (i.e., east–west) circulation shows one westerly jet core in each hemisphere. However, there are actually two westerly jets (Fig. 4.21). They are usually distinct in winter (Fig. 4.22), but they tend to merge in summer. These jets result from the latitudinal temperature gradient, which has two maxima in winter, one near the Arctic and one near the area of continent–ocean contrast in the subtropics. The former is associated with the strong temperature gradients in the vicinity of the polar front and hence is often called the polar-front jet (PFJ). The latter is termed the subtropical jet. Both features migrate latitudinally, poleward in summer, equatorward in winter.

The polar-front jet (PFJ) is a strongly meandering jet with a core at about 300 mb. Core speeds are generally on the order of 150 m/s (60 m/s is the monthly mean). A PFJ also exists in the Southern Hemisphere and its core speeds are comparable to those in the Northern Hemisphere. The PFJ provides a mechanism of instability that gives rise to disturbances (the transient low-pressure systems of the mid-latitudes); it also plays a role in steering these disturbances.

The subtropical jet (Fig. 4.23) shows three maxima over the Northern Hemisphere, corresponding to areas of strong thermal contrast between land and water (Krishnamurti 1961). Although linked to areas of strong temperature contrast, the subtropical jet actually results from a number of factors, including outflow aloft from regions of convection and from the Hadley circulation

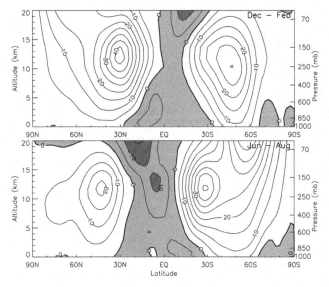

Fig. 4.20 Mean zonal winds for December–February and March–May, based on NCEP data. Areas of easterly winds are shaded. Units: m s⁻¹.

cells (Webster and Yang 1992). The subtropical jet is roughly co-located with the surface position of its subsiding branch. The jet's core is at about 200 mb, with speeds of about 40–50 m/s. Like the polar-front jet, it plays a role in the development and steering of mid-latitude cyclones. It also brings them into the northern subtropics in winter when it is displaced equatorward. The subtropical jet of the Southern Hemisphere exists over Australia, where its core lies at roughly 27° S in winter and its mean speed exceeds 50 m/s (Adams 1983), and over South America, where its latitudinal displacements influence convection over tropical parts of the continent (Gonzalez and Barros 1998).

The upper-level easterly jet core evident at low latitudes in Fig. 4.20 is associated with the tropical easterly jet (TEJ) (Raman *et al.* 2009). This jet is a prominent summer feature with maximum winds of about 30–40 m/s near 150 mb. It has a thermal origin, the high-level temperature contrast between the strongly heated Tibetan plateau and the air overlying the Indian Ocean to the south. The TEJ extends from eastern Asia to West Africa, centered at about 0° of latitude (Fig. 4.24). It may play a role in the formation of wave disturbances over Africa (Nicholson *et al.* 2007, 2008) and the Indian Ocean (Mishra 1993); it definitely provides a steering mechanism for these systems. Over India the TEJ shows considerable variation within the monsoon season and its fluctuations are highly correlated with convective activity (Sathiyamoorthy *et al.* 2007).

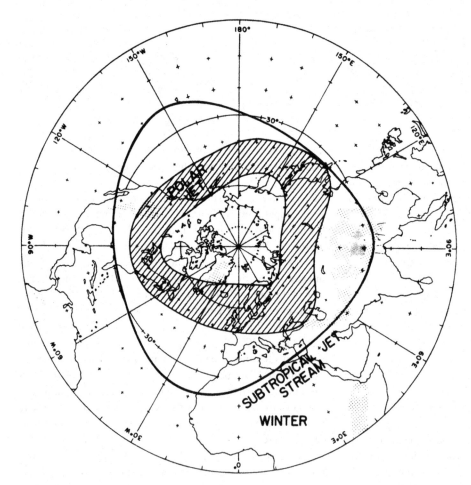

Fig. 4.21 Mean axis of subtropical jet stream during winter and area (shaded) of principal activity of the polar-front jet stream (modified from Riehl 1962).

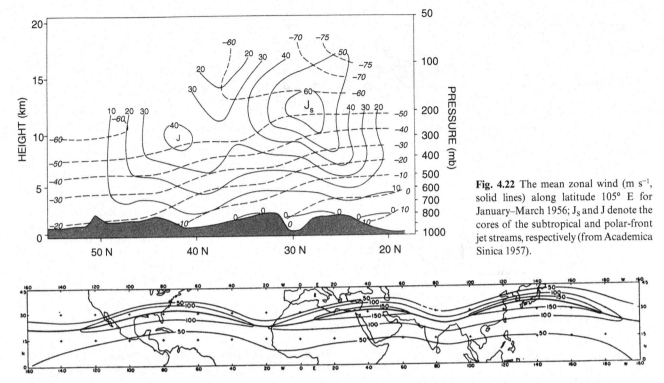

Fig. 4.22 The mean zonal wind (m s^{-1}, solid lines) along latitude 105° E for January–March 1956; J$_s$ and J denote the cores of the subtropical and polar-front jet streams, respectively (from Academica Sinica 1957).

Fig. 4.23 Mean subtropical jet stream for winter 1955/56. Solid line shows mean position of jet stream core and dashed lines are isotachs in knots (after Krishnamurti 1961).

Several other jets are worth mentioning, since they strongly affect regional climates, especially those of many dryland regions. Some reside in the mid-troposphere, others in the lowest levels of the atmosphere. These jet streams variously transport moisture, intensify aridity, create mesoscale circulations, modify the diurnal cycle of precipitation, and trigger or promote convective disturbances. Since they are only regionally important, details about these jet streams and their climatological consequences are found in the relevant chapters dealing with individual continents.

Part of the influence of jet streams relates to a secondary circulation developed around the jet's core. This is known as a "jet streak" circulation (Uccellini and Johnson 1979). Depending on the characteristics of the jet, the acceleration and deceleration of the airstreams entering and exiting the core can produce departures from geostrophic motion that produce a checkerboard pattern of convergence/divergence and vertical motion. This is illustrated in Fig. 4.25. In the Northern Hemisphere, the right/left quadrant of its entrance region is characterized by divergence/convergence and ascent/descent below the level of the jet.

Perhaps the best known mid-level jet is the African easterly jet (AEJ) that lies above North Africa (Parker *et al.* 2005). It is a consequence of the temperature contrast between the warm Saharan air and the cooler air to the south over the coastal forests and the Gulf of Guinea in the Atlantic. It provides the instability for the growth of African easterly waves and provides a steering mechanism for these disturbances. It also transports dust from the African continent across the Atlantic into the Caribbean.

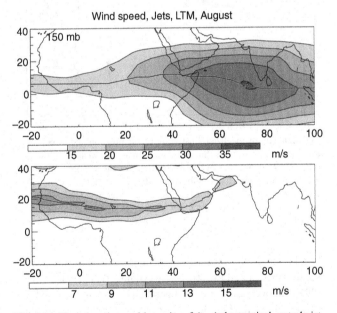

Fig. 4.24 Mean location and intensity of (top) the tropical easterly jet (TEJ) at 150 mb and (bottom) the African easterly jet (AEJ) at 650 mb during August (from Nicholson 2008).

A second mid-tropospheric easterly jet appears in the southern subtropics of Africa during some seasons and this appears to influence the rainy season in these latitudes (Jackson *et al.* 2009). Nicholson and Grist (2002) distinguish these jets by utilizing the terms AEJ-N and AEJ-S for those in the respective hemispheres.

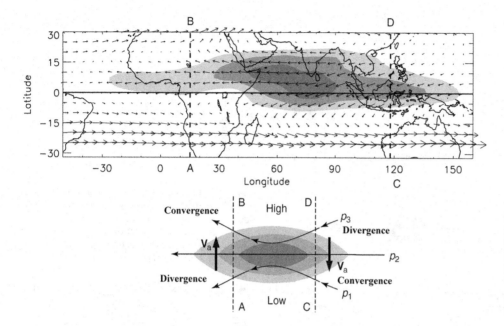

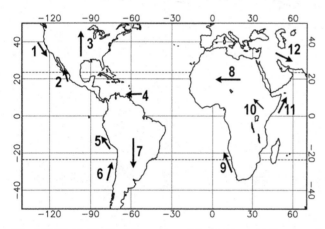

Fig. 4.25 Secondary circulation associated with the entrance and exit regions of the Tropical Easterly Jet (from Webster and Fasullo 2003). The arrows labeled V_a represent the departures from geostrophic motion that produce the secondary circulation.

Fig. 4.26 Highly idealized illustration of the position and direction of the low-level jet streams that influence climate and weather in the drylands. A transient, low-level northerly jet (not shown) also exists over northern Australia. 1 = California coastal jet, 2 = Baja jet, 3 = Great Plains LLJ, 4 = Caribbean LLJ, 5 = Peruvian jet, 6 = Chilean coastal jet, 7 = South American LLJ, 8 = Bodélé jet, 9 = Benguela jet, 10 = Turkana jet, 11 = Somali jet, 12 = Iranian jet.

Mid-level easterly jets exist also near the equator over western South America and the far eastern Atlantic (Hastenrath 1998b, 1999) and over northwestern Australia (Krishnamurti 1979).

Low-level jets (Fig. 4.26), with cores at elevations ranging from 500 to 2000 m, are apparent along several coasts and in some continental interiors (Stensrud 1996). The best known are those along the Peruvian and Somali coasts (Lettau 1978; Hart *et al.* 1978), the Great Plains low-level jet of the central USA (Walters *et al.* 2008), the Turkana jet of East Africa (Indeje *et al.* 2001), and the South American low-level jet (SALLJ) east of the Andes (Marengo *et al.* 2004). The SALLJ has a significant influence on rainfall (Liebmann *et al.* 2004; Salio *et al.* 2007). Various studies have also demonstrated low-level jets along the

Benguela coast (Nicholson 2010), in the Iranian desert (Liu *et al.* 2000), over Australia (Brook 1985; Riley 1989), in the equatorial eastern Pacific (Hastenrath 1998b), along the California coast (Parish 2000), in the Gulf of California (Douglas 1995), in the Caribbean (Muñoz *et al.* 2008), along the coast of north-central Chile (Muñoz and Garreaud 2005), and near the Bodélé Depression of North Africa (Washington *et al.* 2006).

REFERENCES

Academica Sinica, 1957: On the general circulation over eastern Asia. *Tellus,* **9**, 432–446.

Adams, M., 1983: *The Subtropical Jet in the Australian Region.* Meteorological Note 148. Bureau of Meteorology, Melbourne, 37 pp.

Barry, R. G., and Chorley, R. J., 1999: *Atmosphere, Weather and Climate.* Routledge, London, 409 pp.

Brook, R. R., 1985: The Koorin nocturnal low level jet. *Boundary-Layer Meteorology,* **32**, 133–154.

Douglas, M. W., 1995: The summer low-level jet over the Gulf of California. *Monthly Weather Review,* **123**, 2334–2347.

Douglas, H. A., P. J. Mason, and R. Hide, 1972: Investigation of structure of baroclinic waves using 3-level streak photography. *Quarterly Journal of the Royal Meteorological Society.,* **98**, 247–263.

Flohn, H., 1965: Equatorial westerlies over Africa, their extension and significance. *Bonner Meteorologische Abhandlungen,* **5**, 36–48.

Gonzalez, M., and V. Barros, 1998: The relationship between tropical convection in South America and the end of the dry period in subtropical Argentina. *International Journal of Climatology,* **18**, 1669–1685.

Hart, J. E., G. V. Rao, H. van de Boogard, J. A. Young, and J. Findlater, 1978: Aerial observations of the East African low-level jet stream. *Monthly Weather Review,* **106**, 1714–1724.

Hastenrath, S., 1988a: *Climate and Circulation of the Tropics.* Reidel, Dordrecht, The Netherlands, 455 pp.

Hastenrath, S., 1998b: Contribution to the circulation climatology of the eastern equatorial Pacific: lower-atmospheric jets. *Journal of Geophysical Research,* **103**, 19443–19451.

Hastenrath, S., 1999: Equatorial mid-tropospheric easterly jet over the eastern Pacific. *Journal of the Meteorological Society of Japan*, **77**, 701–709.

Hastenrath, S., 2001: Equatorial zonal circulations from the NCEP-NCA reanalysis. *Journal of the Meteorological Society of Japan*, **79**, 719–728.

Hastenrath, S., 2007: Equatorial zonal circulations: historical perspectives. *Dynamics of Atmospheres and Oceans*, **43**, 16–24.

Hastenrath, S., D. Polzin, and L. Greischar, 2002: Annual cycle of equatorial zonal circulations from the ECMWF Reanalysis. *Journal of the Meteorological Society of Japan*, **80**, 755–766.

Holton, J. T., J. M. Wallace and J. A. Young, 1971: Boundary layer dynamics and the ITCZ. *Journal of the Atmospheric Sciences*, **28**, 275–280.

Hoyos, C., and P. J. Webster, 2007: The role of intraseasonal variability in the nature of Asian monsoon precipitation. *Journal of Climate*, **20**, 4402–4424.

Indeje, M., F. H. M. Semazzi, L. Xie, and L. J. Ogallo, 2001: Mechanistic model simulations of the East African climate using NCAR regional climate model: influence of large-scale orography on the Turkana low-level jet. *Journal of Climate*, **14**, 2710–2724.

Jackson, B., S. E. Nicholson, and D. Klotter, 2009: Mesoscale convective systems over western equatorial Africa and their relationship to large-scale circulation. *Monthly Weather Review*, **137**, 1272–1294.

Krishnamurti, T. N., 1961: The subtropical jet stream of winter. *Journal of Meteorology*, **18**, 172–191.

Krishnamurti, T. N., 1979: *Tropical Meteorology: Compendium of Meteorology*. Vol. II, Part 4. World Meteorological Organization, Geneva, 428 pp.

Lettau, H. H., 1978: Explaining the world's driest climate. In *Exploring the World's Driest Climate* (H. H. Lettau and K. Lettau, eds.), Institute for Environmental Studies, University of Wisconsin, Madison, WI, pp. 182–248.

Liebmann, B., G. N. Kiladis, C. Vera, A. C. Saulo, and L. M. V. Carvalho, 2004: Subseasonal variations of rainfall in South America in the vicinity of the low-level jet east of the Andes and comparison to those in the South Atlantic convergence zone. *Journal of Climate*, **17**, 3829–3842.

Liu, M., D. L. Westphal, T. R. Holt, and Q. Xu, 2000: Numerical simulation of a low-level jet over complex terrain in southern Iran. *Monthly Weather Review*, **128**, 1309–1327.

Marengo, J. A., W. R. Soares, S. Saulo, and M. Nicolini, 2004: Climatology of the low-level jet east of the Andes as derived from the NCEP-NCAR Reanalyses: characteristics and temporal variability. *Journal of Climate*, **17**, 2261–2280.

Mintz, Y., and G. Dean, 1952: *The Observed Mean Field of Motion of the Atmosphere*. Geophysics Research Paper No. 17, Air Force Cambridge Research Center, Cambridge, MA.

Mishra, S. K., 1993: Non-linear barotropic instability of upper-tropospheric tropical easterly jet on the sphere. *Journal of the Atmospheric Sciences*, **50**, 3541–3552.

Muñoz, E., A. J. Busalacchi, S. Nigam, and A. Ruiz-Barradas, 2008: Winter and summer structure of the Caribbean low-level jet. *Journal of Climate*, **21**, 1260–1276.

Muñoz, R. C., and R. D. Garreaud, 2005: Dynamics of the low-level jet off the west coast of subtropical South America. *Monthly Weather Review*, **133**, 3661–3677.

Newell, R. E., Vincent, D. G., Dopplick, T. G., Ferruza, D., and Kidson, J. W., 1969: The energy balance of the global atmosphere. In *The Global Circulation of the Atmosphere* (Corby, G. A., ed.). Royal Meteorological Society, London.

Nicholson, S. E., 2008: The intensity, location, and structure of the tropical rainbelt over west Africa as factors in interannual variability. *International Journal of Climatology*, **28**, 1775–1785.

Nicholson, S. E., 2009: A revised picture of the structure of the "monsoon" and land ITCZ over West Africa. *Climate Dynamics*, **32**, 1155–1171.

Nicholson, S. E., 2010: A low-level jet along the Benguela coast, an integral part of the Benguela current ecosystem. *Climatic Change*, **33**, 313–330.

Nicholson, S. E., and J. P. Grist, 2002: On the seasonal evolution of atmospheric circulation over West Africa and Equatorial Africa. *Journal of Climate*, **16**, 1013–1030.

Nicholson S. E., and P. J. Webster, 2007: A physical basis for the interannual variability of rainfall in the Sahel. *Quarterly Journal of the Royal Meteorological Society*, **133**, 2065–2084.

Nicholson, S. E., A. I. Barcilon, M. Challa, and J. Baum, 2007: Wave activity on the Tropical Easterly Jet. *Journal of the Atmospheric Sciences*, **64**, 2756–2763.

Nicholson, S. E., A. I. Barcilon, and M. Challa, 2008: An analysis of West African dynamics using a linearized GCM. *Journal of the Atmospheric Sciences*, **65**, 1182–1203.

Parish, T., 2000: Forcing of the summer low-level jet along the California coast. *Journal of Applied Meteorology*, **39**, 2421–2433.

Parker, D. J., C. D. Thorncroft, R. R. Burton, and A. Diongue-Niang, 2005: Analysis of the African easterly jet, using aircraft observations from the JET2000 experiment. *Quarterly Journal of the Royal Meteorological Society*, **131**, 1461–1482.

Ramage, C. S., 1971: *Monsoon Meteorology*. Academic Press, New York, 296 pp.

Raman, M. R., and Coauthors, 2009: Characteristics of the Tropical Easterly Jet: long-term trends and their features during active and break monsoon. *Journal of Geophysical Research–Atmospheres*, **114**, D19105.

Riehl, H., 1962: *Jet Streams of the Atmosphere*. Department of Atmospheric Science, Colorado State University, Fort Collins, Technical Report No. 32, 117 pp.

Riehl, H., 1979: *Climate and Weather in the Tropics*. Academic Press, London, 611 pp.

Riley, P. A., 1989: *The Diurnal Variation of the Low-Level Jet over the Northern Territory*. Meteorological Note 188, Bureau of Meteorology, Melbourne, 31 pp.

Salio, P., M. Nicolini, and E. J. Zipser, 2007: Mesoscale convective systems over southeastern South America and their relationship with the South American Low-Level Jet. *Monthly Weather Review*, **135**, 1290–1309.

Sathiyamoorthy, V., P. K. Pal, and P. C. Joshi, 2007: Intraseasonal variability of the Tropical Easterly Jet. *Meteorology and Atmospheric Physics*, **96**, 305–316.

Satyamurty, P., C. A. Nobre, and P. L. Silva Dias, 1998: South America. In *Meteorology of the Southern Hemisphere* (D. J. Karoly and D. G. Vincent, eds.), American Meteorological Society, Boston, MA, pp. 119–139.

Stensrud, D. J., 1996: Importance of low-level jets to climate: A review. *Journal of Climate*, **10**, 1698–1711.

Tomas, R. A., and P. J. Webster, 1997: The role of inertial instability in determining the location and strength of near-equatorial convection. *Quarterly Journal of the Royal Meteorological Society*, **123**, 1445–1482.

Trewartha, G. T., and L. H. Horn 1980: *An Introduction to Climate*. McGraw-Hill Book Company, New York, 416 pp.

Uccellini, L. W., and D. R. Johnson, 1979: The coupling of upper and lower tropospheric jet streaks and implications for the development of severe convective storms. *Monthly Weather Review*, **107**, 682–703.

Vera, C., and Coauthors, 2006: Toward a unified view of the American monsoon systems. *Journal of Climate*, **19**, 4977–5000.

Walters, C. K., J. A. Winkler, R. P. Shadbolt, J. van Ravensway, and G. D. Bierly, 2008: A long-term climatology of southerly and northerly low-level jets for the central United States. *Annals of the Association of American Geographers*, **98**, 521–552.

Warner, T. T., 2004: *Desert Meteorology*. Cambridge University Press, Cambridge, UK, 595 pp.

Washington, R., M. C. Todd, S. Engelstaedter, S. Mbainayel, and F. Mitchell, 2006: Dust and the low-level circulation over the Bodélé Depression, Chad: observations from BoDEx 2005. *Journal of Geophysical Research*, **111**, D03201, doi:10.1029/2005JD006502.

Webster, P. J., 1983: Large-scale structure of the tropical atmosphere. In *Large-Scale Dynamical Processes in the Atmosphere* (B. J. Hoskins and R. P. Pearce, eds.), Academic Press, London, pp. 235–275.

Webster, P. J., 2004: The elementary Hadley circulation. In *The Hadley Circulation: Past, Present and Future* (H. Diaz and R. Bradley, eds.), Cambridge University Press, Cambridge, pp. 9–60.

Webster, P. J., 2006: The coupled monsoon system. In *The Asian Monsoon* (B. Wang, ed.), Springer and Praxis, Berlin, pp. 3–66.

Webster, P. J., and J. Fasullo, 2003: *Monsoon: Dynamical Theory*. In *Encyclopedia of Atmospheric Sciences* (J. Holton and J. A. Curry, eds.), Academic Press, London, pp. 1370–1385.

Webster, P. J., and C. D. Hoyos, 2004: Forecasting monsoon rainfall and river discharge variability on 20–25 day time scales. *Bulletin of the American Meteorological Society*, **85**, 1745–1765.

Webster, P. J., and S. Yang, 1992: Monsoon and ENSO: selectively interacting systems. *Quarterly Journal of the Royal Meteorological Society*, **118**, 877–926.

Zhou, J. Y., and K. M. Lau, 1998: Does a monsoon climate exist over South America? *Journal of Climate*, **11**, 1020–1040.

5 The global distribution of arid climates and rainfall

5.1 PRECIPITATION AND ARIDITY

The key elements necessary for the formation of rainfall are moist, unstable air and ascent, which is often produced by a convergent pattern of airflow. Such convergence is associated with surface low-pressure systems. The factors that are classically considered to promote aridity are the converse: a lack of atmospheric moisture, stable air, subsidence, and divergent airflow. A fifth factor that must be considered is an absence of rain-bearing disturbances.

The first factor, a lack of sufficient moisture, is generally not the overriding influence in most arid climates. The case of the Sahara Desert clearly demonstrates this. Maps of precipitable water show that in summer the atmosphere above the Sahara contains about as much moisture as that over the wet southeastern USA. Likewise, the Namib Desert is a humid environment, although it is one of the "driest" deserts on earth in terms of rainfall. With the exception of the polar deserts, where a lack of available moisture plays a major role in the origin of the dry climate, an absence of moist air masses is a major factor in aridity only in the deep continental interior of Asia. Even there, however, the lee rain shadow effects of high terrain also play a major role.

The next three factors, stable air, subsidence, and divergence, are common to two meteorological situations linked to the occurrence of dry climates: the subtropical highs and the lee rain shadows of mountain barriers (see Section 5.2). Such rain shadow effects also play an important role in the aridity of the most extreme deserts of the Asian interior.

Collectively these factors produce a global mean pattern of high rainfall in the equatorial regions and low rainfall in the subtropical latitudes where the subtropical highs prevail (see Fig. 5.1). Rainfall increases in the mid-latitudes, where the influence of the highs gives way to that of the migrating mid-latitude lows. In high latitudes, rainfall is again very low because the low temperatures result in low atmospheric moisture content. In low latitudes, rainfall is predominantly in the high-sun season (the summer season of the respective hemisphere).

5.2 GLOBAL DISTRIBUTION AND ORIGIN OF THE DRYLAND CLIMATES

The earth's major deserts include the Sahara, Kalahari, Namib, and Somali-Chalbi in Africa, the Australian desert, the Monte-Patagonian and Peruvian-Atacama deserts of South America, the Arabian and Iranian deserts of the Middle East, the Thar or Rajasthan desert of India and Pakistan, the Turkestan, Taklamakan, and Gobi deserts of Central Asia, and the Sonoran, Mojave, and Chihuahuan deserts of North America. Their distribution is depicted in Fig. 1.2. These deserts are concentrated in the subtropics, where the aridifying influence of the subtropical highs plays a major role in their existence. Grasslands and woodlands with semi-arid climates often border these deserts, especially along their poleward or equatorward margins. Other semi-arid regions, particularly the grasslands of North America or Central Asia, are in somewhat higher latitudes where other factors play a role, the most common being the lee rain shadow of mountains.

The subtropical highs (Fig. 5.2) are primary factors in the development of dryland climates because four of the five main factors promoting aridity are associated with them. These include divergent air flow, subsidence, thermal stability (see Section 4.3), and blocking of disturbances by these stationary features.

The stability of the subtropical highs and the patterns of ascent and descent associated with the vertical Hadley and Walker cells produce the major areas of dryland climates in the mid- and low latitudes. There is thus a general tendency for aridity in subtropical latitudes, a tendency that is enhanced on the western sides of continents in these latitudes. The major dryland regions that can be linked to the subtropical highs include the Australian desert, the Peruvian-Atacama of South America, the southwestern USA, the Namib and Kalahari of southern Africa, and the western Sahara. However, this factor does not suffice to explain the full extent of the Sahara and its continuation into Arabia and the Middle East. More localized influences that also contribute are discussed in Section 5.5.

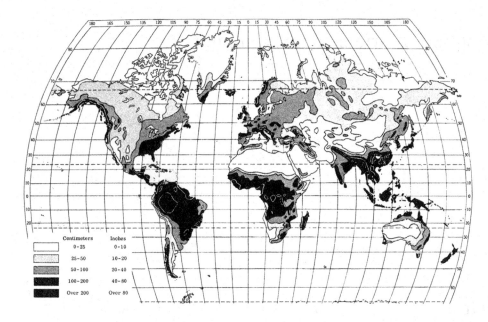

Fig. 5.1 Global patterns of annual rainfall (from Wallace and Hobbs 1977, copyright Elsevier).

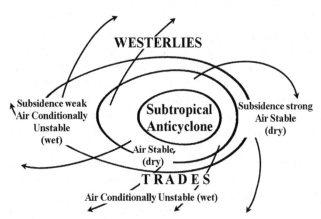

Fig. 5.2 Schematic east–west transect of the subtropical high (modified from Trewartha and Horn 1980).

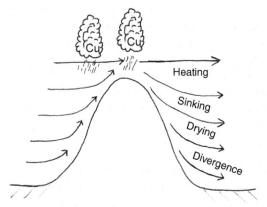

Fig. 5.3 The lee rain shadow can be explained by orographic effects on airflow. As air rises on the windward side, it cools and converges, often triggering clouds and precipitation. The result is drier air that is also heated via the latent heat released upon condensation. As the air subsides on the lee side, it warms adiabatically, further reducing its relative humidity. The sinking motion itself and divergence of the stream lines further contribute to aridity. The result is a "rain shadow," an area of dry climate, in the lee of large mountain barriers.

Another major site of arid climates is on the lee side of mountain ranges. The extent of the aridifying influence can be vast, as the dry climates of the western US Great Plains prove. The high terrain either effectively blocks airflow across it or vastly alters the characteristics of the airstream as it traverses the mountain barrier (Fig. 5.3). In the first case, major rain-bearing systems are deflected away from the lee side. They penetrate to the lee side only through breaks in the barrier or where the large-scale winds are particularly strong. Air masses that do rise above the barrier and penetrate to the lee side are greatly transformed, generally losing their moisture on the windward side where they rise and produce rainfall. The relatively dry air sinks and warms adiabatically via compression, thereby further reducing relative humidity. The subsiding airflow on the lee side is also divergent. Each of these factors effectively promotes aridity.

The major dryland locations where aridity is related to a lee rain shadow include the Patagonian Desert of South America, the intermontane region of the western USA, and the US Great

Plains. Topography also plays a role in the creation of deserts in Ethiopia and the Horn of Africa and, perhaps to some extent, the Thar Desert of India and Pakistan. However, in these cases the aridity cannot be explained by topography alone. In some of these situations, local low-level jet streams appear to complement the aridifying influence of topography, especially in the US Great Plains and in the Somali-Chalbi desert system in the Horn of Africa.

Elsewhere, such as Asia and the Middle East, more complex factors are important. The argument can be made that topography probably plays a significant role in the Iranian desert, where high terrain serves to produce rain shadows and to block the passage of rain-bearing disturbances. Similarly, the most extreme deserts of eastern Asia – the Gobi and the Taklamakan – are surrounded by high terrain. Here and elsewhere in Asia the distance from major sources of moisture and prevailing storm tracks appear to play some role. The latter factor seems to be fundamental in explaining the dry climates in the Mediterranean areas of the Middle East and in the Sahara.

The major point of this discussion is that, while many factors promote arid climates and one or more of these factors can be linked to most dryland regions, the causes of dryland climates are quite complex. In most cases, regional influences must be taken into account to thoroughly explain the existence of true deserts. These are described in detail in Section 5.5. The complexity and the regionalization of these factors give a nearly unique climatic identity to each of the earth's major dryland regions. They also share many common characteristics, climatically and environmentally, but these are consequences of the aridity and not the cause of it.

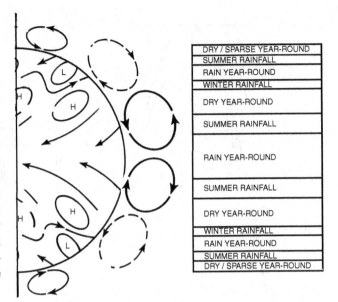

Fig. 5.4 Climatic zones corresponding to the major features of the general atmospheric circulation, as described in Chapter 4.

5.3 GLOBAL PATTERNS OF RAINFALL SEASONALITY

Figure 5.4 illustrates the relationship between major climate zones and the general circulation described in Chapter 4. This might be termed a zonal view of global climate, one in which the major climate elements are basically a function of latitude. The shortcomings of this scenario are already apparent from the discussions in Chapter 4 and earlier in this chapter. Nevertheless, it provides a useful basis for developing the general global picture of climate, especially the seasonality of rainfall. In this simplistic scenario, rainfall occurs where low pressure prevails, and dry conditions where high pressure prevails.

This zonal view, together with the traditional Intertropical Convergence Zone (ITCZ) concept, is particularly useful in describing the patterns of rainfall and winds that prevail in the subtropics, where aridity is linked to the subtropical highs. The poleward and equatorward extents of these drylands are limited by the year-round presence of the mid-latitude low-pressure belt and the ITCZ, respectively (Fig. 5.4). The true deserts lie where the subtropical high is the dominant influence year-round. Where the prevailing influence shifts seasonally between the subtropical highs and the mid-latitude lows or the ITCZ, climate is semi-arid. The seasonal shift of the general circulation explains the rainfall seasonality. The poleward flank of the subtropical high receives winter rainfall when the mid-latitude lows are displaced equatorward. The equatorward flank receives summer rainfall when the ITCZ is displaced poleward, further into the summer hemisphere.

The major wind systems are inherently linked to the global pressure systems (see Chapter 4). Thus, the prevailing wind patterns shift seasonally with the pressure features of the general atmospheric circulation.

An overview of wind and rainfall as a function of latitude is presented in Fig. 5.5. It should be kept in mind, however, that

this simple pattern is interrupted by the secondary factors producing global patterns of climate. These are imposed largely by the presence of continents, large topographic features, and maritime effects in regions reasonably close to the oceans, but also by more local influences.

In the lower latitudes, those where seasonality is largely a function of the changing influence of the subtropical highs and the ITCZ, this zonal picture can be carried a step further (Fig. 5.5). In the equatorial region, rainfall occurs all year round but its seasonal distribution shows two maxima. Beyond this zone is a region in both hemispheres where two wet seasons and two dry seasons occur during the year. The wet seasons again correspond to the transition seasons, when the ITCZ passes through the region twice on its excursion between the two hemispheres. The winter dry season is the more extreme of the two, since the ITCZ is far into the other hemisphere. Further equatorward lie zones with a single wet season and a single dry season. Each zone receives rainfall only when the ITCZ is at its extreme position in the respective hemisphere, which is the summer or high-sun season in the hemisphere.

Although highly idealized, the pattern described above bears a strong resemblance to the seasonal cycle of rainfall in the tropics and subtropics of Africa (see Chapter 16). A similar pattern is evident over Australia, but it is much less evident over the Americas in the Western Hemisphere. The global distribution of the regions with a double and single rainy season is depicted in Fig. 5.6.

5.4 RAIN-BEARING SYSTEMS

The older literature on deserts describes the infrequent rains that occur as isolated, local disturbances. The German literature even

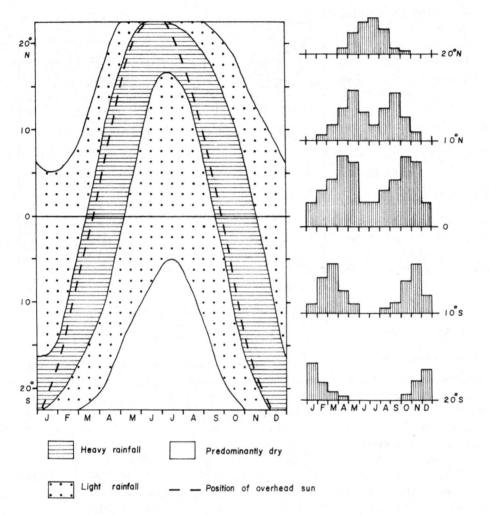

Heavy rainfall   Predominantly dry

Light rainfall   — — Position of overhead sun

Fig. 5.5 Idealized pattern of rainfall seasonality versus latitude (from Nieuwolt 1977).

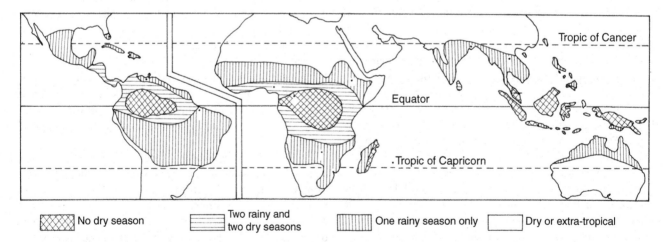

No dry season   Two rainy and two dry seasons   One rainy season only   Dry or extra-tropical

Fig. 5.6 Areas with one and two rainfall maxima in the seasonal cycle (from Nieuwolt 1977).

used the term "*Platzregen*," roughly translated as "rain in one place." It is now known that the occasional rains falling in deserts are generally associated with large-scale weather systems. In contrast to wetter regions, the deserts are affected only infrequently, when these systems are abnormally large or depart from their usual tracks. In the dry regions, the large-scale systems also tend to produce less rainfall than elsewhere, in view of the factors producing the overall atmospheric stability in the region.

The dryland regions often mark the transition between the meteorological regimes of the tropical and temperate latitudes. As a consequence, these areas are generally influenced by rain-bearing disturbances of both tropical and mid-latitude origin. An understanding of the origin and nature of the systems, and how they differ in the two regimes, requires consideration of the major meteorological differences between the tropics and extra-tropics.

The tropics differ fundamentally from the extra-tropics in that the radiation received is relatively uniform in space and time and the Coriolis force is relatively weak. As a result of the uniform distribution of radiation, temperature gradients are small in the low latitudes, and the mid-latitude concept of warm and cold fronts is generally not applicable. Because the Coriolis force is weak, winds tend to blow directly toward low pressure, inhibiting the build-up of large pressure gradients.

This contrast between the tropics and mid-latitudes manifests itself in the mechanisms by which rain-bearing disturbances are produced and in their resultant characteristics. In the higher latitudes, both temperature and pressure change rapidly with latitude, a condition called "baroclinicity." These intense latitudinal gradients provide the energy and instability for the creation of synoptic-scale lows, the major rain-producing system of the extra-tropics. These gradients likewise produce the jet streams and the relatively high mean wind speeds of the mid-latitudes.

Temperature and pressure gradients are weak in the tropics and consequently tropical systems frequently lack a distinct low-pressure center. This is especially true in the equatorial latitudes, where the Coriolis force is weak (it becomes zero at the equator itself). Instead, minor disturbances in the tropical pressure field may become sites of intense convective activity. These are the wave disturbances described later. As a consequence of the thermal uniformity of most of the tropics (a condition termed "barotropy"), the mechanisms that produce tropical disturbances are different from those producing mid-latitude disturbances. Of importance in the tropics are various types of instability associated with the presence of warm, moist air at the surface and the latent heat released when the humid air rises and cools and its moisture condenses aloft.

The most important rain-bearing system of the extra-tropical latitudes is a low-pressure system referred to as an extra-tropical or frontal cyclone. Smaller-scale systems called mesoscale convective systems (MCSs) are also important in some regions, such as the central USA. In the higher latitudes MCSs are most common in summer. They are the dominant rain feature of the tropics. Other tropical disturbances include easterly waves, tropical cyclones and hurricanes, and isolated thunderstorms. In many dryland regions, hybrid systems with both tropical and mid-latitude characteristics are important sources of precipitation. These often take the form of diagonal cloud bands extending from the low to high latitudes or cut-off upper-level lows. Local systems such as mountain–valley winds and sea breezes also result in precipitation.

5.4.1 THE EXTRA-TROPICAL CYCLONE

The basic structure of the extra-tropical cyclone was described early in the twentieth century by Norwegian meteorologists. An idealized picture is shown in Fig. 5.7. The overall characteristics of individual systems may differ from this general picture, but in all cases the major features are a pronounced low-pressure center and a cold front and a warm front separated by a region of warm and generally moist air called the warm sector. The major areas of rainfall are along the two fronts and throughout the poleward flank of the system. Isolated convective rains, often of great intensity, are also quite common in the warm sector. Tornadoes occasionally develop in this region.

The cyclones form from waves along the region of high latitudinal temperature contrast known as the polar front. The areas of formation are regions of maximum thermal contrast or gradient. These regions, shown in Fig. 5.8, are generally where cold land masses meet warm ocean currents. Hence, in the Northern Hemisphere, lows tend to form in the Gulf of Alaska and the Bering Sea or near Iceland. Their frequent occurrence in these locations produces the mean low-pressure features of the general circulation known as the Aleutian Low and Icelandic Low, respectively. These low-pressure systems are steered by the winds aloft, in the mid- and upper troposphere. Since the high-level flow is generally westerly, they tend to move eastward, traversing the USA or Europe in 3–5 days.

These systems shift latitudinally with the seasons, along with the planetary wind belts. When they are displaced equatorward in winter, they produce the winter rainfall of the semi-arid Mediterranean climates of the subtropics. Such a winter rainfall regime prevails in Africa north of the Sahara, the Near East, southern Europe, northern Australia, southern California and parts of the southwestern USA, and the Baja California region of Mexico.

5.4.2 MESOSCALE CONVECTIVE SYSTEMS

The traditional picture of tropical convective rainfall is that it is generally associated with relatively small and isolated cumulus and cumulonimbus clouds. Satellite imagery has completely changed this picture, showing that most precipitation is coupled with intermediate systems that are larger than individual cumulus clouds but smaller than synoptic systems. Such systems have various names, such as linear disturbances, line squalls, disturbance lines, sumatras, cloud clusters (Fig. 5.9), mesoscale convective clusters or complexes (MCCs), and mesoscale convective systems (MCSs). Here the term MCS will be utilized and it applies only to "cold" systems, i.e., those in which ice is present in the upper levels of the cloud (Nesbitt and Zipser 2003). This distinction is important because the presence of ice enhances the efficiency of the precipitation process, generally producing more intense precipitation.

An MCS (Fig. 5.10) is essentially a large, continuous area of deep cloud. Definitions vary but most require a contiguous area

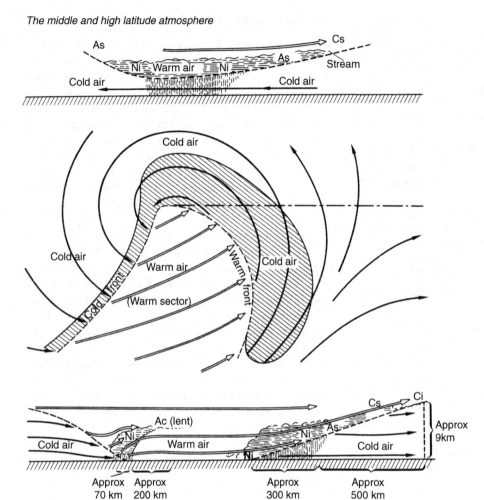

The middle and high latitude atmosphere

Fig. 5.7 The extra-tropical cyclone. The top and bottom diagrams show the typical cloud patterns across the northern and southern peripheries of the system (from Lockwood 1974). As = Altostratus, Ni = Nimbostratus, Cs = Cirrostratus, Ac (lent) = Altocumulus lenticularis, Ci = Cirrus.

of cloud greater than 2000 km². The typical MCS is topped by a large anvil of stratiform cloud (Fig. 5.11). Despite the intense convection associated with these systems, stratiform precipitation accounts for 73% of the rain area and contributes roughly 40% of the total rainfall for the tropics as a whole (Schumacher and Houze 2003). MCSs may be at least partially generated by the aggregation of smaller systems as they grow during the course of the day, but they appear to be linked to larger-scale systems as well. In the tropics, wave disturbances appear to organize the convection into mesoscale systems. Many systems are associated with squall lines.

Mesoscale convective systems were initially studied in the central USA, where they produce 30–70% of the rain falling during the months of April–September (Fritsch *et al.* 1986). The mesoscale systems were initially identified with visible and infrared satellite imagery that provides little information about the precipitation produced by the system. More advanced satellite techniques allow for better descriptions of features, such as the presence of ice, that are more closely correlated with precipitation (Mohr *et al.* 1999). Over the tropical land areas MCSs average roughly 10,000 km² in size and comprise only about 2% of the precipitation-bearing features. However, they provide

about 50% of total rainfall tropics-wide (Fig. 5.12) and up to 90% of the rainfall in areas over select land areas (e.g., Nesbitt *et al.* 2006). By comparison, small cold-cloud precipitation features provide some 30% of the rainfall over land.

MCSs are in a constant state of evolution. These systems produce intense convective rainfall mainly during the afternoon, with most convective events lasting 3 hours or less over land (Ricciardulli and Sardeshmukh 2002). As the system evolves in the later hours of the day and into the night, the cloud anvil that tops the system spreads and produces a large area of stratiform cloud (Nesbitt and Zipser 2003). Thus, the stratiform rain typically occurs at night and for a longer period of time, but the rain rate is roughly one-fourth the rain rate associated with convective clouds.

5.4.3 DIAGONAL CLOUD BANDS

These features, also called tropical plumes, are essentially waves in the upper-level subtropical westerly jet that become elongated and extend equatorward, often merging with tropical low-pressure systems at low levels in the atmosphere. Rubin *et al.* (2007) provide an excellent review of these systems, and Iskenderian

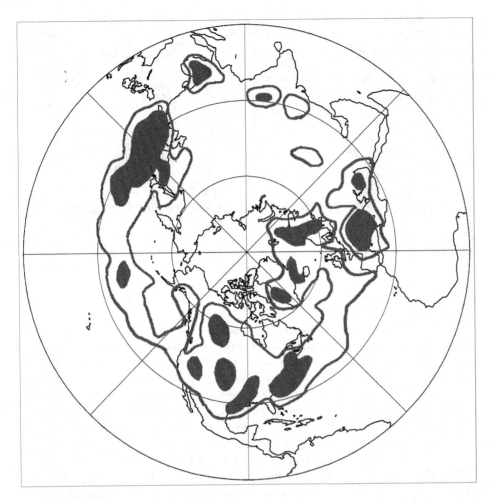

Fig. 5.8 Main regions of cyclogenesis in winter (based on Petterssen 1969).

Fig. 5.9 Satellite photo of two cloud clusters over West Africa, near the Atlantic coast.

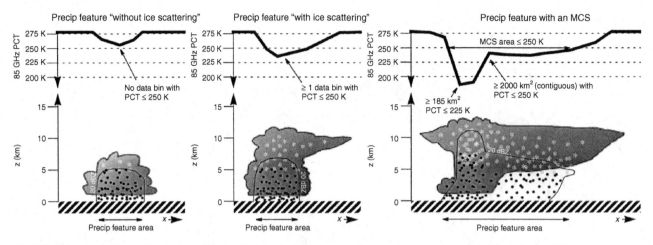

Fig. 5.10 Schematic of typical tropical precipitation features, including mesoscale convective complexes (from Nesbitt *et al.* 2000).

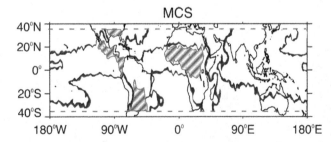

Fig. 5.11 Anvil cloud over West Africa (on right), viewed from the air. A second anvil is in the process of forming on the left.

Fig. 5.12 Fraction of rainfall produced by mesoscale convective complexes (right) and by small cold clouds (left) (from Nesbitt *et al.* 2006). The dark, irregular line represents areas with more than 50%; striped areas, more than 70%.

(1995) presents their global climatologies. Because they span the relatively arid locations between the tropics and subtropics, the cloud bands can be important aspects of the rainfall regime in many dryland areas. For example, cloud bands are a major feature of the summer rainfall regime over southern Africa,

having extensive influence throughout the semi-arid Kalahari (Taljaard 1990).

Similar cloud bands have been described in the eastern Pacific (McGuirk *et al.* 1988), over North Africa (Nicholson 1981; Knippertz 2005), Australia (Wright 1997), and over parts of South America (Satyamurty *et al.* 1998) and the Middle East (Ziv 2001). This system is responsible for incidents of precipitation during the dry season in some areas, such as West Africa (Knippertz and Fink 2008). When the cloud bands interact with other synoptic-scale disturbances, they bring usually high amounts of rainfall to relatively arid regions, such as Israel, the Sinai Peninsula, and the western Sahara (e.g., Ziv 2001; Knippertz and Martin 2005). Rainstorms occur most frequently in the exit region of the trough. In many cases, the system persists for 5–10 days (Satyamurty *et al.* 1998). The result is prolonged rainfall and the persistent cloud cover that reduces evaporative losses. The cloud bands over South America are associated with a quasi-stationary feature called the South Atlantic Convergence Zone (SACZ). In southeast South America, 65% of the cases of extreme precipitation occur when convection in the SACZ is intense and expansive (Carvalho *et al.* 2002).

A cloud band over southern Africa is illustrated in Fig. 5.13. The cloud band emanates from mid-latitude cold fronts associated with wave troughs in the upper-level westerlies. This synoptic situation is typically referred to as a "tropical–temperate trough" (Todd and Washington 1999). They extend diagonally, in a northwesterly direction, into the low latitudes when a tropical cyclonic vortex is situated over Central Africa at about 20° S. They influence rainfall throughout the semi-arid regions of southern Africa, at least as far north as Zambia and Zimbabwe. In many cases, they lie off the coast of southern Africa.

Flohn (1971) describes a tropical–temperate trough over North Africa. A wave in the mid-latitude upper-level westerlies extends diagonally across the Sahara into equatorial latitudes. If the wave overrides a disturbance in the mid-level easterlies (Fig. 5.14), an intense band of cloud and rain can form and extend southwestward from the westerly flow. Since the development

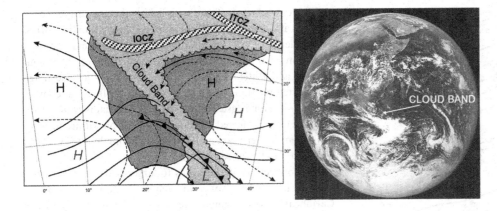

Fig. 5.13 (left) Schematic of a diagonal cloud band over southern Africa, linked to a cold front (from van Heerden and Taljaard 1998). (right) Satellite photo of such a cloud band.

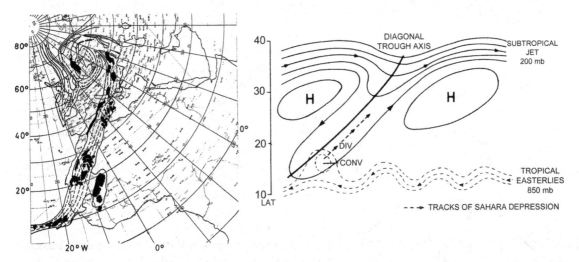

Fig. 5.14 Development of Soudano-Saharan depressions and diagonal cloud bands over West Africa (from Nicholson 1981 and Flohn 1975). Diagram on the left depicts a satellite view of a diagonal cloud band and areas of heavy rain within it (shaded areas) in September 1969, a period of tremendous rainfall and flooding in parts of North Africa. Diagram on the right shows the typical circulation pattern that leads to its development.

of such a system requires westerlies aloft, it generally does not occur during the summer season of June–August. However, this system can produce rainfall at any other time of year as far south as 10° N. In the Sahara and areas to the south of it, this system seldom produces more than 25 mm of rainfall per day, but it can persist for several days. When this occurs in the cool season, the rainfall can be effective for agriculture in parts of Mauritania, Mali, and Algeria (Nicholson 1980). North of the Sahara, these systems produce much more intense rainfall. Such a system was responsible for the extreme floods of September 1969, in Algeria and Tunisia, which are described in Chapter 20. At Biskra, Algeria, 299 mm fell in two days. This is roughly 18 times the September mean of 17 mm.

These westerly troughs also promote the development of Atlas depressions, low-pressure systems forming in the lee of the Atlas Mountains north of the Sahara (Flohn 1975). These are the major rain-producing systems for much of Algeria, Tunisia, and Morocco (Dubief 1947; Knippertz et al. 2003). They are particularly common in those areas in the northern

and central Sahara with rainfall maxima in the transition seasons of March–May and October–November.

5.4.4 TROPICAL WAVE DISTURBANCES

Two major categories of tropical systems are distinguished on the basis of whether or not a distinct, quasi-circular low-pressure core exists. Systems with a well-developed core, such that at least one or more closed isobars can be defined, are called tropical depressions or storms. These are discussed in the following section. Those lacking the low-pressure core are termed wave disturbances. These are irregular perturbations in the pressure field that have a well-defined structure. Wave disturbances are important because they organize convection and probably enhance it. They also give rise to tropical cyclones (Hopsch et al. 2007).

The best known are easterly waves. These disturbances are particularly common over Africa and the Atlantic Ocean. Each year some 60 waves traverse West Africa and cross into the Atlantic,

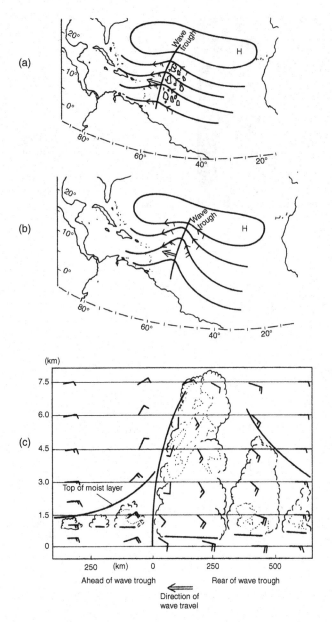

Fig. 5.15 A Caribbean easterly wave (top) cloudiness, (middle) wind directions, and (bottom) vertical view (from Lockwood 1974).

where they sometimes spawn hurricanes. There are distinct patterns of precipitation associated with wave disturbances (Serra *et al.* 2008). There is some question as to what role these play in actually producing precipitation events, as opposed to merely organizing local convection into larger-scale systems. However, over West Africa, wet years are clearly marked by more waves than dry years and the atmospheric conditions characteristic of the wet years are much more conducive to wave generation and growth than those of dry years (Nicholson *et al.* 2008).

The first type to be intensively studied was the Caribbean easterly wave (Fig. 5.15). Ahead of the low-pressure wave trough is fair weather and scattered cumulus; near the trough are well-developed cumulus and occasional showers (Riehl 1979). The

major area of disturbed weather is behind the trough axis: heavy cumulus and cumulonimbus, moderate to heavy thunderstorms, and cooler air. Over Africa and in the eastern Pacific, convection is preferentially in the northerly flow ahead of the trough (Pereira and Rutledge 2006). However, there is a wide range of reported relationships, complicated by the fact that convection may, in some case, trigger waves (e.g., Kiladis *et al.* 2006). Moreover, the preferred location of convection within the wave is different in wet years and dry years (Baum 2006) and varies with latitude (Lavaysse *et al.* 2006).

The connection of the waves to larger-scale aspects of the general circulation is not completely understood. Many of the Caribbean waves originate over Africa, but their structure is modified in passing over the Atlantic and many of their characteristics differ from those of the African waves (also called African easterly waves). The African waves are associated with a jet stream in the mid-troposphere (Grist 2002). The primary wave track lies equatorward of the African easterly jet and the jet provides the energy for their propagation and maintenance. A second wave track exists to the north of the mid-level jet stream, near the surface position of the ITCZ, but little convective activity is associated with these waves (Thorncroft and Hodges 2001; Nicholson 2009). The upper-tropospheric tropical easterly jet might also be important in their generation (Nicholson *et al.* 2008). Waves on the tropical easterly jet itself can also trigger rainfall events (Nicholson *et al.* 2007).

The instabilities associated with the wind shear of the jet are considered to be responsible for the generation of the African waves (e.g., Burpee 1972), but the Pacific easterly waves appear to be generated primarily by latent heat release associated with cloud formation (Serra *et al.* 2008). However, recent research has suggested that topography or latent heating may also trigger African easterly waves (Mohr and Thorncroft 2006). The waves often develop near highlands such as Darfur (e.g., Mekonnen *et al.* 2006), Tibesti, Aîr (Baum 2006), or the Hoggar (Reed *et al.* 1988).

Another common tropical wave disturbance is the Kelvin wave. These systems propagate eastward, rather than westward and, like the easterly wave, they also have a strong impact on precipitation, sometimes organizing it into mesoscale convective complexes (Straub and Kiladis 2002). Convective disturbances linked to Kelvin waves are a major feature of the tropics, but they are not a particularly important rain-bearing system for the drylands. However, they can influence some drier regions, such as West Africa (Mounier *et al.* 2007), Northeast Brazil (Wang and Fu 2007), and other parts of South America (Liebmann *et al.* 2009).

5.4.5 TROPICAL DEPRESSIONS AND STORMS

Tropical depressions and storms differ from tropical waves in that the former are closed circulation systems with a distinct low-pressure core. Various categories are distinguished on the basis of intensity of the pressure and wind fields. The most common is the tropical depression; the low-pressure region is enclosed within a few isobars and winds are less than 17 m s^{-1}. Tropical cyclones have winds above 32 m s^{-1} and have developed a warm core.

The principal regions of tropical cyclone formation include the South, NE and NW Pacific, the Bay of Bengal, the Arabian Sea, the South Indian Ocean (off the NW Australian coast), the western Caribbean, and the Gulf of Mexico. The semi-arid regions that are occasionally (but rarely) affected by them include northern Australia, the arid zones of eastern and southern Africa, the American Southwest, parts of Central America, the Rajasthan desert, and some arid regions of the Middle East.

5.4.6 CUT-OFF LOWS

Some of the most torrential rainfall in drylands is associated with stationary low-pressure systems that develop from waves in the upper-level westerlies. A variety of systems develop in this way and they go by various names. The distinctions between them are not very clear, but all have the capability of bringing tremendous floods to the regions impacted. This is particularly true for certain dryland locations in South America, Australia, and Africa.

Satyamurty and Seluchi (2007) describe what they call a subtropical cold-core vortex and provide a detailed review of its occurrence over South America. They contrast it with semi-stationary tropical upper cold-core vortices that likewise affect parts of South America. The subtropical systems tend to occur in winter and seldom bring severe weather. They are also transient features.

Very similar systems commonly occur over Africa and Australia, where they are most commonly termed cut-off lows. These are also cold core, upper-level systems that have become cut off from the westerlies. Those over southern Africa have been described by Taljaard (1985, 1996) and others. They are frequently associated with flash floods and torrential rains, especially in autumn and spring (van Heerden and Taljaard 1998). Knippertz and Martin (2007) describe cut-off lows over northwest Africa. These can bring torrential rains to parts of North Africa and often bring unseasonal rainfall to sub-Saharan Africa.

Fuenzalida et al. (2005) have created a climatology of cut-off lows in the Southern Hemisphere. These occur particularly frequently over all three Southern Hemisphere continents, but 48% were found to occur over Australia and only 10% over Africa. These systems produce over 50% of the rainfall in parts of southeastern Australia (Pook et al. 2006).

5.5 MESOSCALE INFLUENCES ON RAINFALL

Rainfall is influenced by local or mesoscale factors such as topography and proximity to an ocean or lake. Large-scale orographic (i.e., topographic) effects have been described in the context of the origin of dryland climates in Section 5.2, but orography also influences rainfall on a smaller scale. Topography and water bodies both modify the large-scale airflow and induce local flow patterns. In doing so, they influence the distribution and amount of precipitation and its diurnal cycle. In extreme cases even the seasonal cycle is affected. Local flow patterns are

particularly important when the large-scale flow is weak and insolation is strong; they are especially effective in regions of predominantly convective rainfall. These conditions tend to be met in the low-latitude drylands. Hence, orographic and shoreline effects can be quite significant in these regions, producing, for example, small, arid coastal strips or the mountain "oases" of the Hoggar and Tibesti in the central Sahara. Mean annual rainfall over both regions exceeds at least 100 mm (Dubief 1963). Mean annual rainfall exceeds 200 mm at optimal elevations on the Hoggar, and probably on the Tibesti massif as well (Winiger 1975). By comparison, mean annual rainfall is roughly 10–20 mm in the surrounding lowlands. In some cases, mountains such as the Atlas actually trigger major disturbances that bring rain to the drylands.

The mechanisms by which topography influences rainfall include orographic uplift of the large-scale winds and ascent and convergence produced by mountain–valley wind systems. The latter rise upslope as the highlands heat by day, and flow downslope as the cold air drains at night. The uplift, if sufficiently strong, elevates surface air parcels to the lifting condensation level, the altitude at which clouds form. The result is convective clouds and rainfall. Typically the interaction with the large-scale flow enhances rainfall on the windward slopes and creates rain shadows in the lee. This effect can produce isolated pockets of desert in what is otherwise a humid mountain region. This mosaic pattern is well developed over the Olympic and Cascade Mountains of Washington State, over the highlands of Ethiopia, and over the mountains of Asia.

The more regular mountain–valley winds produce quasi-permanent mist belts at preferential elevations, with intense fogwater precipitation. This is the case on the high East African peaks, such as Mts. Kenya and Kilimanjaro. The altitude of the mists is so constant that the mist belts correspond to dense stands of jungle-like vegetation quite limited in vertical extent. On Mt. Kenya and other East African mountains, the trees are buried beneath dense vines and deep moss covers the ground (Fig. 5.16). Throughout East Africa the peaks of the Rift Valley markedly enhance rainfall, and contrast sharply with the surrounding regions of desert and savanna.

There also exists a pronounced altitudinal zonation of precipitation that is produced by complex factors such as humidity and thermal stability of the air mass, the prevailing large-scale wind patterns, and local orographic effects. Thus, only a few rough generalizations can be made. A common one is that, at least on the windward slopes, rainfall tends to increase with elevation in the extra-tropical latitudes, while in tropical highlands there is a rainfall maximum at intermediate levels (Taylor 1996). This may relate to the contrast between the tropics and extra-tropics with respect to the vertical stratification of winds. Wind speed tends to continuously increase with elevation in the higher latitudes, but reaches a maximum in the lower troposphere in the tropics. However, in the tropics the orographic uplift can be limited by an inversion layer. In such a case, orographic rainfall enhancement is limited to lower elevations. In other cases, the higher elevations are dry because they lie above cloud tops.

Fig. 5.16 Vegetation in the mist belt of Mt. Kenya.

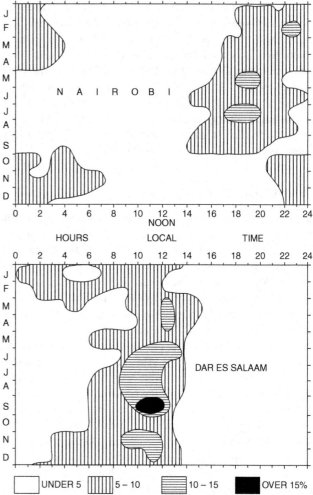

Fig. 5.17 The diurnal cycle of rainfall at Nairobi (top) and Dar es Salaam (bottom) (from Nieuwolt 1977).

Even the seasonal cycle can be affected by topographic factors. Near the Red Sea Trench, Asmara, at an elevation of 2370 m, has a summer maximum, like most of the region. Massawa, a low-level coastal station only 63 km away, has a winter maximum (Flohn 1965).

A survey of the available literature suggests that the above generalizations do not hold in dryland regions. In the Pamiro-Alay Mountains of Middle Asia, for example, where the summits top 5000 m, the precipitation maximum lies at about 2500 m (Walter and Box 1983). In the drier highlands of Africa the rainfall maximum generally lies considerably higher, with various estimates placing it between 1700 and 2200 m (Rohr and Killingtveit 2003). On some windward slopes (e.g., Mt. Cameroon) the maximum lies near the surface (Lauer 1975). On the Hoggar of the central Sahara it lies at roughly 2400 m, but the altitude of the maximum and degree of enhancement vary greatly from year to year and season to season.

The mountain–valley winds that contribute to these maxima have a regular diurnal cycle. As a result, they tend to produce rainfall at particular times of the day. The timing of the rainfall is also affected by the diurnal cycle of convective activity because the orographically-induced rains are generally convective in nature. This typically leads to a rainfall maximum in early or late afternoon. In lower elevations surrounding areas of high relief, the timing of the rainfall will generally be much later. In some cases, the disturbance systems are advected into the lower regions by the downslope valley breezes. This tends to occur in the early evening hours or at night. Afternoon convection in the lower regions may also be suppressed by subsidence compensating for the ascent over the higher terrain. A case in point is the anomalous evening rainfall maximum over the western Sahel. Reed and Jaffe (1981) speculate that this is related to systems enhanced in the highlands of Guinea and advected northwestward in the late evening, and to compensatory subsidence around the highlands in the afternoon.

A case study of the altitudinal distribution of rainfall over Mt. Kilimanjaro, in the East African highlands, demonstrates

several of the above effects (Coutts 1969). During the long rains of March–May, when southerly trades prevail, the rainfall maximum lies on the south and southeast slopes. The northeasterly trades of the short rain period (November and December) produce a rainfall maximum on the northeast slope. During the months of April–September (see Jackson 1989) an inversion between 3965 and 4575 m limits the height of the cloud tops, restricting the development of convection. At other times, convective disturbances build up and move into surrounding lowlands. Such disturbances building up over Mt. Kenya account for the evening and nocturnal rains at Nairobi (Fig. 5.17). At Dar es Salaam, a coastal location some 650 km away, morning rains associated with land breezes prevail (Thompson 1968).

As with topography, shoreline effects can both modify the large-scale flow and induce a local land–sea breeze circulation, as shown in Fig. 5.18. By day, the land is generally warmer than the water, particularly in desert locations, and low pressure builds up over the land. At the sea-breeze front, marking the inland edge of the maritime air, forced ascent leads to cloud formation and frequently rainfall. At night the land breeze,

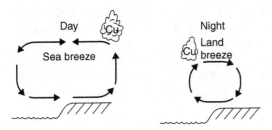

Fig. 5.18 Typical patterns of sea breeze (left) and land breeze (right) circulations.

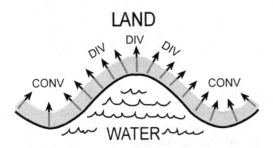

Fig. 5.19 Effects of convex and concave shorelines on airflow and divergence (DIV=divergence, CON= convergence).

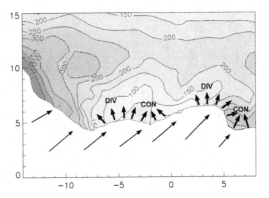

Fig. 5.20 Mean rainfall during July and August and prevailing summer wind direction along the Guinea coast. Thin arrows indicate large-scale wind; shorter thick arrows indicate the prevailing direction of sea breezes. DIV and CON indicate areas of divergence and convergence of sea breezes. The latter tends to enhance rainfall, such as over southwestern Nigeria (c. 5° E).

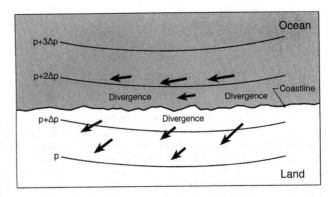

Fig. 5.21 Frictional divergence along a shoreline, a common cause of coastal aridity (from Warner 2004).

generally weaker and shallower, blows from the land to the warmer water. Then, the convection is over the water but the clouds are brought inland in the early morning with the onset of the sea breeze. This generally results in a morning rainfall maximum at coastal locations.

The land–sea breeze circulation is strongest when the land–water contrast is large, insolation high, the nights short, and large-scale pressure gradients and winds weak. Thus, the coastal drylands are ideal for its development. In very dry regions, such as the Namib Desert, the sea breeze seldom produces rainfall inland, but it does in less arid locations such as southern California.

The effects of the sea breeze are modified by coastline configuration. In the case of a convex shoreline, the sea breezes converge, reinforcing the ascent produced at the sea-breeze front (Fig. 5.19). In the arid region of Northeast Brazil rainfall is about 400 mm about 100–200 km inland, but along the shoreline (which has a convex configuration) mean annual rainfall exceeds 2000 mm (Ratisbona 1976). Concave coastlines, on the other hand, reduce rainfall because the sea breezes diverge, producing subsidence inland. This appears to be one factor in the anomalous dry zone along the Guinea coast of West Africa (Trewartha 1970) (Fig. 5.20) and in two small regions of coastal desert in Venezuela.

The effects on the large-scale flow depend on the direction of the wind with respect to the coastline. When winds blow perpendicular to the shore, the increase in friction at the coast causes air to converge and ascend, thus promoting rainfall. This factor markedly enhances rainfall in parts of the Guinea coast of West Africa.

Winds blowing parallel to the coast tend to diminish rainfall because of frictionally-induced wind divergence (Fig. 5.21). The essence of the phenomenon is the frictional contrast between land and water. Because friction is lower over the water than over the land, winds parallel to the shore tend to develop a slight inland component over the land (Bryson and Kuhn 1961). The resulting divergence of the flow reduces rainfall by producing subsidence and suppressing ascent. Frictionally-induced divergence appears to be a factor in the development of the extreme desert along the west coast of South America, the anomalous dry zone of the Guinea coast (Fig. 5.20), and semi-arid coastal strips in the Yucatan Peninsula of Mexico, and along the northern Caribbean coast of South America.

The combined influence of shoreline and topographic effects can produce interesting and complex diurnal cycles of rainfall. A classic case is that of Lake Victoria. During the afternoon, rainfall is suppressed by subsidence as the lake breezes diverge. At night the land breezes converge, with the flow being enhanced by downslope valley breezes from the highlands. The consequence is nocturnal rains, concentrated in the lake's NW quadrant, and an enhancement of rainfall of 30% compared with surrounding regions (Nicholson and Yin 2002). There

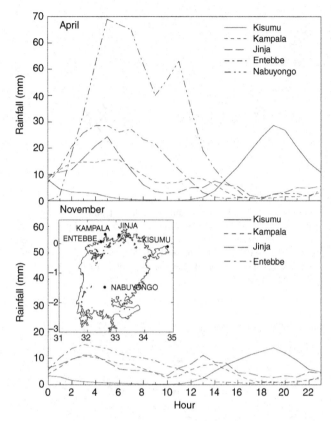

Fig. 5.22 Diurnal cycle of rainfall (mm) along the shore of and at an island in Lake Victoria (from Nicholson and Yin 2002). (top) April, (bottom) November. Location of stations is indicated on the map. Rainfall is markedly higher at the island Nabuyongo, where there is both regional-scale convergence and a convergence of lake breezes. Rainfall occurs at night over the western half of the lake, and during the day over the eastern half.

is an abrupt shift between a nocturnal maximum at Jinja and a diurnal maximum at Kisumu, just some 150 km to the east (Fig. 5.22).

5.6 QUASI-GLOBAL PHENOMENA THAT INFLUENCE PRECIPITATION

Climatic elements, most of all rainfall, fluctuate on a variety of time scales. Day-to-day variability is essentially weather. Systematic fluctuations within a season are termed intraseasonal fluctuations. Longer-term fluctuations include interannual, interdecadal, and multidecadal variations; the time scales of these are roughly less than 10 years, 10–30 or so years, and longer than 20–30 years. These are often termed "oscillations" because of their temporally recurring nature. However, the fluctuations do not occur as regularly as the term "oscillation" would imply.

These fluctuations tend to occur simultaneously over large parts of the globe, indicating a climatic relationship between distant locations. Such relationships are referred to as "teleconnections." Several quasi-global patterns of teleconnections

in surface temperature or pressure have been identified that act on a variety of time scales and produce known spatial patterns of anomalies in temperature, precipitation, or other climatic elements over large portions of the globe. The most widely known is that of El Niño. But several other patterns of climatic variability exist and these influence many of the global drylands.

5.6.1 THE EL NIÑO/SOUTHERN OSCILLATION

A now well-known global-scale phenomenon that modulates rainfall in many of the global drylands is the El Niño/Southern Oscillation (ENSO). This was originally known as a localized meteorological event that brought intense rains to the arid coast of Peru, on the heels of unusual warming of the ocean just offshore.

Some time in the 1970s it was realized that El Niño is not simply a local occurrence, but part of a tropical meteorological/ oceanographic phenomenon that has almost worldwide influence on rainfall and temperature. Some of the most strongly affected regions are the drylands. El Niño occurs in conjunction with large-scale warmings in the Pacific that, in turn, are linked to pressure adjustments over the tropical Pacific (Fig. 5.23). These adjustments are out of phase in eastern and western portions of the Pacific Basin and are particularly strong in the Southern Hemisphere, hence leading to the term "Southern Oscillation" (SO). A low index of the Southern Oscillation (SOI) is associated with its warm, or El Niño, phase; a high index is associated with its cold, or La Niña, phase. Global climate repercussions tend to be the opposite in these two phases. Further details, particularly regarding the influence of ENSO on dryland climate, are presented in Chapter 25.

5.6.2 OTHER QUASI-GLOBAL PATTERNS OF VARIABILITY

Some of the other well-known patterns of climatic variability include the Arctic Oscillation or AO (Thompson and Wallace 1998), the North Atlantic Oscillation or NAO (Barnston and Livezey 1987), the Pacific/North American pattern or PNA (Wallace and Gutzler 1981), the Pacific Decadal Oscillation or PDO (Latif and Barnett 1996), the Atlantic Multidecadal Oscillation or AMO (Enfield *et al.* 2001), the Antarctic Oscillation or AAO (Thompson and Wallace 2000), and the Madden–Julian Oscillation or MJO (Madden and Julian 1994). These act on different time scales, are defined by different parameters, and affect climate in different ways. However, many of these phenomena are interactive, so that the climatic fluctuations in a given location may reflect the simultaneous operation of two or more of these phenomena.

The Pacific/North American pattern and the North Atlantic Oscillation were among the first to be identified. The PNA is low-frequency variability in pressure over extra-tropical regions of the Pacific. Its defining pattern is an out-of-phase

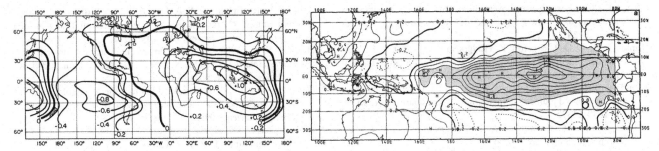

Fig. 5.23 The El Niño/Southern Oscillation. (left) Typical pattern of sea-level pressure anomalies associated with the Southern Oscillation (from Enfield 1992). (right) Sea-surface temperature anomalies (°C) during the mature phase of El Niño (from Rasmusson and Carpenter 1982).

relationship between surface pressure in the northeast Pacific and the central Pacific near Hawaii. Pressure variations in the intermontane and southeastern regions of the USA tend to be part of the pattern as well. The negative phase of the PNA pattern is linked to a westward retraction of the East Asian jet stream and blocking over high latitudes of the Pacific. In the positive phase, this jet stream is enhanced and the exit region of the jet is shifted eastward toward the USA. The Atlantic analog to the PNA is the NAO, a seesaw relationship between the intensity of the Bermuda High over the Atlantic and the Icelandic Low. In its positive (negative) phase, both pressure features are stronger (weaker) than normal. The NAO occurs in all seasons, but it is the dominant mode in winter climate variability.

The NAO and PNA are generally associated with interannual variability. The Pacific Decadal Oscillation and Atlantic Multidecadal Oscillation act on much longer time scales, on the order of 20–30 years. The PDO is a pattern of cooling or warming north of roughly 20° N and, unlike ENSO, it is an extra-tropical pattern. However, its pattern of global teleconnections is very similar to that of ENSO. The AMO is likewise a pattern of sea-surface temperatures, the average temperature of the North Atlantic. It is linked to droughts in many parts of the world.

The two polar oscillations are instead patterns of atmospheric pressure. The AO is defined by anomalies of the opposite sign in the Arctic and in a region centered at roughly 37°–45° N. It is the dominant pattern of non-seasonal sea-level pressure variations poleward of 20° N. The Antarctic Oscillation (AAO) is defined instead by the circulation at 850 mb (Thompson and Wallace 2000). It is characterized by an inverse relationship in pressure anomalies between the Antarctic and the mid-latitudes around 40°–50° S. The terms Southern Annular Mode and Northern Annular Mode are also used for these phenomena.

The Madden–Julian Oscillation is very different from those previously discussed. It is an intraseasonal propagation of rainfall anomalies that traverse equatorial latitudes (Zhang 2005). It is also called the 30–60 day oscillation, the 30–60 day wave, the 40–50 day oscillation and the intraseasonal oscillation. The convection associated with the MJO moves eastward, but some westward-moving anomalies are often triggered at the same time.

REFERENCES

Barnston, A. G., and R. E. Livezey, 1987: Classification, seasonality and persistence of low-frequency atmospheric circulation patterns. *Monthly Weather Review*, **115**, 1083–1126.

Baum, J. D., 2006: African Easterly Waves and their relationship to rainfall on a daily timescale. MS thesis, Department of Meteorology, Florida State University, 170 pp.

Bryson, R. A., and P. M. Kuhn, 1961: Stress-differential induced divergence with application to littoral precipitation. *Erdkunde*, **15**, 287–294.

Burpee, R. W., 1972: The origin and structure of Easterly Waves in the lower troposphere of North Africa. *Journal of the Atmospheric Sciences*, **29**, 77–90.

Carvalho, L. M. V., C. Jones, and B. Liebmann, 2002: Extreme precipitation events in southeastern South America and large-scale convective patterns in the South Atlantic Convergence Zone. *Journal of Climate*, **15**, 2377–2394.

Coutts, H. H., 1969: Rain of the Kilimanjaro area. *Weather*, **24**, 66–70.

Dubief, J., 1947: Les pluies au Sahara central. *Travaux de L'Institut de Recherches Sahariennes*, **IV**, 7–23.

Dubief, J., 1963: *Le Climat du Sahara*. Vol. II. Institut de Recherches Sahariennes, Université d'Alger, Algiers, 275 pp.

Enfield, D. B., 1992: Historical and prehistorical overview of El Niño/Southern Oscillation. In *El Nino: historical and paleoclimatic aspects of the Southern Oscillation* (H. F. Diaz and V. Markgraf, eds.), Cambridge University Press, Cambridge, pp. 95–118.

Enfield, D. B., A. M. Mestas-Nunez, and P. J. Trimble, 2001: The Atlantic multidecadal oscillation and its relation to rainfall and river flows in the continental US. *Geophysical Research Letters*, **28**, 2077–2080.

Flohn, H., 1965: Contributions to a synoptic climatology of the Red Sea Trench and adjacent areas. In *Studies on the Meteorology of Tropical Africa* (H. Flohn, ed.). Bonner Meteorologische Abhandlungen 5, Meteorologischen Institut der Universität Bonn, Bonn, pp. 2–35.

Flohn, H., 1971: *Tropical Circulation Patterns*. Bonner Meteorologische Abhandlungen 15, Meteorologischen Institut der Universität Bonn, Bonn, 55 pp.

Flohn, H., 1975: *Tropische Zirkulationsformen im Lichte der Satellitenaufnahmen*. Bonner Meteorologische Abhandlungen 21, Meteorologischen Institut der Universität Bonn, Bonn, 82 pp.

Fritsch, J. M., R. J. Kane, and C. R. Chelius, 1986: The contribution of mesoscale convective systems to the warm season precipitation in the United States. *Journal of Climate and Applied Meteorology*, **25**, 1333–1345.

Fuenzalida, H. A., R. Sanchez, and R. D. Garreaud, 2005: A climatology of cut off lows in the Southern Hemisphere. *Journal of Geophysical Research – Atmospheres*, **110**, Art. No. D18101.

Grist, J. P., 2002: Easterly waves over Africa. Part I. The seasonal cycle and contrasts between wet and dry years. *Monthly Weather Review*, **130**, 197–211.

Hopsch, S. B., C. D. Thorncroft, K. Hodges, and A. Aiyyer, 2007: West African storm tracks and their relationship to Atlantic tropical cyclones. *Journal of Climate*, **20**, 2468–2483.

Iskenderian, H., 1995: A 10-year climatology of northern hemisphere tropical cloud plumes and their flow patterns. *Journal of Climate*, **8**, 1630–1637.

Jackson, I. J., 1989: *Climate, Water and Agriculture in the Tropics*. Longman, Singapore, 377 pp.

Kiladis, G. N., C. D. Thorncroft, and N. M. J. Hall, 2006: Three-dimensional structure of African easterly waves. Part I. Observations. *Journal of the Atmospheric Sciences*, **63**, 2212–2230.

Knippertz, P., 2005: Tropical–extratropical interactions associated with an Atlantic tropical plume and subtropical jet streak. *Monthly Weather Review*, **33**, 2759–2776.

Knippertz, P., and A. H. Fink, 2008: Dry-season precipitation in tropical West Africa and its relation to forcing from the extratropics. *Monthly Weather Review*, **136**, 3579–3596.

Knippertz, P., and J. E. Martin, 2005: Tropical plumes and extreme precipitation in subtropical and tropical West Africa. *Quarterly Journal of the Royal Meteorological Society*, **112**, 2337–2365.

Knippertz, P., and J. E. Martin, 2007: The role of dynamic and diabatic processes in the generation of cut-off lows over Northwest Africa. *Meteorology and Atmospheric Physics*, **96**, 3–19.

Knippertz, P., A. H. Fink, A. Reiter, and P. Speth, 2003: Three late summer/early autumn cases of tropical–extratropical interactions causing precipitation in northwest Africa. *Monthly Weather Review*, **131**, 116–135.

Latif, M., and T. P. Barnett, 1996: The decadal climate variability over the North Pacific and North America: dynamics and predictability. *Journal of Climate*, **9**, 2407–2423.

Lauer, W., 1975: Klimatische Grundzüge der Hohenstufung tropischer Gebirge. *Deutscher Geographentag Verhandlungen*, **40**, 76–90.

Lavaysse, C., A. Diedhiou, H. Laurent, and T. Lebel, 2006: African easterly waves and convective activity in wet and dry sequences of the West African monsoon. *Climate Dynamics*, **27**, 319–332.

Liebmann, B., and Coauthors, 2009: Origin of convectively coupled Kelvin waves over South America. *Journal of Climate*, **22**, 300–315.

Lockwood, J. G., 1974: *World Climatology: An Environmental Approach*. St. Martins Press, New York, 330 pp.

Madden, R. A., and P. R. Julian, 1994: Observations of the 40–50 day tropical oscillation. *Monthly Weather Review*, **122**, 814–837.

McGuirk, J. P., A. H. Thompson, and J. R. Schaefer, 1988: An eastern Pacific tropical plume. *Monthly Weather Review*, **116**, 2505–2521.

Mekonnen, A., C. D. Thorncroft, and A. Aiyyer, 2006: Analysis of convection and its association with African easterly waves. *Journal of Climate*, **19**, 5405–5421.

Mohr, K. I., and C. D. Thorncroft, 2006: Intense convective systems in West Africa and their relationship to the African easterly jet. *Quarterly Journal of the Royal Meteorological Society*, **132**, 163–176.

Mohr, K. I., J. S. Famiglietti, and E. J. Zipser, 1999: The contribution to tropical rainfall with respect to convective system type, size, and intensity estimated from 85-GHz ice-scattering signature. *Journal of Applied Meteorology*, **38**, 596–606.

Mounier, F., G. N. Kiladis, and S. Janicot, 2007: Analysis of the dominant mode of convectively coupled Kelvin waves in the West African monsoon. *Journal of Climate*, **20**, 1487–1503.

Nesbitt, S. W., and E. J. Zipser, 2003: The diurnal cycle of rainfall and convective intensity according to three years of TRMM measurements. *Journal of Climate*, **16**, 1456–1475.

Nesbitt, S. W., R. Cipelli, and S. A. Rutledge, 2006: Storm morphology and rainfall characteristics of TRMM precipitation features. *Monthly Weather Review*, **134**, 2702–2721.

Nicholson, S. E., 1980: Saharan climates in historic times. In *The Sahara and the Nile*. (M. A. J. Williams and H. Faure, eds.). A. A. Balkema, Lisse, pp. 173–200.

Nicholson, S. E., 1981: Rainfall and atmospheric circulation during drought periods and wetter years in West Africa. *Monthly Weather Review*, **109**, 2191–2208.

Nicholson, S. E., 2009: A revised picture of the structure of the "monsoon" and land ITCZ over West Africa. *Climate Dynamics*, **32**, 1155–1171.

Nicholson, S. E., and X. Yin, 2002: Mesoscale patterns of rainfall, cloudiness and evaporation over the great lakes of East Africa. In *The East African Great Lakes: Limnology, Palaeolimnology and Biodiversity* (E. O. Odada and D. O. Olago, eds.), Kluwer Academic Publishers, Dordrecht, pp. 93–119.

Nicholson, S. E., A. I. Barcilon, M. Challa, and J. Baum, 2007: Wave activity on the Tropical Easterly Jet. *Journal of the Atmospheric Sciences*, **64**, 2756–2763.

Nicholson, S. E., A. I. Barcilon, and M. Challa, 2008: An analysis of West African dynamics using a linearized GCM. *Journal of the Atmospheric Sciences*, **65**, 1182–1203.

Nieuwolt, S., 1977: *Tropical Climatology*. John Wiley and Sons, New York, 297 pp.

Pereira, L. G., and S. A. Rutledge, 2006: Diurnal cycle of shallow and deep convection for a tropical land and an ocean environment and its relationship to synoptic wind regimes. *Monthly Weather Review*, **134**, 2688–2701.

Petterssen, S., 1969: *Introduction to Meteorology*, 3rd edn. McGraw-Hill, New York, 333 pp.

Pook, M. H., P. C. McIntosh, and G. A. Meyers, 2006: The synoptic decomposition of cool-season rainfall in the southeastern Australian cropping region. *Journal of Applied Meteorology and Climatology*, **45**, 1156–1170.

Rasmusson, E. M., and T. H. Carpenter, 1982: Variations in tropical sea-surface temperature and surface wind fields associated with the Southern Oscillation/El Niño. *Monthly. Weather. Review*, **110**, 354–384.

Ratisbona, L. R., 1976: The climate of Brazil. In *Climates of Central and South America* (W. Schwerdtfeger, ed.), World Survey of Climatology 12, Elsevier, Amsterdam, pp. 219–293.

Reed, R. J., and J. D. Jaffe, 1981: Diurnal variation of summer convection over West Africa and the tropical eastern Atlantic during 1974 and 1978. *Monthly Weather Review*, **109**, 2527–2534.

Reed, R. J., A. Hollingsworth, W. Heckley, and F. Delsol, 1988: An evaluation of the performance of the ECMWF operational system in analyzing and forecasting easterly wave disturbances over Africa and the tropical Atlantic. *Monthly Weather Review*, **116**, 824–865.

Ricciardulli, L., and P. D. Sardeshmukh, 2002: Local time- and space scales of organized tropical deep convection. *Journal of Climate*, **15**, 2775–2790.

Riehl, H., 1979: *Climate and Weather in the Tropics*. Academic Press, New York, 611 pp.

Rohr, P. C., and A. Killingtveit, 2003: Rainfall distribution on the slopes of Mt. Kilimanjaro. *Hydrological Sciences Journal*, **48**, 65–77.

Rubin, S., B. Ziv, and N. Paldor, 2007: Tropical plumes over eastern North Africa as a source of rain in the middle east. *Monthly Weather Review*, **135**, 4135–4148.

Satyamurty, P., and M. E. Seluchi, 2007: Characteristics and structure of an upper air cold vortex in the subtropics of South America. *Meteorology and Atmospheric Physics*, **96**, 203–220.

Satyamurty, P., C. A. Nobre, and P. L. Silva Dias, 1998: South America. In *Meteorology of the Southern Hemisphere* (D. J. Karoly and

D. G. Vincent, eds.), American Meteorological Society, Boston, MA, pp. 119–139.

Schumacher, C., and R. A. Houze Jr., 2003: Stratiform rain in the tropics as seen by the TRMM precipitation radar. *Journal of Climate*, **16**, 1739–1756.

Serra, Y., G. N. Kiladis, and M. F. Cronin, 2008: Horizontal and vertical structure of easterly waves in the Pacific ITCZ. *Journal of the Atmospheric Sciences*, **65**, 1266–1284.

Straub, K. H., and G. N. Kiladis, 2002: Observations of a convectively coupled Kelvin wave in the eastern Pacific ITCZ. *Journal of the Atmospheric Sciences*, **59**, 30–53.

Taljaard, J. J., 1985: *Cut-off Lows in the South African Region*. Technical Paper No. 14, South African Weather Bureau, 153 pp.

Taljaard, J. J., 1990: The cloud bands of South Africa. *South African Weather Bureau Newsletter*, **493**, 6–8.

Taljaard, J. J., 1996: *Atmospheric Circulation Systems, Synoptic Climatology and Weather Systems of South Africa. Part 6. Rainfall in South Africa*. Technical Report No. 32, South African Weather Service, Pretoria, 100 pp.

Taylor, D., 1996: Mountains. In *The Physical Geography of Africa* (W. M. Adams, A. S. Goudie, and A. R. Orme, eds.), Oxford University Press, Oxford, pp. 287–306.

Thompson, B. W., 1968: Tables showing the diurnal variation of precipitation in East Africa and Seychelles. *E. A. M. D. Techn. Memorandum* No. 10, 49 pp.

Thompson, D. W. J., and J. M. Wallace, 1998: The Arctic Oscillation signature in the wintertime geopotential height and temperature fields. *Geophysical Research Letters*, **25**, 1297–1300.

Thompson, D. W. J., and J. M. Wallace, 2000: Annular modes in the extratropical circulation. Part I. Month-to-month variability. *Journal of Climate*, **13**, 1000–1016.

Thorncroft, C., and K. Hodges, 2001: African easterly wave variability and its relationship to Atlantic tropical cyclone activity. *Journal of Climate*, **14**, 1166–1179.

Todd, M., and R. Washington, 1999: Circulation anomalies associated with tropical-temperature troughs in southern Africa and the southwest Indian Ocean. *Climate Dynamics*, **15**, 937–951.

Trewartha, G. T., 1970: *The Earth's Problem Climates*. University of Wisconsin Press, Madison, 334 pp.

Trewartha, G. T., and L. H. Horn 1980: *An Introduction to Climate*. McGraw-Hill Book Company, New York, 416 pp.

van Heerden, J., and J. J. Taljaard, 1998: Africa and the surrounding waters. In *Meteorology of the Southern Hemisphere*, American Meteorological Society, Boston, MA, pp. 141–174.

Wallace, J. M., and D. S. Gutzler, 1981: Teleconnections in the geopotential height field during the Northern Hemisphere winter. *Monthly Weather Review*, **109**, 784–812.

Wallace, J. M., and P. V. Hobbs, 1977: *Atmospheric Science: An Introductory Survey*. Academic Press, New York, 467 pp.

Walter, H., and Box, E. O., 1983: Orobiomes of Middle Asia. In *Temperate Deserts and Semi-Deserts* (N. West, ed.), Ecosystems of the World 5, Elsevier, Amsterdam, pp. 161–191.

Wang, H., and R. Fu, 2007: The influence of Amazon rainfall on the Atlantic ITCZ through convectively coupled Kelvin waves. *Journal of Climate*, **20**, 1188–1201.

Warner, T. T., 2004: *Desert Meteorology*. Cambridge University Press, Cambridge, 595 pp.

Winiger, M., 1975: *Bewölkungsuntersuchungen über der Sahara mit Wettersatellitenbildern*. Geographica Bernensia G1, Geographisches Institut der Universität Bern, Bern, 149 pp.

Wright, W. J., 1997: Tropical–extratropical cloudbands and Australian rainfall. I. Climatology. *International Journal of Climatology*, **17**, 807–829.

Zhang, C., 2005: Madden–Julian oscillation. *Reviews of Geophysics*, **43**, RG2003, 2004RG000158.

Ziv, B., 2001: A subtropical rainstorm associated with a tropical plume over Africa and the Middle East. *Theoretical and Applied Climatology*, **69**, 91–102.

6 Radiation, heat, and surface exchange processes

6.1 INTRODUCTION

The heating of the earth–atmosphere system is accomplished via the processes of radiation, conduction, convection, and latent heat release. Radiation is energy in the form of electromagnetic waves that propagate through space; the waves travel at the speed of light and require no medium for propagation. Conduction is heat transfer via molecular collision, the essence of the process being an exchange of kinetic energy between molecules. Thus, it is most effective in a solid medium, in which molecules are relatively closely packed. Convection is heat transfer via mixing of parcels in fluids; it is much more rapid than conduction and acts on a larger scale. Since it requires translation of molecules, the convection process is limited to liquids and gases. Latent heat involves the molecular energy related to phase changes of water. For each gram of water, one calorie of heat is required to break the intermolecular bonds in the solid phase and vaporize the water. When water vapor condenses, this amount of heat is released. Thus, heat is transferred from the ground, where water is evaporated, to the atmosphere, where it condenses.

Over the earth as a whole and over time, the radiation received from the sun is largely equivalent to that emitted by the earth and atmosphere. This situation, termed the radiation balance, is required if the earth is to maintain a thermal equilibrium. For any given location, the radiation received at the surface is balanced by the remaining processes of heat transfer. This is referred to as the surface heat or energy balance. Atmospheric and surface characteristics determine the values of each term in the balance. The surface plays a particularly important role because the nature of the ground and the characteristics of the air layer just above it govern the fluxes that determine the heat balance. Latent heating intimately links the heat and water balances.

6.2 RADIATIVE PROCESSES

Radiative processes in the atmosphere are governed by numerous physical laws. The most basic are the Stefan–Boltzmann law and Wien's law, which respectively govern the influence of temperature on the amount and the predominant wavelengths of radiation emitted by a body. The Stefan–Boltzmann law states that the amount of radiation emitted by a black body (a perfectly absorbing and radiating body) is proportional to the fourth power of its absolute temperature. This dictates that, as a result of high temperatures, most dryland regions lose much energy via radiation from the surface. Wien's law says that the wavelength of maximum emission of radiation is inversely proportional to temperature. The consequence is that solar radiation is "shortwave," primarily visible and ultraviolet, and the earth's radiation is "longwave," primarily infrared. These principles are important in the remote sensing methods described in Chapter 1.

The net amount of radiant energy available at the surface to drive the climate system, termed net radiation, is the difference between the solar radiation absorbed and the net longwave radiation. The latter is the difference between the surface emission of longwave radiation and its heat gain via longwave radiation from the atmosphere. Net radiation can be expressed as

$$R_{net} = R_{sw}\downarrow - R_{sw}\uparrow + R_{lw}\downarrow - R_{lw}\uparrow = R_{sw}\downarrow(1-a_s) + R_{lw}\downarrow - R_{lw}\uparrow \quad (6.1)$$

where the subscript "sw" refers to shortwave or solar radiation and the subscript "lw" refers to the longwave radiation emitted by the earth, atmosphere and clouds. $R_{lw}\uparrow$ is radiation emitted spaceward from the earth's surface. $R_{lw}\downarrow$ is that emitted earthward by clouds and the atmosphere. $R_{sw}\downarrow$ is the solar radiation received at the earth's surface; $R_{sw}\uparrow$ is that reflected back toward space.

Solar radiation reaching the ground generally decreases with latitude. However, the amount of radiation entering the heat balance at various locations is quite diverse because much of the solar radiation is reflected back to space by the earth's surface. The proportion that is reflected is a_s, the surface albedo. Table 6.1 gives typical values for a variety of surfaces. Note that

Table 6.1. *Typical albedo values for natural surfaces.*

Surface	Albedo (%)
Fresh snow	70–95
Old snow	50–75
Desert sands	25–40
Other desert surfaces	35–50
Grass	15–25
Dry soil	15–25
Wet soil	10
Forest – mid-latitude	15–20
Tropical forest	7–12
Water (low sun angle near horizon)	50–90
Water (sun near zenith)	3–5
Thick cloud	70–80
Thin cloud	25–50
Earth + Atmosphere	30

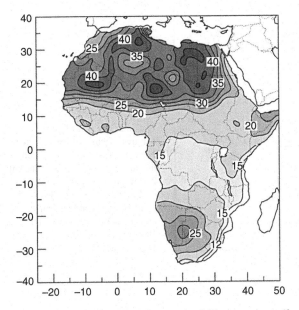

Fig. 6.1 Surface albedo (%) in July over the African continent (from Ba *et al.* 2001).

the actual range of values for any type of surface is very large. The albedo of water averages about 5%, but this is strongly dependent on the angle at which the solar radiation hits the earth's surface. The albedo of vegetated surfaces ranges from about 7% to 12% for tropical rainforests and from 15% to 25% for grasslands.

Soil albedo is, to a large extent, a function of soil moisture and is hence quite variable. The albedo of a wet soil is typically 10%, but 15–25% is more common for a dry soil. Surface albedo can be as high as 45–50% in extreme deserts with light-colored sandy or rock surfaces. This is illustrated in Fig. 6.1, using the surface albedo over Africa in July. Surface albedo reaches a minimum in the equatorial latitudes of the Southern Hemisphere,

where it is less than 15%. In the semi-arid regions it is on the order of 20–35%. Surface albedo attains values exceeding 50% in the central Sahara (Ba *et al.* 2001). In many desert locations, especially saline playas, surface albedo can be higher. See Warner (2004) for a review of measurements in arid regions.

Because of the diversity of surface types, the albedo of the world's deserts varies substantially. This diversity is clearly seen over the Sahara. A comparison of Figs. 6.1 and 6.2 shows that the highest values correspond to the major sand seas and the lowest values appear over the desert highlands (Fig. 6.2). It is also dependent on the type of rock or rock surface covering the desert (Table 6.2).

In part because of the high albedo, the dryland regions tend to be areas of low surface net radiation. Other contributing factors are the high radiant emission from the hot surface and the clear skies that reduce the longwave radiation received from the atmosphere. The contrast with humid regions is more evident in the radiation balance (i.e., net radiation) at the top of the atmosphere (Fig. 6.3). This parameter represents the total amount of energy gained by the earth–atmosphere system via radiation. Net radiation is actually negative in some desert areas, such as the Saharan–Arabian desert belt. The overall effect of this is paradoxical: "hot" deserts, with intensely high surface temperatures, are actually "cool" spots radiatively and do not contribute greatly to the overall atmospheric heating.

The importance of net radiation is apparent when the following fact is considered: for a thermal equilibrium to exist, with no long-term change of mean temperature, the net radiation of the earth–atmosphere system in a given region must be balanced by other forms of heat transfer. When the net radiation is negative, as over the Sahara, maintaining thermal equilibrium requires a gain of heat in other ways. Normally this is accomplished in the atmosphere by advection or latent heat release, but this is not the case over the Sahara. Because of the region's large size, it cannot receive heat by advection; since few clouds develop, there is likewise little gain through latent heat release. Instead, an equilibrium is maintained via large-scale subsidence, which heats the air adiabatically by compression. The subsidence reinforces the aridity by suppressing upward motion and convection. This feedback process may have contributed to the extreme size and intensity of the Saharan–Arabian desert zone.

6.3 THE HEAT BALANCE

A balance must likewise exist at the earth's surface, so that at a given location the net radiative heat gain, expressed as the net radiation, must be balanced by other forms of heat transfer. This heat balance is expressed as

$$R_{\text{net}} = LE + S \tag{6.2}$$

where S is sensible heat transfer (conduction and convection) and LE is latent heat exchange. In the latent heat term, L is the latent heat of condensation/vaporization and E is evaporation.

Table 6.2. *Measured values of surface albedo for various surface types (including three sites in the serir or stone desert) in the central Sahara (from Tetzlaff 1974).*

Surface	Basalt	Sandstone	Serir I	Serir II	Serir III	Sand	River sand
Albedo	0.12	0.18	0.36	0.38	0.36	0.49	0.28

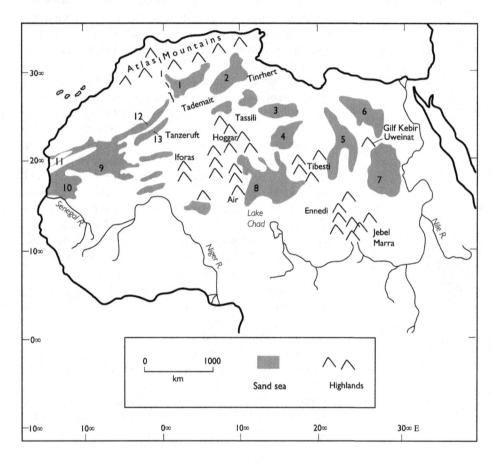

Fig. 6.2 The major physiographic features of the Sahara (courtesy of Nick Lancaster). Major sand seas include 1. Grand Erg Occidental; 2. Grand Erg Oriental; 3. Ubari; 4. Murzuk; 5. Calanscio; 6. Great Sand Sea; 7. Selima; 8. Fachi-Bilma and Ténéré; 9. Majabat al Koubra; 10. Aouker; 11. Akchar; 12. Iguidi; 13: Erg Chech.

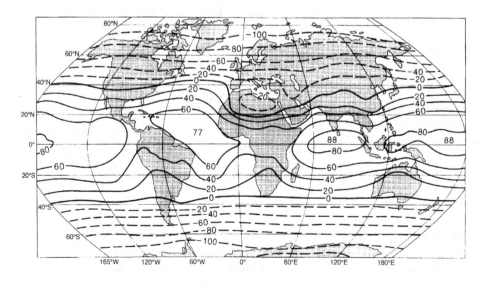

Fig. 6.3 Net radiation at the top of the atmosphere (from Raschke *et al.* 1973).

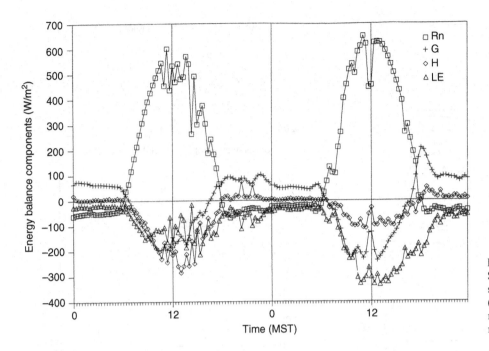

Fig. 6.4 Surface heat balance in the Sonoran Desert (Walnut Gulch watershed, southern Arizona) on two days (Kustas *et al.* 1991). (left) Surface soil moisture is 5%. (right) Surface soil moisture is 10%.

The surface conducts heat to and from both the atmosphere and ground, so that Eq. (6.2) may also be expressed as

$$R_{net} = LE + G + H \qquad (6.3)$$

where G is conduction to or from the subsurface and H is conduction and convection between the surface and atmosphere. The term H is predominantly convection, because conduction is relatively insignificant in a fluid.

The partitioning of the surface contribution to atmospheric heating into sensible and latent heat depends on the nature of the surface. The availability of surface moisture has a decisive influence on this partitioning because of the characteristic properties of water. Its specific heat (the amount of heat required to change the temperature of 1 gram by 1° Celsius) is the highest of any natural substance. Nevertheless, the amount required to vaporize 1 gram of water is 600 times greater, or 600 cal/gram. Water's relatively high transparency also helps to moderate its temperature, as does the mixing that takes place in open water bodies such as the ocean or lakes.

Even small differences in soil moisture can have a very large impact on the surface heat balance. Figure 6.4 shows this balance in the Walnut Gulch watershed of the Sonoran Desert for surface soil moisture contents of 5% and 10%. Latent heat flux is increased by roughly a factor of three with the small addition of moisture. There is a corresponding decrease in sensible heat flux to the atmosphere and a small increase in heat transfer to the soil. Small and Kurc (2003) found that in a semi-arid grassland and a semi-arid shrubland, a wet soil decreased surface temperature by more than 10°C and increased the available energy by some 20%. However, the changes were short-lived, with the soil remaining wet for only a few days at a time.

Figure 6.5 shows simplified representations of the heat balance for eight locations characteristic of various climate types.

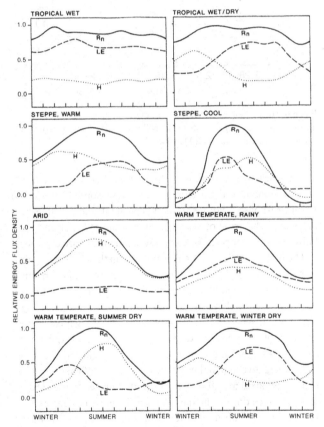

Fig. 6.5 Characteristic annual energy balance for selected climate types (based on Kraus and Alkhalaf 1995).

Except in the tropical locations, the net radiation has a pronounced seasonal cycle, with a maximum in summer and a minimum in winter. The contribution of latent and sensible heat transfer to the balance is quite different at the various locations,

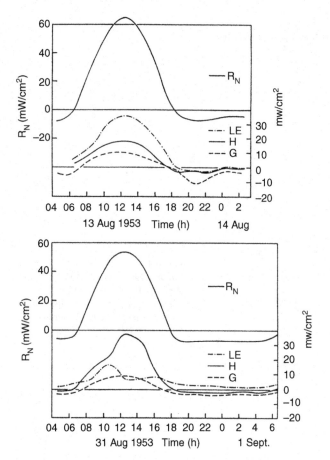

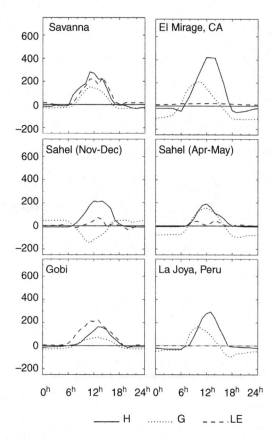

Fig. 6.6 Heat balance over the prairie at O'Neil, Nebraska, shortly after a rain (top) and after a 15-day dry spell (bottom) (from Deacon 1969, based on data in Lettau and Davidson 1957).

Fig. 6.7 Heat balance at various dryland locations during the course of the day, based on experimental measurements: Ivory Coast savanna; El Mirage, Mojave Desert, California; the Sahel near Niamey, Niger in Nov–Dec and Apr–May; Gobi Desert of China; and La Joya, Atacama Desert, Peru. The two curves for the Sahel derive from different experimental programs: ECLATS in Nov–Dec 1980 and Yantala in Apr–May 1984, at the beginning and end of the dry season, respectively.

as is the seasonal variation of these terms. Sensible heat transfer is dominant in the drier climates and seasons. There is also an out-of-phase relationship between latent and sensible heat transfer in the areas of seasonal climate in the lower and mid-latitudes (i.e., steppe climates and warm temperature climates).

The prairie site near O'Neill, Nebraska, in the US Great Plains, provides an example of the diurnal variation of the heat balance for wet and dry conditions in a semi-arid region (Fig. 6.6). Net radiation is strongly positive during the day, but becomes negative at night. When the surface is wet, the remaining fluxes (latent and sensible) are positive during the day (i.e., directed from the surface to the air and ground). At night they are near zero or slightly negative. Latent heat is the largest flux. When the surface is dry, the fluxes are all directed from the surface to the air and/or subsurface during the day and sensible heat flux far exceeds latent. At night all fluxes are small but the latent heat flux is negative (directed from the surface to the atmosphere) and the surface receives sensible heat from both the ground and the air.

The heat balance of several other arid and semi-arid locations is shown in Fig. 6.7: an African tree savanna in the northern Ivory Coast (Coulibaly and Boutin 1983); the Mojave Desert near El Mirage, California (Vehrencamp 1953); the African Sahel (semi-desert shrub grassland) near Niamey, Niger (e.g.,

Durand *et al.* 1988; Frangi *et al.* 1992; see also Miller *et al.* 2009); the Gobi Desert of China (Smith *et al.* 1986); and the Atacama of Peru (Stearns 1969; Lettau and Lettau 1978). Net radiation is not given because its diurnal cycle varies little among these low-latitude sites, except for the magnitude of the midday maximum. Some trends in heat balance are evident among these sites. In general, sensible heat transfer to the air is roughly one-third to half of the net radiation. The degree to which heat transfer to the soil and latent heat contribute to the balance is a function of aridity. In the most arid locations, latent heat transfer is minimal and transfer from the soil to the surface helps to balance the nocturnal losses via net radiation.

The picture is much more complex than that illustrated in the previous examples. Individual vegetation formations can differ substantially in their contribution to the heat balance and its temporal course (Kabat *et al.* 1997; Timouk *et al.* 2009). This is illustrated with an example from the Sahel of West Africa, taken from a detailed study by Verhoef (1995). That study is based on the results of the HAPEX-Sahel and SEBEX experiments, conducted in Niger in 1991/92 and 1989/90, respectively.

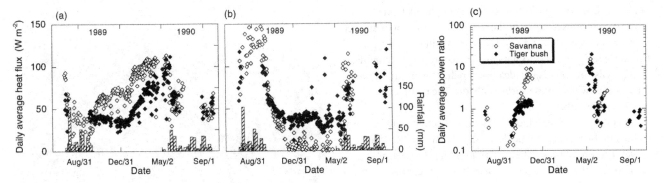

Fig. 6.8 Annual courses of (a) sensible and (b) latent heat for savanna and tiger bush surfaces during the course of 1989 and 1990. Open diamonds are savanna and closed diamonds are tiger bush. Rainfall at the savanna site is also plotted at the bottom (bars). (c) Bowen ratio for savanna and tiger bush during the same period (Verhoef *et al.* 1999).

The vegetation in the region consists of mostly Gueira bush and grass savanna. In some areas there is a formation called "tiger bush" (see Fig. 3.5), that consists of dense thickets surrounded by bare soil in varying proportions. Figure 6.8 contrasts the savanna and tiger bush during the course of one year; precipitation is given at the bottom of the two diagrams and the Bowen ratio is indicated on the right.

During the wet season, the latent and sensible heat flux is roughly the same in the two units but during the dry season the picture is very different. Then the savanna makes a greater contribution to sensible heat flux but the tiger bush makes a greater contribution to the latent heat flux. It appears that the tiger bush formation continues to transpire long after the rains end; this is confirmed by a Bowen ratio of roughly 1 for several subsequent months. This could be a result of soil moisture being retained beneath the dense thicket or deeper root systems of the species in the tiger bush formation (Verhoef 1995). The Bowen ratio for the region as a whole increases rapidly by two orders of magnitude when the rains end, and decreases by more than an order of magnitude at the beginning of the rainy season.

6.4 THE COMPLEXITIES OF SURFACE ALBEDO

The concept of albedo has received tremendous attention in the context of climatic change in dryland regions. Human-induced changes of surface albedo were long ago suggested as possible causes of drought or climatic change (Otterman 1974; Charney 1975). The origin of this theory lies in satellite photos depicting dramatic albedo contrasts along anthropogenically created boundaries. A prominent example is shown in Fig. 6.9a, a satellite image of the border between the Negev and the Sinai. The border is clearly seen as a straight line and the contrast has been attributed to differences in grazing policies on the two sides of the border. The more heavily grazed Sinai, which was then under Egyptian control, appears much lighter in color and is several degrees warmer, a consequence mainly of the contrasting albedo between the sands over the Sinai and biological crusts on the Negev soil (Qin *et al.* 2002).

Fig. 6.9 (a) Satellite photo showing strong albedo contrast along the Negev/Sinai border (photo courtesy of NASA). (b) Albedo contrast between agricultural fields (light squares) in Botswana and the darker, surrounding land.

This issue served well to focus attention on the problem of human impact in the dryland regions and potential human impact on their climates. However, the importance of the so-called "albedo feedback" has been exaggerated. This issue is discussed more fully in Chapter 22. The reason for the exaggerated importance is to some extent the oversimplified treatment of albedo in most discussions of the theory. The concept of surface albedo is actually a very complex one, and failure to understand this complexity has led to numerous, inappropriate generalizations. A case in point is the idea that a reduction in surface vegetation cover will increase surface albedo. The complexity of the concept is described here.

Albedo is generally defined as "surface reflectivity." This in turn is the ratio of reflected to incident solar radiation. "Reflectivity," however, more appropriately refers to a single wavelength (r_λ). Albedo, on the other hand, is reflectivity integrated over all wavelengths of the spectrum. Thus it is defined as:

$$A = \frac{\int_0^\infty r_\lambda m_\lambda d\lambda}{I_0} \tag{6.4}$$

where r_λ is the reflectivity for the wavelength λ, m_λ is incident solar radiation at this wavelength and I_0 is the total incident solar radiation. Theoretically, albedo refers to the whole spectrum but most measuring devices generally see only a portion of the spectrum. Therefore relatively few actual measurements of true albedo exist and it is necessary to extrapolate what the full "spectral" albedo would be from the wavelengths that are seen by the satellite or other measuring device. In many cases, only what is termed "visible albedo" (reflectivity in the visible wavelengths) is observed. This is true for aerial photos and some satellite photos, such as those shown in Fig. 6.9. For a vegetated surface, reflectivity is much higher in the infrared than in the visible (see Chapter 1). Thus, visible images will tend to exaggerate the albedo contrast between vegetation and non-vegetated areas.

The albedo of a particular surface is a function of both surface composition and geometry. The composition determines what wavelengths are absorbed and, to a lesser extent, the degree of absorption. The degree of absorption is largely a function of the transparency of a material. Water illustrates this very well. Because water transmits a large percentage of incident radiation to greater depths, the albedo is only a few percent when the sun is nearly overhead. The geometry, in particular the number of reflecting surfaces and their orientations, determines to a large degree the overall amount of radiation reflected by an opaque surface.

The albedo of a vegetated surface has two components: vegetation albedo and soil albedo. When the vegetation cover approaches 100%, the contribution of soil albedo is negligible. The soil albedo is determined mainly by soil moisture but texture, porosity, color, and structural features such as clustering have some influence (Fig. 6.10). The albedo of vegetation is determined mainly by its geometry, structure, and density, but water and chlorophyll content can also be significant determinants. For both surfaces, albedo also varies with the angle of the sun, the viewing angle, and the ratio of direct to diffuse radiation. Because diffuse radiation is incident upon a surface at many angles, it is also reflected diffusely, thus contributing little to the albedo.

Fig. 6.10 Sharp demarcation between two soil types in Botswana, one highly reflective and the other much darker.

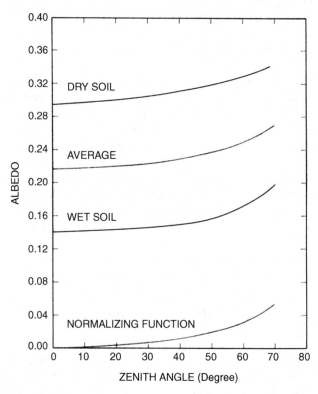

Fig. 6.11 Albedo versus sun angle for wet and dry sand and for water (from Idso *et al.* 1975a).

The influence of sun angle on albedo is shown in Fig. 6.11. At lower solar elevation angles (high zenith angles), the albedo increases very rapidly with solar angle. The rate of increase varies with the type of surface and the degree of wetness. It is particularly strong with water surfaces. At sunset or sunrise, when the sun is near the horizon, the albedo of water approaches 100% and the water becomes a virtual mirror (Fig. 6.12). The implication of this and other factors influencing albedo is that a simple comparison from photos in the visible spectrum does not provide

Fig. 6.12 Photo showing specular (mirror-like) reflection of a water surface at sunset. The image shows a flood of the normally dry surface of the Bonneville salt flats of Utah, USA.

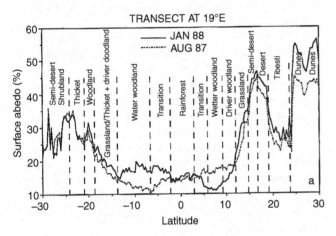

Fig. 6.13 Latitudinal variations in surface albedo over Africa between the wet and dry seasons (from Ba *et al.* 2001). January/August is the wet/dry season for Southern Hemisphere locations; August/January is the wet/dry season for Northern Hemisphere locations. Note that in the woodlands just poleward of roughly 5° of latitude, albedo decreases in the dry season, when vegetation cover is reduced.

reliable information on albedo contrasts. In the case of the international borders mentioned earlier, a detailed assessment generally shows much smaller contrasts (e.g., Michalek *et al.* 2001).

For vegetation cover, the effects of geometry and structure are very complex. This is illustrated by satellite-estimated surface albedo along a transect through Africa (Fig. 6.13). In a surface cover of intermediate density, such as a woodland, the albedo might increase with the number of leaves. Hence, the albedo decreases during the dry season when the trees lose their leaves and the underlying grass cover is seen by the satellite. In the grasslands, the albedo is largely a function of the wetness of the underlying soil, thus the albedo increases rapidly toward the Sahara Desert, where it reaches 56% (Ba *et al.* 2001). In the Taklamakan Desert, values as high as 49% have been measured (Aoki *et al.* 2005). In the true forest, with very dense canopy,

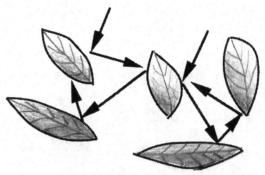

Fig. 6.14 Diagram illustrating the "trapping" of radiation by a canopy of leaves. Each time radiation is incident upon a leaf surface, a portion is absorbed and a portion is reflected back into the canopy at the angle of incidence.

another effect occurs in which the leaves redirect the reflected radiation toward each other and each leaf surface both absorbs and re-reflects the radiation. This is termed "canopy trapping" (Fig. 6.14). As a result, the albedo of a tropical rainforest can be as low as about 5%.

Even the viewing angle of a surface can affect the measured albedo. This is illustrated in Fig. 6.15, showing two views of a sunflower field. The bright face of the sunflowers is always turned toward the sun, so that a view taken toward the sun shows the darker backside of the extensive fields of flowers.

6.5 MICROCLIMATE: THE CLIMATE NEAR THE GROUND

Conditions near the earth's surface are markedly different from those in the free atmosphere. Atmospheric properties such as temperature, moisture content, and wind speed change especially rapidly within the first few meters above the ground because the surface acts as a source or sink of each of these. Temperature and moisture also vary with distance below the surface. This section focuses on variation within the first few meters of the lower atmosphere; the microclimate of the subsurface is dealt with in Section 6.6. The discussion commences with the general and simple case of conditions over a bare soil surface, then continues with the microclimate of a layer of vegetation. More detailed information about microclimates in the drylands is given in Chapter 10.

Figure 6.16 shows typical profiles of temperature during the day and at night. The depicted profiles are primarily applicable to days with high solar insolation and little advection of unseasonably warm or cold air. The typical daytime situation is that of temperature decreasing rapidly with height z above the surface. Since the atmosphere is relatively transparent to solar radiation, it receives most of its heat from the earth's surface, which readily absorbs solar radiation. Transfer of heat to the lowest layers of the atmosphere is accomplished through both radiative and convective processes. At night, the surface cools more rapidly than the air and the temperature stratification is

Fig. 6.15 Two views of a sunflower field illustrating the impact of viewing angle on albedo.

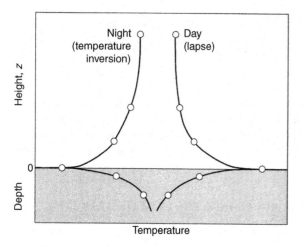

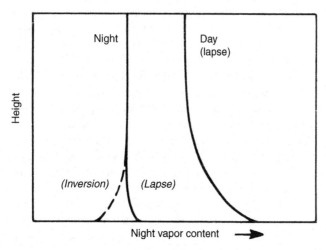

Fig. 6.16 Idealized vertical profiles of soil and air temperature (left) and atmospheric water vapor (right) near the surface on a clear summer day (modified from Oke 1987).

reversed: temperature increases with height above the surface. This situation is referred to as a temperature inversion. In the early evening and morning hours, a nearly isothermal stratification prevails. The temperature pattern in the ground is almost

the inverse: temperature decreases with depth during the day and increases with depth at night. The surface itself is the site of maximum daytime temperatures and minimum nighttime temperatures.

The surface is the source of atmospheric moisture, so the atmospheric water vapor content (expressed as vapor pressure, or the pressure exerted by water molecules) tends to decrease with height. This is the case both during the day and at night (Fig. 6.16). The exception is when condensation occurs at the surface in the form of dew or frost. In such a case, vapor pressure increases with height, at least very close to the surface. The atmospheric moisture content near the surface tends to be higher during the day than at night because the factors critical in the transfer to the atmosphere – energy and wind – are greatly reduced at night. Relative humidity, on the other hand, tends to be higher at night because the cooler night air is more readily saturated.

The vertical profile and diurnal cycle of carbon dioxide are more complex because several competing processes occur simultaneously. Over the land these include photosynthesis, respiration, and decay of organic matter. During the day, when photosynthesis is occurring, plants are sinks of carbon dioxide. Sources are the atmosphere, decaying plant material, and respiration from the soil, plants, and animals. At night, plants, through respiration, become sources of carbon dioxide, along with soil and animals. The atmosphere is a sink. As a result, the concentration of carbon dioxide near the surface tends to be highest at night and lowest during the day.

In contrast to its influence on other atmospheric properties, the earth's surface is a constant sink of momentum. Therefore the wind speed increases rapidly with height in the lowest layers. A bare soil surface is relatively smooth aerodynamically and the wind profile is described by the log law:

$$u_z = \frac{u^*}{\kappa} \ln \frac{z + z_0}{z_0}. \tag{6.5}$$

In the above equation, u_z is mean wind speed u at height z, u^* is a parameter called friction velocity and is a function of

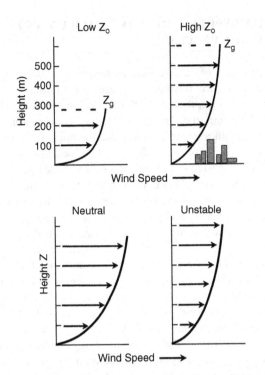

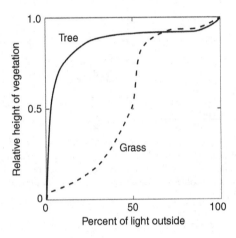

Table 6.3. *Typical aerodynamic roughness lengths for various surface types (based on Sellers 1965; Oke 1987; Warner 2004).*

Surface type	z_0 (cm)
Desert sand	0.001–0.07
Playa, bare and smooth	0.01
Sand with small shrubs	0.17
Dry lake bed	0.003
Stone (0.7 cm) pavement	0.028
Water, still	0.0001–0.01
Ice, smooth	0.001
Grass, 0.02–0.1 m	0.3–1
Grass, 0.25–1.0 m	4–10
Forests	100–600

Fig. 6.17 Simplified representative of u_z in cases of low and high surface roughness (z_0) and varying atmospheric thermal stability. The height at which the wind becomes geostrophic is represented by z_g (based on Oke 1987).

Fig. 6.18 Schematic illustrating radiation attenuation in vegetation canopies. The x-axis is the amount of light reaching a given height in the canopy, expressed as a percentage of the sunlight above the canopy.

the wind regime, z_0 is surface roughness length, and κ is the Karman constant, which roughly equals 0.41. Roughness is low for non-vegetated surfaces, with z_0 rarely exceeding about 0.01. The shape of the wind profile is modified by stability, since the rate of turbulent transfer is reduced under stable conditions. In general, u will increase most rapidly with height when z_0 is low and instability is high (Fig. 6.17). The log law has been adjusted in many ways to account for this effect. Other forms have been produced to account for very tall surface "obstacles," as in the case of a forest. Generally a parameter called the "zero-plane displacement" is introduced; this represents the height above the surface at which the log law becomes valid.

The microclimate of a vegetated surface is much more complex (see Chapter 10). Plant cover creates a situation very different than that over bare soil because (1) there are two active surfaces (ground and canopy) with respect to the transfer of energy, mass, and momentum, (2) the canopy, unlike the ground, is partly transparent to solar radiation, and (3) the vegetation cover is a relatively inhomogeneous surface (Raupach 1994). A layer of vegetation can be a source or sink of heat and carbon dioxide, a sink of momentum, and a source of water vapor. Aerodynamically it is considerably rougher than a bare soil (Table 6.3) and its albedo is generally lower because radiation is trapped in the canopy rather than reflected spaceward. In general, albedo decreases with vegetation height, while z_0 and zero-plane displacement and aerodynamic roughness increase. However, the vegetation/

roughness relationship is complex, as it also depends on other factors, such as the silhouette area of the vegetation and its density (Lettau 1969).

The penetration of solar radiation through a layer of vegetation depends on the density distribution of the foliage (Fig. 6.18). In a relatively thin cover, with foliage evenly distributed with height, like a grass cover, radiation decreases almost linearly from the top of the vegetation layer to the surface. For a dense cover with a distinct layer of dense foliage, like a forest crown, the radiation penetrating through the layer decreases exponentially. The attenuation of solar radiation is described by an equation analogous to Beer's law for a transparent medium. The intensity of radiation E_λ at wavelength λ and some level z in the canopy is expressed as

$$E_\lambda = E_{\lambda,0}\, e^{-az} \tag{6.6}$$

in which z is the level at which E_λ is measured, $E_{\lambda,0}$ is the intensity of solar radiation at the top of the canopy, and a is an

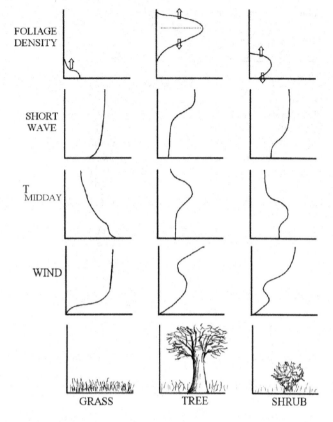

Fig. 6.19 Idealized vertical profiles of shortwave solar radiation, midday temperature and wind speed for grasses, trees and shrubs. Corresponding profile of foliage density is also shown (modified from Oke 1987).

attenuation factor. This factor is often assumed to be equal to 0.6, but in reality the degree of attenuation varies with height and density of the vegetation and with wind. Clearly, a varies with the season in areas of deciduous vegetation. E_λ can also be expressed as

$$E_\lambda = E_{\lambda,0}\, e^{-a(\text{LAI})} = E_{\lambda,0}\, e^{-a(h-z)} \qquad (6.7)$$

where h is the height of the canopy top and LAI is the leaf area index and is defined as

$$\text{LAI} = \Sigma A(z)\, dz \qquad (6.8)$$

with $A(z)dz$ being the leaf area between z and $(z + dz)$ per unit ground (leaf area per unit ground) (Dickinson 1983).

Figure 6.19 summarizes the effect of a plant canopy on conditions near the surface. The canopy retains heat both day and night, but it also insulates the surface below from strong heating or cooling. Both at the surface and in the canopy, vapor pressure attains a maximum and wind speed reaches a minimum. These profiles are highly idealized and strongly dependent on the density distribution of the foliage and other characteristics of the vegetation layer, as well as on ambient meteorological conditions.

6.6 ATMOSPHERIC THERMAL STABILITY AND ADIABATIC PROCESSES

The processes of surface heat exchange are clearly dependent on the vertical distribution of atmospheric characteristics. At the same time, they shape this vertical distribution. This is particularly important in the case of temperature, because its vertical stratification is a critical determinant of the various heat exchange processes, which also influence temperature. Temperature variation with height is called "lapse rate."

The importance of lapse rate is best understood in the context of adiabatic processes. The term adiabatic technically means "without the addition of heat" and refers to processes within a closed system to which no heat is added externally. In such a case, the well-known universal gas law dictates that changes of temperature in a gas are accomplished by changes in volume. Thus, when air encounters a region of lower pressure (as when rising in the atmosphere), it expands. In doing so, it expends energy (it does work) and therefore cools. When air sinks (encountering higher pressure) it is compressed. In this case, work is done on the air, adding energy and hence leading to warming. As long as no condensation or evaporation takes place in the air mass (modifying latent heating), the rate of heating and cooling upon ascent or descent is 1°C/100 m. This is referred to as the dry adiabatic lapse rate.

The tendency for air to rise by convection depends on the relative values of the environmental lapse rate and the adiabatic lapse rate. If the environmental lapse rate is greater than the adiabatic, a rising air parcel will become warmer (and hence lighter) than its environment. The parcel will continue to rise as a result of buoyancy. In this case, the atmosphere is thermally unstable. If the environmental lapse rate is less than the adiabatic, a rising parcel will be colder and hence denser than its surroundings and will sink once the force pushing it upward is removed. In this case, the atmosphere is thermally stable. A temperature inversion is a case of extreme stability.

Stability is also affected by atmospheric moisture content. If a rising air parcel becomes saturated as it cools, condensation takes place (i.e., clouds form) and latent heat is released. This reduces the rate at which it continues to cool as it rises. The rate is a complex function of atmospheric temperature and moisture. However, in the lower atmosphere, dry adiabatic processes prevail, especially in drylands. The issue of stability and adiabatic processes is important not only in heat and moisture transfer near the surface, but also in the development of the surface wind regime and dust devils (see Chapter 13).

6.7 CONDUCTION OF HEAT IN THE SOIL

During periods of high insolation the ground surface heats up and transmits heat to the subsurface via conduction. At times when the surface cools intensely, it becomes cooler than the ground below, and the lower layers then become a reservoir of heat for the surface. The typical patterns of ground temperature

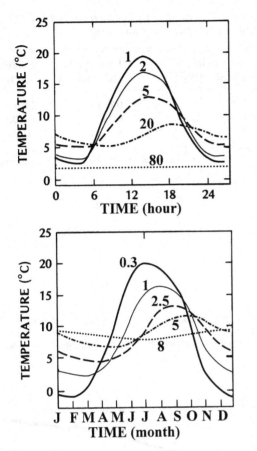

Fig. 6.20 Generalized diurnal and annual cycles of soil temperature at various depths (cm for diurnal, m for annual), based on Geiger 1965, Oke 1987 and others).

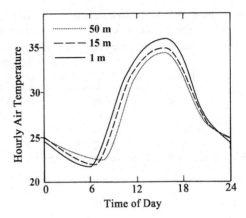

Fig. 6.21 Generalized diurnal cycle of air temperature near the ground (based on Geiger 1965, Oke 1987 and others).

during the course of the day and year are depicted in Fig. 6.20. Several generalizations can be made from these. The first is that the surface tends to be warmer than the subsurface during the day and during the warm season; it tends to be cooler than the subsurface at night and during the cool season. As depth increases, the diurnal range becomes progressively lower and the times of minimum and maximum temperature progressively later. The temperature gradients are largest close to the surface and decrease rapidly with depth below the surface. The layer of maximum temperature gradually moves down from the surface until approximately sunrise. The annual temperature wave strongly resembles the daily pattern, but penetrates to a greater depth and the time lag is on the order of months.

Heat transfer from the surface to the soil is accomplished by conduction, with the relevant flux of heat S being proportional to the temperature gradient, so that

$$S = -k \, dT/dz. \tag{6.9}$$

The thermal conductivity of soil is several orders of magnitude smaller than the equivalent transfer coefficient for air, so that the transfer of heat in the atmosphere is much more rapid and occurs over much larger distances. As a result, the temperature gradients in the lower atmosphere (Fig. 6.21) are considerably

smaller than within the soil and the diurnal and annual waves penetrate to a much greater distance from the surface.

The temperature changes of the surface and subsurface are controlled by a number of thermal properties of the soil, in addition to thermal conductivity. One is defined by relating the time rate of change of temperature to the gradient of its flux:

$$dS/dz = -C_v \, dT/dt \tag{6.10}$$

where C_v is the heat capacity of the soil at constant volume, or the amount of heat needed to raise the temperature of 1 cm^3 by 1°C. This is related to the specific heat (amount required to raise the temperature of 1 gram by 1°C) by the density of the soil. By combining the above two equations, a new equation results which relates the time rate of temperature change to the temperature gradient, such that

$$dT/dt = k/C_v \, dT^2/dz^2. \tag{6.11}$$

In the above expression, k/C_v is called the thermal diffusivity.

The thermal conductivity determines the rate of heat transfer, while the heat capacity or specific heat determines the amount of temperature change corresponding to a given flux of heat. The thermal diffusivity, symbolized by α, determines the time lag of the maximum and minimum temperatures, the temperature range at various depths, and the depth of penetration of the annual and diurnal temperature waves.

Table 6.4 gives typical values of select thermal properties for a variety of natural materials. For soil, these properties are determined by a number of characteristics including moisture content, porosity, organic matter content, type or texture and, to a much lesser extent, the composition of the soil particles. However, soil moisture content is by far the most important factor determining the thermal properties of the soil and it affects the soil in a variety of ways. Figure 6.22 shows the variation of the quantities defined above with soil moisture content. In general, increasing the soil moisture content will decrease its albedo, raise its specific heat and thermal conductivity, and increase the expenditure of incoming energy for latent heat. The increase of

Table 6.4. *Typical values of surface thermal properties (albedo, emissivity e, specific heat, thermal conductivity, thermal diffusivity, and penetration depth of solar radiation) (based on data in Munn 1966; Sellers 1965; Oke 1987; Rosenberg 1974).*

	Albedo	e	Specific heat (cal/g, °C)	Conductivity (10^{-3} cal cm^{-1} s^{-1} °C^{-1})	Diffusivity (10^{-2} cm^2 s^{-1})	Penetration depth
Land	0.1–0.4	0.90	0.3–0.8	0.4–8	0.8–1	1 mm
Water	0.06	0.95	1.00	1.4	0.14	1–100 m
Ice	0.30	0.96	0.50	5.5	1	10 m
Snow	0.80–0.95	0.82–0.95	0.50	0.2–5	–	1 m

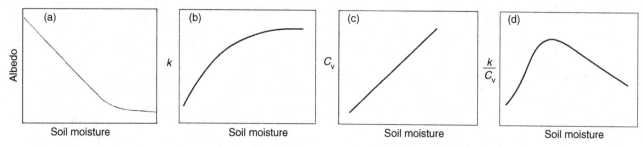

Fig. 6.22 Idealized variation of soil thermal characteristics with soil moisture: (a) albedo a_s, (b) conductivity k, (c) heat capacity C_v, and (d) diffusivity α (based on data in Geiger 1965; Sellers 1965; Idso *et al.* 1975a, and others)

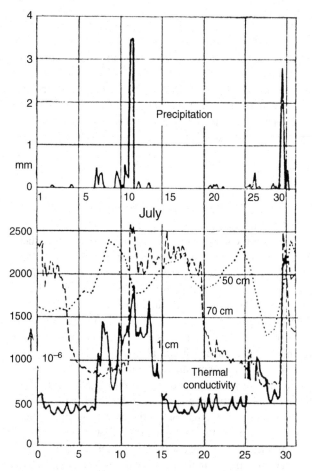

Fig. 6.23 Change in thermal conductivity after rain (from Geiger 1965).

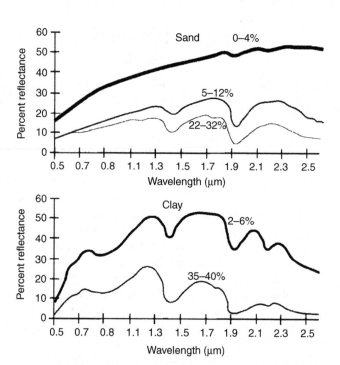

Fig. 6.24 Impact of soil moisture (in %) on visible and near-infrared albedo of (top) sand and (bottom) clay (from Dobos 2006, courtesy of the Taylor & Francis Group, http://www.informaworld.com).

thermal conductivity is readily apparent after rain falls on a dry soil surface (Fig. 6.23). The effect of soil moisture on albedo is particularly high throughout the visible and infrared portions of the spectrum, with that of dry soil being 2–3 times greater than that of wet soil (Fig. 6.24).

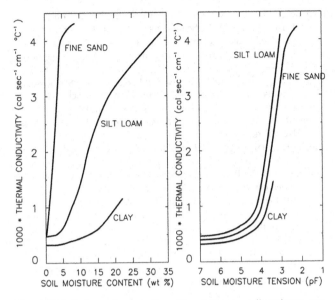

Fig. 6.25 Soil thermal properties with relation to soil moisture tension and soil moisture content for three soil types (modified from Al Nakshabandi and Kohnke 1965).

The relationship between soil moisture content and thermal

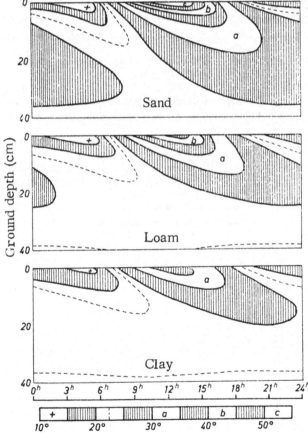

Fig. 6.26 Isotherms of soil temperature as a function of depth and time for three soil types (from Geiger 1965).

Table 6.5. *Soil temperature (°C) on a clear summer day for three soil types with different thermal diffusivities (from Geiger 1965).*

	Sand	Loam	Clay
Surface	40	33	21
5 cm depth	20	19	14
15 cm depth	7	6	4

diffusivity α is more complicated, because the latter is a trade-off between the soil's ability to transmit heat (k) and the amount of heat required to produce a temperature change (C_v). Thus, α increases sharply in going from a dry to wet soil because pore space is rapidly reduced. Above about 15% of field capacity α decreases with increasing soil moisture because C_v continues to increase rapidly but k remains constant.

The effect of soil texture is illustrated in Fig. 6.25. In general, all four thermal properties shown are higher for sand than for silt or clay at a given soil moisture content, and the properties change much more rapidly in sand as the soil is wetted. This is because of the high porosity of sands, which is readily reduced with only a little addition of moisture. The one exception is that the specific heat of clay exceeds that of sand at higher moisture contents. When soil moisture content is expressed as soil moisture tension instead of a percent water content, the differences disappear between soil types (Fig. 6.25). However, since more water is required in sandy and silty soils than in clay to achieve a given moisture tension, the textural differences have a real physical effect on the thermal characteristics of the soil.

This point is illustrated both in Table 6.5, which shows soil temperatures for three soil types with different thermal diffusivities, and in Fig. 6.26 , which illustrates soil temperature as a function of depth and time for the three major types. Clearly, the transfer of heat by conduction is most efficient in sand, less so with a loam, and lowest with a clay soil. As a result, the sand exhibits the greatest diurnal range, highest temperature gradients, and greatest depth of penetration of the diurnal wave.

All deserts are characterized by low soil moisture content and vast areas of desert are characterized by a sandy soil or surface. These factors act together to enhance the temperature extremes imposed on many desert or dryland surfaces by the low surface moisture content.

REFERENCES

Al Nakshabandi, G., and H. Kohnke, 1965: Thermal conductivity and diffusivity of soils as related to moisture tension and other physical properties. *Agricultural Meteorology*, **2**, 271–279.

Aoki, T., and Coauthors, 2005: Spectral albedo of desert surfaces measured in western and central China. *Journal of the Meteorological Society of Japan*, **83A**, 279–290.

Ba, M. B., S. E. Nicholson, and R. Frouin, 2001: Temporal and spatial variability of surface radiation budget over the African continent as derived from METEOSAT. Part II. Temporal and spatial variability of surface global solar irradiance, albedo and net radiation. *Journal of Climate*, **14**, 60–76.

Charney, J. G., 1975: Dynamics of deserts and drought in the Sahel. *Quarterly Journal of the Royal Meteorological Society*, **54**, 642–646.

Coulibaly, Y., and C. Boutin, 1983: Estimation des flux de chaleurs sensible et latente en zone ce savane (Cote d'Ivoire). *Annales de l'Université d'Abidjan*, **XIX**, 79–98.

Deacon, E. L., 1969: Physical processes near the surface of the earth. In *World Survey of Climatology*, (H. Flohn, ed.), Elsevier Amsterdam, vol. 2, pp. 39–104.

Dickinson, R. E., 1983: Land surface processes and climate: surface albedos and energy balance. *Advances in Geophysics*, **25**, 305–353.

Dobos, E., 2006: Albedo. In *Encyclopedia of Soil Science*. Taylor & Francis, London.

Durand, P., J.-P. Frangi, and A. Druilhet, 1988: Energy budget for Sahel surface layer during the ECLATS Experiment. *Boundary-Layer Meteorology*, **42**, 27–42.

Frangi, J. - P., A. Druilhet, P. Durand, H. Ide, J. P. Pages, and A. Ringa, 1992: Energy budget of the Sahelian surface layer. *Annales Geophysicae*, **10**, 25–33.

Geiger, R., 1965: *The Climate Near the Ground*. Harvard University Press, Cambridge, MA, 611 pp.

Idso, S. B., R. D. Jackson, R. J. Reginato, B. A. Kimball, and F. S. Nakayama, 1975: The dependence of bare soil albedo on soil water content. *Journal of Applied Meteorology*, **14**, 109–113.

Kabat, P., A. J. Dolman, and J. A. Elbers, 1997: Evaporation, sensible heat and canopy conductance of fallow savannah and patterned woodland in the Sahel. *Journal of Hydrology*, **188/189**, 494–515.

Kaimal, J. C., and J. J. Finnigan, 1994: *Atmospheric Boundary Layer Flows*. Oxford University Press, New York, 289 pp.

Kraus, H., and A. Alkhalaf, 1995: Characteristic surface energy balances for different climate types. *International Journal of Climatology*, **15**, 275–284.

Kustas, W. P., and Coauthors, 1991: An interdisciplinary field study of the energy and water fluxes in the atmosphere–biosphere system over semiarid rangelands: description and some preliminary results. *Bulletin of the American Meteorological Society*, **72**, 1683–1705.

Lancaster, N., 1996: Desert environments. In *The Physical Geography of Africa* (W. M. Adams, A. S. Goudie, and A. R. Orme, eds.), Oxford University Press, Oxford, pp. 34–59.

Lettau, H. H., 1969: Note on aerodynamic roughness-parameter estimation on the basis of roughness element description. *Journal of Applied Meteorology*, **8**, 828–832.

Lettau, H. H., and B. Davidson, 1957: *Exploring the Atmosphere's First Mile*. Pergamon Press, New York, 578 pp.

Lettau, H. H., and K. Lettau, 1978: Energy budget climatology of the arid region along South America's west coast. In *Exploring the World's Driest Climate* (H. H. Lettau and K. Lettau, eds.), IES Report 101, Institute for Environmental Studies, University of Wisconsin, Madison, WI, pp. 148–155.

Menenti, M., 1984: *Physical Aspects and Determination of Evaporation in Deserts*. Report 10 (special issue), Institute of Land and Water Management Research (ICW), Wageningen, The Netherlands, 202 pp.

Michalek, J. L., J. E. Colwell, N. E. G. Roller, N. A. Miller, E. S. Kasischke, and W. H. Schlesinger, 2001: Satellite measurements of albedo and radiant temperature from semi-desert grassland along the Arizona/Sonora border. *Climatic Change*, **48**, 417–425.

Miller, R. L., A. Slingo, J. C. Barnard, and E. Kassianov, 2009: Seasonal contrast in the surface energy balance in the Sahel. *Journal of Geophysical Research–Atmospheres*, **114**, D00E05.

Munn, R. E., 1966: *Descriptive Micrometeorology*. Academic Press, New York, 245 pp.

Oke, T. R., 1987: *Boundary Layer Climates*, 2nd edn. Halsted, New York, 435 pp.

Otterman, J., 1974: Baring high-albedo soils by overgrazing: a hypothesized desertification mechanism. *Science*, **186**, 531–533.

Qin, Z., P. Berliner, and A. Karnieli, 2002: Micrometeorological modeling to understand the thermal anomaly in the sand dunes across the Israel–Egypt border. *Journal of Arid Environments*, **51**, 281–318.

Raschke, E., T. H. Vonder Haar, W. R. Bandeen, and M. Pasternak, 1973: The annual radiation balance of the earth–atmosphere system during 1969–1970 from Nimbus-3 measurements. *Journal of the Atmospheric Sciences*, **30**, 341–364.

Raupach, M. R., 1994: Simplified expressions for vegetation roughness length and zero-plane displacement as functions of canopy height and area index. *Boundary-Layer Meteorology*, **71**, 211–216.

Rosenberg, N. J., 1974: *Microclimate: The Biological Environment*. Wiley-Interscience, New York, 315 pp.

Sellers, W. D., 1965: *Physical Climatology*, University of Chicago Press, Chicago, 272 pp.

Small, E. E., and S. A. Kurc, 2003: Tight coupling between soil moisture and the surface radiation budget in semiarid environments: implications for land–atmosphere interactions. *Water Resources Research*, **39**, doi:10.1029/2002WR001297.

Smith, E. A., E. R. Reiter, and Y. X. Gao, 1986: Transition of surface-energy budget in the Gobi Desert between spring and summer seasons. *Journal of Climate and Applied Meteorology*, **25**, 1725–1740.

Stearns, C. R., 1969: Surface heat budget of the Pampa de la Joya, Peru. *Monthly Weather Review*, **97**, 860–866.

Tetzlaff, G., 1974: *Die Wärmehaushalt in der zentralen Sahara*. Berichte des Institut für Meteorologie und Klimatologie der TU Hannover No, 13, Hanover, Germany, 113 pp.

Timouk, F., and Coauthors, 2009: Response of surface energy balance to water regime and vegetation development in a Sahelian landscape. *Journal of Hydrology*, **375**, 178–189.

Vehrencamp, J. E., 1953: Experimental investigation of heat transfer at an air–earth interface. *Transactions of the American Geophysical Union*, **34**, 22–30.

Verhoef A., Allen, S. J., and Lloyd, C. R., 1999: Seasonal variation of surface energy balance over two Sahelian surfaces. *International Journal of Climatology*, **19**, 1267–1277.

Warner, T. T., 2004: *Desert Meteorology*. Cambridge University Press, Cambridge, 595 pp.

7 Water balance

7.1 INTRODUCTION

Latent heat exchange links the heat balance to another important component of the climate system, the hydrological cycle. As with energy, a balance must exist if the system is in equilibrium. Thus, in the long term the precipitation (P) must be balanced by the processes of evaporation (E) and runoff (N). On shorter time scales, these processes do not match the precipitation and the imbalance is manifested as changes in moisture storage in the earth–atmosphere system. In its most simplistic form, the surface water balance is expressed as

$$P = E + N + dw/dt \qquad (7.1)$$

where w is available moisture stored in the root zone of the soil and dw/dt represents changes in storage.

Latent heat exchange also provides a fundamental link between vegetation and climate. Distinguishing between the vapor exchange that takes place as transpiration in plants (t) and the direct surficial evaporation from land or water (e), the water balance can be written as

$$P = e + t + N + dw/dt. \qquad (7.2)$$

In this case, e is controlled by atmospheric factors and soil characteristics, including soil moisture, while t is additionally influenced by characteristics of the vegetation (see Chapter 8). It is particularly useful to distinguish between e and t in dryland regions because a large proportion of the surface is bare ground, and surficial evaporation is dominant. Because the two processes are parameterized differently, good estimates of the overall evapotranspiration in dryland regions require separate consideration of the two processes, as described in Section 7.4.

7.2 GLOBAL WATER AND HEAT BALANCE

The latitudinal distribution of precipitation, evapotranspiration, and runoff is shown in Fig. 7.1. Rainfall is highest in the equatorial and mid-latitudes and lowest in the subtropics and high latitudes. Evaporation is highest in the equatorial regions and decreases with increasing latitude. There is a small minimum near the equator because of high cloud cover, cool ocean temperatures, and low wind speeds (Seager et al. 2003). There is no minimum in the subtropics, despite the minimum in rainfall, because evaporation from the oceans compensates for the low rates of evaporation over the relatively dry land areas in these latitudes. Runoff shows minima and maxima that generally match those of rainfall. However, in the subtropics, runoff becomes negative because evaporation exceeds precipitation and the excess loss over the oceans is replenished by runoff from other latitudes, which is transported by ocean circulation.

The relationships between the various components of the water and heat balance are described using a series of ratios. These include the Bowen ratio defined as the ratio of sensible to latent heat; the runoff ratio N/P; the evaporation ratio E/P; and the Budyko dryness ratio. Because evaporation plus runoff equals precipitation, the evaporation and runoff ratios add up to 1. The dryness ratio is defined as R_{net}/LP, where L is the latent heat of vaporization, approximately 600 cal/g. The denominator represents the energy required to evaporate the mean annual rainfall, and the numerator represents essentially a bulk measure of the energy available for evaporation. These ratios are systematic functions of the heat balance and take on characteristic values for particular types of climate (Table 7.1). In general, the runoff ratio tends to be less than 0.1 in arid climates and less than 0.03 in deserts.

Budyko's dryness ratio concept and its major biomes were described in Chapter 3 (Fig. 3.3). A global map of this ratio (Fig. 7.2) shows that it differs by two orders of magnitude between deserts like the Sahara and the Sonoran. Thus, it provides a way to distinguish the degree of aridity within the true deserts. It should be pointed out, however, that these are not absolute limits because of the uncertainties in measuring net radiation. Budyko's relationships were based on a very

Table 7.1. *Annual values of water balance parameters for various biomes.*

	Tundra	Forest	Steppe	Semi-desert	Desert
P	200–2000	200–4000	250–1500	200–400	<200
E/P	<0.3	0.3–0.7	0.7–0.9	0.9–0.97	>0.97
$R_{net}/(LP)$	<0.33	0.33–1.0	1.0–2.0	2.0–3.0	>3.0
N/P	>0.7	0.3–0.7	0.1–0.3	0.03–0.3	<0.03
E/E_P	>0.9	0.7–0.9	0.45–0.7	0.32–0.45	<0.32
P-E_P	0–2000	0–3000	−750–0	−800–−200	<−200

Key: P = annual precipitation; E = actual evapotranspiration; E_P = potential evapotranspiration; R_{net} = net radiation; L = latent heat of vaporization; N = runoff (all parameters expressed in mm, except for R_{net}, which is in J m^{-2}, and L, which is in J m^{-2} mm^{-1}). Based on values in Ripley (1992) and others.

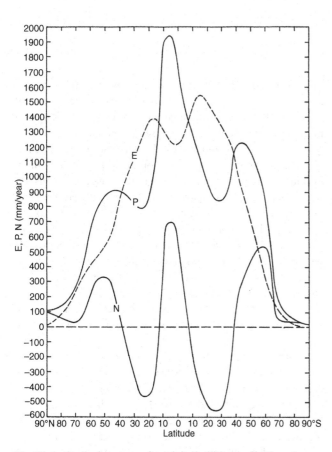

Fig. 7.1 Latitudinal averages of precipitation (P), runoff (N), and evaporation (E) (mm/year) (from Sellers 1965).

limited sample and on values of net radiation calculated from an empirical formula derived from observations in central Europe (Henning 1970). Better net radiation values derived from satellites suggest other limits. Over Africa (see Fig. 3.4) the dryness ratio that distinguishes between forest and woodland lies roughly between 1 and 2, and that delineating the true deserts lies at roughly 15. Nevertheless, the broad associations between vegetation type and the dryness ratio, which combines both energy and water balance, are conceptually valid.

7.3 QUANTIFICATION OF THE WATER BALANCE

The key to assessing the water balance lies in accurately characterizing the soil moisture regime. In order to do this, it is necessary to define a number of terms that characterize the soil moisture reservoir. These include volumetric soil water, saturation, soil moisture tension, exchangeable soil moisture, soil moisture deficit, and the thresholds of field capacity and permanent wilting point.

The amount of water contained in the soil is generally expressed as either the actual water content (in percent by weight or volume) or the soil moisture tension. The *volumetric water content* is the percentage of a wet soil that is occupied by water. *Saturation* occurs when this is equal to the porosity of the soil, i.e., the percent soil occupied by pore space. *Soil moisture tension* is the energy necessary to extract water from the soil matrix. It is indicated in units of pressure, to facilitate comparison with the forces that retain water in the soil against the force of gravity: *surface tension* (at the water surface itself), *adhesion, cohesion,* and *ionic attraction* (clays only). The actual water content is generally used in water balance studies and the soil moisture tension is used in studies of surface heat balance, water availability, and water movement. The relationship between the two depends strongly on soil characteristics, as illustrated in Fig. 7.3.

These forces result in two significant moisture thresholds in the soil. The first, *field capacity*, is the amount of water remaining in a well-drained soil when the velocity of downward flow into unsaturated soil has become small. An alternative definition is the amount of moisture that can be retained against the influence of gravity. The second threshold is *permanent wilting point* and this is related to water removal by plants via suction. Plants cannot exert more than 15,000 cm of pressure. If the water content is so low that soil moisture tension exceeds this, plants cannot remove water. The amount of water available to plants is the difference between field capacity and wilting point and is termed *available* or *exchangeable soil moisture*. The difference between field capacity and the actual water content is the *soil moisture deficit*.

Both field capacity and wilting point are generally expressed as a percentage of soil volume. When so expressed, the amount

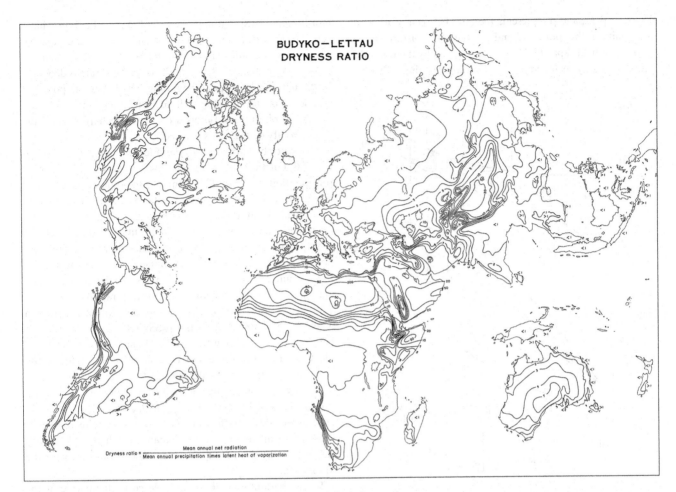

Fig. 7.2 Global map of the Budyko–Lettau dryness ratio (from Henning 1970).

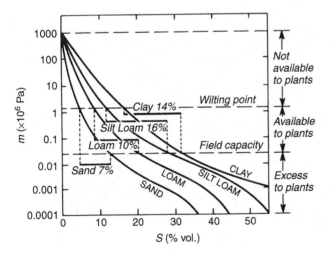

Fig. 7.3 Soil moisture tension (m) in relation to soil water content (S) (from Oke 1987). The water available to plants lies between the two horizontal bars (wilting point and field capacity).

of soil water corresponding to these thresholds varies greatly with soil porosity and texture. In general, both thresholds are lowest for pure sands and increase with increasing clay content

(Fig. 7.4). These thresholds are approximately constant when expressed in units of soil moisture tension (Fig. 7.3).

The classical approach to quantifying the water balance is that developed initially by Thornthwaite (1948) and later modified by Mather (Thornthwaite and Mather 1957). A typical balance diagram and appropriate equations are illustrated in Fig. 7.5. The first step is to calculate a quantity termed *potential evapotranspiration* (E_p). This is essentially the amount of evapotranspiration that can take place under "ideal" conditions, i.e., when moisture is readily available at the surface (see Chapter 8). E_p represents water need, while actual evapotranspiration E_a represents water use; the difference between E_a and E_p is called the *water deficit*. In this approach, E_p is calculated from surface temperature and duration of sunshine and E_a is a function of moisture availability. Soil moisture is calculated from an initial monthly soil moisture value, and precipitation is compared with potential evapotranspiration during that month.

Evapotranspiration is assumed to take place at the potential rate as long as precipitation exceeds E_p. With lower precipitation, evapotranspiration is assumed to be proportional to water availability. It is equated to precipitation and a quantity β, such that $E_a = \beta E_p$, and β represents the ratio between w, moisture availability in the root zone, and w^*, the soil moisture storage capacity or field capacity. Thus, E_p is prorated by the water deficit. In actuality, as long as the soil surface remains sufficiently

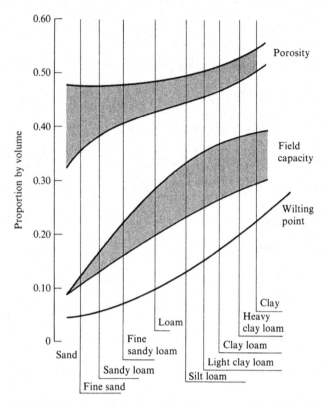

Fig. 7.4 Soil moisture thresholds vs. soil type and water content (%) (modified from Dunne and Leopold 1978). Shading indicates range of values.

moist, evapotranspiration proceeds at the potential rate. This is generally the case as long as soil moisture is above field capacity, which is on the order of 80% saturation. Above field capacity, any excess rainfall is lost as runoff. As soil moisture is depleted, the actual rate of evapotranspiration falls below the potential and soil moisture is being utilized. Whenever rainfall exceeds evapotranspiration, soil moisture is being recharged, until the field capacity of the soil is reached.

The water balance at six dryland locations is represented in Fig. 7.6. For the year as a whole, evaporation nearly balances precipitation, and runoff is minimal. In some cases, evaporation is sustained not only by precipitation but by the extraction of moisture from the air on cool mornings (Malek *et al.* 1997) or by dew on the soil surface. The aridity at a given location is a function of the relative magnitudes of rainfall and evapotranspiration throughout the course of the year. In the dry tropical and subtropical desert climates of Khartoum, Sudan, and Parker, Arizona, potential evapotranspiration exceeds precipitation almost continually, and actual evapotranspiration is essentially equivalent to the precipitation. At the remaining stations, which are wetter and at higher latitudes (with less incoming radiation), the soil is sufficiently moist to sustain some evapotranspiration during most of the year, even during the dry season. In the mid-latitude locations (Los Angeles, California, and Medicine Hat, Canada) the season of water deficit coincides with the summer season of high insolation. In the tropical climates (semi-arid Ndjamena, Chad, and wet–dry Rajpur, India) the period of water deficit coincides with the dry season.

It is important to note the role played by the nature of the precipitation regime in determining the individual balances. This is illustrated by the water balance at two savanna stations in

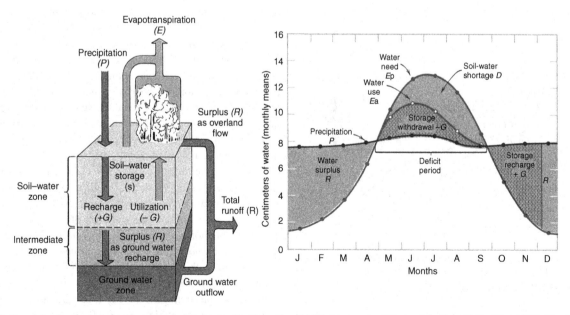

Fig. 7.5 Example of water balance diagram calculated using the Thornthwaite–Mather approach (copyright © A. N. Strahler, 1987). The figure on the left illustrates the losses and gains of soil moisture, including recharge, soil water utilization, and surplus. The figure on the right illustrates the balance, based on monthly values of precipitation, potential evapotranspiration, and actual evapotranspiration.

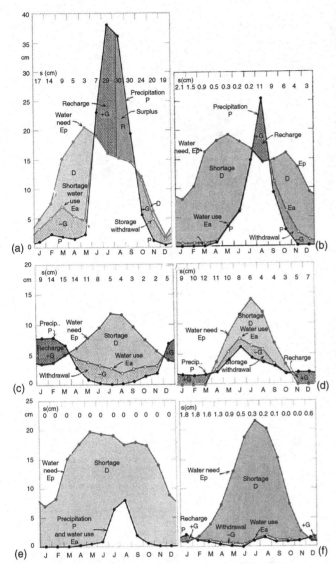

Fig. 7.6 Water balance for typical climate types: (a) tropical – wet–dry (Rajpur, India), (b) dry tropical – semi-arid (Ndjamena, Chad), (c) Mediterranean – semi-arid (Los Angeles, California), (d) dry mid-latitude (Medicine Hat, Alberta), (e) dry tropical – desert (Khartoum, Sudan), (f) dry subtropical – desert (Parker, Arizona) (copyright © A. N. Strahler, 1987).

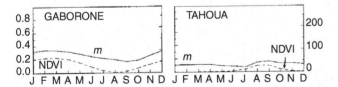

Fig. 7.7 Water balance for two savanna locations (Gaborone, Botswana, and Tahoua, Niger) (from Nicholson *et al.* 1997).

Africa: Gaborone in Botswana and Tahoua in Niger (Fig. 7.7). These were calculated by Nicholson *et al.* (1997), using a variant of the approach developed by the late Heinz Lettau and termed "climatonomy." Mean annual rainfall is roughly comparable at

the two locations (531 mm vs. 381 mm), but the monthly and daily partitioning is quite different. At Gaborone the rainy season lasts 5–6 months, but mean monthly rainfall does not exceed 100 mm; in comparison, there is a 3-month season at Tahoua and rainfall exceeds 150 mm in the wettest month. Differences are strongly apparent in the amount and seasonal cycle of soil moisture (Fig. 7.7). There is markedly more soil moisture at Gaborone and it sustains vegetation growth over most of the year. At Tahoua there is minimal vegetation growth and it is limited to one or two months. This contrast in the rainfall regime and corresponding water balance may help to account for the much more efficient rate of vegetation growth in the Kalahari, as noted by Nicholson and Farrar (1994).

The global water balance is illustrated in Figs. 7.8 through 7.11. The values of E_p in low latitudes range from around 1000 mm/year in regions of frequent cloud cover to nearly 2000 mm/year in relatively cloudless areas like the Sahara. Values generally decrease with latitude, being around 700 mm in the mid-latitudes and decreasing to 200 or 300 mm in the polar latitudes (Fig. 7.8). Actual evapotranspiration (Fig. 7.9) reflects latitudinal effects as well as precipitation; it exceeds 1000 mm/year in the humid tropics and is generally less than about 300 mm/year in arid regions, where it is roughly equivalent to rainfall. In the semi-arid regions, it is generally on the order of 300–900 mm/year. Runoff is primarily a function of rainfall (Fig. 7.10). It ranges from over 1000 mm/year in humid regions to less than 50 mm/year in arid and semi-arid climates. Annual mean soil moisture ranges from less than 25 mm in the deserts to over 125 mm in areas of humid climate (Fig. 7.11).

The approach of Thornthwaite and Mather has many shortcomings, particularly the mismatch between the monthly averages it generally utilizes as input and the time and space scales of the hydrological variables it calculates. For example, runoff and evapotranspiration can occur even if the soil moisture is well below field capacity, because a given rain event can have an intensity that easily exceeds the short-term infiltration capacity of the soil. Likewise, even if rainfall exceeds E_p, the actual evaporation may be less than E_p if the soil is sufficiently dry. Moreover, the rainfall data used as input are assumed to represent areal averages. Since the precipitation field is not uniform in space, especially in dryland regions, runoff will occur in areas where the point rainfall exceeds the infiltration capacity, even if the average values suggest that runoff will not occur.

Despite the shortcomings noted above, the use of time and space averages and prorating potential evapotranspiration by the soil moisture capacity provide the basis for the most common approach to water balance modeling, the so-called "bucket approach." This is the parameterization of runoff used in most atmospheric general circulation models. However, with the high-speed computers now available to do detailed calculations, more sophisticated approaches that consider various parts of the soil–plant system in greater detail are being developed. Some of the relevant concepts are described in Chapter 8.

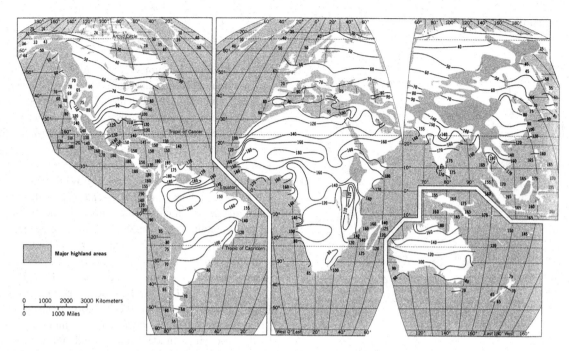

Fig. 7.8 Map of potential evapotranspiration (cm/year) E_p (copyright © A. N. Strahler, 1978).

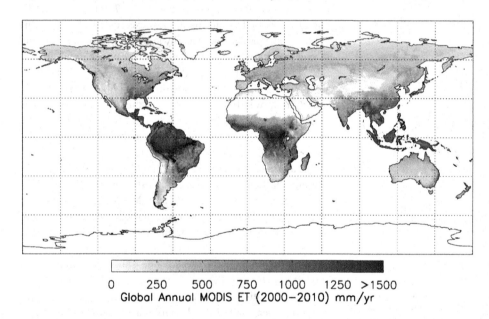

Global Annual MODIS ET (2000–2010) mm/yr

Fig. 7.9 Mean annual evapotranspiration (mm) based on MODIS data (courtesy of Steve Running and Qiaozhen Mu, University of Montana).

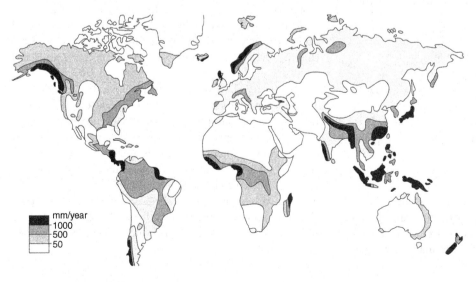

mm/year
1000
500
50

Fig. 7.10 Mean annual runoff (mm) (from Gregory and Walling 1973).

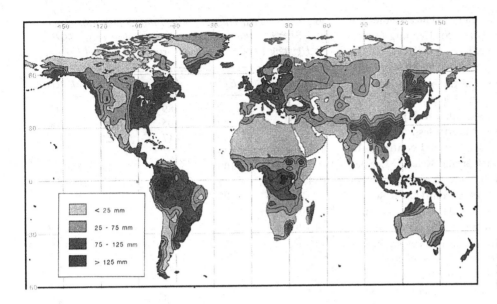

Fig. 7.11 Mean annual soil moisture (mm) (from Willmott *et al.* 1985).

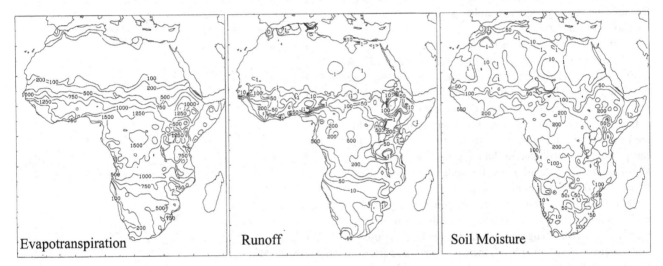

Fig. 7.12 Annual mean evapotranspiration (left), runoff (center), and soil moisture (right) over Africa (from Nicholson *et al.* 1997). Units are mm.

7.4 A CASE STUDY OF WATER BALANCE OVER AFRICA

7.4.1 CLIMATOLOGICAL ESTIMATES OF EVAPOTRANSPIRATION, RUNOFF, AND SOIL MOISTURE

Annual average evapotranspiration is shown in the left panel of Fig. 7.12. The 100 mm contour for evapotranspiration roughly bounds the Sahara in the north, delineating a region that extends latitudinally from about 20° N to 35° N. The 500 mm contour is situated near 15° N, running through Lake Chad. The position of both contours is essentially identical to that of the corresponding rainfall isohyets (not shown), indicating that runoff is very low throughout this region.

Runoff (center panel of Fig. 7.12) is less than 1 mm/year in the desert area extending from about 15° N to 35° N. In tropical latitudes, excluding eastern equatorial Africa, runoff is on the order of 200–500 mm. In semi-arid subtropical latitudes, runoff ranges from about 10 to 200 mm, but there is a strong gradient between 10° S and 15° S, so that it is below 50 mm throughout most of Southern Hemisphere Africa. It approaches zero in the Namib Desert.

Annual mean values of soil moisture (right panel of Fig. 7.12) vary from about 10 mm in the Sahelo-Saharan region to over 200 mm in an equatorial core extending from roughly 10° N to 5° S. Soil moisture ranges from about 10 to 50 mm in the semi-arid Kalahari and karoo of southern Africa, but falls well below 10 mm in the Sahara and the coastal deserts of southern Africa.

The most comprehensive global water balance studies are those of Baumgartner and Reichel (1975), Henning (1989), Legates and McCabe (2005), and Mintz and Walker (1993). These are not strictly comparable to each other because differences exist in the years upon which the balance is based, the stations used in the calculation, the degree of spatial resolution, and various assumptions concerning such parameters as the soil water-holding capacity. Nevertheless, a comparison provides some interesting similarities and contrasts. In general, the estimates of Baumgartner and Reichel, Legates and McCabe. and Mintz and Walker are in excellent agreement with each other, as all rely on data from Thornthwaite–Mather. The differences between these and the "climatonomy" estimates of evapotranspiration in Fig. 7.12 are reasonably small, less than 30 mm/year. Differences are somewhat higher in individual months, reaching 50–70 mm in some areas, but there seems to be some compensation during the course of the year. There is also good agreement for continental estimates of runoff. The disparities can generally be accounted for by differences in the precipitation fields used as input and by differences in the assumed water-holding capacity of the soils (Nicholson *et al.* 1997).

In contrast, the calculations by Henning (1989) are very different from those of any of the other studies (Table 7.2). At most latitudes, Henning's estimates are considerably lower than the others. The likely cause of the differences lies in the net radiation calculations by the various studies. The exception is the arid latitudes of North Africa, where Henning's values are roughly twice those used by Nicholson *et al.* (1997). This can probably be attributed to the unrealistically low values of albedo assumed by Henning over the dryland regions of North Africa (e.g., 0.27 throughout the Sahara).

These comparisons show that the climatological values of the water balance found in the literature can differ substantially. This fact must be recognized when these values are used in other studies. On the other hand, there is substantial agreement between calculations based on many diverse methods, with differences attributed to input data rather than methodological constraints.

A more detailed comparison of three methods of assessment ("climatonomy," Thornthwaite, and Penman) is presented in Fig. 7.13 for individual locations. For evapotranspiration, there are notable differences, particularly for the wetter stations, but these do not appear to be systematic. The range of values can exceed 100 mm/month. There is more agreement between calculations of soil moisture. The difference between the estimates based on "climatonomy" and on Thornthwaite is seldom more than 25 mm/month (and often less) in the wetter stations. However, at the driest stations, such as Gaborone, "climatonomy" predicts significant soil moisture in several months, but the Thornthwaite approach predicts a dry soil. A comparison with vegetation growth suggests that the "climatonomy" (Lettau 1969) estimates are more reasonable in the drier regions (Nicholson *et al.* 1997).

Table 7.2. *Zonal averages of evaporation* E *and evaporation ratio (*E/P*) from three sources (N = Nicholson* et al. *1997; H = Henning 1989; BR = Baumgartner and Reichel 1975).*

Latitude band	E (N)	E (H)	E/P (N)	E/P (BR)
35°–40° N	90	89	594	355
30°–35° N	99	105	215	186
25°–30° N	100	117	23	55
20°–25° N	100	97	35	58
15°–20° N	100	97	244	198
10°–15° N	95	93	727	517
5°–10° N	86	81	1193	773
Equator–5° N	87	74	1144	846
Equator–5° S	88	77	1178	958
5°–10° S	91	84	1058	880
10°–15° S	88	86	1023	772
15°–20° S	95	85	730	715
20°–25° S	98	91	446	559
25°–30° S	97	93	450	428
30°–35° S	97	92	441	423

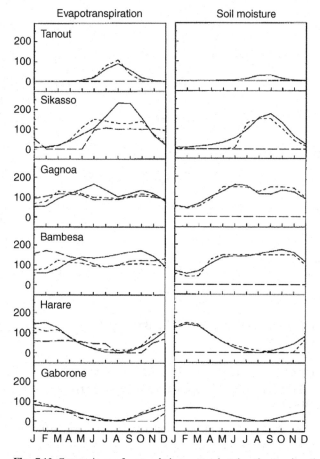

Fig. 7.13 Comparison of water balance at select locations, using the revised version of Lettau's climatonomy model (solid line) and the methods of Thornthwaite (dotted line) and Penman (dashed line) (Nicholson *et al.* 1997). The left-hand diagrams indicate calculated evapotranspiration (mm) and the right-hand diagrams indicate calculated soil moisture (mm). Calculations are based on an assumed water-holding capacity of 150 mm.

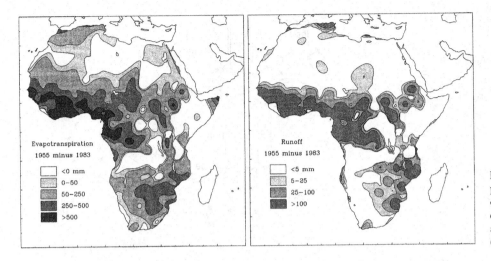

Fig. 7.14 Differences in annual evapotranspiration and runoff (in mm) over Africa between 1955 and 1983, continentally an exceedingly wet and an exceedingly dry year, respectively (from Nicholson *et al.* 1997).

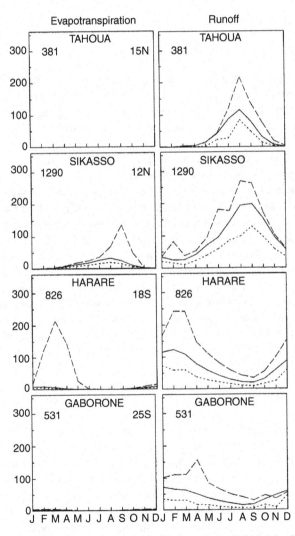

Fig. 7.15 Monthly evapotranspiration (left) and runoff (right) for four locations in wet and dry years. The dashed line indicates the mean for the 10 wettest years during the 1930–1990 period, the dotted line indicates the mean for the 10 driest years during this period, and the solid line indicates the mean for the entire period (from Nicholson *et al.* 1997). Mean annual rainfall (mm) is indicated in the upper left corner of each graph.

7.4.2 INTERANNUAL VARIABILITY OF THE WATER BALANCE

The water balance estimates described in Section 7.4.1 correspond to conditions of mean rainfall over the continent of Africa. It is critically important to be aware of the magnitude of the changes in these conditions from year to year or even from decade to decade. Two extreme years over Africa, 1955 and 1983, serve to illustrate this (Fig. 7.14). Most of Africa was abnormally dry during 1983, while rainfall in 1955 exceeded that in 1983 by over 500 mm in much of the continent. The excess rainfall during 1955 was compensated by nearly the same increase in evapotranspiration throughout most of Africa, indicating that evapotranspiration can vary by hundreds of millimeters from year to year over large areas of Africa. Large differences in runoff are apparent mainly in tropical latitudes and in the eastern half of the southern subcontinent, where fairly humid climates prevail. In these areas, the interannual variability of runoff is on the order of 25 mm to over 100 mm. The overall pattern is that a change to wetter conditions is manifested as increased runoff in humid areas, but as increased evapotranspiration in the dryland regions.

The asymmetry in the variations of runoff and evapotranspiration in wet and dry years is apparent also at individual stations (Fig. 7.15). Evapotranspiration can fluctuate by 100–200 mm/month and tends to increase and decrease with rainfall. Runoff tends to increase in wet years, but shows little change in the dry years. This happens because, once precipitation equals potential evapotranspiration and the soil is saturated, all the additional rainfall becomes runoff. This seldom occurs in the drier locations.

7.5 WATER STORAGE AND MOVEMENT IN AN ECOSYSTEM

Water is stored at five locations in an ecosystem (Fig. 7.16). In the vegetation layer it is held within plant material and on the surfaces of plants. These reservoirs are respectively termed *plant*

Water in the ecosystem

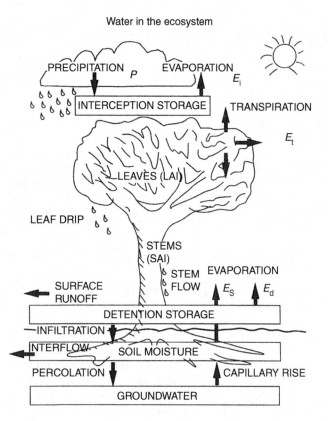

Fig. 7.16 Locations where water is stored in an ecosystem (from Nicholson *et al.* 1996).

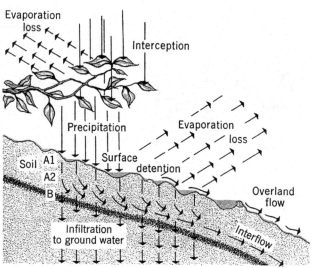

Fig. 7.17 Pathways of water gain and water loss in an ecosystem (copyright © A. N. Strahler, 1987).

water and *interception water. Detention storage* is retained on the ground surface itself, in a thin film or as snow or puddles. Two subsurface reservoirs are the *soil moisture*, the layer where there is pore space and unsaturated flow, and the *groundwater*, the lower layer where the ground is completely saturated and there is no pore space.

Sources of interception water include rain and snow, fog droplets impacted on the surface, condensation, and *guttation* (water exuded onto a leaf surface through the leaf cuticle). Condensation is called *dewfall* if the water is derived from the air, but *distillation* if its source is the soil. The efficiency of interception depends on the nature and amount of precipitation and vegetation characteristics, such as stand architecture and density or surface area of foliage. Many plants in dryland regions have characteristics that tend to maximize the interception of water. Interception is initially high for a dry vegetation layer, but a saturation threshold of storage is reached and efficiency then declines. The excess water falls to the ground as leaf drip or by running down stems (stemflow).

As with interception storage, the sources of detention storage and soil moisture include direct rain and snow, condensation and fog, but they also include leaf drip and stemflow. In drier regions with little ground cover, detention storage is an important source of evapotranspiration. If the penetration of water into the soil is slow, the detention film deepens and promotes surface runoff.

There are three major paths of removal of water from the ecosystem (Fig. 7.16): evapotranspiration to the atmosphere, runoff, and downward flow through the soil layer to the groundwater. The total evapotranspiration is the summation of evaporation of detention water (E_d), interception water (E_i), and soil moisture plus transpiration (E_t) through plants. Runoff is the lateral movement of water on the surface or in the soil.

Runoff takes place at several locations within the ecosystem and results from various mechanisms (Fig. 7.17). Surface runoff, or overland flow, takes place in the porous upper few centimeters of the ground or on the surface itself. Infiltration-excess overland flow (also called Hortonian overland flow) occurs as a result of saturation from above: the detention film becomes thick when the infiltration capacity of the soil is exceeded. Saturation-excess (or Dunne) overland flow occurs when the upper few centimeters of the soil become saturated (i.e., saturation occurs from below). In this case, it is the water-holding capacity of the soil and not the infiltration capacity that is exceeded. Most of the runoff in dryland regions is Hortonian overland flow. Other forms of runoff include throughflow (also called interflow) and base flow. Throughflow occurs in the unsaturated soil moisture zone, as a result of gravity or hydraulic gradients. Base flow is the flow of groundwater.

The infiltration of water to the ground takes place in two modes, depending on the size of soil pores. Infiltration via small pores brings water to the soil moisture layer, where it is extracted by roots or direct evaporation. Movement results from capillary action because the air pressure in the pores is lower than that in the atmosphere. Water moves through large pores directly to the groundwater zone by gravitation, a process called *percolation*. Percolation occurs with greater time lag after rainfall than does infiltration to the soil. After a rain event the rise of the water table can occur several months after the initial jump in soil moisture.

Table 7.3. *Hydrologic parameters associated with various climate types, indicating the proportion of soil moisture that eventually reaches the groundwater (from Miller 1977).*

Zone	Infiltration (mm/year)	Percolation (mm/year)	Ratio
Tundra	300	90	0.30
Taiga/Boreal forest	350	70	0.20
Mixed Conifer/ Deciduous	380	55	0.13
Forest steppe	380	15	0.04
Steppe	290	3	0.01
Semi-desert	200	1	0.005

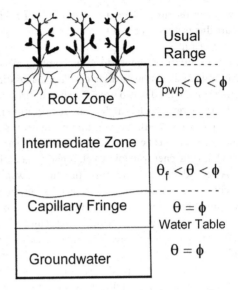

Fig. 7.18 Hydrologic soil profile, showing the range of water content within the various zones. In the root zone (modified from Dingman 1994).

Percolation can be regarded as runoff because the percolated water is of little use to the ecosystem. The relative amount of percolation is a function of climate type. As much as 90 mm/year percolates to the groundwater in the northern tundra and taiga climates, where it comprises about 30% of the infiltration (Table 7.3). In contrast, percolation totals only a few millimeters per year in arid and semi-arid regions and thus comprises, at most, only a few percent of the infiltration.

The distinction between soil moisture and groundwater is an important one because the utilization of these sources of water is quite different. The soil moisture layer essentially constitutes the root zone of plants and the water therein is available to plants. The water content of this zone fluctuates rapidly in space and time. Groundwater is long-term storage and the pathway by which atmospheric water reaches the oceans and lakes, from which it is returned to the atmosphere by evaporation. Groundwater storage fluctuates, but much more slowly than soil moisture. It is also a spatially uniform surface.

In practical terms, the distinction is not as clear cut because there is no tangible separation between the soil moisture and groundwater, and rapid exchange between the two reservoirs takes place. To understand the forces involved in the movement of water, it is necessary to describe the transition zones between the soil moisture and groundwater and the concept of the water table (Fig. 7.18).

The water table is the upper surface of unconfined groundwater. It is the depth at which the ground would be saturated (i.e., the water surface that would result) if the system were under the influence of gravity alone. That is not the case, as water is retained also by forces of adhesion, cohesion, surface tension and, in clays, ionic attraction. These retain water above the level that gravity alone would produce; thus, the "water table" is to some extent a theoretical construct. However, it is also a practical concept, as it provides a representative areal average for the depth of ground saturation.

Water is retained above the water table by capillary action, i.e., the action of those other forces against the downward pull of gravity. This layer is called the capillary fringe. It is saturated, and water is in liquid form only. Above the capillary fringe is the capillary water. That layer is unsaturated and consists of water in both

the liquid and vapor states. Thus, the water table is a dynamic surface of constant exchange between the groundwater and soil layer. Water percolates through large soil pores to the groundwater and back to the soil moisture and root zone by capillary rise.

Movement of soil moisture is accomplished in four ways. Two of them have been discussed: capillary action in response to pressure, and gravity, which produces infiltration and percolation. The other two are responses to the hydraulic gradient (i.e., the gradient of water content) in the soil, and to the temperature gradient in the soil. Because these gradients can be extreme in dryland regions, these are important methods of water movement in dryland regions.

The flux of water in response to a hydraulic gradient is described by Darcy's law. This relationship has two forms, one for the flux of liquid water M_L and the other for water vapor M_V. The former is

$$M_L = K_h \, dm/dz \qquad (7.3)$$

where K_h is the hydraulic conductivity and m is the soil moisture tension. The latter is a measure of the energy required to extract water from the soil matrix (Oke 1987) and is only an indirect measure of moisture content (Fig. 7.3). The tension is determined by a combination of soil porosity and water content; flux is from regions of low to high soil moisture tension. For a given volume of soil moisture, soil moisture tension is lowest in sand, highest in clays.

In the case of water vapor, the flux is given by

$$M_V = -\rho \, k_w \, dq/dz \qquad (7.4)$$

where ρ is the air density, k_w is the molecular diffusivity for water vapor, and q is the specific humidity, defined as the ratio of the mass of water in the air to the mass of moist air. In this case, the moisture flux is directed down the gradient, from regions of high to low moisture.

The moisture flux in response to a temperature gradient is also down-gradient, i.e., from warm to cold areas. This direction of flow can be understood in two ways. The first utilizes the concept of kinetic energy, which is high where temperature is high, low in colder regions. Thus, the moisture movement is analogous to the process of heat conduction, with water flowing from areas of high kinetic energy to low kinetic energy via molecular collision. A second way to understand the flux of water from warm to cold areas involves the concept of saturation. The pore space in soils is close to saturation and the vapor pressure at saturation is a function of temperature. Thus, it is higher in warmer areas and a vapor pressure gradient exists between warm and cold regions. Vapor again flows down the gradient to areas of lower temperature (and lower vapor pressure). As a result, soil water vapor tends to flow upward during the night, when the temperature minimum is generally near the surface, and downward during the day, when soil temperature decreases with depth. The vapor is deposited on the surface by condensation if the surface is sufficiently cool. This process, called distillation, is often confused with dew formation. The vapor flow corresponding to a temperature gradient of 1°C/cm can be as much as 0.4–2 mm/day. This is relatively large compared with evaporation, which is generally on the order of 4–12 mm/day, and it can help to offset the evaporative loss, because the vapor flow is downward if temperature decreases downward. Such is generally the case when evaporation is taking place.

The rate of soil water movement is influenced by soil type, ground cover, and topography. Soil type plays a major role because the capillary action and the tension forces retarding movement are functions of porosity, texture, and water content of the soil. The adhesive forces are weakest for large particles, soils of low organic matter content, and wet soils with an open structure. They are greatest for dry, compact soils with small particles and high organic matter. Thus, sandy soils retain water the least, clay the most. Some clays (montmorillonite) also expand when they absorb water, often forming, with the first post–dry season rainfall, a solid impermeable layer that promotes runoff.

REFERENCES

Baumgartner, A., and E. Reichel, 1975: *The World Water Balance*. R. Oldenbourg, Munich/Vienna, 179 pp.

Dingman, S. L., 1994: *Physical Hydrology*. Macmillan, New York, 575 pp.

Dunne, T., and L. B. Leopold : 1978: *Water in Environmental Planning*. Freeman, New York, 818 pp.

Gregory, K. J., and D. E. Walling, 1973: *Drainage Basin Form and Process*. John Wiley and Sons, New York.

Henning, D., 1970: Comparative heat balance calculations. In *Symposium on the World Water Balance* (Reading), Vol. II, IAHS Publication 93, 361–375, IAHS Press, Wallingford, UK, http://www.iahs.info/redbooks/092.htm.

Henning, D., 1989: *Atlas of the Surface Heat Balance of the Continents*. Gebrüder Borntraeger, Berlin, 402 pp.

Hidore, J. J., and J. E. Oliver, 1993: *Climatology: An Atmospheric Science*. Macmillan, New York, 423 pp.

Legates, D. R., and G. J. McCabe, Jr, 2005: A re-evaluation of the average annual global water balance. *Phys. Geography*, **26**, 467–479.

Lettau, H. H., 1969: Evapotranspiration climatonomy. I. A new approach to numerical prediction of monthly evapotranspiration, runoff and soil moisture storage. *Monthly Weather Review*, **97**, 691–699.

Malek, E., G. E. Bingham, and G. D. McCurdy, 1997: Annual mesoscale study of water balance in a Great Basin heterogeneous desert valley. *Journal of Hydrology*, **191**, 223–244.

Miller, D. H., 1977: *Water at the Surface of the Earth: An Introduction to Ecosystem Hydrodynamics*. International Geophysics Series 21, Academic Press, 557 pp.

Mintz, Y., and G. K. Walker, 1993: Global fields of soil moisture and land surface evapotranspiration derived from observed precipitation and surface air temperature. *Journal of Applied Meteorology*, **32**, 1305–1334.

Nicholson, S. E., and T. J. Farrar 1994: The influence of soil type on the relationships between NDVI, rainfall and soil moisture in semi-arid Botswana. Part I. Response to rainfall. *Remote Sensing of Environment*, **50**, 107–120.

Nicholson, S. E., A. R. Lare, J. A. Marengo, and P. Santos 1996: A revised version of Lettau's evapoclimatonomy model. *Journal of Applied Meteorology*, **35**, 549–561.

Nicholson, S. E., J. Kim, M. B. Ba, and A. R. Lare, 1997: The mean surface water balance over Africa and its interannual variability. *Journal of Climate*, **10**, 2981–3002.

Oke, T. R., 1987: *Boundary Layer Climates*, 2nd edn. Halsted, New York, 435 pp.

Ripley, E. A., 1992: Water flow. In *Natural Grasslands* (R. T. Coupland, ed.), Ecosystems of the World 8A, Elsevier, Amsterdam, pp. 55–73.

Seager, R., R. Murtugudde, A. Clement, and C. Herweijer, 2003: Why is there an evaporation minimum at the equator? *Journal of Climate*, **16**, 3793–3802.

Sellers, W. D., 1965: *Physical Climatology*, University of Chicago Press, Chicago, 272 pp.

Strahler, A. N., and A. H. Strahler, 1978: *Modern Physical Geography*, 1st edn. Wiley, New York.

Strahler, A. N., and A. H. Strahler, 1987: *Modern Physical Geography*. John Wiley and Sons, New York, 544 pp.

Thornthwaite, C. W., 1948: An approach toward a rational classification of climate. *Geographical Review*, **38**, 55–94.

Thornthwaite, C. W., and J. R. Mather, 1957: *Instructions and Tables for Computing Potential Evapotranspiration and the Water Balance*. Publications in Climatology 10. The Johns Hopkins University Laboratory of Climatology, Baltimore, MD, pp. 181–311.

Willmott, C. J., C. N. Rowe and Y. Mintz, 1985: Climatology of the terrestrial seasonal water cycle. *Journal of Climatology*, **5**, 589–606.

8 Evaporation

8.1 INTRODUCTION

The question of assessing surface evaporation has been considered since at least the time of Dalton in the early nineteenth century. Literally hundreds of methods have been described in the literature. These differ greatly with respect to the essence of the method, the assumptions utilized, the conditions when applicable, the units of the constants and variables, and the degree of empiricism involved. This section gives only a broad overview of the methods in widespread use. For the details, the reader is referred to summaries in sources such as Sellers (1965), Rosenberg (1974), Oke (1987), Dunne and Leopold (1978), Brutsaert (1982), Jackson (1989), Lowry and Lowry (1989), Schmugge and André (1991), Dingman (1994), and Raupach (2001).

The total evapotranspiration from the earth's surface is a composite of water flux derived from five types of surfaces: open water, bare soil, vegetation, and snow and ice. The factors and processes in the exchange of vapor with the atmosphere are different in each case and hence the basic surface types will be examined separately in this chapter. However, snow and ice will not be considered, since these surfaces are rare in most of the dryland regions considered in this text.

Three dynamic processes occurring simultaneously contribute to net evapotranspiration: absorption of radiation, advection, and turbulent and molecular exchange. These processes are controlled, respectively, by available energy, wind, and water vapor. The relevant meteorological variables are net radiation or temperature, the atmospheric saturation deficit or soil moisture content, and wind speed or turbulence. The role of wind is to maintain high vapor pressure gradients between the surface and the ground. When surface water is sufficiently available, evapotranspiration is controlled by the atmospheric variables. This situation is sometimes referred to as radiation-limited, but the term atmosphere-controlled is probably more appropriate, because wind also plays a determining role. When the surface is relatively dry, evapotranspiration is largely a function of soil moisture. This is a water-limited or soil-controlled situation.

Assessment of evapotranspiration involves a two-pronged approach, one for the atmosphere-controlled case and the other for the soil-controlled case. This essentially amounts to the assessment of potential evapotranspiration or water demand in the former situation, and the assessment of actual evapotranspiration or water use in the latter. Evaporation over a water surface is predominantly atmosphere-controlled, although some water characteristics can play a role (see Section 8.4.1). Over land the process is generally atmosphere-controlled when the soil moisture is close to or above field capacity and actual evapotranspiration equals potential evapotranspiration to a first approximation. The calculation of potential evapotranspiration is considered in Section 8.2. Methods of assessing actual evapotranspiration in the soil-controlled case are considered in Section 8.3. As the processes controlling evapotranspiration are somewhat different for vegetation and bare soil, these cases are discussed separately in Sections 8.4.2 and 8.4.3.

8.2 ASSESSMENT OF POTENTIAL EVAPOTRANSPIRATION

8.2.1 THE CONCEPT OF POTENTIAL EVAPOTRANSPIRATION

Potential evapotranspiration E_p is commonly defined as the rate at which evapotranspiration can be sustained from a complete and uniform vegetation cover when water availability is unlimited. Although this definition is conceptually appropriate, it is problematic in practice because this rate depends not only on the atmosphere, but also on the surface conditions of soil and vegetation type. Dingman (1994) provides a useful discussion that follows the development of this concept and its appropriate applicability. Much of that discussion is summarized here.

Thornthwaite (1948) introduced the concept of potential evapotranspiration E_p as part of a climate classification scheme. He intended this to be a measure of the maximum amount of evaporation that could be sustained by given atmospheric conditions, as a way of distinguishing between climate types. Thornthwaite calculated it from an empirical formula based on mean monthly temperature and day length, a factor that is related to energy input. His empirical formula is

$$E_p = 1.6[10T_a/I]^a \qquad (8.1)$$

where T_a is mean monthly air temperature (°C), I is an annual heat index based on mean monthly air temperature for all 12 months, and the exponent a is dependent on the heat index. Others have developed analogous formulae using the saturation absolute humidity corresponding to the mean temperature.

Clearly, E_p is influenced not only by the state of the atmosphere but also by such factors as surface albedo, the capacity of the prevailing vegetation to transpire water vapor to the atmosphere, and the presence or absence of interception water. Recognizing this, Penman (1956) modified the definition to "the amount of water transpired … by a short green crop, completely shading the ground, of uniform height and never short of water."

The above definitions relate to what Federer et al. (1996) term "reference surface potential evapotranspiration." They distinguish two other classes of definition: surface-dependent potential evapotranspiration, and potential interception. The former relates to a designated land surface type, as opposed to a specific reference crop, and the latter refers to the maximum sustainable evapotranspiration from externally wetted surfaces, in effect eliminating the impact of transpiration on the process. The range of definitions means that there is considerable ambiguity to the term "potential evapotranspiration" and a variety of assessment methods. Consequently, it is important that reference to this concept be accompanied by specific information regarding the reference surface and method of calculation. A large number of methods are utilized. The following section describes some of the most common.

In must be kept in mind that "potential evapotranspiration" is a theoretical concept and as such has many shortcomings (Brutsaert 1982). This is especially true in water-limited environments, where the meteorological conditions used to assess E_p are different from those that would occur under conditions of unlimited availability (see Bouchet's hypothesis, Section 8.6). In other words, the absence of water and vegetation alters the factors of net radiation, temperature, and wind that are used in various assessments of E_p. Nevertheless, potential evapotranspiration is a useful concept in quantifying the "drying power" of an environment. It has applications in such areas as agriculture, climate assessment, and climate modeling.

8.2.2 OTHER METHODS OF ASSESSING POTENTIAL EVAPOTRANSPIRATION

Potential evapotranspiration is assessed through physical measurements via evaporation pans and or by empirical formulae. Jensen et al. (1990) classified the empirical methods based on data requirements and listed temperature-based, radiation-based, and combination methods as the most common. In more standard usage, the term "combination methods" actually refers to methods that are based on a combination of mass transfer (see below) and energy balance (see Section 8.3) considerations. In the direct measurement approach, the volume of water lost from the pans is recorded. Although this sounds like a simple and robust calculation, in reality the value of E_p obtained depends a great deal on the characteristics of the pan. Thus, no absolute value of E_p can be determined.

The most common empirical approaches to estimating potential evapotranspiration are those of Thornthwaite (1948), Priestley and Taylor (1972), and Penman–Monteith (Monteith 1965). The Priestley–Taylor approach uses only radiation and is an adaptation of the Penman (1948) formula for evaporation from a free water surface. Thornthwaite's calculation of potential evapotranspiration, as described in Section 8.2.1, requires only temperature information. Like Penman's equation, the Penman–Monteith "combination" method utilizes concepts of mass transfer and energy balance. It thus requires several data types: net radiation, air temperature, wind speed, and relative humidity.

Early approaches to calculating evaporation from a free water surface included mainly the mass transfer concept developed by Dalton in 1802 and the energy balance. Dalton reasoned that evaporation would be proportional to the difference between actual and saturation vapor pressure and would occur at a rate that is dependent on wind speed. Equations based on this concept are of the form

$$E_p = (b_1 + b_2 u)(e_{sat} - e_a) \qquad (8.2)$$

where u is wind speed, e_a is the atmospheric vapor pressure, and e_{sat} is the saturation vapor pressure at the appropriate temperature. The empirical constants b_1 and b_2 generally depend on the height at which wind speed and vapor pressure are measured.

Penman (1948) combined these concepts with Dalton's mass transfer concept to produce a formula for potential evapotranspiration. Penman's formula, also called the Penman combination equation, assumes that evaporation from a free water surface is a function of net radiation, saturation deficit, and wind speed, so that E is calculated as:

$$E = \frac{R_n \Delta + \gamma E_a}{\Delta + \gamma} \qquad (8.3)$$

where R_n is the net radiation, Δ is the slope of the curve relating saturation vapor pressure to temperature and γ is the psychrometric constant, equal to 0.622. E_a is a term describing the contribution of mass transfer to evaporation and is in the form shown in Eq. (8.2).

Monteith (1965) revised the Penman equation so that it would be valid in the water-limited case and could incorporate the effects of vegetation on evapotranspiration. The development of this approach was based on biophysical concepts related to water passage through vegetation and this is described in Section 8.4.3, dealing with a vegetated surface.

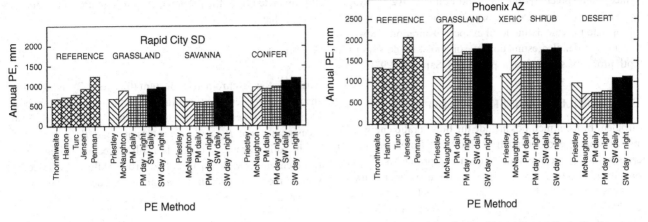

Fig. 8.1 Estimates of annual reference surface and surface-dependent potential evapotranspiration, calculated by various methods, for three cover types at two dryland locations (from Federer *et al.* 1996).

The choice of an assessment method depends on the definition of potential evapotranspiration utilized, the available input data, the purpose of the calculation, and surface characteristics. Dingman (1994) suggests that Penman–Monteith is the best approach to estimating potential evapotranspiration over vegetated surfaces. A study by Jensen *et al.* (1990) compared the results of this method with lysimeter data and found good agreement. However, no method is universally accepted as the best.

Ideally, a definition of potential evaporation should be applicable at all times and places if the goal is to assess regional or global hydrology. In seeking this goal, Federer *et al.* (1996) conducted an intercomparison of 11 methods of assessment and utilized a variety of locations, various cover types, and various time resolutions for input data. Unfortunately, the differences between the various estimates were often on the order of hundreds of millimeters per year, with the greatest range being in arid locations (Fig. 8.1). No method was consistently high or low.

In terms of time resolution, Federer *et al.* (1996) found that the use of five-day or monthly data, instead of daily, did not seriously degrade the estimates of potential evapotranspiration. In general, use of five-day or monthly data tended to give lower values, on the order of 10–15 mm/month for five-day data and 20–40 mm/month for monthly data. In this study, separate calculations for daytime and nighttime had little impact. However, this is not the case in locations where there is a strong diurnal cycle in cloudiness (e.g., Yin *et al.* 2000).

8.3 METHODS OF ASSESSMENT OF ACTUAL EVAPOTRANSPIRATION

Evapotranspiration is a difficult quantity to determine. An informal survey of participants at a NATO meeting on hydrology (Michaud 1997) indicated strong dissatisfaction with the availability of data on evapotranspiration (an overall satisfaction index of −22, compared with −11 for soil moisture, +10 for rainfall, and +14 for runoff). The reason is that few actual measurements exist and that alternative methods require numerous assumptions and approximations, as well as data for other quantities that are equally hard to measure. Because of the complexity of the process and the number of degrees of freedom (numerous characteristics of the wind, soil, soil moisture, and vegetation cover all play a role), a tremendous number of methods exist. The reader is referred to Brutsaert (1982), Shuttleworth (1991), Milly (1991), Burman and Pochop (1994), Dingman (1994), and other sources for details.

The most common methods of assessing actual evaporation are based on turbulent transfer, potential evapotranspiration, and water balance. In climate studies the first two classes are most often used. Some direct measurements can be made via instruments such as lysimeters. The resistance approach of Penman–Monteith (PM) is becoming the most common one utilized for vegetated surfaces, especially at less than climatological time scales. Cleugh *et al.* (2007) compared this method with the aerodynamic resistance/turbulent transfer method for regional evapotranspiration estimates in a tropical savanna in Australia and found the PM method to be superior in estimating both the magnitude and seasonal variation of evapotranspiration.

8.3.1 WATER BALANCE METHODS

The water balance methods require a tally of inputs and outputs, calculating evapotranspiration as a residual. Using the atmospheric water balance, evapotranspiration can be equated to the difference between precipitation, runoff, and changes in soil moisture storage w:

$$E = P - N - \mathrm{d}w/\mathrm{d}t. \tag{8.4}$$

This method is most useful in calculating evapotranspiration for large catchments; it is of limited use in calculating local evapotranspiration. Of the terms required for its calculation, only precipitation is readily available.

The soil water balance can likewise be used, by measuring the change in soil moisture in a soil column and estimating surface input (precipitation, etc.) and drainage. Again, evapotranspiration is calculated as a residual. This approach is more appropriate for calculating local evapotranspiration. However, it requires a detailed estimation of soil moisture flow and the depth profile of soil hydraulic properties (Salvucci 1997).

8.3.2 TURBULENT TRANSFER METHODS

Turbulent transfer methods include the Bowen ratio energy balance approach, eddy correlation, and what Rosenberg (1974) terms aerodynamic methods (the terms "profile" and "mass transfer" methods are also used). All of these require the assessment of gradients and fluxes of quantities such as momentum, sensible heat, and water vapor. Thus, the calculations are most reliable when local measurements are made.

The essence of the aerodynamic approach lies in the definitions of the fluxes of water vapor, sensible heat, and momentum:

$$E = -\rho_a K_v \, dq/dz \tag{8.5}$$

$$H = -\rho_a C_p K_h \, dT/dz \tag{8.6}$$

$$\tau = \rho_a K_m \, du/dz \tag{8.7}$$

where ρ_a is the density of air, z is height above ground, and K_v, K_h, and K_m are the exchange coefficients for water vapor, heat, and momentum, respectively, and the corresponding vertical profile terms are calculated for specific humidity, temperature, and wind. In this approach, the exchange coefficients are assumed to be equal (the "similarity assumption"). Evapotranspiration can then be calculated from the gradient of specific humidity if the simultaneous gradient of temperature or wind and the flux of sensible heat or momentum are known.

Rosenberg (1974) points out that it is difficult to obtain independent measurements of sensible heat or momentum. Furthermore, the assumption of equal exchange coefficients, termed the similarity assumption, is not strictly valid and the values of the exchange coefficients are stability-dependent in ways not well understood. A number of variants of the basic method have been used to deal with this problem.

Like the aerodynamic methods, the Bowen ratio approach relies on near-surface profiles and E is calculated as a residual. It essentially involves a more or less complete assessment of surface energy balance:

$$R_{net} = LE + G + H \tag{8.8}$$

where G and H represent sensible transfer of heat to the ground and atmosphere, respectively. The Bowen ratio B is H/LE, so that the above equation can be rewritten as

$$LE = (R_{net} - S)/(1 + B) \tag{8.9}$$

where S, sensible heat transfer, is $G + H$. Using the similarity assumption and neglecting G, the Bowen ratio can be approximated from the profiles of vapor pressure and temperature, so that

$$B = \gamma \, dT/de \tag{8.10}$$

where γ is the psychrometric constant (0.66 mb/°C) and dT/de is the ratio of the temperature gradient to the vapor pressure gradient in the air near the surface. LE can thus be calculated from measurements of R_{net} and the profiles of e and T.

The eddy correlation methods involve direct measurement of turbulent fluxes at the surface. The turbulent flux is achieved by vertical motion of air. The instantaneous vertical velocity ω can be expressed as the mean vertical velocity $\bar{\omega}$ plus a departure from the mean ω':

$$\omega = \bar{\omega} + \omega'. \tag{8.11}$$

Since the mean vertical velocity $\bar{\omega}$ of air is near zero, the turbulent flux is produced by the rapidly fluctuating ω' (Fig. 8.2). The vertical currents bring with them the momentum, heat, and moisture of the level from which they originated. Thus, since wind speed increases with height near the ground, an updraft is characterized by an instantaneous reduction in speed, and a downdraft is characterized by acceleration; properties of the air, such as temperature and moisture content, depend on the vertical gradients of these properties.

The magnitude of this flux can be expressed as the product of ω' with the simultaneous departure of the property Q from its mean value, so that

$$E = \rho_a \omega' q' \tag{8.12}$$

where E is the flux of water vapor and q' is the instantaneous departure of specific humidity from its mean value. This approach is not limited by assumptions concerning the values of the exchange coefficients, but requires rather sophisticated instrumentation.

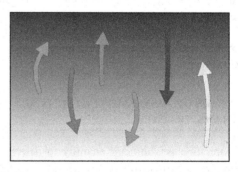

Fig. 8.2 Schematic of turbulent fluxes. The shading indicates the magnitude of wind at a given level, with darker shading indicating greater speeds. Air parcels moving upward and downward, represented by arrows, arrive at a new location with the same velocity as their origin. These create the turbulent fluxes represented as departures from the mean.

8.3.3 METHODS BASED ON POTENTIAL EVAPOTRANSPIRATION

Several approaches to the calculation of actual evaporation are based on potential evapotranspiration and express the actual rate in terms of the proportion of E_p that is realized. The most common of these are based on the degree of soil saturation and are expressed in the form

$$E_0 = f(\theta)E_p \qquad (8.13)$$

where θ is water in the root zone of plants and $f(\theta)$ is commonly based on the relative water content θ_{rel} defined as

$$\theta_{rel} = (\theta - \theta_{pwp})/(\theta_{fc} - \theta_{pwp}) \qquad (8.14)$$

where θ_{fc} and θ_{pwp} are the field capacity and permanent wilting point water contents of the vegetation. These concepts are defined in Chapter 7. The term $(\theta - \theta_{pwp})$ is referred to as available soil moisture and the term $(\theta_{fc} - \theta_{pwp})$ is referred to as the soil water storage capacity. Both are generally expressed as a percent volumetric water content, so that the depth of soil moisture in centimeters requires multiplication by the root depth of the soil.

The above relationship is often written in the form

$$E_0/E_p = \beta \qquad (8.15)$$

where β is the ratio of the actual exchange coefficient for water vapor to the exchange coefficient corresponding to surface resistance $r_s = 0$ (i.e., the case of a wet surface). The parameter β is also termed the evapotranspiration function. The class of methods using this approach is generically termed "beta methods" and various forms of the function β have been utilized. Useful reviews of these are provided by Ye and Pielke (1993), Mahfouf and Noilhan (1991), and Kondo and Saigusa (1990). The β methods have difficulties in the case of very dry surfaces and therefore they are not particularly well suited for

evapotranspiration calculations in dryland regions, despite their common use therein.

Several models have been proposed to describe the functional relationship between β and available soil moisture (Fig. 8.2). It is generally assumed that the relationship is linear, as proposed by Thornthwaite, so that

$$E_0/E_p = w/w^* \qquad (8.16)$$

where w and w^* are the available soil moisture ($\theta - \theta_{pwp}$) and the soil moisture storage capacity ($\theta_{fc} - \theta_{pwp}$). The latter is equivalent to field capacity, but the term soil water storage capacity and the symbols w and w^* are more common in the climatological and meteorological literature. Forms of the equation using θ are more common in the hydrologic and soil literature, where field capacity is most commonly used.

The above equation is the basis for the so-called "bucket method" used in calculating evaporation in many climate models. However, the form $E_0/E_p = \beta$ is much more complex (Dunne and Leopold 1978) and depends greatly on soil type (Fig. 8.3). The linear form is fairly accurate for clay soils but very inadequate for sandy soils. Models proposed by Pierce (1958) and Veihmeyer (1964) appear to be valid for sand and loam soils, respectively.

The Thornthwaite method calculates actual evapotranspiration through a comparison between potential evapotranspiration and precipitation. Willmott et al. (1985) provide an excellent explanation of the method. Some of the details are described in Section 7.2. This method is applied to monthly time scales and appears to provide reasonable values of E_0 that can be verified on regional scales from precipitation and runoff data (Dingman 1994). Nicholson et al. (1997a) use a more physically based approach for calculations over Africa. The results of this method are compared with those of Thornthwaite–Mather and Penman in Section 7.4.1. There is particularly good agreement with calculations based on Thornthwaite–Mather. The exception was in the more arid regions of the continent, where the

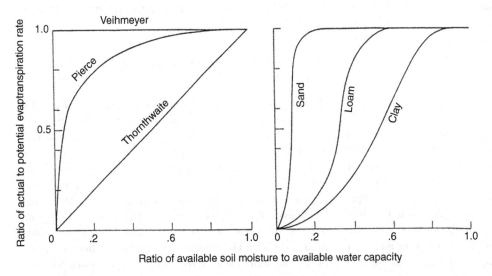

Fig. 8.3 (left) Three different models of the variation of actual evaporation with soil moisture content, expressed as relative water content. (right) Relationship between ratio of actual to potential evapotranspiration and soil moisture availability, represented by the relative water content (Eq. 8.21) and approximately equal to β, in three different soils (from Dunne and Leopold 1978).

presence of significant vegetation indicates the presence of soil moisture and evapotranspiration, but Thornthwaite's method indicates negligible values of both. Thus, the Thornthwaite–Mather approach must be used with caution in arid and some semi-arid locations.

8.4 EVAPORATION FROM SPECIFIC SURFACE TYPES

8.4.1 FREE WATER AND LAKE EVAPORATION

Evaporation over a water surface is the easiest case to deal with, because there is a constant supply of water and the process is almost solely controlled by atmospheric factors. In this case, actual evaporation E_0 is generally assumed to be equal to potential evapotranspiration E_p. The limiting factors include the saturation deficit $(e_s - e)$, wind speed, and net radiation or surface temperature. E_p can be assessed using a variety of methods, as discussed in Sections 8.2 and 8.3.

Particularly useful are Dalton's concept of mass transfer (Eq. 8.2) and the Penman equation (Eq. 8.3). These assume that there is no energy for evaporation available from storage in the subsurface of the water medium and that there is no available energy advected by the water. These assumptions are generally valid over what is termed a "free water" surface, such as an evaporation pan. Over a large lake these assumptions are questionable and sometimes the energy available from storage and advection via water must be included in the calculations.

Despite the simplicity of the open water case, compared with soil or vegetation, the various methods of calculating evaporation can produce markedly different results. Estimates of mean annual evaporation over Lake Victoria range from 1370 to over 1700 mm/year. Yin and Nicholson (1998) compared the results of Penman and energy balance methods for Lake Victoria, using the same input parameters for both. The Penman formula gave 1743 mm/year, while the energy balance approach gave 1551 mm/year.

Evaporation of interception water on the surface of plant materials and detention water on the surface of soils can be treated as evaporation from a free water surface. Evaporation in these cases takes place at the potential rate, as long as the canopy or soil surface remains wet. Under these conditions the formula for potential evapotranspiration is applicable.

8.4.2 BARE SOIL

There are three phases in the process of evaporation over bare soil. These vary with the degree of wetness of the soil. The first phase is that of a wet soil surface; in this case, evaporation is atmosphere-limited. The process is controlled by the same three meteorological variables controlling the process over open water, and actual evapotranspiration is equivalent to potential. The soil is considered "wet" as long as its moisture content is above field capacity. This threshold is defined as the maximum amount of water that the soil can hold against the force of

gravity. It ranges from about 25% of saturation in sand to about 90% in clay.

In the second phase, soil moisture is below field capacity. As moisture declines so does hydraulic conductivity, so that in this phase evaporation is water-limited, i.e., controlled by the soil moisture content. The rate of evaporation depends primarily on the vertical distribution of soil moisture and the rate at which water evaporated from the surface is replenished from below. Water movement is slow below the first few centimeters. The surface dries out first. The layer at which energy and moisture converge and maximum evaporation takes place goes deeper and deeper into the soil as drying continues. As indicated earlier, the rate can be roughly approximated from the parameter β, the ratio of actual soil moisture w to field capacity w^*, and calculating potential evapotranspiration from meteorological variables.

The form of this function is discussed in Section 8.3.3, with typical forms for three soil types illustrated in Fig. 8.3. Since soil moisture is progressively depleted as evaporation proceeds, the rate of evaporation decreases rapidly over time and the cumulative evaporation increases approximately at a rate equal to the square-root of time (Kurc and Small 2004; Salvucci 1997). Eventually evaporation ceases, as soil moisture is depleted, and the cumulative curve becomes asymptotic.

The third phase occurs when the surface is very dry and a second critical level of moisture content is reached. At this point the soil is so dry that liquid water movement nearly ceases and evaporation consists mostly of vapor from soil pores. In such a case, the water available for evaporation lies well below the source of energy (radiation or surface warmth) and evaporation is small. Since vapor movement is from areas of high molecular energy (i.e., warm areas) to areas of low molecular energy (cold areas), in this phase the temperature gradient is of primary importance and the conditions leading to high evaporation are essentially the opposite of those in the wet phase. Surface heating, which enhances evaporation from plants and surficial water, suppresses this form of vapor transfer to the atmosphere. Thus in winter, when the surface is cooler than the ground below, this process can be significant, but it is generally negligible in summer.

8.4.3 EVAPOTRANSPIRATION OVER A VEGETATED SURFACE

Water is stored in five locations in an ecosystem (see Fig. 7.16): on the surface of plants, in plants, on the soil surface, in the soil, and in the groundwater. All reservoirs sustain the evaporative flux (Fig. 8.4). The total evapotranspiration is the summation of evaporation from detention water on the surface, out of soil pores, off the surfaces of plant materials, and via transpiration through the stomata. The dominant mode of water loss is transpiration except when the canopy is wet. One major distinction between vegetation and bare soil is that transpiration short-circuits the normal channels of vertical soil moisture transfer by deriving water from the root zone, sometimes well below the surface. Thus, evaporative loss can be greater than over bare soil, especially when the soil is relatively dry. In this case there are also three phases in the process of evapotranspiration and these are similar to the phases in the case of bare soil.

During the wet phase, evapotranspiration is atmosphere-limited and $E_0 = E_p$. As the soil dries out, a point is reached where the soil moisture instead becomes the limiting factor. Beyond this point, evapotranspiration is a function of soil moisture and is roughly proportional to it, as in the water-limited soil case. This critical threshold may not be reached until the soil moisture

Fig. 8.4 Streamers of evaporative flux coming from the rainforest canopy in Cameroon.

is well below field capacity, because the roots extend downward to the moister lower layers of the surface or even to the water table. Thus E_0 may continue at the potential rate long after field capacity is reached, especially when the dominant vegetation has long roots, as is often the case in dryland regions.

A second critical level in the drying process is the wilting point, the point at which plants no longer remain turgid. The two critical levels of field capacity and wilting point differ greatly between various soil types and are strongly dependent on soil texture. In general, the higher the sand content (and the lower the clay content), the lower the wilting point and field capacity.

For a vegetated surface, transpiration is regulated by a series of plant, soil, and air resistances, as depicted in Fig. 8.5. The rate of water flow through the soil to the atmosphere depends on the vapor pressure difference between the air and the leaf, and the water potential difference between the leaf and the soil. Biophysical characteristics of the soil, plant, and atmosphere govern the rate of transfer corresponding to a given pressure or potential difference.

The resistance is the inverse of the conductance and it has been adopted in many formulations of evaporation because of its mathematical convenience. This is illustrated by considering the concepts of resistances acting in series versus those acting in parallel. In the diagram in Fig. 8.5, the soil, root, and stem

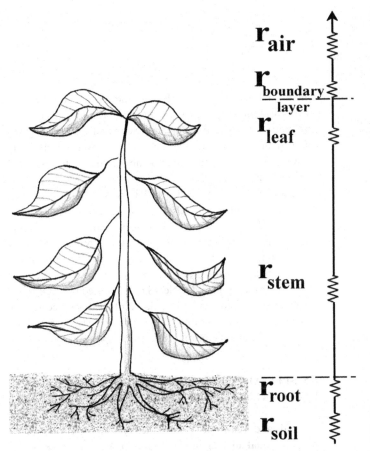

Fig. 8.5 Resistances to water flow by soil, plant and atmosphere. This is an analogy to an electric circuit with resistances acting in series.

resistances act in series, i.e., sequentially. In this case, the resistances are additive, so that r, the total resistance, can be calculated as the sum of all three.

In general, the resistances within the plant (root and stem resistances) are small compared with soil resistance. The largest resistance, between plant and atmosphere, comes from the leaf because here water changes from the liquid to the vapor phase. The individual leaf resistances collectively determine the resistance offered by the canopy (canopy resistance r_c), but these act in parallel. In this case, the bulk resistance by the canopy does not equal the sum of the resistances, but rather must be calculated as

$$1/r_c = 1/r_1 + 1/r_2 + 1/r_3 \dots + 1/r_n \qquad (8.17)$$

where r_1 and so on are the individual leaf resistances for n leaves.

Both the leaf epidermis and stomata offer resistance to the passage of water that also act in parallel, so that

$$1/r_1 = 1/r_e + 1/r_s \qquad (8.18)$$

where r_1 is leaf resistance, r_e is cuticular resistance and r_s is stomatal resistance. The stomata (Fig. 8.6) are essentially "guard cells" that act as valves to regulate the transfer of water vapor and carbon dioxide by controlling the aperture to the stomatal pores within the leaf. Stomatal closure is tied to light intensity and water loss, with stomata closing at night and reducing the aperture to the pores as available moisture decreases (i.e., as water stress increases). The stomata act to oppose the hydraulic conductivity of the soil. Minimum stomatal resistance occurs when conditions for transpiration are optimal; this is usually an order of magnitude smaller than r_e and under such conditions is the main contributor to r_1. When the stomata are completely closed (e.g., when no soil moisture is available), the leaf resistance is equal to r_e.

The transfer of moisture E from the leaf to the boundary layer can thus be calculated as

$$E = \frac{\rho C_p}{\gamma L} \frac{\delta e}{r_1 + r_a} \qquad (8.19)$$

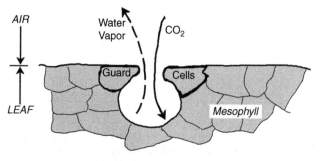

Fig. 8.6 Diagram of plant stomata (modified from Oke 1987).

Table 8.1. *Aerodynamic properties of various surfaces: aerodynamic resistance* ra, *canopy resistance* rc, *and total resistance* rtotal *(from Kaimal and Finnigan 1994).*

Surface	r_a (s m^{-1})	r_c (s m^{-1})	r_{total} (s m^{-1})
Open water	200	0	200
Grass	50	50	100
Forests	20	200	200

where δe is the difference between the saturation vapor pressure at leaf temperature and vapor pressure of the outside air, γ is the psychrometric constant, ρ is the density of moist air, C_p is the specific heat of air at constant pressure, and r_a and r_1 are aerodynamic and leaf resistances, respectively. Generally, a bulk stomatal resistance r_s or canopy resistance r_c is substituted for r_1 and used to characterize the regulation of water vapor transfer by plants. The canopy resistance can be calculated from leaf resistance by the formula

$$r_c = r_1 / \text{LAI} \qquad (8.20)$$

where LAI is leaf area index, as defined by Eq. (6.8) (Chapter 6).

Monteith (1963) combined the above with Penman's equation to produce a form of Eq. (8.3) based on resistances. Monteith initially used the concept of stomatal resistance but later proposed the use of canopy resistance r_c. The resultant equation is then

$$E_0 = \frac{1}{L} \frac{\frac{\Delta}{\gamma}(R_{net} - G) + \frac{\rho L \varepsilon}{P r_a} \cdot \delta e}{\frac{\Delta}{\gamma} + 1 + \frac{r_c}{r_a}} \qquad (8.21)$$

where G is sensible heat transfer into the surface, Δ is the slope of the saturation vapor pressure/temperature curve, γ is the psychrometric constant, ρ is atmospheric pressure and ε is a constant equal to 0.622. The resistances change over time and vary between ecosystems, but the remaining terms of the equation are relatively constant.

Canopy resistance decreases with increasing solar irradiance, but increases with soil water stress. It appears to be independent of wind speed, but tends to increase with size and density of the canopy (see Table 8.1). Aerodynamic resistance, on the other hand, is a function of surface roughness and turbulence and it is relatively small in open canopy space but quite large over open water or grass (Table 8.1). In a wetted leaf, $r_c = 0$ and the stomata play no regulatory role. At other times, because the resistances act in series rather than in parallel, canopy resistance exerts the dominant control on evapotranspiration.

8.5 SPECIAL CONSIDERATIONS CONCERNING EVAPOTRANSPIRATION IN DRYLANDS

In arid and semi-arid environments, evapotranspiration is roughly equal in magnitude to precipitation on time scales

Table 8.2. *Measurements of evapotranspiration (ET) in semi-arid environments (from Kurc and Small (2004).*

Dugas *et al.* (1996)	Jornada, New Mexico	ET, soil evaporation
Gash *et al.* (1991)	Sahel, western Niger	ET, soil moisture
Kabat *et al.* (1997)	Sahel, western Niger	ET
Malek *et al.* (1997)	Great Basin, Nevada	ET, soil moisture
Stannard *et al.* (1994)	Walnut Gulch, Arizona	ET
Taylor (2000)	Sahel, western Niger	ET, soil moisture, soil evaporation
Tuzet *et al.* (1997)	Sahel, western Niger	ET, shrub transpiration
Unland *et al.* (1996)	Sonoran Desert, Arizona	ET

longer than seasons. Its variability on shorter time scales is less directly constrained by precipitation, being limited by soil moisture most of the time (Rodriguez-Iturbe 2000). This shorter-term variability is critical for the coupled cycling of water, energy, and carbon in dryland environments (Kurc and Small 2004). Evapotranspiration is a critical determinant of the partitioning of moisture between runoff, infiltration, and recharge (Laio *et al.* 2001); of ecological processes such as plant productivity, soil respiration, and biogeochemical cycling (Porporato *et al.* 2003); and of the potential impact of land–atmosphere interactions on weather and climate (Pielke 2001). Despite this broad role in the functioning of dryland ecosystems, there are surprisingly few ecosystem-level observations of evapotranspiration in semi-arid environments. The major ones are summarized in Table 8.2. These include a limited range of soil and vegetation types.

Unfortunately, little knowledge of evapotranspiration in drylands can be gleaned from the more extensive studies in humid environments because numerous aspects of the process are significantly different in the drylands. One of the most fundamental differences is the spatial heterogeneity of the surface. This includes both the patchiness of the environment in terms of soils and vegetation, and the very irregular distribution of soil moisture owing to the randomness of the distribution of the rainfall. Very generally, the drier the environment, the greater the degree of spatial heterogeneity. Special approaches are necessary to deal with the combination of vegetation and bare soil (Shuttleworth and Wallace 1985) and with the extreme spatial variability of the physical properties of the land surface (Humes *et al.* 1994). A second difference is the extreme contrast between wet periods and dry periods (the occurrence of rainfall in distinct pulses and seasons) and its influence on biophysical processes within the vegetation and soil (Schwinning and Sala 2004; Austin *et al.* 2004). A third difference has to do with the physics of the evaporation process. Because of the extreme dryness of the soil during all or part of the year, vapor flow within the soil becomes a significant component of evapotranspiration (Katata *et al.* 2007). This problem is discussed in great detail by Menenti (1984). Specific aspects related to playas are discussed in Chapter 10. The contribution of transpiration to evapotranspiration is also more complex than in humid environments because of the strong contribution of direct evaporation from

soil and because of the great diversity in water-use efficiency and adaptive strategies of dryland vegetation.

These aspects of the drylands are illustrated using results from extensive field studies conducted primarily in the USA and Africa. However, the knowledge that emerges is consistent with that gained from studies in many other parts of the world. Some notable field programs include Monsoon '90 (Kustas *et al.* 1991), Walnut Gulch '92, (Moran *et al.* 1994), JORNEX (Ritchie *et al.* 1998), and SALSA (Goodrich *et al.* 2000) in the American Southwest, FIFE (Sellers *et al.* 1988) in the US Great Plains, and HAPEX-Sahel in West Africa (Goutourbe 1997). The results described in this section underscore several aspects of evapotranspiration in drylands that pose a challenge for both modeling and remote sensing. These include the large contrasts between individual species and between the individual components of a vegetation formation (canopy, understory, soil); the different responses of these components and of different species to changes in moisture availability; and the impact of the photosynthetic pathway (C3, C4, CAM; see Section 3.2.2) on evapotranspiration and response to fluctuating moisture supply. The response of below-ground processes to these fluctuations is similarly important, but has received little attention.

8.5.1 SPATIAL HETEROGENEITY

The spatial heterogeneity of soil and vegetation that typifies drylands poses challenges in the assessment of regional scale water balance parameters (Malek *et al.* 1997). Examples from the West African Sahel, where plant–water–climate relationships have been studied in two major field experiments, HAPEX-Sahel (1992) and SEBEX (1989/90), illustrate some of these challenges. The surface types that have been extensively monitored include fallow agricultural fields (millet), a bush savanna, and tiger bush (also called patterned woodland). The latter two formations, which are discussed in greater detail in Chapter 10, have been studied by Verhoef (1995) and Kabat *et al.* (1997). These differ in the microclimatic conditions that determine temperature, soil moisture, and available energy, as well as in the biophysical characteristics of the dominant vegetation species.

In the tiger bush formation, 67% of the surface is bare soil. The albedo contrast between the bushy thickets and the surrounding bare soil patches is strong: roughly 0.15 in the thicket

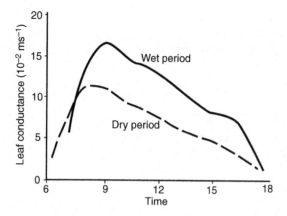

Fig. 8.7 Average diurnal course of *Guiera senegalensis* leaf conductance during wet and dry periods (roughly August/September and October, respectively) (modified from Verhoef *et al.* 1996).

versus 0.25 for the soil. Within the savanna, however, the albedo contrast between the bushes and understory of grasses and herbs is an order of magnitude smaller. The understory or underlying soil layers can be more than 10°C cooler than the bush canopies, and the dense thicket of the tiger bush can locally retard the loss of soil moisture. Thus the conditions of radiation, temperature, and soil moisture contrast strongly within the different components of these formations.

The most important biophysical characteristic is leaf conductance. This varies tremendously during the course of the day and between the wet and dry seasons (Fig. 8.7). Soil thermal properties (diffusivity, conductivity, heat capacity) also change dramatically in response to soil moisture, but remain fairly constant during the dry season (Fig. 8.8). Leaf conductance is quite different for the savanna and tiger bush formations, both in the bush and understory. Canopy conductance, which reflects leaf conductance, structure, and leaf area index (LAI), is markedly higher and is much more sensitive to moisture availability for the tiger bush than is that of the savanna (Fig. 8.9).

Likewise, the impact of changes in soil moisture on leaf conductance varies by about a factor of two among the three dominant savanna species (Verhoef 1995). The leaf conductance of *Mitracarpus scaber* and *Jacquemontia tamnifolia*, both undergrowth species, shows a greater response to changes in moisture availability than does that of *Guiera senegalensis*, the overstory species (Fig. 8.10). The sensitivity to moisture is reflected in the diurnal cycle of conductance, which is greatest in the early morning, when humidity is generally relatively high.

The canopy structure also distinguishes the savanna and tiger bush formations. The contrast imposes differences between these regions in solar irradiance, soil heat flux, and sensible and latent heat flux at the surface (Tuzet *et al.* 1997). Both "shading" and "wake" effects play a role. The sensible heat flux is greater in the tiger bush by roughly a factor of two (Fig. 8.11).

The net result is that the savanna and tiger bush differ markedly in their contribution to surface evapotranspiration and in the importance of the individual surface components within them (Figs. 8.11 and 8.12). In general, the soil contribution to

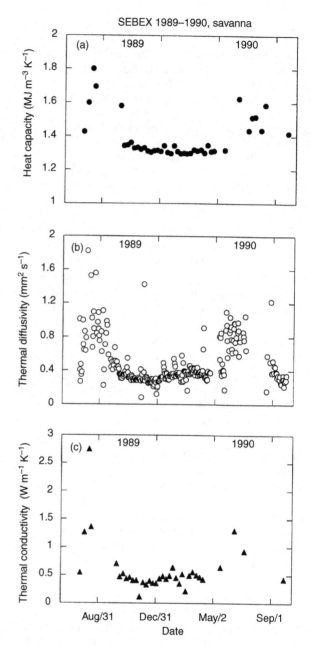

Fig. 8.8 Thermal properties for the SEBEX savanna site during 1989/90: (a) heat capacity, (b) thermal diffusivity, and (c) thermal conductivity (modified from Verhoef *et al.* 1996).

evapotranspiration is the smallest, even in the tiger bush formation with extensive patches of bare soil (Fig. 8.12). The greatest contribution comes from the dense tiger bush thicket. Over the savanna the bush species contribute substantially more than the herb and grass species (Verhoef 1995), but less than for the tiger bush. In the savanna, the partitioning of evapotranspiration between the bush and the surrounding grassland depends on moisture availability (Tuzet *et al.* 1997).

Evapotranspiration can reach 4–5 mm/day in both the savanna and the tiger bush, declining to roughly 2 mm/day after the rains have ceased (Fig. 8.11). However, the cumulative result

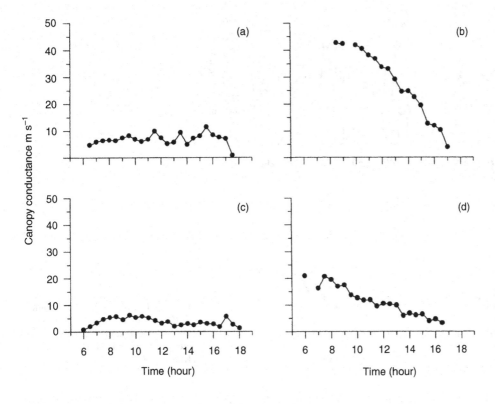

Fig. 8.9 Hourly values of canopy conductance for tiger bush (a, b) and a fallow savanna (c, d) in the Sahel. On the left (a, c), the data are for a day that was cloudier and with higher soil moisture than the day on the right (b, d) (from Kabat *et al.* 1997).

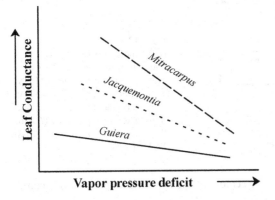

Fig. 8.10 Relationship between daily average leaf conductance and vapor pressure deficit for three species in a Sahelian savanna site (from Verhoef 1995).

(Fig. 8.12) is substantially more evaporation from the tiger bush formation. This contrast is partly due to the differences in sensible heating that produce differences in the evaporative fraction. Differences in vegetation conductance and the response of conductance to vapor pressure deficit also play a role. This can be attributed to the differences in the C3 and C4 species composition of the two formations (Kabat *et al.* 1997), a result that has implications for modeling dryland ecosystems.

Giambelluca *et al.* (2009) examined evapotranspiration in two savannas with contrasting tree density but similar understory. One consisted of tall trees (8–10 m) with a cover density of 50%, the other included shorter trees (3–4 m) and 5% cover. Evapotranspiration was markedly similar in the

two cases, suggesting that LAI is a strong determinant of evapotranspiration.

Kremer and Running (1996), observing dryland ecosystems in the USA, underscore many of these same points. They examined sagebrush/bunchgrass, bunchgrass, and cheatgrass communities in a western steppe. Each community was shown to have a different impact on water balance and soil moisture availability, and each required different model parameterizations for soil moisture availability and carbon accumulation. The implications of the study are that for modeling ecosystem processes in arid lands, the relationships between soil water availability, transpiring leaf area, water-use efficiency, and the respiration costs of substantial below-ground productivity are key ecophysiological considerations. This makes the modeling (and understanding) of vegetation–atmosphere interactions in dryland regions more difficult than in more humid environments.

8.5.2 THE CONTRIBUTION OF TRANSPIRATION IN DRYLANDS

The amount of water available to the ecosystem is determined not only by the spatial and temporal distribution of rainfall, but also by how the rainfall is redistributed via interception, stemflow, infiltration, percolation, evaporation, and runoff (Newman *et al.* 2006). Most studies lump soil evaporation, transpiration, and the evaporation of canopy interception into a single term, evapotranspiration (Loik *et al.* 2004; Huxman *et al.* 2005). However, the partitioning of transpiration and evaporation is important because generally different moisture reservoirs are

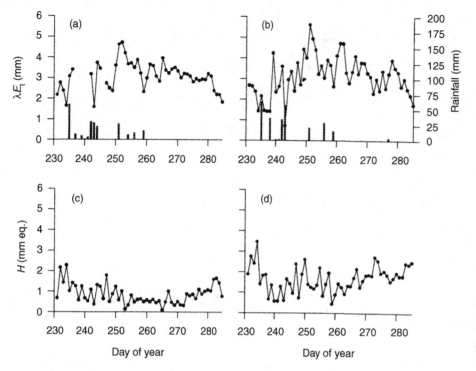

Fig. 8.11 Daily total evaporation (λE_t) and sensible heat (H) for fallow savanna (a, c) and tiger bush (b, d) measured during HAPEX-Sahel (from Kabat *et al.* 1997).

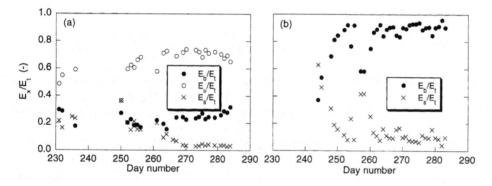

Fig. 8.12 Contribution of (a) bush (E_b), understory herb layer (E_u), and soil evaporation (E_s) to total evaporation (E_t) for the HAPEX-Sahel savanna site; (b) contribution of bush (E_b) and soil (E_s) evaporation to total evaporation (E_t) for the tiger bush formation (from Verhoef 1995).

involved (i.e., near surface vs. deeper soil layers), the two components behave differently with respect to variations in rainfall, and the degree of temporal variability is quite different, especially on daily time scales.

The transpirational component is of interest because it is the main pathway of water removal from lower soil horizons of dryland environments (Schlesinger *et al.* 1987) and it therefore affects the rate and depth of $CaCO_3$ (caliche) deposition. Precipitation of carbonate begins at a depth correlated with the mean annual wetting of the soil and the process proceeds upwards as a layer is deposited. The carbonate layer impedes downward water movement and is impenetrable to most roots, so that vegetation growth is constrained to higher layers of the soil. Hence, there is considerable feedback between plant transpiration in dryland regions, vertical distribution of vegetation and soil moisture, and soil hydraulic conductivity. Distinguishing between transpiration and soil evaporation is thus important in assessing these feedbacks (Newman *et al.* 2006) and in determining the potential impact of vegetation change in dryland

regions on both ecological and hydrological processes (Huxman *et al.* 2005; Scott *et al.* 2006).

Despite the importance of transpiration, relatively few studies have attempted to quantify its contribution to evapotranspiration in the drylands. Kemp *et al.* (1997), Reynolds *et al.* (2000), and Huxman *et al.* (2005) summarize the few available studies and demonstrate a very large range of values.

Reynolds *et al.* (2000) examined those for semi-arid shrublands and found transpiration to total evapotranspiration ratios varying from 7% to 80%. Yepez *et al.* (2003) found a ratio of 85% for a mesquite community in a semi-arid region of the Sonoran Desert during the post-monsoon season. Tree transpiration contributed 70% of the evapotranspiration, the grass understory 15%. Grasses contributed 50% of the total understory evapotranspiration. In the Walnut Gulch watershed, in the transition between the Sonoran and Chihuahuan deserts, Nagler *et al.* (2007) found transpiration to evapotranspiration ratios varying from 75% to 100%, with an average of 87% for a shrub site and 82% for a grassland site.

Empirical studies in desert environments have also produced a wide range of results (Kemp *et al*. 1997), with some studies suggesting that transpiration is important only in regions of cold-season rainfall. Ross (1977) found only a very small contribution of transpiration in the Australian desert, while Lane *et al*. (1984) found that transpiration accounted for 27% of soil moisture loss at a site in the Mojave Desert with only 25% plant cover. Other studies report quite substantial contributions of transpiration. In Great Basin shrub communities, where much of the precipitation falls in winter, Caldwell *et al*. (1977) found that transpiration roughly equaled soil evaporation. Scott *et al*. (2006) reported an overall contribution of 58% in the Chihuahuan Desert during the July–October rainy season. The ratio was only 21% early in the season, but increased to roughly 70% during the three months when the plants are active (August–October).

The range of reported values is large even for individual plant communities. For creosote-dominated communities, the reported ratios of transpiration to total evapotranspiration, based on field studies, are 72% in the Chihuahuan Desert (Schlesinger *et al*. 1987), 7% in the Sonoran Desert (Sammis and Gay 1979), and 27% in the Mojave Desert (Lane *et al*. 1984). In contrast, the modeling assessment of Liu *et al*. (1995) suggested values as high as 80% for creosote in the Sonoran Desert. Kemp *et al*. (1997), using a soil-water model, derived a value of 40% for creosote at the Jornada research site in New Mexico.

The large range of reported values is due to several factors, including variations in the relative abundance of individual plant types and the total percent ground cover. The Kemp *et al*. study found the latter factor to be particularly important, with the dominance of transpiration increasing almost linearly with ground cover. Of five communities studied, the lowest transpiration/evapotranspiration ratio was 40% and corresponded to 30% cover; the highest ratio, 69%, corresponded to 70% cover. That study also found that the vertical distribution of soil moisture and the surface energy budget are very important, with soil texture having only minor influence. The importance of individual plant types is a function of root depth and distribution, as well as water-use efficiency.

Micrometeorological (e.g., Bowen ratio or eddy covariance systems), hydrological (e.g., microlysimeters), and ecophysiological techniques (e.g., sap-flow measurements, stable isotopes) allow for quantification of plant transpiration in the field (Williams *et al*. 2004). Soil water and surface hydrological models have also been used to estimate the contribution of transpiration. The different approaches to assessing transpiration and the fact that most studies cover a brief time span, generally days to weeks (Scott *et al*. 2006), account for some of the range in transpiration/evapotranspiration ratios. To reduce the influence of these factors, Reynolds *et al*. (2000) used a mechanistic model to determine the transpiration to evapotranspiration ratio for the Chihuahuan Desert for a 100-year period. The year-to-year variations were large, with annual ratios for a mixed shrub community ranging from 6% to 60%

as a result of interannual fluctuations in rainfall, corresponding changes in the growth of annuals, and differential water use by various plant types in the wet and dry years. For the C4 grassland in this desert, the model predicted a range from 1% to 58%.

8.5.3 TEMPORAL VARIABILITY: THE IMPACT OF MOISTURE PULSES ON EVAPOTRANSPIRATION

Evapotranspiration and the relative importance of its individual components vary both seasonally and intraseasonally, with the latter variations reflecting the episodic pulses of precipitation. The typical course of hydrologic parameters during and following a rain event in dryland regions is shown in Fig. 8.13. During the rain event, runoff occurs at roughly the same time as the rainfall. Total evapotranspiration generally peaks during the rain event or just afterwards, then decreases rapidly. The decrease is commonly exponential (Hunt *et al*. 2002; Kurc and Small 2004). During the rain event, most of the evapotranspiration is from soil or interception water, with transpiration generally peaking after the rain event and persisting during the interstorm period.

Kurc and Small (2004) examined intraseasonal variations of evapotranspiration and evaporative fraction in semi-arid grassland and shrubland ecosystems in New Mexico (Fig. 8.14). The maximum values of both variables, roughly 4 mm/day and 70%, were observed directly following rain events. Minimum values, roughly 0.5 mm/day and 10%, occurred within a few days of a rain event. The drydown following the event showed the expected exponential form and occurred more quickly in the shrubland. Time constants that roughly represent drying time were 1.4 days for the shrubland and 2.6 days for the grassland.

Because transpiration draws on moisture in the deeper soil layers, a more temporally stable water reservoir than surface water, its daily variations are much smaller than those of soil evaporation or evaporation of canopy interception. It is also the dominant component of total evapotranspiration during the dry season. Soil evaporation is important during the rainy season, especially during and immediately following rain events, but generally becomes negligible during the dry season.

An observational study of a shrub community in the Chihuahuan Desert (Scott *et al*. 2006) illustrates the temporal variations of the two components (Fig. 8.13). At the onset of the rainy season, evapotranspiration is primarily soil evaporation because the dominant shrub community takes about 10 days to respond to soil moisture flux. Within the season, short-duration rain events (less than 2 days) are still dominated by soil evaporation. During interstorm periods soil evaporation is generally negligible, while transpiration peaks. The peak in transpiration is generally several days after the rain event. Transpiration then decays quickly, indicating that optimum moisture conditions are never met. Nevertheless, transpiration is generally on the order of 1 mm/day even at the end of the interstorm period, indicating continued vegetation growth. The magnitude of variability between the storm and interstorm periods is comparable to that

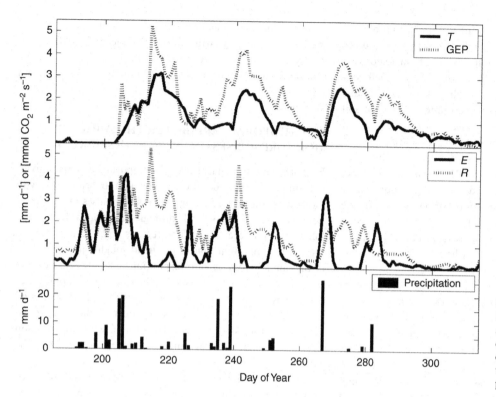

Fig. 8.13 Schematic of the hydrologic processes associated with a rain pulse. E is the evaporation component, T is the transpiration component, GEP the gross ecosystem productivity and R is the respiration (from Scott *et al.* 2006).

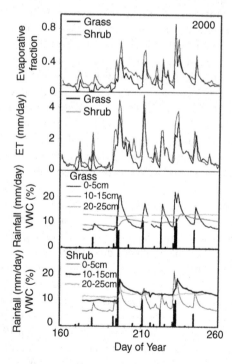

Fig. 8.14 Daily time series of evaporative fraction (midday), evapotranspiration (daily total), precipitation, and volumetric water content in three soil layers (0–5 cm, 10–15 cm, and 20–25 cm) at grassland and shrubland sites in New Mexico (from Kurc and Small 2004).

found by Kurc and Small (2004), with maxima on the order of 4–5 mm/day and minima on the order of 0.5–1 mm/day.

Nicholson *et al.* (1997b) examined the temporal course of soil evaporation and transpiration, applying a water balance

model to data for several HAPEX-Sahel experimental sites in West Africa. The results suggested a one-month lag between peak soil evaporation, which occurs during the month of maximum rainfall, and transpiration (Fig. 8.15). For individual rain events (Fig. 8.16), the lag was generally one or two days, and the model suggests that transpiration was sustained at fairly high levels between rain events (Marengo *et al.* 1996). This contrast with the exponential decrease noted in other studies is probably a result of the relatively high soil moisture content, as the simulation dealt with the late rainy season when soil moisture peaked. This suggests that soil moisture saturation also plays a significant role in determining the relative importance of transpiration.

8.5.4 CONTROLS ON EVAPOTRANSPIRATION IN DRYLANDS

The physical factors controlling evapotranspiration and the influence of vegetation on this process depend to a first approximation on prevailing climate. In humid regions, where surface vegetation cover is nearly complete, transpiration comprises nearly all of the vapor flux to the atmosphere. In arid regions, plants have relatively little influence because most of the precipitation is evaporated regardless of vegetation cover (Wilcox *et al.* 2003). The controls are most complex in semi-arid regions, where vegetation cover is partial and both soil evaporation and transpiration contribute significantly to the total evapotranspiration.

In dryland ecosystems, evapotranspiration is water-limited most of the time, so that water availability is assumed to be the main determinant of evapotranspiration (Noy-Meir 1973; Rodriguez-

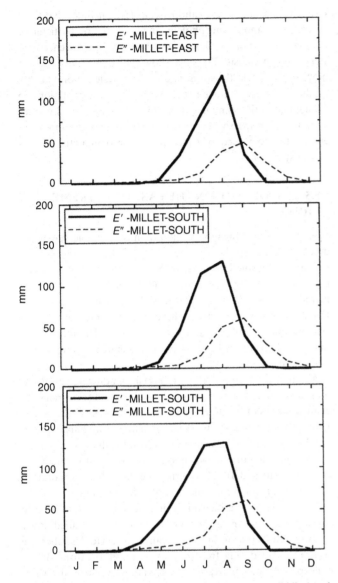

Fig. 8.15 Model estimates (mm/day) of evaporation (E', solid line) and transpiration (E'', dashed line) at three millet sites in the West African Sahel near Niamey, Niger (from Marengo et al. (1996).

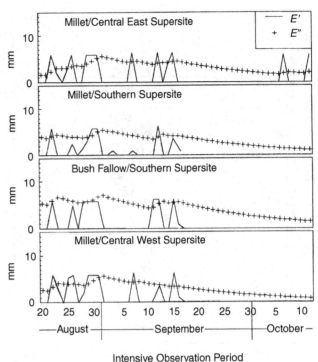

Fig. 8.16 Model estimates (mm/day) of evaporation (E', line) and transpiration (E'', +) in the Sahel during the HAPEX-Sahel intense observation period (IOP) during 1992 (from Nicholson et al. 1997b).

Iturbe 2000). Consequently, evapotranspiration is generally estimated via assumed relationships to soil moisture. However, vegetation and soil characteristics also play a role. Relevant factors include species composition and cover, phenology, stomatal response, root depth and distribution, and soil texture (Kemp et al. 1997). Soil texture does not appear to exert direct control, but instead influences the amount and type of plant cover which, in turn, influence both evaporation and transpiration.

Various forms of the soil moisture/evapotranspiration relationship have been proposed, as discussed in Section 8.3. Generally, a very simple form of the relationship is assumed in studies of soil moisture dynamics, plant productivity, and climate modeling (e.g., D'Odorico et al. 2000; Rodriguez-Iturbe 2000; Fernandez-Illescas et al. 2001; Laio et al. 2001; Ridolfi et al. 2000). The relationship is as follows: evapotranspiration

increases linearly with root zone water content between the wilting point and a soil moisture content that sustains potential evapotranspiration (often on the order of 80% of field capacity). Modeling studies have suggested that the relationship is more complex (Guswa et al. 2002), as roots generally draw moisture from particular layers of the soil.

Considerable debate exists on this point, whether plant transpiration is directly sensitive to the wettest layers (e.g., Waring and Schlesinger 1985) or responds to the average soil moisture content, as assumed by most models of transpiration (Kemp et al. 1997). The first case allows for water to be extracted from one layer even when the rest of the soil is dry. The second implies that water extraction/transpiration slows down when any portion of the soil dries out.

Studies of a grassland site and a shrubland site in central New Mexico (Small and Kurc 2003; Kurc and Small 2004) showed that evapotranspiration is closely tied to surface moisture conditions. For both environments the midday evaporative fraction and daily evapotranspiration correlated well with water content in the first 5 cm of soil, but correlated poorly with water content at greater depths or averaged throughout the entire root zone. This suggests that, in these ecosystems, soil evaporation is a large component of the total evapotranspiration and that predicting ET based on root-zone-averaged soil moisture is inappropriate in semi-arid environments. This may also be true for other semi-arid environments.

Nagler et al. (2007) likewise examined grassland and shrubland sites in the southwestern USA and came to a different

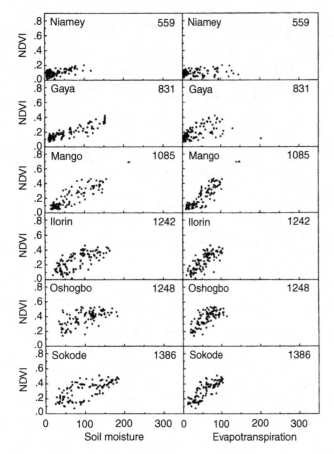

Fig. 8.17 Normalized difference vegetation index (NDVI) versus model-calculated soil moisture and evapotranspiration at six locations along an aridity gradient in West Africa. Mean annual rainfall is indicated in the upper right corner for each location (from Nicholson *et al.* 1996).

conclusion. The strong correlation of evapotranspiration with vegetation index and its lack of correlation with rainfall suggest that transpiration contributed the vast majority of the total vapor flux to the atmosphere. Nicholson *et al.* (1996) similarly used a remotely sensed vegetation index to examine these correlations in the West African Sahel and found that the control switches from rainfall to vegetation at the point along the aridity gradient where water becomes the limiting factor in growth (Fig. 8.17). This suggests a progressive switch from the dominance of soil evaporation to transpiration along the gradient. The transition from water-limited to radiation-limited environments occurs where annual rainfall is on the order of 1000 mm.

The second important debate relates to the importance of ecophysiological versus atmospheric controls on transpiration. Specifically, in the case of well-watered vegetation, do the stomata play a dominant regulatory role, as plant physiologists contend, or is net radiation the primary factor, as meteorologists assume? Meinzer (1993) suggests a compromise viewpoint: evapotranspiration can be predicted if leaf area, plant water status, and atmospheric characteristics are known, but for a given leaf area, different canopy structures may result in contrasting

transpiration rates at the canopy scale. He concludes that three aspects of canopy morphology/architecture are particularly important in drylands: leaf orientation, leaf pubescence, and stem photosynthesis. The data show that these traits become more common, both within and across species, along aridity gradients. These characteristics may represent adaptations to drought conditions and may have direct consequences for transpiration. For sparse canopies, these traits may be important enough that they need to be incorporated into predictive models of transpiration.

8.5.5 EVAPORATION IN THE CASE OF EXTREME ARIDITY

It is generally assumed that evaporation is negligible in extreme deserts and during the extreme aridity of the dry season, barring any unseasonal rain events. Moreover, the most common methods of assessment are not suitable in such cases and special methodologies must be developed (Kobayashi and Nagai 1995). For these reasons, few studies have ever attempted to measure evaporation in these cases of extreme aridity. The recent ones that have been carried out have found significant amounts of evaporation.

Agam (Ninari) and Berliner (2004) found diurnal variations of water content in the uppermost soil layers and corresponding changes in latent heat flux during the dry season in the Negev Desert. A follow-up study (Agam (Ninari) *et al.* 2004) showed that latent heat flux plays a major role in the dissipation of net radiation during the season, amounting to 20% of the net radiation at night and 10–15% during the day. Thus, evaporation can be significant even in the case of extreme aridity.

Menenti (1984) verifies this, using a combination of remote sensing and field measurements to produce a detailed study of evapotranspiration in the Libyan Desert. He describes the concepts of a drying front and an evaporation front. These concepts are particularly applicable in the case where the water table is fairly close to the surface, such as in many playas. In this case, evaporation can be sustained by the groundwater (Malek 2003). However, it is strongly inhibited by salt crusts (Kampf *et al.* 2005). A study of an aquifer in the Algerian/Tunisian Sahara indicated values of evaporation of 315×10^6 m³/acre for the year 1950 and 280×10^6 m³/acre for the 1970 (Menenti 1984). This may explain a feature noted long ago by Starr and Peixoto (1958): the western Sahara appears to be a source of moisture to the atmosphere. Pike (1970) similarly found significant evaporation sustained by groundwater in playas in the Arabian Gulf.

At some depth below the desert surface the soil moisture content rapidly increases as the water table is approached (Fig. 8.18). This is termed the drying front. The diffusivities of water vapor and liquid water both change rapidly with soil moisture content. That for liquid water steadily decreases with decreasing soil moisture, but the diffusivity of the vapor increases with decreasing moisture availability to some maximum value. At very low moisture contents the vapor diffusivity exceeds the

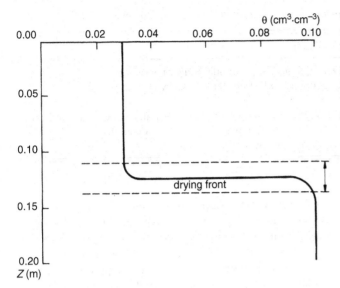

Fig. 8.18 Profile of soil moisture content (θ) versus soil depth (Z, in m) in the Libyan Desert (from Menenti 1984).

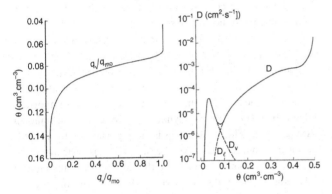

Fig. 8.19 (a) Ratio of vapor flux (q_v) to moisture flux (q_{mo}) as a function of soil moisture (θ). (b) Variation of liquid diffusivity (D_l), vapor diffusivity (D_v), and total diffusivity D of water with soil moisture content (θ) (Menenti 1984).

liquid diffusivity. This occurs at approximately the moisture content θ_e corresponding to that representative of the drying front at some depth z_e. As soil moisture decreases to θ_e, the contribution of vapor flux to moisture flux in the soil steadily increases and the overall amount of evaporation increases. When θ_e is reached, net evaporation attains a maximum, then decreases rapidly with decreasing soil moisture (Fig. 8.19). This maximum is termed the evaporation front.

8.6 ESTIMATION OF REGIONAL EVAPOTRANSPIRATION

The classical approaches to the evaluation of evaporation were developed primarily in the context of agricultural studies and management and climate classification. These traditional approaches are based on the assumption that site-specific meteorological parameters provide reliable input data. This assumption is not true when the goal is regional evaluation. Similarly, biophysicists have generally used the "big leaf" model that simply translates leaf processes to canopy scale. One of the biggest challenges in understanding and quantifying land–atmosphere interactions is extrapolating from these smaller scales to the larger regional scales, a process termed "upscaling." It involves capturing land–atmosphere feedbacks, the effects of land surface heterogeneity on surface fluxes, and the atmosphere boundary layer dynamics that become operative at progressively larger scales (Anderson *et al.* 2003). These challenges are at the heart of the remote sensing of regional-to-continental scale fluxes, such as evapotranspiration, and of representing the land surface in global climate models.

To understand the interactions involved in upscaling, the concept of the soil–plant–atmosphere continuum is useful. This is the pathway of water vapor and other materials (e.g., carbon dioxide) from the soil to the atmosphere, as illustrated in Fig. 8.5. At each point along the way, there is a resistance to passage from one part of the continuum to the next. The value of the resistances depends on the nature of the soil and plants and on environmental conditions. The relative values of the resistances determine the degree of coupling between the surface and the atmosphere.

Several factors render difficult the upscaling along the soil–plant–atmosphere continuum and from local to regional scales. One is the heterogeneity of the surface, which is particularly pronounced in dryland regions. This is well illustrated by the Sahelian studies described in Section 8.5. Because the processes controlling the fluxes between the surface and atmosphere are non-linear, a simple averaging of the surface properties is inadequate. The second factor is the modification of the atmosphere near the surface by the surface itself. Finally, the strength of the interactions between soil, plants, and atmosphere (i.e., along the continuum) is a function of scale, so that extrapolating processes at the leaf scale to the atmosphere scale is not straightforward. In other words, the landscape behaves differently from the individual leaves or plants.

An excellent illustration involves the concept of potential evapotranspiration. As described in Section 8.3, the ratio of actual to potential evapotranspiration is often assumed to be proportional to soil moisture. However, when water availability is limited, as in dryland environments, land–atmosphere feedbacks at the regional scale actually modify the potential evapotranspiration. This is referred to as the Bouchet (1963) hypothesis. As explained by Ramírez *et al.* (2005), the hypothesis deals with the situation in which actual evaporation E_0 falls below E_p for the wet environment with no water limitations. When this happens, energy that would have otherwise been used for evaporation instead contributes to sensible heating and drying of the boundary layer. This essentially increases E_p. When E_0 increases, less energy is available for boundary layer heating and E_p is reduced. Bouchet hypothesized that a complementary relationship exists in which the sum of actual evaporation plus the modified potential evapotranspiration in the water-limited environment is equal to twice the potential

Table 8.3. *Scale dependence of feedback and controls in the soil–plant–atmosphere continuum (from Anderson* et al. *2003).*

Scale	Dimension	Control surface	Feedback
Leaf	a few cm	Leaf boundary layer	Decoupling by leaf boundary-layer resistance
Plant	cm–m	Compound leaf/cluster boundary layer	Decoupling by leaf cluster boundary-layer resistance
Canopy	100 m–km	Surface layer	Decoupling by canopy boundary-layer and aerodynamic resistance
Landscape	several km	Planetary boundary layer	PBL growth entrainment; large eddy circulations
Mesoscale	10–100 km	Mid-troposphere	Organized mesoscale circulations

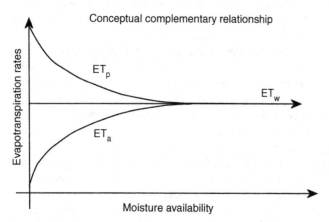

Fig. 8.20 Schematic representation of the complementary relationship between potential (ET_p) and actual (ET_a) evapotranspiration in water-limited environments (ET_w) predicted by Bouchet's hypothesis (from Ramírez *et al.* 2005).

evapotranspiration in the wet environment under the same atmospheric conditions (Fig. 8.20). This idea has been incorporated into regional models of evapotranspiration (Hobbins *et al.* 2004; Kim and Entekhabi 1998) and has been verified through field observations at a broad range of sites in the USA (Ramírez *et al.* 2005).

Anderson *et al.* (2003) provide a very clear explanation of upscaling and the feedbacks and controls at various scales along the continuum, starting at the leaf scale. Table 8.3 illustrates the sequence. The leaf modifies forcing fields in its immediate vicinity. The assemblage of leaves into plants or canopies means that the environment of each leaf is affected by all other leaves. These modifications can affect bulk behavior in such a way as to modify canopy-level states of physical variables, fluxes, and sensitivity to environmental changes (Jacobs and De Bruin 1992; Jones 1998; Gottschalck *et al.* 2001). The spatial scale over which plant behavior is organized and uniform determines the degree of modification and the strength of the thereby imposed feedbacks (e.g., Raupach 1998). Feedback exists between the various scales (Scheffer *et al.* 2005).

There is some debate as to whether stomatal (i.e., leaf) resistance or radiation is most important in determining the passage of water from the leaf to the atmosphere (McNaughton and Jarvis 1991). The relative importance of stomatal versus

atmosphere control depends in part on scale. The strength of the plant–atmosphere feedback depends on how decoupled the leaf surface is from the external atmosphere (Jarvis and McNaughton 1986). The larger the leaf boundary-layer resistance, the more effectively water vapor is trapped near the leaf surface – and the less control stomatal resistance has on transpiration (Anderson *et al.* 2003).

The boundary-layer resistance becomes greater at the plant scale than the leaf scale, progressively decreasing the degree to which the stomata control transpiration. At the canopy scale, aerodynamic resistance in the surface layer controls the passage of water vapor into the free atmosphere. Thus, aerodynamically rough canopies will be more closely coupled with the atmosphere than will be smoother canopies, and they will exert more control over transpiration. Although grasslands are generally smoother than forests, the spatial structure of the vegetation is important and shrublands can be aerodynamically rougher than uniform forest canopies. Thus, patchiness is an important factor in feedback at the landscape scale. At the same time, as scale increases, radiation increasingly dominates the control of transpiration (Albertson *et al.* 2001). At larger scales (e.g., the mesoscale), large surface gradients of heat and moisture, associated with different surface types, exert their influence on the atmosphere by creating mesoscale circulations (Pielke *et al.* 1991).

Both remote sensing and mathematical models of biophysical processes are applied to assessing regional evapotranspiration (e.g., Cleugh *et al.* 2007; Zhang *et al.* 2008; Guerschman *et al.* 2009). For producing a spatial picture, parameters and fluxes are extrapolated to grid-scales of various sizes. For satellite imagery, these grid-points are termed pixels. Climate models are similarly built upon regular spatial grids upon which atmospheric parameters are calculated. A typical climate model has grids with a spatial resolution on the order of one or two degrees of latitude and longitude. Many satellite data sets have similar resolution (see Chapter 1). Clearly, the land surface is not uniform at this scale. Even at the higher resolution of tens of meters associated with some remote sensing products and regional models, the grid boxes contain significant spatial heterogeneity. This is especially true for dryland regions.

A current challenge in modeling and remote sensing deals with subpixel or subgrid-scale heterogeneity. A number of field experiments, such as SALSA (Goodrich *et al.* 2000), HAPEX-Sahel (Goutourbe 1997), and EVA-GRIPS (Mengelkamp *et al.*

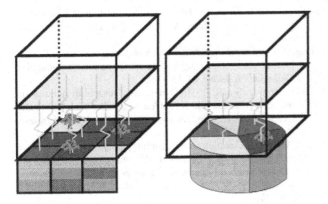

Fig. 8.21 Sketch of (left) mosaic and (right) tile approaches to describing subgrid-scale land-surface heterogeneity (from Mengelkamp *et al.* 2006).

2006) have dealt with this problem. As grid scale increases, so does the probability that a single grid box will contain several vastly different surfaces. In modeling, the central question is how to assign appropriate values to bulk parameters that are associated with strongly heterogeneous pixels (Anderson *et al.* 2003). This problem involves not only a census of the various surfaces within a grid box but also their specific spatial arrangement. In remote sensing, the satellite provides a bulk parameter, which must be "disaggregated" to produce estimates on a smaller scale that are more compatible with ground-based observations (Kustas *et al.* 2003; Anderson *et al.* 2004). In view of the large spatial heterogeneity in dryland regions, these issues are of particular concern in studies related to arid and semi-arid environments.

Numerous approaches have been used to describe subgrid-scale land-surface heterogeneities in climate models. These involve three basic approaches (Mengelkamp 2006). Two calculate fluxes for specific, uniform surface types within a grid cell. In the "tile" approach, the landscape in the grid cell consists of uniform patches, each constituting a certain fractional area of the cell. In contrast, the "mosaic approach" uses a regular subgrid, with each homogeneous surface type comprising a certain number of subgrid cells (Fig. 8.21). The third approach involves utilizing a carefully chosen "effective parameter" to represent the combined influence of all surfaces within a grid cell (Arain *et al.* 1996). Models using subgrid-scale parameterization instead of relying on the dominant vegetation/soil type in a pixel produce greater evaporation estimates by a factor of two or three (Mölders and Raabe 1996; Heinemann and Kerschgens 2005).

REFERENCES

Agam, N. (Ninari), and P. R. Berliner, 2004: Diurnal water content changes in the bare soil of a coastal desert. *Journal of Hydrometeorology*, **5**, 922–933.

Agam, N. (Ninari), P. R. Berliner, and A. Zangvil, 2004: Soil water evaporation during the dry season in an arid zone. *Journal of Geophysical Research*, **109**, doi:10.1029/2004JD004802.

Albertson, J. D., G. G. Katul, and P. Wiberg, 2001: Relative importance of local and regional controls on coupled water, carbon, and energy fluxes. *Advances in Water Resources*, **24**, 1103–1118.

Anderson, M. C., W. P. Kustas, and J. M. Norman, 2003: Upscaling and downscaling: a regional view of the soil–plant–atmosphere continuum. *Agronomy Journal*, **95**, 1408–1423.

Anderson, M. C., J. M. Norman, J. R. Mecikalski, R. D. Torn, W. P. Kustas, and J. B. Basara, 2004: A multi-scale remote sensing model for disaggregating regional fluxes to micrometeorological scales. *Journal of Hydrometeorology*, **5**, 343–363.

Arain, A. M., J. D. Michaud, W. J. Shuttleworth, and A. J. Dolman, 1996: Testing of vegetation parameter aggregation rules applicable to the Biosphere–Atmosphere Transfer Scheme (BATS) at the FIFE site. *Journal of Hydrology*, **177**, 1–22.

Austin, A. T., and Coauthors, 2004: Water pulses and biogeochemical cycles in arid and semiarid ecosystems. *Oecologia*, **141**, 221–235.

Bouchet, R. J., 1963: Evapotranspiration réelle evapotranspiration potentielle, signification climatique. in *General Assembly of Berkeley, Transactions, Vol. 2: Evaporation.* International Association of Scientific Hydrology, Berkeley, CA, pp. 134–142.

Brutsaert, W. H., 1982: *Evaporation into the Atmosphere: Theory, History and Applications.* D. Reidel, Dordrecht, 316 pp.

Burman, R., and L. O. Pochop, 1994: *Evaporation, Evapotranspiration and Climatic Data.* Developments in Atmospheric Science 22, Elsevier, Amsterdam, 278 pp.

Caldwell, M. M., R. S. White, R. T. Moore, and L. B. Camp, 1977: Carbon balance, productivity; water use of cold-winter desert shrub communities dominated by C3 and C4 species. *Oecologia*, **29**, 275–300.

Cleugh, H. A., R. Leuning, Q. Mu, and S. W. Running, 2007: Regional evaporation estimates from flux tower and MODIS satellite data. *Remote Sensing of Environment*, **106**, 285–304.

Dingman, S. L., 1994: *Physical Hydrology.* Macmillan, New York, 575 pp.

D'Odorico, P., L. Ridolfi, A. Porporato, and I. Rodriguez-Iturbe, 2000: Preferential states of seasonal soil moisture: the impact of climate fluctuations. *Water Resources Research*, **36**, 2209–2219.

Dugas, W. A., R. A. Hicks, and R. P. Gibbens, 1996: Structure and function of C3 and C4 Chihuahuan desert plant communities: energy balance components. *Journal of Arid Environments*, **34**, 63–79.

Dunne, T., and L. B. Leopold : 1978: *Water in Environmental Planning.* Freeman, New York, 818 pp.

Federer, C. A., C. Vörösmarty, and B. Fekete, 1996: Intercomparison of methods for calculating potential evaporation in regional and global water balance models. *Water Resources Research*, **32**, 2315–2321.

Fernandez-Illescas, C., A. Porporato, F. Laio, and I. Rodriguez-Iturbe, 2001: The ecohydrological role of soil texture in a water-limited system. *Water Resources Research*, **37**, 2863–2872.

Gash, J., J. Wallace, C. Lloyd, A. Dolman, M. Sivakumar, and C. Renard, 1991: Measurements of evaporation from fallow Sahelian savannah at the start of the dry season. *Quarterly Journal of the Royal Meteorological Society*, **117**, 749–760.

Giambelluca, T. W., and Coauthors, 2009: Evapotranspiration and energy balance of Brazilian savannas with contrasting tree density. *Agricultural and Forest Meteorology*, **149**, 1365–1376.

Goodrich, D., A. Chehbouni, B. Goff, A. Bégué, D. Luquet, Y. Nouvellon, and D. Lo Seen, 2000: Preface paper to the semi-arid land–surface–atmosphere (SALSA) program. *Agricultural and Forest Meteorology*, **105**, 3–20.

Gottschalck, J. C., R. R. Gillies, and T. N. Carlson, 2001: The simulation of canopy transpiration under doubled CO2: the evidence and impact of feedbacks on transpiration in two 1-D soil–vegetation–atmosphere transfer models. *Agricultural and Forest Meteorology*, **106**, 1–21.

Goutourbe, J. P. (ed.), 1997: *HAPEX-Sahel*. Elsevier, 1088 pp.

Guerschman, J. P., and Coauthors, 2009: Scaling of potential evapotranspiration with MODIS data reproduces flux observations and catchment water balance observations across Australia. *Journal of Hydrology*, **369**, 107–119.

Guswa, A. H., M. A. Celia, and I. Rodriguez-Iturbe, 2002: Models of soil moisture dynamics in ecohydrology: a comparative study. *Water Resources Research*, **38**, doi:10.1029/2001WR000826.

Heinemann, G., and M. Kerschgens, 2005: Comparison of methods for area-averaging surface energy fluxes over heterogeneous land surfaces using high-resolution non-hydrostatic simulations. *International Journal of Climatology*, **25**, 379–403.

Hobbins, M. T., J. A. Ramírez, and T. C. Brown, 2004: Trends in pan evaporation and actual evaporation across the conterminous U.S.: paradoxical or complementary? *Geophysical Research Letters*, **31**, L13503, doi:10.1029/2004GL019846.

Humes, K. S., W. P. Kustas, M. S. Moran, W. D. Nichols, and M. A. Weltz, 1994: Variability of emissivity and surface temperature over a sparsely vegetated surface. *Water Resources Research*, **30**, 1299–1310.

Hunt, J., F. Kelliher, T. McSeveny, and J. Byers, 2002: Evapotranspiration and carbon dioxide exchange between the atmosphere and a tussock grassland during a summer drought. *Agricultural and Forest Meteorology*, **111**, 65–82.

Huxman, T., B. Wilcox, D. Breshears, R. Scott, K. Snyder, E. Small, K. Hultine, W. Pockman, and R. Jackson, 2005: Ecohydrological implications of woody plant encroachment. *Ecology*, **86**, 308–319.

Jackson, I. J., 1989: *Climate, Water and Agriculture in the Tropics*. Longman, New York, 377 pp.

Jacobs, C. M. J., and H. A. R. De Bruin, 1992: The sensitivity of regional transpiration to land-surface characteristics: significance of feedback. *Journal of Climate*, **5**, 683–698.

Jarvis, P. G., and K. G. McNaughton, 1986: Stomatal control of transpiration: scaling up from leaf to region. *Advances in Ecological Research*, **15**, 1–49.

Jensen, M. E., R. D. Burman, and R. G. Allen, 1990: Evapotranspiration ad irrigation water requirements. *ASCE Manuals and Reports on Engineering Practice No. 70*, 332 pp.

Jones, H. G., 1998: Stomatal control of photosynthesis and transpiration. *Journal of Experimental Botany*, **49**, 387–398.

Kabat, P., A. J. Dolman, and J. A. Elbers, 1997: Evaporation, sensible heat and canopy conductance of fallow savannah and patterned woodland in the Sahel. *Journal of Hydrology*, **188–189**, 494–515.

Kampf, S. K., S. W. Tyler, C. A. Ortiz, J. F. Muñoz, and P. L. Adkins, 2005: Evaporation and land surface energy budget at the Salar de Atacama, Northern Chile. *Journal of Hydrology*, **310**, 236–252.

Katata, G., H. Nagai, H. Ueda, N. Agam, and P. R. Berliner, 2007: Development of a land surface model including evaporation and adsorption processes in the soil for the land–air exchange in arid regions. *Journal of Hydrometeorology*, **8**, 1307–1324.

Kemp, P. R., J. F. Reynolds, Y. Pachepsky, and J. L. Chen, 1997: A comparative modeling study of soil water dynamics in a desert ecosystem. *Water Resources Research*, **33**, 73–90.

Kim, C. P., and D. Entekhabi, 1998: Feedbacks in the land-surface and mixed-layer energy budgets. *Boundary-Layer Meteorology*, **88**, 1–21.

Kobayashi, T., and H. Nagai, 1995: Measuring the evaporation from a sand surface at the Heife Desert station by the dry surface-layer (DSL) method. *Journal of the Meteorological Society of Japan*, **73**, 937–945.

Kondo, J., and N. Saigusa, 1990: A parameterization of evaporation from bare soil surfaces. *Journal of Applied Meteorology*, **29**, 385–389.

Kremer, G., and S. W. Running, 1996: Simulating seasonal soil water balance in contrasting semi-arid vegetation communities. *Ecological Modelling*, **84**, 151–162.

Kurc, S., and E. Small, 2004: Dynamics of evapotranspiration in semiarid grassland and shrubland ecosystems during the summer monsoon season, central New Mexico. *Water Resources Research*, **40**, WO9305.

Kustas, W. P., and Coauthors, 1991: An interdisciplinary field study of the energy and water fluxes in the atmosphere-biosphere system over semiarid rangelands: description and some preliminary results. *Bulletin of the American Meteorological Society*, **72**, 1683–1706.

Kustas, W. P., M. C. Anderson, J. N. Norman, and A. N. French, 2003: Estimating subpixel surface temperatures and energy fluxes from the vegetation index–radiometric temperature relationship. *Remote Sensing of Environment*, **85**, 429–440.

Laio, F., A. Porporato, L. Ridolfi, and I. Rodriguez-Iturbe, 2001: Plants in water-controlled ecosystems: active role in hydrologic processes and response to water stress. II. Probabilistic soil moisture dynamics. *Advances in Water Resources*, **24**, 707–724.

Lane, L. J., E. M. Romney, and T. E. Hakonson, 1984: Water balance calculations and net production of perennial vegetation in the northern Mojave Desert. *Journal of Range Management*, **37**, 12–18.

Liu, B. L., F. Phillips, S. Hoines, A. R. Campbell, and P. Sharma, 1995: Water movement in a desert soil traced by hydrogen and oxygen isotopes, chloride, and chlorine-36, southern Arizona. *Journal of Hydrology*, **168**, 92–110.

Loik, M., D. Breshears, W. Lauenroth, and K. Belnap, 2004: A multiscale perspective of water pulses in dryland ecosystems: climatology and ecohydrology in the western USA. *Oecologia*, **141**, 181–269.

Lowry, W. P., and P. R. Lowry II, 1989: *Fundamentals of Biometeorology: Interactions of Organisms and the Atmosphere. Vol. I. The Physical Environment*. Peavine, McMinnville, OR, 310 pp.

Mahfouf, J. F., and J. Noilhan, 1991: Comparative study of various formulations of evaporation from bare soil using in situ data. *Journal of Applied Meteorology*, **30**, 1354–1365.

Malek, E., 2003: Microclimate of a desert playa: evaluation of annual radiation, energy, and water budgets components. *International Journal of Climatology*, **23**, 333–345.

Malek, E., G. E. Bingham, and G. McCurdy, 1997: Annual mesoscale study of water balance in a Great Basin heterogeneous desert valley. *Journal of Hydrology*, **191**, 223–244.

Marengo, J. A., S. E. Nicholson, A. R. Lare, B. A. Monteny, and S. Galle, 1996: Application of evapoclimatonomy to monthly surface water balance calculations at the HAPEX-Sahel supersites. *Journal of Applied Meteorology*, **35**, 562–573.

McNaughton, K. G., and P. G. Jarvis, 1991: Effects of spatial scale on stomatal control of transpiration. *Agricultural and Forest Meteorology*, **54**, 269–301.

Meinzer, F. C., 1993: Stomatal control of transpiration. *Trends in Ecology and Evolution*, **8**, 289–294.

Menenti, M., 1984: *Physical Aspects and Determination of Evaporation in Deserts Applying Remote Sensing Techniques*. Institute for Land and Water Management, Wageningen, The Netherlands, 202 pp.

Mengelkamp, H.-T., and Co-authers 2006: Evaporation over a heterogeneous land surface: the EVA-GRIPS Project. *Bulletin of the American Meteorological Society*, **87**, 775–786.

Michaud, J., 1997: Participant opinion survey: how satisfied are we with hydrologic data? In *Land Surface Processes in Hydrology: Trials and Tribulations of Modeling and Measuring* (S. Sorooshian, H. V. Gupta, and J. C. Rodda, eds.), Springer, Berlin, 497 pp.

Milly, P. C. D., 1991: A refinement of the combination equations for evaporation. *Surveys in Geophysics*, **12**, 145–154.

Mölders, N., and A. Raabe, 1996: Numerical investigations on the influence of subgrid-scale surface heterogeneity on evapotranspiration and cloud processes. *Journal of Applied Meteorology*, **35**, 782–795.

Monteith, J. L., 1963: Gas exchange in plant communities. In (L. T. Evans, ed.) *Environmental Control of Plant Growth*. Academic Press, New York, pp. 95–111.

Monteith, J. L., 1965: Evaporation and the environment. *Symposia of the Society for Experimental Biology*, **19**, 205–234.

Moran, M. S., D. C. Goodrich, and W. P. Kustas, 1994: Integration of remote sensing and hydrologic modeling through multidisciplinary semi-arid campaigns: Monsoon '90, Walnut Gulch '92 and SALSA-MEX. In *Proceedings of the Sixth International Symposium: Physical Measurements and Signatures in Remote Sensing*, Val d'Isere, France, 17–21 January.

Nagler, P. L., E. P. Glenn, H. Kim, W. Emmerich, R. L. Scott, T. E. Huxman, and A. R. Huete, 2007: Relationship between evapotranspiration and precipitation pulses in a semiarid rangeland estimated by moisture flux towers and MODIS vegetation indices. *Journal of Arid Environments*, **70**, 443–462.

Newman, B. D., and Coauthors, 2006: Ecohydrology of water-limited environments: a scientific vision. *Water Resources Research*, **42**, W06302, doi:10.1029/2005WR004141.

Nicholson, S. E., A. R. Lare, J. A Marengo, and P. Santos, 1996: A revised version of Lettau's evapoclimatonomy model. *Journal of Applied Meteorology*, **35**, 549–561.

Nicholson, S. E., J. Kim, M. B. Ba, and A. R. Lare, 1997a: The mean surface water balance over Africa and its interannual variability. *Journal of Climate*, **10**, 2981–3002.

Nicholson, S. E., J. A. Marengo, J. Kim, A. R. Lare, S. Galle, and Y. H. Kerr, 1997b: A daily resolution evapoclimatonomy model applied to surface water balance calculations at the HAPEX-Sahel supersites. *Journal of Hydrology*, **188–189**, 946–964.

Noy-Meir, I., 1973: Desert ecosystems: environment and producers. *Annual Review of Ecology, Evolution and Systematics*, **4**, 25–51.

Oke, T. R., 1987: *Boundary Layer Climates*, 2nd edn. Halsted, New York, 435 pp.

Penman, H. L., 1948: Natural evaporation from open water, bare soil, and grass. *Proceedings of the Royal Society of London, Series A*, **193**, 120–145.

Penman, H. L., 1956: Evaporation: an introductory survey. *Netherlands Journal of Agricultural Science*, **4**, 7–29.

Pielke, R., Sr, 2001: Influence of the spatial distribution of vegetation and soils on the prediction of cumulus convective rainfall. *Reviews of Geophysics*, **39**, 151–177.

Pielke, R. A., G. Dalu, J. S. Snook, T. J. Lee, and T. G. F. Kittel, 1991: Nonlinear influence of mesoscale land use on weather and climate. *Journal of Climate*, **4**, 1053–1069.

Pierce, L. T., 1958: Estimating seasonal and short-term fluctuations in evapotranspiration from meadow crops. *Bulletin of the American Meteorological Society*, **39**, 73–78.

Pike, J. G., 1970: Evaporation of groundwater from coastal playa (sabkhah) in the Arabian Gulf. *Journal of Hydrology*, **11**, 79–88.

Porporato, A., P. D'Odorico, F. Laio, and I. Rodriguez-Iturbe, 2003: Hydrologic controls on soil carbon and nitrogen cycles. I. Modeling scheme. *Advances in Water Resources*, **26**, 45–58.

Priestley, C. H. B., and R. J. Taylor, 1972: On the assessment of surface heat flux and evaporation using large-scale parameters. *Monthly Weather Review*, **100**, 81–92.

Ramírez, J. A., M. T. Hobbins, and T. C. Brown, 2005: Observational evidence of the complementary relationship in regional evaporation lends strong support for Bouchet's hypothesis. *Geophysical Research Letters*, **32**, L15401, doi:10.1029/2005GL023549.

Raupach, M. R., 1998: Influences of local feedbacks on land–air exchanges of energy and carbon. *Global Change Biology*, **4**, 477–494.

Raupach, M. R., 2001: Combination theory and equilibrium evaporation. *Quarterly Journal of the Royal Meteorological Society*, **127**, 1149–1181.

Reynolds, J., P. Kemp, and J. Tenhunen, 2000: Effects of long-term rainfall variability on evapotranspiration and soil water distribution in the Chihuahuan desert: a modeling analysis. *Plant Ecology*, **150**, 145–159.

Ridolfi, L., P. D'Odorico, A. Porporato, and I. Rodriguez-Iturbe, 2000: Impact of climate variability on the vegetation water stress. *Journal of Geophysical Research*, **105**, 18013–18025.

Ritchie, J. C., and Coauthors, 1998: JORNEX: A remote sensing campaign to study plant community response to hydrologic fluxes in desert grasslands. *Rangeland Management and Water Resources* (D. F Potts, ed.), American Water Resources Association, Middleburg, VA, pp. 65–74.

Rodriguez-Iturbe, I., 2000: Ecohydrology: A hydrologic perspective of climate–soil–vegetation dynamics. *Water Resources Research*, **36**, 3–9.

Rosenberg, N. J., 1974: *Microclimate: The Biological Environment*. John Wiley and Sons, New York, 315 pp.

Ross, M. A., 1977: Concurrent changes in plant weight and soil water regimes in herbaceous communities in Central Australia. *Australian Journal of Ecology*, **2**, 257–268.

Salvucci, G. D., 1997: Soil moisture independent estimation of stage-two evaporation from potential evaporation and albedo or surface temperature. *Water Resources Research*, **33**, 111–122.

Sammis, T. W., and L. Y. Gay, 1979: Evapotranspiration from an arid zone plant community. *Journal of Arid Environments*, **2**, 313–321.

Scheffer, M., M. Holgren, V. Brovkin, and M. Claussen, 2005: Synergy between small- and large-scale feedbacks of vegetation on the water cycle. *Global Change Biology*, **11**, 1003–1012.

Schlesinger, W. H., P. J. Fonteyn, and G. M. Marion, 1987: Soil moisture content and plant transpiration in the Chihuahuan Desert of New Mexico. *Journal of Arid Environments*, **12**, 119–126.

Schmugge, T. J., and J. André, eds., 1991: *Land Surface Evaporation*. Springer, New York, 424 pp.

Schwinning, S., and O. E. Sala, 2004: Hierarchical organization of resource pulse responses in arid and semiarid ecosystems. *Oecologia*, **141**, 221–220.

Scott, R., T. Huyman, W. Cable, and W. Emmerich, 2006: Partitioning of evapotranspiration and its relation to carbon dioxide exchange in a Chihuahuan Desert shrubland. *Hydrological Processes*, **20**, 3227–3243.

Sellers, W. D., 1965: *Physical Climatology*, University of Chicago Press, Chicago.

Sellers, P. J., F. G. Hall, G. Asrar, D. E. Strebel, and R. E. Murphy, 1988: The first ISLSCP field experiment (FIFE). *Bulletin of the American Meteorological Society*, **69**, 22–27.

Shuttleworth, W. J., 1991: Evaporation models in hydrology. In *Land Surface Evaporation* (T.J. Schmugge and J. André, eds.), Springer, New York, pp. 93–120.

Shuttleworth, W. J., and J. S. Wallace, 1985: Evaporation from sparse crops-an energy combination theory. *Quarterly Journal of the Royal Meteorological Society*, **111**, 839–855.

Small, E., and S. Kurc, 2003: Tight coupling between soil moisture and the surface radiation budget in semiarid environments: implications for land–atmosphere interactions. *Water Resources Research*, **39**, WR001297.

Stannard, D., and Coauthors, 1994: Interpretation of surface flux measurements in heterogeneous terrain during the Monsoon '90 experiment. *Water Resources Research*, **30**, 1227–1229.

Starr, V. P., and J. P. Peixoto, 1958: On the global balance of water vapor and the hydrology of deserts. *Tellus*, **10**, 189–194.

Taylor, C. M., 2000: The influence of antecedent rainfall on Sahelian surface evaporation. *Hydrological. Processes*, **14**, 1245–1259.

Thornthwaite, C. W., 1948: An approach towards a rational classification of climate. *Geographical Review*, **38**, 55–94.

Tuzet, A., J.-F. Castell, A. Perrier, and O. Zurfluh, 1997: Flux heterogeneity and evapotranspiration partitioning in a sparse canopy: the fallow savanna. *Journal of Hydrology*, **188–189**, 482–493.

Unland, H. E., P. R. Houser, W. J. Shuttleworth, and Z.-L. Yang, 1996: Surface flux measurement and modeling at a semi-arid Sonoran desert site. *Agricultural and Forest Meteorology*, **82**, 119–152.

Veihmeyer, F. J., 1964: Evapotranspiration. In *Handbook of Applied Hydrology* (V. T. Chow, ed.), McGraw-Hill, New York.

Verhoef, A., Van den Hurk, B. J. J. M, Jacobs, A. F. G. and Heusinkveld, B. G., 1996: Thermal soil properties for a vineyard (EFEDA-I) and a savanna (HAPEX-Sahel) site. *Agricultural and Forest Meteorology*, **78**, 1–18.

Waring, R. H., and W. H. Schlesinger, 1985: *Forest Ecosystems*, Academic Press, San Diego, CA, 340 pp.

Wilcox, B., D. Breshears, and C. Allen, 2003: Ecohydrology of a resource-conserving semiarid woodland: effects of scale and disturbance. *Ecological Monographs*, **73**, 223–239.

Williams, D. G., and Coauthors, 2004: Evapotranspiration components determined by stable isotope, sap flow and eddy covariance techniques. *Agricultural and Forest Meteorology*, **123**, 79–96.

Willmott, C. J., C. N. Rowe and Y. Mintz, 1985: Climatology of the terrestrial seasonal water cycle. *Journal of Climatology*, **5**, 589–606.

Ye, Z., and R. A. Pielke, 1993: Atmospheric parameterization of evaporation from non-plant-covered surfaces. *Journal of Applied Meteorology*, **32**, 1248–1258.

Yepez, E., D. Williams, R. Scott, and G. Lin, 2003: Partitioning overstory and understory evapotranspiration in a semiarid savanna woodland from the isotopic composition of water vapor. *Agricultural and Forest Meteorology*, **119**, 53–68.

Yin, X., and S. E. Nicholson, 1998: The water balance of Lake Victoria. *Hydrological Sciences Journal*, **43**, 789–811.

Yin, X., and S. E. Nicholson, and M. B. Ba, 2000: On the diurnal cycle of cloudiness over Lake Victoria and its influence on evaporation from the lake. *Hydrological Sciences Journal*, **45**, 407–424.

Zhang, Y. A., F. H. S. Chiew, L. Zhang, R. Leuning, and H. A. Cleugh, 2008: Estimating catchment evaporation and runoff using MODIS leaf area index and the Penman–Monteith equation. *Water Resources Research*, **44**, W1020.

Part III The climatic environment of drylands

9 Defining aridity: the classification and character of dryland climates

9.1 INTRODUCTION

In this book the term dryland is used to designate those regions that have a pronounced dry season during at least part of the year. This encompasses vastly diverse regions that, in the literature, are variously termed arid, extremely arid, hyper-arid, desert, semi-desert, semi-arid, semi-dry, and subhumid. The essence of the problem of classification is ultimately the definition – and quantification – of aridity. There are at least as many definitions as there are deserts. Most definitions of aridity relate to climate, but people who are interested in vegetation, landforms, soils, or hydrology are likely to have quite different concepts of aridity and drylands. The best aridity index or climate classification scheme is that which is most appropriate for the task at hand, be it understanding vegetation growth, modeling climates, or assessing agricultural potential or human comfort. In this chapter, an overview of the most common aridity indices and climate classification schemes is provided. The ways in which these various systems define and distinguish the various dryland climates are emphasized.

9.2 DEFINITION AND INDICES OF ARIDITY

In his classic book Petrov (1976) equates aridity with a deficiency of rainy days and ground moisture. Meigs (1953), whose classification of arid lands is the one most widely used today, writes that "many distinctive environmental traits distinguish arid lands, but one is essential – lack of precipitation." Nevertheless, rainfall alone is clearly insufficient for defining climatic boundaries. In Australia, for example, the generally accepted limit for an arid zone is 250 mm, but some areas of the tropical north-west are clearly arid, although mean annual rainfall exceeds 500 mm (Mabbutt 1979). For the Sahara, Dubief (1959) suggests that the 50 mm and 100 mm isohyets, respectively, be used to delineate its extra-tropical northern border and its tropical southern border.

The inadequacy of mean rainfall as a delineating criterion is further underscored by a comparison of four locations: the central Sahel, the northeastern edge of the Kalahari, the central Great Plains of the USA, and central Europe (Fig. 9.1). In each case, mean annual rainfall is on the order of 500–600 mm, but photos of these environments show that the ground cover ranges from grassland and shrubland to mid-latitude forest.

The fundamental commonality in most climatic definitions of aridity is moisture availability, which reflects the balance between precipitation and evapotranspiration. Thornthwaite (1931) was probably the first to quantify aridity based on water balance. His initial work utilized four criteria to distinguish degrees of aridity: precipitation effectiveness (based on precipitation and evaporation); its seasonal variation; temperature efficiency; and its concentration in the summer period. Thornthwaite (1948) modified the first criterion to develop a "moisture index" and focused more directly on water balance. This index I_m is defined as

$$I_m = [(S - 0.6d)/e] \times 100 \tag{9.1}$$

where S is the sum of monthly surpluses of precipitation over potential evapotranspiration, d is the sum of monthly water deficits, and e is annual evapotranspiration estimated from mean monthly temperature with an adjustment for the seasonality of rainfall and soil moisture storage. The water deficit is calculated as the difference between potential evapotranspiration and rainfall plus available soil moisture. Thornthwaite and Mather (1962) published a revised index I_m', defined as

$$I_m' = [(S - d)/e] \times 100 \tag{9.2}$$

and this is now the more commonly used of the two indices.

Many other approaches to quantifying aridity have been proposed. Stadler (2005) lists 41 different indices of aridity. One of the better known, that of Budyko (1986) is based on radiation balance. Budyko defined what he termed a "radiative index of

(a)

(b)

(c)

(d)

Fig. 9.1 Vegetation in four diverse locations, all of which receive roughly 500 mm of precipitation per year: (a) the central Sahel near Niamey, (b) the northeastern edge of the Kalahari north of Gaborone, (c) the central Great Plains of the USA in eastern South Dakota, and (d) central Europe, in the forest northwest of Prague.

dryness" or the dryness ratio (see Chapter 7). This is the ratio between annual net radiation and the latent heat of condensation times mean annual precipitation. Conceptually, this is the ratio between the energy available and the energy required to evaporate the mean annual rainfall.

9.3 CLIMATE CLASSIFICATION

Climate classification has evolved considerably from empirical systems that were essentially vegetation classes interpreted in climatic terms. More recent systems are based on the water and radiation balance or on the meteorological causes of climate variation. In this section, some of the most common global climate classification schemes are summarized and compared. For more complete reviews of other global and regional classifications and indices of aridity, see Jackson (1989), Strahler and Strahler (1987), and Barry and Chorley (2004). A particularly comprehensive review is provided by Oliver and Hidore (2002).

The classic scheme is the one developed by Koeppen (1918) and later revised by his students Geiger and Pohl (1954). Koeppen attempted to assign empirical climate limits to recognized vegetation classifications. In this scheme, dry climates are those in which evaporation exceeds precipitation on average throughout the year. There is no water surplus, and therefore there are no permanent streams. For the climates characterized as "dry" in the Koeppen–Geiger scheme, there are two degrees of aridity: "steppe" and "desert." Hot and cold deserts and steppes are also distinguished. The boundaries of the dry climates are determined by a combination of temperature and precipitation, with the limits depending on the seasonality of the precipitation. This is illustrated in Fig. 9.2.

This empirical approach was followed by many others, often in the context of generalizing the climatic limits of ecosystems. Ripley (1992), for example, showed that grasslands and savannas tend to lie where mean annual rainfall ranges from 50 to 150 mm, but the limits fall considerably lower in cooler locations (see Fig. 3.2). Another example is the model of Box (1981), which predicts the distribution of vegetation types from eight climatic variables. Since the time of that study, numerous other models have been developed to predict vegetation distribution from climate variables. Even the quantitative approach of Thornthwaite entails a certain amount of empiricism in deriving its boundaries for climatic types. In his 1948 paper, locations where I_m (Eq. 9.1) fell between −20 and −40 are designed as semi-arid, and those with I_m below −40 as arid. For the revised index (Eq. 9.2), the values of −67, −33, and 0 separate arid, semi-arid, dry subhumid, and wet subhumid climates.

The weakness of this approach, besides the empiricism, is that there is as much disagreement about vegetation classification as there is about climate. Stark examples of this are the controversies concerning the definitions of grasslands (Coupland 1992) and savannas (Adams 1996). This is illustrated in Fig. 9.3, which shows the limits of the Mojave, Chihuahuan, and

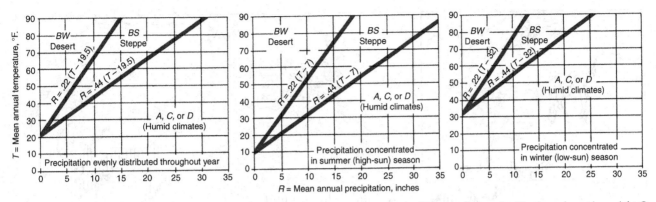

Fig. 9.2 Temperature and precipitation limits of desert, steppe and humid climates in the Koeppen–Geiger classification scheme (copyright © A. N. Strahler, 1987). The limits are different for areas of year-round, summer, and winter precipitation.

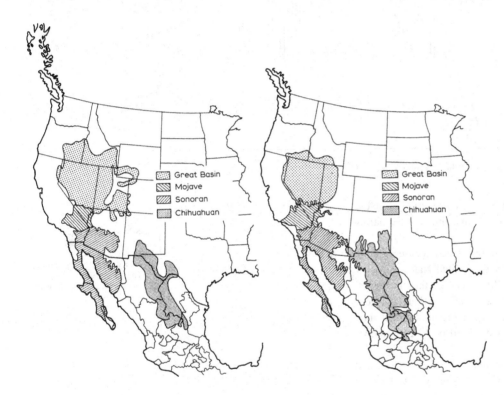

Fig. 9.3 Disparate boundaries of the Mojave, Chihuahuan, and Sonoran deserts (from MacMahon and Wagner 1985); (left) as defined by Shreve (1942); (right) as defined by MacMahon 1979.

Sonoran deserts according to the classifications of two different authors. The root of the problem, of course, is that neither climate nor vegetation falls into discrete classes, but rather they exist along a continuum in which boundaries are actually transition zones. Nevertheless, for many problems and applications, some degree of classification or distinction is meaningful.

Flohn (1957) proposed an alternative "genetic" approach (from "genesis"). It is based on such causative factors as the air-mass source regions, the prevailing movements of air masses, and the synoptic and general circulation features that produce the seasonal cycle of weather. Trewartha and Horn (1980) used a similar approach, emphasizing climate in relationship to general circulation features. This approach is consistent with the discussions in this book, which emphasize the dynamic causes of regional patterns of aridity and how

they relate to the variability of the precipitation regime in the drylands.

Strahler (1978) followed Thornthwaite's water balance approach and classified world climates into seven precipitation regions; one of which is the tropical deserts and one of which is the mid-latitude deserts and steppes. Strahler's scheme, depicted in Fig. 9.4, establishes climatic limits for the various zones, based on the calculated soil moisture (Strahler and Strahler 1987). Dry climates are defined as those in which the total annual soil-water deficit is at least 15 cm and in which there is no water surplus. Desert subtypes are climates with soil moisture storage not exceeding 2 cm in any month. Semi-desert climates are those with soil moisture exceeding 2 cm in at least one month, but with fewer than two months when it exceeds 6 cm. The remainder of the dry locations are classified as semi-arid or steppe.

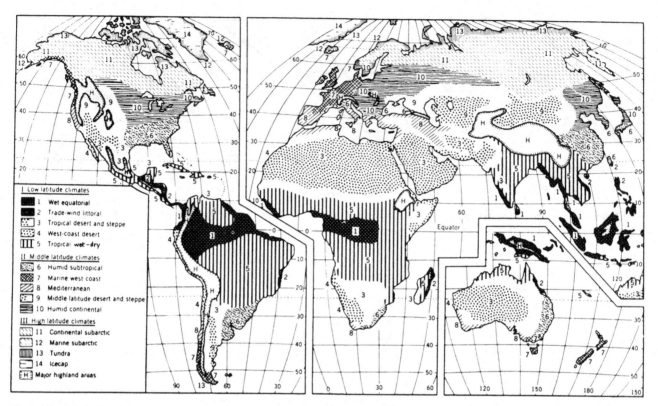

Fig. 9.4 Strahler's climate classification (copyright © A. N. Strahler, 1978).

This system has the advantage of directly considering available water in the form of soil moisture, but the climatic boundaries are somewhat arbitrary.

The classification scheme of Meigs (1953) focuses specifically on aridity. At the request of the United Nations, Meigs expanded on Thornthwaite's (1948) system, adding a class of "extremely arid." This was defined as locations where I_m is less than -57, with no seasonal preference in the rainfall distribution and with at least one recorded period of 12 consecutive months without precipitation.

The dryness ratio is also well suited for classifying arid climates. Budyko's (1986) formulation conveniently distinguishes the drylands as those environments in which the available energy exceeds that needed to return all of the precipitation to the atmosphere via evaporation. He indicated that desert climates are those in which the dryness ratio exceeds 3 and that a ratio of 1 marks the border of semi-arid climates (Fig. 3.3). Semi-deserts are regions where the ratio ranges from 2 to 3, and steppes or savannas have dryness ratios between 1 and 2. Ripley (1992) shows that the dryness ratio, combined with temperature or latitude information, can even delineate various types of grasslands (see Fig. 9.5).

Thus Budyko's dryness ratio has many merits but also some shortcomings. Two important advantages are its ability to depict the degree of aridity and aridity gradients. Another is that this ratio facilitates the calculation of related water balance indices that, in turn, permit the quantification of evaporation and runoff (Hare 1977, 1980; Nemec and Rodier 1979).

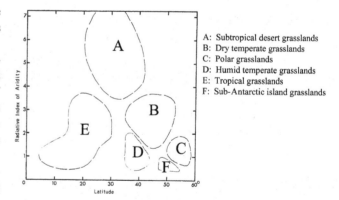

Fig. 9.5 Cluster diagram of world grassland sites, in terms of Budyko's (1986) radiative index of dryness and latitude (i.e., moisture availability decreasing upward and temperature decreasing from left to right) (from Ripley 1992). The enclosed areas labeled with letters represent moisture–climate groups, but strongly resemble the grassland types classified by Walter (1985).

Budyko's (1986) calculations of runoff ratio as a function of net radiation and dryness ratio are shown in Fig. 9.6. On the negative side, the delineating values of the index are somewhat arbitrary and may not be universally applicable. Budyko's study was weighted toward central Europe. Hare (1977) pointed out that Budyko's scheme would classify as desert many regions that have significant perennial vegetation cover. He suggested that a value of 7 might be a more appropriate lower limit for desert climates. Also, over Africa, quite different limits are apparent

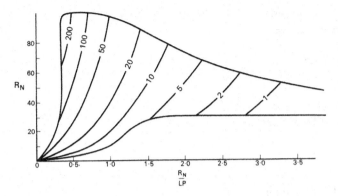

Fig. 9.6 Runoff (cm) as a function of net radiation and dryness ratio (from Budyko 1986.)

(see Chapter 7). At least part of the discrepancy lies in the determination of net radiation, with traditional methods of estimation often producing values that are very different from those derived from satellites.

The dryland boundaries, as established by the various classification schemes, show good agreement as to the location and extent of dryland regions, but tend to show some disagreement on the subclasses of aridity. The advantage of the dryness ratio (Fig. 7.2) is readily apparent because it uniquely depicts two important characteristics of the drylands: the degree of aridity and the spatial gradients. The degree of aridity helps to distinguish and characterize the various dryland regions, while the spatial gradient gives some indication of the long-term variability of the precipitation regime.

The relative agreement of the most common systems at the continental and global scale indicates that any of these systems are probably appropriate for studies at this scale. At the regional scale, it becomes important to look at more complex climatic factors. This is shown in a study by Ellery *et al.* (1991) that examined temperatures on days when soil moisture was adequate for plant growth and on days when it was not. The delineation of South African biomes was strongly influenced by both, and by the seasonality of rainfall.

9.4 LIMITATIONS OF CLIMATE CLASSIFICATION SCHEMES IN DRYLAND REGIONS

The requisite characteristics of any classification scheme for the dryland climates include a sound physical basis, a minimum of empiricism, quantification and subclassification of aridity, and accessibility of data. The commonly used classifications of Meigs and Budyko best satisfy these requirements. There are inherent shortcomings in all of the classification schemes, but most are imposed by the difficulties associated with the lack of available data (especially a problem in dryland regions), the problems involved in estimating evapotranspiration, and the complexity of the environmental systems which the schemes attempt to simplify.

One weakness in most schemes is the overly simplistic treatment of rainfall and the associated processes of evapotranspiration and

runoff. The UNESCO system, like most others, utilizes monthly or annual means in calculating aridity. Characteristics such as the daily distribution (French: "repartition"), the duration and frequency of dry spells, rainfall intensity, reliability, and the diurnal cycle of rainfall are ignored. The importance of these factors is illustrated by studies dealing with the concept of "effective rainfall" (e.g., Noy-Meir 1985; Scholes 1990; LeHouérou 1984). The essence of this concept is that plant growth does not respond to rainfall below a certain threshold of intensity (e.g., 3 mm/day).

A recent trend in hydrology is to parameterize the water balance by assuming that a certain probabilistic distribution of precipitation in time and space variability corresponds to a given annual or monthly rainfall amount (e.g., Rodriguez-Iturbe *et al.* 1987; Eagleson *et al.* 1987). Evaporation and runoff are calculated based on these time–space fields of rainfall. The idea is that rainfall is concentrated at certain points in time and space and that, at some of these points, critical thresholds of rainfall intensity are surpassed that permit evapotranspiration or runoff to occur. This concept is being integrated into numerical modeling of the atmospheric general circulation and climate and was utilized in the recent global water balance calculations of Willmott *et al.* (1985). Incorporating such concepts into climate classification would be desirable, particularly for arid lands where the time and space variation of precipitation is extreme. Current high-speed computers make this possible, but limitations are imposed by the availability of data on the time–space distribution of rainfall.

With respect to the needs of numerical models, a more pragmatic classification might be one that distinguishes water-limited and temperature- or radiation-limited environments. Because the parameterization of exchange processes differs so fundamentally in these types of environments (see Chapter 8), a broad determination of this distinction can serve many useful purposes in assessing and modeling climate. Churkina and Running (1998) categorized global environments on this basis and found that while C4 grasslands are water-limited, C3 grasses are limited by both water and temperature. They also found some degree of water limitation in the more humid forest environments. A distinction based on water limitation is also compatible with some vegetation classification schemes, such as those distinguishing the wet and dry savanna (also termed moist eutrophic and dry eutrophic) (Nicholson 1996).

One major limitation in all of these is that they do not account for climate as a dynamic variable, with transient boundaries between the various regimes. This problem is magnified in arid regions, where interannual variability is large. Analyses by Kendall (1935) and Trewartha and Horn (1980) show the fluctuations of the boundary between dry and humid climates in the central USA (Fig. 9.8). In one 5-year period the boundary shifted by more than 800 km. The case of the West African Sahel provides a further fitting illustration. In this region 30-year means of precipitation changed by 30–40% during the course of the twentieth century (Nicholson *et al.* 2000). Clearly, areas defined as semi-desert would have changed to desert and many semi-arid regions would have become arid. The best way

Table 9.1. *Common climatic characteristics of arid and semi-arid environments.*

Typical climatic characteristics of dryland regions
    Low rainfall, which is episodic in time and highly variable in space
    Severe moisture stress during some or all of the year
    Rainfall events of short duration but frequently high intensity
    High temperature, extreme diurnal or annual temperature range
    High insolation and low humidity (except for fog-shrouded coastal deserts)
    High potential evaporation
    Strong, turbulent and probably dusty winds
Additional characteristics of semi-arid regions
    Pronounced rainfall seasonality
    Marked and abrupt alternation between extremes of dry and wet conditions
    Relatively short "wet" season
    High sensitivity to climatic fluctuations and climatic change
    High risk of drought and flood

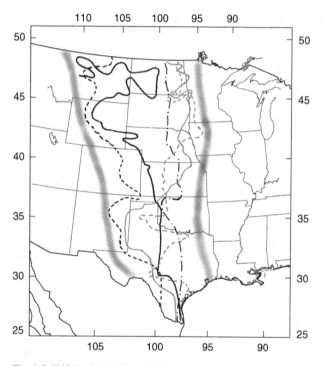

Fig. 9.7 Shift in dryland boundaries over the USA: indicated are the boundaries between 'B' and 'C' or 'D' climates in five successive years and the extremes over a roughly 10-year period (based on Trewartha and Horn 1980, and Kendall 1935).

of dealing with this weakness is to recognize it and any implications it has for the problem at hand, and also to focus on process and genesis, rather than the delineation of artificial boundaries. For this reason, the quantifiable moisture indices of Thornthwaite and Budyko can be particularly useful.

9.5 GENERAL CLIMATIC ENVIRONMENT OF DRYLANDS

Despite the great diversity of the earth's drylands, these regions share many common climatological characteristics. Most of

these common characteristics are consequences of the aridity or of the factors that produce it. Some are consequences of the low-latitude location of most drylands. These common characteristics, which are evident to some degree in both the prevailing thermal and moisture regimes of most dryland regions, are summarized in this section. Chapters 11 and 12 contain more detailed discussion of the hydrologic regime. The most general characteristics of the dryland hydrologic regime are summarized in Table 9.1.

The most common climatic characteristic of the drylands is scant rainfall that is discontinuously available in both time and space. The high temporal and spatial variability is a consequence of the randomness of the convective disturbances producing rainfall. In most semi-arid regions, rainfall is concentrated in a brief season, often during summer when potential evapotranspiration is high, 2000–2500 mm/year in the Australian desert and up to 6000 mm/year in the Sahara (Cooke and Warren 1973). Long dry spells occur within the rainy season and a dry season prevails for most of the year. When the rainy season arrives, the dormant vegetation and barren ground are rapidly transformed to a field of green (Fig. 9.8a). In the arid regions there is generally no preferred seasonality and sometimes years can go by with little or no rainfall. The dryland vegetation is well adapted to these erratic patterns and rapidly responds to moisture availability (Fig. 9.8b).

The rainfall distribution in drylands tends to be skewed such that there is an overabundance of subnormal years, with the long-term average being inflated by a few years with exceedingly high rainfall. Local factors, such as topography, have a great influence on rainfall, creating large spatial gradients in mean rainfall. Annual rainfall totals are determined by a small number of rain events, during which rainfall is usually of high intensity and short duration. Runoff is high during such events, but overall the proportion of rainfall that generates runoff is quite low.

In any one year, there are usually a number of areas where rainfall is abnormally low, even if good rainfall prevails on a regional basis. This happens because the disturbances that bring

(a)

(b)

Fig. 9.8 Dryland vegetation responds effectively to periods of moisture availability. (a) The dry and wet season in the woodland south of the Sahel. (b) Vegetation in the bed of the Kuiseb (Namibia), a response to a flood 10 years earlier.

rainfall have limited areas in which rain actually occurs. The spatial variability is particularly high in the wet years. During the dry years, rainfall is more uniformly suppressed, because the large-scale environment is simply not conducive to the development of precipitation.

The dryland regions are also subjected to extreme, albeit infrequent, weather events such as snowfall and floods (Goudie and Wilkinson 1980). Chicama, Peru, where mean annual precipitation is 4 mm, received 394 mm in 1925. Lima, Peru, where mean annual precipitation is 46 mm, received 1524 mm that same year. The North African flood of September 1969 brought 210 mm to Biskra in two days and 319 mm to El Djem in three days. These totals are well in excess of the annual means at these stations. In 1934 extreme rainfall affected both the northern and southern extremes of Africa: 370 mm in three days in the central Sahara and 110 mm during March (mean = 5 mm) to Swakopmund, Namibia. These were not local events, but instead were linked to large-scale systems. For example, 1925 had a major El Niño event, 1934 saw extremely high rainfall

throughout southern Africa, and the 1969 floods were linked to a major synoptic system that extended across northern Africa into the southern Sahel.

Two of the most common thermal characteristics are high incoming solar radiation and, often, thermal instability in the near-surface atmosphere. Ground temperature can be exceedingly high, especially in arid regions; the annual and diurnal temperature ranges are often extreme. Most dryland regions also have a relatively high surface albedo but low net radiation. Many of the thermal characteristics are related to the low thermal conductivity of the dry desert soils. This limits heat transfer to the subsurface, creating a large temperature gradient just beneath the surface, and produces extreme diurnal and annual temperature ranges.

The high insolation is largely a function of the subtropical location of many dryland regions and the relatively low cloud cover typical of most of them. The low atmospheric moisture content in some drylands also enhances the amount of solar radiation reaching the surface, because water vapor is a strong

Fig. 9.9 (a) As a vehicle approaches a highway mirage, (b) its image is reflected. (c) An Australian highway appears sand-covered, but it is absolutely clear. (d) Mirage in the Namib Desert; a lake appears to be surrounding the distant topography.

absorber in some solar wavelengths. The relatively sparse vegetation and the low ground moisture further enhance surface temperatures. The net result of these conditions is that little solar energy is expended for evaporation or stored within the vegetation layer; it is used instead to heat the ground surface, which in turn conducts very little heat to the subsurface. Consequently, the temperature gradient near the surface can be extremely large. These temperature gradients are the cause of the mirages frequently seen in deserts (Fig. 9.9). The sandy soils of many dryland regions further inhibit the retention of soil moisture and hence enhance these characteristics.

9.6 THERMAL EXTREMES

Excellent summaries of thermal extremes in deserts are provided by Goudie and Wilkinson (1980), Lettau (1978), Dubief (1959), Petrov (1976), and Chang (1958). Table 9.2 provides examples of many of the extremes. Lettau further gives a detailed discussion of the atmospheric and surface factors that promote both high and low temperatures in deserts, stressing the radiative regime, the nature of the surface, and local climatic effects, such as cold air drainage.

In Australia and Asia the absolute maximum air temperatures are on the order of 48°–50°C, but temperatures of 57°C and 58°C have been recorded in the Sahara. In Death Valley, California, an air temperature of 65.8°C was recorded over a hot, basalt surface (Warke and Smith 1998). Ground temperature is even more extreme, with temperatures in excess of 70°C having been recorded at numerous locations around the world. The basalt surface in Death Valley has reached 72.6°C. In the Sahara, Augiéras measured a surface temperature of 78°C and estimated that temperatures can reach 80°C or more, a value Dupuis claimed to have observed at Edjeleh (Dubief 1959). At

Table 9.2. *Examples of extreme thermal conditions in deserts (compiled from Lettau and Lettau (1978); Petrov (1976); Dubief (1959), and other sources).*

Location	Extreme air temperature (°C)
Abdaly, Kuwait	52
Abqaiq, Saudia Arabia	52
Abukmal, Syria	49
Buraimi, Oman	51
Death Valley, California	65
Lake Assal, Djibouti	52
Mexicali, Mexico	50
Oodnadatta, South Australia	51
Phoenix, Arizona	50
Repetek, Turkmenistan	50
Sahara Desert	58
Tucson, Arizona	56

Location	Ground temperature (°C)
Central Sahara	80
Chinguetti, Mauritania	72
Death Valley, California	88
Fada, Chad	70
Karakum dunes, Turkmenistan	70
Loango, Gabon	82
Namib Desert	78
Peruvian dunes	67
Poona, India	75
Port Sudan, Sudan	84
Repetek, Turkmenistan	79
Tamanrasset, Algeria	70
Tucson, Arizona	74

Location	Extreme diurnal range of air temperature (°C)
Central Asia	60
Gobabeb, Namibia	50
Pampa de la Joya, Peru	63
Sahara Desert	56
Tucson, Arizona	56

Location	Absolute range of air temperature (°C)
Gobabeb, Namibia	60
Mashad, Iran	72
Mazar al Sharif, Iran	72
Milford, Utah	79
Ouargla, Algeria	57
Repetek, Turkmenistan	81
Sarmiento, Argentina	71
Tsabong, Botswana	58
Winslow, Arizona	71

Port Sudan on the Red Sea, a sand temperature of 83.5°C has been recorded (Petrov 1976). Monod obtained a ground temperature of 72°C in the Ahnet and the Adrar of Mauritania. Quezel obtained an even higher temperature of 75°C in Borkou, northern Chad (Rognon 1989). These extreme temperatures

Table 9.3. *Diurnal versus annual temperature range (mean, °C) for selected locations.*

Station	Location	Latitude (°)	Diurnal	Annual
Abadan, Iran	Inland	30 N	15	22
Alice Springs, Australia	Inland	24 S	15	16
Calcao, Peru	Coastal	11 S	11	5
Khartoum, Sudan	Inland	16 N	15	12
Ouarghla, Algeria	Inland	32 N	16	24
Phoenix, Arizona, USA	Inland	33 N	16	23
San Diego, California, USA	Coastal	32 N	10	10
Walvis Bay, Namibia	Coastal	23 S	11	5

contribute to a rapid turbulent transfer of heat to the atmosphere. These temperatures are still quite far from the theoretical maximum of roughly 90°C predicted by Garratt (1992).

The above examples of extreme temperatures are measurements from highly localized sites. Mildrexler *et al.* (2006) used satellite imagery to assess maximum surface temperatures. The data represent averages for areas of roughly 25–30 km². During the years 2003–2005 the hottest locations were in Queensland, Australia, with a temperature of 69.3°C, and the Lut Desert of Iran, with a temperature reaching 70.7°C. The conditions in the latter desert are so extreme that even bacteria cannot survive.

The factors that contribute to the extreme temperatures reached in the drylands also contribute to other "typical" thermal characteristics of these regions, such as a large diurnal and annual temperature range. Since little heat is transmitted to deeper layers of the soil, there is no thermal reservoir to prevent rapid cooling at night or during the cool season. The lack of clouds and the absence of dense vegetation cover to retain heat further enhances the cooling. Nocturnal cooling is enhanced by the relatively low atmospheric moisture content in the lower atmosphere of most drylands. Water vapor readily absorbs the longwave radiation emitted by the surface, thus reducing nocturnal cooling. For these reasons, the annual temperature range can be exceedingly high in dryland regions in the higher subtropical latitudes. In the low latitudes there is no cold season, so the annual temperature range is small and is generally exceeded by the diurnal range. The contrast between annual and diurnal range is particularly large in the drylands.

Typical annual and diurnal ranges of air temperature are illustrated in Table 9.3 for eight stations representing different latitudes and both coastal and inland locations. At the two low-latitude coastal stations (Walvis Bay, Namibia, and Callao, Peru) the annual and diurnal ranges are on the order of 5° and 11°C, respectively. Inland at Khartoum (in the Sudan, at 16° N) and Alice Springs (in Australia, at 24° S) the annual range is two to three times higher and the mean diurnal range, 15°C at both locations, is also greater. At the inland stations at higher latitudes (Ouarghla, Algeria, at 32° N, Abadan, Iran, at 30° N, and Phoenix, Arizona at 33° N) the annual temperature range is on the order of 23°C and exceeds the diurnal, which is on the

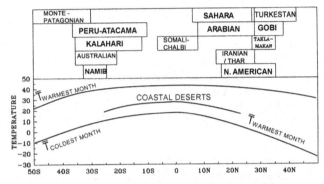

Fig. 9.10 Annual temperature range at inland and coastal deserts as a function of latitude. The mean temperatures of the warmest and coldest months are plotted for the indicated deserts. The contrast between coastal and interior deserts is apparent in the mean temperature of the warmest month, but not in temperatures for the coldest month.

order of 15°C. In comparison, at San Diego, on the Pacific west coast of the USA, both the annual and diurnal range are on the order of 10°C.

An extreme example comes from Bir Milrha in the Libyan Desert, where on one day the nocturnal minimum was −0.6°C, while the maximum temperature exceeded 37°C (Thompson 1977). At Death Valley (California), an even greater range, 41°C, was recorded on a day in August 1891 (Goudie and Wilkinson 1980).

The annual range is considerably greater at the surface itself. In the Saharan highlands of Algeria, the annual range of surface ground temperature is about 60°C; it reaches 65°C in some places at lower elevation. The absolute annual range at Tamanrasset reached 81.9°C in 1935; over sand the range was even higher, 83.9°C. At nearby Asekrem, the absolute annual range of ground temperature reached 78.5°C in 1957 (Dubief 1959).

Figure 9.10 summarizes the effects of latitude and coastal versus inland location. The mean maximum and minimum monthly temperatures for dryland locations worldwide are plotted. Both temperature and annual range increase with latitude, but much more so in the Northern Hemisphere. In the Southern Hemisphere, which is 90% water, the temperature

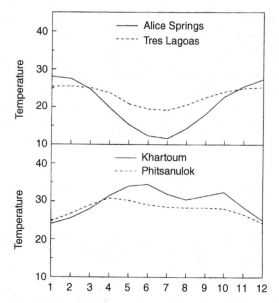

Fig. 9.11 Seasonal cycle of monthly mean temperatures (°C) at Khartoum (Sudan), Alice Springs (Australia), Phitsanulok (Thailand), and Tres Lagoas (Brazil).

regime is markedly more moderate. The coastal deserts also stand out both in terms of lower annual range and lower summer temperatures.

Annual range is influenced not only by latitude and location with respect of water, but also by the degree of aridity. This is illustrated using pairs of stations at roughly the same latitude, one arid and one humid (Fig. 9.11). One pair, Alice Springs, Australia, and Tres Lagoas, Brazil, is in the Southern Hemisphere. Khartoum, Sudan, and Phitsanulok, Thailand, are in the Northern Hemisphere. The arid locations (Khartoum and Alice Springs) have a much lower annual range. Surprisingly, the contrasts are greatest in summer in the Northern Hemisphere stations and in winter in the Southern Hemisphere stations.

REFERENCES

Adams, M. E., 1996: Savanna environments. In *The Physical Geography of Africa* (W. M. Adams, A. S. Goudie, and A. R. Orme, eds.), Oxford University Press, Oxford, pp. 196–210.

Barry, R. G., and R. J. Chorley, 2004: *Atmosphere, Weather and Climate.* Routledge, London, 421 pp.

Box, E. O., 1981: Predicting physiognomic vegetation types with climate. *Vegetatio*, **45**, 127–139.

Budyko, M. I., 1986: *The Evolution of the Biosphere.* Reidel, Dordrecht, 423 pp.

Chang, J. H., 1958: *Ground Temperature.* Blue Hill Observatory, Boston, MA, 496 pp.

Churkina, G., and S. W. Running, 1998: Contrasting climatic controls on the estimated productivity of global terrestrial biomes. *Ecosystems*, **1**, 206–215.

Cooke, R. U., and A. Warren, 1973: *Geomorphology in Deserts.* Batsford, London.

Coupland, R. T., 1992: The mixed prairie. In *Ecosystems of the World*, 8A, Natural Grasslands. Introduction and Western Hemisphere (R. T. Coupland, ed.), Elsevier, Amsterdam, pp. 151–182.

Dubief, J., 1959: *Le Climat du Sahara.* Vol. 1. Mémoires de l'Institut de Recherche Saharienne, Algiers, 312 pp.

Eagleson, P. S., N. M. Fennessey, W. Qinliang, and I. Rodriguez-Iturbe, 1987: Application of spatial Poisson models to air mass thunderstorm rainfall. *Journal of Geophysical Research*, **92**, 9661–9678.

Ellery, W. N., R. J. Scholes, and M. T. Mennis, 1991: An initial approach to predicting the sensitivity of the South African grassland biome to climate change. *South African Journal of Science*, **87**, 499–503.

Flohn, H., 1957: Zur Frage der Einteilung der Klimazonen. *Erdkunde*, **11**, 161–175.

Garratt, J. R., 1992: Extreme maximum land surface temperatures. *Journal of Applied Meteorology*, **31**, 1096–1105.

Geiger, R., and W. Pohl, 1954: Revision of the Köppen-Geiger *Klimakarte der Erde. Erdkunde*, **8**, 58–61.

Goudie, A., and J. Wilkinson, 1980: *The Warm Desert Environment.* Cambridge University Press, New York, 88 pp.

Hare, F. K., 1977: Climate and desertification. In *Desertification: Its Causes and Consequences.* UN Conference on Desertification, Pergamon, Oxford, pp. 63–120.

Hare, F. K., 1980: Long-term annual surface heat and water balances over Canada and the United States South of 60°N: reconciliation of precipitation, run-off and temperature fields. *Atomsphere-Ocean*, **18**, 127–153.

Jackson, I. J., 1989: *Climate, Water and Agriculture in the Tropics.* Longman, New York, 377 pp.

Kendall, H. M., 1935: Notes on climatic boundaries in the eastern United States. *Geographical Review*, **25**, 117–124.

Koeppen, W., 1918: Klassifikation der Klimate nach Temperatur, Niederschlag und Jahreslauf. *Petermanns Geographische Mitteilungen*, **64**, 193–203.

LeHouérou, H. N., 1984: Rain use efficiency: a unifying concept in arid land ecology. *Journal of Arid Environment*, **7**, 1–15.

Lettau, H. H., 1978: Extremes of diurnal temperature ranges. In *Exploring the World's Driest Climate* (H. H. Lettau and K. Lettau, eds.), IES Report 101, Institute for Environmental Studies, University of Wisconsin, Madison, WI, pp. 67–73.

Lettau, H. H., and K. Lettau, 1978: *Exploring the World's Driest Climate.* Institute for Environmental Studies, University of Wisconsin, Madison, WI.

Mabbutt, J. A., 1979: *Desert Landforms.* MIT Press, Cambridge, MA, 340 pp.

MacMahon, J. A., 1979: North American deserts: their floral and faunal components. *Arid-Land Ecosystems: Structure, Functioning and Management* (D. Goodall and R. A. Perry, eds.), Cambridge University Press, Cambridge, pp. 21–82.

MacMahon, J. A., and F. A. Wagner, 1985: The Mojave, Sonoran and Chihuahuan deserts of North America. In *Hot Deserts and Arid Shrublands* (M. Evenari, I. Noy-Meir, and D. W. Goodall, eds.), *Ecosystems of the World*, 12A, Elsevier, Amsterdam, pp. 105–202.

Meigs, P., 1953: The distribution of arid and semi-arid homoclimates. In *UNESCO, Arid Zone Research*, Series 1, *Reviews of Research on Arid Zone Hydrology*, pp. 203–210.

Mildrexler, D. J., M. Zhao, and S. W. Running, 2006: Where are the hottest spots on earth? *EOS Transactions*, **87**, 461–467.

Nemec, J., and J. A. Rodier, 1979: Streamflow characteristics in areas of low precipitation (with special reference to low and high flows). *The Hydrology of Areas of Low Precipitation.* IAHS Publication No. 128, 125–140.

Nicholson, S. E., 1996: Savanna climate. In *Encyclopedia of Climate and Weather* (S. H. Schneider, ed.), Simon and Schuster, New York, pp. 657–660.

Nicholson, S. E., B. Some, and B. Kone, 2000: A note on recent rainfall conditions in West Africa, including the rainy seasons of the 1997 El Niño and the 1998 La Niña years. *Journal of Climate*, **13**, 2628–2640.

Noy-Meir, I., 1985: Desert ecosystem structure and function. In *Hot Deserts and Arid Shrublands* (M. Evenari, I. Noy-Meir, and D. W. Goodall, eds.), *Ecosystems of the World*, 12A, Elsevier, Amsterdam, pp. 93–103.

Oliver, J. E., and J. J. Hidore, 2002: *Climatology: An Atmospheric Science*. 2nd edn, Prentice-Hall, 410 pp.

Petrov, M. P., 1976: *Deserts of the World*. Wiley, New York, 447 pp.

Ripley, E. A., 1992: Water flow. In *Natural Grasslands* (R. T. Coupland, ed.), *Ecosystems of the World*, 8A, Elsevier, Amsterdam, pp. 55–73.

Rodriguez-Iturbe, I., B. Febres de Power, and J. B. Valdès: 1987: Rectangular pulses point process models for rainfall: analysis of empirical data. *Journal of Geophysical Research*, **92**, 9645–9656.

Rognon, P., 1989: *Biographie d'un Désert*. Plon, Paris, 347 pp.

Scholes, R. J., 1990: The influence of soil fertility on the ecology of southern African dry savannas. *Journal of Biogeography*, **17**, 415–419.

Shreve, F., 1942: The desert vegetation of North America. *Botanical Review*, **8**, 195–246.

Stadler, S. J., 2005: Aridity indices. In *Encyclopedia of World Climatology* (J. E. Oliver, ed.), Springer, Dordrecht, pp. 89–94.

Strahler, A. N., 1978: *Physical Geography*. 3rd edn. John Wiley and Sons, New York.

Strahler, A. N., and A. H. Strahler, 1987: *Modern Physical Geography*. John Wiley and Sons, New York, 544 pp.

Thompson, R. D., 1977: *The Climatology of the Arid World*. Reading Geographical Papers, University of Reading, 37 pp.

Thornthwaite, C. W., 1931: The climate of North America according to a new classification. *Geographical Review*, **21**, 633–655.

Thornthwaite, C. W., 1948: An approach towards a rational classification of climate. *Geographical Review*, **38**, 55–94.

Thornthwaite, C. W., and J. R. Mather, 1962–65: *Average Climatic Water Balance Data of the Continents*. Eight volumes. Centerton, NJ: C. W. Thornthwaite and Associates.

Trewartha, G. T., and L. H. Horn 1980: *An Introduction to Climate*. McGraw-Hill Book Company, New York, 416 pp.

Walter, H., 1985: *Vegetation of the Earth and Ecological Systems of the Geo-Biosphere* (3rd edition). Springer, New York, 318 pp.

Warke, P. A., and B. J. Smith, 1998: Effects of direct and indirect heating on the validity of rock weathering simulation studies and durability tests. *Geomorphology*, **22**, 347–357.

Willmott, C. J., C. N. Rowe and Y. Mintz, 1985: Climatology of the terrestrial seasonal water cycle. *Journal of Climatology*, **5**, 589–606.

10 Desert microclimate

10.1 INTRODUCTION

In this chapter, "microclimate" will be used in two ways, one to designate the climatic conditions near the surface (both in the atmosphere and in the ground just below the surface) and another to designate the particular climatic conditions of microhabitats within the drylands. The latter include dunes, oases, desert depressions, riverine environments within deserts, and beneath the canopy of the savanna vegetation. This chapter begins with an examination of typical conditions of temperature, moisture, and wind near the surface of a sand or stone desert. Then the microclimates of select habitats are considered: the desert dunes, the more extreme conditions of desert depressions, and the more moderate environments of the riparian valleys, oases, desert lakes and irrigated fields, and within a vegetation layer are described.

The research that is available includes many classic studies from several desert research institutes. These include the Repetek Desert Research Center in Turkmenistan; the Desert Ecological Research Unit at Gobabeb, Namibia; the University of Arizona and the Arid Lands Office, in Tucson; the Nevada Desert Research Center; the Egyptian Meteorological Authority; the Lanzhou Institute of Desert Research in China; the Chihuahuan Desert Research Institute in Mexico; CSIRO in Alice Springs, Australia; the Arid Ecosystems Research Center at the Hebrew University in Jerusalem, the Blaustein Institute for Desert Research at Ben-Gurion University in Tel Aviv, and the Institute for Arid Zone Research at the University of the Negev in Beer Sheva. Most of the discussion in this chapter will be based on a selection of available examples, many from important but little-known studies of half a century ago. Books by Dubief (1959) and Menenti (1984) also contain a wealth of information about desert microclimates.

The purpose of this extensive chapter is both to provide an overview of desert microclimates and to create an archive of some of these forgotten or little-known studies. Although discussion in based on only a few examples, in most cases generalizations can be made about other regions. This is particularly true for the thermal regime. It is more difficult to generalize in the case of subsurface hydrology, because this is strongly dependent on many local factors. Even so, the material presented here has broader implications.

10.2 MICROCLIMATE OF DESERT SURFACES

Relatively few studies of desert microclimates have been conducted, because special instrumentation is required to measure conditions very near the surface and below it. Figure 10.1 shows an example of a field set-up at a station in the Sudan.

The thermal conditions near the surface of a sand desert are extreme. One reason is that most of the radiant energy received is utilized for heating the surface; very little is expended as latent heat to evaporate surface moisture. Also the specific heat of the dry surface is low and the sand heats rapidly. The extreme heating of the surface is enhanced by the low thermal conductivity of the dry sand; very little heat is conducted to the subsurface. Consequentially, the subsurface provides no reservoir of heat and the nocturnal and winter cooling are extreme. The result is a large annual temperature range and large day/night differences at the surface, and extreme temperature gradients below ground.

10.2.1 GROUND TEMPERATURES

The diurnal pattern of heating and cooling is illustrated in Fig. 10.2 for a typical day at several locations. At Repetek and Tucson and in the Oregon desert the surface heats to over 50°C by day, but cools to around 20°C at night. In the Sahara, nighttime surface temperature drops to near freezing, despite a daytime high of 43°C. At Tucson, even more extreme cases have been recorded. On one June day, the temperature just below the surface (0.4 cm depth) ranged from 71.5°C in the day to 15°C at night.

In many of the extreme deserts a diurnal range of between 40°C and 45°C is typical, but the diurnal range can be much greater. In the Namib, temperatures on a given day can range from 10°C to nearly 70°C (Robinson and Seely 1980). The world's most extreme range may be that at Pampa de la Joya in the coastal desert of Peru. A diurnal range of 67°C was recorded during a field expedition in July 1964 (Lettau 1978).

The temperature conditions rapidly become more moderate with depth in the ground. At Repetek the diurnal range approaches zero just 10 cm below the surface. At that soil depth, the daily mean is 15°C lower than at the surface. At night the surface temperature is nearly 15°C cooler than that just a few centimeters below. In the Sahara the diurnal wave extends about 25 cm below the surface, where the temperature is nearly 25°C lower than at the surface. In the pumice desert of Oregon the temperature can drop some 30°C within the first few centimeters (Gay 1970). At Tucson the afternoon maximum temperature drops at least 40°C in less than 2 cm and the diurnal range at 0.4 cm in the ground is 58.5°C.

In the deserts of the higher latitudes, the temperature gradients in the ground vary greatly with the season, as these gradients are so strongly dependent on solar radiation. This is shown by extensive measurements at Repetek (Fig. 10.3). The most extreme conditions are clearly seen to occur in the winter and summer seasons of maximum and minimum incoming solar radiation. A rain event can also disrupt the temperature patterns because moistened sand conducts heat about 10 times faster than dry sand. After a rain, the temperature gradient is markedly reduced. Changes in cloud cover, i.e., insolation, also have a marked effect on the temperature profile. This is evident in March, when the top layers become nearly isothermal on many days.

Ground surface temperatures are strongly influenced by the nature of the surface material, especially by contrasts in albedo and thermal conductivity. Figure 10.4 shows temperature variation in the Sahara over basalt (albedo ~0.2), sandstone, and a sand dune (albedo ~0.5). Ground temperatures reach nearly 80°C over sandstone and basalt, compared with 64°C over the dune. The stone conducts heat to a greater depth than the sand because the pore spaces in the sand inhibit heat conduction. The diurnal wave may penetrate 1 m or more in basalt compared with 10–30 cm in the sand (Fig. 10.4). The effect of the pore space is evident even in the sandstone, which has somewhat lower temperatures and subsurface temperature gradients than the basalt.

In Death Valley, California, the dark-colored basalts heat up much more quickly and efficiently than lighter-colored sandstones and limestones (Warke and Smith 1998). On one summer day in 1992, when the air temperature was 65.8°C, the temperature of the basalt surface reached 72.65°C, while the temperature only reached 57.65°C over the sandstone (not shown) and limestone (Fig. 10.5). The basalt was notably hotter than the air, by nearly 7°C during the day, while the limestone was some 8°C cooler than the air.

10.2.2 NEAR-SURFACE THERMAL REGIME OF THE AIR

The thermal patterns in the air layer just above the desert surface are generally more moderate than on the ground, but they

Fig. 10.1 Micrometeorological instrumentation at Wad Medani, Sudan: rows of thermistors measure temperature at various depths.

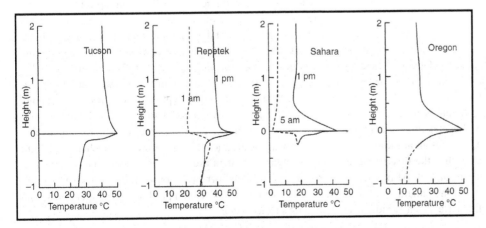

Fig. 10.2 Diurnal pattern of heating and cooling for Tucson (SW USA), Repetek (the Karakum Desert of Central Asia), the Sahara, and an Oregon (USA) pumice desert (based on Sinclair 1922; Petrov 1976; Dubief 1959; Gay 1970).

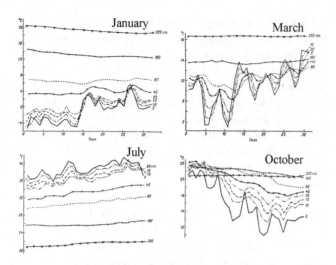

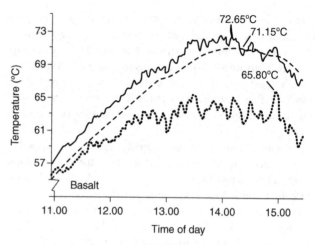

Fig. 10.3 March average daily sand temperatures at various depths in Repetek during October, January, March, and July of 1933 (from Petrov 1976).

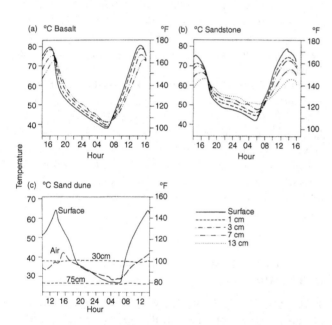

Fig. 10.4 Temperatures over (a) basalt, (b) sandstone, and (c) a sand dune in the Sahara at Tibesti (from Peel 1974).

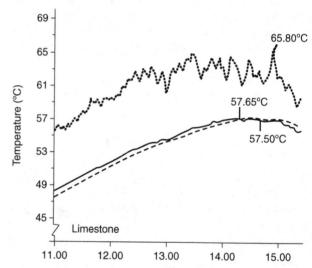

Fig. 10.5 Surface (solid line) and subsurface (dashed line) temperatures over basalt and limestone in Death Valley, CA, on August 3, 1992 (from Warke and Smith 1998). Air temperature (65.8°C, dotted line) is indicated for comparison.

are still extreme. In the Sahara, air temperatures in the shade have reached 58°C at Azizia (Libya) and Tindouf (Algeria) (Dubief 1959; Rognon 1989). At Death Valley, 64.8°C has been recorded (Roof and Callagan 2003).

Various deserts exhibit marked differences in near-surface temperature gradients (Fig. 10.2). At Repetek, air temperature 10 cm above the surface is about 5°C lower than at the surface, so that the air temperature gradient within the first 10 cm is roughly half that within the first 10 cm of soil. At night, the surface air layer is nearly isothermal. In the Sahara, the air at 30 cm is nearly 30°C cooler than the surface. The gradient is

more gradual at Tucson; air temperatures at 10 cm are only about 2°C lower than at the surface.

The temperature gradients near the ground produce the characteristic wind regime of deserts: high wind speeds and turbulence during the day and a dramatic change to calm conditions at night. The temperature lapse rate within the first few centimeters above the surface may exceed 1°C/2 mm during the day, which is 50,000 times greater than the dry adiabatic lapse rate of 1°C/100 m. This is roughly equivalent to a Richardson number on the order of 5, compared with a value of 1, which marks the onset of instability. Hence extraordinary thermal instability prevails near the ground and in the lower atmosphere during the day. At night, a temperature inversion equivalent to several hundred degrees Celcius per meter may prevail a few centimeters above the surface. The inversion produces extreme stability, suppressing surface winds.

10.2.3 ANNUAL TEMPERATURE RANGE

The absolute annual temperature range is the difference between the absolute maximum and absolute minimum temperatures. It generally increases with increasing distance into a continental interior. At Repetek, deep in the continental interior, the highest sand surface temperature ever recorded is 79.4°C and the minimum is −40°C; thus the absolute annual range is nearly 120°C. In deserts with a less continental location the range is less extreme. For example, it is only 66°C at Death Valley (Felton 1965), and 60°C in the Namib, where sand temperatures frequently exceed 70°C (Robinson and Seely 1980). In the Sahara, surface temperatures as high as 80°C and as low as −8.6°C have been reported (Dubief 1959).

The annual range also diminishes rapidly with depth below the surface, especially in the high-latitude deserts where the incoming radiation fluctuates dramatically during the course of the year. At Repetek, just 5 cm below the surface, the highest and lowest recorded temperatures are 55.5° and −25°C. Thus, the annual range is reduced by roughly one-third just 5 cm below the surface. By comparison, the dampening effect is much reduced in the Sahara, at roughly 30° N; data for Tamanrasset and el Oued show that the annual range is not reduced by one-third until about 70–80 cm below the surface (Dubief 1959).

At Wad Medani, Sudan, micrometeorological measurements were made on a routine basis for many years. Figure 10.6 shows mean monthly temperatures at five depths from March 1960 to February 1961. This illustrates both the effect of lower latitude (14° N) and of rainfall. Here the annual cycle is more complex than at Repetek or in the Sahara, where insolation reaches a maximum at the summer equinox. At Wad Medani, insolation peaks twice as the sun migrates northward and southward. The large drop in temperature in July and August results both from this factor (the overhead sun is far to the north) and from rainfall.

10.2.4 MOISTURE CONDITIONS

The depth profile of moisture is considerably more complex than that of temperature because it is influenced by such factors as grain size, composition and compaction, and depth of the water table. Generally the subsurface is wetted from above, unless the water table is near the surface. Assuming that 1 mm of rainfall can wet about 1 cm of sand, Petrov (1976) suggests that 100 mm of rainfall will moisten the sand to a depth of about 100 cm. Infiltration in the larger pores allows for more rapid penetration to greater depths than capillary action in the small pores. The relative importance of the two is strongly dictated by grain size. Thus, the depth to which water penetrates varies markedly with soil texture.

After a rain event, the moisture profile generally shows a decrease from the surface to the lower layers. When drying commences, evaporation takes place from the surface itself; thus, the surface is the first to be moistened but is also the first to dry out. Evaporation takes place within a depth of 50–100 cm; during the dry season this layer will be desiccated, but moisture may be found in the layers below. In many desert sands there is a layer of desiccation commencing somewhere beneath the surface, of variable vertical extent and depth; it may range from a few centimeters to a few meters. Soil water will be found beneath the layer of desiccation.

The cycle of desiccation and moistening is illustrated in Fig. 10.7 for Tengri Sirte in the eastern Ala Shan Desert of Central Asia (near Shap'o-t'ou), where a summer monsoon rainfall regime prevails. The data are for barchan sand, hence the grain distribution is uniform, the sand highly porous, and the field capacity may therefore be as low as 5% (Petrov 1976). Because of the rapid infiltration in sands, rainfall can penetrate to a depth of nearly 20 cm within 2 minutes, within 5 minutes to 35 cm, and within 30 minutes to 120 cm (Orlov 1928). The drying generally extends to only about 50 cm, but the rains penetrate to well over 300 cm. Subsurface layers of desiccation are apparent at some times, usually during the rainy season.

This pattern will often differ in areas where the groundwater is relatively close to the surface and soil moisture results from capillary action in the capillary fringe. The conditions in playas (Section 10.5) illustrate this. The pattern will also differ

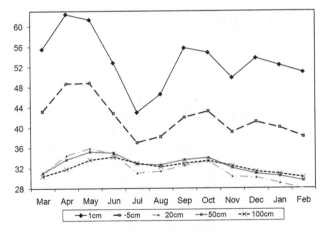

Fig. 10.6 Mean monthly soil temperatures (°C) for the year from March 1960 to February 1961 at five soil depths, ranging from 1 cm to 100 cm, at Wad Medani, Sudan.

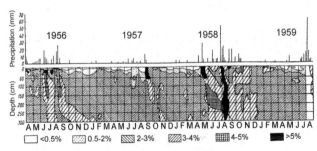

Fig. 10.7 Cycles of desiccation and moistening, as indicated by moisture in the upper 3-meter layer of sand at Tengri Sirte in the eastern Ala Shan of Central Asia (near Shap'o-t'ou) (from Petrov 1976).

depending on the temperature and moisture conditions at the surface itself, on the soil texture, and on plant cover.

The desert sands receive moisture not only from precipitation but also from condensation. The strong radiative cooling at night is conducive to dew formation. The dew can be a significant component of the water balance and can provide a means of survival for animal and plant life, especially when dew formation occurs during the dry season.

10.3 THE MICROCLIMATE OF DUNES

The sand dune is a peculiar micro-environment because its topography plays a major role in modifying airflow and distributing radiation, heat, and moisture. Consequently, conditions differ markedly on various parts of the dune, so much in fact that various habitats within the dune have their own unique biota (Seely and Louw 1980; Krasnov *et al.* 1996). The contrast is greatest during the afternoon hours, when incoming solar radiation and wind are at their peak. The overall microclimate of the dune also contrasts with that of the surrounding flat plains. When the sun is high overhead (as in tropical deserts) the incoming radiation is spread over a larger surface area, and surface temperatures of dunes are often lower than in the plains. When the sun is near the horizon (as in the cold deserts of the higher latitudes), the dune surface may receive stronger insolation than the plain, thus warming more rapidly.

10.3.1 WINDS AND TRANSPORT

A dune acts as an obstruction in an airstream: wind is strong at the dune crest (Fig. 10.8), but moderate on the dune slopes

(see Chapter 2). This produces the characteristic silhouette of a dune (Fig. 10.9), with a slowly rising plinth beginning at the windward dune base, a strongly inclined slipface just over the crest on the leeward slope, and a gently inclined plinth continuing to the leeward dune base. Winds deflate material from the windward plinth and transport it across the crest to the slipface and leeward plinth, which are zones of deposition of sand, dust, and detritus (Fig. 10.10).

The airflow and sand transport around the dune play at least some role in regulating its micro-environments (Hastenrath 1978). Because the lighter materials are transported more readily, the sand becomes coarser, denser, and darker from the windward crest toward the horns. The lowest bulk density, the lightest color, and possibly the finest grains, are found on the slipface. These sand characteristics influence both the thermal and moisture conductivity of the various parts of the dunes, and hence moisture and temperature patterns. Surface albedo is lower on the darker windward side than on the slipface (for the Peruvian dunes, dark gray and albedo of 0.14 vs. light gray and 0.25 on the slipface).

10.3.2 THERMAL ENVIRONMENT

Dune temperatures are influenced by the albedo contrast, as well as by aspect with respect to the sun. Figure 10.11 shows diurnal temperature variation on the NNW-facing slipface of a barchan in Peru and contrasts it with variation on the WNW- and NNE-facing leading edges of the horns. All three sites heat up rapidly at sunrise, with the temperature climbing by 58°–63°C within six hours. Maximum temperatures are up to 5°C warmer on the slipface than at the leading edges. The cooling is somewhat less abrupt

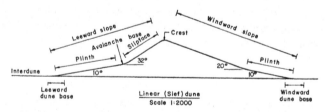

Fig. 10.9 Topographically distinct regions of a linear dune (from Robinson and Seely 1980).

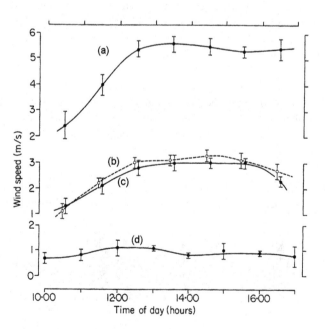

Fig. 10.8 Wind speed measured at the dune top (a), at the top (b) and bottom (c) half of the slipface, and at the avalanche base (d) of a Namib dune (from Seely *et al.* 1990).

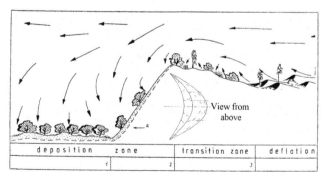

Fig. 10.10 Winds and sand transport around a typical barchan dune (based on Walter and Box 1983 and Bryson 1978).

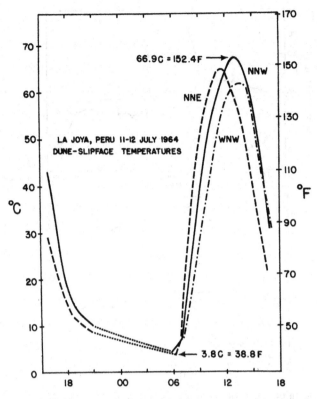

Fig. 10.11 Diurnal variation of sand temperature at 2 mm depth on the slipface of a barchan dune in Peru (Lettau 1978).

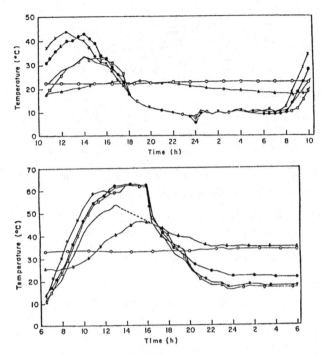

Fig. 10.12 Diurnal cycle of temperature over dunes in the Namib Desert during winter (top) and summer (bottom) in 1977 (from Robinson and Seely 1980). Temperatures are shown for six microhabitats: surface, 10 cm and 20 cm on a slipface, the avalanche base, the surface of the windward slope 8 m from the crest, and the dune base. Symbols for the surface are (starting from the lowest to highest line at 12 to 1300 hrs): open circles (almost straight line) = 20 cm depth, triangular symbols = 10 cm depth, solid line = dune base, open squares = avalanche base, solid circles = surface of windward slope, x = dune surface. Reproduced with permission from Springer.

and minimum temperatures, reached around sunrise, show little contrast. Thus, the maximum diurnal range is on the slipface.

In the deserts of the higher latitudes, where the radiation regime changes greatly with the seasons, the situation is more complex. Temperature measurements on a barchan in the Karakum Desert on a summer day show that in the morning and late afternoon, temperatures are rather uniform across the dune surface. However, the slipface of the dune heats up rapidly and by 08:00 hours it is about 7°C warmer than the windward slope or the inter-barchan depression. By 13:00 hours the slipface and the windward side are several degrees cooler than the depression and nearly 13°C cooler than the crest. In general, the slipface has the greatest diurnal range, the greatest near-surface temperature gradient, and the highest temperatures. The most moderate environment is the crest, where higher wind speeds enhance sensible heat flux to the atmosphere during daytime hours.

The above example provides a general idea of the range of variability within the dune, but the patterns will differ from one location to another and from one season to another. In this example, the slipface lies to the SE and faces the morning and afternoon sun; the summer season accentuates the effect. The orientation of the dune with respect to solar geometry will modify the pattern, as will location in the tropics, where the diurnal and seasonal cycles of radiation are less extreme.

An interesting example of dune microclimates comes from the Namib Desert. Here the combination of microclimatic dune habitats and large variations in macroclimatic conditions has

created a virtual mosaic of habitats that may help to account for the evolution of the desert's extremely diverse endemic fauna. Robinson and Seely (1980) have investigated the climatic conditions and the biotic communities on four sections of a dune shown in Fig. 10.9: the interdune, dune base, plinth, and slipface. They identified four distinct biotic communities in the various habitats. In a companion paper, Seely and Louw (1980) examined the impact of rainfall on the ecology and energetics of the various habitats.

The variation in dune surface temperature is about 20°C between winter and summer, with a diurnal amplitude of about 35°–40°C in winter, but nearly 70°C in summer (Fig. 10.12). At 10 cm the annual amplitude is reduced to less than 15°C; the diurnal amplitude is about 20°C in summer, but only about 5°C in winter. By 20 cm depth, the diurnal essentially disappears, and the annual wave is reduced to an amplitude of about 10°C. The slipface generally heats up first; it is the warmest habitat from early morning to midday, but it experiences temperatures similar to those on the plinth and windward slope later in the day. The coolest niche is the dune base. On foggy days, heating of the dune surface is delayed, but the fog has little effect on subsurface temperatures.

Most dune animals are most active when surface temperatures are between 20° and 45°C. In winter this generally occurs

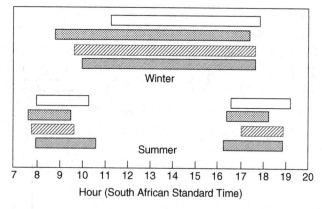

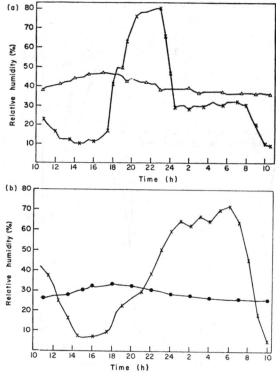

Fig. 10.13 Winter and summer contrast in the time during the day when dune surface temperature on the windward slope (blank), slipface (stippled), avalanche base (wide hatching), and dune base (narrow hatching) is between 20°C and 45°C (from Robinson and Seely 1980).

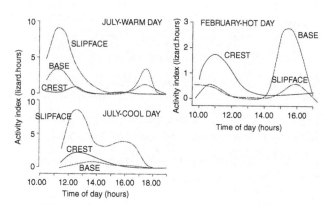

Fig. 10.14 Mean surface activity of the desert lizard *Angolosaurus skoogi* on the dune crest (solid line), slipface (dot-dashed line), and avalanche base and adjacent plain (dashed line) on a warm winter day (top left) and a cool winter day in July (lower left) and on a typical summer day in February (top right) (from Seely *et al*. 1990).

Fig. 10.15 Relative humidity in winter (a) and summer (b) at the slipface surface (x) and at 10 cm depth (•) (from Robinson and Seely 1980).

10.3.3 MOISTURE CONDITIONS

Moisture also varies between the dune habitats. In the Namib, fog-water precipitation provides as much or more moisture than rainfall in most locations; it occurs on the coastal dunes about

between the hours of 10:00 and 18:00, and in summer between about 08:00 and 10:00 and from 17:00 to 19:00 hours. (Fig. 10.13). In winter, the slipface is within this range most often, and the exposed windward slope least often. In summer the dune base is most favorable, and the avalanche base of the plinth least favorable. The fauna respond to these seasonal and diurnal shifts in the thermal environment of the dune. The Namib lizard *Angolosaurus skoogi*, for example, seeks the slipface in winter, being more active and spending more time there on the warm winter days than the cool ones (Fig. 10.14). In summer, the lizard's active period is much more restricted and morning activity is mostly on the cool dune crest and afternoon activity is mostly on the cool base and interdune depression.

120 days per year, but only about 36 days per year some 50 km inland. Nearly 70% of the fog-water is precipitated on the windward slope of the dunes, and nearly half of that on the upper third. The lower third of the leeward slope receives almost no fog-precipitation. Fog penetrates the top 1–2 cm, but usually evaporates within a few hours. Rainfall penetrates deeper and is therefore retained longer; it creates a wet subsurface horizon that may remain for several weeks (Robinson and Seely 1980). On the slipface (Fig. 10.15) the relative humidity varies from 100% at night to a midday low of about 10%; at 10 cm the range is about 45–25%. Rains of 3–14 mm suffice to produce a moisture content in the sand of 2–8%. Mann *et al*. (1976) have reported comparable results for dunes in the Rajasthan Desert of India.

In the Negev, where rainfall rather than fog is the main moisture source, considerable differences in soil moisture were noted from the dune crest, plinth, and interdune corridor (Kadmon and Leschner 1995). The highest moisture content was in the interdune corridor down to a depth of 75–80 cm, and the lowest was on the crest. At 75–80 cm, moisture content was highest on the plinth. Below that the pattern was reversed, compared with that at the surface, but the differences were quite small. Contrasts in the deposition of dew have also been observed (Jacobs *et al*. 2000), with 50% less deposition on the slopes than within the interdune area.

These conditions of temperature and humidity determine the character of the biotic community, especially the plants,

Fig. 10.16 In this field of transverse dunes in the Namib, vegetation is confined to the interdune depressions, with coverage particularly high near the base of the dunes.

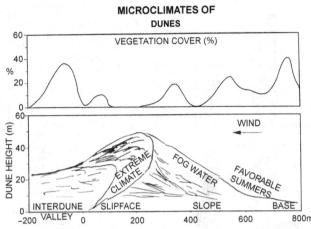

Fig. 10.17 Microclimatic conditions on a typical Namib dune and percent vegetation cover along a transect across the dune (from Robinson and Seely 1980).

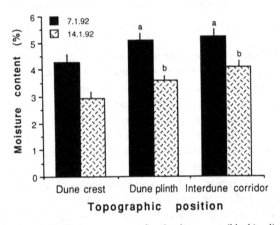

Fig. 10.18 Moisture content for the dune crest (black), plinth (gray), and interdune corridor (white) three days and ten days after a rainstorm of 49 mm (from Kadmon and Leschner 1995).

but the distribution of wind-blown detritus is an equally important limiting factor for the fauna of the dunes. The detritus derives from surrounding savannas, in which considerable biomass thrives on as little as 20 mm of rainfall. It accumulates in the wind-shadow of the slipface, particularly in summer. Because it absorbs large quantities of water from the fogs (Tschinkel 1973), it may provide moisture to the faunal communities of the dune during the fogs, as it does in the Negev (Broza 1979).

10.3.4 BIOTA AND HABITATS

In the Namib, the interdune community is mostly ephemeral grasses, which are absent in rainless years. Beetles and a few species of reptiles thrive here. The vegetation cover is usually greatest at the dune base (Fig. 10.16) and is quite diverse, including tubers, grasses, a lily, a leaf succulent, and the nara melon. The animal community is equally diverse, with species of beetles, reptiles, and small mammals. Oryx typically graze at the dune base. On the plinth, the plant cover is relatively sparse overall and generally absent on the upper third (Fig. 10.17). Two grasses, *Stipagrostis sabulicola* and *Trianthema hereroensis* are found here, but the latter is usually present only during wetter years. They tend to form hummocks 1–2 m high. Many invertebrates inhabit the plinth. The slipface, with its steep angle, is usually devoid of live vegetation, but it is where the detritus accumulates. Here the animals that thrive, like the beetles, are sand-swimmers rather than the burrowers that inhabit the other niches.

A study by Nechayeva *et al.* (1973) in the Karakum Desert found comparable results for plant communities. The greatest number of species was found in the stable environments of the interdune depressions and immobile sands, such as at the dune base. The contrast was particularly large for herb species, where shallow roots make them susceptible to sand movement and other environmental disturbances. There were

72 species of herbs (mostly annuals) in the immobile sands, and 80 in the depressions, comprising 82% and 79% of the species in the respective locations. By comparison, only 10 species of herbs were found in the areas of mobile sands. There was less contrast in tree and shrub species (7, 11, and 13 for the mobile and immobile sands and the depressions, respectively). Presumably the deeper roots provide stability in the areas of mobile sand.

Studies on linear dunes in the western Negev (Kadmon and Leschner (1995) similarly concluded that stability was the most important factor determining the composition and abundance of annual plants. In this desert most of the moisture is derived from rainfall. Moisture content, organic matter, texture, and stability were compared for the crest, plinth, and interdune depressions. Near the surface, moisture content was greatest in the interdune depressions and lowest on the crest (Fig. 10.18); the interdune corridor retained moisture the longest, while the crest lost it most quickly. The interdune depressions also had

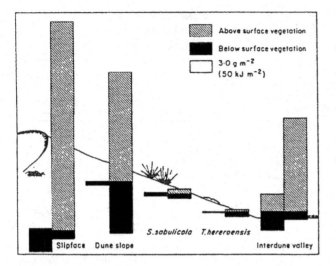

Table 10.1. *Change in biomass after rain in five microhabitats in the Namib (from Seely and Louw 1980).*

Location	Dry year	Wet year
Fauna	0.01	0.06
Surface vegetation	2.68	23.98
Surface detritus	0.01	2.08
Subsurface vegetation	2.99	17.32
Subsurface detritus	0.38	0.62

Fig. 10.19 Change in biomass after rain in five microhabitats in the Namib (from Seely and Louw 1980). Left-hand column represents biomass during a dry year and right-hand column represents biomass during a wet year. Note high above-surface biomass within the slipface habitat.

the highest concentration of fines (silt and clay) and organic matter, and the greatest stability over time. The crest was the most unstable region and had the lowest amount of fines and organic matter. These characteristics made the interdune corridor the preferred habitat for annual plants and the crest the most barren. Of 30 annual plant species in the region, 27 were found in the interdune corridor, 18 were found on the plinth, and none were found on the crest.

Other factors appear to determine the patterns of animal communities within the desert environment. Krasnov *et al.* (1996) studied the distribution of rodent communities in five habitats in the Negev: dunes, gravel plains, limestone cliffs, wadis on loess hills, and wadis on gravel plains. They found that communities were arranged along three gradients: soil hardness (from rock to sand), relief (from cliffs to flat plains), and vegetation density. Species diversity and richness were highest in the limestone cliffs and lowest in wadis on the gravel plains. Rodent biomass was highest in wadis among the loess hills and lowest in the gravel plains.

The contrast in dune habitats becomes even greater after substantial rains. Remarkable changes in the biota, especially the amount of biomass, are observed after heavy rains. Seely and Louw (1980) conducted ecological surveys in 1975 during a prolonged dry period, which fortuitously occurred just prior to the extremely rainy year 1976. Mean annual rainfall for the 13-year period ending in 1975 was 14 mm, with a range of 30 mm; but during the period January–March 1976, 114 mm fell at Gobabeb in the central Namib. Figure 10.19 shows the plant biomass during the two periods for five dune microhabitats. Rainfall produced the most dramatic increase on the slipface, where live vegetation increased from 0 to 25 g/m^3 and detritus increased over sevenfold. The most stable habitat was the interdune valley. The dune slope showed a marked increase in plant

biomass, but little of it was in the perennial grasses that inhabit the slope. The largest increase overall was in the surface vegetation cover. In the dry period, the subsurface vegetation exceeded surface vegetation; after the rains, most of the biomass was on the surface (Table 10.1).

10.4 MICROCLIMATIC STUDIES IN THE SAHARA

Dubief (1959) carried out some rather remarkable microclimatic studies in the Saharan desert of Algeria. These were conducted in the context of the possibility of inhabiting the desert, and the studies included experiments carried out by painting iron plates and boxes various colors and measuring temperatures on sides with different aspects. Dubief included an excellent comparison between various desert locations as well, emphasizing such environmental variables as minimum and maximum temperatures, and annual and diurnal range.

Figure 10.20 shows the diurnal variation of temperature for the air, sand, and lead plates colored black, white, and metallic gray. In the diagram on the left, the plates were placed directly on the sand; on the right, they were placed on planks of wood, thus preventing conduction to the subsurface. The white plaque, with an exceedingly high albedo, was a few degrees cooler than the sand but roughly 8°–10°C warmer than the air temperature. In contrast, the black plaque reached a temperature of 73.2°C when placed on the sand, and 81.3°C when placed on the planks. The albedo contrast between the black and white plaques evoked a temperature difference of about 20°C.

The results of a two-day experiment with a black iron box are shown in Fig. 10.21. The box, both inside and out, warmed up much more rapidly than the air, a consequence of its low albedo and low specific heat. By midday, the air temperature had only reached about 47°C, while the box heated up to about 80°C; the air inside was about 66°C. On both days the east-facing side heated up first, considerably before the horizontal top face or the air, but it began to cool by early morning. The west-facing side heated up several hours later and attained its maximum temperature in mid-afternoon. The temperature on the west-facing side reached 84°C on the second day. The occurrence of sand storms markedly decreased the external temperatures of the box but not the air temperature.

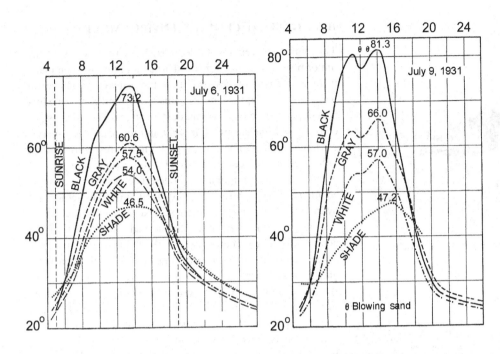

Fig. 10.20 Diurnal variations in temperature on iron plaques painted different colors and placed on the sand (middle unlabeled curve) at Touggourt, Algeria, on two July days. Air temperature in the shade is given for comparison (from Dubief 1959).

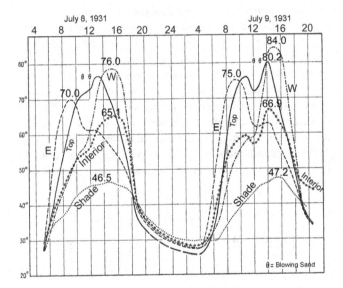

Fig. 10.21 Diurnal temperature variations on the top and the east and west faces of an iron box, painted black and placed on the sand, at Touggourt, Algeria, on two July days (from Dubief 1959). Temperature inside the box and air temperature in the shade are given for comparison.

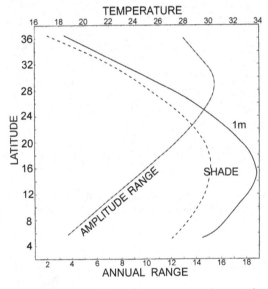

Fig. 10.22 Latitudinal variation of temperature at the ground surface and at 1 m depth (from Dubief 1959). Annual temperature range is also shown.

10.5 THE MOST EXTREME NICHES: DESERT DEPRESSIONS

Desert depressions are variously called playas, pans, flats, sebkhas, dry lakes, and many other local names. They provide a surface environment that contrasts sharply with that of the surrounding desert, be it dunes or a sparsely vegetated surface. The environment of a typical playa is extreme because its geometric form and surface characteristics enhance heating and because abrupt floods can suddenly transform the arid surface (see Fig. 13.26).

Depressions that are subject to flooding or with sufficient subsurface water may sustain a sparse vegetation cover (Figs. 10.22 and 10.23). More typical, though, are the vast, light-colored plains that characterize Etosha Pan in northern Namibia (Fig. 10.24), Sua Pan (within the Makgadikgadi pan system of Botswana) (Fig. 10.25), or the Bonneville salt flats of Utah (Fig. 13.1). Sediments are usually fine-grained or evaporite deposits and the surface is generally exceedingly flat. Such depressions are highly reflective and aerodynamically smooth.

Fig. 10.23 Sossusvlei, in the central Namib Desert, supports a partial vegetation cover.

Fig. 10.24 Vast white plains of the Etosha Pan, northern Namibia. Numerous dust devils are just visible at the edge of the pan.

Fig. 10.25 Sua Pan, a part of the Makgakgadi pan system of Botswana.

10.5.1 GENERAL CLIMATIC ENVIRONMENT

The common perception of a desert is a barren area with spatially uniform surface characteristics. This assumption is often made in many climate studies, especially those involving numerical models. Desert depressions belie this picture and contribute to a mosaic of surface types within desert regions. They show stark contrast with the surrounding sand and rock surfaces in terms of climate and, in some cases, vegetation. The resultant contrast influences the surface thermal and hydrologic environment.

Despite many commonalities of desert depressions in terms of geometry and moisture availability, their surface materials and hydrologic characteristics are markedly diverse. Consequentially it is hard to generalize about the thermal environment of the depressions. The higher albedo of the light-colored materials will moderate the thermal environment, but the finer particles may be less efficient than sands in transferring heat to the sub-surface. If flooding has left a residual layer of moisture near the surface, this will also ameliorate the thermal extremes. The hard crusts of some depressions inhibit heat and moisture flux.

In many depressions groundwater is an active force (Fig. 10.26); capillary action pulls water from the depths upward toward the surface. This movement produces patterns of surface relief that increase roughness and reduce reflectivity. Such depressions are often significant sources of surface evaporation and latent heat transfer to the atmosphere.

One commonality of most desert depressions is a fast and turbulent wind regime, particularly in the afternoon, because friction is low over the smooth surface. Over the dry surface of El Mirage Lake in California (Vehrencamp 1953), wind speeds at a height of 2 m range from near 0 in the early morning to 8 m/s in the late afternoon, the time of maximum surface heating. Turbulence is also intense in the unstable air over the pans and

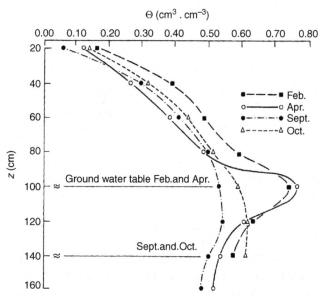

Fig. 10.26 Soil moisture content (Θ) and groundwater table of Idri playa during February–October 1978 (from Menenti 1984).

dry lakes, so that dust devils are a common occurrence. Over some desert surfaces, the high-velocity winds create a constant roar.

Sellers (1965) gives a surface roughness value of 0.003 for a dry pan in California and 0.001 for smooth mud flats. Over the exceedingly smooth surface of dry lakes, vehicles are tested for high-speed performance. A long-standing record of 630 miles per hour (mph) was set at Lake Bonneville, Utah in 1970. The current record, 760 mph, was set in 1997 over a dry lake bed in the Black Rock Desert of Nevada. That speed broke the sound barrier.

10.5.2 SURFACE THERMAL AND HYDROLOGIC REGIME: MICROCLIMATIC ASPECTS

The microclimatic environment of a desert depression depends primarily on its surface characteristics and the quantity and vertical distribution of surface and subsurface water. The geomorphic processes that produced the depression determine the composition and texture of the surface materials, the roughness and the stability of the surface. The available surface and subsurface moisture determines whether vegetation cover is present. Together these characteristics determine the radiative, thermal, and hydrologic environments of the depression and its contrast with other areas of surrounding desert.

Extensive microclimatic measurements have been made over a playa in northwestern Utah (Malek *et al.* 1990, 2002; Malek 2003). These have concentrated on evaluation of the radiation, energy, and water balances. One interesting finding is that, despite the region's prevailing aridity, 25% of the available net radiation was utilized for evaporation. The total evaporation during the year 2000 was 168 mm, while precipitation amounted to 108 mm. This indicated that considerable water was extracted from the shallow water table of the playa, a finding confirmed by studies of other playas (e.g., Tyler *et al.* 1997). However, saline crusts can inhibit the extraction of subsurface water via evaporation, even when the water table is near the surface (Kampf *et al.* 2005). The measurements in Utah also indicated that the playa receives at least 10% more solar radiation than typical urban meteorological stations, where wind speeds were some 30–50% lower than over the playa.

One of the most detailed studies of desert depressions is that of Menenti (1984). He utilized remote sensing and field measurements to study these contrasts in the playas of the Idri region of western Libya. The NNW playa has a water surface, the NE playa is dry soil, and the W playa is wet soil. Clear differences are seen between the various playa surfaces. At high sun angles, reflectivity is roughly 0.18 for a playa with a water surface, 0.25 for a playa with wet soil, and 0.5 for a dry playa. The differences between the three playa surfaces are greater than the seasonal changes in reflectivity of the surfaces. This produces small-scale variations in the surface reflectance across the region (Fig. 10.27).

Other notable contrasts are in the surface thermal conductivity and latent heat flux. The rocks of the Qarqaf highlands

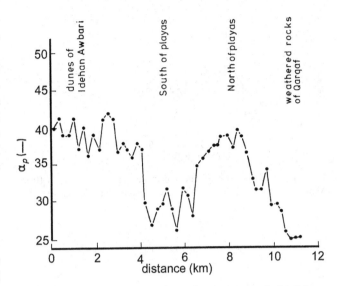

Fig. 10.27 Cross-section of a surface reflectance map calculated from Landsat data; the cross-section extends from the dunes across the playas toward the Qarqaf highlands (from Menenti 1984).

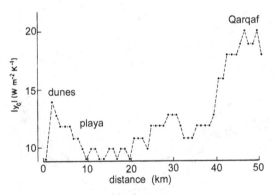

Fig. 10.28 Cross-section of thermal admittance (y_0) of the surface in the Idri region in September 1978 (from Menenti 1984).

transfer a considerable amount of heat to the subsurface, but the dry playas do not (Fig. 10.28). The playas lose heat to the atmosphere via evaporation and latent heat flux (Fig. 10.29). There is some loss from the region of dunes, probably from vegetation on the dunes, and almost none over the rocks of the Qarqaf.

The surface temperature integrates the various effects of reflectivity, thermal conductivity, and latent heat flux. It also shows large variations across the region. On a single day in winter, satellite estimates of surface radiative temperature across the region varied from 298°K in the dunes to 304°K in the playas and over the rock surface. Ground measurements showed even greater contrasts.

Sua Pan, in Botswana, may be the only playa with a permanent first-class meteorological station. Extensive studies of surface reflectivity have been made here (e.g., Abdou *et al.* 2006). Figure 10.30 gives recent climate statistics for Sua Pan and two nearby stations: Letlhakane, to the west and surrounded by

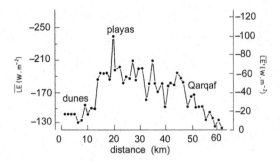

Fig. 10.29 Cross-section of latent heat flux in the Idri region (from Menenti 1984). The absolute values are imprecise, but the relative trends are reliable.

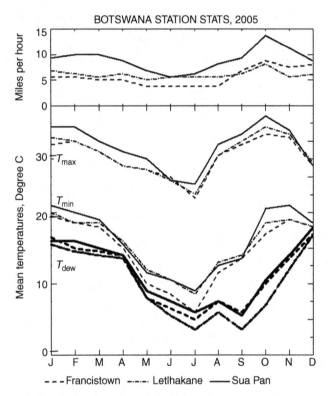

Fig. 10.30 Meteorological conditions at Sua Pan, Botswana, compared with conditions at Francistown and Letlhakane (top, wind; bottom, maximum, minimum, and dew-point temperatures).

desert vegetation, and Francistown, to the east and surrounded by woodland vegetation. Monthly means for the year 2005 are shown for daily maximum and minimum temperatures, dew-point temperature, and wind speed.

The largest contrast is in wind speed, but contrasts are also apparent in temperature and humidity. Mean speeds are roughly twice as high over the aerodynamically smooth surface of the pan. On many days, wind at Sua Pan is 15–20 m/s, while it is calm or light at the other locations. Sua Pan is consistently warmer than the other stations by several degrees, with maximum differences in summer. The contrast with Francistown is particularly striking, roughly 3°C for both the maximum and

minimum temperatures. For Letlhakane the differences are much greater for maximum temperature than for minimum. The warmer maximum temperatures can be understood in terms of the dry and barren surface. An explanation for the higher minimum temperatures is not readily apparent. One possible factor is subsurface moisture. When the author visited the region in 1992 the pan was completely dry, but in recent years there has been standing water in some areas, particularly the southeast. Flooding and subsequent surface evaporation would have left a subsurface moist layer, which could keep temperatures warmer at night, compared with the dry surface of the other stations. The standing water is also consistent with the higher dew points at Sua Pan, but advection from the more humid east might also play a role.

10.5.3 THERMAL EXTREMES

Dry desert depressions are the hottest environments on earth. Excellent examples are Death Valley in the western USA, the Qattara Depression of Egypt, and the chotts of Tunisia and northeastern Algeria. At El Arfiane, Algeria, on the edge of the chotts (elevation 23 m) an air temperature of 49.5°C has been recorded and mean monthly temperature (day and night average) reaches 40°C in July and August. These values are relatively high for its latitude, 34° N (Dubief 1959). In Death Valley, where air temperatures as high as 56.7°C have been recorded, the mean maximum temperature exceeds 46°C during July (Roof and Callagan 2003).

Death Valley (elevation −59 m) and the Qattara Depression of Egypt (elevation −133 m) are desert depressions only in a morphological sense. Unlike true playas, they are not drainage features, although playas exist within them. They are included here to illustrate the impact of geometry on the thermal extremes. Both have remarkably steep sides (Fig. 10.31) and lie well below sea-level. With the forbidding summer heat, temperature records for the Qattara Depression are unavailable, but

Fig. 10.31 Photo of the Qattara Depression of Egypt. The steep walls of the depression, seen in the upper left, sharply drop to the depression floor, which is 300 m below sea-level.

Fig. 10.32 Vehicles attempting to ascend the valley walls of the Qattara Depression. The steep sides and soft sand make the ascent extremely challenging. In this case, each made numerous unsuccessful attempts, so that the caravan spent several hours trying to exit the depression.

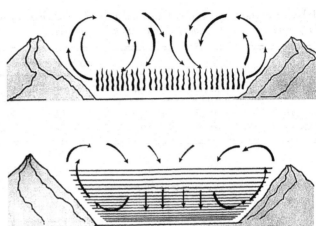

Fig. 10.34 Schematic illustrating the mechanism of extreme heating of a desert depression.

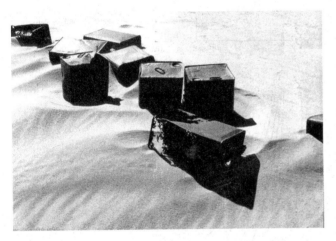

Fig. 10.33 Photo of gas canisters left over from World War II in regions surrounding the Qattara Depression.

Fig. 10.35 Photo showing the sand piled up along the walls of the Qattara Depression by high winds moving upward from the hot valley floor.

summer temperatures would probably exceed 58°C, the record set at Azizia, Libya. Much of the region is impassable by vehicles of any kind (Fig. 10.32). Thus, during World War II, the Qattara Depression provided natural protection for the British troops isolated on the coast at El Alamein (Fig. 10.33)

The extreme heat results primarily from the radiative geometry of the basin (Fig. 10.34), but other factors can enhance this effect. In Death Valley, for example, heating is intensified by a subsidence inversion created as air passes over the surrounding mountains and sinks in the lee (Roof and Callagan 2003). Also, the prevailing winds are perpendicular to the orientation of the valley, keeping the warm air trapped within the valley.

During the day, the valley bottom and sides become extremely hot (Fig. 10.34). In Death Valley, rock surface temperatures have reached 93°C. The hot light air is constrained by the basin walls and rises along them. It is replaced by cooler air descending over the center of the basin. Closed circulation cells develop, with the descending air heating up adiabatically. As air continues to descend over the central basin, it is further heated by the warm basin floor and again rises in the ascending branch of the cell. Its velocity

can be sufficient to transport sand (Fig. 10.35). Heat is continually added as it passes over the basin walls and further heats up during the subsequent descent. In this way a cycle of heating continues during the day, causing extreme air temperatures.

At night, the basin floor and walls cool as efficiently as they warmed up during the day. The cold air drains to the bottom, creating a temperature inversion within the basin at night. This further stabilizes the air in the depression and keeps the cold air from mixing with the warmer overlying air. In this way, the diurnal temperature fluctuations are magnified in the desert depressions.

10.6 MODERATE ENVIRONMENTS: OASES, IRRIGATED FIELDS, DESERT LAKES, AND RIPARIAN VALLEYS

The presence of water near or at the surface substantially modifies the microclimate of the surface and overlying air layer. In the case of desert lakes, the water itself is the primary factor, but

Table 10.2. *Annual energy and water balances of Tunisian oases (after Flohn and Ketata 1971). Radiative fluxes are in units of MJ/ m² per day; evaporation and precipitation are in mm/year.*

Surface	Albedo	Radiative fluxes			Bowen ratio	Water balance	
		Net	Sensible	Latent		Evaporation	Precipitation
Desert	0.20	6.9	5.8	1.0	5.8	150	150
Oases	0.15	8.6	−3.1	11.8	−0.26	1680	150

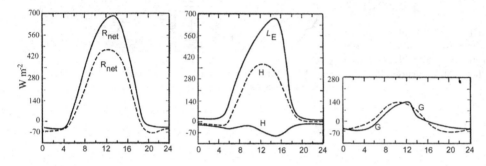

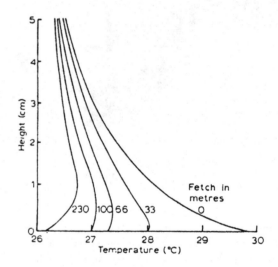

Fig. **10.36** Diurnal variations of the heat balance in the semi-desert of Turkestan (dashed line) and an adjacent, irrigated oasis (solid line) (modified from Flohn 1969 and Budyko 1956). Left: net radiation; center: latent (L_E) and sensible (H) heat transfer to the air; right: sensible heat transfer G to the soil. E is zero for the semi-desert.

in other situations, such as oases, irrigated fields, and riparian valleys, surface vegetation cover further ameliorates the microclimate. In some cases, these changes might suffice to produce local boundary-layer circulation systems that can, in turn, modify cloudiness.

10.6.1 DESERT OASES AND IRRIGATED FIELDS

A desert oasis extracts heat and energy from the warmer overlying air. If moisture supply in the oasis is adequate, the daily evaporation is sustained at the potential rate and net radiation is roughly balanced by latent heat transfer (Fig. 10.36). During the daytime hours the overall energy balance (the net radiation) is positive, and sensible heat is transferred from the surface to the soil below. At night, there is a negative balance and the soil conducts heat to the surface. Throughout the entire day the sensible heat transfer is directed from the overlying air to the surface. As a result, the air cools markedly as it passes over the oasis. This is the so-called "oasis effect." The contrast with the surrounding desert is clear: in the desert, net radiation is almost completely balanced by the sensible heat transfer to the air.

A study in Tunisia (Table 10.2) similarly showed that the oasis, on average, gained heat from the atmosphere. The Bowen ratio was negative (−0.26), while in the surrounding semi-desert, with a mean annual rainfall on the order of 150 mm, the Bowen ratio was 5.8. The contrast in albedo between the two areas was not particularly large, 0.20 for the semi-desert, compared with 0.15 for the oasis.

Figure 10.37 shows the temperature change with the passage of desert air across an oasis (an irrigated field of rye grass in the middle of a desert). Evaporation from the oasis surface augments the moisture content in the overlying air. The depth of the moist air deepens continually as the air traverses the oasis.

Fig. **10.37** Cooling and modification of the temperature profile of warm air as it passes over an irrigated field of ryegrass (from Dyer and Crawford 1965).

At the same time, the overlying air cools as it loses heat to the cooler oasis surface. This creates an internal boundary layer that is cooler and moister than that over the desert. This boundary layer continuously diminishes in intensity and depth downstream, as evaporation ceases and heat transfer is again from the surface to the air. The moisture gained as the air passes over the oasis can have an impact on the temperature and humidity conditions in desert areas downwind (Zhang *et al.* 2003).

Omara (1971) measured the ground temperatures at the Giza Agrometeorological Station near Cairo, comparing a dry desert site with a bare soil kept near field capacity and a site with short vegetation cover (also kept near field capacity). The latter two

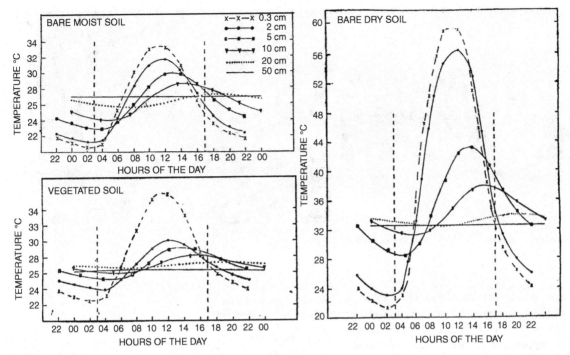

Fig. 10.38 Diurnal variation in temperature on a typical late summer day as a function of depth at three sites in the Egyptian desert near Giza: bare dry soil, bare moist soil, vegetation (from Omara 1971).

sites are analogous to oases. Figure 10.38 shows the temperatures at the three sites as a function of depth. Over the dry soil, temperatures at 0.3 cm depth climb to nearly 60°C during the day but plummet to 22°C at night. Daytime temperature drops by over 25°C in the first 10 cm. In contrast, daytime temperature reaches about 36°C over the vegetated plot, and 34°C over the bare moist soil. The difference is presumably attributable to heat stored in the vegetation. In both cases it drops at night by only some 12°–13°C. The annual patterns, however, are almost identical for the vegetation and moist soil, but the bare soil heats up much more intensely in summer (Fig. 10.39). The degree of winter cooling is comparable at all three sites.

10.6.2 DESERT LAKES

Large permanent lakes, such as Lake Chad and the Aral Sea, are sustained in arid environments via runoff from surrounding humid areas. As in the case of an oasis, these provide a comparatively mild climate for some distance away from the shores. The contrast between a bare desert surface and a desert lake is extreme.

When the Aswan Dam was built, the Egyptian Meteorological Authority (el-Bakry and Metwally 1982) set up a station to study the impact of the resulting Lake Nasser on climate. Comparisons were made between a land station (the Aswan Airport) and a station on a raft over the lake. Temperature over the lake was only about 1°–2°C cooler than over the land on average; the lake was warmer than the land during some winter months. The differences were occasionally as large as about

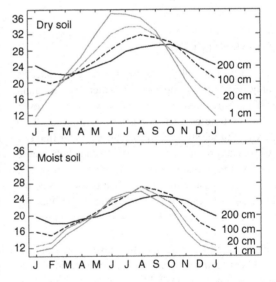

Fig. 10.39 Annual variation of temperature at the bare soil sites (dry and moist) shown in Fig. 10.38 (from Omara 1971).

5°C. Wind speeds were generally up to 1 m/s lower over the lake, despite the decrease in aerodynamic roughness. Relative humidity also increased substantially over the lake from an annual average of roughly 25% to about 35%.

Segal *et al.* (1983) used a mesoscale meteorological model to simulate the effects of flooding the Qattara Depression of Egypt. Their model suggests that the microclimate would be affected for some 200 km distance from the lake, but that the

Fig. 10.40 Vegetation in dry stream beds in Namibia (left) and Australia (right).

effect was not always a moderating one. The simulation showed that on one side of the lake, winds at 2 m above the surface would increase by up to 3 m/s and temperature would decrease by up to 5°C. On the opposite side, where the lake breeze converges with the sea breeze from the Mediterranean, the model indicated that wind speed would decrease and temperatures would become more extreme. Thus, the net effect of a desert lake depends to some extent on the larger-scale meteorological situation.

10.6.3 RIPARIAN VALLEYS

The case of a riparian valley is quite different, in that the surface is only occasionally flooded with water and the main contrast with the surrounding desert is the vegetation cover. In many cases, the time between floods may be around 10–20 years. This creates, in the stream bed, subsurface water that can support lush vegetation growth. Figure 10.40 shows the dense vegetation in dry stream channels in the Namib and Australian deserts. In the former case, only 10 years had elapsed since the last flood but a dense canopy of trees had developed. Once the tree canopy develops, the ground retains more moisture from subsequent rains to help sustain the vegetation during drier years. The deep-rooted trees also tap the subsurface water. In some cases, the dry channels contain fossil water from former streams. This is evident in the Dallol Bosso of Niger, where lush agriculture is supported on the former channels of the Niger. Other microclimatic effects include the results of floods spilling onto lands surrounding the river. A classic example is that of the Nile, which supports lush vegetation along its banks in what is climatically an extreme desert. The daounas of Mali, depressions that receive the overflow from the Niger in wet years, also support extensive agriculture.

The channel geometry influences the thermal regime of a riparian valley (Kassas and Imam 1957). Soil temperatures in a runnel (water channel) of the Nile valley near Cairo were on average about 10°C cooler than on the slopes above (Fig. 10.41).

Sa = Soil temperature at the top of the South facing slope

Sb = Bottom of the South facing slope

Na = Top of the North facing slope

Nb = Bottom of the North facing slope

Fig. 10.41 Schematic illustrating a water channel in which micrometeorological measurements were made near Cairo, Egypt, in 1957 (modified from Kassas and Imam 1957).

The south-facing slope was also consistently warmer than the north-facing slope throughout the year (Fig. 10.42). The contrast was markedly greater during the day than at night (Fig. 10.43).

Williams (1954) measured soil and air temperatures on a flat wadi bed near Cairo. A comparison with the above illustrates the overall impact of slope (Fig. 10.44). Surface temperatures are higher over the flat wadi bed during the day (56°C vs. 47°–50.5°C at various locations on the slopes). This probably represents the diminished intensity of the solar beam when received over a slanted surface. At night the wadi bed is colder by 5°–8°C. In this case, some of the radiation emitted at night by the slopes is probably retained within the valley, reducing the net nocturnal cooling. This is analogous to the desert depression illustrated in Fig. 10.34.

The geometry also affects moisture conditions, including the penetration of moisture into the soil, evaporative loss, and runoff. Evaporative loss depends on the facing and location on the slope because of the varying conditions of insolation and temperature at these sites. Table 10.3 shows soil moisture in the runnel near Cairo at various depths at two slope locations on December 21 and February 17. In December water has only begun to percolate

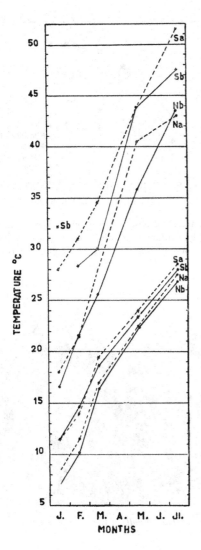

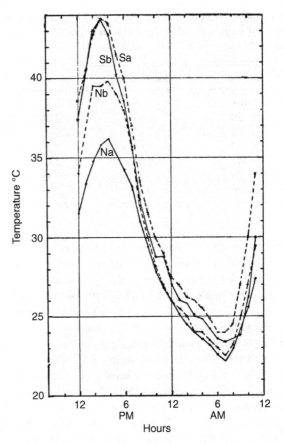

Fig. 10.43 Diurnal cycle of soil temperature in May at the top and bottom of a north- and south-facing slope of a riparian valley in Egypt (see Fig. 10.41, from Kassas and Imam 1957).

Fig. 10.42 Maximum and minimum temperatures (monthly means) at the top and bottom of north- and south-facing slopes of a riparian valley near Cairo (see Fig. 10.41, based on data in Kassas and Imam 1957).

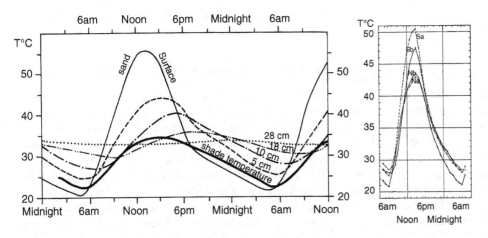

Fig. 10.44 Diurnal cycle of summer soil temperatures over a sand patch in a wadi bed (left, from Williams 1954) and in a riparian valley near Cairo (right, from Kassas and Imam 1957).

Table 10.3. *Soil moisture at different depths (cm) at two points on the slopes of a riparian valley near Cairo on December 21 and February 17 (data from Kassas and Imam 1957).*

Site	Depth	December 21	February 17
K2	0–10	3.86	2.86
	10–20	4.97	2.76
	20–30	3.22	5.94
	30–40	1.60	9.60
L8	0–10	2.41	2.03
	10–20	3.52	2.19
	20–30	0.42	2.93

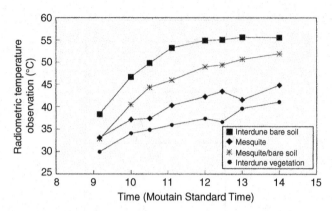

Fig. 10.45 Temperature observations from handheld radiometers for bare soil in the interdune flats, mesquite, a mixture of mesquite and bare soil, and interdune vegetation at the Jornada experimental range of New Mexico (from Havstad *et al.* 2000).

to greater depths and the maximum is in the 10–20 cm layer. By February, soil moisture clearly increases downward as evaporation and percolation reduce moisture in the layers near the surface. By then soil moisture is notably higher in the lower portion of the slope. At the surface itself (i.e., 0–5 cm layer) the higher elevations always have a lower soil moisture content.

The impact of slope on microclimatic conditions is apparent not only in riparian valleys, but in any region of significant terrain. This strongly impacts the distribution of vegetation. In the Sonoran Desert of North America (Nobel and Linton 1997) the percentage ground cover of the C3 shrub *Encelia farinosa* was eight times higher on a 20° south-facing slope than on a similar north-facing slope at 820 m elevation. This was apparently a consequence of higher temperatures impacting rooting depth. Growth of a C4 bunchgrass and a CAM leaf succulent was also more prolific on north-facing slopes, but probably because of greater moisture availability.

10.7 MODERATE ENVIRONMENTS: MICROCLIMATES IN VEGETATION AND UNDER THE SAVANNA CANOPY

Vegetation cover strongly moderates the surface environment, particularly temperature and wind. Its influence is manifold. It acts as a radiative surface that both emits and absorbs, and hence can act as either a sink of radiation and heat (as at night) or a source (as during the day). The canopy captures and re-evaporates water and helps to retain soil moisture. Transpiration from the leaves enhances the humidity of the air in a plant cover. Thus, vegetation serves as a source of moisture and a sink of momentum.

In arid regions individual plants ameliorate the microclimate. In semi-arid regions, the microclimate is controlled by the canopy structure and relative abundance of the three functional elements: grasses, shrubs, and trees. However, the ambient climate still has an overriding influence. The effect of a canopy is generally greater in the harsher climate of the tropics than in the more moderate regime of the mid-latitudes. Advection puts much heat into the energy budget of the subcanopy area. Transpiration and leaf temperatures may therefore remain high in the subcanopy zone, despite shading and the resultant reduction in soil temperature.

Fig. 10.46 *Welwitschia mirabilis* of the Namib Desert.

The microclimatic impact of vegetation is illustrated in Fig. 10.45, which compares the temperatures in four locations in the Jornada experimental range of New Mexico: bare soil in the interdune flats, mesquite, a mixture of mesquite and bare soil, and interdune vegetation (Havstad *et al.* 2000). Throughout the day temperatures are highest in the bare interdune soil and lowest in the interdune vegetation. Maximum contrast is obtained around noon, when the temperature range between the sites is nearly 20°C. Early in the morning the range is less than half that.

10.7.1 PLANT TEMPERATURES

An extreme example of the moderating influence of vegetation is the *Welwitschia* plant, endemic to the Namib Desert (Marsh 1990). A typical plant (Fig. 10.46) can be 1–2 m across. Figure 10.47 shows temperature measurements made near a *Welwitschia* plant in January and April. In January, surface temperatures reached 78°C, compared with a relatively cool air temperature of 39°C. Subsurface temperature was 60°C near the plant, but only 37°C under the leaves, where the surface temperature in the leaf litter was 41°C. In the fairly well-shaded stem of the plant, temperature was as low as 32°C. In April, when the air temperature was only 8°C cooler, ground temperature had decreased by 24°C to make it 54°C. Thus, the plant and the air layer within its leaves maintain temperatures comparable to the ambient air, creating a much more favorable thermal environment for desert fauna. The plant is home to beetles, lizards, and other small animals.

An analogous habitat is that in and around the Saguaro cactus of the Sonoran Desert in the southwestern USA. Its more moderate temperature and higher moisture content make it hospitable to birds, small animals, ants, flies, and other insects. Birds and packrats build nests in Saguaro cacti (Fig. 10.48). Gibbs *et al.* (2003) studied the microclimate of the Saguaro and two other columnar cactus species of the Sonoran Desert, the Senita and the organ pipe cactus (Fig. 10.49). The temperature of air inside the cacti and inside the rotting tissue of damaged plants (Fig. 10.50) was compared with the temperature of the outside air. The Senita heats and cools markedly more rapidly than the

other two species (Fig. 10.51); heating and cooling is least rapid in the necrotic tissue, making it a "comfortable" habitat for flies.

The interiors of all three cactus species have mean temperatures similar to those of the outside air, but they maintain lower

Fig. 10.48 Saguaro cactus with numerous openings for birds and small animals to nest in.

Fig. 10.47 The microclimate around a *Welwitschia* plant in April and January (based on Marsh 1990).

Fig. 10.49 Sonoran cacti: Saguaro (left), Senita (center), and organ pipe (right).

Fig. 10.50 Necrotic (rotting) tissue in dying Saguaro cacti. The rot caused the arms on the Saguaro on the right to collapse.

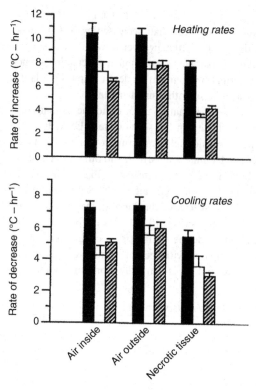

Fig. 10.51 Heating and cooling rates of the cacti shown in Fig. 10.49: Saguaro (black), Senita (white), and organ pipe (diagonals) (from Gibbs *et al.* 2003). Rates are given for air inside, air outside, and necrotic tissue.

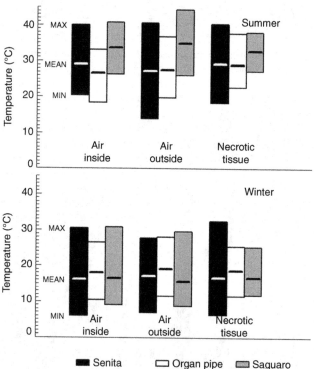

Fig. 10.52 Mean, maximum and minimum temperature in winter and summer for the cacti shown in Fig. 10.49 (from Gibbs *et al.* 2003). Temperatures are given for air inside, air outside, and necrotic tissue.

maximum temperatures in summer and higher minimum temperatures in winter (Fig. 10.52). For the Saguaro, the interior air is about 4°C cooler than outside on summer afternoons and about 1°C warmer at night in winter. The rotting tissue further minimizes the environmental extremes, with maximum temperatures some 6°–8°C cooler than the air in summer and minimum temperatures 2°–3°C higher than the air in winter. Conditions are fairly similar in the organ pipe cactus but the Senita cactus moderates the environment somewhat less (Fig. 10.52).

Even considerably smaller plants have thermal regimes that differ from that of the ambient air. Each plant surface acts to absorb and emit radiation, but its "efficiency" as a thermal surface depends on both the orientation and geometry of the surface and on its composition. Temperature contrasts between the four cactus species shown in Fig. 10.53 are presented in Fig. 10.54 for spring (March) and summer (June). In general, the greatest temperature contrast between vegetation and air occurs during the colder season; the difference can reach almost 20°C. For two species of *Opuntia*, plant temperature is lower than air temperature during the hot season and higher than air temperature

Fig. 10.53 Four cactus species: (a) *Echinocereus engelmannii* (http://cactiguide.com), (b) *Opuntia acanthocarpa*, (c) *Opuntia bigelovii*, and (d) *Opuntia engelmannii*.

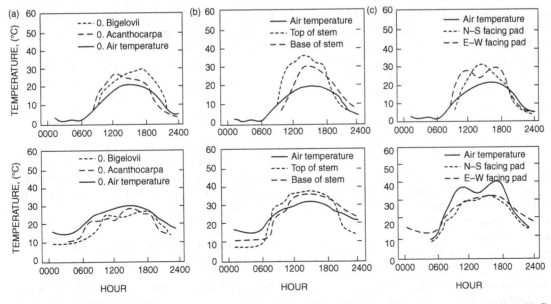

Fig. 10.54 Diurnal march of temperatures of air and of four cactus species: (top row) March; (bottom row) June. (a) Comparison of *Opuntia bigelovii* and *Opuntia acanthocarpa* (from Gibbs and Patten 1970), (b) comparison of top and base of stem of *Echinocereus engelmannii*, (c) comparison of N–S and E–W facing pads of *Opuntia engelmannii*.

during the cold season. This reflects the absorption of solar radiation by the plant surfaces. For *Opuntia engelmannii* the orientation of the pad modifies temperature by up to 10°C. For *Echinocereus engelmannii* the top of the stem is warmer than its more sheltered base.

Rauh (1985) reports on a cactus in the central Peruvian desert with microclimatic adaptations to extreme aridity that involve both temperature and moisture content. *Melocactus peruvianus*, a small barrel-like cactus, survives surface temperatures of 60°–70°C. Under such conditions, the temperature of its central water tissue can be as low as 45°C. This species also tolerates extreme desiccation by shrinking completely, then expanding after the sporadic rains, restoring the water reserves in its tissues.

10.7.2 MICROCLIMATE OF THE TROPICAL SAVANNA

Vegetation influences temperature, humidity, and wind in the layer between the canopy top and the bare soil, but the resultant near-surface profiles of these properties depend greatly on the structure of the vegetation layer. Three distinct structures are those of grass, shrubs, and trees. Some savannas include all three elements. The active surface in the exchange process is that where foliage is maximized. Each element has a characteristic pattern of temperature, humidity, and wind (see Fig. 6.19). The fluxes of heat and moisture are directed away from the canopy or grass layer, while the flux of momentum is toward them. At night, the flux of heat is toward the canopy, which cools more rapidly than the ambient air or the subsurface.

Tree leaf area index (LAI) in savannas is low compared with that of forests or grasslands. LAI in the subcanopy zone may be 2–6 times higher than the average for the savanna as a whole because the tree leaves are typically concentrated into a tree cover fraction of 15–50%. A LAI of 0.5–1.0 is typical for dry savannas (Scholes and Archer 1997).

The canopy of the tree or shrub layer shields the grass layer from the full intensity of solar radiation (Fig. 10.55), thereby reducing evaporation of soil moisture, and it also prevents extensive heat loss from the surface at night (Kidron 2009). The degree to which the canopy reduces solar radiation depends on the cover, height, and spatial pattern of the canopies – characteristics that vary along the grassland/forest continuum (Martens *et al.* 2000). The reduction of both direct and indirect solar radiation can range from 25% to 90%. Since the semi-arid subtropics are some of the sunniest areas in the world, the amount of radiation reaching grasses below the canopy may still be sufficient for a relatively high rate of photosynthesis. The canopy also blocks the force of the wind and transpires moisture into the subcanopy air layer. This ameliorates the ambient environment under the canopy, moderating the thermal and moisture regimes and thereby facilitating the growth of the grass layer. The grass layer shields the bare soil from the sun.

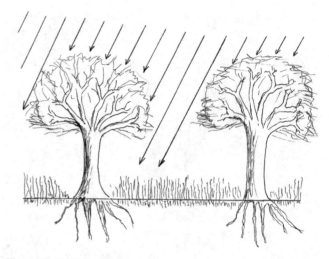

Fig. 10.55 Shading effect of trees in a savanna.

Measurements taken in a West African savanna illustrate the moderating influence of the vegetation on the ambient environment near the surface (Lewis and Berry 1988; d'Hoore 1954). In the early afternoon, when the air temperature reached 36°C, soil temperature at a depth of 1 cm reached nearly 53.5°C in bare soil and 50.4°C in the soil under the savanna grass cover. In contrast, the temperature under the shrubs was only 26.9°C. Even at a depth of 10 cm, temperatures were markedly lower under the shrubs (25.4°C vs. 36.8°C under the bare soil).

Another good example of the microclimate produced by vegetation comes from the HAPEX-Sahel and SEBEX experiments, conducted in Niger in 1991/92 and 1989/90, respectively. The vegetation in the region consists of bush and grass savanna (Fig. 10.56a) and some areas with a formation called "tiger bush" or "*brousse tigre*." The latter is a patterned formation of dense thickets surrounded by bare soil in varying proportions; the resultant effect resembles the irregular pattern of stripes on a tiger (Fig. 10.56b). An experimental savanna plot, from which the bush elements had been cleared, was used to represent the grass and herb understory and create an environment typical of the Sahel grassland further north. Radiation, temperature, and other properties were measured in the vegetation layers and in the underlying soil.

Contrasts are clearly evident in the thermal regimes (Verhoef 1995). Temperature is generally lowest in the bush canopies and highest in the soil. Even at a depth of 10 cm, soil temperatures are lower beneath the bushes than beneath the understory or on the bare soil surface (Fig. 10.57).

10.7.3 MICROCLIMATE OF A MID-LATITUDE SEMI-ARID WOODLAND

A mid-latitude woodland is structurally similar to a tropical savanna, with a mixture of tree canopy and intercanopy grass. The main differences are in the size and spacing of the elements, the solar elevation angle, and the seasonal contrasts in the

(a)   (b)

Fig. 10.56 Vegetation formations in the West African Sahel of Niger: (a) savanna, (b) tiger bush.

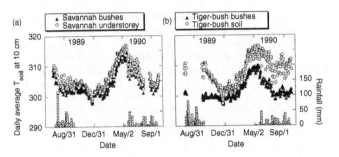

Fig. 10.57 Daily averages of soil temperature at 10 cm depth beneath the SEBEX savanna (a) and tiger bush (b) (from Verhoef 1995). Rainfall at both sites is indicated by bars.

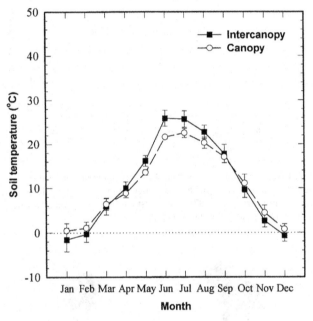

Fig. 10.58 Average monthly soil temperature (°C) in canopy and intercanopy patches at a depth of 2 cm in a semi-arid mid-latitude woodland (from Breshears *et al.* 1998).

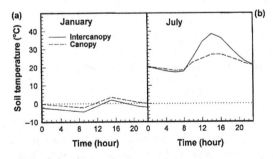

Fig. 10.59 Average diurnal soil temperature (°C) in canopy (dashed line) and intercanopy (solid line) patches at a depth of 2 cm in a semi-arid mid-latitude woodland during (a) January and (b) July (from Breshears *et al.* 1998).

radiation regime. The generally lower elevation angles allow more radiation to penetrate under the canopy. This reduces the contrast between canopy and intercanopy space in the mid-latitudes, compared with the tropics.

Figure 10.58 shows this for a semi-arid woodland in piñion–juniper woodland in northern New Mexico (Breshears *et al.* 1998). Soil temperatures at 2 cm depth were higher in intercanopy patches in summer, with differences in monthly averages on the order of 5°C; in winter, canopy patches are warmer, but the difference is only about 2°C. The contrast is greatest when the sun is most intense. Thus, in summer, the midday soil temperature difference between canopy and intercanopy can be nearly 15°C. In winter, when the controlling influence is longwave cooling, this is reduced to about 2°C (Fig. 10.59).

The canopy and intercanopy patches also differ with respect to the soil moisture regime (Breshears *et al.* 1997). This is a consequence both of interception of precipitation by the canopy, reducing understory moisture, and increased evaporation over the warmer intercanopy patches. The intercanopy patches generally have significantly greater soil moisture, especially during periods of precipitation (Fig. 10.60). In summer, when insolation is higher, the soil in intercanopy patches can support potential evaporation equivalent to the maximum for its soil

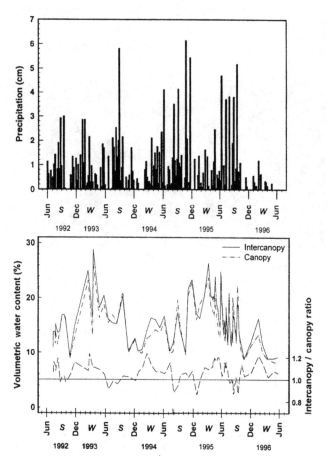

Fig. 10.60 (top) Precipitation and (bottom) volumetric soil water content for canopy (dashed line) and intercanopy (solid line) locations (from Breshears *et al.* 1997). The lowest curve gives the ratio of intercanopy to canopy soil water content.

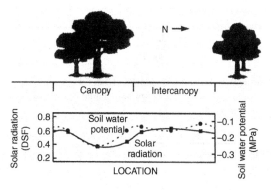

Fig. 10.61 Mean values for gradients in solar radiation and soil water potential along a north–south transect through a canopy and an intercanopy patch (Breshears *et al.* 1997).

texture, but values under the canopy are less than half that. Consequently, soil moisture under the canopy may exceed that in the intercanopy patches in summer.

The contrast is summarized in Fig. 10.61, which shows typical values of solar radiation along a north–south transect between

canopy and intercanopy patches. Radiation is relatively uniform within intercanopy space, but decreases sharply toward the center of the canopy patches. Soil moisture likewise decreases toward the center of canopy patches.

REFERENCES

Abdou, W. A., and Coauthors, 2006: Sua Pan surface bidirectional reflectance: a case study to evaluate the effect of atmospheric correction on the surface products of the Multi-angle Imaging SpectroRadiometer (MISR) during SAFARI 2000. *IEEE Transactions on Geoscience and Remote Sensing*, **44**, 1699–1706.

Breshears, D. D., P. M. Rich, F. J. Barnes, and K. Campbell, 1997: Overstory-imposed heterogeneity in solar radiation and soil moisture in a semiarid woodland. *Ecological Applications*, **7**, 1201–1215.

Breshears, D. D., J. W. Nyhan, C. E. Heil, and B. P. Wilcox, 1998: Effects of woody plants on microclimate in a semiarid woodland: soil temperature and evaporation in canopy and intercanopy patches. *International Journal of Plant Sciences*, **159**, 1010–1017.

Broza, M., 1979: Dew, fog and hygroscopic food as a source of water for desert arthropods. *Journal of Arid Environments*, **2**, 43–39.

Bryson, R. A., 1978: The albedo pattern of the La Joya dunes. In *Exploring the World's Driest Climate* (H. H. Lettau and K. Lettau, eds.), IES Report 101, Institute for Environmental Studies, University of Wisconsin, Madison, WI, pp. 54–56.

Budyko, M. J., 1956: *Teplowoj Balans Zemnoj Powerchnosti.* Gidrometeoizdat, Leningrad, 255 pp.

d'Hoore, J. L., 1954: *L'accumulation des sesquioxides libres dans les sols tropicaux.* INEAC (Institut National pour l'Etude Agronomique du Congo) Serie scientifique 62, Brussels, 132 pp.

Dubief, J., 1959: *Le Climat du Sahara.* Vol. 1. Mémoires de l'Institut de Recherche Saharienne, Algiers, 312 pp.

Dyer, A. J., and Crawford, T. V., 1965: Observations of the modification of the microclimate at a leading edge. *Quarterly Journal of the Royal Meteorological Society*, **91**, 345–348.

el-Bakry, M. M., and Z. Metwally, 1982: Relations between some parameters measured over land and lake stations at Aswan. *Meteorological Research Bulletin*, **14**, Egyptian Meteorological Authority, Cairo, 61–74.

Felton, E. L., 1965: *California's Many Climates.* Pacific Books, Palo Alto, CA, 169 pp.

Flohn, H., 1969: Local wind systems. In *General Climatology* (H. Flohn, ed.), World Survey of Climatology 2, Elsevier, Amsterdam, pp. 139–171.

Flohn, H., and M. Ketata, 1971: *Investigations on the Climatic Conditions of the Advancement of the Tunisian Sahara.* WMO Technical Note 116, World Meteorological Organization, Geneva.

Gay, L. W., 1970: *Energy Balance Estimates of Evapotranspiration.* Water Studies in Oregon, Oregon Water Resources Institute, Corvallis, OR.

Gibbs, A. G., M. C. Perkins, and T. A. Markow, 2003: No place to hide: microclimates of Sonoran Desert *Drosophila. Journal of Thermal Biology*, **28**, 353–362.

Gibbs, J. G., and D. T. Patten, 1970: Plant temperatures and heat flux in a Sonoran Desert ecosystem. *Oecologia*, **5**, 165–184.

Hastenrath, S. L., 1978: Physical properties of dune sand. In *Exploring the World's Driest Climate* (H. H. Lettau and K. Lettau, eds.), IES Report 101, Institute for Environmental Studies, University of Wisconsin, Madison, WI, pp. 89–103.

Havstad, K. M., W. P. Kustas, A. Rango, J. C. Ritchie, and T. J. Schmugge, 2000: Jornada experiment range: a unique arid land location for experiments to validate satellite systems. *Remote Sensing of Environment*, **74**, 13–25.

Jacobs, A. F. G., B. G. Heusinkveld, and S. M. Berkowicz, 2000: Dew measurements along a longitudinal sand dune transect, Negev Desert, Israel. *International Journal of Biometeorology*, **43**, 184–190.

Kadmon, R., and H. Leschner, 1995: Ecology of linear dunes: effect of surface stability on the distribution and abundance of annual plants. In *Arid Ecosystems* (H.-P. Blume and S. M. Berkowicz, eds.), *Advances in Geoecology*, **28**, 125–144.

Kampf, S. K., S. W. Tyler, C. A. Ortiz, J. F. Muñoz, and P. L. Adkins, 2005: Evaporation and land surface energy budget at the Salar de Atacama, Northern Chile. *Journal of Hydrology*, **310**, 236–252.

Kassas, M., and M. Imam, 1957: Climate and microclimate in the Cairo desert. *Bulletin de la Société de Géographie d'Egypte*, **30**, 25–52.

Kidron, G. J., 2009: The effect of shrub canopy upon surface temperatures and evaporation in the Negev Desert. *Earth Surface Processes and Landforms*, **34**, 123–132.

Krasnov, B., G. Shenbrot, L. Khokhlova, and E. Ivanitskaya, 1996: Spatial patterns of rodent communities in the Ramon erosion cirque, Negev Highlands, Israel. *Journal of Arid Environments*, **32**, 319–327.

Lettau, H. H., 1978: Extremes of diurnal temperature ranges. In *Exploring the World's Driest Climate* (H. H. Lettau and K. Lettau, eds.), IES Report 101, Institute for Environmental Studies, University of Wisconsin, Madison, WI, pp. 67–73.

Lewis, L. A., and L. Berry, 1988: *African Environment and Resources*. Unwin Hyman, Boston, MA, 404 pp.

Malek, E., 2003: Microclimate of a desert playa: evaluation of annual radiation, energy, and water budgets components. *International Journal of Climatology*, **23**, 333–345.

Malek, E., G. E. Bingham, and G. D. McCurdy, 1990: Evapotranspiration from the margin and moist playa of a closed desert valley. *Journal of Hydrology*, **120**, 15–34.

Malek, E., C. Biltoft, J. Klewicki, and B. Giles, 2002: Evaluation of annual radiation and windiness over a playa: possibility of harvesting the solar and wind energies. *Journal of Arid Environments*, **52**, 555–564

Mann, H. S., A. N. Lahiri, and O. P. Pareek, 1976: A study on the moisture availability and other conditions of unstabilised dunes in the context of present land use and the future prospects of diversification. *Annals of Arid Zones*, **15**, 270–284.

Marsh, B. A., 1990: The microenvironment associated with *Welwitschia mirabilis* in the Namib Desert. In *Namib Ecology: 25 Years of Namib Research* (M. K. Seely, ed.), Transvaal Museum Monograph No. 7, Pretoria, pp. 149–154.

Martens, S. N., D. D. Breshears, and C. W. Meyer, 2000: Spatial distributions of understory light along the grassland/forest continuum: effects of cover, height and spatial pattern of tree canopies. *Ecological Modelling*, **126**, 79–93.

Menenti, M., 1984: *Physical Aspects and Determination of Evaporation in Deserts*. Report 10 (special issue), Institute of Land and Water Management Research (ICW), Wageningen, The Netherlands, 202 pp.

Nechayeva, N. T., V. K. Vasilevskaya, and K. G. Antonova, 1973: *Lifeforms of the Plants in the Kara-Kum Desert*. Akademii Nauk SSSR, Moscow, 243 pp. (in Russian).

Nobel, P. S., and M. J. Linton, 1997: Frequencies, microclimate and root properties for three codominant perennials in the northwestern Sonoran Desert on north- vs. south-facing slopes. *Annals of Botany*, **80**, 731–739.

Orlov, B. P., 1928: On the study of ecological conditions in the southeastern part of Transcaspian Kara-Kum. *Trudy po Prikl. Bot.*, **19**, Leningrad (in Russian).

Omara, A. A., 1971: Diurnal and annual temperature patterns in a soil at Giza subjected to three treatments. *Meteorological Research Bulletin*, **3**, Meteorological Authority of the Arab Republic of Egypt, Cairo, 185–202.

Peel, R. F, 1974: Insolation and weathering: some measures of diurnal temperature changes in exposed rocks in the Tibesti region, central Sahara. *Zeitschrift für Geomorphologie*, Supp., **21**, 19–28. Schweizerbart Publishers, www.schweizerbart.de.

Petrov, M. P., 1976: *Deserts of the World*. Wiley, New York, 447 pp.

Rauh, W., 1985: The Peruvian-Chilean deserts. In *Hot Deserts and Arid Shrublands* (M. Evenari, I. Noy-Meir, and D.W. Goodall, eds.), Ecosystems of the World 12A, Elsevier, Amsterdam, pp. 239–268.

Robinson, M. D., and M. K. Seely, 1980: Physical and biotic environments of the southern Namib dune ecosystem. *Journal of Arid Environments*, **3**, 183–203.

Rognon, P., 1989: *Biographie d'un Désert*. Plon, Paris, 347 pp.

Roof, S., and C. Callagan, 2003: The climate of Death Valley, California. *Bulletin of the American Meteorological Society*, **84**, 1725–1739.

Scholes, R. J., and S. R. Archer, 1997: Tree-grass interactions in savannas. *Annual Review of Ecology and Systematics*, **28**, 517–544.

Seely, M.K., and G. N. Louw, 1980: First approximation of the effects of rainfall on the ecology and energetics of a Namib Desert dune ecosystem. *Journal of Arid Environments*, **3**, 25–54.

Seely, M. K., D. Mitchell, and K. Goelst, 1990: Boundary layer microclimate and *Angolosaurus skoogi* (Sauria: Cordylidae) activity on a northern Namib dune. In *Namib Ecology: 25 Years of Namib Research* (M. K. Seely, ed.), Transvaal Museum Monograph No. 7, Pretoria, pp. 155–162.

Segal, M., R. A. Pielke, and Y. Mahrar, 1983: On climatic changes due to a deliberate flooding of the Qattara Depression (Egypt). *Climatic Change*, **5**, 73–84.

Sellers, W. D., 1965: *Physical Climatology*. University of Chicago Press, Chicago, 272 pp.

Sinclair, J. G., 1922: Temperature of the soil and air in a desert. *Monthly Weather Review*, **50**, 142–144.

Tschinkel, W. R., 1973: The sorption of water vapor by windborne plant debris in the Namib Desert. *Madoqua Series* II, **2**, 21–24.

Tyler, S. W., and Coauthors, 1997: Estimation of groundwater evaporation and salt flux from Owens Lake, California, USA. *Journal of Hydrology*, **200**, 110–135.

Vehrencamp, J. E., 1953: Experimental investigation of heat transfer at an air–earth interface. *Transactions – American Geophysical Union*, **34**, 22–30.

Verhoef, A., 1995: Surface energy balance of the shrub vegetation in the Sahel. Thesis, Department of Meteorology, Wageningen Agricultural University, The Netherlands, 247 pp.

Walter, H., and Box, E. O., 1983: The Karakum Desert, an example of a well-studied eu-biome. In *Ecosystems of the World*, **5**, *Temperate Deserts and Arid Semi-Deserts* (N. E. West, ed.). Elsevier, Amsterdam, pp. 105–159.

Warke, P. A. S., and B. J. Smith, 1998: Effects of direct and indirect heating on the validity of rock weathering simulation studies and durability tests. *Geomorphology*, **22**, 347–357.

Williams, C. B., 1954: Some bioclimatic observations in the Egyptian desert. In *Biology of Deserts* (J. L. Cloudsley-Thompson, ed.), Institute of Biology, London, pp. 18–27.

Zhang, Q., L. Song, R. Huang, G. Wei, S. Wang, and H. Tian, 2003: Characteristics of hydrologic transfer between soil and atmosphere over Gobi near Oasis at the end of summer. *Advances in Atmospheric Sciences*, **20**, 442–452.

11 Precipitation in the drylands

11.1 THE NATURE OF RAINFALL IN THE DRYLANDS

The major characteristics of the precipitation regime of dryland regions have been summarized in Chapter 9. Some of the most universal are the skewed distribution of rainfall, the extreme spatial and temporal variability, the relationship between variability and mean rainfall, and the concentration of rainfall in a few large events. These characteristics, and the extent to which they, and other axioms about dryland climates are valid, are briefly reviewed in this chapter. Characteristics of the rainy season and rainfall occurrences and their quantification are also dealt with. Since many of the points are related to the issue of frequency distributions, this topic is treated extensively.

These characteristics of the dryland rainfall regime have been known for some time from analyses of gauge data, but the physical reasons were never explained in detail until remote sensing techniques allowed for a more thorough look at the physical characteristics of storms. The most important work has resulted from the Tropical Rainfall Measuring Mission (TRMM) satellite, which was launched in 1997. It provides coverage from roughly 35° N to 35° S – an area that includes most of the dryland regions of Africa, Australia, South America, and the Middle East, India, Mexico, and a small sector of the southwestern deserts of the USA.

Two extremely important results that have emerged from analysis of TRMM data are that a small number of very large features, the mesoscale convective systems (MCSs) (see Chapter 5), produce most of the rainfall in the low latitudes, and that both convective and stratiform rainfall is associated with these features. These systems develop in the afternoon, but persist through the night. Stratiform rain accounts for roughly 40% of the rain associated with MCSs and occurs primarily at night or in the early morning (Schumacher and Houze 2003). The convective rainfall peaks during the afternoon hours. MCSs are generally in the size range of 1000–30,000 km^2, with an average size over land of roughly 10,000 km^2 (Nesbitt et al. 2006), thus a horizontal scale on the order of 100–150 km. Overall, the total area of rain within the storm is only about 30–50% (Liu et al. 2007).

On a regional basis, MCSs comprise around 10–20% of the storms, but produce about 70% of the rainfall. Very intense MCSs, which comprise only about 1% of the storms, produce about 35% of the rainfall (Mohr et al. 1999). These characteristics, together with the random distribution of the rain area within the systems, result in a highly skewed distribution of daily rainfall and tremendous spatial variability.

The contribution of MCSs varies greatly among the dryland regions of the low latitudes (see Fig. 5.12). They occur most frequently in the drylands of South America and West Africa; they bring up to 90% of the rainfall in the La Plata Basin and about 70% of the total rainfall in the Sahel (Nesbitt et al. 2006). MCSs occur less frequently in southern Africa and Australia, and their occurrence is negligible in the deserts of the Middle East, Mexico and the southwestern USA. The most intense category of MCSs is an important contributor to dryland rainfall only in South America and West Africa. These facts provide much of the explanation for the contrasts between the daily precipitation regimes of the various drylands.

In making generalizations about dryland rainfall, it is important to keep in mind the strong contrasts between convective rainfall and stratiform or frontal rains (Fig. 11.1). Convective rainfall tends to be brief but intense and relatively localized. Stratiform rain and frontal rain (which is primarily associated with stratiform clouds) both tend to be less intense but of longer duration and relatively widespread. Frontal and convective rainfall events have vastly different statistical properties and different impacts on runoff, erosion (Fig. 11.2), and soil–water balance. Intense convective storms often result in rapid flooding in dryland environments devoid of vegetation (Fig. 11.3).

Most of the generalizations made about dryland precipitation relate to convective rainfall; however, the contribution of convective rainfall varies tremendously among the drylands. Frontal rainfall prevails in the higher latitudes and in mid-latitudes in winter. Convective rainfall prevails in the low latitudes and in

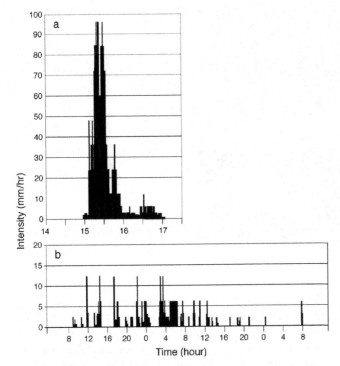

Fig. 11.1 Comparison of convective and frontal rainfall at Nizzana, Israel (from Sharon *et al.* 2002). The convective rainfall (top) fell on November 2, 1994, and was brief (2 hours) and intense. The frontal rain (bottom) fell on December 5, 1994, and was prolonged (2 days) and less intense.

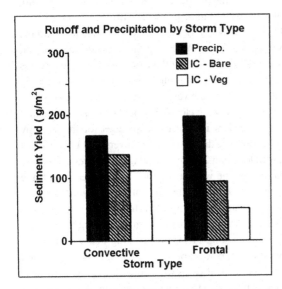

Fig. 11.2 Precipitation, runoff and erosion from convective (summer) vs. frontal (winter) rains (adapted from Wilcox *et al.* 2003). Data are indicated for both bare ground and for a vegetated surface.

summer in many mid-latitude locations, especially in the central USA. In the low latitudes the local contribution of stratiform rain is proportional to the overall contribution of mesoscale convective systems to total rainfall. This also varies markedly across the global drylands. Consequently, the character of the rainfall regime is very diverse in the world's drylands, despite certain commonalities.

11.2 AVAILABILITY OF PRECIPITATION DATA

The density of precipitation measuring stations in the drylands is generally quite low, so that our understanding of dryland precipitation and its variability is limited by data availability. Notable exceptions are dense gauge networks in the central Sahel near Niamey (Lebel and Le Barbé 1997) and in the Walnut Gulch area of Arizona (Goodrich *et al.* 2008). These were established for hydrological experiments. Figure 1.7 shows a map of precipitation reporting stations represented in the Global Historical Climatology Network (GHCN); the major deserts distinctly emerge as areas with few stations. Although the GHCN by no means includes all possible stations, the relative distribution is representative. Figure 11.4 shows more detailed maps of reporting stations in West Africa and Australia. The map for Africa is based on data collected by the author and by the Agence pour la Sécurité de la Navigation Aérienne en Afrique et à Madagascar (ASECNA). The Australian map shows stations of the Australian Bureau of Meteorology. Although hundreds or thousands more stations are shown in the more humid regions, the networks in the deserts are still extremely sparse and the density is relatively low in the semi-arid regions.

Even in areas with relatively good coverage, the data may not be easily accessible. The global precipitation archives available for research are described in Chapter 1. To go beyond this and obtain the much denser networks of stations, as shown in Fig. 11.4, one usually has to turn to the meteorological services of the individual countries. Some countries, such as the USA and Australia, provide the data in digitized format at relatively low cost. Other countries, especially developing countries, do not routinely publish the extensive precipitation data collected, although many have them available in digitized archives. Unfortunately, in most African countries and some South American countries the cost of purchasing even monthly data is prohibitive.

Although daily and hourly data are acquired at most reporting stations, these are even more difficult and more costly to obtain. For some countries, historical archives of such data are available in miscellaneous publications that may be available in meteorological libraries. These were generally published decades ago, but nonetheless contain useful statistics on daily or hourly rainfall. A few examples are studies of rainfall intensity–duration–frequency in East Africa (Taylor and Lawes 1971), Ghana (Dankwa 1974), and Zimbabwe (Department of Meteorological Services 1975).

11.3 THE FREQUENCY DISTRIBUTION OF RAINFALL IN DRYLANDS

11.3.1 CHARACTERIZING THE FREQUENCY DISTRIBUTION OF RAINFALL IN DRYLANDS

The magnitude of values in a time series of some variable is often represented by a frequency distribution, which shows the likelihood of the variable taking on specific values within its range of variation. This is illustrated in Fig. 11.5, which shows three types

Fig. 11.3 A rapid downpour inundated White Sands, Mexico, within minutes.

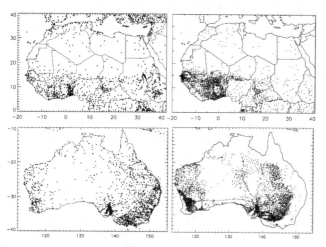

Fig. 11.4 Rainfall stations in the drylands of West Africa and Australia (mean annual rainfall below 500 mm). (left) Stations available in the Global Historical Climatology Network (GHCN); (right) stations based on locally archived data sets (ASECNA and select meteorological services for West Africa, Bureau of Meteorology for Australia).

of frequency distributions: Gaussian or normal, positive skew, and negative skew. These distributions are defined using various measures of central tendency (most common values) and scatter around the central tendency. The mean is the arithmetic average of all observations, the median is the value on either side of which 50% of the observations fall, and the mode is the most common magnitude of the variable. In a normal distribution, all three of these measurements of central tendency are equivalent. In a distribution that is positively skewed, the value of the mean exceeds that of both the median and the mode.

It is generally assumed that in arid and semi-arid regions the frequency distribution of rainfall amounts for years, months, or shorter periods is positively skewed. That is, there is a preponderance of rainfall below the mean, which is inflated by a handful of extreme occurrences of rainfall well above the mean. One example of the extremes comes from Iquique, Chile (mean annual rainfall 3 mm). A 4-year rainless period was absolutely dry, but

on one occasion a single rain brought 64 mm. At Helwan (Egypt) seven convective storms produced a quarter of the rain that fell during an entire 20-year period (Trewartha and Horn 1980). In Cairo, where mean annual rainfall is 25 mm, rain fell in only 13 of the 30 years between 1890 and 1919, but 43 mm fell on one day in 1919 (Gautier 1970). In such regions, the skewed distribution is probably the appropriate model. In semi-arid regions, however, annual rainfall is not necessarily strongly positively skewed.

Figure 11.6 shows the frequency distribution of annual rainfall and rainfall in the wettest month at several arid and semi-arid stations in Mali (West Africa), Botswana (southern Africa), and the southwestern United States. All are in regions of summer rainfall, and in most cases 70–100 years of data are available. In most cases a positive skew is apparent for annual data but it is not necessarily strong. More evident are the very diverse shapes of the distributions, with some even being clearly bimodal (e.g., Timbuktu and Francistown). The skew is, in general, stronger for the monthly data but, even there, the distributions are diverse. At Phoenix and Las Vegas, with less than 25 mm on average in the wettest month, the distribution for the wettest month (August) peaks with the smallest events and decreases steadily with event size. The distributions for Tucson and Mopti are close to normal.

The degree of skewness increases with decreasing time scales, so that daily rainfall (Fig. 11.7) is very strongly skewed. In most cases, daily rainfall is below about 20–30 mm. However, at each station a small number of cases with rainfall ranging from 40 to 120 mm occur. This is consistent with a small number of exceedingly large disturbances producing the vast majority of rainfall (see Section 11.1).

A number of statistical tests can help to quantitatively assess the degree to which data are normally distributed. O'Brien and Griffiths (1967) suggest that the Cornu criterion is the most appropriate for small samples on the order of 100 observations or fewer. The criterion utilizes the ratio between the average and the standard deviation. When this ratio lies between certain limits, dictated by the number of observations, it can be assumed that the distribution is approximately normal. Figure 11.8 shows

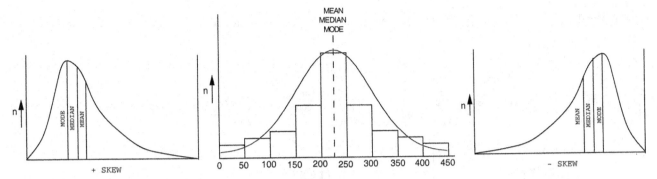

Fig. 11.5 Diagram showing three types of frequency distribution: positive skew, Gaussian or normal, and negative skew.

the Cornu criterion for some 854 African stations with 50 years of data. Values are indicated for annual rainfall, for January totals at Southern Hemisphere stations, for August at Northern Hemisphere stations, and for March totals at all stations. These are the rainy season months in southern, northern, and equatorial Africa, respectively. Annual data for some 80% of the stations tested meet the criterion at the 5% significance level. For March rainfall, the number is 54%; for January and August it is 64%.

Overall, the assumption of a normal distribution, which provides the basis for numerous statistical tests and analyses in climatology, is inappropriate for annual rainfall in most arid regions but may be appropriate for semi-arid regions. The assumption is probably inappropriate in dry season months because rainfall will generally be near zero or will be quite significant. Alternate frequency distributions are therefore commonly used for analyzing dryland precipitation.

One, the gamma distribution (Groisman *et al.* 1999), is particularly applicable when rainfall is zero or near zero in a large proportion of the years or months. Thus it is appropriate for daily totals. The gamma distribution is defined by a shape parameter and a scale parameter, the combination of which gives the mean and variance of the distribution (Fig. 11.9).

The skew of the rainfall distribution in dryland regions and the prevalence of many dry months pose a challenge to many types of statistical analyses. Various approaches have been utilized to deal with the problem. One approach is to utilize rank order statistics, assigning each observation a value related to its order in the sequence from driest to wettest. Another is to "normalize" the data set by a square root or logarithmic transformation (e.g., Nicholson 1986).

At some locations, the issue is complicated by a bimodal distribution, in which rainfall totals are concentrated in two distinct ranges. This generally implies that two distinctly different rainfall regimes are operative and that the dominant regime fluctuates from year to year (Nicholson and Webster 2007). Such is the case for the Sahel region as a whole (Fig. 11.10). A wet mode is associated with a more northerly position of the tropical rainbelt and a dry mode with a more southerly position. Bimodality is evident to some extent in the August or January totals for several stations shown in Fig. 11.6. This is the case for Timbuktu (a Sahelian station), Francistown, and Las Vegas.

Most studies dealing with rainfall distribution evaluate records for individual stations, which reflect a combination of the interannual variability of large-scale atmospheric processes and the erratic distribution of convective rainfall (see Section 11.4.4). For many applications, including most meteorological studies, areal averages are more useful, as the spatial variability is removed. This also reduces the skew of distribution. Of the 84 African rainfall regions evaluated by Nicholson and Entekhabi (1986), the time series for all but 10 arid regions met the Cornu criterion, suggesting that the distribution is normal when mean areally averaged rainfall is as low as 200 mm.

11.3.2 FREQUENCY DISTRIBUTIONS AND DAILY EVENT SIZE

One commonality of daily rainfall in dryland regions is the predominance of small daily events, with the frequency of events rapidly decreasing with the magnitude of the event. The large events produce a disproportionate amount of rainfall, but the small events have considerable ecological significance. Other than these generalizations, there are few commonalities among the drylands concerning daily rainfall. The contribution of small events, the typical size of daily events, and the daily totals associated with the most extreme events all vary greatly.

Figure 11.11 shows the frequency distribution of daily rainfall totals for three deserts of the American Southwest (Huxman *et al.* 2004). The total number of small events (daily rainfall less than 5 mm) decreases from 22 per year in the Sonoran Desert to 10 in the Mojave. The vast majority of events in the Mojave are less than 11 mm/day. Events of this size produce most of the rainfall in the Sonoran Desert, but events up to 11 mm/day are common in the Chihuahuan Desert. The contribution of the small events to annual rainfall ranges from 76% in the Chihuahuan Desert to 85% in the Sonoran to 95% in the Mojave, thus increasing with the degree of aridity. In the Chihuahuan Desert, roughly 7% of the rainfall comes from events with more than 10 mm/day; in the other two deserts, these large events contribute less than 1%.

The distribution of small events has been examined at a few other locations. At 213 stations in the western USA small events

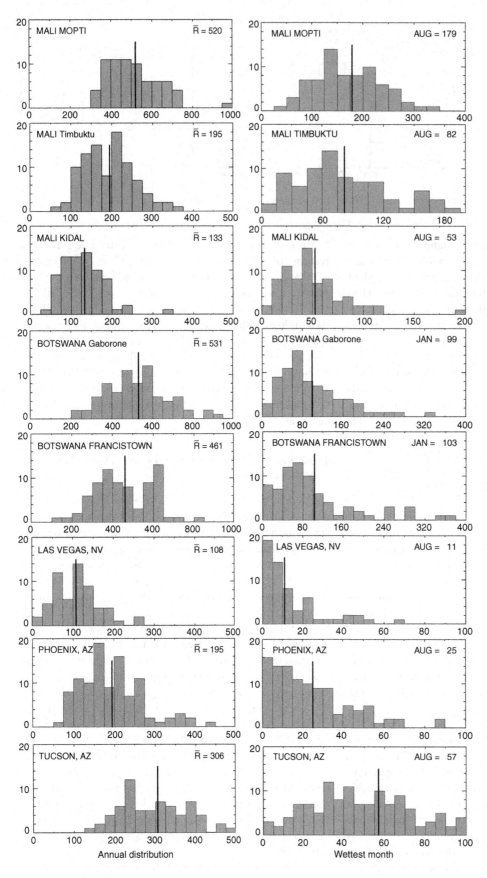

Fig. 11.6 Frequency distribution of annual rainfall (left) and rainfall in the wettest month (right) for stations in Mali (West Africa), Botswana (southern Africa), and the southwestern USA. The annual or monthly means (mm) are indicated with vertical bars and are also indicated in the upper right of each graph. The frequency is expressed as a percent of all years of record.

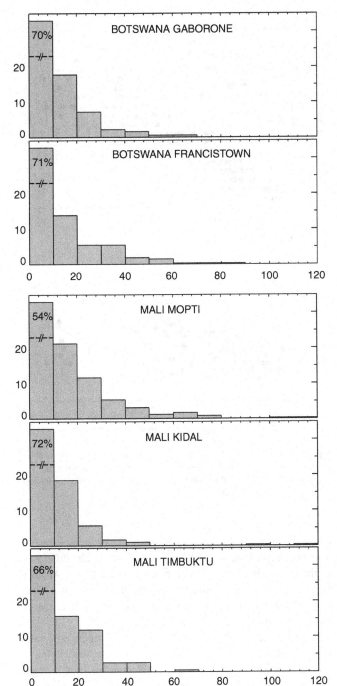

Fig. 11.7 Frequency distribution of daily rainfall at two stations in Botswana in January and three stations in Mali in August. The frequency is expressed as a percent of all days when rainfall occurred and the calculation is based on all available years.

on average comprised 47% of the total rain events, with the range at individual stations extending from 24% to 65% (Loik *et al.* 2004). On the Colorado Plateau (Fig. 11.12), 73% of the events were 5 mm or less, but these contribute some 25% of the rainfall (Sala and Lauenroth 1982). In the Ordos region of China, 55% of the rain events bring less than 5 mm (Cheng *et al.* 2006).

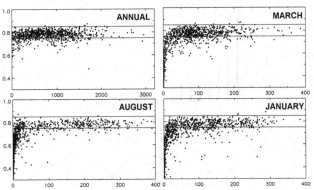

Fig. 11.8 Scatter diagram of the ratio of average value to standard deviation vs. mean annual rainfall for African stations and for August at Northern Hemisphere stations, January at Southern Hemisphere stations, and March for all stations. Each station record is 50 years in length. The Cornu criterion gives the value of this ratio that can be expected for normally distributed variables and is a function of the number of observations (i.e., the number of years of record). For any station falling outside these limits, annual rainfall is not normally distributed; it is probably normally distributed for stations falling within the limits.

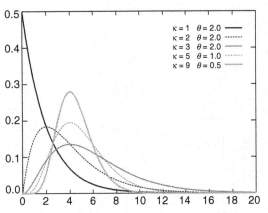

Fig. 11.9 Schematic of the gamma distribution, an applicable model for many arid regions and dry months. The various curves reflect different combinations of shape and scale parameters.

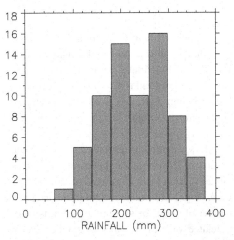

Fig. 11.10 Frequency distribution for regionally averaged annual rainfall in the Sahel from 1950 to 1997.

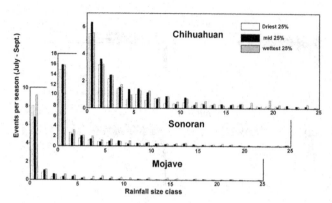

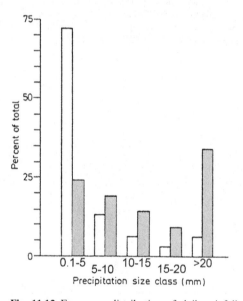

Fig. 11.11 Frequency distribution of daily rainfall totals for stations in the Chihuahuan, Sonoran, and Mojave deserts (from Huxman *et al.* 2004). Data are from Las Vegas, Tucson, and the Jornada experimental site, respectively.

Fig. 11.12 Frequency distribution of daily rainfall on the Colorado Plateau (from Sala and Lauenroth 1982). White columns = percent of total number of events; gray columns = percent of total rainfall.

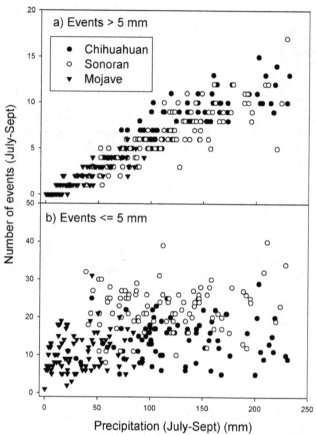

Fig. 11.13 The relationship between total seasonal precipitation (July–September) and (a) the number of large (>5 mm) versus (b) small (≤5 mm) rainfall events for the Chihuahuan, Sonoran, and Mojave deserts (from Huxman *et al.* 2004).

The contribution of small events to total rainfall is similarly diverse at the African stations shown in Fig. 11.7. Daily events with rainfall less than 5 mm contribute 7% of August rainfall at Mopti, where mean annual rainfall is 520 mm. At Gaborone, where annual rainfall (531 mm) is similar to that at Mopti, the contribution of small events to rainfall in January, the wettest month, is 21%.

At the other end of the spectrum, the largest events contribute a disproportionate share of the rainfall and generally the lion's share of the interannual variability. In all three desert regions of the southwestern USA (Fig. 11.13), annual rainfall is correlated linearly with the number of events of more than 5 mm/day, but is completely uncorrelated with the number of smaller events (Huxman *et al.* 2004). Moreover, the amount of rainfall from small events is relatively invariant from year to year (Fig. 11.14)

(Schwinning and Sala 2004), at least in the American Southwest. This is also the case in the West African Sahel. The difference in the daily rainfall distribution between wet and dry Augusts is generally one or two events with rainfall in excess of 40 mm (Fig. 11.15). Lamb *et al.* (1988) even found that there are more weak events in the drier years.

In some locations daily rainfall can be very extreme, especially in the tropics. Often several times the annual average will fall in one day. In Gabes, Tunisia, a flood in September 1969 brought 400 mm in one day, which is about four times the mean annual rainfall (Thornes 1994). The same event brought nearly 800 mm to Sidi bou Zid, Tunisia, during September and October, months in which the mean rainfall is on the order of 10–20 mm.

The difference between areas with relatively low daily totals, such as the American Southwest, and those where daily rainfall not uncommonly exceeds 100 mm appears to be in the character of the rainfall. One factor is whether or not the rainfall is convective or frontal. This is illustrated in Fig. 11.1 with data from Nizzana, Israel. The intensity of the convective rainfall exceeds 100 mm/hour and the rain event persists for only 2 hours. The intensity of the frontal rain barely exceeds 10 mm/hour, but the event persists for 2 days. For Sahelian Africa, where rainfall is

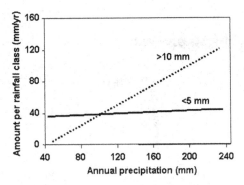

Fig. 11.14 The amount of rainfall occurring in events greater than 10 mm/day (dotted line) and less than 5 mm/day (solid line), as a function of annual rainfall (from Schwinning and Sala 2004).

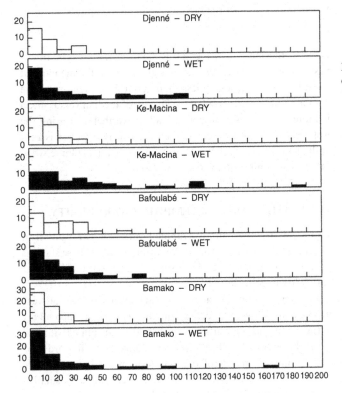

Fig. 11.15 Frequency distribution of daily rainfall totals at four Sahelian stations in Mali during three wet Augusts (1952, 1954, 1961) and three dry Augusts (1937, 1941, 1951) (from Nicholson 2000).

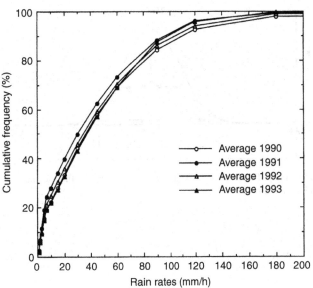

Fig. 11.16 Average rain rate distributions (equivalent hourly rate) for each of the four HAPEX-Sahel years 1990–1993 (from Lebel *et al.* 1997).

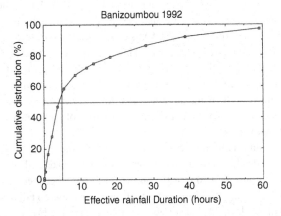

Fig. 11.17 Effective rainfall duration at Banizoumbou, Niger (from Lebel *et al.* 1997).

almost exclusively convective, the average rain rate exceeds 100 mm/hour roughly 10% of the time (Fig. 11.16), although not all events last a full hour. The duration of some 80% of rain events is less than two hours; 70% are less than one hour (Fig. 11.17).

11.4 TEMPORAL AND SPATIAL VARIABILITY

11.4.1 QUANTIFYING TEMPORAL VARIABILITY

Long-term planning of agriculture, subsistence, and water supply requires knowledge of the overall character of the rainy season and the range of variability of its major characteristics.

This entails establishing a "mean" or "normal" year and quantifying the likelihood that a given year falls within the range of "normal." This is a formidable task and many studies have shown that "mean" and "typical" are in no way synonymous.

The most general characteristics of the rainy season are its length, seasonal pattern of occurrence, and amount of rainfall. The first characteristic is quite problematic, as it requires a definition of the onset and end of the season or a definition of dry and wet months. The seasonality is sometimes quantified by such measures as the percent concentration in a certain number of months, the number of months with precipitation exceeding some threshold value, or by harmonic analysis (e.g., Horn and Bryson 1959).

Variability is most often described using various measures of scatter about the mean. One of these, the standard deviation σ is calculated as

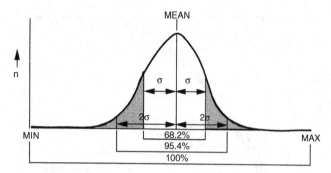

Fig. 11.18 Illustration of statistical properties of the normal or Gaussian distribution.

$$\sigma = \sqrt{\frac{1}{N} \sum_{i=1}^{N} (x_i - \mu)^2}. \qquad (11.1)$$

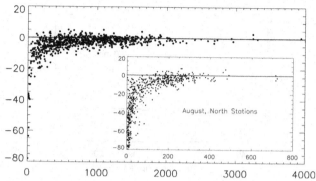

Fig. 11.19 Scatter diagram comparing the difference between mean and median rainfall (as % of mean). Annual values are indicated for all African stations with at least 50 years of record. August values, typical of wet season months, are indicated for Northern Hemisphere stations with at least 50 years of records.

where μ is the mean of all observations, x_i is observed rainfall at a given time (e.g., month or year), and N is the number of observations. In the case of a normal distribution, 68.3% of occurrences fall within one standard deviation of the mean, and 95.5% within two standard deviations (Fig. 11.18). The standard deviation is an absolute quantity; a useful relative quantity, the coefficient of variation (CV), is expressed as the ratio of the standard deviation to the mean. These are the most commonly used parameters for describing rainfall variability. Their interpretation is clear if the frequency distribution is Gaussian. They lose much of their meaning if it is not, as is often the case in dryland regions. In that case, other measures of variability are more appropriate.

A common alternative is the use of probabilities, such as percentiles, deciles, and quartiles. These are calculated directly from observations, with 10% of the observations lying between adjacent deciles and 25% of the observations lying between adjacent quartiles. Thus, 90% of all observations exceed the first decile, 10% exceed the ninth decile, and so on. Similarly the upper and lower quartiles delineate the highest and lowest quarters of the values, respectively. Variability can be quantified in a manner analogous to the standard deviation, using the quartile deviation (half of the difference between the upper and lower quartile) and expressing the relative variability as the quartile deviation divided by the median. Comparable measures of variability can also be calculated for the gamma distribution, which is defined by a shape parameter κ and a scale parameter θ. Its mean is equal to the product of these two parameters $\kappa\theta$ and its variance is equal to $\kappa\theta^2$ (Fig. 11.9).

A number of studies have suggested that, in subhumid regions, rainfall is better expressed as a departure from the median rather than the mean value (e.g., Gibbs and Maher 1967). The calculation of the median is complicated, so that the mean is often used for convenience; also mean values are routinely reported in climatic records, while median values are not.

Data for Africa, shown in Fig. 11.19 for approximately 1400 stations, suggest that, at least for annual rainfall, the difference

between the mean and median is small in most cases. In the vast majority of cases, the mean is greater than the median but the difference is generally less than about 5% in regions where annual rainfall exceeds 400 mm. The use of median versus mean is more critical for monthly rainfall. Nevertheless, during the wet season, the difference between the two parameters is generally less than 10% although it becomes relatively large when monthly rainfall is on the order of 250 mm or less.

11.4.2 MAGNITUDE OF TEMPORAL VARIABILITY

Figure 11.20 shows the interannual variability of rainfall at select dryland stations around the world. At all of these stations the interannual variability is reasonably large, but the degree of variability is quite diverse.

In many regions decadal (10-year) means or the 30-year means used routinely to define climatic normals are also highly variable. In the semi-arid Sahel, where mean annual rainfall is on the order of 200–800 mm, the range is on the order of 50% to more than 75% of the long-term mean. In comparable regions of southern Africa, the range is considerably smaller, on the order of 25–50%. In the Sahel, even 30-year climatological means show substantial variation. That difference between mean annual rainfall for 1931–1960 and 1968–1997 is on the order of 30–40% of the long-term mean.

It is often assumed that variability is greater in regions of summer rainfall than in regions of winter rainfall and that it is greater in arid regions than semi-arid. The few stations shown in Fig. 11.20 do not support these generalizations. Some winter rainfall stations, such as Biskra and Nouadhibou, show tremendous year-to-year fluctuations. Variability is greater at Phoenix, with winter rainfall, than at nearby Tucson, with summer rainfall. Rainfall is markedly more irregular at Santiago than at Oudtshoorn, with a similar annual mean, although both are regions of winter rainfall.

Figure 11.21, showing the global distribution of the coefficient of variation (CV) of annual rainfall, illustrates this

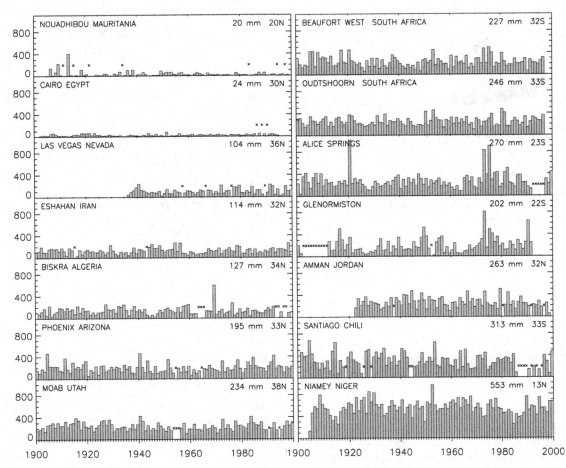

Fig. 11.20 Time series representing the interannual variability of rainfall at select arid and semi-arid stations worldwide. Mean annual rainfall and latitude are indicated.

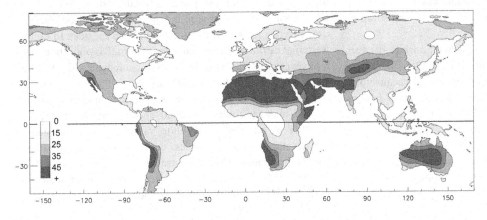

Fig. 11.21 Map of the coefficient of variation of annual rainfall (based on data in Global Historical Climatology Network).

diversity. A comparison with mean annual rainfall (see Fig. 5.1) shows that, in general, the CV ranges from about 5% or 10% in humid regions to over 50% in arid regions. In a very general sense, the map in Fig. 11.21 suggests that (1) variability is large in dryland regions and (2) there is a strong inverse relationship between the magnitude of variability and mean rainfall.

Figure 11.22 examines this for equatorial, Southern Hemisphere, and Northern Hemisphere stations in Africa. The inverse relationship is evident and the variability initially decreases

rapidly with increasing rainfall. It is on the order of 100–300% at the driest stations. It levels off at a rainfall of about 800–1000 mm/year, ultimately approaching about 15% in all three regions. Notably, there is a much greater spread in this parameter at the Southern Hemisphere stations.

Figure 11.23 shows the relationship between mean annual rainfall and the variability for dryland stations globally, utilizing all stations in the Global Historical Climatology Network with mean annual rainfall below 800 mm and with at least 50

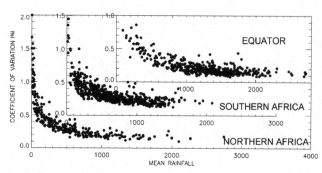

Fig. 11.22 Coefficient of variation (CV) of annual rainfall versus mean annual rainfall for African stations in the equatorial region (10° N to 10° S), Southern Hemisphere (poleward of 10° S), and Northern Hemisphere (poleward of 10° N).

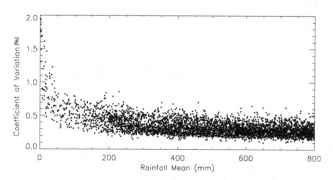

Fig. 11.23 Coefficient of variation (CV) of annual rainfall vs. mean annual rainfall for global dryland stations with at least 50 years of records in NOAA's Global Historical Climatology Network archive.

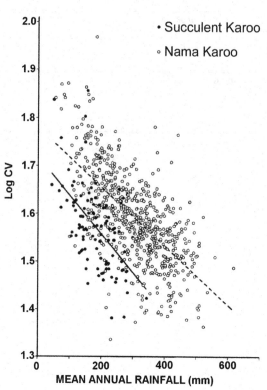

Fig. 11.24 Relationship between variability and mean annual rainfall for stations in the succulent karoo of South Africa, a winter rainfall region (●), and in the nama karoo, a summer rainfall region (○) (from Desmet and Cowling 1999).

years of records. The same inverse relationship is evident, but the spread is extremely large. When rainfall is as low as 200 mm, the CV ranges from about 25% to 75%. At 800 mm, it ranges from about 15% to 40%.

This suggests that the exponential relationship between variability and mean annual rainfall is valid globally only in a very general sense. There is significant spread because of a dependence on latitude, regional controls, and other factors. The seasonality of rainfall is a large factor (e.g., Desmet and Cowling 1999). For example, variability is much greater at all levels of rainfall in the nama karoo of South Africa, a summer rainfall region, than in the succulent karoo, a winter rainfall region (Fig. 11.24). Variability is also anomalously large where El Niño has a large impact on rainfall (Nicholls and Wong 1990).

Individual stations may deviate from expectation because of local factors that enhance or reduce variability. For example, at Mayumba, Gabon, year-to-year changes in the coastal currents and water temperatures (Nicholson and Entekhabi 1987) and local wind patterns dramatically enhance variability. February and March rainfall at Mayumba was 16 and 58 mm, respectively, in 1958, but 607 and 413 mm during the same months of the following year, when the coastal water temperature was anomalously high. Daily totals in February reached 352 mm in 1959, but only 11 mm in 1958.

Many other dryland regions are also atypical in terms of interannual variability. Throughout Northeast Brazil, where mean annual rainfall everywhere exceeds about 500 mm, the CV exceeds 50%. The region is known for the episodic severe droughts that frequently accompany El Niño episodes. In southwestern Australia, where mean annual rainfall is only about 400 mm/year, variability is roughly 25%, compared with 40% for similar regions elsewhere in Australia.

It is useful to point out that the relationship between mean rainfall and variability described above relates to a *relative* measure of variability. Absolute variability, i.e., the standard deviation, shows much less contrast between humid and subhumid regions. This is illustrated in Fig. 11.25 using data for Africa. In most cases when mean annual rainfall ranges between 500 and 1800 mm, the standard deviation is on the order of 125–275 mm.

The limited range in absolute variability has implications for interpreting and predicting hydrologic changes in semi-arid regions in which exogenous rivers make a large contribution to local water supply. It is often assumed that rainfall variability in the humid source regions controls the variability of this supply. However, the absolute year-to-year variability of the region's rainfall may be roughly as large as that in the humid source region (Nicholson 1981). Despite the major contribution of river flow from equatorial regions, local rainfall has a significant impact on lake levels.

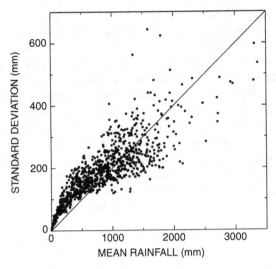

Fig. 11.25 Comparison of annual mean and standard deviation of rainfall for African stations with at least 50 years of record.

Fig. 11.26 Twenty years after this Moroccan dam was constructed, it had never contained any water, because its planning was based on an inadequate climatic record.

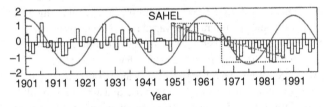

Fig. 11.27 Rainfall variability in the Sahel since 1901, illustrating the description of recent fluctuations as either a downward trend or a step function reduction in precipitation. A curve with a 30-year periodicity is superimposed upon the time series.

Clearly it is difficult to make any generalizations concerning the reliability of rainfall means and standard deviations in semi-arid regions. Perhaps the most important point to be made is that short-term means may be quite unrepresentative of normal conditions. If only a 10-year record is available, it may happen to correspond to an unusually long wet or dry episode.

There may even be disastrous consequences if agriculture and water management are based on short-period means. An example comes from Morocco where the Mansour Eddabhi Dam (Fig. 11.26) was constructed in the 1970s, relying on brief rainfall records for an estimate of its potential. By 1984 the reservoir created by the dam had never contained water because rainfall had never been sufficient to fill it. Likewise, plans for a dredging operation in the Okavango Delta of Botswana, proposed by an engineering firm (SMEC 1987), were based on the hydrologic conditions in one very abnormal year in the 1930s (Peter Smith, Botswana Ministry of Water Affairs, personal communication).

11.4.3 TEMPORAL STRUCTURE OF VARIABILITY

Three descriptors of the temporal variations of rainfall include trends, cycles, and persistence. The first term relates to progressive, longer-term changes; the second to quasi-regular intervals between wetter or drier conditions; and the third to the duration of anomalous conditions. These statistics are applicable to such questions as to whether rainfall in one season or year can be predicted from the rainfall of the previous season or year, or how long a drought or wet period is likely to persist.

The analysis of trends and of cycles is often problematic, and both of these statistical approaches should be used and interpreted with caution. A trend is not an absolute, but rather must be established for some arbitrarily defined time period. A number of studies have described the long, downward trend in Sahel rainfall since about 1950. Although a statistically significant

trend clearly exists since that time, the change in rainfall can as readily be described by a step function (Fig. 11.27), with a wet period persisting until some time in the 1960s and a dry episode since about 1968.

Cycles in rainfall variability are perhaps more properly referred to as "quasi-periodicities," since the term "cycle" implies a regularity that is not evident in such phenomena. These are generally detected using the much-abused statistical technique of spectral analysis. A thorough survey of the literature would probably find reference to cycles of every conceivable length in tropical rainfall variability, some of which are established only from a subjective look at plots of rainfall time series. These are commonly used in studies of drought occurrence in semi-arid regions.

One example is the study by Faure and Gac (1981), which used a purported 30-year cycle to predict that the Sahel drought would end in 1985. In fact, that year was only slightly wetter than the two previous years, which were some of the driest on record. Wet conditions did not return until some 20 years later. Figure 11.27 shows a 30-year cycle imposed upon a time series of rainfall anomalies in the Sahel. Clearly there are major departures from the 30-year curve. Winstanley (1973), using a time series of roughly 70 years, claimed to identify a 200-year

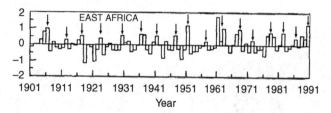

Fig. 11.28 Time series of regionally averaged annual rainfall in East Africa, illustrating the prominent 5–6 year quasi-periodicity of the fluctuations. Arrows show maxima, with 16 occurring within 90 years, but the timing is irregular.

cycle and predicted that a severe drought would occur in the Sahel and in India around the year 2030.

Studies of the periodic behavior of rainfall can be useful when care is taken to rigorously establish the periodicities and their reliability, when the periodicities explain a high percentage of the variance, and when underlying dynamic causes can be identified. A case in point is East African rainfall. A study by Rodhe and Virji (1976) established three pronounced cycles of about 2.3, 3.5, and 5–6 years, with the latter explaining 25% of the variance of interannual variability. The 5–6 year cycle is readily apparent in a simple plot of rainfall totals, although it has tended to disappear during the last few decades (Fig. 11.28). All three cycles have been clearly demonstrated in dynamic causal factors, such as the atmospheric quasi-biennial oscillation (QBO), El Niño, and sea-surface temperature (SST) variability. Hence, the established cycles may have some very limited forecast potential. Nevertheless, the three significant cycles collectively explain just over 40% percent of the variance of annual totals (Nicholson and Entekhabi 1986) and appear very irregularly. Thus, it is more useful to use the cycles in a diagnostic fashion to establish links to physical causes of variability, rather than a prognostic one.

A few studies have looked at the question of interannual persistence, or how long conditions of anomalously high or low rainfall are likely to last. This question is of interest because the consequences of a succession of dry years are likely to be more severe than those of the same number of dry years interspersed with wet ones.

This is also highly variable among the drylands. The West African Sahel (Fig. 11.27) is an area where year-to-year persistence is abnormally high. The number of multi-year sequences of wet or dry conditions departs significantly from random expectations. Lag 1 autocorrelations for rainfall (the correlation between successive pairs of consecutive years) are on the order of 0.4–0.5 in the central Sahel (Nicholson and Palao 1993). Rainfall in the southern Kalahari, the Sahel's Southern Hemisphere equivalent, shows no such persistence, nor does rainfall in East Africa (Fig. 11.28). Mooley and Parthasarathy (1984) likewise found no significant interannual persistence in Indian monsoon rainfall during the period 1871–1978.

The concept of persistence often comes up in the context of predictability. Attempts are commonly made to predict seasonal rainfall variations on the basis of rainfall in previous months

or seasons. Nicholson and Entekhabi (1986) examined persistence on intra-annual time scales by calculating the correlation between consecutive seasons and months. They concluded that in few cases was the correlation statistically significant, let alone large enough to have forecast potential. For the Sahel and nearby regions there is a small positive correlation between the early rainy season (June–July) and the late rainy season (August–September). Only in a limited area of the northwestern Sahel is the correlation significant. Even there, June–July rainfall explains only 16–25% of the variance of August–September rainfall (Nicholson and Palao 1993). This is similarly true for Indian summer monsoon rainfall: the first and second halves of the season are completely independent (Mooley and Appa Rao 1971).

In a small number of cases, persistence may be large enough to have forecast potential. In southern Africa (Nicholson and Entekhabi 1986), March and April rainfall anomalies are highly correlated, as are April and May anomalies. Also, rainfall in the December–March season correlates well with rainfall in the April–May season. The correlation coefficients reach 0.59 in some regions. Here, the physical reasons for the persistence are relatively clear. Sea-surface temperature (SST) fluctuations along the coast are a major factor in rainfall variability and SST anomalies persist for many months. Likewise, near the Guinea coast of West Africa (southern Ghana, Ivory Coast, Togo, Benin), the correlation between July and August rainfall is 0.72. This relationship might also be linked to the controlling influence of SST, in this case in the Gulf of Guinea.

11.4.4 SPATIAL VARIABILITY

It is well established that rainfall in dryland regions is erratic both in time and space. However, spatial variability has been systematically examined in only a few studies. Spatial variability is evident both in mean conditions and during individual rain events.

A dramatic example of the spatial variability of means is precipitation in the state of Washington in the USA. Aridity is largely topographically induced, with rain shadows in the lee of higher terrain. Spatial gradients are as large as 400 mm/10 km. Humid environments in which annual rainfall exceeds 2500 mm lie adjacent to deserts with less than 200 mm. Such patterns are commonly associated with mountains, but extreme gradients result from other factors as well. In the West African Sahel, where the prevailing circulation changes from high pressure to the Intertropical Convergence Zone (ITCZ) over a short distance, the rainfall gradient is 1 mm/km. The gradient between the Peruvian desert and the rainforest of Ecuador to the north is even steeper, with mean annual rainfall changing from less than 200 to over 2000 mm within some 5° of latitude, or roughly 500 km. These gradients are one factor in the large interannual variability in dryland regions.

Even in areas with similar long-term means, the precipitation falling on a given day, or within a month or year, may vary tremendously over quite short distances. This is particularly

Table 11.1. *Annual rainfall (mm) during the years 1957, 1958, and 1963 at four stations in Niger, West Africa. The long-term annual mean is also given for each station. Data are from ORSTOM (1976).*

Station	Distance from Niamey City	Annual mean	1957	1958	1963
Niamey City	0	594 mm	608	622	558
Niamey Airport	5	614 mm	732	523	474
Kolo	20	602 mm	542	409	903
Say	40	688 mm	876	728	604

Fig. 11.29 (left) Two widely separated rain cells in a storm in New Mexico. (right) The line between the rain and non-rain areas is very sharp, leading to a high degree of spatial variability, especially in stationary systems.

true of convective rainfall, and is a result of the relatively small spatial scales and short time scales of convective systems. Table 11.1 shows the annual rainfall in the years 1957, 1958, and 1963 at four stations in Niger. The stations are only 5–40 km apart. In 1958, the driest of the three years, annual rainfall varied from 409 to 728 mm. The values for 1963, which ranged from 474 to 903 mm, show that the variation is random rather than a systematic result of local effects. Kolo, the wettest station in 1963, was the driest in the other two years.

For individual days or rain events, the spatial variability is considerably greater. Rain cells are localized in storms and many storms have more than one rain cell (Fig. 11.29). In two storms in the Walnut Gulch catchment of southern Arizona, the cells with intense rain are on the order of 5 km (Fig. 11.30). Daily rainfall at five stations in the Jornada long term ecological research (LTER) site in southern New Mexico (Kemp *et al.* 1997) shows comparable spatial variability (Fig. 11.31). The stations lie along a transect at roughly 500 m intervals. The differences between the stations are random and are often as great as 5–10 mm, or roughly on the order of 20–50%. Interestingly,

the largest event (with roughly 40 mm/day) had relatively little variability, with the differences between four of the five stations being on the order of 2 mm. This suggests that the system producing this event may have been larger, so that peak rainfall occurred over a larger area.

In Africa, where the storm intensity is markedly greater than in the southwestern USA, the contrasts are remarkable. Table 11.2 shows rainfall at two stations 3.2 km apart near Dar es Salaam, Tanzania (Jackson 1969). Daily totals at the two stations differ by factors of 2–20. The magnitude of spatial variability is similar in West Africa. Table 11.3 gives daily totals at five locations in and around Bamako, Mali. Distances between the stations are on the order of 2–6 km. On August 27, one location received 163 mm, and two others received over 100 mm, but the remaining stations registered 30 and 38 mm. On September 11, one location received 127 mm, while another received only 17 mm.

Spatial variability can be illustrated using correlation–distance relationships. These have been evaluated in locations as diverse as the Netherlands, the central SA, the southwestern USA, Namibia, Israel, Jordan, and Tanzania (Fig. 11.32a). The

Table 11.2. *Daily rainfall at two stations near Dar es Salaam, Tanzania, 3.2 km apart, during April 1967 (from Jackson 1969).*

	6th	7–8th	10th	12th	13th	24th	25th
Station A	54.9	33.3	2.3	13.5	13.5	14.2	21.1
Station B	100.3	3.3	31.8	64.8	5.1	8.4	0.8

Table 11.3. *Rainfall (mm) at sites around Bamako, Mali, on two days in 1954. Distance between the various stations ranges from approximately 2 to 6 km. Data are from ORSTOM (1974).*

	City	Airport	Zoo	Koulouba	Dungfing
August 27	101	163	121	38	30
September 11*	52	55	127	44	17

*Rain at the zoo site was registered on September 12.

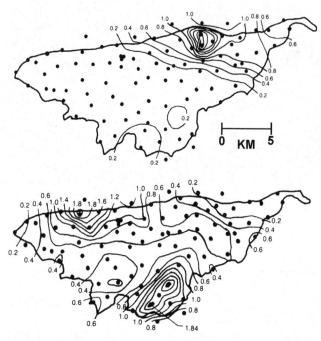

Fig. 11.30 Distribution of rainfall (inches) from thunderstorm cells over Walnut Gulch, Arizona, an area of roughly 100 km². (top) A single-cell event. (bottom) A multiple-cell event. Dots represent rain gauges (after Renard 1970, courtesy of US Department of Agriculture).

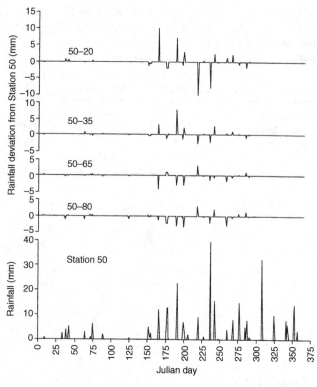

Fig. 11.31 Jornada daily rain at five stations (from Kemp *et al.* 1997).

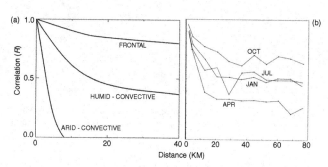

Fig. 11.32 Correlation vs. distance (km) between gauges: (a) schematic illustration for various climate types; (b) four different months at Ruvu, Tanzania (mean annual rainfall 1000 mm) (based on Jackson 1974).

correlation as a function of the distance between gauges depends strongly on the nature of the rainfall regime. This correlation tends to be extremely high for frontal rainfall in either arid or humid regions, lower for convective rainfall in humid regions, and extremely low for convective rainfall in arid regions. In the tropical and semi-arid locations, the correlation falls below 0.2 at distances of about 5–20 km, but remains in excess of 0.6 for tens of kilometers (or greater) in humid mid-latitude locations over the USA and Europe.

These relationships are also seasonally dependent. In northern Tanzania, which experiences two rainy seasons each year, inter-station correlation is markedly higher during the briefer and weaker "short rains" of the austral spring than during the "long rains" of the austral autumn. In October the correlation falls to roughly 0.7 at a distance of 20 km and remains at this level at inter-station distances up to 80 km (Fig. 11.32b). In April, the heart of the "long rains," it falls to 0.4 at about 10 km and is on the order of 0.3 at distances up to 80 km. The interannual variability of the short rains is largely controlled by large-scale processes, such as El Niño, while more local effects are important during the long rains (see Chapter 16). The seasonal differences can be even larger in mid-latitude locations, where convection prevails in summer and frontal rains occur in winter.

In some cases the correlation increases at larger inter-station distances that roughly represent the size of storm cells. In the southwestern USA, typical sizes of the cores of the storms are on the order of tens of kilometers. In the Namib Desert, correlation is maximized at both 40–50 km and 80–100 km (Sharon 1981).

Several characteristics of rain-bearing systems in the tropics are responsible for this variability. One is that rainfall tends to be concentrated in deep convection cells imbedded within non-rainy regions of massive cloud areas (e.g., MCSs). Also, continuity of mass requires that the updrafts producing clouds and rainfall must be balanced by compensating subsidence, which suppresses precipitation in the surroundings. Thus, cumulus or cumulonimbus cells are in a constant state of flux, with the descending raindrops signaling the decay of the cloud, and with the downdrafts spreading out at the surface and triggering uplift and convection ahead of the decaying cell. The life cycle of a single cloud may be on the order of minutes in extreme cases. Another factor is inherent to drylands. The disturbances are few in number and specific locations are randomly hit or missed by the major disturbance regions or the regions of intense convection within them.

The spatial variability of precipitation is thus inherently large in convective systems, which prevail most of the year in the tropics and in summer in higher latitudes. It is considerably smaller in typical mid-latitude synoptic systems, such as those that produce most of the rainfall in the cooler seasons. Therefore, it is impossible to generalize about the spatial variability in a given region without having information on the dynamical character of the rainfall regime.

This is illustrated with the stations Gourma-Rharous and Timbuktu in Mali, which lie 115 km apart, and Griquatown and Postmasburg in South Africa, which lie 60 km apart. All are in semi-arid regions with predominantly summer rainfall. Time series of annual rainfall from 1926 to 1991 and scatter plots of rainfall at the two station pairs for the year are shown in Fig. 11.33. The latter are given for annual rainfall and for rainfall in the wettest month. In Mali, where the systems are purely tropical in origin and primarily convective, the two stations show very dissimilar fluctuations. Most of the rainfall at the South African stations is brought by hybrid tropical/extra-tropical systems and is frequently associated with frontal systems. The year-to-year patterns are very similar at these two stations. This contrast is evident at the monthly scale also, although both sets of stations show greater differences at the monthly scale.

11.4.5 INTRASEASONAL VARIABILITY

The total rainfall during a season or year is not a good predictor of water availability for the ecosystem, agriculture, or water resources. The sequence of events is very significant (Miao *et al.* 2009). Several characteristics of the rainy season are relevant: its onset, end and length; its reliability; the occurrence of breaks in the season; the length and frequency of dry spells; and the timing of drought. These characteristics, though important, are

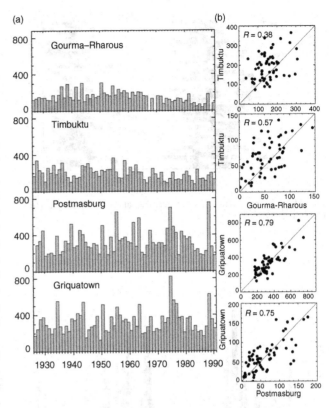

Fig. 11.33 Rainfall at two South African stations (Postmasburg and Griquatown) and two stations in Mali, West Africa (Timbuktu and Gourma-Rharous), 1926–1991. The South African stations lie 60 km apart; the Mali stations lie 115 km apart. (a) Time series of annual rainfall. (b) Scatter plots of the station pairs for annual totals and for rainfall in the wettest month.

difficult to evaluate because they all rely on arbitrary definitions and thresholds. Unfortunately, there is no universal agreement on what signals the onset or end of the rainy season, nor on what constitutes a dry month or dry spell.

Figure 11.34 illustrates the intraseasonal variability and variations of the start and end of the rainy season at Nairobi, Kenya, based on wet and dry pentads (5-day periods). Numerous dry spells occur within the season, the length and number varying from year to year. The onset and end of the season is also quite variable, especially for the long rains. Lengthy intraseasonal breaks are a regular feature of the Indian monsoon (Fig. 11.35). These tend to occur at intervals of 20–30 days and last as long as 40 days. In the drylands of the western USA dry spells of 10–30 days in duration are fairly common (Fig. 11.36).

The intraseasonal dry spells can be fairly regular or random in occurrence. In Botswana and parts of South Africa breaks in the season tend to occur in late December/early January (Fig. 11.37). This appears to reflect a change from a prevailing mid-latitude rainfall regime to a tropical rainfall regime.

Figure 11.38 illustrates intraseasonal variability at dryland locations in Africa and the USA with varying degrees of aridity. Most stations receive on average a few hundred millimeters per year. For the African stations both a wet year and a dry year

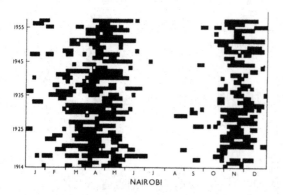

Fig. 11.34 Rainy pentads (5-day periods) at Nairobi, Kenya (from Griffiths 1959).

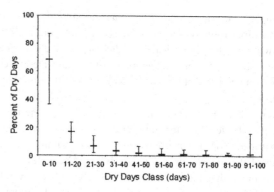

Fig. 11.36 Frequency distribution of dry spells of various duration (days between precipitation events) (from Loik *et al.* 2004).

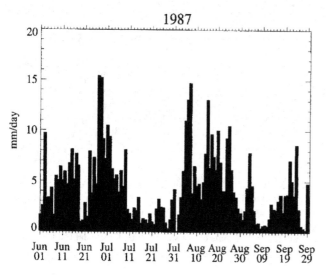

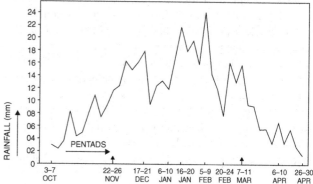

Fig. 11.37 Mean pentad (5-day) rainfall for Francistown, Botswana, 1922–1980 (from Bhalotra 1984).

Kalamare contrasts starkly, with few days receiving more than 40 mm. Daily rainfall is even lower at the stations in the USA. At Walnut Gulch maximum daily rainfall is on the order of 10 mm.

The large rainfall events are "pulses" of moisture availability to the ecosystem. The rapid availability of moisture produces a greater ecosystem response than a gradual increase in moisture availability to the same level (Fig. 11.39) and has a different impact on water balance (Scheffer *et al.* 2008).

There is considerable debate about the best way to quantify intraseasonal variability. One approach is to derive probability estimates of such parameters as the onset, end, and duration, number and length of dry spells, or sequences of various amounts of rainfall. Currently, considerable effort is being directed toward an alternative statistical approach to modeling the occurrence of wet and dry days or spells. These models often use a first- or second-order Markov process and consider the dependence of the occurrence of rainfall on whether or not rain has fallen during the previous day or days. For a review of some of these models, see Rodriguez-Iturbe (1986) and Ali *et al.* (2003).

11.4.6 CHARACTERIZING RAINFALL EVENTS

Intensity, duration, frequency, size, and spatial distribution of rainfall are important characteristics of rain events, especially

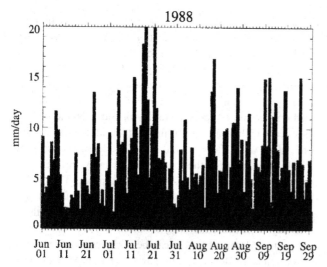

Fig. 11.35 Satellite estimates of daily rainfall averaged for the sector 10–15° N, 75–80° E for June–September, 1987 and 1988, illustrating breaks in the Indian summer monsoon (from Lawrence and Webster 2001).

are shown. The patterns are very diverse among the stations. At Niafunke, daily totals are typically 20 mm or less, but a handful of events brought 60–150 mm. The daily rainfall regime at

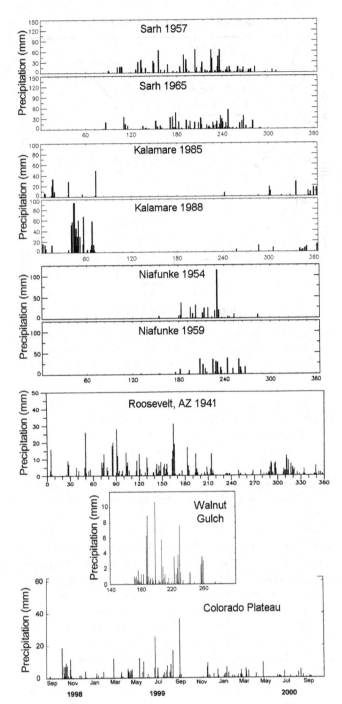

Fig. 11.38 Time series of daily rainfall at locations in Africa and the USA: Sarh (Chad), Kalamare (Botswana), Niafunke (Mali), Roosevelt (Arizona), Walnut Gulch (Arizona), and Colorado Plateau (Utah). For the African locations, a wet and a dry year are given.

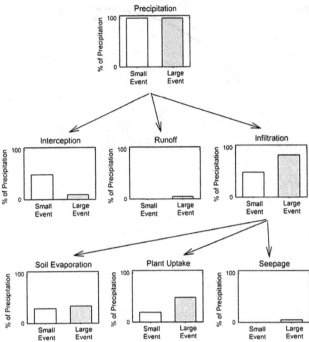

Fig. 11.39 Water budget components for large and small pulses of precipitation. All panels are normalized relative to 100% of the event; hence the figure emphasizes the relative size of each component for small versus large events, not the actual amount (from Loik et al. 2004).

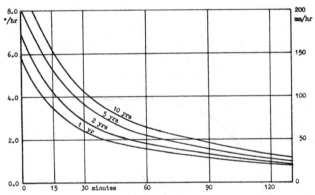

Fig. 11.40 Intensity–duration–frequency relations for rainfall at four return periods (from Department of Meteorological Services 1975).

in arid and semi-arid regions. In general, there is a strong relationship between mean intensity and duration (Fig. 11.40). The more intense systems occur less frequently, comprising a very small percentage of rainfall events, but they bring a large proportion of the total rainfall.

Figure 11.41 demonstrates this using Niamey, Niger, as an example (mean annual rainfall = 553 mm). The cumulative rainfall amounts from the largest 10 rain events are shown for September, a month during which rainfall generally occurs on 5–10 days. Separate tabulations are indicated for the wettest 10 years, the driest 10 years, and the long-term mean. During the wet years, approximately 75% of the monthly total falls on the wettest four days; the wettest day typically brings over 40% of the monthly total. During the dry years, over 90% falls in the wettest four days; the wettest day accounts for nearly 60% of the monthly total.

Another characteristic of interest is the relationship between point and areal rainfall. In drylands, rainfall is patchy and station density is likely to be low, so that the representativeness of

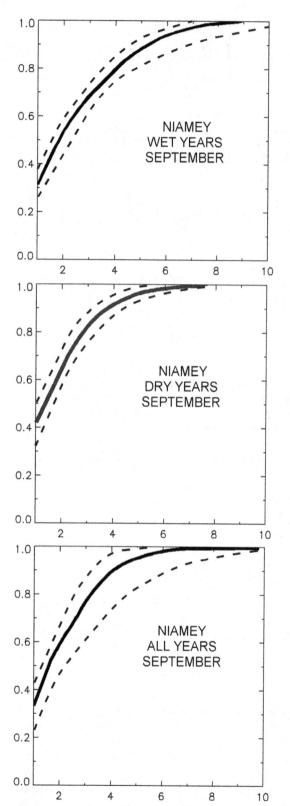

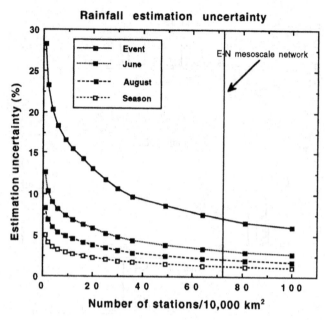

Fig. 11.42 Average estimation uncertainty of areal rainfall estimation over the 1° × 1° square (in percent of the areal rainfall). Values are computed for a "typical" event with 15.5 mm of rainfall (from Lebel and Le Barbé 1997).

Fig. 11.41 Cumulative % of monthly rainfall at Niamey, Niger, brought by individual rainfall events, ordered from largest to smallest daily totals. The mean number of rain events during the month is indicated. Curves are shown for the 10 driest years, the 10 wettest years, and for all of the 30 years from 1931 to 1960. Thick lines indicate an average of all years, and thin lines indicate the limits between which two-thirds of all years fall.

rainfall at a single gauge is questionable. A number of studies (e.g., Rodriguez-Iturbe *et al.* 1998) have established the probable concentration of rainfall in time and space associated with measured "point" (i.e., station) rainfall amounts. This is critically important for estimating runoff and evapotranspiration – quantities that are particularly difficult to estimate in dryland regions.

Lebel and Le Barbé (1997) examined the relationship between point and areal rainfall in the West African Sahel. They found that for a 1° × 1° area the estimation uncertainty for spatial averages reaches a minimum when the station density reaches eight stations per 10,000 km². At this density, the error is on the order of 1–3% for monthly and seasonal rainfall, but about 7% for event rainfall (Fig. 11.42). They concluded that, for a 1° × 1° area, two stations can provide an estimate with less than 2% uncertainty for annual or seasonal rainfall and less than 5% uncertainty for monthly rainfall. Xie and Arkin (1995), in a global study, similarly concluded that the error of gauge estimates of areal rainfall is generally less than 10% if five or more gauges are available within an area of 2.5° × 2.5° of latitude and longitude. A statistical analysis of gauge data also showed that errors in spatial estimates stabilize when five stations or more are available (Nicholson 1986).

The results of the studies cited above cannot be directly extrapolated to all dryland regions. It is relevant that three independent analyses provide results that are roughly comparable. Thus, they suggest at least an order of magnitude estimate of the error introduced by extrapolating areal estimates from gauge data and this is probably valid for all but the most arid regions.

11.5 FOG, DEW, AND ADSORPTION AS WATER SOURCES IN DRYLANDS

In many deserts the lack of precipitation is partially compensated by the direct acquisition of moisture from the air. The available moisture comes in three forms: fog, dew, and adsorption water (Agam and Berliner 2006). Fog is water droplet condensation in the atmosphere. Dew is water that condenses on the surface and may later infiltrate into the soil. Adsorption water enters the soil matrix as gas and condenses in the pores. In many deserts, these sources provide more water than does rainfall, and many desert plants and animals can readily utilize these sources.

Unfortunately, relatively little quantitative data of these phenomena are available because measurement is more complex than for rainfall. Assessment is made via a variety of field measurement techniques (e.g., Agam and Berliner 2006; Heusinkveld *et al.* 2006) and modeling (e.g., Jacobs *et al.* 2002). Both water volume and duration of dew are quantified. The models have potential applications for satellite assessment of dew. Remote sensing methods already exist for detecting the presence of fog (Cermak and Bendix 2008).

11.5.1 FOG

Fog provides a significant moisture source in some desert regions, particularly the deserts located along cold-water coasts. These include the Namib (Henschel and Seely 2008), the Chilean-Atacama Desert (Munoz-Schick *et al.* 2001; Larrain *et al.* 2002), the western Sahara, and Baja California (Martorell and Ezcurra 2002). In these coastal deserts, fog is generally of the advection type, a result of coastal stratus clouds penetrating inland. In the interior deserts, fog is primarily a result of radiative cooling or orographic effects.

Many desert plants, animals, and microorganisms utilize fog-water to survive the harsh conditions in their environment. Certain plants can directly absorb fog-water through their leaves or thorns. Others collect fog-water and transmit it downward by stemflow to their roots. Growth patterns of some plant species reflect the distribution of fogs. Organisms can absorb water vapor, drink the fog-water deposited on surfaces, or collect the fog-water on the body.

The amount of fog-water can be considerable, especially during the dry seasons. In the Atacama Desert, an average of 2.5 L/m^2 was measured during a 1-year period (Westbeld *et al.* 2009). At Anaga, on Tenerife in the Canary Islands, hourly fog-water deposition can reach 6.7 L/m^2 (Marzol 2008). The daily maximum is 63 L/m^2.

The frequency and intensity of fog is seasonally dependent and varies greatly from year to year. On Tenerife the maximum occurs during the summer dry season, with the frequency approaching 80% from evening to early morning. During the same hours, the frequency is roughly 30% in winter. In the hyper-arid Atacama Desert of Chile, the fog maximum is in winter and in the Namib the maximum occurs during the dry spring months (September–November) (Cereceda *et al.* 2008;

Henschel and Seely 2008). Fog occurrence also tends to have a diurnal cycle, with a maximum at night and a minimum at mid-day. Consequently, fog duration is generally less than 24 hours.

The occurrence of fog is strongly dependent on atmospheric conditions. It requires only that the water vapor concentration in the atmosphere reaches saturation (Agam and Berliner 2006). Conditions at the surface play only an indirect role, in that they may enhance nocturnal cooling, which favors the formation of radiation fogs. Mixing by wind inhibits it. The deposition of fog on a given surface is mainly a manifestation of settling and interception (Jacobs *et al.* 2002). In general, fog interception increases with wind speed (Martorell and Ezcurra 2002), but fog amount does not necessarily correlate with wind speed (Marzol 2008) because humidity conditions also play a role. Fog interception decreases with the slope of the surface (Kidron 2005).

Fog is essentially a cloud that intercepts the ground. Because saturation is a function of temperature, the condensation that produces fog or cloud droplets is altitude-dependent. In advection fog resulting from coastal stratus, the location and extent of the fogs and the fog-water collected depend on the base and thickness of the coastal stratus. In areas of higher terrain, mountain–valley winds can promote the fog generated when air flows upslope during the day and cools upon ascent. The result is a cloud layer that intercepts the higher terrain.

Thus, the occurrence of orographically induced fog is altitude-dependent. On a regional basis, the altitude depends on prevailing meteorological conditions, specifically the mean temperature and moisture variations in the lower layers of the atmosphere. In the central fog desert of the South American coast, the coastal stratus and fog occur in a layer some 250 m thick with a base between 500 and 800 m ASL (Lettau 1978). Further north, near Lima, the base generally lies between 150 and 300 m ASL and has a distinct upper boundary around 500–700 m ASL. Further south, in northern Chile, the cloud base generally lies above 650 m ASL and the fog maximum lies at 700–800 m ASL. In the Namib coastal desert, fog occurs most frequently between 300 and 600 m ASL (Lancaster *et al.* 1984). In the Negev, fog and dew both increase steadily with altitude from sea-level up to at least 1000 m ASL (Kidron 1999a).

When condensation occurs at a particular elevation, dense stands of vegetation develop. These locations are variously termed "fog or cloud forests" or "mist or cloud belts" or "fog oases." In some regions as many as three such altitudinal zones exist. Here the transition from the desert vegetation of the plains to the lush vegetation of the mist belts is abrupt. Examples of this arid–humid juxtaposition are in the Chilean-Atacama Desert of South America (the lomas formations) and the desert of Baja California (see Chapter 20).

The altitude of the mist belts in the arid mountains of Mexico varies from around 1200 m ASL in the Sierra de San Francisco of Baja California, which receives Pacific coastal fogs, to 2200 m ASL in the Sierra de El Doctor, in the rain shadow desert of central-southern Mexico (Martorell and Ezcurra 2002). At San Francisco a second layer of thin coastal fogs exists at 400 m ASL.

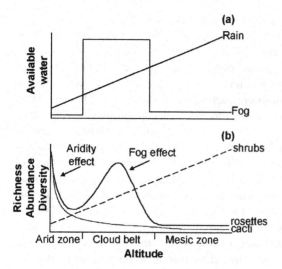

Fig. 11.43 Schematic illustration of three life forms (shrubs, cacti, and rosette scrub) in relation to rainfall and fog in the Tehuacán Valley of Mexico (from Martorell and Ezcurra 2002).

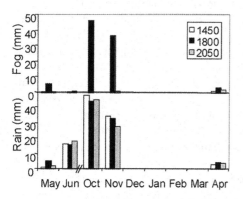

Fig. 11.44 Fog and rain recorded monthly at three altitudes (in m) in the Tehuacán Valley of Mexico (from Martorell and Ezcurra 2002). Fog measurements are shown for 2 m above the ground.

In the arid Tehuacán Valley, both rainfall and fog influence vegetation distribution. Rainfall increases steadily with increasing elevation, but fog is concentrated around 1800 m ASL (Fig. 11.43). At this altitude the fog creates a cloud belt in which rosettes and cacti thrive, while the abundance of shrubs continues to increase with altitude, as does rainfall. Cacti, with the best survival mechanisms, are most abundant in the low-altitude arid zone and their abundance decreases rapidly with altitude. Fog occurs here most frequently during April–June and October/November (Fig. 11.44). Fog precipitation increases with height above the surface, ranging from less than 10 mm at 0.3 m in height to nearly 200 mm/year at 2 m above the surface (Fig. 11.45). Overall, fog-water increases moisture availability by 10–100%.

11.5.2 DEW

Some of the most extensive work on the accumulation of dew has been carried out in Israel. Pioneering work on dew

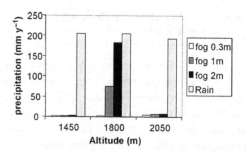

Fig. 11.45 Total annual rainfall and fog at three altitudes and three heights above ground in the Tehuacán Valley, Mexico (from Martorell and Ezcurra et al. 2002).

measurement began in 1936. The National Meteorological Service began routine dew measurements at a large number of stations in 1945 and published the data regularly in annual summaries (Zangvil 1996).

Dew is an important source of moisture for plants, insects, and small animals in desert environments. Desert arthropods, such as ants and beetles, and soil fauna, such as nematodes, rely on dew (Broza 1979; Moffett 1985; Steinberger et al. 1989). Dew helps plants recover from water stress and in some deserts it can be the sole source of water for plants.

Dew also influences geomorphologic processes, such as weathering (Evenari et al. 1982), enhances the development of karst formations in arid environments (Castellani and Dragoni 1987), and helps to consolidate mineral dust deposited on the surface (Goossens and Offer 1995). Where vegetation is sparse, it can stabilize sand dunes and surface soil (Jacobs et al. 1999). It is a source of moisture for biological crusts that likewise stabilize soil and dunes (Kidron 1999b; Danin et al. 1989). The biological crusts, in turn, exert considerable feedback on dew deposition (Liu et al. 2006; Rao et al. 2009; Zhang et al. 2009).

The cycle of deposition and evaporation is accompanied by changes in latent heat flux. Therefore, dew affects surface energy balance, being a major sink of early-morning energy availability (Pitacco et al. 1992) but supplying latent heat to the surface at night (Katata et al. 2007). Consequently, the presence of dew can reduce diurnal temperature variations.

Dew is much less dependent than fog on large-scale atmospheric conditions. However, its formation is influenced by radiation energy fluxes at the surface and by turbulent transport of heat and moisture in the boundary layer. The dew formation process is controlled by the available energy (net radiation minus soil heat flux) and by the vapor pressure gradient between the atmosphere and soil.

There are two phases in the process of dew formation: nucleation of the liquid phase (the initial contact) and droplet growth (Beysens 1995). Characteristics of the surface, including slope angle and aspect, wetness, and soil texture are important in determining growth (Heusinkveld et al. 2006; Jacobs et al. 2000; Kidron 2005). In the Negev, dew formation was found to be twice as great on playas, which have a high silt and clay content, than on the sandy dunes, and twice as great in interdune areas than on dune slopes. Dew formation requires that surface temperature be lower than or equal to dew-point temperature. When the soil is dry, dew formation can begin at much higher temperatures than when it is wet.

Dew can occur frequently, whenever the air is not too dry and wind speeds are low (Jacobs *et al.* 2002). In some areas, such as the deserts of Nevada, it occurs almost on a daily basis, even occasionally during dry season months (Malek *et al.* 1999). Annually about 14 mm is accumulated on average. Deposition is on the order of 5–10 mm/year in parts of the Andes (Kalthoff *et al.* 2006). In the northwestern Negev and in the Negev highlands, dew forms on some 200 nights per year on average (Evenari *et al.* 1982; Zangvil 1996). An equivalent of 30 mm accumulates on average, with peak accumulation in spring and autumn. Dew accumulation often exceeds annual rainfall in drought years.

Dew accumulates during the night and begins to evaporate after sunrise (Jacobs *et al.* 1999). Theoretically, up to 1 mm per night can occur (Jacobs *et al.* 1990). However, the nightly dew formation is extremely variable. In the Negev highlands it is in the range of 0.06–0.1 mm per night (Zangvil 1996). Up to 0.2 mm per night accumulates in the northwestern Negev (Jacobs *et al.* 2000). In some locations in Israel as much as 0.8 mm can accumulate on a given day.

11.5.3 ADSORPTION WATER

Fewer studies have considered adsorption water. Most measurement techniques do not distinguish between adsorption water and dew, making assessment difficult. The source of much water assumed to be dewfall is probably adsorption water, especially during the dry season (Agam and Berliner 2006). Water adsorption is an important component of the water cycle in dryland regions and, like dew, influences the energy balance via its role in evaporation and latent heating.

Two conditions are required for water vapor adsorption to occur. The surface temperature must be higher than the dew-point temperature, and the humidity of the soil's pores must be lower than the relative humidity of the air. The gaseous vapor in the air condenses on solid surfaces in the soil, creating a surficial water layer that has the same appearance as dew. A distinction can be made on the basis of temperature. If it is not below the dew point, the mechanism is adsorption.

Water is held in the soil matrix by adsorption to particle surfaces or in the poles by capillary forces. Distinguishing between these two reservoirs is difficult, so that measurements generally reflect the combined retention. The relative humidity of the soil pores determines which mechanism is dominant. Retention is primarily via capillary forces at high relative humidities (greater than about 60%), but by physical adsorption at lower relative humidities. The dominance of adsorption increases with increasing clay content as a result of greater surface-to-volume ratios in clay particles and because of ionic attraction.

REFERENCES

Agam, N., and P. R. Berliner, 2006: Dew formation and water vapor adsorption in semi-arid environments: a review. *Journal of Arid Environments*, **65**, 572–590.

Ali, A., T. Lebel, and A. Amani, 2003: Invariance in the spatial structure of Sahelian rainfall fields at climatological scales. *Journal of Hydrometeorology*, **4**, 996–1011.

Beysens, D., 1995: The formation of dew. *Atmospheric Research*, **39**, 215–237.

Bhalotra, Y. P. R., 1984: *Climate of Botswana. Part I. Climatic Controls.* Department of Meteorological Services, Gaborone, 68 pp.

Broza, M., 1979: Dew, fog and hygroscopic food as a source of water for desert anthropods. *Journal of Arid Environments*, **2**, 43–49.

Castellani, V., and W. Dragoni, 1987: Some considerations regarding karstic evolution of desert limestone plateaus. In *International Geomorphology 1986: Proceedings of the First International Conference on Geomorphology* (V. Gardner, ed.), Vol. 2, Wiley, Manchester, pp. 1199–1206.

Cereceda, P., H. Larrain, P. Osses, M. Farías, and I. Egaña, 2008: The spatial and temporal variability of fog and its relation to fog oases in the Atacama Desert, Chile. *Atmospheric Research*, **87**, 312–323.

Cermak, J., and J. Bendix, 2008: A novel approach to fog/low stratus detection using Meteosat 8 data. *Atmospheric Research*, **87**, 279–292.

Cheng, X., and Coauthors, 2006: Summer rain pulse size and rainwater uptake by three dominant desert plants in a desertified grassland ecosystem in northwestern China. *Plant Ecology*, **184**, 1–121.

Danin, A., Y. Bar-Or, I. Dor, and T. Yisraeli, 1989: The role of cyanobacteria in stabilization of sand dunes in southern Israel. *Ecologia Mediterranea*, **XV**, 55–64.

Dankwa, J. B., 1974: *Maximum Rainfall Intensity-Duration Frequencies in Ghana.* Departmental Note No. 23, Ghana Meteorological Services Department, Legon, 38 pp.

Department of Meteorological Services, 1975: *Rainfall Intensity in Rhodesia.* Rainfall Handbook Supplement No. 7, Salisbury, Rhodesia, 64 pp.

Desmet, P. G., and R. M. Cowling, 1999: The climate of the karoo: a functional approach. *The Karoo: Ecological Patterns and Processes* (W. R. J. Dean and S. J. Milton, eds.), Cambridge University Press, Cambridge, pp. 3–16.

Evenari, M., L. Shahan, and N. Tadmor, 1982: *The Negev: The Challenge of a Desert.* Harvard University Press, Cambridge, MA, 345 pp.

Faure, H., and J. Y. Gac, 1981: Will the Sahelian drought end in 1985? *Nature*, **291**, 475–478.

Gautier, E.-F., 1970: *The Sahara: The Great Desert.* Cass, London, 281 pp.

Gibbs, W. J., and J. V. Maher, 1967: *Rainfall Deciles as Drought Indicators.* Bulletin 48, Bureau of Meteorology, Commonwealth of Australia, Melbourne, 84 pp.

Goodrich, D. C., and Coauthors, 2008: Long-term precipitation database, Walnut Gulch Experimental Watershed, Arizona, United States. *Water Resources Research*, **44**, doi:10.1029/2006WR005782.

Goossens, D., and Z. Y. Offer, 1995: Comparisons of day-time and night-time dust accumulation in a desert region. *Journal of Arid Environments*, **31**, 253–281.

Graf, W. L., 1988: *Fluvial Processes in Dryland Rivers.* Springer-Verlag, Berlin, 346 pp.

Griffiths, J. F., 1959: Bioclimatology and the meteorological services. In *Proceedings of the Symposium on Tropical Meteorology in Africa.* Munitalp Foundation and World Meteorological Organization, Nairobi, pp. 282–300.

Groisman, P. Y., and Coauthors, 1999: Changes in the probability of heavy precipitation: important indicators of climatic change. *Climatic Change*, **42**, 243–283.

11 Precipitation in the drylands 210

Henschel, J. R., and M. K. Seely, 2008: Ecophysiology of atmospheric moisture in the Namib Desert. *Atmospheric Research*, **87**, 362–368.

Heusinkveld, B. G., S. M. Berkowicz, A. F. G. Jacobs, and W. C. A. M. Hillen, 2006: An automated microlysimeter to study dew formation and evaporation in arid and semiarid regions. *Journal of Hydrometeorology*, **7**, 825–832.

Horn, L. H., and R. A. Bryson, 1959: Harmonic analysis of the annual march of precipitation over the United States. *Annals of the Association of American Geographers*, **50**, 157–171.

Huxman, T. E., and Coauthors, 2004: Precipitation pulses and carbon fluxes in semiarid and arid ecosystems. *Oecologia*, **141**, 254–268.

Jackson, I. J., 1969: The persistence of rainfall gradients over small areas of uniform relief. *East African Geographical Review*, **7**, 37–43.

Jackson, I. J., 1974: Inter-station rainfall correlation under tropical conditions. *Catena*, **1**, 235–236.

Jacobs, A. F. G., W. A. J. Van Pul, and A. Van Dijken, 1990: Similarity moisture dew profiles within a corn canopy. *Journal of Applied Meteorology*, **29**, 1300–1306.

Jacobs, A. F. G., B. G. Heusinkveld, and S. M. Berkowicz, 1999: Dew deposition and drying in a desert system: a simple simulation model. *Journal of Arid Environments*, **42**, 211–222.

Jacobs, A. F. G., B. G. Heusinkveld, and S. M. Berkowicz, 2000: Force-restore technique for ground surface temperature and moisture content in a dry desert system. *Water Resources Research*, **36**, 1261–1268.

Jacobs, A. F. G., B. G. Heusinkveld, and S. M. Berkowicz, 2002: A simple model for potential dewfall in an arid region. *Atmospheric Research*, **64**, 285–295.

Kalthoff, N., M. Fiebig-Wittmaack, C. Meissner, M. Kohler, M. Uriarte, K. Bischoff-Gauss, and E. Gonzales, 2006: The energy balance, evapotranspiration and nocturnal dew deposition of an arid valley in the Andes. *Journal of Arid Environments*, **65**, 420–443.

Katata, G., H. Nagai, H. Hueda, N. Agam, and P. R. Berliner, 2007: Development of a land surface model including evaporation and adsorption processes in the soil for the land–air exchange in arid regions. *Journal of Hydrometeorology*, **8**, 1307–1324.

Kemp, P. R., J. F. Reynolds, Y. Pachepsky, and J. L. Chen, 1997: A comparative modeling study of soil water dynamics in a desert ecosystem. *Water Resources Research*, **33**, 73–90.

Kidron, G. J., 1999a: Altitude-dependent dew and fog in the Negev Desert, Israel. *Agricultural and Forest Meteorology*, **96**, 1–8.

Kidron, G. J., 1999b: Differential water distribution over dune slopes as affected by slope position and microbiotic crust, Negev Desert, Israel. *Hydrological Processes*, **13**, 1665–1682.

Kidron, G. J., 2005: Angle and aspect dependent dew and fog precipitation in the Negev desert. *Journal of Hydrology*, **301**, 66–74.

Lamb, P. J., M. A. Bell, and J. D. Finch, 1988: Variability of Sahelian disturbance lines and rainfall during 1951–1987. In *Water Resources Variability in Africa during the XXth Century* (E. Servat, D. Hughes, J.-M. Fritsch, and M. Hulme, eds.), International Association of Hydrological Sciences, Gentbrugge, Belgium, pp. 19–26.

Lancaster, J., N. Lancaster, and M. K. Seely, 1984: Climate of the Central Namib Desert. *Madoqua*, **14**, 5–61.

Larrain, H., and Coauthors, 2002: Fog measurements at the site "Falda Verde" north of Chanaral compared with other fog stations of Chile. *Atomspheric Research*, **64**, 273–284.

Lawrence, D. M., and P. J. Webster, 2001: Interannual variations of the intraseasonal oscillation in the South Asian summer monsoon region. *Journal of Climate*, **14**, 2910–2922.

Lebel, T., and L. Le Barbé, 1997: Rainfall monitoring during HAPEX-Sahel. 2. Point and areal estimation at the event and seasonal scales. *Journal of Hydrology*, **188/189**, 97–122.

Lebel, T., J. D. Taupin, and N. D'Amato, 1997: Rainfall monitoring during HAPEX-Sahel. 1. General conditions and climatology. *Journal of Hydrology*, **188/189**, 74–96.

Lettau, H. H., 1978: Explaining the World's Driest Climate. In *Exploring the World's Driest Climate* (Lettau, H. H. and K. Lettau, eds.). Institute for Environmental Studies, University of Wisconsin, Madison, WI, pp. 182–248.

Liu, C., E. J. Zipser, and S. W. Nesbitt, 2007: Global distribution of tropical deep convection: different perspectives from TRMM infrared and radar data. *Journal of Climate*, **20**, 489–503.

Liu, L. C., S. Z. Li, Z. H. Duan, T. Wang, Z. S. Zhang, and X. R. Li, 2006: Effects of microbiotic crusts on dew deposition in the restored vegetation area of Shapotou, northwest China. *Journal of Hydrology*, **328**, 331–337.

Loik, M., D. Breshears, W. Lauenroth, and K. Belnap, 2004: A multi-scale perspective of water pulses in dryland ecosystems: climatology and ecohydrology in the western USA. *Oecologia*, **141**, 181–269.

Malek, E., G. McCurdy, and B. Giles, 1999: Dew contribution to the annual water balances in semi-arid desert valleys. *Journal of Arid Environments*, **42**, 71–80.

Martorell, C., and E. Ezcurra, 2002: Rosette scrub occurrence and fog availability in arid mountains of Mexico. *Journal of Vegetation Science*, **13**, 651–662.

Marzol, M. V., 2008: Temporal characteristics of fog water collection during summer in Tenerife (Canary Islands, Spain). *Atmospheric Research*, **87**, 352–361.

Miao, S. L., C. B. Zou, and D. D. Breshears, 2009: Vegetation responses to extreme hydrological events: sequence matters. *American Naturalist*, **173**, 113–118.

Moffett, M. W., 1985: An Indian ant's novel method for obtaining water. *National Geographic Research*, **1**, 146–149.

Mohr, K. I., J. S. Famiglietti, and E. J. Zipser, 1999: The contribution to tropical rainfall with respect to convective system type, size, and intensity estimated from the 85-GHz ice-scattering signature. *Journal of Applied Meteorology*, **38**, 596–606.

Mooley, D. A., and G. Appa Rao, 1971: Distribution function for seasonal and annual rainfall over India. *Monthly Weather Review*, **99**, 796–799.

Mooley, D. A., and B. Parthasarathy, 1984: Fluctuation in all-India summer monsoon rainfall during 1871–1965. *Climatic Change*, **6**, 287–301.

Munoz-Schick, M., R. Pinto, A. Mesa, and A. Moreira-Munoz, 2001: Fog oases during the El Niño Southern Oscillation 1997–1998, in the coastal hills south of Iquique, Tarapaca region, Chile. *Revista Chilena de Historia Natural*, **74**, 389–405.

Nesbitt, S. W., R. Cifelli, and S. A. Rutledge, 2006: Storm morphology and rainfall characteristics of TRMM precipitation features. *Monthly Weather Review*, **134**, 2702–2721.

Nicholls, N., and K. K. Wong, 1990: Dependence of rainfall variability on mean rainfall, latitude, and the Southern Oscillation. *Journal of Climate*, **3**, 163–170.

Nicholson, S. E., 1981: Lake Chad and its relation to Sahelian climate history. In *The Sahara: Ecological Change and Early Economic History* (J. A. Allen, ed.), Menas Monograph No. 1, Westview Press, Boulder, CO, pp. 35–60.

Nicholson, S. E., 1986: The spatial coherence of African rainfall anomalies: interhemispheric teleconnections. *Journal of Climate and Applied Meteorology*, **25**, 1365–1381.

Nicholson, S. E., 2000: Land surface processes and Sahel climate. *Reviews of Geophysics*, **38**, 117–139.

Nicholson, S. E. and D. Entekhabi, 1986: The quasi-periodic behavior of rainfall variability in Africa and its relationship to the Southern Oscillation. *Archives for Meteorology, Geophysics, and Bioclimatology, Series A*, **34**, 311–348.

Nicholson, S. E. and D. Entekhabi, 1987: Rainfall variability in Equatorial and Southern Africa: relationships with sea-surface temperatures along the southwestern coast of Africa. *Journal of Climate and Applied Meteorology*, **26**, 561–578.

Nicholson S. E., and I. M. Palao, 1993. A re-evaluation of rainfall variability in the Sahel. Part I. Characteristics of rainfall fluctuations. *International Journal of Climatology*, **13**, 371–389.

Nicholson S. E., and P. J. Webster, 2007: A physical basis for the interannual variability of rainfall in the Sahel. *Quarterly Journal of the Royal Meteorological Society*, **133**, 2065–2084.

O'Brien, J. J., and J. F Griffiths, 1967: Choosing a test of normality for small samples. *Archiv fur Meteorologic, Geophysik und Bioklimatologie*, **16**, 267–272.

ORSTOM (Office de la recherché scientifique et technique outr é-mer), 1974: *Précipitations journalières de l'origine des stations à 1965: République du Mali*. Etienne Julienne, Paris, 1081 pp.

ORSTOM (Office de la recherché scientifique et technique d'outre-mer), 1976: *Précipitations journalières de l'origine des stations à 1965: République du Niger*. Etienne Julienne, Paris, 505 pp.

Pitacco, A., N. Gallinaro, and C. Giulivo, 1992: Evaluation of actual evapotranspiration of a *Quercus ilex L.* stand by the Bowen ratio-energy budget method. *Vegetation*, **99/100**, 163–168.

Renard, K. G., 1970: The hydrology of semiarid rangeland watersheds. Rep. ARS-41-162, Agric. Res. Service, US Dept. Agric., Washington DC.

Rao., B. Q., Y. D. Liu, W. B. Wang, C. X. Hu, L. Dunhai, and S. B. Lan, 2009: Influence of dew on biomass and photosystem II activity of cyanobacterial crusts in the Hopq Desert, northwest China. *Soil Biology and Biochemistry*, **41**, 2387–2393.

Rodriguez-Iturbe, I., 1986: Scale of fluctuations of rainfall models. *Water Resources Research*, **22**, 15S–37S.

Rodriguez-Iturbe, I., M. Marani, P. D'Odorico, and A. Rinaldo, 1998: On space–time scaling of cumulated rainfall fields. *Water Resources Research*, **34**, 3461–3469.

Rodhe, H., and H. Virji, 1976: Trends and periodicities in East African rainfall. *Monthly Weather Review*, **104**, 307–315.

Sala, O. E., and W. K. Lauenroth, 1982: Small rainfall events: an ecological role in semiarid regions. *Oecologia*, **53**, 301–304.

Scheffer, M., E. H. van Nes, M. Holmgren, and T. Hughes, 2008: Pulse-driven loss of top-down control: the critical-rate hypothesis. *Ecosystems*, **11**, 226–237.

Schumacher, C., and R. A. Houze Jr., 2003: Stratiform rain in the tropics as seen by the TRMM precipitation radar. *Journal of Climate*, **16**, 1739–1756.

Schwinning, S., and O. E. Sala, 2004: Hierarchy of responses to resource pulses in arid and semi-arid ecosystems. *Oecologia*, **141**, 211–220.

Sharon, D., 1981: The distribution in space of local rainfall in the Namib desert. *Journal of Climatology*, **1**, 69–75.

Sharon, D., A. Margalit, and S. M. Berkowicz, 2002: Locally modified surface winds on linear dunes as derived from directional raingauges. *Earth Surface Processes and Landforms*, **27**, 867–889.

SMEC, 1987: *Study of Open Water Evaporation in Botswana*. Snowy Mountains Engineering Corporation, Ministry of Local Government and Lands, Gaborone, Botswana.

Steinberger, Y., I. Loboda, and W. Gamer, 1989: The influence of autumn dewfall on spatial and temporal distribution of nematodes in the desert ecosystem. *Journal of Arid Environments*, **16**, 177–183.

Taylor, C. M., and E. F. Lawes, 1971: *Rainfall Intensity-Duration Frequency Data for Stations in East Africa*. East African Meteorological Department, Nairobi, 30 pp.

Thornes, J. B., 1994: Catchment and channel hydrology. In *Geomorphology of Desert Environments* (A. D. Abrahams and A. J. Parson, eds.), Chapman and Hall, London, pp. 257–287.

Trewartha, G. T., and L. H. Horn, 1980: *An Introduction to Climate*, 5th edn. McGraw-Hill, New York.

Westbeld, A., and Coauthors, 2009: Fog deposition to a *Tillandsia* carpet in the Atacama Desert. *Annales Geophysicae*, **27**, 3571–3576.

Wilcox, B., D. Breshears, and C. Allen, 2003: Ecohydrology of a resource-conserving semiarid woodland: effects of scale and disturbance. *Ecological Monographs*, **73**, 223–239.

Winstanley, D., 1973: Recent rainfall trends in Africa, the Middle East and India. *Nature*, **243**, 464–465.

Xie, P., and P. A. Arkin, 1995: An intercomparison of gauge observations and satellite estimates of monthly precipitation. *Journal of Applied Meteorology*, **34**, 1143–1160.

Zangvil, A., 1996: Six years of dew observations in the Negev Desert, Israel. *Journal of Arid Environments*, **32**, 361–371.

Zhang, J., Y. M. Zhang, A. Downing, J. H. Cheng, X. B. Zhou, and B. C. Zhang, 2009: The influence of biological crusts on dew deposition in Gurbantunggut Desert, Northwestern China. *Journal of Hydrology*, **379**, 220–228.

12 Hydrologic processes in the drylands

12.1 INTRODUCTION

Compared with humid environments, relatively little is known about surface hydrological processes in dryland regions. Over the last decade or so there has been a broad effort to improve our understanding of runoff and other hydrologic aspects of drylands (Wheater *et al.* 2007). Some generally accepted tenets, such as the absence of groundwater recharge, are being contradicted by new studies (Scanlon *et al.* 2006). Much of this is in the context of research in a new interdisciplinary area called ecohydrology (Newman *et al.* 2006a). Consequently, the hydrologic processes associated with precipitation events are fairly well understood (Weltzin *et al.* 2003). Much less is understood about the processes that influence the rate of soil moisture change between precipitation events.

Much has been learned about the impact of the characteristics of rainfall on surface hydrological processes and how surface vegetation cover influences these relationships. This chapter goes into detail on these concepts, because they are critical in understanding how climate translates into water availability (or the lack of it) in dryland regions and in projecting the impact of future climate change.

12.1.1 WATER SUPPLY IN DRYLANDS

The availability of precipitation is only one aspect of the water supply in dryland regions. The availability of water for use by humans, plants, and animals is also a function of the disbursement of the precipitation into evaporation and runoff and the presence of water in various reservoirs. These reservoirs can be characterized as endogenous or exogenous. Endogenous sources are concentrations of water originating inside the dryland region itself, including streams and surface or subsurface reservoirs. The reservoirs are water that has accumulated in sands (which have high infiltration, but dry out only at the surface), aquifers in alluvial fans or occasionally in limestone rocks (karst features), playas, and freshwater accumulations along some coasts. A much greater volume of water is exogenous (also termed exotic), originating in more humid environments outside of the dryland regions; this includes groundwater and rivers. A prime example is the Nile, which flows out of equatorial highlands across hyper-arid Egypt and serves as a lifeline for the country

For a given location in the drylands, various geographic factors determine the overall availability of water. These include soil type and the nature of other surface materials, slope and topography, runoff and the nature of the drainage system, proximity to groundwater (depth of the water table), and proximity to seasonal or exotic streams. Goudie and Wilkinson (1980) identify several niches in arid regions, with differences in supply as a result of geographical differences or microclimatic environments. The driest of these are rocky slopes and the wettest are exogenous stream valleys. The contrasts among the various niches may be remarkable, with lush vegetation thriving in dry streambeds or along their banks (Fig. 10.40). Favorable environments include highlands (such as Tibesti, in North Africa), where rainfall is higher than on the plains; the foot of slopes or inselbergs or in depressions, where runoff collects; and oases, which may be springfed from groundwater. Under sand dunes, the water table tends to be elevated and is often of use to plants; sands also accentuate changes between wetter and drier conditions.

12.1.2 CONTRASTS WITH HUMID REGIONS

The hydrologic regime in arid and semi-arid regions differs fundamentally from that in wetter regions. The differences result from a large number of factors, but three of them are most basic: the relative sparseness of the vegetation cover, the low amount of rainfall, and its episodic and localized nature. Vegetation utilizes water and facilitates infiltration, promoting storage and delayed runoff. It also returns moisture to the atmosphere by evapotranspiration. When vegetation cover is sparse, runoff is more immediate and less moisture infiltrates into the soil. Consequently, precipitation events have a more direct and immediate impact on streamflow, and extreme rainfall events reach the stream with little moderation during passage through the ground. Thus, the intermittency of rainfall in time and space is reflected in streamflow, which is also episodic or strongly seasonal and, in arid regions, spatially intermittent and variable.

The critical hydrological differences between dryland and humid environments are the components of the water cycle that are most important, and the degree of spatial and temporal variability of these components (Thornes 1994). These create characteristic patterns of runoff, streamflow, drainage, and soil moisture in drylands that are markedly different from those in humid regions.

In temperate environments, percolation to deep soil layers, throughflow, saturation-excess overland flow, and groundwater are the most significant components of the water cycle. Drainage occurs in well-defined and well-connected channels. Stream discharge increases downstream, as channels continue to drain into the stream.

When precipitation is sparse, percolation is limited, infiltration is relatively shallow, baseflow (lateral groundwater flow) is usually absent, and Hortonian (infiltration-excess) overland flow is the primary form of runoff. In the drylands, both the temporal and spatial variability of water availability are large, the former because of the episodic nature of rainfall events, the latter because of the erratic distribution of rainfall and the spatial heterogeneity of the landscape. The drainage patterns are diffuse and often discontinuous. The rain falls with high intensity on ground that is barren or covered with sparse vegetation. Evaporation is high, and surface characteristics serve as major controls on infiltration. Infiltration is also strongly dependent on the size of the rainfall event. Most streams are ephemeral and flash floods are common. As a result of high transmission losses (infiltration plus evaporation), discharge decreases downstream. Most dryland streams terminate within the drainage basin.

12.2 RUNOFF

Runoff in dryland regions is difficult to measure because its observed depth is often less than the measuring accuracy of gauges, giving a small signal-to-noise ratio, and because the channels are ephemeral and their configuration changes over time. Moreover, because runoff events can be brief, important, and localized, the time required for adequate characterization can be very long. Much of the traditional knowledge is based on small plots with simulated rainfall, modeling studies, or point measurements or field measurements from two semi-arid watersheds in Israel (Nahal Yael) and Arizona (Walnut Gulch). For these locations, long-term records are available, commencing with the work of Schick (1970) and Renard and Keppel (1966). Extensive field measurements of runoff were also carried out in the Sahel (Rodier 1975).

What is known about runoff in dryland landscapes is thus based on little or no field data (Wilcox et al. 1997, 2003), and some of our classical knowledge about the hydrology of arid zones is oversimplified or even wrong. Generalized conclusions often fall apart when more detailed questions are asked about specific dryland landscapes. Simple questions remain unanswered. How much water runs off? What type of runoff provides the water and where does it go? How does runoff change with scale? Overall, there is considerable uncertainty in the sources and amounts of runoff in dryland regions because work with small plots or point measurements provides little insight at the landscape scale (Beven 2002).

Extensive research during the last 10 years has helped to change this situation. Two important new ideas involve the importance of scale in hydrological processes and the interactions between soil, vegetation, and runoff. When processes are examined at various scales, many of the classic generalizations about runoff in drylands are found to be questionable or just plain wrong. This section attempts to succinctly summarize much of the recent research, including the concept of vegetation patchiness and its impact on runoff. This concept highlights both the impact of scale and the relationships with soil and vegetation.

12.2.1 THE NATURE OF RUNOFF IN DRYLAND REGIONS

In drylands, runoff is characterized by the same attributes as rainfall: brief, infrequent, localized, often of high intensity, and varying greatly from year to year. Also like rainfall, a few strong events make up most of the total. Daily runoff in two Texas watersheds during a 3-year period illustrates these characteristics (Fig. 12.1). Each region experienced only three major runoff events, each lasting at most a few days. At North Concho, near Carlsbad, runoff was nearly absent in 1985 and 1987; the three major runoff events occurred within a period of a few months. Typical is the frequency distribution of runoff on the Pajarito Plateau in northern New Mexico (Fig. 12.2). Of 44 events during an 8-year period, the 10 largest runoff events produced almost 90% of the total runoff (Wilcox et al. 2003).

Another characteristic of runoff in dryland regions is that it exhibits a strong non-linear decrease with increasing drainage area (Fig. 12.3). It tends toward zero as the scale under consideration increases. As some point, mean annual precipitation is completely balanced by evapotranspiration, and runoff occurs only locally. The vast Sahelian region of West Africa is a case in point (Leduc et al. 1997). This region lacks an integrated drainage network and the main hydrologic units are endoreic areas, pools on the order of a few hectares to a few square kilometers that collect surface runoff. In contrast, in humid regions the ratio of runoff to drainage area reaches an asymptotic value with increasing drainage area.

Four types of runoff are generated: saturation-excess overland (Dunne) flow, infiltration-excess (Hortonian) overland

(a)

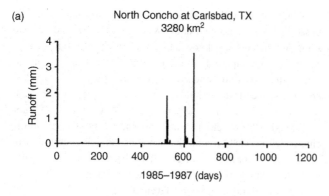

(b)

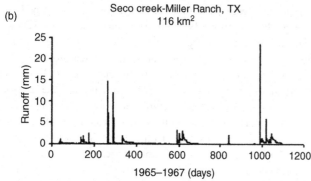

Fig. 12.1 Daily runoff (mm) in two Texas watersheds (Wilcox 2002).

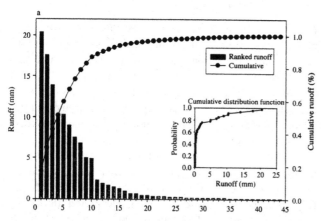

Fig. 12.2 Ranked runoff events and cumulative runoff on the Pajarito Plateau, northern New Mexico (Wilcox et al. 2003).

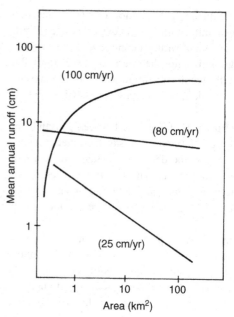

Fig. 12.3 Mean annual runoff (cm) in relation to drainage area (km²) (Graf 1988).

flow, lateral subsurface flow (interflow), and groundwater flow (see Chapter 7). Overland flow and interflow are generally brief and rapid, while groundwater flow is slow and more even. Precipitation is the source of overland flow in most drylands. Ground moisture generates the other types of runoff.

Infiltration-excess overland flow is the dominant source of runoff in drylands, while saturation-excess overland flow is uncommon in semi-arid settings (Graf 1988). Exceptions are the piñon–juniper and ponderosa pine woodlands of Arizona, where prolonged frontal rains can saturate the shallow, low-permeability soils (Wilcox et al. 1997). Lateral subsurface flow is generally considered to be unimportant, although it can be significant in certain dryland regions, especially in wet years (Wilcox et al. 1997; Guntner and Bronstert 2004).

12.2.2 OVERLAND FLOW

Runoff is essentially water moving downhill under the influence of gravity. Overland flow is generated when the precipitation arriving at the surface cannot locally penetrate into the ground or adhere to soils and vegetation. The spatial pattern of generation is diffuse, because of the heterogeneity of the surface characteristics controlling it and because of the high spatial variability of rainfall. Along its path, the water encounters "obstacles" that deflect its flow: vegetation, stones, and higher topographic relief. These obstacles concentrate the flow into rills and eventually gullies (Abrahams et al. 1986). The flow terminates when "transmission" losses via evapotranspiration and infiltration into the ground total 100% of the runoff.

The character of the runoff's path depends largely on its intensity and the magnitude of transmission losses. In humid environments, where the total volume of runoff is high, runoff is channeled through a well-defined drainage system of rivers, tributaries, and smaller flows and eventually reaches the ocean. In semi-arid regions, the volume of runoff is significantly lower and transmission losses are promoted by the sparse vegetation cover (Fig. 12.4). The runoff typically reaches a stream that may be ephemeral or episodic rather than perennial. The stream frequently terminates within the drainage basin (endoreic drainage). In some cases, such as Sahelian West Africa, the runoff gathers instead in small pools (Fig. 12.5) (Desconnets et al. 1997).

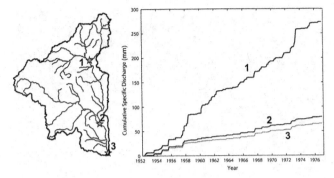

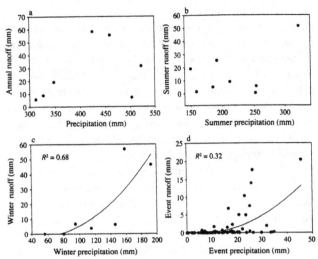

Fig. 12.4 Cumulative discharge along the Rio Puerco, New Mexico, illustrating transmission losses (from Newman *et al.* 2006b).

Fig. 12.6 Runoff versus (a) annual, (b) summer, (c) winter, and (d) event rainfall in a piñon–juniper woodland on the Pajarito Plateau of northern New Mexico (Wilcox *et al.* 2003).

12.2.3 CONTROLS ON RUNOFF

It was long assumed that rainfall was the main control on runoff in dryland regions, so that seasonal or annual rainfall could serve as a proxy for runoff. A recent study that sought to confirm this (Wilcox *et al.* 2003) found little relationship (Fig. 12.6). There was no correlation on either event or annual time scales or for the summer season. There was a moderate correlation in winter, when saturation-excess runoff is common and snowpack may occur. The few generalizations that can be made concerning the rainfall–runoff relationship in drylands are (1) that runoff is more strongly dependent on rainfall intensity than on duration, and (2) other factors being equal, the longer the duration of rainfall and the greater its intensity, the greater is the runoff generation.

The primary factors that influence the amount and nature of runoff are the percentage of vegetation cover and bare ground, and the intensity, duration, amount, and seasonal distribution of rainfall. In the drylands, runoff is also strongly influenced by other surface characteristics, such as micro- and macrotopography, soil hydraulic properties (including soil texture and condition), surface drainage patterns, soil chemistry, stone cover, biological crusts, surface cracks, snowpack, and antecedent soil moisture. This myriad of factors results in highly localized patterns of runoff.

The Texas watersheds in Fig. 12.1 illustrate how strong one of these factors can be. The two regions have similar conditions of rainfall but vastly contrasting soil types. The maximum runoff per unit area is on the order of 2–4 mm in the North Concho watershed, but on the order of 15–20 mm in the Seco River watershed.

The controls on runoff act in diverse ways to modify its path, determine the water volume available for runoff, and determine the losses by evaporation and infiltration. Rainfall not only affects

Fig. 12.5 Runoff pool in the West African Sahel.

Runoff generation is very low in desert lowlands, where sands and stone pavements promote infiltration, clay pans inhibit it, and drainage is disorganized. In desert uplands, where much rock is exposed and rainfall is often higher and more intense than on the plains, runoff generation can be quite high. The response of streams in these regions is particularly rapid and intense; runoff and even floods can result from very little rainfall.

The threshold for runoff generation can be quite small, but runoff does not guarantee streamflow. Thresholds for initiating streamflow were found to be 7.5 mm of rainfall with a mean intensity of 0.5 mm/min in the Sinai and 5mm with an intensity of 0.5 mm/min in the Hoggar of the Sahara (Mabbutt 1979).

the water available for runoff; its characteristics also significantly influence infiltration, interception, evaporation, and soil moisture storage. Organic matter and leaf litter promote infiltration; soil crusts inhibit it. Vegetation, stones, and crusts concentrate the flow into rills and gullies, creating the characteristic reticular or diffuse pattern of drainage channels in most dryland regions. The reticular flow contrasts strongly with the sheet flow common in more humid regions. Its major impact on runoff is that it restricts the infiltration of runoff to relatively small areas (Thornes 1994). Microscale topography has a similar impact.

The prevailing spatial pattern of vegetation in drylands is largely a result of extensive feedbacks between plants and surface hydrology. The characteristics of vegetation that influence runoff include its density, leaf area index, root depth, seasonal growth cycle, and water-use efficiency. Woody vegetation modifies runoff in several ways (Wilcox 2002). It alters soil infiltration characteristics via root penetration and the addition of organic matter. It preserves soil moisture through shading and mulching, but also draws off soil moisture through transpiration and interception. It directly impacts runoff by both deflecting and capturing flow (runon). Its roots create macropores that alter subsurface water flow both horizontally and vertically (Guntner and Bronstert 2004).

Studies in the southwestern USA have indicated that the greatest impact of woody vegetation on runoff magnitude and streamflow is via evapotranspiration (Wilcox 2002). Much of the evapotranspiration consists of water that was intercepted by the vegetation. A portion of this generally becomes stemflow that once again becomes available for runoff. The degree of interception and stemflow is markedly different for shrubs and herbaceous vegetation and for individual species, so that few generalizations can be made about their overall importance in dryland regions (Wilcox 2002).

Stones and crusts likewise intercept precipitation; the intercepted water is then redistributed via evaporation and runoff. This serves to increase runoff. Unfortunately, there is little quantitative knowledge of this process and, even qualitatively, the few studies have produced quite different answers. Stones embedded in the surface interact with precipitation differently from stones lying on the surface (Thornes 1994). For example, stones lying on the surface may inhibit crusting. Because the biological material in crusts fills the soil pores, inhibiting crust formation, this will increase infiltration, thereby reducing runoff. The crusts (Fig. 12.7) also influence soil temperature, soil moisture profile (Qin *et al.* 2002), and soil roughness, which influences runoff path length, tortuosity, velocity, and residence time of overland flow on both the micro- and macroscales (Belnap and Eldridge 2001).

12.2.4 INFLUENCE OF VEGETATION PATCHINESS ON RUNOFF

In semi-arid shrublands and grasslands, infiltration and runoff are strongly influenced by surface characteristics, the most important being vegetation, microtopography, and soil

Fig. 12.7 A biological crust (light-colored area in foreground) on soil on the Bandelier Plateau of New Mexico.

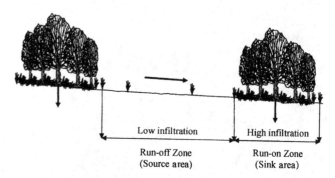

Fig. 12.8 Schematic shows the intercanopy and canopy patches as sources and sinks of runoff, with the maxima of infiltration along the edges of the canopy plots and on the vegetated intercanopy plots.

hydraulic properties, including biological crusts. In grasslands the primary control on infiltration and runoff is microtopography. In shrublands the primary control is plant size, structure, and spacing (Bhark and Small 2003). The patchiness of the shrublands plays a major role in the redistribution of precipitation within the ecosystem.

Studies from semi-arid regions in Australia, Niger, Mexico, and the United States have all shown that vegetation patterns influence the spatial distribution of surface runoff both by obstructing and by intercepting overland flow (Tongway and Ludwig 1997). Vegetation functions to redistribute water at small scales, generally less than 10 m² (Fig. 12.8). At these scales, vegetation patches may function as either sources or sinks of water. Source areas, generally areas that are bare or have only sparse vegetation, produce runoff. Sink areas, which are downslope of the sources, receive and store the runoff and thereby become enriched and relatively productive. The transfer of water at these small scales is both frequent and substantial enough to be hydrologically and ecologically significant (Reid *et al.* 1999).

A dichotomy exists between canopies and intercanopy areas (Breshears and Barnes 1999). Characteristics that differentiate

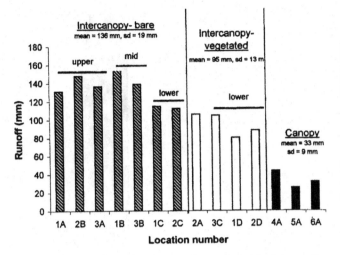

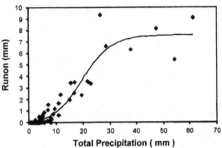

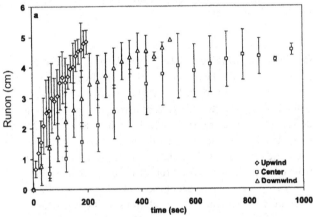

Fig. 12.9 Runoff from three types of experimental plots on the Bandelier Plateau, New Mexico (from Reid *et al.* 1999, reprinted with permission from Soil Science Society of America).

Fig. 12.10 Runon to lower, vegetated intercanopy patches as a function of the precipitation event (from Reid *et al.* 1999, reprinted with permission from Soil Science Society of America).

Fig. 12.11 Runon rates at upwind (diamonds), central (squares), and downwind (triangles) locations of shrub mounds in the Jornada experimental range in the Chihuahuan Desert of New Mexico (from Ravi *et al.* 2007). Vertical bars indicate the range of values.

these regions include (1) the presence of the canopy, which modifies the intensity and amount of precipitation reaching the soil and soil microclimate (Breshears *et al.* 1998), (2) the presence of litter, which increases soil organic matter, and thus enhances infiltration capacity (Schlesinger *et al.* 1999), and (3) soil morphology and soil faunal activity (Bromley *et al.* 1997a). The result is the "islands of fertility" described in detail in Chapter 3 and an enhancement of the patchiness of the surface cover.

Reid *et al.* (1999) assessed the impact of vegetation patchiness on runoff in a piñon–juniper woodland on the Bandelier Plateau of northern New Mexico. Measurements were made on three patch types: canopy patches (beneath woody plants), vegetated patches in intercanopy areas, and bare patches in intercanopy areas. The patches were markedly different in terms of both runoff and erosion (Fig. 12.9), with intercanopy patches generating markedly more runoff than the canopy patches. Highest average runoff was from the bare patches: 136 mm, compared with 33 mm for the canopy patches.

This study also contrasted the intercanopy plot types for three types of precipitation events: convective (high intensity) and frontal rainfall (duration longer than 5 hours) events and minor precipitation occurrences (less than 15 mm). For all three types, the bare intercanopy patches generated more runoff than the vegetated intercanopy patches (Fig. 12.9). The minor events comprised 79 rainfall occurrences and the largest share of total rainfall (45%). Only 16 larger events occurred. Intense convective events generated the most runoff, roughly 70–90% of the rainfall from these events. The minor events produced little runoff but a much greater proportion of the runoff was captured and stored in the soil from this type of event. The total capture, or runon, was comparable for convective and frontal rains, suggesting that the longer duration of the latter compensated for the lower intensity. Overall, the runon is markedly greater for the larger events, especially those bringing more than 25 mm, but the higher frequency

of minor events resulted in a comparable overall contribution (Fig. 12.10). This suggests that the minor events are ecologically significant.

Ravi *et al.* (2007) examined an even finer scale, comparing runoff generation and infiltration in various parts of the canopy patches in the Chihuahuan Desert, near Las Cruces, New Mexico. The infiltration was markedly lower in the center of the patch compared with the edges of the canopy (Fig. 12.11). This suggests a transfer of both water and nutrients from the patch center to the edge, with the water carrying nutrients amassed underneath the canopy. This increases the fertility of the edges, allowing the patches to grow.

Landscape ecologists evaluate the health or functionality of semi-arid rangelands on the basis of interactions between runoff and vegetation (Ludwig *et al.* 2005). A dysfunctional ecosystem is one from which a significant portion of water and nutrients are being lost, generally because the network of vegetation patches is too spotty to trap surface runoff. This can lead to the catastrophic change discussed in Chapter 3.

For this reason, a current challenge is to develop a unifying concept to assess this loss in diverse ecosystems that incorporates the effects of vegetation patchiness on runoff. One useful

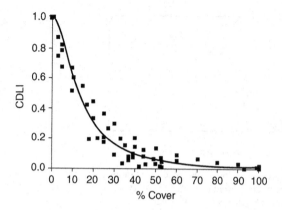

Fig. 12.12 Curvilinear relationship between the cover-based directional leakiness index (CDLI) and vegetation cover (%), based on plots in Australia (Ludwig *et al.* 2005b).

metric is the "leakiness index" formulated by Ludwig *et al.* (2007) to quantify the degree to which a landscape loses potential soil water via runoff. It exploits the connectivity between patch types demonstrated by the redistribution of precipitation downslope from bare to vegetated patches (Reid *et al.* 1999) by quantifying the losses each time water flowing downslope encounters a vegetated patch. The index ranges from 0 to 1, from complete water loss via runoff to complete retention, and utilizes remotely sensed data for calculations.

There is a general inverse relationship between leakiness and vegetation cover (Fig. 12.12). However, the index is extremely sensitive to the spatial configuration of the vegetation cover, particularly the degree to which vegetated patches are contiguous. This suggests that the spatial distribution of vegetation is critical in maintaining a healthy ecosystem. Figure 12.13 illustrates this for two patches: A, with 34% total cover and B with 47% total cover but a large bare spot toward the bottom of the hillslope. The leakiness of site B was 0.71, compared with 0.21 for site A, despite the much higher degree of overall cover.

12.3 DRYLAND STREAMS

Streams fall into three classes: perennial, ephemeral, and intermittent. Perennial rivers flow year-round; ephemeral rivers flow for only short periods, seasonally or in response to episodic rainfall. Some perennial streams may become ephemeral in particularly dry years (Fig. 12.14). Intermittent rivers are discontinuous in space; often much of the river will dry up, leaving only ponds in the depressions in the channel. Endogenous streams in arid regions are predominantly ephemeral; both ephemeral and intermittent streams are common in semi-arid regions. The water level in these streams is highly localized in time and space, reflecting the characteristics of the dryland rainfall regime. Most of the perennial rivers in these regions are exogenous (also called exotic or allogenic). Since they tend to originate in subhumid regions, these rivers,

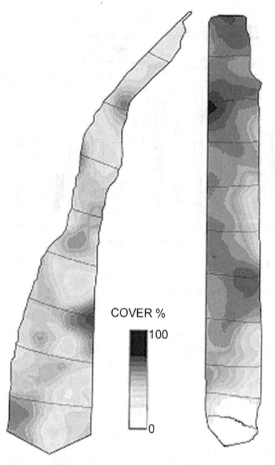

Fig. 12.13 Ground cover of two patches in on a hillslope in Queensland, Australia. Average cover is 34% (Site A, on the left) and 47% (Site B, on the right), but the leakiness indices are 0.21 and 0.71, respectively (from Ludwig *et al.* 2005b).

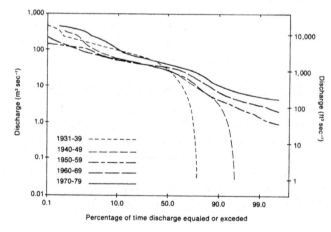

Fig. 12.14 Flow duration curves for the Platte River, Nebraska, for wet and dry years. The stream becomes ephemeral in the dry years (from Kirchner and Karlinger 1981).

such as the Nile or Niger, tend to have a strongly seasonal regime (see Section 12.4). For much of the year, water is well below the level of the stream bank and is accessible only with pumps.

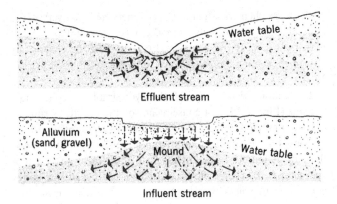

Fig. 12.15 The relationship of effluent and influent streams to the water table (copyright © A. N. Strahler, 1987).

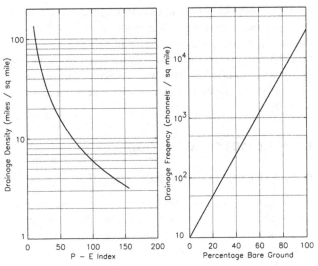

Fig. 12.16 Drainage density versus aridity (PE index) and percent ground cover in the southwestern USA (modified from Melton 1957).

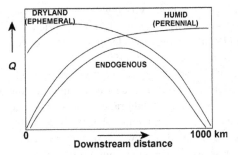

Fig. 12.17 Discharge (Q) versus downstream distance for streams of humid and dryland regions.

The streams of dryland regions generally have a different relationship with the water table than those of humid regions. They are less dependent on groundwater, being sustained primarily by overland flow. Consequently, streams in dry climates generally lose water through seepage and recharge the groundwater below. These are termed influent streams, in contrast to the effluent streams of moist climates, which receive flow from the groundwater (Fig. 12.15). Thus, near the endogenous streams of arid regions, the water table may be locally determined by the river, and vegetation growth will be extremely lush along the stream banks or even in its dry bed. In contrast, the water table determines the streamflow in humid regions.

The ephemeral streams in dryland regions generally commence in the uplands of the basin as overland flow (Bull 1997). The flow becomes organized on the hillslopes, where it moves through deep, permanent channels with steep sides. These are generally termed wadis or arroyos, but have numerous local names. Drainage density is thus high on the hillslopes, where it is proportional to the degree of aridity and inversely related to the percent of bare ground (Fig. 12.16).

In the lowlands, the streamflow is moderated by the basin it traverses. The ground soaks up both flow from the uplands and the meager rains that fall. Both infiltration and evaporation can be high, so that generally only localized runoff occurs in the lowlands. It often collects in pools, where a large portion evaporates or infiltrates into the ground. The net result is that in the lowlands much of the flow is lost to infiltration or evaporation before reaching stream channels. For that reason, the channels then tend to disperse and disintegrate in the lowlands and the network becomes disorganized. This is the typical "areic" drainage pattern, in which runoff occurs over multiple restricted areas of gently sloping terrain.

These processes are responsible for the different patterns of stream discharge in arid and humid regions (Fig. 12.17). In the perennial streams of humid regions, discharge increases more or less continuously downstream. In the ephemeral streams of drylands, discharge initially increases downstream as the water collects in the channel, but after a point it diminishes downstream as the network becomes dispersed in the lowlands. This

is especially true for sand, which has a high infiltration capacity, but clays will swell and absorb water, likewise leaving little available for runoff and streamflow. Evaporation also contributes to the diminution of the discharge.

The pattern of discharge versus downstream distance in exogenous streams has some commonalities with both perennial and ephemeral streams (Fig. 12.17). In the channel segments in the humid source region, discharge increases downstream. After the stream enters the more arid environment, discharge usually diminishes downstream. Like ephemeral streams, they constantly lose water through evaporation and infiltration, so that their response to episodic floods becomes increasingly moderated with distance downstream.

The largest exotic streams, such as the Colorado River or the Nile, eventually reach the ocean. For most ephemeral streams, drainage ceases long before the flow can reach the ocean (i.e., the drainage is endoreic). The situation is often different in extremely wet years. In Namibia, for example, long qualitative records exist of the Kuiseb and Swakop Rivers "coming down" from the highlands and reaching the Atlantic. This phenomenon is generally limited to coastal deserts. The Kuiseb reached the Atlantic in 1934, 1962/63, and 1985 (Lancaster 1996). A more recent flooding occurred in 2000, quickly transforming the vegetated stream bed into a raging river over 3 m deep (Sisamu and Siteketa 2006). This was only the seventeenth time the river had reached the Atlantic since 1837. Another major flood occurred in 2006. More typically, the streams of arid regions terminate in a playa or dry lake bed and do not reach the ocean.

12.4 STREAMFLOW AND FLOODS

12.4.1 STREAMFLOW CHARACTERISTICS

The streams of dryland regions contrast with those in humid environments in several ways, but the most obvious are a tendency for episodic flow and abrupt and intense floods. Where rainfall is episodic, streamflow is also typically brief and infrequent (Fig. 12.18). The lag between peak rainfall and peak flow is also much shorter than in humid regions. Floods at Nahal Yael, in the Gulf of Aquaba (Schick 1970) peak within 5–20 minutes and usually last just 1–4 hours (the longest persisted for 24 hours). Flooding can occur very rapidly, but terminates equally rapidly. Flow velocity is strong. In several studies of streams in arid and semi-arid regions, velocities on the order of 3–10 m/s have been measured (Thornes 1994).

Peak flow (Fig. 12.19) and frequency of flooding generally diminish progressively downstream. The reduction in peak flow represents transmission losses via infiltration/seepage into the banks and streambed and via evaporation. For the brief floods in many streams, the duration of flow is too short to sustain much evaporative loss, so seepage is the main transmission loss. This loss can account for as much as 30% of the flow (Graf

1988). For ephemeral streams the channels are often very wide and increase very rapidly with drainage area, unlike perennial streams. Reid and Frostick (1989) describe the ephemeral stream as seeming "to have grown too large for its basin" (Fig. 12.20).

The characteristic shape of the hydrograph of ephemeral streams (Fig. 12.19) is a consequence of the nature of runoff in dryland regions. The response of streams to runoff depends not only on the total amount of runoff generated, but also on the path it follows from watershed to the stream and on the characteristics of the catchment and on antecedent moisture conditions. In general, overland flow is the fastest and most direct, and thus promotes high and immediate runoff generation, and a quick and peaked response in streamflow. Groundwater is the slowest and least direct, but most reliable, source of streamflow.

12.4.2 DRYLAND FLOODS

Graf (1988) identifies four types of floods in dryland streams: flash floods, single peak events, multiple peak events, and seasonal floods. Flash floods are limited to basins of 100 km^2 or less, and flow goes from zero to a maximum within a few minutes, or at most a few hours. Single peak floods are longer and range from a few hours to several days in duration. These often result from frontal rains. Multiple peak floods are generally associated either with multiple precipitation events within one frontal system or storm, or from the contributions of various tributaries (Fig. 12.21). Seasonal floods are typical of exogenous streams flowing through a dryland region.

Figure 12.22 shows the hydrograph for a flash flood that occurred in Eldorado Canyon, Nevada, on September 14, 1974. The flow increased nearly instantaneously from zero to 2000 m^3/s. Flash floods are generally associated with an advancing

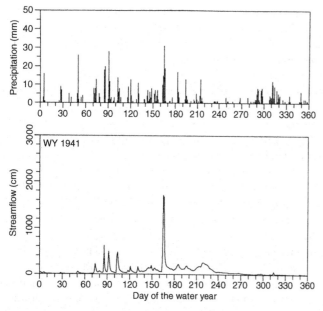

Fig. 12.18 Daily precipitation and daily streamflow in the Salt River at Roosevelt, Arizona, during the water year 1941 (October 1940 to September 1941) (from Cayan and Webb 1992).

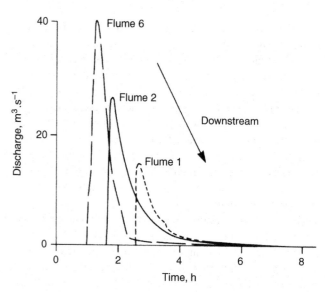

Fig. 12.19 Peak flow versus downstream distance (Reid and Frostick 1989).

Fig. 12.20 (left, center) Two wide stream channels in northern Arizona. (right) Small but steep-sided stream channel in the West African Sahel.

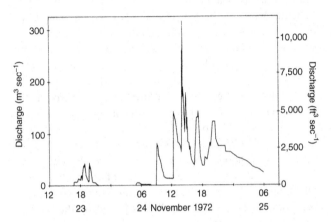

Fig. 12.21 Multiple peak flood, from Wadi Watir in the southeastern Sinai (from Graf 1988).

wave of water, termed a "bore." The bore advances rapidly downstream, in some cases as much as 50–100 m/h (Thornes 1994).

Examples of seasonally flooding streams are the Nile and Niger (Fig. 12.23) and the Ganges. Seasonal streams may or may not sustain flow during the dry season. The Niger becomes extremely low at the end of the dry season, but the Nile does not. These rivers often overflow their banks, providing moisture and nutrients in narrow zones along the river. These zones are important for agriculture. Egyptian agriculture thrived on the Nile floods prior to the construction of the Aswan Dam. In the region of Timbuktu, the Niger overflows in very wet years and floods depressions called douanas. At the end of the nineteenth century, this was a regular enough occurrence that wheat was grown in the douanas and exported from this very arid region.

The data for the Niger shown in Fig. 12.23a illustrate not only the seasonal nature of the discharge, but also the complexity of the flow as it moves downstream from a humid to an arid region (Fig. 12.23b). The flow is slight near the headwaters (Kouroussa, Guinea, c. 11° N) where rainfall averages over 2000 mm/year. Peak discharge is reached near Koulikoro, Mali (c. 13° N), where the rainfall regime becomes semi-arid and mean annual rainfall is on the order of 800 mm/year. Flow wanes as the river traverses the lowlands of the Niger Inland Delta, where

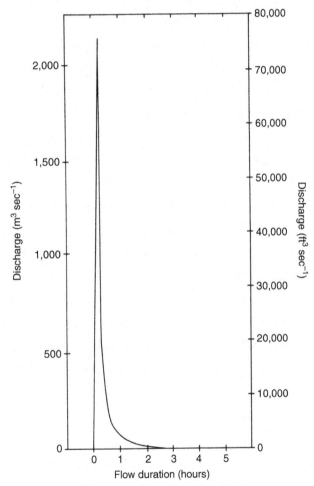

Fig. 12.22 Hydrograph for flash flood in the Eldorado Canyon (Nevada) in 1974 (from Graf 1988).

the river spreads into a network of braided streams, marshes, and lakes. Discharge is reduced by more than 50% when the river reaches Mopti, Mali (mean annual rainfall c. 500 mm), but increases somewhat downstream (Diré, Mali, c. 16° N) with the input of the Bani tributary. From there, the river's flow continually wanes as it traverses the arid and semi-arid Sahel in the upper reaches of the Niger Bend. By the time it reaches Niamey, Niger (c. 13° N, mean annual rainfall roughly 500 mm), peak

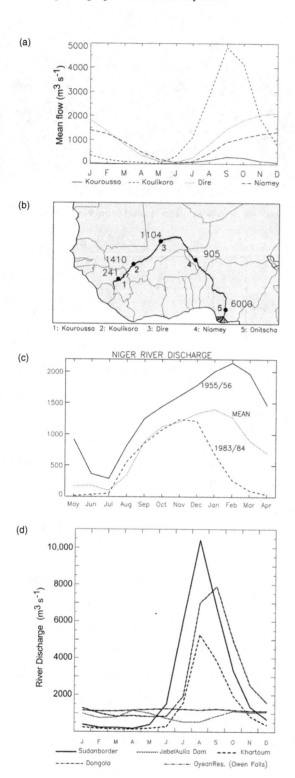

Fig. 12.23 Discharge of the Niger and Nile Rivers. (a) Mean flow (m³ s⁻¹) at several gauging stations along the Niger, (b) map of the course of the two rivers through Africa, select gauging stations, and mean annual discharge at these stations, (c) flow of the Niger at Niamey in wet and dry years (average is also indicated), and (d) discharge of the White and Blue Nile and the main Nile north of Khartoum. Data are from the Center for Sustainability and the Global Environment website: www.sage.wisc.edu/riverdata/.

flow is roughly one-quarter that at Koulikoro. Further south, the river gains flow both from higher rainfall and additional tributaries. When it reaches the Atlantic just south of Onitsha, Nigeria, average discharge is some 6000 m³/s¹, which is roughly 6–7 times greater than the discharge at Niamey.

There is also considerable lag in peak flow downstream. This occurs in September at Koulikoro, but its timing is progressively later downstream, occurring on average in January at Niamey. Downstream from Koulikoro the lag is strongly affected by rainfall conditions (Fig. 12.23c). In the very dry year 1973/74, it shifted progressively downstream from September to December. In the very wet year 1953/54 the peak at Niamey was delayed until February or March.

The discharge of the Nile is similarly complex. The White Nile commences in the humid equatorial rainfall regime at Lake Victoria. The lake's discharge is relatively even throughout the year, on the order of 1100–1200 m³/s. When it reaches Khartoum (Fig. 12.23d), the flow is somewhat lower after its passage through the swamps of the Sudd and through the Sahelian latitudes. The Blue Nile originates at Lake Tana in the highlands of Ethiopia. Peak discharge in August averages nearly 11,000 m³/s. At Khartoum, where it joins the White Nile, the flood is roughly half that, a result of evaporative losses as it traverses the arid lands of the Sudan. From Khartoum, the Nile continues across the Sahara, continually losing flow to evaporation until it is joined by the Atbara, which brings flow off the Ethiopian highlands during the summer.

12.4.3 RELATIONSHIP OF DISCHARGE TO RAINFALL

This complexity makes it difficult to assess rainfall conditions from streamflow in exogenous rivers, despite many attempts to do so (including those of the author). In the case of the Niger, the interannual variability of rainfall in its headwaters (Fig. 12.23c) is more often than not out of phase with those in the semi-arid Sahel, through which it flows. The very wet decade of the 1950s in the Sahel (Fig. 12.24) appears only as part of a sustained period of high Niger flows in the 1950s and 1960s. The period of peak flows in the 1920s is not evident in the rainfall time series. These differences reflect not only the fact that the river flow integrates the humid and dryland regions, but also the existence of a memory of past conditions and possible influences of land-use/land-cover change.

The Nile flow downstream from Khartoum integrates the contributions of the Blue Nile and White Nile, posing a challenge to interpret the rainfall conditions producing the flow. To some extent the summer flow regime is most strongly influenced by the White Nile, and the subsequent flood is a fairly good indicator of Blue Nile flow, and hence rainfall over the Ethiopian plateau. The flow of the White Nile is further complicated by the fact that flow in its headwaters is controlled by the level of Lake Victoria (Yin and Nicholson 2002). Its flow is

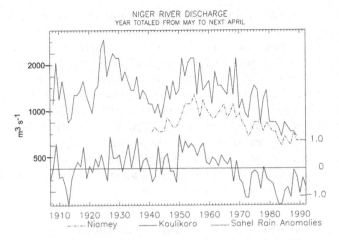

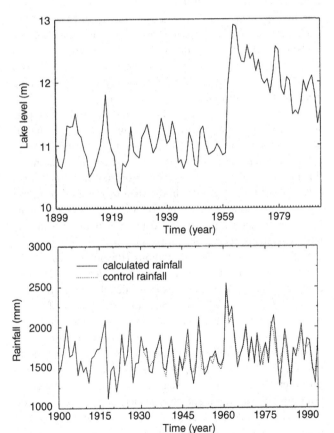

Fig. 12.24 Interannual variations in the flow (m³ s⁻¹) of the Niger at Koulikoro and Niamey, compared with rainfall variations in the Sahel (see Fig. 12.23b for location). Data are from the Center for Sustainability and the Global Environment website: www.sage.wis.edu/riverdata/.

Fig. 12.25 The level of Lake Victoria (top) and mean annual rainfall over its catchment as calculated from the lake level (bottom) (from Yin and Nicholson 2002).

not indicative of equatorial rainfall conditions, because rainfall is not directly correlated with the level of Lake Victoria. Rainfall is reflected instead by the year-to-year changes in lake level (Fig. 12.25).

12.5 GROUNDWATER AND ITS CONNECTION TO SURFACE PROCESSES

The groundwater aquifers in dryland regions also have many characteristics that distinguish them from aquifers of humid regions and limit the degree to which they can be exploited for irrigation. Their thickness is quite variable over short distances, as is their water quality. They are also frequently well below the surface, and tapping them requires expensive, deep wells. Finally, most aquifers in dryland regions are predominantly or entirely fossil, hence recharge is almost negligible and their potential for irrigation is short-range. There is, however, evidence of recharge in regions of the northeastern Sahara, the Sahel, parts of Arabia, segments of the Kalahari, and within the southwestern USA (Goudie and Wilkinson 1980; Verhagen 1990; Desconnets et al. 1997; Goodrich et al. 1997).

12.5.1 SURFACE WATER–GROUNDWATER INTERACTIONS

In the drylands, surface water–groundwater interactions strongly control the availability of the region's limited natural resources (Sophocleous 2002). Groundwater often provides opportunistic conditions for vegetation growth (Scott et al. 2006). In the southwestern USA, for example, mesquite thrives on the desert floor only where recent rain has occurred, but it thrives on riparian terraces where groundwater or deep soil water is available. Deep-rooted trees in the drylands rely primarily on groundwater, while the understory grasses and herbs generally rely on precipitation. Areas of surface water–groundwater interactions are also the loci of "hot spots" or "hot moments" of biochemical cycling, another factor that markedly influences the ecosystem (McLain et al. 2003)

In dryland regions these interactions are restricted in both time and space. They typically occur on small and distinct parts of the landscape, the locations dictated by geomorphic, topographic, and geologic factors. These interactions are also episodic and variable in time. A few studies have attempted to relate basin shape and network geometry to the location, duration, and frequency of surface water–groundwater interactions (Loik et al. 2004; McLain et al. 2003). In this regard, an important descriptor is drainage density (Tucker et al. 2001).

12.5.2 GROUNDWATER RECHARGE

One of the most important interactions, groundwater recharge, is probably negligible or non-existent over much of the semi-arid landscape. It is generally restricted to uplands and topographic lows such as drainages, occurring seldom in the inter-drainage regions. For drainages in closed basins, contributing streams can lead to surface water–groundwater interactions over broad, flat regions that temporarily flood. Thus playa lakes are areas of significant recharge (Menenti 1984). Canyons, valleys, run-off pools, arroyos, and wadis are potential sites of recharge (Newman et al. 2006b; Desconnets et al. 1997).

Fig. 12.26 Rapid inundation of a playa in White Sands, New Mexico, during a brief rain. The albedo of the water surface is extremely high because the sun is near the horizon.

The 1991–1993 HAPEX-Sahel experiment provided some of the most detailed information available about recharge. This region of West Africa lacks an integrated drainage network, and the main hydrologic units are endoreic areas, pools on the order of a few hectares to a few square kilometers that collect surface runoff. Regionally averaged, the recharge is about 50–60 mm/year, or about 10% of the rainfall (Leduc *et al.* 1997). Most of that comes from runoff pools.

Beneath the Wankama pool the infiltration reaches to at least 30 m after flood events. On average, only about 10 mm of the *in situ* rainfall infiltrates to the aquifer (Desconnets *et al.* 1997). Under the tiger bush formation there is a recharge of 13 mm/year, on average, in the upper 70 m of the soil profile. The measured range of recharge beneath this formation is 10–19 mm (Bromley *et al.* 1997b).

The proportion of rainfall that eventually infiltrates depends not only on total rainfall but also on its temporal distribution. The years 1991 and 1992 had similar rainfall amounts but infiltration was 20% in 1992, but only 5% in 1991. The difference was that the rain events of 1992 were clustered into a brief period, minimizing evaporation.

12.6 PLAYAS

The hydrologic regime of the playas is largely dependent on the relative importance of surface water inflow (precipitation and runoff) and groundwater discharge. Moist playas are largely dependent on groundwater, dry playas on surface sources. The greater the importance of groundwater, the more reliable the moisture supply and the longer the moist period in the basin. Those entirely dependent on groundwater are often perennial. The variability of water on the playa surface is lower when groundwater supplements the surface inflow.

The dry playas, which are primarily dependent on episodic rainfall, are often closed basins (those without an outlet).

These strongly reflect local or regional rainfall fluctuations. The response time to rainfall events is dependent on the ratio of the area of the drainage basin to that of the playa surface. The estimate for the world's largest playa, Lake Eyre, is 1.9 years (basin/playa ratio of 100). This was close to the length of the filling and drying cycle following the flood of 1974 (Tetzlaff and Bye 1978), but the cycle following a flood in 1949 lasted 3–4 years (Mabbutt 1979). In many cases, the response to rainfall is almost immediate because the surface is impenetrable hardpan or the surface clays may expand and seal when water is absorbed. In such cases, a brief shower can bring a rapid and spectacular inundation of the surface (Fig. 12.26).

12.7 MOISTURE PULSES AND ECOLOGICAL PROCESSES

12.7.1 MOISTURE PULSES

For generations two ecological paradigms have been utilized to conceptualize the plant–water relationships in dryland regions. The first, termed "pulse–reserve" was put forth by Noy-Meir (1973). At its core is the episodic nature of precipitation in the drylands: water is discontinuously available over time, with water availability occurring in discrete rainfall events (Katul *et al.* 2007). This has long been recognized as a controlling aspect of biophysical processes in arid lands.

The second paradigm is Walter's two-layer soil water partitioning model, that considers bulk rainfall (e.g., seasonal or annual) and distinguishes between plant water utilization in a surface layer and a deeper layer of soil. In applications of this model (Eagleson and Segarra 1985; Walker 1987) grasses are assumed to utilize water in the shallow soil layer (e.g., top 10 cm) and arboreal species are assumed to utilize water in the deeper layer, allowing for the coexistence of both types in dryland ecosystems, such as the savanna.

Both models have their limitations (Reynolds *et al.* 2004). Those of the pulse–reserve model of Noy-Meir are mainly its conceptual nature and lack of specific detail. However, recent research has unequivocally shown that the pulses of moisture availability control the biophysical processes in water-limited ecosystems (Huxman *et al.* 2004a) and the use of the "pulse model" paradigm is rapidly growing. It is illustrated in Fig. 12.27, based on Ludwig *et al.* (2005a) and others. A "trigger" event (a precipitation event) is spatially redistributed or "transferred" via runon and runoff, then transformed into soil moisture available for biological processes (the "reserve," or water storage). When adequate reserve is available, a "pulse" event in plant growth or microbial activity occurs.

Reynolds and colleagues (Reynolds *et al.* 2004) have modified the pulse–reserve concept to make it more widely applicable. The modifications include explicit consideration of (1) what constitutes a biologically significant rainfall event, (2) how rainfall pulses are translated into usable soil moisture pulses, and (3) how different plant functional types utilize the pulses. A revised "threshold–delay" model explicitly includes such factors

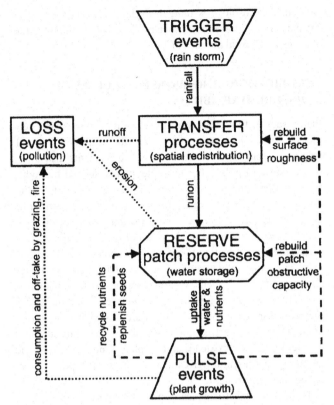

Fig. 12.27 The trigger–transfer–reserve–pulse (TTRP) framework linking "trigger" events, such as rainstorm inputs of water, through spatial "transfer" (runoff–runon) and "reserve" (patch) processes, to "pulse" events, such as plant growth (Ludwig *et al.* 2005a). These linkages are denoted with solid arrows. Feedbacks and flows out of the system are indicated with dashed or dotted arrows.

as thresholds for establishing a "significant" event, potential delayed (i.e., lagged) responses of plants to rainfall, plant phenology, and various aspects of rooting.

The response threshold reflects the ability of organisms and plants to utilize soil moisture pulses of different infiltration depths or duration. The thresholds vary not only for different functional types but also for different biophysical processes. Plant functional types can be distinguished by threshold responses to soil moisture and lag times between pulse and response. This creates a non-linear response of plants to precipitation pulses

There is a hierarchy of events and associated biophysical processes (Schwinning and Sala 204), so that small and large events have significantly different consequences (Loik *et al.* 2004). Small events trigger a limited number of minor ecological events, such as microbial activity. Larger pulses trigger a wider range of minor events and some larger events. Longer and deeper pulses or a series of small pulses are needed to trigger a photosynthetic response in higher-order plants (Huxman *et al.* 2004a, 2004b).

To better understand the role of pulses, ecological studies are beginning to focus on intraseasonal precipitation and evaluating the impact of the magnitude, frequency, and timing of rainfall on biological processes in water-limited ecosystems (e.g., Schwimming *et al.* 2004). Associated with the episodic pulses of rainfall are drying and wetting cycles on a variety of time scales, some as short as hours or days. These cycles are most pronounced in drylands with strongly seasonal precipitation (Austin *et al.* 2004).

The drying and wetting cycles produce a two-phase resource availability, the pulse and inter-pulse periods. At the beginning of a pulse, there is low uptake because there are few plants, but there is abundant water (Cheeson *et al.* 2004). As plants grow, each individual plant reduces what is available to its neighbor. At some point, evapotranspiration reduces soil moisture to a level at which active growth of some species ceases. The inter-pulse period commences for that species. This allows for coexistence of species despite the competition for the scarce moisture reserve. Surviving the inter-pulse period is a critical biological challenge.

12.7.2 FACTORS INFLUENCING THE TRANSFORMING OF PRECIPITATION PULSES TO SOIL WATER

Understanding ecosystem structure and function requires knowledge of where, when, and how much precipitation is converted into available soil moisture. The amount of water in a pulse, its infiltration depth, and the subsequent pattern and duration of the moisture pulse varies in time and space. This reflects not only the stochastic nature of precipitation, but also the heterogeneity of such factors as soil type and characteristics and plant cover (Cheeson *et al.* 2004). The deeper a moisture pulse infiltrates, the longer it takes to deplete. With increasing depth, the fraction of water lost via evaporation decreases and transpiration increases (if plant roots are present). Root densities also drop off sharply with depth. The net result is that soil moisture pulses in shallow soil layers are frequent and brief, but those in the lower layers, although slower to recharge, remain useful for plant growth for a longer time.

Loik *et al.* (2004) summarized the biophysical factors that produce the spatio-temporal heterogeneity of soil moisture that characterizes dryland environments and controls ecological processes within them. These factors include soil depth, texture, hard pan layers, parent material, organic matter, snow pack depth, vegetation type, LAI, and soil surface characteristics such as the presence of stones and biological crusts. These factors affect both the horizontal and vertical heterogeneity of soil water availability through their impact on infiltration capacity, hydraulic conductivity, water-holding capacity, precipitation interception, and runoff. Soil texture and porosity are particularly important, as they contribute to water infiltration rate and surface runoff (Laio *et al.* 2006; Huxman *et al.* 2004a). The temporal pattern of rainfall dictates the temporal heterogeneity in availability, especially in the upper layers of the soil.

The prime influence on horizontal heterogeneity is the dichotomy between canopy and intercanopy space. Although the moisture flux is generally from intercanopy to canopy space, with moisture accumulating beneath the canopy, canopy patches can be drier than intercanopy spaces during small events with significant canopy inception and subsequent evaporation. Soil crusts influence these same parameters, but they can also have a significant impact on water loss through evaporation. The crusts can reduce it by capping the soil surface or increase it by increasing surface temperatures. Crusts can be as much as 14°C warmer than adjacent bare soil surfaces (Belnap and Lange 2001).

Superimposed upon these factors are the rainfall events that provide pulses of moisture to the ecosystem. Water availability, and its heterogeneity vertically and horizontally, is constrained by the nature and timing of the rain events (Austin *et al.* 2004). The frequency and periodicity of the moisture pulses and their control on ecosystem processes can be examined at a number of scales, such as intraseasonal events, seasonal distribution, and interannual variability (Schwinning and Sala 2004). Both the large, infrequent pulses and the small but prolonged events are important for maintenance of the ecosystem (Wiegand *et al.* 2004).

The characteristics of the rainfall events that influence ecosystem processes include their class-size distribution, frequency, sequence, and season of occurrence (Loik *et al.* 2004). Small events generally wet only the upper layers of the soil; large events penetrate to deeper layers. However, small events are significant in that they generally comprise a large fraction of the total rainfall in dryland regions (Sala and Lauenroth 1982). Seasonal influences are due to contrasts in the nature of events and in ambient temperature. These factors control the magnitude of loss to surface runoff and evaporation. Hence spring and summer precipitation, much of which is lost to runoff and evaporation, may only reach only shallow depths of the soil, while the gentler autumn and winter rainfall penetrates more deeply into the soil.

The sequence of events and, especially, the number of days between events also play a role by influencing the accumulation of soil moisture. Hence each event has some residual memory, since runoff, evapotranspiration, and seepage to lower layers all depend on antecedent soil moisture. A succession of events generally has an additive effect on soil moisture, since the influx via rainfall has a shorter time scale than the processes that reduce the soil moisture (Loik *et al.* 2004). Thus, a cluster of rainfall events is markedly more important than a single event (Reynolds *et al.* 2004). The frequency of events plays a similar role. The number of days between events is particularly important for evapotranspiration because of the two-stage process (water-limited vs. radiation-limited conditions). It is also important because it has a major impact on the intensity and length of the critical inter-pulse period that challenges plant survival.

Meteorologists generally refer to the temporal distribution of rain events within the season as "intraseasonal variability." Interest in this topic has been slow to develop, but during the last decade numerous studies have begun to focus on this (e.g., Hoyos and Webster 2007). One incentive for such work is the interest in global change – the projected changes in the earth's environment due to global warming. Implications of the recent work in ecohydrology is that, in dryland regions, changes in the pattern of intraseasonal variability can be as important as changes in the total amount of rainfall.

12.7.3 BIOLOGICAL RESPONSE TO PULSES OF MOISTURE AVAILABILITY

Plants vary in their response to precipitation pulses via phenology, morphology (e.g., root distribution), and growth responses. Temperature and nutrient requirements also affect the response to pulses. For example, plants favored by higher temperatures will respond more efficiently to rainfall during warm seasons than to cool-season rainfall (Cheeson *et al.* 2004). Species may emerge from a dormant state at different rates following a rain pulse. The result of these variations is that in dryland regions water utilization by plants is partitioned spatially and temporally (Schwinning *et al.* 2005a, 2005b). Plants vary with respect to when they utilize soil moisture, the extent to which they utilize water in the upper soil layers, and how readily they can respond to pulses of moisture availability (Reynolds *et al.* 2004). This partitioning is important in facilitating both competition and coexistence among plant functional types and among species, and may give advantage to some types and species in times of drought.

There is also a vertical partitioning of water use, with grasses tending to utilize surface moisture while woody plants, with deeper roots, access water in lower layers of the soil. The depth of utilization generally increases from soil crust biota to herbaceous plants to shrubs then to trees. As a consequence, grasses tend to respond to small precipitation events, shrubs to large ones (Cheng *et al.* 2006). However, this sequence is not universal. In some dryland environments, rooting depth and soil water depletion are similar for grasses and trees (see Chapter 3).

Partitioning of water use over time may be equally important. This partitioning includes preferential use of water in different seasons as well as lagged responses to precipitation events. Many plants can utilize moisture during only one season, even if moisture is available in other seasons. Many woody plant species use only cold-season rain for growth (Snyder *et al.* 2004). Schwinning *et al.* (2005a, 2005b) studied water uptake and growth for three species of grasses and shrubs on the Colorado Plateau, a cold desert. For all three species, water uptake is more sensitive to summer rainfall, but the plants do not grow in summer, so that growth is controlled by winter rains. Winter rain is the primary driver for plant productivity throughout the North American warm deserts, despite differences in the seasonal distribution of precipitation between the various desert regions (Reynolds *et al.* 2004).

In the Great Basin Desert of southern Utah, where precipitation is equally distributed through winter and summer, most species utilize winter rainfall, but the use of summer rainfall is strongly life-form dependent (Ehleringer *et al.* 1991). The desert's annuals and succulent perennials depend solely on summer rainfall but most herbaceous and woody perennials utilize both summer and winter precipitation. The differential response to winter and summer rainfall is related to root distribution. Summer rainfall is usually stored in the uppermost soil layers, so that shallow-rooted plants preferentially utilize it. However, during very rainy summers, water can penetrate to greater depths, where deeper-rooted plants can use it.

Temporal partitioning of water utilization is achieved via lagged response to moisture pulses. The lags are a consequence of threshold values of water availability required for various biophysical processes (Gao and Reynolds 2003). It is often assumed that smaller events, which wet shallow soil layers, favor grasses, while larger events favor deeper-rooted woody plants. The few studies that have actually examined this hypothesis (e.g., Schwinning et al. 2003; Cheng et al. 2006) have reached diverse conclusions, suggesting that partitioning by event size does not appear to be a universal characteristic of dryland ecosystems.

The lagged response also reflects a "memory" in the ecosystem that resides in antecedent moisture conditions (Potts et al. 2006) and factors influenced by these conditions, such as soil organic matter and relative population of grasses, herbs, and woody plants. This memory constrains the ability of a system to respond to a pulse (Wiegand et al. 2004), be it from individual events or wet or dry seasons. As an example, the recovery from drought may be delayed by a season or more (Prince et al. 1998). Hence the precipitation history is of considerable importance in understanding ecosystem-level responses to moisture pulses.

REFERENCES

Abrahams, A. D., A. J. Parsons, and S. H. Luk, 1986: Field measurement of the velocity of overland flow using dye tracing. *Earth Surface Process and Landforms*, **11**, 653–657.

Austin, A. T., and Coauthors, 2004: Water pulses and biogeochemical cycles in arid and semiarid ecosystems. *Oecologia*, **141**, 221–235.

Belnap, J., and D. Eldridge, 2001: Disturbance and recovery of biological soil crusts. In *Biological Soil Crusts: Structure, Function, and Management* (J. Belnap and O. L. Lange, eds.), Springer-Verlag, Berlin, pp. 363–384.

Belnap, J., and O. L. Lange (eds.), 2001: *Biological Soil Crusts: Structure, Function, and Management*. Springer-Verlag, Berlin.

Beven, K., 2002: Runoff generation in semi-arid areas. (L. B. Bull and M. J. Kirby, eds.) *Dryland Rivers: Hydrology and Geomorphology of Semi-arid Channels*. John Wiley and Sons, Chichester, pp. 57–105.

Bhark, E. W., and E. E. Small, 2003: Association between plant canopies and the spatial patterns of infiltration in shrubland and grassland of the Chihuahuan desert, New Mexico. *Ecosystems*, **6**, 185–196.

Breshears, D. D., and F. J. Barnes, 1999: Interrelationships between plant functional types and soil moisture heterogeneity for semi-arid landscapes within the grassland/forest continuum: a unified conceptual model. *Landscape Ecology*, **14**, 465–478.

Breshears, D. D., J. W. Nyhan, C. E. Heil, and B. P. Wilcox, 1998: Effects of woody plants on microclimate in a semiarid woodland: soil temperature and soil evaporation in canopy and intercanopy patches. *International Journal of Plant Sciences*, **159**, 1010–1017.

Bromley, J., J. Brouwer, A. P. Barker, S. R. Gaze, and C. Valentin, 1997a: The role of surface water distribution in an area of patterned vegetation in a semi-arid environment, south-west Niger. *Journal of Hydrology*, **188–189**, 1–29.

Bromley, J., and Coauthors, 1997b: Estimation of rainfall inputs and direct recharge to the deep unsaturated zone of southern Niger using the chloride profile methods. *Journal of Hydrology*, **188–189**, 139–154.

Bull, W. B., 1997: Discontinuous ephemeral streams. *Geomorphology*, **19**, 227–276.

Cayan, D. R., and R. H. Webb, 1992: El Niño/Southern Oscillation and streamflow in the western United States. In *El Niño: Historical and Paleoclimatic Aspects of the Southern Oscillation* (H. F. Diaz and V. Markgraf, eds.), Cambridge University Press, Cambridge, pp. 29–68.

Cheeson, P., and Coauthors, 2004: Resource pulses, species interactions, and diversity maintenance in arid and semi-arid environments. *Oecologia*, **141**, 236–253.

Cheng, X., and Coauthors, 2006: Summer rain pulse size and rainwater uptake by three dominant desert plants in a desertified grassland ecosystem in northwestern China. *Plant Ecology*, **184**, 1–12.

Desconnets, J. C., J. D. Taupin, T. Lebel, and C. Leduc, 1997: Hydrology of the HAPEX-Sahel Central Super-Site: surface water drainage and aquifer recharge through the pool systems. *Journal of Hydrology*, **188–189**, 155–178.

Eagleson, P. S., and R. I. Segarra, 1985: Water-limited equilibrium of savanna vegetation systems. *Water Resources Research*, **21**, 1483–1493.

Ehleringer, J. R., S. L. Philips, W. S. F. Schuster, and D. R. Sandquist, 1991: Differential utilization of summer rains by desert plants. *Oecologia*, **88**, 430–434.

Gao, Q., and J. F. Reynolds, 2003: Historical shrub–grass transitions in the northern Chihuahuan Desert: modeling the effects of shifting rainfall seasonality and event size over a landscape gradient. *Global Change Biology*, **9**, 1–19.

Goodrich, D. C., L. J. Lane, R. M. Shillito, S. N. Miller, K. H. Syed, and D. A. Woolhiser, 1997: Linearity of basin response as a function of scale in a semiarid watershed. *Water Resources Research*, **33**, 2951–2965.

Goudie, A., and J. Wilkinson, 1980: *The Warm Desert Environment*. Cambridge University Press, New York, 88 pp.

Graf, W. L., 1988: *Fluvial Processes in Dryland Rivers*. Springer-Verlag, Berlin, 346 pp.

Guntner, A., and A. Bronstert, 2004: Representation of landscape variability and lateral redistribution processes for large-scale hydrological modelling in semi-arid areas. *Journal of Hydrology*, **297**, 136–161.

Hoyos, C. D., and P. J. Webster, 2007: The role of intraseasonal variability in the nature of Asian monsoon precipitation. *Journal of Climate*, **20**, 4402–4424.

Huxman, T. E., and Coauthors, 2004a: Precipitation pulses and carbon fluxes in semiarid and arid ecosystems. *Oecologia*, **141**, 254–268.

Huxman, T. E., and Coauthors, 2004b: Response of net ecosystem gas exchange to a simulated precipitation pulse in a semi-arid grassland: the role of native versus non-native grasses and soil texture. *Oecologia*, **141**, 295–305.

Katul, G., A. Porporato, and R. Oren, 2007: Stochastic dynamics of plant–water interactions. *Annual Review of Ecology, Evolution, and Systematics*, **38**, 767–791.

Kircher, J. E., and Karlinger, M. R., 1983: Effects of water development on surface-water hydrology, Platte River basin in Colorado, Wyoming, and Nebraska upstream from Duncan, Nebraska. US Geological Survey Professional Paper 1277-B, 49 pp.

Laio, F., P. D'Odorico, and L. Ridolfi, 2006: An analytical model to relate the vertical root distribution to climate and soil properties. *Geophysical Research Letters*, **33**, L18401.

Lancaster, N., 1996: Desert environments. In *The Physical Geography of Africa* (W. M. Adams, A. S. Goudie, and A. R. Orme, eds.), Oxford University Press, Oxford, pp. 211–237.

Leduc, C., J. Bromley, and P. Schroeter, 1997: Water table fluctuation and recharge in semi-arid climate: some results of the HAPEX-Sahel hydrodynamic survey (Niger). *Journal of Hydrology*, **188–189**, 123–138.

Loik, M., D. Breshears, W. Lauenroth, and K. Belnap, 2004: A multi-scale perspective of water pulses in dryland ecosystems: climatology and ecohydrology in the western USA. *Oecologia*, **141**, 181–269.

Ludwig, J. A., B. P. Wilcox, D. D. Breshears, D. J. Tongway, and A. C. Imeson, 2005a: Vegetation patches and runoff–erosion as interacting ecohydrological processes in semiarid landscapes. *Ecology*, **86**, 288–297.

Ludwig, J. A., R. Bartley, and A. C. Liedloff, 2005b: Modelling landscape leakiness and sediment yields from savanna hillslopes: the critical role of vegetation configuration. In *MODSIM 2005*, Melbourne, Australia, pp. 361–366.

Ludwig, J. A., G. N. Bastin, V. H. Chewings, R. W. Eager, and A. C. Liedloff, 2007: Leakiness: a new index for monitoring the health of arid and semiarid landscapes using remotely sensed vegetation cover and elevation data. *Ecological Indicators*, **7**, 442–454.

Mabbutt, J. A., 1979: *Desert Landforms*. MIT Press, Cambridge, MA, 340 pp.

McLain, M. E., and Coauthors, 2003: Biogeochemical hotspots and hot moments at the interface of terrestrial and aquatic ecosystems. *Ecosystems*, **6**, 301–312.

Melton, F. A., 1957: *An Analysis of the Relations Among Elements of Climate, Surface Properties, and Geomorphology*. Technical Report 11, Department of Geology, Columbia University, New York.

Menenti, M., 1984: *Physical Aspects and Determination of Evaporation in Deserts*. Report 10, Institute for Land and Water Management (ICW), Wageningen, The Netherlands, 202 pp.

Newman, B. D., and Coauthors, 2006a: Ecohydrology of water-limited environments: a scientific vision. *Water Resources Research*, **42**, W06302, doi:10.1029/2005WR004141.

Newman, B. D., E. R. Vivoni, and A. R. Groffman, 2006b: Surface water–groundwater interactions in semiarid drainages of the American southwest. *Hydrological Processes*, **20**, 3371–3394.

Noy-Meir, I., 1973: Desert ecosystems: environment and producers. *Annual Review of Ecology and Systematics*, **4**, 25–51.

Potts, D. L., and Coauthors, 2006: Antecedent moisture and seasonal precipitation influence the response of canopy-scale carbon and water exchange to rainfall pulses in a semi-arid grassland. *New Phytologist*, doi:10.1111/j.1469-8137.2006.01732.x

Prince, S. D., E. Brown De Colstoun, and L. L. Kravitz, 1998: Evidence from rain-use efficiencies does not indicate extensive Sahelian desertification. *Global Change Biology*, **4**, 359–373.

Qin, Z., P. Berliner, and A. Karnieli, 2002: Micrometeorological modeling to understand the thermal anomaly in the sand dunes across the Israel–Egypt border. *Journal of Arid Environments*, **51**, 281–318.

Ravi, S., P. D'Odorico, and G. S. Okin, 2007: Hydrologic and aeolian controls on vegetation patterns in arid landscapes. *Geophysical Research Letters*, **34**, L24S24, doi:10.1029/2007GL031023.

Reid, I., and L. E. Frostick 1989: Channel forms, flows and sediments in deserts. In *Arid Zone Geomorphology*, (D. S. G. Thomas, ed.), Wiley and Sons, Chichester, UK, pp. 117–135.

Reid, K. D., B. P. Wilcox, D. D. Breshears, and L. MacDonald, 1999: Runoff and erosion in a piñon–juniper woodland: influence of vegetation patches. *Soil Science Society of America Journal*, **63**, 1869–1879.

Renard, K. G., and R. V. Keppel, 1966: Hydrographs of ephemeral streams in the Southwest. *American Society of Civil Engineers, Journal of the Hydraulics Division*, **92**, HY2, 33–52.

Reynolds, J. F., P. R. Kemp, K. Ogle, and R. J. Fernández, 2004: Modifying the "pulse-reserve" paradigm for deserts of North America: precipitation pulses, soil water, and plant responses. *Oecologia*, **141**, 194–210.

Rodier, J., 1975: *Evaluation de l'écoulement annuel dans le Sahel tropical africain*. Collection Travaux et Documents, ORSTOM, Paris, 121 pp.

Sala, O. E., and W. K. Lauenroth, 1982: Small rainfall events: an ecological role in semiarid regions. *Oecologia*, **53**, 301–304.

Scanlon, B. R., K. K. Keese, A. L. Flint, L. E. Flint, C. B. Gaye, S. Simmers, and W. M. Edmunds, 2006: Groundwater recharge in semiarid and arid regions. *Hydrological Processes*, **20**, 3335–3370.

Schick A. P., 1970: Desert floods. In *Symposium on the Results of Research on Representative Experimental Basins*, IAHS/UNESCO, pp. 478–493.

Schlesinger, W. H., A. D. Abrahams, A. J. Parsons, and J. Wainwright, 1999: Nutrient losses in runoff from grassland and shrubland habitats in southern New Mexico. I. Rainfall simulation experiments. *Biogeochemistry*, **45**, 21–34.

Schwinning, S., and O. E. Sala, 2004: Hierarchical organization of resource pulse responses in arid and semiarid ecosystems. *Oecologia*, **141**, 211–220, doi:10.1007/s00442-004-1520-8.

Schwinning, S., O. E. Sala, and J. R. Ehleringer, 2003: Dominant cold desert shrubs do not partition warm season precipitation by event size. *Oecologia*, **136**, 252–260.

Schwinning, S., O. E. Sala, M. E. Loik, and J. R. Ehleringer, 2004: Thresholds, memory, and seasonality: understanding pulse dynamics in arid-semi-arid ecosystems. *Oecologia*, **141**, 191–193.

Schwinning, S., B. I. Starr, and J. R. Ehleringer, 2005a: Summer and winter drought in a cold desert ecosystem (Colorado Plateau). Part I. Effects on soil water and plant water uptake. *Journal of Arid Environments*, **60**, 547–566.

Schwinning, S., B. I. Starr, and J. R. Ehleringer, 2005b: Summer and winter drought in a cold desert ecosystem (Colorado Plateau). Part II. Effects on plant carbon assimilation and growth. *Journal of Arid Environments*, **61**, 61–78.

Scott, R., T. Huxman, W. Cable, and W. Emmerich, 2006: Partitioning of evapotranspiration and its relation to carbon dioxide exchange in a Chihuahuan Desert shrubland. *Hydrological Processes*, **20**, 3227–3243.

Sisamu, C., and V. Siteketa, 2006: Extreme weather at Gobabeb. *Gobabeb Times*, **2**(2), 3.

Snyder, K. A., L. A. Donovan, J. J. James, R. L. Tiller, and J. H. Richards, 2004: Extensive summer water pulses do not necessarily lead to canopy growth of Great Basin and northern Mojave Desert shrubs. *Oecologia*, **141**, 325–334, doi:10.1007/s00442-003-1403-4.

Sophocleous, M., 2002: Interactions between groundwater and surface water: the state of the science. *Hydrogeology Journal*, **10**, 52–67.

Strahler, A. N., and A. H. Strahler, 1987: *Modern Physical Geography*. John Wiley and Sons, New York, 544 pp.

Teztlaff, G., and J. A. T. Bye, 1978: Water balance of Lake Eyre for the flooded period January 1974 to June 1976. *Transactions of the Royal Society of South Australia*, **102**, 91–96.

Thornes, J.B., 1994: Catchment and channel hydrology. In *Geomorphology of Desert Environments* (A. D. Abrahams and A. J. Parsons, eds.), Chapman and Hall, London, pp. 257–287.

Tongway, D. J., and J. A. Ludwig, 1997: The conservation of water and nutrients within landscapes. In *Landscape Ecology, Function and Management: Principles from Australia's Rangelands* (J. A. Ludwig, D. J. Tongway, D. A. Freudenberger, J. C. Noble, and K. C. Hodgkinson, eds.), CSIRO Publishing, Melbourne, Australia, pp. 13–22.

Tucker, G., F. Catani, A. Rinaldo, and R. Bras, 2001: Statistical analysis of drainage density from digital terrain data. *Geomorphology*, **36**, 187–202.

Verhagen, B. T., 1990: Isotope hydrology of the Kalahari: recharge or no recharge? *Palaeoecology of Africa and the Surrounding Islands*, **21**, 143–158.

Walker, B. H., 1987: A general model of savanna structure and function. In *Determinants of Savannas* (B. H. Walker, ed.), IUBS Monograph Series, No. 3, IRL Press, Oxford, pp. 1–12.

Weltzin, J. F., and Coauthors, 2003: Assessing the response of terrestrial ecosystems to potential changes in precipitation. *Bioscience*, **53**, 941–952.

Wheater, H., S. Sorooshian, and K. D. Sharma, 2007: *Hydrological Modelling in Arid and Semi-Arid Areas*, Cambridge University Press, Cambridge, 206 pp.

Wiegand, K., F. Jeltsch, and D. Ward, 2004: Minimum recruitment frequency in plants with episodic recruitment. *Oecologia*, **141**, 363–372.

Wilcox, B. D., 2002: Shrub control and streamflow on rangelands: A process based viewpoint. *Journal of Range Management*, **55**, 318–326.

Wilcox, B. P., B. D. Newman, D. Brandes, D. W. Davenport, and K. Reid, 1997: Runoff from a semiarid ponderosa pine hillslope in New Mexico. *Water Resources Research*, **33**, 2301–2314.

Wilcox, B., D. Breshears, and C. Allen, 2003: Ecohydrology of a resource-conserving semiarid woodland: effects of scale and disturbance. *Ecological Monographs*, **73**, 223–239.

Yin, X., and S. E. Nicholson, 2002: Interpreting annual rainfall from the levels of Lake Victoria. *Journal of Hydrometeorology*, **3**, 406–416.

13 Desert winds and dust

13.1 WIND SPEEDS AND TURBULENCE NEAR THE SURFACE

13.1.1 SURFACE WINDS

Many deserts are characterized by a steady and intense daytime wind that blows with a noisy roar, and utter stillness of the wind in the desert night. This wind regime is a result of two factors, the aerodynamic smoothness of the barren desert surface and the extreme temperature gradients near the surface.

Surface roughness z_0 is small in arid regions, so that wind speeds increase rapidly within the first few meters of the surface. Among the world's deserts, however, roughness varies by about an order of magnitude. One of the smoothest surfaces is the Bonneville salt flats, an ancient lake bed in Utah (Fig. 13.1). The roughness of a similar dry lake bed in California was measured to be 0.003 (Vehrencamp 1951). Its smoothness is indicated by the land-speed records set there: 622 mph (~1000 km/h) for a rocket-powered vehicle in 1970, and 407 mph (655 km/h) for gasoline engines.

During the day, because of intense surface heating, the temperature gradient near the surface is greatly in excess of the adiabatic lapse rate, but at night an intense inversion generally forms. The near-surface wind profile constantly responds to these thermal effects, as illustrated in Fig. 13.2. When extreme surface heating destabilizes the air near the ground, the near-surface wind speed gradient is large because the surface air readily mixes with the faster-moving air above. Consequently, in dryland regions, winds close to the ground will tend to be fast and turbulent during the day. At night, when a surface temperature inversion creates extreme stability, winds are very light. The onset of the calm or turbulent conditions is quite rapid, since the temperature pattern changes quickly between day and night.

The wind regime in dryland regions is a composite of microscale effects, local factors, and prevailing atmospheric circulation regimes, ranging from the regional to the planetary scale.

Where the subtropical highs prevail, the large-scale wind is generally weak and local circulations are therefore well developed. Large-scale winds are typically higher in the mid-latitude drylands, where the westerlies prevail. The regional winds are often reflected in the patterns of dunes and wind erosion. The local influences are generally not sufficiently strong to override the prevailing wind regimes, but when the latter are weak, intense thermal currents and turbulent gusts arise.

13.1.2 TURBULENCE

The turbulence is prescribed by competitive forces involving the wind shear and the temperature gradient at the surface. Its development is best understood in the context of forced and free convection. Forced convection is the vertical motion of air induced by mechanical forces such as the friction within the air, or external factors such as surface roughness. The magnitude of these forces is related to the vertical wind shear. The ensuing vertical currents are relatively small, so that weak, molecular processes are important. Free convection results from density differences in a fluid; in the air the relevant forces are thermal and are proportional to the temperature gradient. Thermal instability is conducive to free convection but stability inhibits it.

When the wind is turbulent, the vertical velocity is irregular, with turbulent elements or eddies exhibiting a broad spectrum of wavelengths and frequencies. This results in rapid fluctuations in wind speed and direction (Fig. 13.3). The onset of turbulence is determined by the relative magnitudes of the thermal and mechanical forces. A dimensionless parameter called the Richardson number Ri represents these magnitudes. This is calculated as the ratio between the vertical temperature gradient, or thermal instability, and the vertical wind shear, or mechanical instability.

The Richardson number represents a balance of the forces that suppress and enhance turbulence. Wind shear and roughness enhance it. The buoyant forces represented by the temperature gradient can either suppress turbulence (thermal stability) or enhance it (thermal instability). Turbulence, or free convection, will occur when particular values of the Richardson number are exceeded.

In general, wind is turbulent, except for an exceedingly small laminar sub-layer very close to the surface (Fig. 13.4). Particles that are small enough to reside within the laminar sub-layer are not prone to erosion. Larger particles that protrude into the turbulent layer are. This is why a mixture of particle sizes makes a surface more prone to erosion than one with uniform particles.

The relevance of these concepts to the desert wind regime is twofold: the thermal stability of the surface affects both the mean wind speed near the ground and the kinetic energy of the short-term eddy or turbulent fluxes. The latter effect is critical in aeolian processes such as dust generation, surface erosion, and dune movement. For a given large-scale wind velocity, the speed will be higher close to the ground over a desert than over a vegetated surface because of thermal instability and lower surface roughness. If the regime is turbulent (in the case of free convection), the mean wind speed will not be affected, but instantaneous departures from the mean speed will be more frequent and more intense than in the laminar (non-turbulent) case (Fig. 13.5). This will mean that the threshold velocities for particle or dune movement are more often exceeded in turbulent conditions.

13.2 LOCAL WINDS

In many arid and semi-arid regions, winds from particular directions will bring a rapid onset of extreme conditions of temperature, dustiness, or aridity. They reflect the ambient conditions in their source region. These winds can result from transient synoptic-scale circulations, or may be persistent manifestations of stable features, like the heat lows over the Sahara or Tibetan plateau, or surface topography. Isolated desert mountains, such as Tibesti or the Hoggar in the Sahara, have intense mountain–valley winds.

Cold-air invasions from higher latitudes affect many deserts, but these are generally short-lived incursions coupled with mid-latitude synoptic-scale systems that extend abnormally far into the tropics. An example is a trough in the mid-latitude westerlies extending into dry regions of the Kalahari, Namibia, southern Australia, Brazil, or even into the Sahara (see Chapter 4). Such troughs occasionally bring a blanket of snow into the Sahara or a sudden onset of freezing temperatures in other regions. Some deserts are occasionally subjected to cool, moist coastal air. This is common in Baja California, Chile, and northern Peru, where sea breezes play a major role in inland sediment transport (Cooke and Warren 1975).

One of the best-known and most persistent desert winds is the hot, dry northeasterly current called the *harmattan* (Fig. 13.6a). It arises on the equatorward side of the semi-permanent Azores subtropical high over the Atlantic. This wind is best developed in winter, when it blows out of the central

Fig. 13.1 Bonneville salt flats, Utah, during partial flooding.

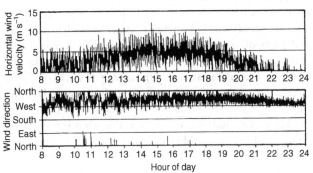

Fig. 13.3 Turbulent fluctuations of wind speed and direction on a summer day at Beer Sheva, in the northern Negev Desert (from Pye and Tsoar 1990, copyright Elsevier).

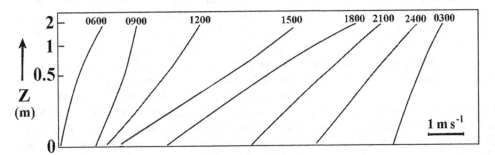

Fig. 13.2 Wind versus height above the surface at various times during the day at Davis, CA, on July 30–31, 1962 (modified from Munn 1966).

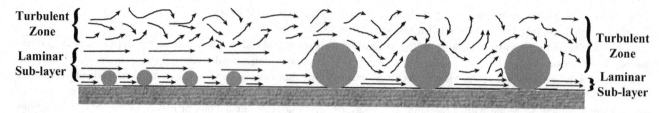

Turbulent Zone

Laminar Sub-layer

Turbulent Zone

Laminar Sub-layer

Fig. 13.4 Near the surface is a very thin sub-layer with laminar airflow, above which the flow quickly becomes turbulent. Small grains lie within the laminar layer and do not create a rough surface, but larger grains reach the turbulent zone and increase surface roughness (modified from Chepil 1958).

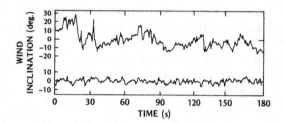

Fig. 13.5 Schematic of wind fluctuations during (bottom) stable and (top) unstable (turbulent) conditions utilizing an index derived from the inclination of the surface wind vector. The y-axis is an indicator of the magnitude of vertical motion, with positive values reflecting upward motion, negative reflecting downward motion (from Priestley 1959, University of Chicago Press, all rights reserved).

Sahara across most of West Africa south of the desert. When the harmattan advances to the Guinea coast it gives relief from the stifling and unhealthy humid conditions, but also brings dust (ben Mohamed *et al.* 1992; Pinker *et al.* 1994) and extreme heat.

Other persistent winds prevail in the eastern Mediterranean and parts of the Middle East in summer, in response to a semi-permanent trough over the Persian Gulf and an intense heat low over Pakistan and Afghanistan (Fig. 13.6b). This circulation produces steady northwesterly or westerly winds that can prevail for 40–50 days at a time. Over the eastern Mediterranean and the Aegean Sea, this wind is termed the *etesian* and prevails in May and early June. There the northwesterly wind traverses water, so that it brings relatively cool weather.

Further east over the Persian Gulf and the lower valleys of the Tigris and Euphrates, these northwesterly winds are called the *shamal* (Fig. 13.7). Here they are of continental origin, so the shamal is dry, very hot, and dusty. Prevailing from June through early July, it brings dust storms to Iraq, Kuwait, and other Gulf states. Some 50% of the dust storms in Baghdad are associated with the shamal (Houseman 1961). The shamal also plays a major role in dune formation in this region (Saqqa and Atallah 2004). This wind is often restricted by topography and surface temperature inversions, occasionally creating speeds in excess of 40 m/s (Membery 1983).

When the nocturnal inversion is particularly strong, a nocturnal jet may form with a core near 200 m (Fig. 13.8). The strong winds are restricted to the layer below the base of the

inversion, around 400–450 m (Membery 1983). The dust is also trapped beneath the inversion layer. A similar jet develops near the Bodélé Depression of West Africa (Washington and Todd 2005; Washington *et al.* 2006a, 2006b), enhancing dust generation in what is probably the world's most intense source region for dust (Prospero *et al.* 2002).

Hot and dry winds also blow northward out of the Sahara along its Mediterranean margin. Such winds are known as the *sirocco* or *scirocco* in the Middle East, parts of North Africa, and the Mediterranean, where they occasionally bring remarkable "dustfalls," lightly covering the surface with red particles. The sirocco generally blows from a southerly or southeasterly direction and also has a large number of local names: *khamsin* in Egypt, *ghibli* in Libya, *chili* in Tunisia, *chergui* in Morocco (Cerveny 1996), and *irifi* in the western Sahara. In Libya its intense heat is exacerbated as it crosses the mountains and is adiabatically warmed upon leeward descent. In the western Sahara the irifi can bring remarkably abrupt temperature changes. In March 1941 this wind brought a sudden temperature increase at Smara from 18.3°C at noon to 42.8°C at 16:00 hours (Goudie and Wilkinson 1980).

Descending air in the lee of mountains also produces marked changes in weather conditions in numerous dryland regions. These hot winds, which are both adiabatically warmed and desiccated, are generically termed *foehn* winds. They also have many local names. East of the Rocky Mountains in the USA, the subsiding westerlies are termed a *chinook*. The temperature can rise 30°–40°C within a few hours, during which time half a meter or more of snow can be removed via melting and evaporation. At Pincher Creek, Alberta, a temperature rise of 21°C occurred within 4 minutes of the onset of a chinook (Barry and Chorley 1999). In Loma, Montana, the temperature on a January day rose from −47°C to 9°C within 24 hours.

On the western slopes of the mountains, especially in California, a similar hot and dry foehn occurs with easterly winds (Fig. 13.9). These typically arise in association with a persistent high-pressure cell over the West Coast (Castro *et al.* 2003). Termed the Santa Ana, this wind's extreme heat and aridity brings relief from the local air pollution, but also causes tremendous wildfires. A popular misconception is that the wind's dryness is a result of passage over desert regions. The Santa Ana winds actually form when the desert is relatively cold, during autumn and early spring, and warm up adiabatically as they

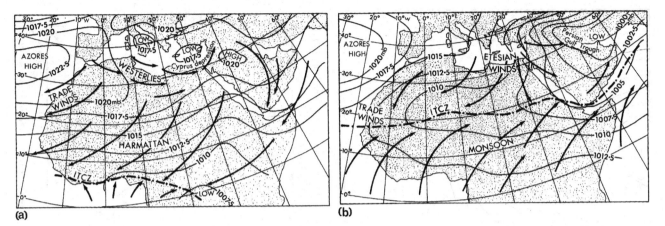

Fig. 13.6 Circulation pattern corresponding to (a) harmattan winds, the NE flow over West Africa that prevails in winter, and (b) etesian winds in the Middle East (from Pye and Tsoar 1990, copyright Elsevier).

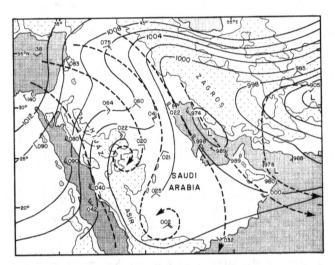

Fig. 13.7 Typical shamal synoptic situation over the Arabian Gulf on June 9, 1982. Surface isobars and 850 mb streamlines (dashed lines) are shown (from Pye 1987; Membery 1983, copyright Elsevier).

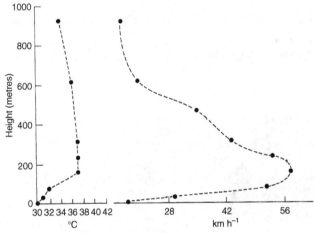

Fig. 13.8 Low-level temperature (°C) and wind speed (km/h) profiles during a Gulf shamal. Peak winds are at the top of the temperature inversion (redrawn from Pye 1987; Membery 1983).

descend the western slopes. Although the speed is typically on the order of 35 mph, the Santa Ana can reach hurricane force when channeled through mountain passes. This and the tinder created by the desiccated chaparral is an explosive mixture that triggers wildfires (Fig. 13.10).

The changing weather conditions associated with foehn winds are illustrated by the so-called "berg" winds of the Namib Desert (Lancaster *et al.* 1984). Figure 13.11 shows temperature and relative humidity at Gobabeb in the central Namib during the course of a typical foggy day with westerly winds from the Atlantic coast (top) and during an easterly berg wind (bottom). During the fog, temperature has a marked diurnal cycle, increasing rapidly to about 25°C in mid-morning when the fog lifts; humidity is relatively high (40–100%) and decreases abruptly in mid-morning as temperature rises. During the berg wind, temperature and humidity are by comparison relatively constant throughout the day. The minimum temperature is about 23°C,

and a steady temperature of about 36°C is maintained throughout the afternoon. Humidity ranges from near zero to a mid-morning maximum of about 40%. The berg wind often brings temperatures in excess of 40°C.

Cerveny (1996) lists around 100 names of local winds, many of which impact dryland regions. Some of the better known winds in Asia include the *afghanets*, a southerly wind from Afghanistan into Uzbekistan; the *sukhovei* (literally, Russian for "dryness"), a dry easterly wind in Russia; and the *simoom*, a dry, dust-laden desert wind in the Middle East. Dry winds in South America are the warm and often dusty *zonda*, a foehn wind that blows out of the Andes into Argentina with speeds up to 33 m/s, and the *pampero*, a cold, strong, southerly wind with a rapid onset that suddenly and dramatically lowers temperatures on the Pampas. In North America, dust-bearing winds include the *tolverane*s in Mexico and the *palouser* in Idaho and Montana (Goudie 1978).

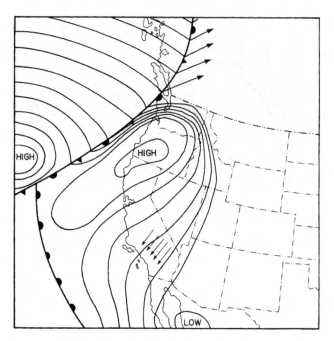

Fig. 13.9 Circulation pattern corresponding to Santa Ana winds (from Pye 1987, copyright Elsevier).

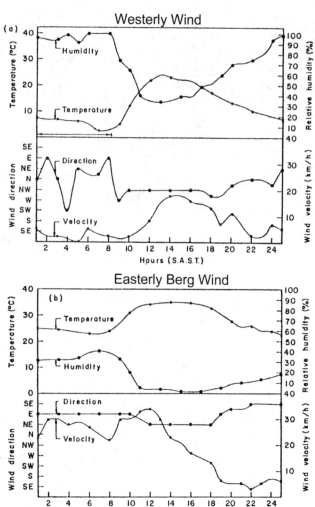

Fig. 13.11 Temperature, humidity, and wind velocity and direction at Gobabeb, Namibia, (a) during a fog situation (September 8, 1971) and (b) with an easterly berg wind (April 27, 1973) (from Lancaster *et al.* 1984).

13.3 DUST MOBILIZATION, TRANSPORT, AND DEPOSITION

The factors that make a surface susceptible to the mobilization of dust are reviewed in Chapter 2, along with the physical processes producing particle movement and uplift. These include particle size/soil texture, soil structure/aggregates, moisture content, and surface roughness. Although airborne dust is primarily silt and clay, a high content of silt and clay does not necessarily make the surface conducive to erosion. The sand content is important, and a clay content greater than 10% decreases erodibility (Gillette 1988). Important sources of dust are sandy, loamy sand, and sandy loam soils. In general, disturbed soils are much more easily eroded than undisturbed.

The environmental factors that determine the potential for dust production include not only the surface erodibility, but also the nature of the vegetation cover and climate. In general,

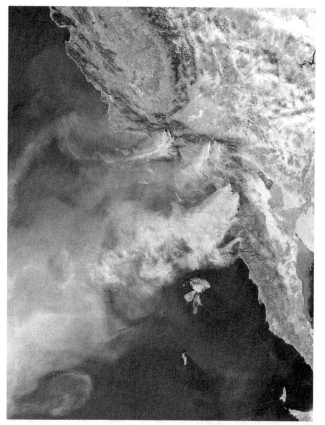

Fig. 13.10 Satellite image of wildfires associated with Santa Ana winds in late 2003 (from nasa.gov/centers/Goddard/news/topstory/2003).

the denser the vegetation cover, the greater its capacity to bond the surface materials. This capacity depends also on vegetation type (e.g., form and structure), percent cover, and root density and structure. The climatic factors include wind speed, turbulence, wind direction, rainfall, humidity, air temperature, etc. The soil and vegetation factors tend to operate on a local scale and the climate factors at a larger scale, although microclimate can also play a role.

A number of studies have shown that dust storm frequency is highest in semi-arid regions, not in the very arid deserts. However, there is also considerable variation in the relationship between rainfall and dust storm frequency, suggesting the influence of other factors as well. This is especially true in regions of higher rainfall, where the mean frequency is low but year-to-year variations in dust storms are quite large. The relationship is complicated by the fact that many regions experience dust transported from afar as well as locally generated dust. Also, wind is necessary to mobilize the dust, even when the surface is very barren and dry.

Table 13.1 examines the relationships between the amount of dust carried in the air and both rainfall and wind for the locations identified by TOMS data as being major source regions of dust (Prospero et al. 2002). The table presents the linear correlation between mean monthly values of the TOMS aerosol index (see Chapter 1) and mean monthly rainfall and wind speed at a meteorological station within each source region. This correlation measures the similarity of the annual cycles of the two variables being compared. It is notable that most stations show a correlation between the TOMS index and either rainfall (a negative correlation) or wind (a positive correlation), suggesting that either can be the limiting factor in dust production. Only a handful of stations show a significant correlation with both variables.

The minimum wind speed needed to generate blowing dust varies greatly with surface characteristics. In the American Southwest (Pye 1987), the highest speeds, on the order of 15–18 m/s, are required over undisturbed playas, crusted alluvial fans, mature desert pavements, and interdune flats. The most easily eroded surfaces are river beds and sand dunes, where a wind speed as low as 5 m/s can mobilize particles. However, the dunes contain few silt- and clay-sized particles that can contribute to the atmospheric dust load. Playas, takyr soils, loess, and humid tropical soils can be less important sources of dust than sandy alluvium and many semi-arid soils. In the western Sahara, mean deflation velocities range from 6.5 to 13.0 m/s (Helgren and Prospero 1987). The frequency of dust events goes up rapidly with speeds in excess of 10 m/s.

Once the dust is mobilized and entrained into the atmosphere, the suspended particles are transported over large distances by the wind. During transport, particles are continually lost from the airstream by deposition. Dry deposition occurs in three ways:

1. the vertical motion of turbulent fluxes falls below the gravitational fall velocity of the particles;
2. the particles form aggregates and settle back to the ground;
3. the particles are captured by collision with rough, moist, or electrically charged surfaces (Pye 1987).

Table 13.1. *Correlation of the seasonal cycle of the TOMS aerosol index (AI) with those of rainfall and wind at meteorological stations within the indicated source region. The correlations are based only on 12-monthly values, hence only very high correlations are meaningful.*

Source region	Rain	Wind
India/Pakistan 1	0.48	0.82
India/Pakistan 2	0.35	0.87
Afghanistan 1	−0.61	0.62
Afghanistan 2	−0.63	0.62
Taklamakan	0.61	0.73
Caspian	−0.76	0.95
Iran 1	−0.62	0.90
Iran 2	−0.37	0.70
Aral Sea	0.04	0.20
Tigris–Euphrates	−0.81	0.70
Tunisian chotts	−0.82	0.11
Danikil 1	−0.25	0.69
Danikil 2	−0.38	0.79
Libyan Desert	−0.14	0.86
Saudi Arabia 1	−0.44	0.33
Saudi Arabia 2	−0.56	−0.38
Bodélé Depression	−0.01	−0.33
Mali	0.52	−0.19
Western Sahara	0.12	−0.53
Sudan 1	0.34	−0.06
Sudan 2	0.25	0.63
Lake Bonneville	0.46	0.70
Salton Sea	−0.72	−0.41
Bolivia	−0.17	0.49
Etosha Pan	−0.11	0.08
Makgadikgadi Pan	−0.05	0.84
Patagonia 1	−0.26	0.68
Patagonia 2	0.39	0.68
Lake Eyre	0.47	0.38

With gravitational settling, the largest particles settle out first. Wet deposition involves removal of particles by rain or snow. This can occur when falling rain or snow captures particles or when cloud droplets capture particles or condense upon them and later precipitate. In general, the further from the source, the greater the ratio of wet to dry deposition. This suggests that small particles are preferentially entrained into cloud droplets or utilized as cloud condensation nuclei. Studies in Israel confirm this: washed-out dust contained 50–65% clay, but dry-deposited dust contained less than 20% clay (Ganor 1975).

Deposition is promoted by reduction of ambient wind speed. This can occur as a result of meteorological conditions, local surface cover, or topography. Wind speed is reduced when roughness suddenly increases, such as when air traversing a smooth surface encounters vegetation. Topography can play a similar role. Obstacles capture dust. Depressions do too, because wind slows down as the airstream traverses a depression (Fig. 13.12).

Over the continents, only sporadic estimates of deposition are available. These suggest deposition rates on the order of 10–200 t/km² per year. Although much of this is dry deposition, rains that occur early in the season can scavenge tremendous quantities of dust. Much of the dust generated over the continents is transported and deposited over the surrounding oceans. Figure 13.13 shows the transport trajectories. The Pacific receives dust from Asia and Australia. The Atlantic receives dust from both Africa and North America, with a small contribution from South America. That over the Indian Ocean comes primarily from southern Africa.

13.4 DUST DEVILS

13.4.1 FORMATION

Dust devils are highly visible examples of surface convection that occur frequently in arid regions (Fig. 13.14). Several may be active at any given moment. Dust devils are essentially low-pressure, warm-core vortices that can move with the prevailing winds or remain stationary. The sand and dust that they entrain creates their visibility. Their formation requires certain geographic and atmospheric conditions:

- relatively flat terrain with loose material
- intense surface heating
- a lapse rate greatly exceeding the dry adiabatic
- ambient winds below a certain critical threshold.

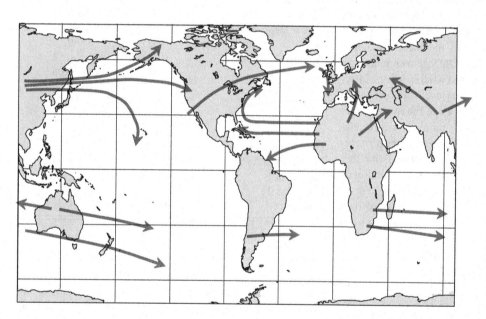

Fig. 13.12 Deposition of dust in the lee of topographic obstacles due to flow divergence and reduction of wind speed (Pye 1987, copyright Elsevier).

Fig. 13.13 Typical trajectories of soil aerosol transport (based on Gillette 1993, Pye 1987, Tsai *et al.* 2008, Gao and Washington 2009, and others). Extensively modified from a figure by Jennings (1993). Original figure (c) 1993 The Arizona Board of Regents.

Fig. 13.14 (left) Australian dust devil, (right) dust devils over Etosha Pan, Namibia.

Over a dry and relatively bare surface, such as a playa, the intense heating creates an extremely high temperature. The atmosphere above is much cooler and denser. This creates a superadiabatic lapse rate and a thermally unstable situation. The convection associated with the dust devil pulls some of the denser air toward the surface, thereby increasing the thermal stability.

An unstable vertical temperature stratification is insufficient to produce a dust devil. Generally, a source of angular momentum for rotation is required, with Kelvin–Helmholz instability possibly providing a trigger for the rotation (Barcilon 1967; Barcilon and Drazin 1972). Therefore, dust devil formation is favored by local roughness elements that can produce wake eddies or concentrate vorticity (Balme and Greeley 2006) and by boundaries between different surface types (e.g., irrigated fields in bare desert regions) where strong thermal gradients develop.

One theory of formation involves local surface "hot spots" that force intense upward motions (Sinclair 1969). This theory is supported by theoretical work on convection and by laboratory experiments. It also helps to explain the existence of spatially preferential sites of formation, but the concentration in the lee of small hills suggests also that the generation of vorticity by terrain irregularities or by surface obstacles also plays a role. Ives (1947) even suggests that disturbances of the surface air layer by small animals can trigger formation.

The formation and frequency of occurrence of dust devils are complicated by many factors. The areas of most frequent occurrence are hot, flat surfaces, especially those near disturbed ground (e.g., freshly plowed or irrigated fields) (Balme and Greeley 2006). Gentle slopes may favor their formation, but mountains and hills do not. They can rise over a gentle slope, but cannot descend one.

13.4.2 MOTION, STRUCTURE AND MAINTENANCE

The typical diameter of a dust devil is 1–50 m and the typical vertical extent is a few meters to a few hundred meters. Measurements near Tucson, Arizona, showed that 55% of the observed dust devils have diameters between 3 and 15 m, and 15% have diameters larger than 15 m. However, large devils can reach well over a hundred meters in diameter. Sailplane flights into dust devils have shown that the thermal updraft in their core can extend up to at least 800 m (Sinclair 1969), but the updrafts generally extend to the top of the convective layer, which can extend to 3–4 km in some desert regions. The height of the dust column, however, rarely exceeds 1 km.

Studies near Tucson, Arizona, and in the Mojave Desert showed that the highest frequency occurred around 13:00–14:00 hours (Fig. 13.15), coinciding with the lowest vertical stability of the atmosphere and not necessarily with the highest surface temperature. The maximum frequency in the Near East was found to occur slightly earlier, around 12:30–13:30 hours (Flower 1936).

In an area near Tucson, the number of dust devils sighted per day in an area of roughly 100 square miles ranged from 20 to 86. In the Arva Valley, west of Tucson, 1663 dust devils were observed in an area roughly the same size but during a 22-day period. On one day, 169 occurred. Within these two areas the distribution was highly variable and the dust devils appeared to be concentrated in dry riverbeds in the lee of small hills.

The pattern of motion associated with a dust devil is similar to that of a helical vortex, with warm air parcels spiraling in toward the center of the vortex and gently rising (Fig. 13.16). There is also a weaker and cooler downdraft in the core of the vortex. The warm air moves in toward the low-pressure center horizontally until it reaches the dust column. There it rises rapidly and within the dust column the radial velocity nearly vanishes, creating a rotating column of air (Fig. 13.16). In the core of the dust devil is descending motion that, together with centrifugal forces, suppresses the presence of dust particles.

Within the dust devil, peak vertical velocity occurs within the region of maximum temperature (Fig. 13.17). The radial velocity (Fig. 13.18) reaches its peak value outside the region of maximum tangential and vertical velocities (Sinclair 1966, 1973). At low levels the radial velocity of the spiraling air is on the order of 5 m/s, but its tangential velocity can reach about

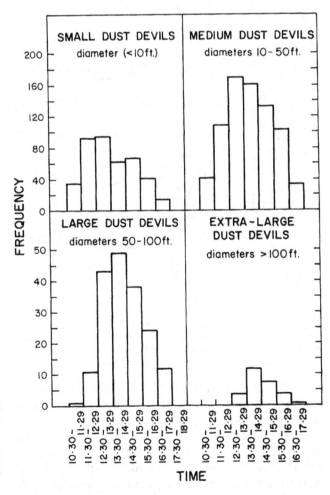

Fig. 13.15 Diurnal frequency of dust devils of various sizes near Tucson, Arizona (from Sinclair 1969).

Fig. 13.16 Helical flow into center of a dust devil vortex. Within the core, radial velocity vanishes, creating a rotating column of air (small arrows). Air rises in the center of the core.

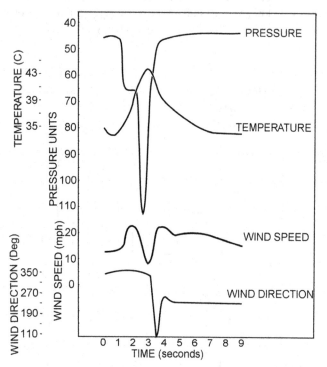

Fig. 13.17 Schematic of temperature, pressure, vertical velocity, and horizontal wind direction with the passage of a dust devil (based on information in Sinclair 1965, 1969; Fitzjarrald 1973; Balme and Greeley 2006).

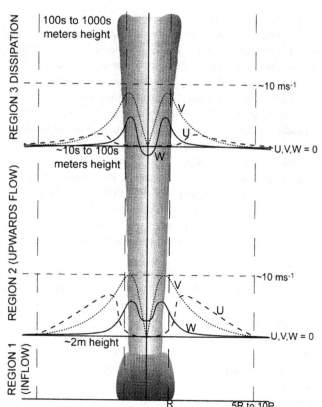

Fig. 13.18 Vertical model of dust devil (from Balme and Greeley 2006).

wind speeds below this. Formation is suppressed by very high values of ambient wind (Rennó *et al.* 1998).

The air within the vortex is generally 4°–8°C warmer than the ambient environment and the pressure drop in the center is roughly 2.5–4.5 mb (Rennó *et al.* 1998). The spiraling air parcels continue to absorb heat from the surface as they move with the dust devil. This provides a continuous supply of warm boundary-layer air to the vortex, thereby sustaining the vertical motion within it. This process also serves to transfer sensible heat from the surface to the boundary layer, and thus dust devils probably play an important role in heat transfer processes in arid regions.

The efficiency at which the warm air is supplied determines the lifetime of a dust devil. The lifetime is generally on the order of a few seconds to 15 or 20 minutes. Some 65% of the dust devils in the Arva Valley of Arizona had a duration less than 1.5 minutes and 92% had a duration less than 3.5 minutes. The lifetime can be considerably longer, especially for stationary dust devils. Ives (1947) reports a case in the Sonora in which a large dust devil persisted for four hours and removed one cubic yard of sand per hour. A similar event was reported near Cairo, Egypt; a dust devil that began on a sand mound remained stationary for nearly two hours then wandered away (Humphreys 1940). On the Bonneville salt flats in western Utah, large dust devils with a lifetime of 8 hours have been observed to travel horizontal distances of some 40 miles. In general, the larger the dust devil, the longer its duration (Sinclair 1969). This is

22 m/s (Schweisow and Cupp 1976). The vertical velocity of the updraft ranges from about 2 to 15 m/s, with the larger velocities being associated with the largest dust devils.

To a first approximation, the dust devil moves in roughly the same direction and at roughly the speed of the ambient wind, typically about 5 m/s, and it slopes vertically in the direction of the shear of the ambient wind. In environments with high ambient wind speeds (greater than 10 m/s), the larger-diameter dust devils tend to form, but the highest frequency of formation is at consistent with the ability of the larger vortices to entrain more warm air for maintenance.

The dust devil can take on many shapes (Balme and Greeley 2006). It can be crooked or sinuous due to wind shear. It can be a tall and thin column, or wider than it is tall. Some are disordered rotating dust clouds. The most common shape is an inverted cone (Metzger 1999). Infrequently the dust appears as a broad rotating mass with little structure but containing dynamic "ropes." The author once observed a similar phenomenon, a vertical sheet of dust in central Botswana. From a distance it appeared to be a wall of fire, but upon approach, it took the form of a tall, ruffling curtain (the "ropes"). It dissipated within a few minutes.

13.5 MONITORING DUST

13.5.1 PRE-SATELLITE ANALYSIS OF DUST

Producing a reliable assessment of global mineral dust concentration and global dust sources has been an elusive problem. Until the 1980s, when satellites began to provide spatially and temporally continuous coverage, dust assessments were based on archival references to dust storms, analysis of meteorological records of visibility, and sporadic surface measurements of radiation attenuation or captured particles. Each approach provides a different quantity and there is no consistent standard. Moreover, each approach is representative of total atmospheric particulate loading, of which mineral dust is only one component.

The most widely available information is on dust storm occurrence. Dust storms are a specific meteorological event in which turbulent wind systems are actively entraining millions of particles into the air. Dust storms are generally short-lived, but are accompanied by extreme reductions in visibility. Dust haze is a different phenomenon. It occurs when aeolian dust particles are suspended in the air and not being actively entrained (Goudie and Middleton 1992). The particles result from a previous event or long-distance transport. Dust haze can be a persistent phenomenon, lasting for days or weeks. In practical terms, however, the distinction is blurred because both dust storms and dust haze are frequently assessed from visibility data.

Monitoring can be accomplished by analysis of meteorological records of visibility. Meteorological records explicitly specify the cause of the visibility reduction. In West Africa, mineral dust can be distinguished by its reddish appearance and even by taste, as the observer breathes in the dust-laden air. Whenever the visibility is less than 1 km, it correlates well with the amount of suspended dust in the atmosphere (Fig. 13.19).

Analysis of meteorological data provides records on dust occurrence going back roughly a century, depending on location. Archival information provides the longest records, but it is also the most ambiguous and the most subjective. Ancient literature, for example, includes references to "dust rain," "dust fog," and "yellow fog" dating back to 1150 BC (Goudie and Middleton 1992). Zhang (1985) compiled records to construct a chronology commencing in 300 AD. The dust outbreaks in the twelfth and thirteenth centuries, the sixteenth and eighteenth

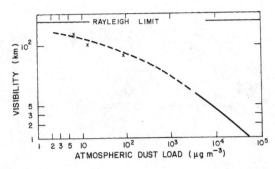

Fig. 13.19 Visibility versus atmospheric dust load (from Hall 1981; Pye 1987, copyright Elsevier).

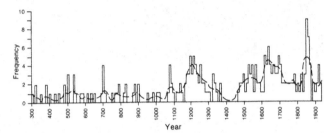

Fig. 13.20 Historical dust outbreaks in earlier centuries (from Zhang 1985).

centuries, and the mid-nineteenth century are very prominent (Fig. 13.20).

The only direct measurements come from dust-collecting filters. Collection is made at only a handful of locations. However, other ground-based instruments can provide reliable assessments by measuring radiation attenuation. These include sun photometers, which measure sunlight at various wavelengths, and LiDAR (laser radar), which measures the return signal off an emitted beam of radiation. These give a good picture of atmospheric concentration.

13.5.2 LONG-TERM MONITORING

Until recently there was no systematic attempt at global, long-term aerosol monitoring. The first global monitoring program began in the 1990s, sponsored by NASA and many other international scientific agencies (Holben et al. 1998). Termed AERONET (Aerosol Robotic Network), this program includes some 100 stations worldwide. They provide continuous measurements of remotely sensed aerosol optical depth. Holben et al. (2001) summarize the data for 9 primary sites and 21 other locations around the world. More recently various satellites have provided the capability of monitoring dust from space (Fig. 13.21).

A number of studies have assessed the interannual variability of dust over longer time frames. Dust deposition at Barbados has been monitored since 1965, using ground-based collectors (e.g., Prospero and Nees 1986; Prospero and Lamb 2003). Chiapello and Moulin (2002) and Torres et al. (2002) have pieced together satellite records from TOMS and Meteosat/VIS to examine dust in the Atlantic over the coast of Senegal over the period

Fig. 13.21 Dust image from SeaWiFS, showing an intense plume off the coast of West Africa.

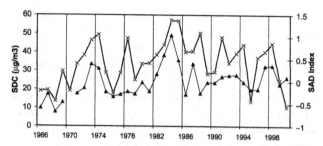

Fig. 13.22 Comparison between the summer dust concentration (SDC, μg m⁻³) in Barbados (solid triangles) and a Sahel drought index (SAD) (crosses, inverse of rainfall) (from Ginoux *et al.* 2004).

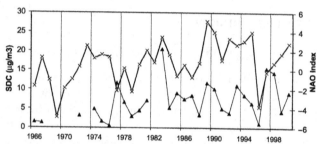

Fig. 13.23 Comparison between the North Atlantic Oscillation (NAO) index (crosses) and the summer dust concentration (SDC, μg m⁻³) in Barbados (solid triangles) (Chiapello and Moulin 2002).

1979–1997. Interesting correlations have been found between the summer dust concentration in Barbados and aridity in the Sahel (Fig. 13.22) (Ginoux *et al.* 2004) and between the summer dust concentration in Barbados and the North Atlantic Oscillation (NAO) (Fig. 13.23) (Chiapello and Moulin (2002). A number of Chinese authors have also produced multidecadal records of dust in Asia (e.g., Natsagdorj *et al.* 2003; Zhou and Zhang 2003).

N'tchayi Mbourou *et al.* (1997) utilized visibility data to track dust haze over West Africa. The frequency and severity of dust haze (Fig. 13.24) increased continually throughout West Africa from the 1950s, when the region experienced good rainfall, through the 1980s, following two periods of protracted drought. The increase was particularly marked in northern Mali, where dust haze increased from 10 or 20 days per year to an almost daily presence (N'tchayi Mbourou *et al.* 1997). A strong relationship with rainfall is apparent (Fig. 13.25a).

Goudie and Middleton (1992) quantified dust storm occurrence at 30 stations throughout the arid world. In most cases the period 1950–1980 was examined. There was no worldwide trend; dust tended to increase in some regions, such as Maiduguri, Nigeria, and decrease or vary randomly in others, such as Mexico City and Choibalsan, Mongolia. At many stations, a relationship with rainfall was apparent, but at other locations, such as Minqin, China, and Zamiin-Uud, Mongolia, other factors appear to govern dust frequency (Fig. 13.25b,c).

13.6 CHARACTERISTICS AND METEOROLOGICAL ASPECTS OF MINERAL DUST

13.6.1 DISTRIBUTION, SOURCES AND SOURCE STRENGTH

Aerosols are an important constituent of the atmosphere. They provide condensation nuclei, trace constituents, soil nutrients, and soil materials. They modify the sediment chemistry of the oceans and the heat balance of the atmosphere. The main sources of aerosols include cosmic dust, volcanic dust, industrial emissions, gases converted to particulates, smoke from fires, sea salt, and mineral materials derived from surface deflation of sediments and soils (Pye 1987). The total emission to the atmosphere is on the order of 6000 Tg/year. Over half of this is sea salt and probably one-third or more is mineral aerosol (Duce 2005). However, estimates of the latter have varied by more than an order of magnitude, ranging from about 130 Tg/year (Joseph *et al.* 1973) to over 5000 Tg/year (Schütz 1980).

Satellite imagery (Husar *et al.* 1997; Prospero *et al.* 2002; Washington *et al.* 2003) has provided the first global view of aerosol distribution (Fig. 13.26). The size and intensity of the Sahara–Arabian desert belt is clearly shown. A number of other regions with high aerosol index (AI) are also apparent. These are listed in Table 13.2, together with the maximum mean aerosol index for the region. Values vary from ~3.4 near the Bodélé Depression to ~0.5 in the Great Basin of the USA.

Three of the four largest sources are in West Africa. The largest, the Bodélé Depression, owes its strength at least partially to a low-level jet in the region that churns up dust and moves it westward (Washington *et al.* 2006b). This source dwarfs all others and is responsible for up to half of the total atmospheric loading of mineral dust.

An interesting finding is that nearly all of the source regions have two commonalities. One is that most were once lakes, in

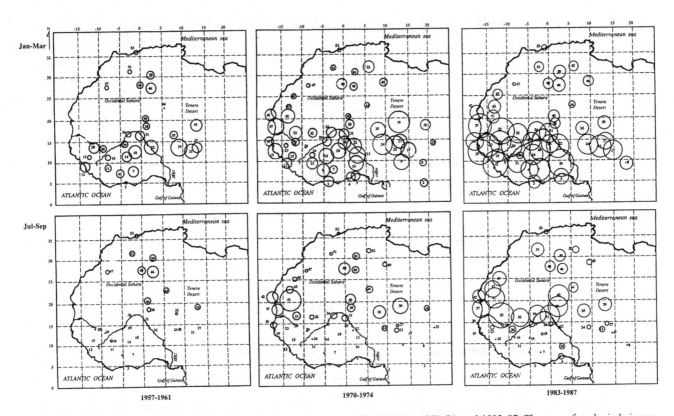

Fig. 13.24 Seasonal means of dust frequency over West Africa for the periods 1957–61, 1970–74, and 1983–87. The area of each circle is proportional to the number of hours with visibility reduced to less than 5 km because of wind-borne dust; numbers correspond to stations (from N'tchayi Mbourou *et al.* 1997). (top) January–March (season of maximum dust frequency). (bottom) July–September (season of minimum dust frequency).

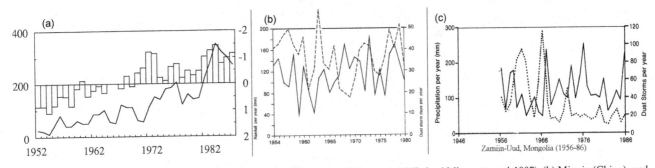

Fig. 13.25 Dust haze or dust storms versus precipitation at (a) Nouakchott (Mauritania) (N'tchayi Mbourou *et al.* 1997), (b) Minqin (China), and (c) Zamiin-Uud (Mongolia) (from Goudie and Middleton 1992). In (b) and (c), the solid line is rainfall and the dashed or dotted line is dust storm days per year.

most cases during the Holocene. The other is that most lie in topographic basins. Figure 13.27 compares the location of these lakes over West Africa and Arabia with the dust source regions indicated by TOMS. The former lakes are repositories of fine sediments and commonly provide dust sources because they have high silt/clay ratios and generally low material bonding. Depressions such as the Bodélé of North Africa also tend to gather aeolian sediments because of the reduction of wind speed as the dust-transporting winds pass over them and spread out into the depression. Consequently, over West Africa neither the large sand ergs of the Sahara nor its

desert mountain massifs are major source regions of mineral dust.

The source strength is determined by the erosion potential of the surface. Several factors control this. One is the nature of the source material, and how readily it is deflated. Texture plays a large role, as do factors such as aggregation, which influences the binding, or the presence of larger materials (e.g., gravel). The larger materials affect both surface roughness and wind speed and in some cases act as an agent of deflation. The surface wind regime is also critical, since it provides the energy for deflation. This energy is proportional to the square of the

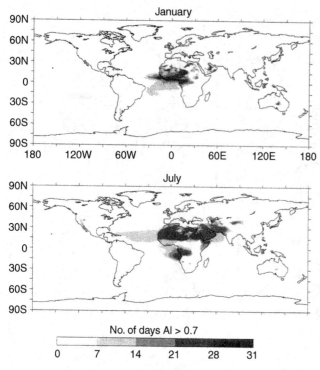

No. of days AI > 0.7

Fig. 13.26 Area average TOMS aerosol index (AI) for major dust source regions in January and July (from Washington *et al.* 2003, courtesy of the Taylor & Francis Group, http://www.informagroup.com). The vast magnitude of the Saharan source is readily seen.

Table 13.2. *TOMS AAI (absorbing aerosol index) for major source regions of mineral dust (data provided by J. Prospero, University of Miami).*

Source region	Absorbing aerosol index
Tunisia	1.4
Libya	1.7
Mauritania	1.7
Mali	2.1
Chad Basin	3.4
Sudan	1.4
Ethiopia	1.0
Arabia	1.0
Oman	1.5
Tigris–Euphrates	1.6
Caspian Sea	0.8
Aral Sea	0.8
Iran–Pakistan	1.0
India	1.1
Tarim Pendi Basin	1.0
Gobi Desert	0.4
Salton Sea	0.5
Bonneville salt flats	0.7
Lake Eyre Basin	0.4
Altiplano	0.7
Patagonia	0.5

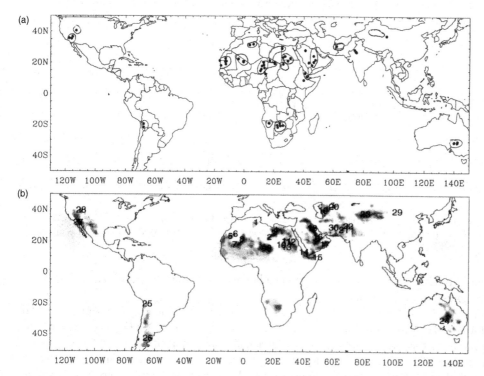

Fig. 13.27 (a) Major dust sources identified by Prospero *et al.* (2002). (b) Major now-dry Holocene lakes, based on data in Street-Perrott and Perrott (1993). The dust sources in (a) are outlined.

wind speed. Therefore the degree of turbulence is very important since it produces short-term variations in excess of threshold velocities, thresholds that the mean wind speed might not exceed.

A second set of factors relates to the condition of the surface, including the nature and amount of vegetation cover and the soil moisture content. Both affect particle binding and reduce erosion. If the depressions and lake beds remain dry or lose their vegetation cover, they are highly prone to wind erosion. A study of depressions that serve as source regions for mineral dust showed an inverse relationship between dust storm frequency and both leaf area index (an index of vegetation density) and net primary productivity (Engelstaedter *et al.* 2003). The grasslands, despite low rainfall, did not generate a large amount of dust because of the large percentage of the surface covered by vegetation.

The vegetation, soil moisture, and wind vary from season to season and from year to year. Thus the source strength is by no means constant. TOMS data from West Africa suggest that dust is at a minimum in November, when vegetation reaches its peak following the wet season. The dust maximum is in May, toward the end of the dry season in most regions (Prospero *et al.* 2002). Visibility data show a somewhat different picture, but confirm the late dry season/early wet season maximum and late wet season/early dry season minimum (N'tchayi Mbourou *et al.* 1997). The seasonality can change over time, especially the length of the dust season. Figure 13.28 shows the seasonal cycle at select stations in West Africa, based on visibility data. These cover three progressively drier periods, 1957–1961, 1970–1974, and 1983–1987. At stations in the Sahel (14°–19° N), the number of months with intense dust haze increased from 6 or 7 in the 1950s to 10 or 12 in the 1970s and 1980s; the frequency of occurrence also increased dramatically.

Interannual variability is also quite pronounced and is generally related to drought occurrence. Case studies for West Africa, Australia, and the United States clearly illustrate the link. The TOMS aerosol index (Fig. 13.29) was much higher and dust was more widespread during 1984, in the middle of a protracted drought in West Africa, than in 1986, when rainfall had increased to just subnormal values. Over Australia, dust storms occurred over roughly half the continent during the extensive drought of 1964, but in only a handful of locations during 1974, when rainfall conditions were good throughout Australia (Fig. 13.30). In the Great Plains of the USA, dust storm occurrence peaked during the droughts of the 1930s and 1950s. During the "Dust Bowl" years of the 1930s, the number of days with blowing dust at Dodge City, Kansas, increased to 120, compared with 0–20 days during more typical years (Fig. 13.31).

13.6.2 GRAIN SIZE AND COMPOSITION

Mineral dust consists mainly of silt- and clay-sized soil particles, i.e., particles less than about 50–70 microns. In general, the ability of a particle to remain in atmospheric suspension depends on the efficiency with which turbulent currents can keep it "buoyant" against the gravitational forces pulling it earthward. This depends on the particle's size, shape, smoothness, and density.

The composition of mineral dust varies greatly over place and time, in response to fluctuations in sources and winds. The dust content of the atmosphere at any given location includes both

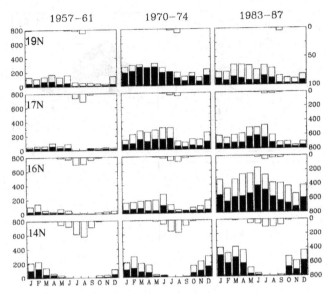

Fig. 13.28 Mean monthly rainfall and number of hours with visibility reduced to less than 5 km (shaded bars) and less than 10 km (unshaded bars) at select West African stations for the periods 1957–61 (wet), 1970–74 (drought), and 1983–87 (severe drought) (from N'tchayi Mbourou *et al.* 1997). The number of hours with reduced visibility is indicated on the left vertical axis. Mean monthly rainfall during each period is indicated at the top of the graph, with the amount indicated on the right vertical axis.

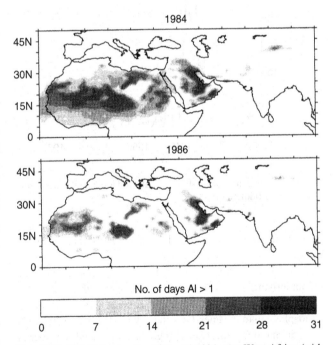

Fig. 13.29 TOMS images for 1984 and 1986 over West Africa (with thanks to Joseph Prospero, University of Miami). Shading indicates number of days with an aerosol index (AI) greater than 1.

a local component and a component that has been transported from distant sources. Its composition is determined by three main factors (Pye 1987): (1) the nature of the source material, (2) the speed and turbulence of the eroding and transporting

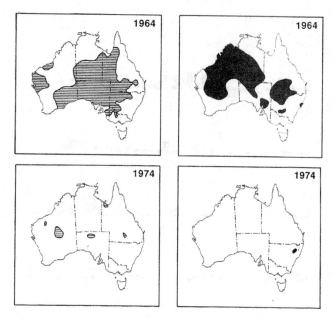

Fig. 13.30 Australian drought area (black) vs. dust storms (horizontal shading) for the years 1964 and 1974 (from McTainsh *et al.* 1989).

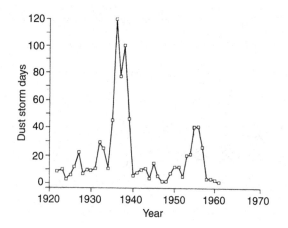

Fig. 13.31 Frequency of dust storms at Dodge City, Kansas (from Goudie and Middleton 1992).

wind, and (3) the vertical and horizontal distance from the source.

Close to the source the composition tends to resemble that of the underlying soil. With distance, the larger and denser particles progressively fall out. They also reside at lower levels in the atmosphere than the fines. Over the Sahara, the bulk of the dust is 30–50 microns. It decreases progressively to roughly 1 micron at 5000 km downwind (Duce 1995). At many locations there is a bimodal size distribution, due to the combination of local and remote sources.

Dust consists of a wide variety of minerals (Pye 1987). Quartz, feldspars, and mica are almost ubiquitous. Dusts that are rich in these minerals tend to be coarse-grained. Numerous clay minerals such as kaolinite, illite and chlorite, and calcareous materials such as calcite, dolomite and gypsum are also common. By

far the most abundant is silicon dioxide; its percent contribution ranges from 33% to 66%. Aluminum oxide is the next most abundant, followed by ferric oxide. The mineralogy and chemical composition is a signature for the source region of the dust. Because the composition varies over relatively small distances, dust can be traced to very specific regions. The Si/Al ratio is a particularly useful "tracer."

13.6.3 THE SAHARAN AIR LAYER

Air mass formation is a well-known concept in mid-latitudes. When air remains stationary over a surface for a sufficiently long period of time, it takes on the characteristics of that surface. This happens most often in the vicinity of high-pressure cells. Air masses are transported with the transient synoptic-scale winds, bringing their ambient characteristics to the regions they enter.

An analogous situation occurs over the Sahara. In winter, high pressure and low wind speeds prevail. A massive layer of hot, dusty air forms aloft. This is the haze layer associated with the northeast harmattan. In the summer, Saharan dust outbreaks linked to easterly waves add to the dust concentration in the layer. Dust concentrations aloft are several times greater than in the marine boundary layer, which undercuts it over the Atlantic. Three or four waves per month propagate westward at a speed of about 8 m/s, transporting the Saharan air layer westward. The Saharan air traverses the Atlantic in 5–6 days and conserves many of its properties en route. Saharan dust outbreaks are well known in the Mediterranean, but they have even reached northern Europe. The dust appears as a red coating on numerous surfaces; in most of the European outbreaks, the dust accompanies precipitation.

The concept of the Saharan air layer (SAL) arose from measurements made during the GARP (Global Atmospheric Research program) Atlantic Tropical Experiment (GATE), which took place over the eastern tropical Atlantic and West Africa during 1974. Conceptual models of the SAL were proposed by Carlson and Prospero (1972), Karyampudi (1979), Swap *et al.* (1996), and others (Fig. 13.32). Numerical models have confirmed many of their features.

In the eastern tropical Atlantic, just off the West African coast, the SAL resides between roughly ~900 mb/1800 m and ~500 mb/5500 m. Over the continent it can extend even higher (Fig. 13.33). The SAL is bound by subsidence inversions at the top and bottom. The lower inversion is associated with the trade winds in the marine layer. Toward its southern edge lies the mid-level African easterly jet (AEJ). The east–west extent of this layer is typically 2000–3000 km, but it can extend some 5000 km and cover an area slightly larger than the contiguous United States (Dunion and Velden 2004).

The SAL is well mixed, with uniform temperature and moisture within. Temperature decreases with height, but potential temperature is relatively constant, so neutral stability prevails. The SAL is characterized by unusually high temperature and low relative humidity. A typical temperature would be 44°C. A

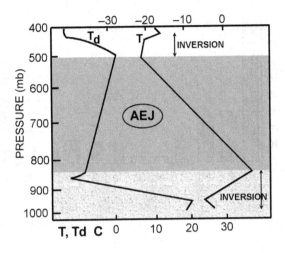

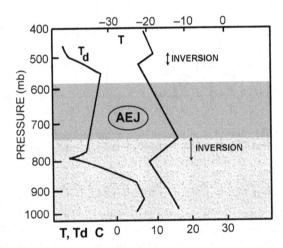

Fig. 13.32 Conceptual model of the Saharan air layer (SAL) (from Carlson and Prospero 1972). The Saharan air layer (dark shading, indicating intense dust layer) is trapped between two inversions. The bottom diagram depicts the Saharan air layer over the eastern Atlantic and the top diagram depicts it over the continent. Dust is evident below the inversion but in lower concentrations.

Fig. 13.33 High clouds over the northern Sahel mixed with red Sahelian dust (photo courtesy of C. Breed).

typical mixing ratio would be 2 g/kg (Diaz *et al.* 1976). This is markedly drier and warmer than conditions in the same region when the SAL is not present. In the layer between 600 and 850 mb, relative humidity is some 25–45% lower and the mixing ratio some 2.5–5.5 g/kg lower in the SAL air than in the non-SAL air (Dunion and Velden 2004). The SAL's unusual characteristics allow it to be readily identified from soundings as far west as the Caribbean, Miami, and eastern Brazil (Prospero 1996). A similar structure is seen during dust outbreaks over the Mediterranean.

The SAL is dome-shaped at the top and bottom. Its base rises and its top sinks westward toward the Caribbean. The layer retains its integrity well into the eastern Caribbean, but it is weaker, elevated, and constricted. Its base is closer to 700 mb/3 km and its top lies closer to 600 mb/4 km. Even this far west

of the African continent, the inversion layers are intact. The lower trade inversion is, however, less intense and commences near 800 mb, instead of 900 mb. The aerosol optical depth in the layer exceeds 1.0 in outbreaks over the continent but is more typically about 0.15 over the Atlantic. Within the layer there is an anticyclonic rotation that draws some of the dust northward then northwestward.

Significantly more detailed information about the SAL was obtained from LiDAR observations collected during the LiDAR In-space Technology Experiment (LITE; Karyampudi *et al.* 1999). Two of the most interesting observations are the existence of two plumes within the layer and the existence of vertical circulations near the AEJ imbedded within the SAL. One of the plumes originates from over the northern Sahara and the other from the Bodélé region of Chad. The vertical circulations are important in determining its interaction with synoptic systems and tropical cyclones (see Section 13.7.5).

13.6.4 LONG-DISTANCE TRANSPORT AND DUST OUTBREAKS

Materials generated at the surface are transported long distances across the oceans from continent to continent. This includes mineral dust, pollutants, and biogenic material from vegetation and biomass burning. Along the way these aerosols are deposited by way of both dry deposition (gravitational settling) and wet deposition (precipitation scavenging). Long-distance transport is the source of most aerosols reaching the oceans. The transport is at higher altitudes, typically in a layer that extends to several kilometers and to 5 or even 6 km on occasion (Prospero 1996).

The major dust sources are apparent from the transport paths shown in Fig. 13.13. Dust from Africa goes to the Atlantic, Europe, and North and South America; on occasion dust from North Africa reaches the Middle East. The United States and South America are additional sources of dust in

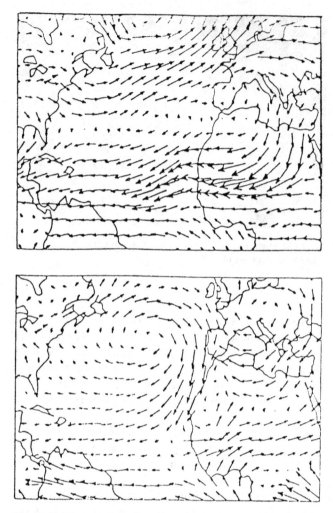

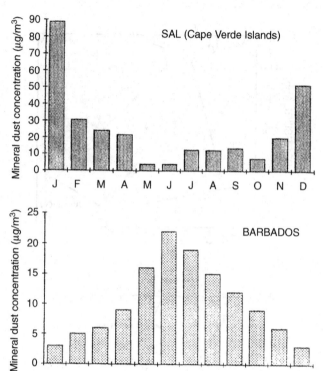

Fig. 13.35 Mineral dust concentration at the surface level: (top) Sal, Cape Verde Islands (1992–94) and (bottom) Barbados, West Indies (1973–92) (from Chiapello *et al.* 1995).

Fig. 13.34 Mean winds in the mid- and upper troposphere, indicating winter (January – top) and summer (July – bottom) circulation patterns over the Atlantic and Euro-African sector (from Chiapello *et al.* 1995).

the Atlantic (Gillette *et al.* 1993). Dust from Asia is transported mainly to the Pacific (Zhang *et al.* 1997). Australian dust makes its way to both the Indian and Pacific Oceans. Dust from the United States is also transported to the Pacific. The prevailing directions of transport are consequences of the prevailing winds, easterly in the tropics and westerly in the mid-latitudes. This is a consequence of the prevailing wind regimes in these regions. Latitudinal transport also occurs, generally in response to the development of synoptic-scale systems.

Dust from West Africa is transported westward by the African easterly jet across the Atlantic to South America and the United States. The transport trajectory varies with the season, as the circulation changes and the AEJ migrates north and south (Fig. 13.34). This source varies greatly from year to year, in parallel with annual rainfall in the semi-arid Sahel. The strongest correlation is with rainfall in the previous year, a factor that determines the erosion potential of the surface

at the end of the subsequent dry season (Prospero and Lamb 2003).

Figures 13.35 and 13.36 illustrate the dust transport. The first of these figures shows the mineral dust concentration at Sal in the Cape Verde Islands, just west of Dakar, Senegal (16°45′ N), and at Barbados (13°10′ N), several thousand kilometers further west. Despite the distance from the source, dust concentration within the Saharan air layer at Barbados can reach 400 µg/m³ during severe outbreaks (Talbot *et al.* 1986). The seasonal cycles are opposite at the two locations, with a maximum in summer at Barbados and in winter at Cape Verde. This is consistent with the winter and summer circulation patterns shown in Fig. 13.34. Figure 13.36 shows the areas of dust fallout over the Atlantic. These similarly illustrate the seasonal migration of the "Saharan plume."

The Saharan dust reaches well into the North and South American continents, into the Great Plains, and even the Amazon. Periodic outbreaks often accompany the summer easterly waves (Jones *et al.* 2003). At Miami, Florida, dust outbreaks lasting several days or longer occur every summer (Prospero 1999). The dust also regularly reaches Tallahassee, Florida, some 900 km further north. The maximum concentration occurs in July, with a monthly mean of between 10 µg/m³ and 100 µg/m³. Similar outbreaks of African dust have also been monitored further west at Fort Myers, Florida. The values of particulates in both cities are high enough to occasionally push

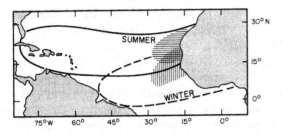

Fig. 13.36 Areal extent of the Saharan dust plume over the Atlantic during winter and summer (from Pye 1987; Schütz 1980, copyright Elsevier). Hatched areas are those of prime dust fallout.

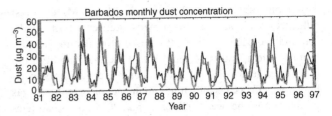

Fig. 13.37 Monthly dust concentrations (μg m^{-3}) in Barbados from January 1981 to December 1996, both observed (gray line) and simulated with the GOCART model (black line) (from Ginoux et al. 2004).

the particulate concentration above the pollution standards set by the Environmental Protection Agency (Prospero et al. 2001). Even as far away as central Illinois, dust deposition episodes of silicon and aluminum can be traced back to the Saharan air layer (Gatz and Prospero 1996).

The arid and semi-arid regions of eastern Asia provide another source for the long-distance transport of mineral dust (Shao and Wang 2003; Qian et al. 2004). The deserts in Mongolia and in western and northern China (the Taklamakan and Badain Juran, respectively) are the major areas of Asian dust storms (Chen et al. 1999; Zhang et al. 2003). The northern part of the Indian subcontinent, including the Tibetan plateau, is also a source region. Dust loading reaches up to more than 500 kg/km^{-2} near the source regions (Zhao et al. 2006).

Unlike the African sources, the Chinese source strengths are relatively constant from year to year, so that variability is linked instead to changes in transport, i.e., weather patterns. However, there has been a long-term increase over the last 100 years, due to anthropogenic changes that degrade the land along the desert margins and that may have expanded the area of desert by about 2–10% (Gong et al. 2006). The dust is transported across the Pacific mainly with the westerlies at elevations of 3–10 km and occasionally produces dust outbreaks in western North America. Notable outbreaks of Asian dust occurred in the Olympic Peninsula of Oregon in 1997 and again in April 1998 (Wilkening et al. 2000). During its transport, deposition shifts from dry processes near the source regions to almost exclusively wet processes over the ocean and western North America.

13.6.5 MODELING AND PREDICTION

A number of models of dust mobilization and transport have been developed for use both in conjunction with and independent of atmospheric general circulation models (AGCM). An excellent review is provided by Ghan and Schwartz (2007). Some of the notable models include a dust tracer within the NASA/GISS AGCM (Tegen and Miller 1998), the Northern Aerosol Regional Climate Model (NARCM) developed by the Meteorological Service of Canada (e.g., Zhao et al. 2006), and the GOCART model of Georgia Tech and NASA Goddard

(Ginoux et al. 2001). The NARCM is a combination of the Canadian Aerosol Model (CAM) and the Canadian Regional Climate Model (RCM) (Gong et al. 2006). The GFDL Coupled Model CM2.1 also includes aerosol distribution and optical depth (Ginoux et al. 2006). All of these models have met with a fair degree of success.

The GOCART (GeorgiaTech/Goddard Ozone Chemistry Aerosol Radiation Transport) model provides an example of their capabilities. It simulates the global distribution of dust, sulfate, carbonaceous, and sea-salt aerosols. Simulations of African dust concentrations at Barbados, in the Caribbean, and at Miami (Fig. 13.37), compare very favorably with observations (Ginoux et al. 2001).

13.7 IMPACTS OF MINERAL DUST

This impact of tropospheric aerosols on climate has been considered for quite some time, mostly in the context of volcanoes and air pollution. Only within the last one or two decades has the global significance of mineral aerosols been fully recognized. The result is a proliferation of national and international programs to study aerosols and their climatic impacts from a combination of field measurements, satellites, and models (e.g., Seinfeld et al. 2004; Diner et al. 2004; Ackerman et al. 2004; Bates et al. 1998).

Atmospheric aerosols include sulfate, black carbon particulates, organic carbon, and sea salt, as well as mineral dust. The effects of aerosols are quite diverse because they depend strongly on physical (e.g., size and shape) and chemical properties. Thus, the effects of mineral dust can be quite different from those of other atmospheric aerosols, especially those resulting from anthropogenic sources. The interaction is complex because some of the component particles primarily absorb radiation and others primarily reflect it.

Mineral dust can modify the characteristics of the atmosphere in several ways. The most obvious is its impact on radiation. Via scattering and absorption, the dust alters the distribution of atmospheric heating, modifies cloud thermal properties, and reduces atmospheric visibility. The dust also modifies atmospheric characteristics by way of its impact on cloud microphysical processes, such as the dust particles' interaction with atmospheric moisture and cloud droplets.

This is one of the less well-understood effects. Mineral dust influences acid rain. It also provides nutrients both to the soil and the ocean. Finally, the dust outbreaks impart conditions of temperature, humidity, and thermal stability from the source region. There is considerable concern about its long-term effects, because it has an impact on greenhouse gas production in the oceans (Jickells *et al.* 2005), on ocean temperatures, and therefore on ocean heat and moisture fluxes. Through its impact on the ocean surface, mineral dust may affect hurricane formation and development (Dunion and Velden 2004).

It has long been known that the African continent produces large quantities of mineral dust. It was, however, assumed that the African dust would have little impact on large-scale climate because of its low light scattering efficiency and low concentration. However, Li *et al.* (1996) have shown that, although the mass scattering efficiency of this mineral dust is only one-quarter that of non–sea salt sulfate, its annual mean concentration over the North Atlantic is 16 times as great. Thus, it is more important in light scattering and in climate.

13.7.1 RADIATIVE EFFECTS

The impact of African dust on the atmosphere's thermal structure is complex because it modifies both the shortwave solar radiation transmitted through to the surface and the longwave infrared radiation emitted to space (Li *et al.* 1996; Tegen *et al.* 1996; Andreae 1996). The dust absorbs and reflects the shortwave radiation and absorbs and emits the longwave. The net effect of these four simultaneous processes depends not only on the characteristics of the dust and its vertical distribution, but also on atmospheric and surface characteristics. Atmospheric stability, cloud cover, convective activity, and surface albedo all modulate the dust's impact. The scattering effect dominates in the visible wavelengths, but evidence suggests that either heating or cooling can result. The dust is an effective absorber of longwave radiation and is likely to evoke a greenhouse-type warming of the atmosphere, at least locally, in the dust layer. This can modify atmospheric dynamics.

Observational studies of the effect of the African dust layer on heat and radiation balance were made during a 1980 field campaign in the Sahel of Niger (the ECLATS experiment). The presence of a dust layer was shown to increase the downward infrared flux to the surface at night and to increase the radiative cooling rate of the atmosphere (Guedalia *et al.* 1984). As a consequence, surface air temperature was on average 3°C cooler during the day and 3°C warmer at night when dust was present than on haze-free nights (Druilhet and Durand 1984). Overall the dust reduced the diurnal temperature range by about 30%. The effect was more pronounced at the surface itself, thereby altering the stability of the boundary layer as well as sensible and latent heat fluxes and convection. The dust also influences the growth of the boundary layer (Goutourbe *et al.* 1997).

Estimates of the overall impact on atmospheric heating were made by Carlson and Benjamin (1980) and by Fouquart *et al.* (1987). Both studies suggested that heating rates for the longwave and shortwave spectrum were on the order of 4–5°K/day. The combined heating for most of the atmosphere beneath the top of the dust layer (500 mb) was about 1°K/day. Carlson and Benjamin found that the maximum heating rates are near the level of the maximum dust concentration (700 mb) and also near the surface beneath the Saharan air layer (i.e., below 900 mb). Fouquart *et al.*, however, found the heating to be relatively evenly distributed within the dust layer, suggesting little impact on vertical structure.

Theoretical calculations of dust's global influence on radiation are generally in accord with observational results. The radiative effect of each particle depends upon its cross-sectional area, so that the smallest particles, which have the longest atmospheric residence times, also have the largest effect per unit area. The net radiative effect of mineral dust is to redistribute the radiative heating from the surface into the dust layer (Miller and Tegen 1998, 1999). This reduces surface heating and enhances heating in the dust layer itself.

Figure 13.38a shows the calculated effect on surface radiation during June, July, and August (JJA) (Miller and Tegen 1998). Most of the impact is over Africa and Asia. There is relatively little influence of the dust aerosols at the top of the atmosphere. The influence is overwhelmingly negative, with local effects as high as −60 W/m². In December–February (not shown) the radiative influence is small (and also negative).

During JJA the dust reduces globally averaged surface net radiation by nearly 3 W/m² on average. This is comparable to the reduction of surface radiation by stratospheric sulfate aerosols one year after the eruption of Mt. Pinatubo (Miller and Tegen 1998). Mineral dust has little influence at the top of the atmosphere (TOA) in contrast to the influence of other types of atmospheric aerosols. The weak thermal absorption at the TOA is balanced by the small amount of reflection of solar radiation.

13.7.2 MODELING THE EFFECTS OF MINERAL DUST

Several general circulation model simulations have evaluated the impact of dust on radiation and the consequences thereof for atmospheric temperature and other phenomena such as cloud formation and rainfall. Figures 13.38b and 13.38c show the impact on temperature and rainfall, as predicted by the NASA/GISS atmospheric GCM. In JJA the largest effects on surface temperature are in the vicinity of the dust cloud. The change beneath the cloud is generally about 0.5°K. The temperature effect is generally small in regions of convection. The impact during the December, January, and February (DJF) season is in roughly the same locations but is much smaller (Miller and Tegen 1998).

(a)

JJA Change in net surface radiation by dust aerosols (Wm^{-2})

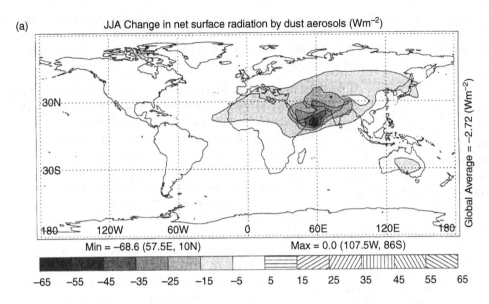

Min = −68.6 (57.5E, 10N)   Max = 0.0 (107.5W, 86S)

−65   −55   −45   −35   −25   −15   −5   5   15   25   35   45   55   65

(b)

JJA change in surface temperature by dust aerosols (K)

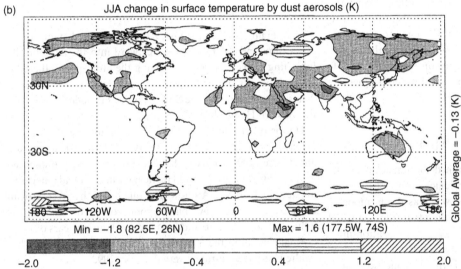

Min = −1.8 (82.5E, 26N)   Max = 1.6 (177.5W, 74S)

−2.0   −1.2   −0.4   0.4   1.2   2.0

(c)

JJA change in precipitation by dust aerosols (mm day^{-1})

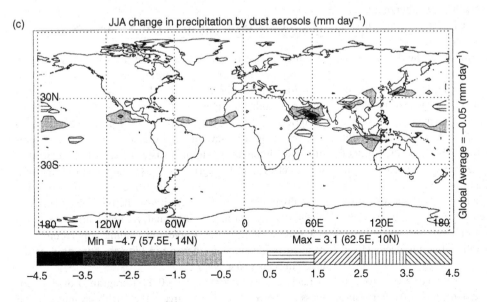

Min = −4.7 (57.5E, 14N)   Max = 3.1 (62.5E, 10N)

−4.5   −3.5   −2.5   −1.5   −0.5   0.5   1.5   2.5   3.5   4.5

Fig. 13.38 Effect of mineral aerosols on climate parameters during June, July, and August (JJA) as predicted by the NASA/GISS atmospheric general circulation model (from Miller and Tegen 1998): (a) net surface radiation (Wm^{-2}), (b) surface temperature (°C), and (c) rainfall.

Model simulations predict little influence on precipitation except in regions encompassed by the dust cloud, and the impact is greater during June–August (Fig. 13.38c) than during December–February. The simulations indicate a general reduction of subsidence, except near the ITCZ (Intertropical Convergence Zone), where ascent is instead reduced. The reduction in precipitation is generally in the vicinity of the ITCZ. The overall effect on tropical circulation is strongly dependent on the optical properties and vertical distribution of the dust (Miller and Tegen 1999). A simulation of the impact of the South Asian haze layer with the Community Climate Model (CCM) suggests a large impact on rainfall within the haze layer (Chung et al. 2002).

Another interesting model result is a negative feedback in the dust–climate interaction. The presence of a dense aerosol layer has an inhibiting effect on the further generation of dust. This may be due in part to the impact of reduced surface temperatures on surface wind. The magnitude of the feedback varies with season and with the optical properties of the aerosol. During JJA the feedback is strongest for dust that strongly absorbs radiation. In DJF it is strongest for more reflective dust (Perlwitz et al. 2001). Overall, the model results underscore the importance of the optical properties of dust in determining its radiative impact and the consequences thereof.

13.7.3 IMPACT OF THE SAHARAN AIR LAYER

Until recently, few studies had examined the impact of West Africa's vast dust layer on meteorological processes in the Sahel. Most of the relevant work was based on models. Various studies showed that regional climate cannot be realistically simulated without introducing dust mobilization into the model. Models of convection and wave development also showed that the dust has an impact on synoptic conditions, and therefore presumably on rainfall as well. The Saharan air layer (SAL) appears to be a critical factor in determining whether an African easterly wave (AEW) develops into a tropical cyclone. This was examined in field studies over the Atlantic during 2006 (Zipser et al. 2009).

Over the continent, the Saharan dust layer is important for the maintenance and possible growth of some wave disturbances (Karyampudi and Carlson 1988). It also acts to strengthen the mid-level African easterly jet (AEJ). A comparison of the atmosphere's thermal structure in the SAL and to the south of it shows that it alters the stability mechanisms and energy conversion processes important in generating waves (Chang 1993). It also alters wave characteristics such as growth rate and wavelength. It also influences the microphysics of clouds (Khain et al. 2005), with dust particles serving both as cloud condensation nuclei (Twohy et al. 2009) and as freezing nuclei (Field et al. 2006).

Over the Atlantic, the SAL tends to suppress tropical cyclone activity (Dunion and Velden 2004). Its pervasive presence throughout the tropical Atlantic may be one reason for the less frequent development of hurricanes in the Atlantic than in the Pacific. The SAL can inhibit tropical cyclones in four ways.

1. By warming the lower troposphere, the dust enhances the pre-existing Atlantic trade wind inversion. This further stabilizes the environment, inhibiting vertical motion and possibly inhibiting the development of convection in weak AEWs.
2. The SAL introduces a dry air intrusion into the cyclone. Although this can promote convection along its western and southern boundary (Chen 1985), the dryness more generally suppresses convection by enhancing evaporatively driven downdrafts (Emanuel 1989; Powell 1990; Khain et al. 2008).
3. The dust layer can reduce Atlantic sea-surface temperatures (Lau and Kim 2007).
4. The strong AEJ associated with it locally enhances the vertical wind shear, a factor that has a destructive influence on tropical cyclones.

This might also help the SAL retain its thermodynamic properties thousands of kilometers westward into the Atlantic. By inhibiting convection and stabilizing the atmosphere, the SAL also inhibits its own modification as it traverses the Atlantic. This is one reason why the SAL maintains its identity well into the Western Hemisphere.

On the other hand, it does not have an adverse affect on all tropical cyclones or hurricanes. Case studies by Karyampudi and Pierce (2002) suggest that the SAL positively influenced Tropical Storm Ernesto and Hurricane Luis, but had a negative impact on Hurricane Andrew. The quality of the West African rainy season might have played some role in determining its influence. Both Ernesto and Luis occurred during years of near-normal rainfall in the Sahel (1994 and 1995, respectively), but Andrew occurred during a very dry year, 1992.

13.8 BIOMASS BURNING IN THE SAVANNAS

Another source of atmospheric aerosols is biomass burning. This includes the burning of living and dead vegetation for land clearing and the burning of wood for heating, cooking, and charcoal production (Fig. 13.39). Burning is a common means of agricultural management that has been practiced for millennia. In the dryland savanna and grassland environments, fires are set to eliminate agricultural waste and stubble after harvesting, to clear land for agricultural use (cultivation or grazing) and shifting cultivation, and to control grass, weeds, and litter. Humans initiate the vast majority of the burning, perhaps as much as 90%; lightning-induced natural fires contribute only a few percent of the total. Rapidly expanding populations have resulted in a significant increase in biomass burning over the last century (Levine 1991).

Each year approximately 8700 Tg (teragrams) of dry biomass is burned. The main source is the earth's savannas, which contribute nearly half the total. Most burning takes place in the savannas that prevail over Africa and South America (Fig. 13.40). The products of biomass burning are the atmospheric trace gases that control its chemistry and heating:

carbon monoxide (CO), methane (CH_4), nonmethane hydrocarbons (NMHCs), and particulate carbon (Levine *et al.* 1995). Burning is now recognized as the source of as much as 40% of gross atmospheric carbon dioxide and 39% of tropospheric ozone.

The AVHRR instrument on the NOAA satellites has been an effective tool in monitoring fires. Figure 13.41 shows the fire count for Africa and South America during September–October 1992. During this season the maximum over Africa is in the Southern Hemisphere in latitudes of 10°–18° S, but a broad band of fire activity extends throughout the savanna from 5° to 20° S. Over South America the maximum is in the savanna of Northeast Brazil's dry zone, but considerable fire activity occurs in the eastern Amazon as well.

(a)

(b)

Fig. 13.39 Anthropogenic contributions to tropospheric chemistry. (a) Wood burning for cooking in highly populous developing countries contributes to atmospheric aerosols. (b) Agricultural burning in the savanna of southern Africa.

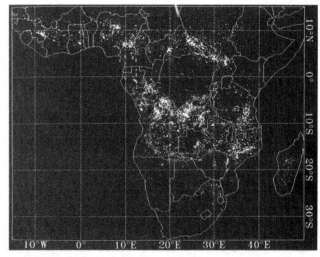

Fig. 13.40 Burning over the African savannas, as seen from nighttime images from the Defense Meteorological Satellite Project (DMSP) (from Cahoon *et al.* 1992). Mapping is based on the entire year 1987.

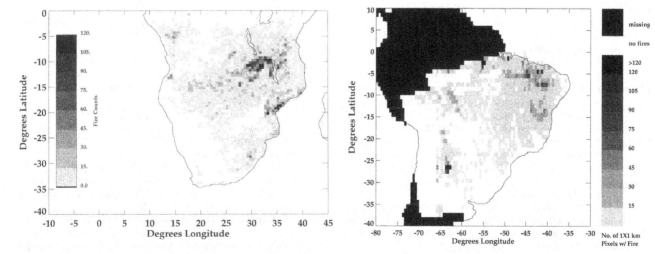

Fig. 13.41 Fire count for (left) Africa and (right) South America during September–October 1992 (from Thompson *et al.* 1996).

REFERENCES

Ackerman, T. P., and Coauthors, 2004: Integrating and interpreting aerosol observations and models within the PARAGON framework. *Bulletin of the American Meteorological Society*, **85**, 1523–1533.

Andreae, M. O., 1996: Raising dust in the greenhouse. *Nature*, **380**, 389–390.

Balme, M., and R. Greeley, 2006: Dust devils on Earth and Mars. *Reviews of Geophysics*, **44**, doi:10.1029/2005RG000188.

Barcilon, A., 1967: A theoretical and experimental model for a dust devil. *Journal of the Atmospheric Sciences*, **24**, 453–466.

Barcilon, A., and P. G. Drazin, 1972: Dust devil formation. *Geophysical and Astrophysical Fluid Dynamics*, **4**, 147–158.

Barry, R. G., and R. J. Chorley, 1999: *Atmosphere, Weather and Climate*. Routledge, London, 409 pp.

Bates, T. S., B. J. Huebert, J. L. Gras, F. B. Griffiths, and P. A. Durkee, 1998: International Global Atmospheric (GAC) project's first aerosol characterization experiment (ACE-1): overview. *Journal of Geophysical Research*, **103**, 16297–16318.

ben Mohamed, A., J. P. Frangi, J. Fontan, and A. Druilhet, 1992: Spatial and temporal variations of atmospheric turbidity and related parameters in Niger. *Journal of Applied Meteorology*, **31**, 1286–1294.

Cahoon, D. R., Jr., B. J. Stocks, J. S. Levine, W. R. Cofer III, and K. P. O'Neill, 1992: Seasonal distribution of African savanna fires. *Nature*, **359**, 812–815.

Carlson, T. N., and S. G. Benjamin, 1980: Radiative heating rates for Saharan dust. *Journal of the Atmospheric Sciences*, **37**, 193–213.

Carlson, T. N., and J. M. Prospero, 1972: The large-scale movement of Saharan air outbreaks over the equatorial North Atlantic. *Journal of Applied Meteorology*, **11**, 283–297.

Castro, R., A. Parés-Sierra, and S. G. Marinone, 2003: Evolution and extension of the Santa Ana winds of February 2002 over the ocean, off California and the Baja California Peninsula. *Ciencias Marinas*, **29**, 275–281.

Cerveny, R. S., 1996: Small-scale or local winds. In *Encyclopedia of Climate and Weather* (S. H. Schneider, ed.), Oxford University Press, New York, pp. 693–697.

Chang, C.-B., 1993: Impact of desert environment on the genesis of African wave disturbances. *Journal of the Atmospheric Sciences*, **50**, 2137–2145.

Chepil, W. S., 1958: *Soil conditions that influence wind erosion*. US Dept. Agri. Tech. Bull., No. 1185.

Chen, W., D. W. Fryrear, and Z. Yang, 1999: Dust fall in the Taklamakan Desert of China. *Physical Geography*, **20**, 189–224.

Chen, Y.-L., 1985: Tropical squall lines over the eastern Atlantic during GATE. *Monthly Weather Review*, **113**, 2015–2022.

Chiapello, I., and C. Moulin, 2002: TOMS and METEOSAT satellite records of the variability of Saharan dust transport over the Atlantic during the last two decades (1979–1997). *Geophysical Research Letters*, **29**, doi:10.1029/2001GL013767.

Chiapello, L., G. Bergametti, F. Dulac, L. Gomes, B. Chatenet, J. Pimeneta, and E. S. Suares, 1995: An additional low layer transport of Sahelian and Saharan dust over the north-eastern tropical Atlantic. *Geophysical Research letters*, **22**, 3191–3174.

Chung, C. E., V. Ramanathan, and J. R. Kiehl, 2002: Effects of the South Asian absorbing haze on the Northeast monsoon and surface–air heat exchange. *Journal of Climate*, **15**, 2462–2476.

Cooke, R. U., and A. Warren, 1975: *Geomorphology of Deserts*. Batsford, London, 394 pp.

Diaz, H. F., T. N. Carlson, and J. M. Prospero, 1976: *A Study of the Structure and Dynamics of the Saharan Air Layer over the Northern Equatorial Atlantic During BOMEX*. NOAA Technical Memorandum, ERL WMPO-32, 61 pp.

Diner, D. J., and Coauthors, 2004: PARAGON: an integrated approach for characterizing aerosol climate impacts and environmental interactions. *Bulletin of the American Meteorological Society*, **85**, 1491–1501.

Druilhet, A., and P. Durand, 1984: Etude de la couche limite convective sahélienne en présence de brumes sèches (Expérience ECLATS). *Boundary-Layer Meteorology*, **28**, 51–77.

Duce, R. A., 1995: Sources, distribution, and fluxes of mineral aerosols and their relationship to climate. In *Dahlem Workshop on Aerosol Forcing of Climate* (R. J. Charlson and J. Heintzenberg, eds.), John Wiley, New York, pp. 43–72.

Duce, R. A., 2005: Aerosols. In *Encyclopedia of World Climates* (J. E. Oliver, ed.), Kluwer, Dordrecht, pp. 4–6.

Dunion, J. P., and C. S. Velden, 2004: The impact of the Saharan air layer on Atlantic tropical cyclone activity. *Bulletin of the American Meteorological Society*, **85**, 353–365.

Emanuel, K. A., 1989: The finite-amplitude nature of tropical cyclogenesis. *Journal of the Atmospheric Sciences*, **46**, 3431–3456.

Engelstaedter, S., K. E. Kohfeld, I. Tegen, and S. P. Harrison, 2003: Controls of dust emissions by vegetation and topographic depressions: an evaluation using dust storm frequency data. *Geophysical Research Letters*, **30**, 1294, doi:10.1029/2002GL016471.

Field, P. R., and Coauthors, 2006: Some ice nucleation characteristics of Asian and Saharan desert dust. *Atmospheric Chemistry and Physics Discussions*, **6**, 2991–3006.

Fitzjarrald, D. E., 1973: A field investigation of dust devils. *Journal of Applied Meteorology.*, **12**, 808–813.

Flower, W. D., 1936: *Sand Devils*. UK Meteorological Office Technical Note 5, No. 71, 16 pp.

Fouquart, Y., B. Bonnell, G. Brogniez, J. C. Buriez, L. Smith, J. J. Morcrette, and A. Cerf, 1987: Observations of Saharan aerosols: results of ECLATS field experiment. II. Broadband radiative characteristics of the aerosols and vertical radiative flux divergence. *Journal of Climate and Applied Meteorology*, **26**, 38–52.

Ganor, E., 1975: Atmospheric dust in Israel: sedimentological and meteorological analysis of dust deposition. PhD thesis, Hebrew University of Jerusalem (in Hebrew).

Gao, H., and R. Washington, 2009: Transport trajectories of dust originating from the Tarim Basin, China. *International Journal of Climatology*, **30**, 291–304.

Gatz, D. F., and J. M. Prospero, 1996: A large silicon aluminum aerosol plume in central Illinois: North African desert dust? *Atmospheric Environment*, **30**, 3789–3799.

Ghan, S. J., and S. E. Schwartz, 2007: Aerosol properties and processes: a path form field and laboratory measurements to global climate models. *Bulletin of the American Meteorological Society*, **88**, 1059–1083.

Gillette, D. A., 1988: Threshold friction velocities for dust production for agricultural soils. *Journal of Geophysical Research*, **93**, D10, 12645–12662.

Gillette, D. A., E. M. Patterson Jr., J. M. Prospero, and M. L. Jackson, 1993: Soil aerosols. In *Aerosol Effects on Climate* (S. G. Jennings, ed.), University of Arizona Press, Tucson, AZ, pp. 73–109.

Ginoux, P., M. Chin, I. Tegen, J. M. Prospero, B. Holben, O. Dubovik, and S. Lin, 2001: Sources and distributions of dust aerosols simulated with the GOCART model. *Journal of Geophysical Research*, **106**, 20255–20273.

Ginoux, P., J. M. Prospero, O. Torres, and M. Chin, 2004: Long-term simulation of global dust distribution with the GOCART model: correlation with North Atlantic Oscillation. *Environmental Modelling and Software*, **19**, 113–128.

Ginoux, P., L. H. Horowitz, V. Ramaswamy, I. V. Geogdzhayev, B. N. Holben, G. Stenchikov, and X. Tie, 2006: Evaluation of aerosol distribution and optical depth in the GFDL coupled model CM2.1 for present climate. *Journal of Geophysical Research*, **111**, D22210, doi:10.1029/2005JD006707.

Gong, S. L., X. Y. Zhang, T. L. Zhao, X. B. Zhang, L. A. Barrie, I. G. McKendry, and C. S. Zhao, 2006: A simulated climatology of Asian dust aerosol and its trans-Pacific transport. Part II. Interannual variability and climate connection. *Journal of Climate*, **19**, 104–122.

Goudie, A. S., 1978: Dust storms and their geomorphological implications. *Journal of Arid Environments*, **1**, 291–310.

Goudie, A. S., and N. J. Middleton, 1992: The changing frequency of dust storms through time. *Climatic Change*, **20**, 197–225.

Goudie, A., and J. Wilkinson, 1980: *The Warm Desert Environment*. Cambridge University Press, New York, 88 pp.

Goutourbe, J. P., J. Noilhan, P. Lacarrere, and I. Braud, 1997: Modelling of the atmospheric column over the Central sites during HAPEX-Sahel. *Journal of Hydrology*, **188/189**, 1017–1039.

Guedelia, D., C. Estournel, and R. Vehil, 1984: Effects of Sahel dust layers upon nocturnal cooling of the atmosphere (ECLATS Experiment). *Journal of Climatology and Applied Meteorology*, **23**, 644–650.

Hall, F. F., 1981: Visibility reductions from soil dust in the western United States. *Atmospheric Environment*, **15**, 1929–1933.

Helgren, D. M., and J. M. Prospero, 1987: Wind velocities associated with dust deflation events in the western Sahara. *Journal of Climate and Applied Meteorology*, **26**, 1147–1151.

Holben, B. N., and Coauthors, 1998: AERONET: A federated instrument network and data archive for aerosol characterization. *Remote Sensing of Environment*, **66**, 1–16.

Holben, B. N., and Coauthors, 2001: An emerging ground-based aerosol climatology: aerosol optical depth from AERONET. *Journal of Geophysical Research*, **106**, 12067–12097.

Houseman, J., 1961: Dust haze in Bahrain. *Meteorological Magazine*, **90**, 50–52.

Humphreys, W. J., 1940: *Physics of the Air*. McGraw-Hill, New York, 676 pp.

Husar, R. B., J. M. Prospero, and L. L. Stowe, 1997: Characterization of tropospheric aerosols over the oceans with the NOAA advanced very high resolution radiometer optical thickness operational product. *Journal of Geophysical Research*, **102**, 16889–16909.

Ives, R. I., 1947: Behavior of the dust devil. *Bulletin of the American Meteorological Society*, **28**, 168–174.

Jennings, S. G., 1993: *Aerosol Effects on Climate*. University of Arizona Press, Tucson, TX.

Jickells, T. D., and Coauthors, 2005: Global iron connections between desert dust, ocean biogeochemistry, and climate. *Science*, **308**, 67–71.

Jones, C., N. Mahowald, and C. Luo, 2003: The role of easterly waves on African desert dust transport. *Journal of Climate*, **16**, 3617–3628.

Joseph, J. H., A. Manes, and D. Ashbel, 1973: Desert aerosols transported by khamsinic depressions and their climatic effects. *Journal of Applied Meteorology*, **12**, 792–797.

Karyampudi, V. M., 1979: A detailed synoptic-scale study of the structure, dynamics, and radiative effects of the Saharan air layer over the eastern tropical Atlantic during GARP Atlantic Tropical Experiment. MS thesis, Department of Meteorology, Pennsylvania State University, 136 pp.

Karyampudi, V. M., and T. N. Carlson, 1988: Analysis and numerical simulations of the Saharan air layer and its effects on easterly wave disturbances. *Journal of the Atmospheric Sciences*, **45**, 3102–3136.

Karyampudi, V. M., and H. F. Pierce, 2002: Synoptic-scale influence of the Saharan air layer on tropical cyclogenesis over the eastern Atlantic. *Monthly Weather Review*, **130**, 3100–3128.

Karyampudi, V. M., and Coauthors, 1999: Validation of the Saharan dust plume conceptual model using Lidar, Meteosat and ECMWF data. *Bulletin of the American Meteorological Society*, **80**, 1045–1075.

Khain, A., D. Rosenfeld, and A. Pokrovsky, 2005: Aerosol impact on the dynamics and microphysics of deep convective clouds. *Quarterly Journal of the Royal Meteorological Society*, **131**, 2639–2663.

Khain, A., P. N. BenMoshe, and A. Pokrovsky, 2008: Factors determining the impact of aerosols on surface precipitation from clouds: an attempt at classification. *Journal of the Atmospheric Sciences*, **65**, 1721–1748.

Lancaster, J., N. Lancaster, and M. K. Seely, 1984: Climate of the central Namib Desert. *Madoqua*, **14**, 5–61.

Lau, K. M., and K. M. Kim, 2007: Cooling of the Atlantic by Saharan dust. *Geophysical Research Letters*, **34**, L23811.

Levine, J. S., 1991. *Global Biomass Burning: Atmospheric, Climatic, and Biospheric Implications*. MIT Press, Cambridge, MA, 569 pp.

Levine, J. S., W. R. Cofer III, D. R. Cahoon Jr., E. L. Winstead, 1995: Biomass burning: a driver for global change. *Environmental Science and Technology*, **29**, 120A–125A.

Li, X., H. Maring, D. Savoie, K. Voss, and J. M. Prospero, 1996: Dominance of mineral dust in aerosol light-scattering in the North Atlantic trade winds. *Nature*, **380**, 416–419.

McTainsh, G. H., R. Burgess, and J. R. Pitblado, 1989: Aridity, drought and dust storms in Australia (1960–84). *Journal Arid Environment*, **16**, 11–22.

Membery, D. A., 1983: Low-level wind profiles during the Gulf Shamal. *Weather*, **38**, 18–24.

Metzger, S. M., 1999: Dust devils as aeolian transport mechanisms in southern Nevada and in the Mars Pathfinder landing site. PhD thesis, University of Nevada, Reno.

Miller, R., and I. Tegen, 1998: Climate response to soil dust aerosols. *Journal of Climate*, **11**, 3247–3267.

Miller, R., and I. Tegen, 1999: Radiative forcing of a tropical direct circulation by soil dust aerosols. *Journal of the Atmospheric Sciences*, **56**, 2403–2433.

Munn, R. E., 1966: *Descriptive Micrometeorology*. Academic Press, New York.

Natsagdorj, L., D. Jugder, and Y. S. Chung, 2003: Analysis of dust storms observed in Mongolia during 1937–1999. *Atmospheric Environment*, **38**, 1401–1411.

N'tchayi Mbourou, G., J. J. Bertrand, and S. E. Nicholson, 1997: The diurnal and seasonal cycles of wind-borne dust over Africa north of the equator. *Journal of Applied Meteorology*, **36**, 868–882.

Perlwitz, J., I. Tegen, and R. L. Miller, 2001: Interactive soil dust aerosol model in the GISS GCM. Part I. Sensitivity of the soil dust cycle to radiative properties of soil dust aerosols. *Journal of Geophysical Research*, **106**, 18167–18192.

Pinker, R. T., G. Idemudia, and T. O Aro, 1994: Characteristic aerosol optical depths during the harmattan season on sub-Sahara Africa. *Geophysical Research Letters*, **21**, 685–688.

Powell, M. D., 1990: Boundary layer structure and dynamics in outer hurricane rainbands. Part II. Downdraft modification and mixed-layer recovery. *Monthly Weather Review*, **118**, 918–938.

Priestley, C. H. B., 1959: *Turbulent Transfer in the Lower Atmosphere*. University of Chicago Press, Chicago.

Prospero, J. M., 1996: Saharan dust transport over the North Atlantic Ocean and Mediterranean. In *Impact of African Dust across the Mediterranean* (S. Guerzoni and R. Chester, eds.), Kluwer, Dordrecht, pp. 133–151.

Prospero, J. M., 1999: Long-term measurements of the transport of African mineral dust to the southeastern United States: implications for regional air quality. *Journal of Geophysical Research*, **104**, 15917–15927.

Prospero, J. M., and P. J. Lamb, 2003: African droughts and dust transport to the Caribbean: climate change implications. *Science*, **302**, 1024–1027.

Prospero, J. M., and R. T. Nees, 1986: Impact of the North African drought and El Niño on mineral dust in the Barbados trade winds. *Nature*, **320**, 735–738.

Prospero, J. M., I. Olmez, and M. Ames, 2001: Al and Fe in PM 2.5 and PM 10 suspended particles in south central Florida: the impact of the long-range transport of African mineral dust. *Water, Air and Soil Pollution*, **125**, 291–317.

Prospero, J. M., P. Ginoux, O. Torres, S. E. Nicholson, and T. E. Gill, 2002: Environmental characterization of global sources of atmospheric soil dust identified with the Nimbus 7 Total Ozone Mapping Spectrometer (TOMS) absorbing aerosol product. *Reviews of Geophysics*, **40**(1), 1002, doi:10.1029/2000RG000095.

Pye, K., 1987: *Aeolian Dust and Dust Deposits*. Academic Press, London, 334 pp.

Pye, K., and H. Tsoar, 1990: *Aeolian Sand and Sand Dunes*. Unwin Hyman, London, 396 pp.

Qian, W. H., X. Tang, and L. S. Quan, 2004: Regional characteristics of dust storms in China. *Atmospheric Environment*, **38**, 4895–4907.

Rennó, N. O., M. L. Burkett, and M. P. Larkin, 1998: A simple thermodynamical theory for dust devils. *Journal of the Atmospheric Sciences*, **55**, 3244–3252.

Saqqa, W., and M. Altallah, 2004: Characterization of the aeolian terrain facies in Wadi Araba Desert, southwestern Jordan. *Geomorphology*, **62**, 63–87.

Schütz, L., 1980: Long range transport of desert dust with special emphasis on the Sahara. *Annals of the New York Academy of Sciences*, **338**, 515–532.

Schweisow, R. L., and R. E. Cupp, 1976: Remote Doppler velocity measurements of atmospheric dust devil vortices. *Applied Optics*, **15**, 1–2.

Seinfeld, J. H., and Coauthors, 2004: ACE-Asia. Regional climatic and atmospheric chemical effects of Asian dust and pollution. *Bulletin of the American Meteorological Society*, **85**, 367–380.

Shao, Y. P., and J. J. Wang, 2003: A climatology of Northeast Asian dust events. *Meteorologische Zeitschrift*, **12**, 187–196.

Sinclair, P. C., 1966: A quantitative analysis of the dust devil. PhD dissertation. The University of Arizona, Nevada, AZ, 292 pp.

Sinclair, P. C., 1969: General characteristics of dust devils. *Journal of Applied Meteorology*, **8**, 32–45.

Sinclair, P. C., 1973: The lower structure of dust devils. *Journal of the Atmospheric Sciences*, **30**, 1599–1619.

Street-Perrott, F. A., and R. A. Perrott, 1993: Holocene vegetation, lake levels and climate of Africa. In *Global Climates since the Last Glacial Maximum* (H. E. Wright *et al.*, eds.), University of Minnesota Press, Minneapolis, Chap. 13, pp. 318–356.

Swap, R., S. Ulanski, M. Cobbett, and M. Garstang, 1996: Temporal and spatial characteristics of Saharan dust outbreaks. *Journal of Geophysical Research*, **101**, 4205–4220.

Talbot, R. W., R. C. Harriss, E. V. Browe, G. L. Gregory, D. I. Sebacher, and S. M. Beck, 1986: Distribution and geochemistry of aerosols in the tropical North Atlantic troposphere: relationship to Saharan dust. *Journal of Geophysical Research*, **91**, 5173–5182.

Tegen, I., and R. Miller, 1998: A general circulation model study on the interannual variability of soil dust aerosol. *Journal of Geophysical Research*, **103**, 25975–25995.

Tegen, I., A. A. Lacis, and I. Fung, 1996: The influence on climate forcing of mineral aerosols from disturbed soils. *Nature*, **380**, 419–422.

Thompson, A.M., and Coauthors, 1996: Ozone over southern Africa during SAFARI-92 TRACE A. *Journal of Geophysical Research*, **101**, 23793–23807.

Torres, O., P. K. Bhartia, J. R. Herman, A. Sinyuk, P. Ginoux, and B. Holben, 2002: A long-term record of aerosol optical depth from TOMS observations and comparison to AERONET measurements. *Journal of the Atmospheric Sciences*, **59**, 398–413.

Tsai, F., G.T-J. Chen, T.-H. Liu, W.D. Lin and J.-Y. Tu, 2008: Characterizing the transport pathways of Asian dust. *Journal Geophysical Research*, **113**, D17311.

Twohy, C. H., and Coauthors, 2009: Saharan dust particles nucleate droplets in eastern Atlantic clouds. *Geophysical Research Letters*, **36**, L01807, doi:10.1029/2008GL035846.

Vehrencamp, J. E., 1951: *An Experimental Investigation of Heat and Momentum Transfer at a Smooth Air-Earth Interface*. Department of Engineering, University of California, Los Angeles.

Washington, R., and M. C. Todd, 2005: Atmospheric controls on mineral dust emission from the Bodélé Depression, Chad: the role of the low-level jet. *Geophysical Research Letters*, **32**, L17701, doi:10.1029/2005GL023597.

Washington, W., M. Todd, N. J. Middleton, and A. S. Goudie, 2003: Dust-storm source areas determined by the total ozone monitoring spectrometer and surface observations. *Annals of the Association of American Geographers*, **93**, 297–313.

Washington, R., and Coauthors, 2006a: Links between topography, wind, deflation, lakes and dust: the case of the Bodélé Depression, Chad. *Geophysical Research Letters*, **33**, L09401, doi:10.1029/2006GL025827.

Washington, R., M. C. Todd, S. Engelstaedter, S. Mbainayel, and F. Mitchell, 2006b: Dust and the low-level circulation over the Bodélé Depression, Chad: observations from BoDEx 2005. *Journal of Geophysical Research*, **111**, D03201, doi:10.1029/2005JD006502.

Wilkening, K. E., L. A. Barrie, and M. Engle, 2000: Trans-Pacific air pollution. *Science*, **290**, 65–67.

Zhang, D. E., 1985: Meteorological characteristics of dust fall in China since historic times. In *Quaternary Geology and Environment of China* (T. S. Liu, ed.). China Ocean Press, Beijing

Zhang, X. R., R. Arimoto, and Z. S. An, 1997: Dust emission from Chinese desert sources linked to variations in atmospheric circulation. *Journal of Geophysical Research*, **102**, 28041–28047.

Zhang, X., S. L. Gong, T. L. Zhao, R. Arimoto, Y. Q. Wang, and Z. J. Zhou, 2003: Sources of Asian dust and role of climate change versus desertification in Asian dust emission. *Geophysical Research Letters*, **30**, doi:10.1029/2003GL018206.

Zhao, T. L., S. L. Gong, X. Y. Zhang, J.-P. Blanchet, I. G. McKendry, and Z. J. Zhou, 2006: A simulated climatology of Asian dust aerosol and its trans-Pacific transport. Part I. Mean climate and validation. *Journal of Climate*, **19**, 88–103.

Zhou, Z. J., and G. C. Zhang, 2003: Typical severe dust storms in northern China during 1954–2002. *Chinese Science Bulletin*, **48**, 2366–2370.

Zipser, E. J., and Coauthors, 2009: The Saharan air layer and the fate of African easterly waves. *Bulletin of the American Meteorological Society*, **90**, 1137–1156.

Part IV The earth's drylands

14 North America

14.1 OVERVIEW OF REGIONAL GEOGRAPHY AND CLIMATE

A number of arid and semi-arid regions are found in North America (Fig. 14.1). The best known are the Sonoran, Mojave, and Chihuahan deserts of the American Southwest and northern Mexico (Fig. 14.2a,b,c); the semi-arid regions along the Pacific coast (which are dealt with in Chapter 20); the grasslands of the Great Plains to the east of the Rockies (Fig. 14.2d,e); and the intermontane Great Basin Desert (Fig. 14.2f). Arid and semi-arid conditions are also found in numerous lee valleys in the western mountains, in lands bordering the western Gulf of Mexico, and in the northwestern Yucatan of Mexico. The three deserts of the Southwest and Mexico are classified as "hot" deserts, that of the Great Basin, a "cold" desert.

The prevailing climates of the contiguous United States and southern Canada are a result of two major geographic factors superimposed on the planetary-scale circulation features of the subtropical and mid-latitudes. The latter include the subtropical highs, the westerlies, and the mid-latitude cyclones associated with them. These produce a pattern of humid climates in northern and eastern North America and relatively dry climates in the West and Southwest. In the tropical sectors far to the south, the trade winds and tropical systems also play a role.

The geographic controls are the north–south oriented high cordillera in the west (Fig. 14.1) and the Great Plains of the interior. These impart several distinctive features of North American climates. The western mountain barrier blocks the Pacific maritime air from most of the interior. Its interaction with the westerly flow creates dry climates in its lee, i.e., in the intermontane basin between the western cordillera (the Sierra Nevada and the Cascade Range) and Rocky Mountains and in the interior east of the Rockies. The absence of an east–west barrier in the interior plains provides an uninterrupted pathway for large latitudinal excursions of both polar and tropical air. It also allows the moisture from the Gulf of Mexico to penetrate far inland.

The most basic consequences of the geographic controls are a very limited development of maritime climates and an extremely continental interior, with great temperature variability from day to day and an extremely high annual temperature range. The aridity of this region, with little reservoir of heat in the ground to dampen temperature fluctuations, enhances the continentality.

The western cordillera is also the primary cause of North America's dryland climates. Some 15% of the continent is classified as arid or semi-arid and most of it lies in the lee of western mountain ranges. Figure 14.3, which shows an east–west transect of surface elevation and rainfall at 40° N, clearly shows the relationship. The arid climates coincide with the low-lying intermontane basins. A secondary cause of the western aridity is the North Pacific subtropical high, the main factor in the aridity of the southwestern deserts.

The basic pressure patterns of winter and summer are shown in Fig. 14.4. At the surface in winter the Aleutian and Icelandic Lows prevail over the Pacific and Atlantic, and high pressure prevails over the continent. Two weak cores of high pressure are apparent, the more western one being part of the Pacific subtropical High. At the surface in summer, the North Pacific and Azores subtropical highs, situated over the relatively cool oceans, are dominant features, while two shallow low-pressure cells are maintained over the Southwest and Great Basin and over northeastern Canada. In the mid-troposphere, the dominant pattern in both seasons is a ridge of high pressure over the Great Basin and Rockies and a trough over eastern North America. The latter appears to be a lee trough, the effect of the mountain barrier on the westerly flow, but strong east–west temperature gradients along the east coast enhance the trough in winter.

The areas of winter cyclogenesis that most directly influence North America are in the North Pacific, along the eastern seaboard, and in the lee of the cordilleran chain. There are three prominent depression tracks across the continent in winter (Fig. 14.5). One is from west to east near 45–50° N, along the

Fig. 14.1 Location of highlands and arid and semi-arid regions of North America.

Canadian border. A second track results from the steering of the large mid-tropospheric trough over the eastern half of the continent: a southward track over the central United States, then turning northeastward further east. A third track is along the eastern seaboard from south to north along the polar front, the mid-tropospheric trough likewise providing the northward steering. In summer all three tracks are displaced northward some 5–10° of latitude. The absence of mid-latitude depression tracks over the Southwest is clearly apparent.

The major air masses that prevail over the North American continent are mainly continental air from the north or maritime air from the Pacific or the Gulf of Mexico (Fig. 14.6). In summer, maritime tropical air masses permeate most of the eastern half of the USA. Maritime polar air affects the northwest and occasionally the upper Great Plains and New England. Continental tropical air prevails in the Southwest. Continental polar air seldom crosses southward across the US/Canadian border in summer. In winter, intensely cold continental Arctic air masses prevail over Canada and affect much of the northern USA and Great Plains. Occasional incursions of maritime polar air stretch across the Rockies into the Midwest or extend from the North Atlantic into New England. Maritime tropical air prevails in the Southwest and Southeast.

The arid regions are clearly evident in the pattern of mean annual rainfall (Fig. 14.7), as is the very pronounced rainfall gradient from the west to the east over the interior Great Plains. The drylands are also evident from the high degree of variability (Fig. 14.8). Rainfall variability is high in all arid and semi-arid regions, but it is particularly so in the American Southwest and Baja California. The coefficient of variation (CV) exceeds 35% in much of the region. The most variable rainfall conditions occur near the boundary between the USA and Mexico east of Baja California and over far southern California and southwestern Arizona. In this region, which is strongly impacted by El Niño, the CV ranges from 40% to over 60%. In contrast, the CV is only on the order of 20–25% in the other North American

desert regions, even though rainfall is lower in those areas than in the areas of coastal California and the Southwest where the variability exceeds 40%.

Despite the arid and semi-arid conditions in the region, there is no strong seasonal concentration of rainfall (Fig. 14.8). In most areas, no more than 40–50% of the precipitation occurs during the wettest quarter of the year. This increases to about 60–70% in southwestern Mexico and the southern extreme of Baja California. In much of the Great Basin and the southwestern deserts, including Baja California, there is no month in which mean rainfall exceeds 25 mm. The number of months in which mean rainfall exceeds 25 mm is between four and six in most of the semi-arid zone, increasing to seven or eight months in the central Great Plains and eastern Mexico.

14.2 PRECIPITATION REGIONS OF NORTH AMERICA

In the North American drylands there are three major precipitation divisions (Fig. 14.9): the predominantly summer precipitation region of the western Great Plains; the Mediterranean winter rains region of southern California and the southwestern deserts; and the intermontane Great Basin with two precipitation seasons, one in winter and one in late spring or summer. This last pattern also characterizes the southern Great Plains.

The configuration of wettest months is more complex (Fig. 14.9). In the interior, there is generally a spring or early summer maximum, progressing northward from March to June. Over the Great Plains, the wettest month is generally May or June. In the areas dominated by the North American monsoon, the wettest month occurs in summer. This is generally August in the Southwest and parts of the Great Basin. Over Mexico the wettest month is quite variable, but it tends to be September in the east and July or August in more western sectors. Over the western United States and northern Baja California, December or January is generally the wettest month. The driest month (not shown) is generally June in the Southwest, shifting to July and August in southern California. Eastward over Arizona, New Mexico, Texas, and Mexico, the driest month is progressively earlier in the year, shifting from May and April to February or January. In the Great Plains it is generally December or February in the north and January elsewhere.

The seasonality is associated with the origin of the precipitation. The warm-season rains of the Great Plains result both from localized convection within the prevailing humid tropical air mass from the Gulf of Mexico and from mesoscale convective systems (MCSs). The winter rainfall of the West Coast is associated with the equatorward migration of the mid-latitude westerlies and cyclones; the region's dry summers are a manifestation of the poleward displacement of the North Pacific subtropical high. The intermontane region is a transition zone between these two regions, so that the infrequent precipitation can be of Pacific or tropical origin. For that reason there are two rainfall maxima in the annual cycle.

Fig. 14.2 Vegetation clearly illustrates the contrasts in the North American deserts. (a) The Joshua tree is an apt symbol of the Mojave. (b) The Sonoran desert is characterized by various cactus species; symbolic of the Sonoran is the tall Saguaro cactus, (c) The Chihuahuan Desert is predominantly shrubland. The Great Plains are predominantly grasslands: (d) short-grass steppe of southern Colorado (Great Sand Dunes in the background), (e) tall grass steppe of the more eastern plains. Vegetation is very diverse in the Great Basin and strongly dependent on soil type. Death Valley (f) is in a transition zone at the western edge of the Basin; the large shrub in the bottom left corner is creosote.

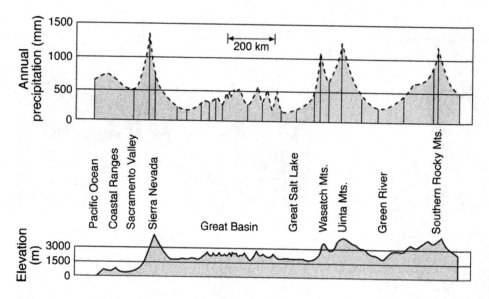

Fig. 14.3 Transect of mean annual precipitation (mm) (top) and elevation (bottom), showing contrast in the rainfall in the mountains, plateaus, and basins (from Bailey 1941).

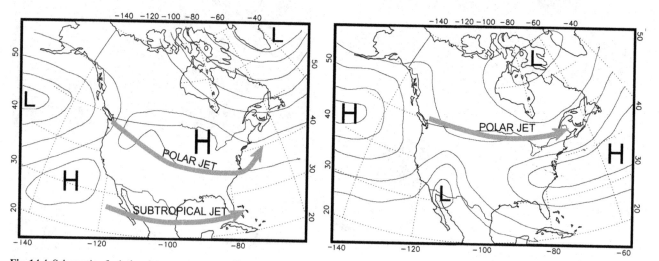

Fig. 14.4 Schematic of wind and pressure systems affecting North America in winter (left) and summer (right) (modified from Trewartha and Horn 1980).

The one in the warm season is due to convective rainfall of tropical or subtropical origin, while that of the cooler season is generally frontal rainfall associated with Pacific moisture and disturbances.

Superimposed on these basic patterns are large spatial variations of the precipitation regime imposed by a spatial hierarchy of controls (Mock 1996). The complexity of controls is such that some stations in mountainous areas of Utah and Colorado have a winter precipitation maximum, while nearby stations have spring and summer maxima (Fig. 14.10). The controls on interannual variability are often different from those prescribing seasonality. An interesting example of this is mesoscale regions of ENSO influence in the mountainous terrain of the northwest, with precipitation anomalies of different sign in very nearby locations (Leung *et al.* 2003). This complexity makes it very difficult to make generalizations about the precipitation

climatology of the western United States and its variability in space and time.

14.3 THE SEMI-ARID INTERIOR: THE AMERICAN GREAT PLAINS

14.3.1 GENERAL GEOGRAPHY AND CLIMATE

East of the Rocky Mountains and extending from 26° N to 52° N is a semi-arid region of short-grass steppe and mixed and tall-grass prairie called the Great Plains (Coupland 1992; Daubenmire 1992; Kucera 1992; Lauenroth and Milchunas 1992). The land slopes eastward from about 1675 m at the foot of the Rocky Mountains to 600 m near its eastern boundary. There are only a few areas of significant relief: the Black Hills, the sandhills of Nebraska, and the badlands of the Dakotas.

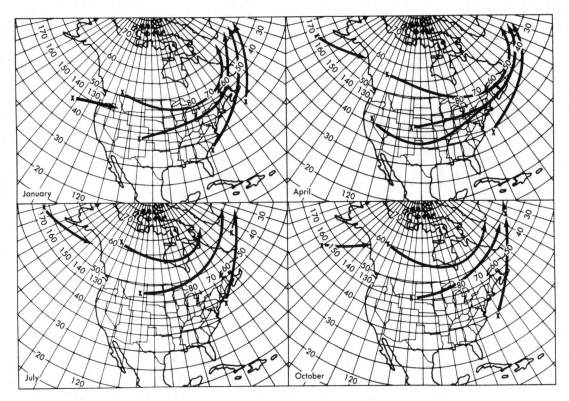

Fig. 14.5 Typical tracks of mid-latitude cyclones over North America during four seasons (from Reitan 1974).

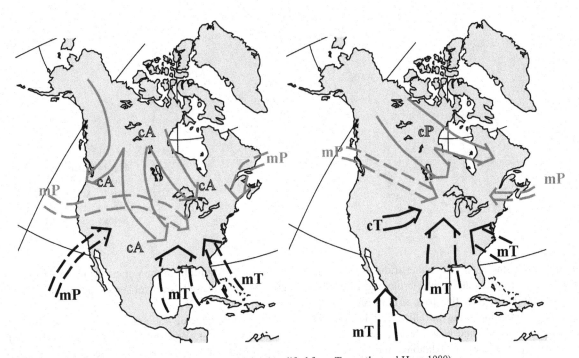

Fig. 14.6 Prevailing air masses in winter (left) and summer (right) (modified from Trewartha and Horn 1980).

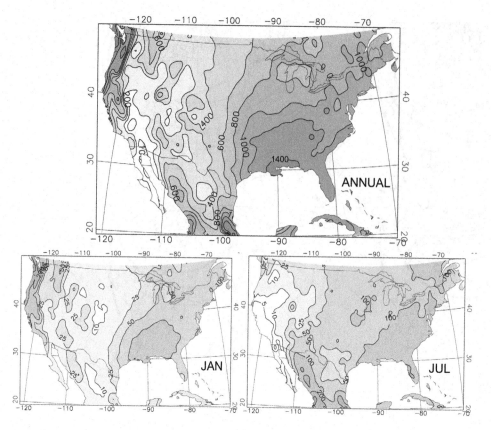

Fig. 14.7 (top) Mean annual precipitation (mm). (bottom) Mean precipitation in January and July.

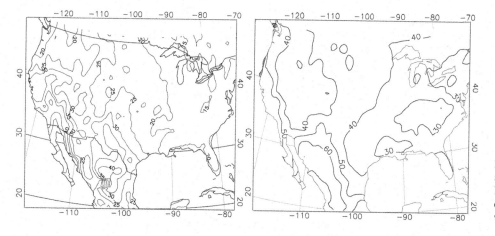

Fig. 14.8 (left) Coefficient of variation (%) of annual rainfall over North America. (right) Percent concentration of rainfall in the wettest quarter of the year.

The soils are mostly mollisols, with some aridisols in the more arid west. These are generally highly fertile and often dark-colored. The steppe occupies the drier, southwestern portion of the plains, commencing at the foothills of the Rockies. This includes eastern Colorado and New Mexico, western Kansas, and the western panhandles of Oklahoma and Texas. To the north and east is the northern mixed prairie; directly to the east is the southern mixed prairie. These are distinguished by the presence of both short- and mid-height grasses. Generally, the more resistant short grasses are dominant during drier periods, while the mid-height grasses abound under more favorable

conditions of rainfall. In the wetter regions to the east are tall-grass prairies, as well as assemblages of forbs and forest species in patches.

The eastern edge of these grasslands, which protrudes eastward into Illinois, Wisconsin, and Indiana, is referred to in some of the literature as the "prairie wedge." This appears to be a vegetation anomaly, situated within surrounding areas of forest to the north, south, and east. This area of prairie was once thought to be anthropogenic, but a number of climatic analyses have revealed coincident patterns of such factors as the intensity of westerly flow, amounts of winter precipitation and snow,

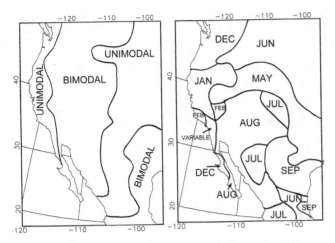

Fig. 14.9 Precipitation seasonality over western North America: (left) areas with unimodal and bimodal seasonal cycles, (right) the wettest month.

frontal positions, the frequency of severe drought, and the frequency of lightning-induced fires. Thus, the origin is clearly climatic (Changnon *et al.* 2002).

The temperature gradient over the Great Plains in winter is one of the strongest in the world (Fig. 14.11). The isotherms run east–west, and mean January temperatures range from −15°C in the Canadian Great Plains to 25°C in southern Texas. This temperature gradient gives much of the character to the region's climate, including the extreme variability of winter temperatures. The summer temperature gradient is much weaker, so that mean July temperatures range from about 20°C in the Great Plains of Canada to 30°C over Texas and Mexico (Fig. 14.11).

The climate is consequently markedly continental. The mean annual range is largely a function of latitude, being between 7° and 12°C in the tropical latitudes of Mexico, 12°–20°C in the

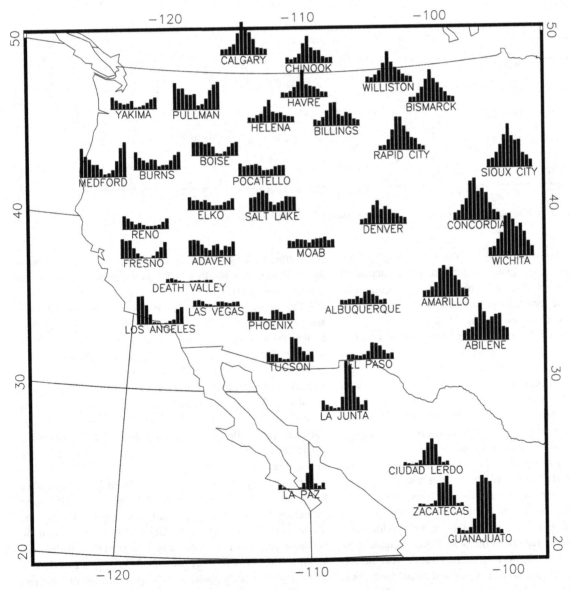

Fig. 14.10 Seasonal cycle of precipitation at select North American stations.

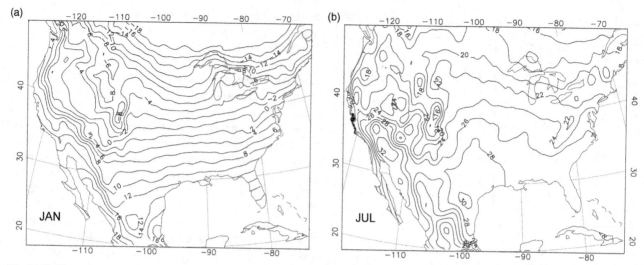

Fig. 14.11 Mean monthly temperatures over North America (°C) (a) January (b) July.

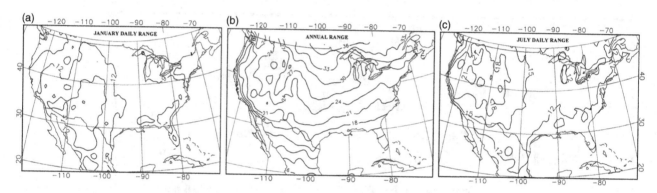

Fig. 14.12 Mean temperature range (°C): (a) diurnal in January, (b) annual, (c) diurnal in July.

subtropical latitudes of the southern plains, and ranging from about 20°C in the central plains to 35°C in the northernmost plains (Fig. 14.12). The diurnal temperature range is quite variable throughout the region, ranging from as low as 10°C in winter to as high as 18°C in summer (Fig. 14.12).

The patterns of wind tend to enhance the regional temperature gradient, particularly in winter, and the annual range. In the southern and central plains of the USA, southerly directions are most prevalent in summer. In winter, the flow tends to be northerly in the northernmost sectors. Surface wind speeds are relatively high throughout the region, generally on the order of 4–8 m/s for much of the year, and the more northern regions tend to experience higher surface winds.

The temperature regime is moderated to some extent by the relatively high cloud amounts prevailing in this region. Typically, monthly cloud cover within the Great Plains ranges from 40–50% during the months with minimum cover to 60–70% during the cloudiest months. However, the region is often cloudless, especially in the drier regions to the west. Relative humidity is generally high in the early morning throughout the year, on the order of 60–80% at most stations. Afternoon humidity shows a strong seasonal cycle, with summer values as low as 20–30% in

the northern plains, 40% in the southern plains. Winter months are more humid, typically 60–70% in the north, 50–60% in the southern and central plains.

Mean annual rainfall, which is concentrated in the warm season, is basically a function of longitude, increasing from about 300 mm/year just east of the mountains near 105° W to about 900 mm on the eastern edge of the plains at 90–95° W (Fig. 14.7). The border between dry and humid climates fluctuates markedly from year to year and lies on average at about 100° W (Fig. 9.8). Snowfall occurs throughout the Great Plains, but its frequency and importance increase with latitude, particularly over the southern plains. From the Texas panhandle northward, a cover of at least 2.5 cm can be expected in two out of three years.

A common occurrence in the Great Plains is severe weather: tornadoes, severe thunderstorms, damaging hail, and strong, damaging winds. Most of the region experiences 40–50 thunderstorm days per year, decreasing to 20–30 per year in the northern regions. Thunderstorms, like rainfall, have a maximum in summer, when they occur on 7–9 days per month. Hail frequency is generally about 2–4 days per year, except in eastern Colorado and Wyoming, where it exceeds 6 days per year. That region of the plains frequently experiences very damaging hail-storms (Changnon 2008).

The area with the most severe weather is centered on the plains of Oklahoma. This area has the greatest frequency of large hail (greater than 19 mm) and the greatest number of tornadoes (from 8 to more than 16 per year on average). Hail frequency reaches 30 days per year in eastern Colorado and western Kansas.

Downbursts and microbursts are also common in parts of the region, particularly along the eastern margins of the Rockies. Severe winds can also occur when deep waves originate over the mountains, bringing fast-moving tropospheric air down to the surface. Such storms in Boulder, Colorado, dubbed "moun-tainadoes," can bring winds in excess of 125 mph (Julian and Julian 1969). One such storm, on January 7, 1969, produced about one million dollars worth of damages. Gusts of 130 mph can be reached within seconds of near-calms.

Severe convective windstorms are also common. The con-vective windstorms include squalls, bow echoes (Davis *et al.* 2004), super-cell thunderstorms, and other strong but less organized convective features. During a 4-year period, two-thirds of the severe, damaging winds over the northern high plains were related to organized convection, rather than to isolated downbursts or microbursts (Klimowski *et al.* 2003). These storms tend to originate at the interface of the Rockies and the Great Plains and have a lifetime of 2–4 hours. The bow echoes, a crescent or bow-shaped line of convection marking the edge of strong wind bursts associated with thunderstorms, were responsible for the largest proportion (29%) of high-wind events. Some 20% were associated with squall lines and 9% with super-cell thunderstorms. Some 86% of all bow ech-oes and 56% of squall lines identified during this period were associated with severe surface winds, including, in some cases, microbursts or downbursts.

14.3.2 FACTORS CONTROLLING ARIDITY AND PRECIPITATION

A number of factors contribute to the overall aridity of the Great Plains. One of these is the lee rain shadow of the Rockies (see Chapter 5). Moreover, the expansive north–south extent of the high terrain effectively blocks the flow of Pacific moist air from penetrating into the region. Another factor in the aridity is the dynamic deformation of the westerlies by the mountains, producing a ridge of persistent high pressure that extends well eastward into the plains.

Precipitation exhibits a marked annual cycle, with the max-imum or maxima coinciding with the warm season and with a single minimum in winter. In southern and western areas, such as New Mexico and Colorado, the maximum occurs in late sum-mer (July–August) and the minimum occurs in mid-winter. The cold season is very dry, with relative humidity as low as 30–40%. There is some trace of a slight maximum in this region in May, a feature of the annual cycle that becomes more conspicuous fur-ther east in Texas and Oklahoma. There the primary maximum

is in September, and a secondary minimum occurs in July and August. This minimum is a consequence of the westward move-ment of the Azores High in July.

Despite the similarities in aridity and warm-season rainfall, the southern and northern Great Plains have notably different precipitation regimes. The most striking contrast is between the May rainfall maximum in the south and the June maxi-mum in the central and northern Great Plains, eastward to Minnesota and Iowa. There are also contrasts in the origin of rainfall. In the southern plains, summer rainfall tends to be of tropical origin, with moisture coming off the Gulf of Mexico. Surges of moisture are associated with the development of a southerly low-level jet stream to the east of the Rockies (see Section 14.3.3). In the central and northern region, rainfall is concentrated in early summer and is generally of extra-tropical origin, associated with the storm tracks imbedded in the westerlies. The June maximum is related to the number of disturbances and their mean storm track. Precipitation drops rapidly in July, when the storm tracks shift northward and fewer disturbances reach as far south as the central and north-ern plains.

Other causes of summer rainfall in the central plains are due to local thunderstorms and MCSs (Ashley *et al.* 2003). These develop in the warm, humid and unstable Gulf air along the southern flank of the Azores High. The MCSs are particularly common in winter and spring, when they bring long-lasting and widespread rainfall and severe weather. These systems last on average about 12 hours and initiate convective activity in the late evening.

14.3.3 THE NOCTURNAL PRECIPITATION MAXIMUM AND THE LOW-LEVEL JET

One peculiar feature of the central Great Plains is a nocturnal rainfall maximum (Fig. 14.13a); 60–65% of the rainfall occurs between about 20:00 and 08:00 hours (Higgins *et al.* 1997a). The region also experiences a high frequency of nocturnal thunderstorms (Pitchford and London 1962). Traditionally, this was considered unusual for a region with convective summer precipitation.

When this nocturnal maximum was first identified, there were several hypotheses concerning its origin. One hypothesis was orographic effects from the Rockies. During the day, a large-scale "mountain"-type circulation prevails (see Chapter 5), with flow from the plains to the mountains and uplift and con-vection over the high terrain, but compensating divergence and subsidence over the plains. The system is reversed at night, with subsidence over the mountains and downward flow to the plains that would produce convergence and uplift over the plains to the east. A related theory, proposed by the late climatologist F. Kenneth Hare, was that the nocturnal rainfall is related to disturbances that are formed in the high terrain during the day but carried eastward to the plains with the downslope flow at night. A similar phenomenon occurs in parts of East Africa (Chapter 16).

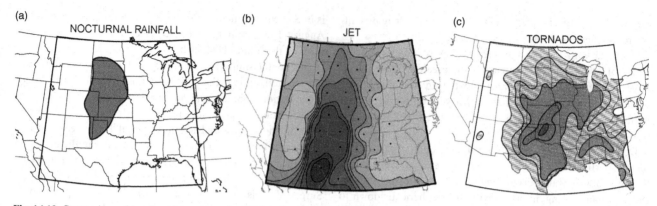

Fig. 14.13 Comparison of low-level jet, nocturnal rainfall and tornado activity: (a) areas with predominantly nocturnal rainfall, (b) frequency of occurrence of a southerly low-level jet, and (c) the location of "tornado alley" (based on Walters *et al.* 2008; Higgins *et al.* 1997a; Eagleman 1980). The shaded area in (a) is where nocturnal rainfall exceeds diurnal by 25% or more.

It is now known that the nocturnal rainfall peak is related to MCSs, which occur with greatest frequency at night. The timing appears to be linked to the nocturnal development of a low-level jet, the axis of which is generally centered over Oklahoma and Kansas, the core region of nocturnal rainfall (Fig. 14.13b) (Augustine and Caracena 1994). As a climatological feature, the Great Plains low-level jet (LLJ), is a southerly to southwesterly flow, commencing around early evening and persisting to early morning. Although it is the mean of LLJs that develop on individual days, a LLJ is evident on about 80% of the soundings taken at 06:00 hours (Bonner 1968). This feature can develop all year round, but occurs most frequently in summer, least frequently in winter (Walters *et al.* 2008).

The axis of the jet also coincides with the plains region described earlier with large hail, severe weather, and frequent tornadoes, termed "tornado alley" (Fig. 14.13c). The jet exhibits a speed maximum between 300 and 1500 m and a magnitude often in excess of 25 m/s. The low-level jet also contributes to instability and tornado formation by providing a constant influx of warm, moist, unstable air from the Gulf of Mexico. Convergence associated with the jet may also help to trigger rain-bearing disturbances (Wang and Chen 2009).

Southerly jets develop over other parts of the central United States, as well. Their characteristics, such as speed and elevation, are regionally distinct (Walters *et al.* 2008). Northerly low-level jets also appear over the Great Plains. However, they also occur much further north. They develop most frequently over Kansas, Nebraska, the Dakotas, and eastern Montana and Wyoming. Occurring most frequently in winter, the northerly low-level jets are often associated with blizzards and other adverse weather conditions.

Both the origin and characteristics of the southerly and northerly jets over central USA are distinctly different (Walters *et al.* 2008). The northerly jets are typically associated with various synoptic situations. The southerly jets tend to be a result of boundary-layer conditions, such as surface thermal gradients (Song *et al.* 2005). Their average speed shows an interesting correlation with the strength of the Azores/Bermuda High (Ting and Wang 2006), indicating significant synoptic influence.

A strong diurnal cycle in speed and elevation is typical of the southerly low-level jets, but not the northerly. The northerly jets are generally larger in scale.

14.3.4 DROUGHT

Long, persistent, and spatially extensive drought episodes are a major characteristic of the North American drylands. An excellent review is found in Woodhouse *et al.* (2009). The most persistent, multi-year droughts generally occur in the Great Plains (particularly west of ~90° W) and over northwest Mexico (Kangas and Brown 2007; Mo and Schemm 2008). The long drought period of the "Dust Bowl" years of the 1930s and a subsequent drought in the 1950s are well known. Over 55% of the United States experienced moderate to extreme drought in 1934, and over 40% of the area was affected during the years 1954–1956 (van der Schrier *et al.* 2006).

Extensive studies of historical drought occurrence, generally based on tree rings, show that severe, multidecadal episodes were an inherent feature in past centuries as well. The historical droughts share the spatial characteristics and causal mechanisms of modern drought. Periods of particularly severe and widespread drought included the end of the sixteenth century and the twelfth century (Meko *et al.* 2007). Some have been more prolonged and severe than any identified in the instrumental record (Stahle *et al.* 2007).

Both the historical and modern analyses underscore two major drought patterns influencing the North American drylands. These are a NW/SW oriented dipole in the western United States and Mexico, with precipitation anomalies of the opposite sign in the Southwest and Northwest, and a similar east–west oriented dipole extending throughout most of the United States (Woodhouse *et al.* 2009). The first pattern is typical of the precipitation response to El Niño and the second is related to extra-tropical factors, such as changes in the position of the jet stream and the upper-atmospheric waves within it. As a consequence of this multiplicity of factors, the characteristics of drought in the region depend on the time and space scales considered (Kangas and Brown 2007).

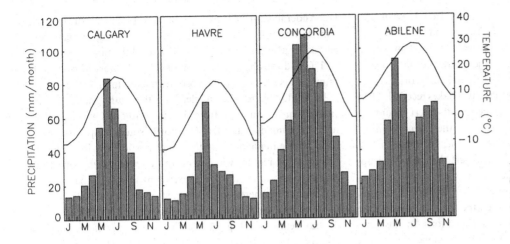

Fig. 14.14 Monthly temperature (°C) and precipitation (mm) at four typical Great Plains stations: Calgary, Havre, Concordia, and Abilene (see Fig. 14.10 for location).

Various studies examining the meteorological origin of Great Plains droughts, as well as floods, have shown that large-scale processes, such as sea-surface temperature (SST) in the Pacific, play a major role. The role of cool and persistent tropical Pacific SSTs in producing periods of extended drought in the Great Plains is very clear (Herweijer *et al.* 2007). However, warm conditions in the Atlantic may reinforce the impact of the Pacific. This was clearly the case in the droughts of the 1930s and the 1950s (Seager *et al.* 2008; Quiring and Goodrich 2008). There is also evidence that abnormally strong westerlies do tend to be associated with widespread drought in the Great Plains and Midwest (Booth *et al.* 2006). This is a long-held climatological axiom. However, numerous counter-examples can be found (Trenberth and Guillemot 1996).

14.3.5 CLIMATIC ELEMENTS AT TYPICAL STATIONS

The expansive Great Plains, extending from southern Canada into Mexico, provide an extreme range of climatic conditions. Several stations are used to show the shifting character from north to south (Fig. 14.14). These include Calgary, Alberta; Havre, Montana; Concordia, Kansas; and Abilene, Texas.

At Calgary, in the Canadian Great Plains, monthly mean temperatures range from −10°C in January to almost 17°C in July. The extremes range from −43°C in January to 36°C in July. Precipitation can occur in any month, but is concentrated in the summer months of May–August. Snow has fallen in every month except July. About one-third of its annual precipitation of 444 mm falls as snow. Cloudiness is high throughout the year, but peaks in spring. It ranges from about 50% in July to 70% in June.

In the northern Great Plains of the USA, conditions are only slightly more moderate. At Havre, monthly temperatures range from −10°C in January to 21°C in July. The extremes range from −49°C in January to 42°C in June and July. Precipitation can occur in any month, but is concentrated in the summer months of May–July. Snow is rare in summer, but it has fallen in every month except July and August. On average, Havre receives about one-third of its annual precipitation of 302 mm as snow.

Cloudiness is high throughout the year, ranging from about 36% in July to 64% in December.

In the central Great Plains, temperatures are markedly warmer in summer and winter. At Concordia the mean temperature of the coldest month, January, is − 2°C; the mean temperature of the warmest, July, is almost 27°C. Snowfall becomes more infrequent, contributing less than 10% of the annual precipitation. Snow is generally limited to the months of October–April. The temperature extremes range from −32°C in January and February to 47°C in August. Precipitation can occur in any month, but is concentrated in the months of April–September. Cloudiness is high throughout the year, ranging from about 40% in late summer and autumn to 60% in winter.

In the southern Great Plains, snow and freezing temperatures are uncommon. At Abilene, the mean temperature of the coldest month, January, is 7°C; the mean temperature of the warmest, July, is over 28°C. Snowfall is generally limited to the months of November–March and accounts for only a small amount of the annual precipitation. The temperature extremes range from −23°C in January to 43°C in August. Precipitation can occur in any month, but is concentrated in the months of April–September. Cloudiness is high throughout the year, ranging from about 40% in late summer and autumn to 60% in winter.

14.4 THE GREAT BASIN

Between the Pacific coast mountains and the Rockies is an extensive intermontane basin with no drainage to the sea. It lies in the rain shadow of both ranges, being on the leeside of the Sierra Nevada in the case of the prevailing westerlies, but leeward of the Rockies when warm, humid air flows northwestward from the Gulf of Mexico. Although referred to as a desert, most parts of this region are classified as steppe or semi-desert. The landscape of the Great Basin was molded by the extensive pluvial lakes that covered the region during the late Pleistocene. Valley floors, with elevations exceeding 1200 m, are interspersed among some 200 mountain ranges that occupy 40% of the region. Most of the rest of the landscape has been formed from loessial, alluvial, or lacustrine fill where pluvial lakes existed.

West (1983a, 1983b) identifies three major ecosystem types in the Great Basin. Their dominance and abundance are strongly dependent on soil type. Many of the soils are halomorphic, remnants of the evaporites left when the pluvial lakes receded. These soils support what is commonly termed a salt-desert shrubland, dominated by shrubs and half-shrubs of the *Chenopodiaceae* family. Where the salinity becomes extreme, the region is virtually devoid of vegetation. The salt-desert shrublands occupy 30% of Utah and 37% of Nevada.

Elsewhere lie the sagebrush semi-desert and the sagebrush steppe, an intermediate type between the shrub desert in drier locations and the grasslands in wetter locations. In the semi-desert *Artemisia* (sagebrush) is dominant, but in the steppe there is equal dominance of *Artemisia* and bunchgrasses. *Artemisia* is favored in more arid microhabitats, dictated by topography or soils. This steppe occupies 24% of Wyoming and over 20% of Idaho and Oregon. Few generalizations can be made about the soils, except that they are not highly saline. Soils types are quite varied in the steppe. In the semi-desert, they grade from aridisols to mollisols and alfisols.

14.4.1 CLIMATIC OVERVIEW

The climate of the Great Basin stands in strong contrast to that of the Great Plains. Although there are some commonalities in the rainfall and temperature regimes, the Great Basin is markedly devoid of severe weather. Although on average 15–25 thunderstorms occur per year, few bring much rainfall. Rainfall exceeds 12 mm generally on only 2–5 days per year. Hail and tornadoes are relatively rare. Large hail events and tornadoes occur less than once per year over most of the region. Fog also occurs infrequently, generally between 3 and 10 days per year in most of the region.

The distinctive character of the region's climate is the cold winters, the warm summers, the high winds, and the extreme annual and diurnal range. The frost-free period ranges from 80 to 120 days. Mean January temperatures are largely a function of elevation and are on the order of 0° to −5°C in the lower elevations (Fig. 14.11). The impact of elevation is evident from a comparison of the stations Boise and Elko in Fig. 14.15. Boise

is considerably further north but has a warmer yet more extreme climate because of its lower elevation. Freezing temperatures can occur at Elko in any month; the absolute minimum there is −21°C and the absolute maximum 40°C. At Boise the absolute minimum is −27°C and the absolute maximum is 44°C.

In winter there also a clear north–south gradient, with January means ranging from 5°C in southern Utah to −10°C near the Canadian border. The north–south temperature gradient is often enhanced by the wind, particularly in winter. In much of the region, winds tend to be southeasterly or northwesterly. The latter are more prevalent in winter, particularly further north. There are, however, strong regional variations.

Mean July temperatures are on the order of 20°–25°C in most of the Great Basin (Fig. 14.11). Thus, the annual range here and elsewhere in the basin is generally 20°–25°C (Fig. 14.12). The diurnal range is equally large in some months, exceeding 25°C in many locations during the summer, when there is intense insolation during the day and intense radiative cooling at night (Fig. 14.12).

14.4.2 THE PRECIPITATION REGIME

Annual precipitation in the intermontane basin is commonly 200–400 mm, except in the arid Southwest (see Section 14.5). High-velocity winds are frequent, promoting high evaporation and thus enhancing the aridity. In summer the cloudiness is as low as 30%, further promoting evaporation. Winter cloudiness can be as high as 75%. The Great Basin lies between the Mediterranean climates of the West Coast and the summer rainfall climates of the Great Plains. Consequently, another distinctive climatic feature is the bimodal character of the seasonal cycle of rainfall.

One maximum occurs consistently in the winter season, the second in the warm season, but varying between spring and late summer. Toward the west, the winter maximum is the primary one, reflecting the closer proximity to the Pacific moisture sources and disturbances. This pattern is apparent at Boise, Elko, and Adaven, typical of the northern, central, and southern Great Basin, respectively (Fig. 14.15). Even further west, in the western cordillera, the warm-season maximum disappears, so that stations like Reno, along the western edge of the Basin, show only a winter precipitation maximum. Further east, at

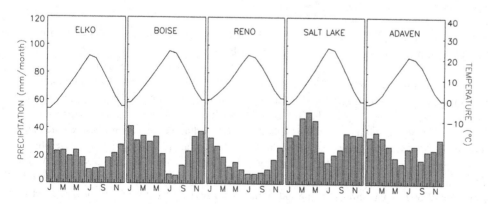

Fig. 14.15 Monthly temperature (°C) and precipitation (mm) at five typical Great Basin stations: Elko, Boise, Reno, Salt Lake City, and Adaven (see Fig. 14.10 for location).

locations such as Salt Lake City, the warm-season maximum associated with air from the Gulf of California and the Gulf of Mexico tends to be the main one. The bimodal distribution is most strongly developed in the far south of the intermontane basin, in the Sonoran Desert. There the primary maximum tends to be July or August, with a secondary maximum in February.

There is also a clear contrast between the precipitation regimes of the northern and southern Great Basin. Going from north to south, there is a shift from a winter to a summer maximum and the contribution of snowfall progressively declines. The timing of the secondary maximum also shifts from June in the north to July–August further south.

The occurrence of the secondary peak in June rather than July–August relates to an abrupt poleward displacement of the North Pacific High during these months, with its center as far north as 40° N. The May–June rainfall is linked to a surface low and an upper-level trough at 500 mb, while the North Pacific High still exerts its aridifying influence in the southern Great Basin, where the summer maximum occurs in July and August.

The winter and summer precipitation in the Great Basin have quite different origins. The winter maximum in the north is a consequence of the position of the jet streams and its associated cyclones. Cyclonic control is stronger and cyclonic activity more frequent with increasing latitude. The rains of July–August are generally linked to the North American monsoon, with the southern Great Basin lying on its northern extreme (Adams and Comrie 1997; Vera *et al.* 2006). In contrast to the frontal precipitation of winter, the rainfall is convective, high intensity, brief, and localized. Summer thunderstorms account for as much as 40–60% of the annual precipitation, but because of their limited spatial extent, they account for only about 15% of the runoff into reservoirs (Trewartha 1970).

14.5 THE MOJAVE, SONORAN, AND CHIHUAHUAN DESERTS

The deserts of the arid Southwest are quite different from those of the Great Basin and the Great Plains. Being further south, these deserts are warmer and they receive less influence from the mid-latitude westerlies and associated disturbances. A consequence of the latter is that winds are generally much weaker

than in the Great Plains, with mean speeds on the order of 2–5 m/s. On the other hand, they experience stronger influences from tropical systems and from the Pacific Ocean.

14.5.1 GENERAL GEOGRAPHY

Three major deserts are delineated in the arid Southwest: the Mojave of southern Utah, Nevada, and California; the Sonoran of Arizona, southeastern California, and northwestern Mexico; and the Chihuahuan of southern New Mexico, southwestern Texas, and north-central Mexico (Fig. 14.1). They are distinguished on the basis of their vegetation (Fig. 14.2), but the underlying climatic factors in vegetation distribution, primarily rainfall and temperature, are equally distinctive.

The annual march of temperature and precipitation for typical stations in these three deserts is shown in Fig. 14.16. Those in the Sonoran Desert are Tucson, Arizona, and La Paz, Mexico, near the southern tip of Baja California. Death Valley represents the Mojave Desert, but its climate is extreme due to the basin topography. El Paso and Ciudad Lerdo represent the Chihuahuan Desert.

The major climatic elements are to a large extent a function of distance from the Pacific coast, major topographic features, and altitude. These are quite diverse among the three deserts. Roughly half of the low-lying Sonoran Desert lies within 100 km of the Pacific Ocean or the Gulf of California. The Mojave and Chihuahuan deserts lie inland and the latter is located on a plateau. The altitudinal extremes range from −86 m in the Mojave to more than 1525 m in the Chihuahuan.

Within the Sonoran Desert there is a marked variation in climate, due in part to its large latitudinal span and its location near the transition between tropical and temperate climates. The contrasting conditions of coastal desert to the south and inland desert to the north accentuate these variations. Conditions are much more uniform within the smaller and more northern Mojave Desert. In the Chihuahuan Desert there is also a relatively large north–south gradient in climate because, like the Sonoran Desert, this region spans more than 10° of latitude.

All three deserts lie within the Basin and Range physiographic province of western North America. A common land surface feature is isolated and short north–south mountains.

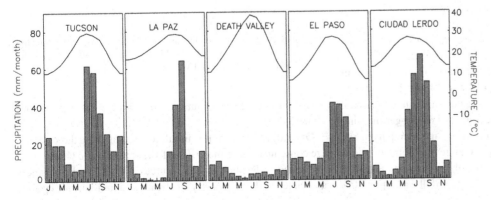

Fig. 14.16 Monthly temperature (°C) and precipitation (mm) at five typical Southwest desert stations: Tucson, La Paz, Death Valley, El Paso, and Ciudad Lerdo (see Fig. 14.10 for location).

The ratio of the desert basin to the total area is generally over 50%, but it is much higher in the low-lying Sonoran Desert, with its coastal location. This mixture of basins and high terrain produces several characteristic surface features of these deserts: the pediments, bajadas or alluvial fans, and playas (see Chapter 2). In recent times, vegetation changes due to both climatic and anthropogenic factors have cut deep arroyos, with some previously insignificant streams being lowered more than 20 m.

The soils of the North American deserts are mainly aridisols, common to most arid locations. The low areas, including channels and playas, tend to include a clay horizon that places them in the suborder termed argids. In some areas, particularly where the soils are calcareous, the argids develop significant areas of hard calcium carbonate, termed "caliche." These are particularly common in the Chihuahuan Desert. The thickness and depth of the caliche vary. Frequently the layer has its surface in the upper meter or so of soil and may range from a few millimeters to 40 cm in thickness. In extreme cases, caliche layers 90 m thick have been found. Saline soils also occur and can form hardpans or crusts. Sand dunes occupy a relatively small part of the North American deserts. Silica dunes are extensive in the northern Sonoran Desert and parts of southern California, such as Death Valley. Rare gypsum dunes (e.g., the white sands of New Mexico) and calcium carbonate dunes are found, and these develop a unique flora.

The Sonoran Desert is the most diverse of the three, containing a variety of cactus types, trees and tall shrubs, and numerous species of perennial shrubs (MacMahon and Wagner 1985). A characteristic vegetation type is the tall Saguaro cactus (Fig 14.2). The Mojave has mostly relatively short perennial shrubs and cacti, though the latter are less diverse than in the Sonoran Desert. Taller (up to 5 m) yuccas, such as the Joshua tree, are found in some regions of higher elevation. Perennial shrubs and cacti characterize the Chihuahuan Desert; yucca and agave species are also common. This desert has a more significant grass cover than the others of the Southwest.

14.5.2 THE PRECIPITATION REGIME

The three desert provinces of the Southwest can to some extent be differentiated on the basis of the seasonality of rainfall. Within this region the percent winter rainfall (November to April) gradually decreases from west to east, while the summer rains (May to October) become increasingly dominant (Fig. 14.17). Consequently, rainfall tends to be concentrated in the winter season in the Mojave Desert (with a weak warm-season peak), to be strongly bimodal (with both winter and summer rains) in the Sonoran, and to occur primarily in summer in the Chihuahuan Desert (Figs. 14.17 and 14.18). These differences manifest themselves in the distribution of annual species: the few annuals in the Mojave are primarily winter-germinating species, both summer and winter annuals are abundant in the Sonoran Desert, and in the Chihuahuan mainly summer annuals are found. An interesting aspect of parts of the Sonoran Desert is a high frequency of nocturnal rainfall with up to 70%

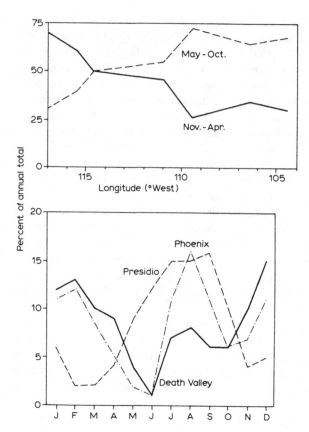

Fig. 14.17 (top) Percentage of mean annual precipitation falling between May and October (dashed line) and between November and April (solid line) along an east–west transect in the southwestern deserts (from MacMahon and Wagner 1985). (bottom) The seasonal cycle at stations typical of the Mojave (Death Valley, CA), Sonoran (Phoenix, AZ), and Chihuahuan (Presidio, TX) deserts.

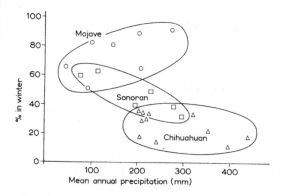

Fig. 14.18 Percentage of winter rainfall plotted against annual precipitation for a variety of sites in the Mojave, Sonoran and Chihuahuan deserts (from MacMahon and Wagner 1985).

occurring between the hours of 20:00 and 08:00 (Adams and Comrie 1997).

Other differences in the rainfall regimes of these three deserts are consequences of the origin of the rainfall. The winter rains are linked to Pacific moisture sources and Pacific disturbances associated with the westerly jet stream and mid-latitude cyclone storms in its path. Rainfall is primarily frontal-type and is therefore relatively widespread, of moderate intensity and generally long duration (several hours to days) (Turner and Hastings 1965). Summer rainfall, which prevails in the Chihuahuan Desert, is part of the so-called "North American monsoon." Convective disturbances develop in the moist unstable air from the Gulf of Mexico and the Gulf of California. Rainfall is generally localized, intense, and of short duration.

The amount of rainfall is more difficult to generalize because the factors that determine it, principally elevation and location with respect to high terrain, are quite diverse and vary greatly within relatively small distances. Within the individual desert provinces, it is mostly a function of elevation. In the Sonoran Desert, annual precipitation typically ranges from 50–100 mm in the far southwest to 300–350 mm further eastward as elevation increases. In the Chihuahuan Desert, there is a similar range but rainfall at a given location is generally somewhat lower than in the Sonoran Desert at a comparable elevation. The Mojave Desert is the most arid, with annual rainfall generally of the order of 50–125 mm/year.

The occurrence of thunderstorms is quite variable. They are relatively common in much of the American portion of all three deserts, with as many as 7 per month during summer at Phoenix, and as many as 15 per year at Las Vegas. They are rare in most of the Mexican regions further south.

Relative humidity shows some systematic contrasts between the three deserts. Monthly means range from about 25% to over 60% in the Chihuahuan Desert, which is relatively humid compared with other interior deserts. In the Sonoran, this ranges from about 20% during the driest summer months to 50% in winter in the areas furthest from the coast. In the coastal desert of Baja California it is quite high, generally 80–90% in the morning but typically 70–80% even in the afternoon. The Mojave Desert is markedly drier: daytime relative humidity ranges from 11% in June to 32% in December. In all three deserts fog is relatively rare, except at higher elevations.

Other aspects of the precipitation regime are fairly similar in the three deserts. Cloudiness is moderate, on the order of 20% in the driest months and 40–50% in wettest months. Snow is relatively rare throughout the region. It does not occur at all in the southernmost sectors. It occurs once every few years in the Mojave Desert, generally in January. In the Chihuahuan Desert, the relatively infrequent winter precipitation events are generally in the form of snow. This occurs when the polar front lies anomalously far to the south.

14.5.3 THE NORTH AMERICAN MONSOON

Over northwest Mexico and the southwestern United States a monsoon system exists during the summer season. This system is analogous to the Asian monsoon in that its origin lies in the seasonal changes of heating over the continent, resulting in a seasonal reversal of wind and pressure patterns (Higgins et al. 2003). The heating occurs over the lowland deserts, creating a thermal low centered over the Arizona/California border.

The North American monsoon (also termed the Arizona monsoon or Mexican monsoon) is responsible for some 60–80% of the annual precipitation in northwest Mexico and 40% of the precipitation in the southwestern USA (Ladwig and Stensrud 2009). It also brings occasional summer rainfall to the Great Basin further north, but there mid-latitude influences become increasingly prevalent. Some of the older literature treats this summer rainfall regime as essentially a spillover from that influencing the Great Plains. This is not the case. The precipitation maximum in the southern Great Plains occurs in May, long before the monsoon season begins over the southwestern USA.

The onset phase of the monsoon occurs in May and June. Locally its onset becomes progressively later with increasing latitude. Over the southwestern USA it commences abruptly in July, with the prevailing weather shifting from hot and dry to cool and rainy (Higgins et al. 1997b). The prevailing winds shift to southeasterly. The monsoon is fully developed during July through early September, when it begins to decay (Vera et al. 2006). In the southwestern USA, maximum monsoonal precipitation occurs during August. During the peak monsoon season in the north (July and August), the southern monsoon regions experience a brief period of reduced rainfall (Higgins et al. 1999).

Several changes in the large-scale atmospheric circulation occur at the onset of the monsoon. A thermal low develops at the surface over the desert Southwest, the North Pacific subtropical high is displaced northward, and the Azores/Bermuda High expands westward and northward. When the summer monsoon rains begin over the Southwest, there is generally a decrease in rainfall over the Great Plains (Douglas et al. 1993; Mock 1996), possibly a consequence of the westward movement of the high.

Low-level moisture in the monsoonal region comes mainly from the Gulf of California, with the Gulf of Mexico contributing moisture in the 700 mb to 200 mb layer. The air from the Gulf of Mexico is blocked at lower levels by the high terrain of north-central Mexico, but is an important moisture source in the mid-troposphere. Moisture transport from the Gulf of California is achieved mainly by way of a low-level, southerly jet with a core near 450 m and with speeds in excess of 7 m/s (Douglas 1995).

In Arizona and New Mexico, about 80% of the precipitation occurs after temperatures in the Gulf of California exceed 28.5°C. The sea-surface temperatures in this region appear to influence the timing, intensity, and extent of the monsoon (Mitchell et al. 2002). The precipitation associated with the monsoon appears to be triggered mainly by surges of the cooler, moist, and unstable Gulf air. The surges may be associated with a variety of synoptic and mesoscale disturbances, including the passage of easterly waves, mesoscale convective systems, and upper-level inverted troughs (Ladwig and Stensrud 2009). The surges are associated with much of the thunderstorm activity in the southwestern USA (McCollum et al. 1995). Further north, on the extreme northern edge of the monsoon, precipitation can be triggered by mid-latitude air mass intrusions (Anderson and

Roads 2002) and by mid-latitude disturbances off the Pacific (Vera *et al.* 2006). Interannual variability is likewise associated with a multitude of factors, such as sea-surface temperature in the Pacific and Gulf of Mexico, the El Niño/Southern Oscillation (ENSO), and land surface conditions (such as winter snowpack).

14.5.4 THE THERMAL REGIME

Temperature throughout the region is mainly a function of elevation (MacMahon and Wagner 1985), although latitude and proximity to the coast play some role. The Sonoran is at the lowest elevation, generally at altitudes below 600 m; consequently it is the warmest of the three deserts; the Chihuahuan is the highest, with about half its area above 1200 m; and the Mojave is intermediate in elevation, with most regions lying between 600 and 1200 m.

Temperature differences between the Mojave and Sonoran are most apparent in the winter season; differences between the Mojave and Chihuahuan are most apparent in summer. In the Sonoran, mean monthly maxima are on the order of 40°C in summer and 20°C in winter; mean monthly minima range from about 4°C in December and January to 24°C in July and August. In the Mojave, typical mean monthly maxima and minima are similar to the Sonoran in the warm season, but winter temperatures are much lower. Winter mean maxima are as low as about 13°C and mean minima approach freezing in December and January.

There is a large diversity of climatic conditions within the Sonoran Desert as a result of varying degrees of continentality. The contrasts are clearly apparent between Phoenix, which lies far inland, and La Paz, which is situated on the coast of a narrow peninsula (Fig. 14.16). At Phoenix the mean temperature of the coldest month, January, is 10.4°C, compared with 18.3°C at La Paz. That of the warmest month is 32.9°C at Phoenix and 29.8°C at La Paz. Thus, the annual range is 23°C at Phoenix, but only 12°C at La Paz. In Phoenix, the temperature extremes range from 48°C in July and September to 10.4°C in January, compared with 40.6° and 30.5°C at La Paz. The diurnal range is generally 8°–12°C at La Paz, compared with 14°–18°C at Phoenix.

The Chihuahuan Desert, being the highest of the three, is also the coolest. Winter temperatures are similar to those of the Mojave, but summer temperatures are cooler than those of the other two deserts. In summer, mean maxima only reach about 35°C, while mean minima are as low as about 20°C. Freezing temperatures can occur in any month from November through April. The region has a very continental climate. As an example, temperature extremes at El Paso range from 43°C in June and July to −21°C in January, giving an absolute range of 64°C. The absolute range is smaller further south.

Monthly means are more uniform across the three deserts. January means are on the order of 5°–10°C in the American Southwest, but on the order of 10°–15°C in Baja California and other drylands of Mexico (Fig. 14.11). Mean July temperatures

are on the order of 20°–30°C throughout the region (Fig. 14.11). The mean annual range is largely a function of latitude, being between 7° and 12°C in the tropical latitudes of Mexico, and 12°–20°C in the American sections (Fig. 14.12). The diurnal range is on the order of 12°C in Baja California and elsewhere in Mexico, but reaches 15°–18°C in the American portions, depending on the season (Fig. 14.12).

14.6 LOCALIZED DRYLAND REGIONS IN NORTH AMERICA

The previous sections describe the major deserts and semi-arid expanses in North America. Elsewhere dry conditions exist as a consequence of topography or coastal effects. These include areas of Washington state, where arid valleys lie within tens of kilometers of wet mountain peaks, coastal areas of Mexico along the western Gulf of Mexico, and the northwestern tip of the Yucatan Peninsula, where coastal effects produce a narrow belt of aridity extending outward into the ocean.

Central Washington is an intermontane region where mean annual rainfall ranges between 200 and 400 mm. In the west there are steep rainfall gradients along the eastern slopes of the Cascade Mountains; mean annual rainfall changes from less than 200 mm near Yakima to over 3000 mm on Mt. Rainier within 100 km of horizontal distance and 4000 m of elevation change. This is an area of predominantly medium-length grasses called the Palouse Prairie. Precipitation occurs during the winter season from mid-September to mid-June and frequently falls in the form of snow. Snow cover can remain in the region for several weeks. Summer temperatures are mild, but winter temperatures can drop to −45°C (Daubenmire 1992).

In northwestern Mexico, in a low-lying area that represents a far south protrusion of the Great Plains, the plains grasslands give way to a semi-arid scrub and thorn forest. This area borders the western Gulf of Mexico and runs from about 25° to 18° N. Rainfall is on the order of 600–800 mm, which is low for an eastern littoral location in the low latitudes. The cause of the relatively dry conditions appears to be related to conditions in summer, when a ridge of high pressure moves into the region. The diffluence of the tropical easterly current over the western Gulf, with one branch turning southward and the other northward, produces subsidence. The concavity of the coastline may further reduce summer rainfall (see Chapter 5). Also, most of the region's rainfall is linked to easterly waves from the Caribbean, and these penetrate less frequently into the western Gulf of Mexico than into areas further east.

Further south is a semi-arid region along the northwestern tip of the Yucatan Peninsula, where mean annual rainfall is below 500 mm at some stations (Trewartha 1970). This region stands in stark contrast to the rest of the peninsula, where annual rainfall is on the order of 1000–1400 mm/year. The meager rainfall is concentrated in the warmer months of May–October. The aridity, which is even greater seaward from the coast, appears to be linked to flow that is offshore or parallel to the shore during the warm season. Frictional divergence as the offshore flow accelerates over the smoother water or as the winds blow along the coast (see Chapter 5) creates subsidence and suppresses rainfall. Relatively cool water probably enhances this effect. There is also a slight reduction in rainfall in mid-summer, probably due to a westward displacement of a prevailing, upper-level trough and a simultaneous northwestward migration of the Subtropical High at the surface.

REFERENCES

Adams, D. K., and A. C. Comrie, 1997: The North American monsoon. *Bulletin of the American Meteorological Society*, **78**, 2197–2213.

Anderson, B. T., and J. O. Roads, 2002: Regional simulation of summertime precipitation over the southwestern United States. *Journal of Climate*, **15**, 3321–3342.

Ashley, W. S., and Coauthors, 2003: Effects of mesoscale convective complex rainfall on the distribution of precipitation in the United States. *Monthly Weather Review*, **131**, 3003–3017.

Augustine, J. A., and F. Caracena, 1994: Lower-tropospheric precursors to nocturnal MCS development over the central United States. *Weather and Forecasting*, **9**, 116–135.

Bailey, R. W., 1941: Climate and settlement of the arid region. In *Climate and Man: Year of Agriculture -1941*. US Dept. of Agriculture, Washington, DC, pp. 188–196.

Bonner, W. D., 1968: Climatology of the low level jet. *Monthly Weather Review*, **96**, 833–850.

Booth, R. K., J. E. Kutzbach, S. C. Hotchkiss, and R. A. Bryson, 2006: A reanalysis of the relationship between strong westerlies and precipitation in the Great Plains and Midwest regions of North America. *Climatic Change*, **76**, 427–441.

Changnon, S., 2008: Temporal and spatial distributions of damaging hail in the continental United States. *Physical Geography*, **29**, 341–350.

Changnon, S. A., K. E. Kunkel, and D. Winstanley, 2002: Climate factors that caused the unique tall grass prairie in the central United States. *Physical Geography*, **23**, 259–280.

Coupland, R. T., 1992: The mixed prairie. In *Natural Grasslands: Introduction and Western Hemisphere* (R. T. Coupland, ed.), Ecosystems of the World 8A, Elsevier, Amsterdam, pp. 151–182.

Daubenmire, R., 1992: The Palouse Prairie. In *Natural Grasslands: Introduction and Western Hemisphere* (R. T. Coupland, ed.), Ecosystems of the World 8A, Elsevier, Amsterdam, pp. 297–312.

Davis, C., and Coauthors, 2004: The bow echo and MCV experiment: observations and opportunities. *Bulletin of the American Meteorological Society*, **85**, 1075–1093.

Douglas, M. W., 1995: The summer low-level jet over the Gulf of California. *Monthly Weather Review*, **123**, 2334–2347.

Douglas, M. W., R. A. Maddox, K. Howard, and S. Reyes, 1993: The Mexican monsoon. *Journal of Climate*, **6**, 1665–1677.

Eagleman, J. R., 1980: *Meteorology: The Atmosphere in Action*. Van Nostrand, New York, 384 pp.

Hasting, J. R., and R. M. Turner, 1965: *The Changing Mile: An Ecological Study of Vegetation Change with Time in the Lower Mile of an Arid and Semiarid Region*. University of Arizona Press, Tucson, TX, 317 pp.

Herweijer, C., R. Seager, E. R. Cook, and J. Emile-Geay, 2007: North American droughts of the last millennium from a gridded network of tree-ring data. *Journal of Climate*, **20**, 1353–1376.

Higgins, R. W., Y. Yao, E. S. Yarosh, J. E. Janowiak, and K. C. Mo, 1997a: Influence of the Great Plains low-level jet on summertime precipitation and moisture transport over the central United States. *Journal of Climate*, **10**, 481–507.

Higgins, R. W., Y. Yao, and X. Wang, 1997b: Influence of the North American monsoon system on the U. S. summer precipitation region. *Journal of Climate*, **10**, 2600–2622.

Higgins, R. W., Y. Chen, and A. V. Douglas, 1999: Interannual variability of the North American warm-season precipitation regime. *Journal of Climate*, **12**, 653–680.

Higgins, R. W., and Coauthors, 2003: Progress in Pan American CLIVAR research: the North American Monsoon System. *Atmosfera*, **16**, 29–65.

Julian, L. T., and P. R. Julian, 1969: Boulder's winds. *Weatherwise*, **22**, 109–112+126.

Kangas, R. S., and T. J. Brown, 2007: Characteristics of U.S. drought and pluvials from a high-resolution spatial dataset. *International Journal of Climatology*, **27**, 1303–1325.

Klimowski, B., M. J. Bunkers, M. R. Hjelmfelt, and J. N. Colvert, 2003: Severe convective windstorms over the northern High Plains of the United States. *Weather and Forecasting*, **18**, 502–519.

Kucera, C. L., 1992: The tall-grass prairie. In *Natural Grasslands: Introduction and Western Hemisphere* (R. T. Coupland, ed.), Ecosystems of the World 8A, Elsevier, Amsterdam, pp. 227–268.

Ladwig, W. C., and D. J. Stensrud, 2009: Relationship between tropical easterly waves and precipitation during the North American monsoon. *Journal of Climate*, **22**, 258–271.

Lauenroth, W. K., and D. G. Milchunas, 1992: The short-grass steppe. In *Natural Grasslands: Introduction and Western Hemisphere* (R. T. Coupland, ed.), Ecosystems of the World 8A, Elsevier, Amsterdam, pp. 183–226.

Leung, L. R., Y. Qian, X. D. Bian, and Hunt, A., 2003: Hydroclimate of the western United States based on observations and regional climate simulation of 1981–2000. Part II. Mesoscale ENSO anomalies. *Journal of Climate*, **16**, 1912–1928.

MacMahon, J. A., and F. A. Wagner, 1985: The Mojave, Sonoran and Chihuahuan deserts of North America. In *Hot Deserts and Arid Shrublands* (M. Evenari, I. Noy-Meir, and D. W. Goodall, eds.), Ecosystems of the World 12A, Elsevier, Amsterdam, pp. 105–202.

McCollum, D. M., R. A. Maddox, and K. W. Howard, 1995: Case study of a severe mesoscale convective system in central Arizona. *Weather and Forecasting*, **10**, 643–665.

Meko, D. M., and Coauthors, 2007: Medieval drought in the upper Colorado River basin. *Geophysical Research Letters*, **34**, doi:10.1029/2007GL029988.

Mitchell, D. L., D. Ivanova, R. Rabin, T. J. Brown, and K. Redmond, 2002: Gulf of California sea surface temperatures and the North American monsoon: mechanistic implication from observations. *Journal of Climate*, **15**, 2261–2281.

Mo, K. C., and J. E. Schemm, 2008: Droughts and persistent wet spells over the United States and Mexico. *Journal of Climate*, **21**, 980–994.

Mock, C. J., 1996: Climatic controls and spatial variations of precipitation in the western United States. *Journal of Climate*, **9**, 1111–1125.

Pitchford, K. L., and J. London, 1962: The low-level jet as related to nocturnal thunderstorms over the Midwest United States. *Journal of Applied Meteorology*, **1**, 43–47.

Quiring, S. M., and G. B. Goodrich, 2008: Nature and causes of the 2002 to 2004 drought in the southwestern United States compared with the historic 1953 to 1957 drought. *Climate Research*, **36**, 41–52.

Reitan, C. H., 1974: Frequencies of cyclones and cyclogenesis for North-America, 1961–1970. *Monthly Weather Review*, **102**, 861–868.

Seager, R., Y. Kushnir, M. F. Ting, M. Cane, N. Naik, and J. Velez, 2008: Would advance knowledge of 1930s SSTs have allowed prediction of the Dust Bowl drought? *Journal of Climate*, **21**, 3261–3281.

Song, J., K. Liao, R. L. Coulter, and B. M. Lesht, 2005: Climatology of the low-level jet at the southern Great Plains atmospheric Boundary Layer Experiments site. *Journal of Applied Meteorology*, **44**, 1593–1606.

Stahle, D. W., F. K. Fye, E. R. Cook, and R. D. Griffin, 2007: Tree-ring reconstructed megadroughts over North America since AD 1300. *Climatic Change*, **83**, 133–149.

Ting, M., and H. Wang, 2006: The role of the North American topography on the maintenance of the Great Plains summer low-level jet. *Journal of the Atmospheric Sciences*, **63**, 1056–1068.

Trenberth, K. E., and C. J. Guillemot, 1996: Physical processes involved in the 1988 drought and 1993 floods in North America. *Journal of Climate*, **9**, 1288–1298.

Trewartha, G.T., 1970: *The Earth's Problem Climates*. University of Wisconsin Press, Madison, WI, 334 pp.

Trewartha, G. T., and L. H. Horn, 1980: *An Introduction to Climate*. McGraw-Hill, New York, 416 pp.

van der Schrier, G., 2006: Summer moisture availability across North America. *Journal of Geophysical Research – Atmospheres*, **111**, doi:10.1029/2005JD006745.

Vera, C., and Coauthors, 2006: Toward a unified view of the American monsoon systems. *Journal of Climate*, **19**, 4977–5000.

Walters, C. K., J. A. Winkler, R. P. Shadbolt, J. van Ravensway, and G. D. Bierly, 2008: A long-term climatology of southerly and northerly low-level jets for the central United States. *Annals of the Association of American Geographers*, **98**, 521–552.

Wang, S.-Y., and T. C. Chen, 2009: The late-spring maximum of rainfall over the U.S. central plains and the role of the low-level jet. *Journal of Climate*, **22**, 4696–4709.

West, N. E., 1983a: Overview of North American temperate deserts and semi-deserts. In *Temperate Deserts and Semi-Deserts* (N. E. West, ed.), Ecosystems of the World 5, Elsevier, Amsterdam, pp. 321–330.

West, N. E., 1983b: Great Basin–Colorado Plateau sagebrush semi-desert. In *Temperate Deserts and Semi-Deserts* (N. E. West, ed.), Ecosystems of the World 5, Elsevier, Amsterdam, pp. 331–350.

Woodhouse, C. A., J. L. Russell, and E. R. Cook, 2009: Two modes of North American drought from instrumental and paleoclimatic data. *Journal of Climate*, **22**, 4336–4347.

15 South America

15.1 OVERVIEW OF SOUTH AMERICAN CLIMATE

South America extends from equatorial to subpolar latitudes and consequently experiences a wide diversity of climate, governed by remarkably complex meteorological phenomena. Most of the continent, however, is within the tropics. On the continent are three arid regions, the Atacama-Peruvian Desert along the west coast and the Monte and Patagonian deserts east of the Andes in the south (Fig. 15.1). A number of other regions are classified as semi-arid, including areas surrounding the Monte and Patagonian deserts, northeastern Brazil, the high-altitude Puna, the woodlands and thorn forest of the Espinal and Chaco, and the northern Caribbean coast of Venezuela and Columbia. Subhumid environments include the Pampas grassland and parts of the Brazilian savanna.

The major characteristics of South American climate are a function not only of the general atmospheric circulation, but also of the steep and narrow Andean cordillera spanning its full north–south extent and of the geographic characteristics of the Southern Hemisphere (see below). The Andes have a strong impact on transient disturbances and low-level atmospheric circulation (Seluchi *et al.* 2006). As in North America, this north–south mountain barrier precludes the strong development of true maritime climates, except in the south. It also produces a mosaic of climates, with isolated dry valleys in enclosed basins only some 100 km away from subtropical forests (Mares *et al.* 1985). Because of the relatively small latitudinal expanse of most of South America, both the Pacific and Atlantic Oceans have a strong influence on the continent. The maritime influence is particularly strong in the mid- and high latitudes.

This influence is clearly reflected in the prevailing thermal regime over South America. Latitudinal gradients and annual and diurnal range are relatively small compared with those over other continents. In the extra-tropical latitudes, mean annual temperatures range from about 5°C at the southern tip near 55° S to about 24°C near 20° S. Figure 15.2 shows mean temperature for July, which is the coldest month over most of the continent,

and January, which is the warmest month south of roughly 20° S and also in parts of northern South America, especially along the Caribbean coast.

In July, temperatures are fairly uniform through most of this region, ranging from 5°C near 45° S to 15°C near 25° S. In January the surface gradient becomes strong, with temperatures increasing from about 10°C near 45° S to 24°C near 35° S. Throughout this region the annual temperature range is only 10°–15°C (Fig. 15.3). Owing to the increasingly strong maritime influence southward, the range is greatest in the mid-latitudes. The diurnal range (Fig. 15.3) increases from the equatorial to the high latitudes in January, ranging from about 9°C to over 15°C. In July, it is greatest in the southern equatorial latitudes, where it reaches 18°C, and in the coastal desert.

The prevailing general circulation features of interest with respect to South American climate are the South Pacific and South Atlantic Highs, the Intertropical Convergence Zone (ITCZ) over the oceans, the equatorial trough over the continent (which merges with the marine ITCZ), and the mid-latitude westerlies and associated disturbances (Figs. 15.4 and 15.5). The subtropical highs are quasi-permanent features that extend very close to the equator and show little seasonal variability. Consequently, the seasonal excursion of the equatorial trough over South America is relatively limited, barely reaching the equator as it recedes from the Northern Hemisphere in the austral summer. There is also a zone of weak and shallow easterlies near the equator that intensify during the boreal summer, and weak remnants of the tropical easterly jet aloft (Fig. 15.6). Though a weak system, expansion of this easterly upper-tropospheric flow during December–February correlates with the seasonal change from dry winter to wet summer (Garreaud *et al.* 2003).

The continent is also influenced by a low-level jet that is part of the South American monsoon system (see Section 15.3.1) and by several coastal jets. These include the southerly Peruvian jet (Lettau 1978), a southerly jet off the coast of central Chile (Garreaud and Muñoz 2005), and the easterly Caribbean low-level jet (Muñoz *et al.* 2008). A mid-level easterly jet in the

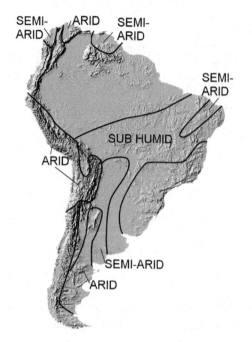

Fig. 15.1 Approximate location of arid and semi-arid regions shown on a relief map of South America.

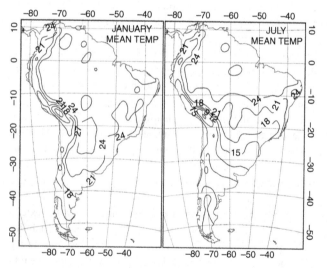

Fig. 15.2 Mean maximum and mean minimum temperature (°C) for January and July.

equatorial Pacific might also influence continental weather patterns (Hastenrath 1999).

The South Pacific High is particularly stable and intense, its eastern flank being fixed in position by the Andes. In contrast, the western flank of the South Atlantic High is weaker and its position less stable, moving westward into the interior of Brazil in the austral winter and seaward during the austral summer. Its core shifts from 15° W, 27° S in August to 5° W, 33° S in

February. The ITCZ occupies its northernmost position in the late austral winter (10° N over the Atlantic, 13° N over the Pacific). As both subtropical highs move away from the continent in summer, the ITCZ moves southward, to 3° N over the Atlantic and 5° N over the Pacific. Its seasonal migration over the Atlantic determines the rainfall over semi-arid Northeast Brazil (Hastenrath 1991).

As a result of the extreme contrast between the Antarctic continent and the polar ocean and the paucity of land in the mid- and subpolar latitudes, the Southern Hemisphere westerly flow is intense and strongly zonal, with little meridional exchange of warm and cold air. As in the Northern Hemisphere, both a polar jet and subtropical jet exist. The latitudinal location of the latter influences both the location of tropical convection (Gonzalez and Barros 1998) and the equatorward penetration of cold fronts (Vera and Vigliarolo 2000).

Only the tip of Chile is far enough poleward to disrupt the strongly zonal westerly flow and it is here, in a corridor between the two subtropical highs, that cold fronts occasionally surge equatorward, affecting the weather of areas east of the Andes and south of the equator (Vera *et al.* 2002). Such transient incursions of mid-latitude air from east of the Andes into subtropical and tropical regions of the continent are a distinctive feature of the synoptic climatology of South America; they are year-round features, but their strongest influence on precipitation occurs in the austral summer.

15.2 THE RAINFALL REGIME

The continental distribution of rainfall is illustrated in Fig. 15.7. The coastal desert and the drylands of the southern latitudes are readily apparent, with rainfall well under 50 mm/year in the former and on the order of 100–300 mm in the latter. In the semi-arid regions of the interior, rainfall is on the order of 300–1200 mm. In the dry zones of Northeast Brazil and the northern Caribbean coast it again drops to values below 1000 mm, reaching as low as 300 mm in some places.

The aridity along the Pacific coast is primarily a function of its location on the eastern equatorial flank of the South Pacific High. Further south, where prevailing westerlies place the region in the lee of the Andes, aridity is primarily a result of rain-shadow effects. However, this effect is probably insufficient to explain the full development of dryland climates over southern South America; other factors, such as the weakly developed cyclonic systems and the presence of anticyclonic flow for much of the year, probably play a role. Explanations for the arid climates elsewhere in South America are more complex. A strong South Atlantic High certainly plays a role in Northeast Brazil (as it does over central and southern South America) but surface divergence and consequent subsidence also play a role. The very small arid region along the Caribbean coast appears, however, to be caused primarily by divergent airflow that results from the frictional stress-differential between the coast and the adjacent sea.

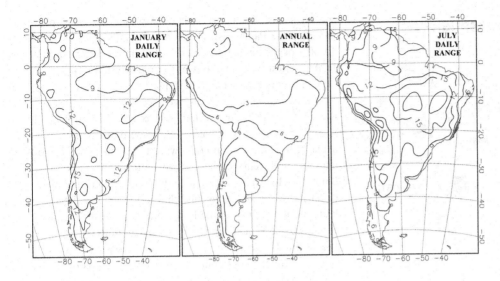

Fig. 15.3 Mean temperature range (°C): (left) diurnal in January, (center) annual, (right) diurnal in July.

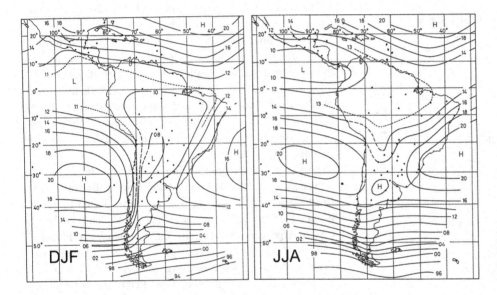

Fig. 15.4 Surface wind and pressure near South America in January (DJF) and July (JJA) (from Schwerdtfeger 1976).

The patterns of rainfall seasonality over South America are extremely diverse (Figs. 15.8 and 15.9) and are not what might be expected from the simple scenario of seasonal displacement of circulation features toward the summer hemisphere (see Chapter 5, Section 5.3). The coastal Atacama-Peruvian Desert (see Chapter 20) west of the cordillera separates summer and winter precipitation regimes, but it is peculiarly extensive. Its southern tip experiences winter precipitation and June is generally the wettest month (Fig. 15.10). There is little seasonal preference from there northward until the summer rainfall regime is reached near 18° S. There the wettest month is February, shifting to March in the latitudes from roughly 14° S to the equator.

To the east of the cordillera there is neither a well-developed dry zone nor a distinct shift from tropical summer rainfall to subtropical winter rainfall. Instead the summer rainfall regime of the equatorial latitudes extends into the subtropical Monte Desert near 30–35° S and surrounding drylands. Over much of eastern South America precipitation reaches its maximum in the transition seasons. The pattern becomes particularly complex in Northeast Brazil. In the dry zone to the north, autumn is the rainy season, with a maximum in March, shifting progressively later toward the eastern tip of the continent. From about 2° S to 18° S the coast has year-round rainfall, but within these latitudes there is a sharp change from autumn rainfall, with a maximum in April or May, to summer rainfall with a December maximum. Further south the maximum shifts to winter, then to autumn in the eastern grasslands, where March is the wettest month. The subpolar Patagonian Desert around 40–50° S has predominantly cold-season rainfall, with May or June generally being the wettest month. However, the pattern is complex, and rainfall occurs during most months.

In much of the world, the patterns of precipitation tend to be pronounced dry seasons in the arid, semi-arid, and subhumid climates, and year-round rainfall in the humid regions. Roughly the opposite is true over South America (Fig. 15.11). In the dryland regions, precipitation is not strongly seasonal except in the

LOWER TROPOSPHERE

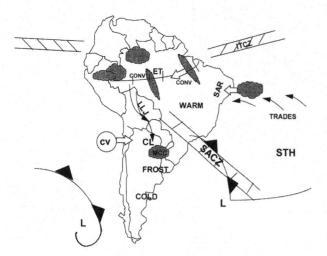

UPPER TROPOSPHERE

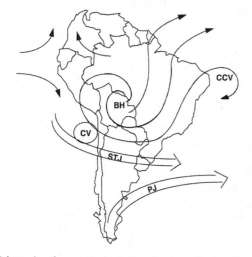

Fig. 15.5 Schematic of general circulation features affecting South America (from Satyamurty *et al.* 1998). CL = Chaco Low; LLJ = low-level jet; ET = equatorial trough; MCC = mesoscale convective complex; CONV = convective activity; STH = subtropical high; EA = extra-tropical anticyclone; L = low-pressure center; CCV = cold-core vortex; CV = cyclonic vortex; STJ = subtropical jet; PJ = polar jet.

coastal desert. In the drylands of the Southern Hemisphere, including the Patagonian Desert and the grasslands, generally 30–40% of the annual precipitation falls in the wettest quarter. In the wettest parts of the Amazon rainforest, 40–50% of the rainfall occurs during the wettest quarter.

Consequently, few months are exceedingly dry. Except in the true deserts, mean precipitation exceeds 25 mm in 10–12 months over most of the continent. The driest month shows a markedly complex pattern, owing to the influence of topography and remarkably complex circulation patterns. In Southern Hemisphere regions with warm-season rainfall, the driest month tends to be July, shifting to June in the higher latitudes and August in the lower latitudes and the northeast. It tends to be September, October, or November in Northeast Brazil, but January, February, or March along the Caribbean coast and in the southernmost latitudes with predominantly winter precipitation.

For most of South America's drylands, the variability of precipitation is low compared with regions on other continents that receive similar amounts of precipitation. This, like the temperature pattern, is a result of the strong, stabilizing maritime influence over much of South America. The coefficient of variation (CV) (Fig. 15.12) lies between 20% and 25% in most areas. However, it exceeds 40% in the dry northeastern part of Brazil, and in the coastal desert it ranges from 40% in the wetter areas on the slopes of the Andes to over 100% in the driest regions in the core of the desert.

These areas where precipitation is most variable are strongly influenced by the Pacific El Niño/Southern Oscillation (ENSO) (see Chapter 25). The areas most strongly affected by ENSO include the Peruvian-Atacama Desert (see Chapter 20), northeastern Argentina, southeastern Brazil, Uruguay and Paraguay, and Northeast Brazil. While ENSO influences rainfall over much of the continent, competing factors in variability include the Pacific Interdecadal Oscillation (PDO) and temperatures in the tropical South Atlantic (Andreoli and Kayano 2006). El Niño episodes tend to reduce rainfall over the continent, while La Niña episodes tend to increase it (Grimm *et al.* 2000). However, the opposite is true for the dryland regions lying just south of Northeast Brazil (see Section 15.3.2).

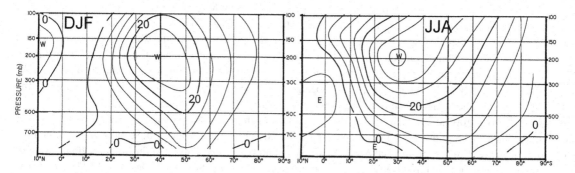

Fig. 15.6 Mean zonal winds (m s⁻¹) in the South American sector in December–February and June–August (from Schwerdtfeger 1976).

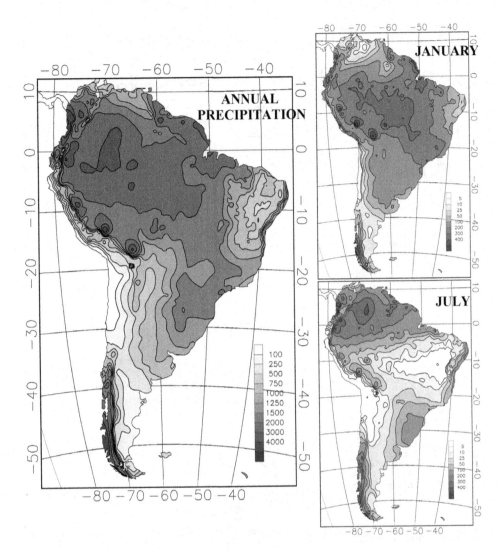

Fig. 15.7 Precipitation over South America: (left) annual mean (mm), (right) monthly means for January and July.

15.3 REGIONAL-SCALE METEOROLOGICAL CIRCULATION SYSTEMS AND CONVECTIVE DISTURBANCES

15.3.1 THE SOUTH AMERICAN MONSOON

Perhaps the most interesting regional-scale feature of South American climate is what is termed the South American monsoon. This system includes many elements common to the North American and Asian monsoons. These include the role of land–sea contrast in evoking a seasonal circulation, a thermally direct circulation with rising motion over the continent, and a progressive advance and retreat of the rainfall linked to the monsoon (Vera *et al.* 2006a). However, many of the typical "monsoonal" elements are lacking, such as a seasonal wind shift and a pronounced seasonal cycle of precipitation in much of the region (Zhou and Lau 1998).

Major elements of the South American monsoon (Fig. 15.6) include the Bolivian High, the South Atlantic Convergence Zone (SACZ), the Chaco Low, the Northeast Brazil (or Nordeste) Low, and a low-level northerly jet east of the Andes. The

Bolivian High at 200 mb is the dominant upper-air feature in the austral summer (Satyamurty *et al.* 1998). Its climatological importance lies in a linked upper-tropospheric system, the Nordeste Low, which is centered just off the coast of Northeast Brazil. The low results when strong southerly flow on the eastern flank of the Bolivian High crosses the equator and becomes cyclonic. Convective activity frequently occurs on the periphery of the low, so that its zonal and meridional movements cause various regions to become excessively wet or dry.

The SACZ is another dominant summer feature (it is usually absent in winter). This is an elongated band of strong cloudiness and precipitation stretching diagonally from northwest to southeast, from the Amazon into the Atlantic (Liebmann *et al.* 2004). The SACZ undergoes very large variations in extent, location, and intensity. It has active and break phases analogous to those of the Asian monsoon (Carvalho *et al.* 2004). When it forms, it dissects the South Atlantic High so that, unlike other subtropical highs, the South Atlantic High actually weakens in summer. The SACZ appears to be associated with an upper-level trough in the westerlies and low-level convergence. This situation, with

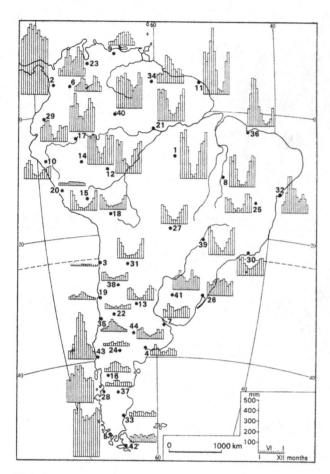

Fig. 15.8 Annual cycle of precipitation at select stations in South America (from Schwerdtfeger 1976).

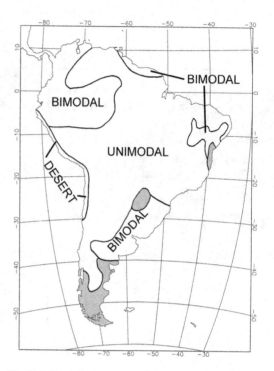

Fig. 15.9 Precipitation seasonality over South America: areas with unimodal and bimodal seasonal cycles are delineated.

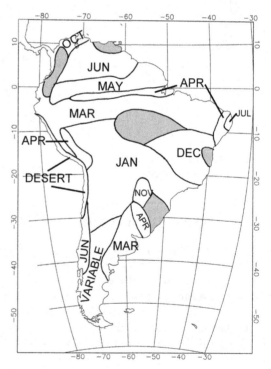

Fig. 15.10 Precipitation seasonality over South America, showing the wettest month of the year.

a westerly trough overriding a surface low, is analogous to that conducive to the formation of diagonal cloud bands over South Africa, so that the low may play a role in the development of the convective activity.

Changes in the Bolivian High and the SACZ are important factors in rainfall variability. The high exhibits strong changes in intensity, as well as latitudinal displacements in response to ENSO. These changes, which are modified by Rossby waves from the Pacific, modulate intraseasonal rainfall variability over central South America (Garreaud *et al.* 2003). Fluctuations of the SACZ and the convection within it are affected by transient wave disturbances, changes in sea-surface temperatures, and cold-frontal incursions from the higher latitudes. Its intensity and location modulate both interannual and intraseasonal variability in the dryland regions of central South America, east of the Andes (Carvalho *et al.* 2004).

Another regional feature over central South America, east of the Andes, is persistent, northerly, low-level flow. This regime is present year-round but it is strongest in the austral spring. On some occasions a low-level jet develops within the northerly flow. This appears to be forced dynamically (as opposed to thermally) by the presence of the Andes and is hence a year-round feature. Termed the South American low-level jet (SALLJ), its

core is at about 850 mb (Fig. 15.13) and typical speeds are on the order of 12–15 m/s (Marengo *et al.* 2004). This jet is responsible for the transport of water vapor and heat to the regions of Paraguay and northern Argentina from the Amazon (Vera

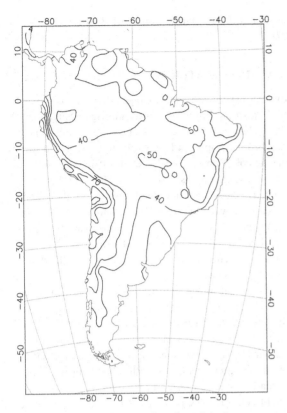

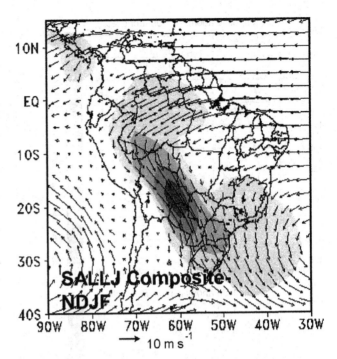

Fig. 15.13 The South American low-level jet (SALLJ) (m s⁻¹) during the warm season (from Marengo *et al.* 2004). This represents a composite of well-developed jet cases. Shading indicates moisture transport.

Fig. 15.11 Degree of seasonality of precipitation: percent concentration of rainfall in the wettest quarter of the year.

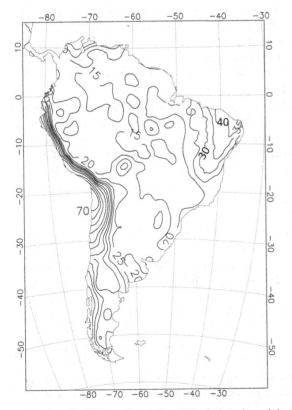

Fig. 15.12 Coefficient of variation (%) of annual precipitation over South America.

et al. 2006b). The SALLJ has been shown to be a critical factor in the development of intense disturbances, such as mesoscale convective systems (see below) and hence affects rainfall (Salio *et al.* 2007). It was also a factor in the extreme heat wave that plagued Argentina during 2002/03 (Cerne *et al.* 2007). On about 17% of summer days the SALLJ extends as far south as the Chaco, around 25–30° S (Nicolini and Saulo 2006). In such cases (termed Chaco low-level jet events) significant rainfall occurs over Uruguay, Paraguay, and northern Argentina.

15.3.2 THE RAINFALL "SEE-SAW"

A very distinctive pattern of intraseasonal variability characterizes a large sector of South America that includes several dryland regions. It takes the form of a rainfall dipole, termed a "see-saw" by Nogués-Paegle and Mo (1997). The dipole is modulated by ENSO, so that it also characterizes interannual variability. El Niño years tend to produce negative rainfall anomalies in the northern half of the monsoon region, but positive anomalies south of 25° S. The dipole represents the interplay between the SACZ and the SALLJ. The northwestern segment of the SACZ is relatively invariant, but the southeastern extension varies markedly in intensity, inversely with the intensity of the jet. Its variations bear some relationship to the Madden–Julian Oscillation.

When the jet is weak (Fig. 15.14), the SACZ is intense and high rainfall occurs over southeastern Brazil. When the jet is strong, the SACZ is anomalously weak and peak rainfall occurs over the La Plata Basin, downstream of the jet.

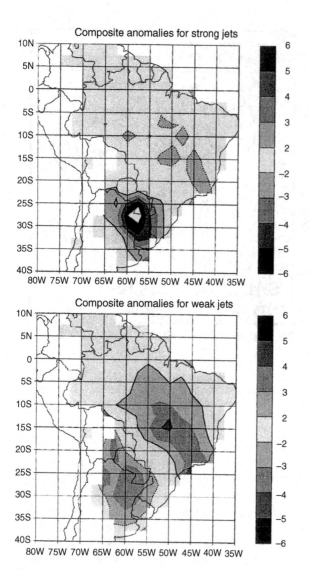

Fig. 15.14 Rainfall dipole for strong and weak jet cases (from Liebmann *et al.* 2004). Rainfall anomalies (mm/day) for the two cases are indicated via shading and contours.

This includes parts of the Pampas, but its margins extend over the Chaco, xerophyll forest, *pampas* and *espinal* regions. Convection in either region is triggered by waves coming off the Pacific.

In the phase of the dipole with a rainfall maximum in the SACZ, rainfall tends to occur during the day. In the opposite phase, with the rainfall maximum downstream of the SALLJ, rainfall tends to be nocturnal (Nogués-Paegle and Vera 2009). Consequently, there is a nocturnal maximum in rainfall over the La Plata Basin and elsewhere in subtropical South America, and a diurnal maximum to the north of it, in southeastern Brazil. The diurnal variability of the jet, which has a late afternoon–evening maximum, might explain the prevalence of nocturnal convection in this region (Zipser *et al.* 2004; Nicolini *et al.* 2004). Moreover, convective systems that occur on days when the jet is well developed tend to have a nocturnal maximum in

precipitation, but those occurring on other days do not (Salio *et al.* 2007).

15.3.3 TRANSIENT FEATURES

South America experiences various types of transient disturbances of both tropical and extra-tropical origin (Satyamurty *et al.* 1998). These influence day-to-day weather and appear on all scales, from synoptic and mesoscale to local. The disturbances include cold fronts, coastal and cut-off lows, mesoscale convective complexes, squalls and linear systems, and convection associated with Kelvin and Rossby waves off the Pacific or African easterly waves (Garreaud and Rutllant 2003; Salio *et al.* 2007; Satyamurty and Seluchi 2007; Siqueira and Marques 2008; Liebmann *et al.* 2009). As over southern Africa, tropical/extra-tropical interaction plays an important role in the development of these systems.

Intense storms that influence other tropical continents, hurricanes or typhoons and tropical cyclones, are rare or absent over South America. The reason for this is that such storms seldom occur in the eastern tropical Pacific or the western South Atlantic. Their absence may be a consequence of the relative coolness of these waters, compared with other tropical oceans. Virulent tornadoes are likewise absent. Severe weather events are often related to cut-off lows, particularly the mid-tropospheric subtropical cold-core vortices that generally lie in the latitudes between the subtropical and tropical westerly jets (Fuenzalida *et al.* 2005). These systems, also called subtropical cyclones, are distinct from tropical cold-core vortices that influence the low latitudes of South America.

Cold fronts (Fig. 15.15) are the most common transient weather event over the continent (Siqueira and Machado 2004). They are often triggered by mid-latitude cyclones from the Pacific. These systems, termed "cold surges," occur in all seasons, but their strongest influence is on summer precipitation (Garreaud and Wallace 1998). They occur in all latitude bands, but are most frequent between 35° S and 40° S. In winter, they often trigger severe weather events in southern parts of South America (Vera and Vigliarolo 2000). They are responsible for a large part of the rainfall in northern Argentina, Uruguay, Paraguay, southern and western Brazil, Bolivia, and southern Peru. The cold fronts or their remnants produce a significant proportion of the rainfall in Northeast Brazil during the months of November through February and May through July. The cold fronts cause intense frost south of 30° S and occasional, moderate frost in parts of Brazil up to 20° N, where they generally dissipate. One particularly intense variant of the cold front, termed the *friagem*, occurs in winter (Marengo *et al.* 2002). It penetrates across the equator and can cause dew-point temperature to drop by 10°C in one day, so that the *friagem* is extremely dry as well as cold.

Another common, transient disturbance in South America is the mesoscale convective system (MCS). Several factors appear to operate in tandem to produce them: a mountain breeze in an unstable atmosphere, the SALLJ, and the upper-level subtropical

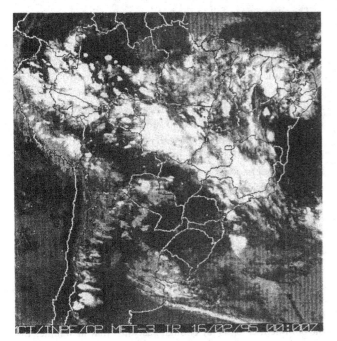

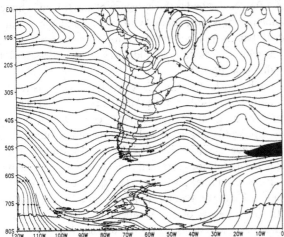

Fig. 15.15 Cold front, as evident in infrared satellite imagery (top) and corresponding 250 mb streamline analysis (bottom) for February 16, 1995 (from Satyamurty *et al.* 1998).

MCSs can develop suddenly and are particularly common in the months of November–April. The SALLJ appears to play a large role in organizing mesoscale convective systems in its exit region over the La Plata Basin (Zipser *et al.* 2004; Salio *et al.* 2007). This basin, in fact, experiences some of the highest frequencies of MCSs worldwide (Zipser *et al.* 2006). On 41% of the days when the jet is present, at least one MCS develops, compared with 12% of the days when the jet is not present (Salio *et al.* 2007). Those linked to the jet also tend to have a nocturnal precipitation maximum.

Mesoscale and small-scale systems embedded in the trade winds approach the northern parts of the coast of Northeast Brazil from the equatorial Atlantic (Satyamurty *et al.* 1998). These do not have quite the same characteristics as Caribbean easterly waves, but they are important features of day-to-day weather. The coasts are also influenced by a number of convective systems, including squalls. The trades over the Atlantic significantly influence precipitation along the eastern coasts, but their influence varies seasonally, depending on the orientation of the coast with respect to the direction of the trades.

15.4 THE DRYLAND BIOMES OF SOUTHERN SOUTH AMERICA

The dryland biomes of southern South America, stretching from 22° S to nearly 50° S, include most of Argentina and small parts of Bolivia, Paraguay, Uruguay, and Brazil (Fig. 15.16). Far to the west, the extreme coastal desert all falls within the northern sector of this latitudinal zone; this is discussed in Chapter 20. There are three other distinct arid biomes, all semi-desert regions of Argentina, and three somewhat more humid biomes of xerophyllous forest, woodland, and savanna grassland to the north and east. Far to the south, in the high temperate latitudes, are small areas of grass steppe. These biomes are discussed individually in the remainder of this chapter, using the stations shown in Fig. 15.17 to illustrate the climatic conditions within them.

15.4.1 ARID BIOMES

The three arid biomes are the Patagonian, the Puna, and the Monte semi-deserts. The most arid, the Monte, is a subtropical region in north-central Argentina that is predominantly scrubland. To the north and east, semi-arid xerophyllous thorn forests and woodlands separate the Monte from the subhumid savanna grasslands further east. To the west is the Puna, a high-altitude semi-desert extending some 10° of latitude north–south at elevations of 3400–4500 m ASL in the Andes. This area is predominantly a high Andean grass and shrub steppe, but many of the dominant genera are also abundant in Patagonia. Patagonia is a cold semi-desert region that is predominantly grassland or scrub; it lies south of the Monte and extends from the Andes to the coast.

jet. The juxtaposition of the two jets provides shear instability to trigger convective activity, while the moisture transport out of the Amazon via the SALLJ provides moisture for the formation and maintenance of cloud and precipitation complexes. The MCSs move from their source region in northern Argentina and Paraguay to affect southwestern Brazil and Uruguay with intense rainfall. They bring around 70–90% of the rainfall in regions influenced by the SACZ and SALLJ, while accounting for only about 10% of all rainfall events (Nesbitt *et al.* 2006; Salio *et al.* 2007). A larger and less frequently occurring class of system, the mesoscale convective complex (MCC), is another important feature of this region. These systems, extending over at least 50,000 km², account for as much as 60% of the rainfall in parts of this region (Durkee *et al.* 2009).

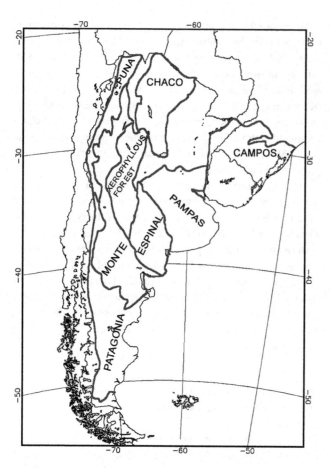

Fig. 15.16 The arid and semi-arid biomes of southern South America.

The main reason for the region's aridity is its location in the lee rain shadow east of the Andes. The westerly current loses its moisture in the higher elevations, then descends, diverges, and warms adiabatically in the lee, increasing dynamic stability and further reducing relative humidity. As in the western USA, the mountain barrier also deforms the westerlies, producing a quasi-permanent ridge over the mountains and in the lee, but a trough further eastward. This ridge further suppresses rainfall.

From north to south across this region (Fig. 15.10), the precipitation regime changes from predominantly summer rainfall to winter rain and snowfall. Within the Monte Desert most stations receive mainly summer rainfall. Annual means range between 85 and 200 mm, with 250–300 mm at a few locations (Mares *et al.* 1985). The driest area is the north, where the Monte blends into the Atacama Desert of Chile. Patagonia receives mainly winter precipitation and strong winds all year round. Mean annual rainfall throughout the region is about 200–300 mm. It is roughly separated from the Monte Desert by the 13°C mean annual isotherm. Frost generally occurs on at least 250 days per year and snowfall is common. The Puna is also a cold, windy semi-desert but at higher elevations and lower latitudes than Patagonia. This region receives both summer and winter precipitation, with a high frequency of frost, winter snow and summer hail.

The summer rainfall to the north is convective in nature and linked to the subtropical regimes further north. Most of the rainfall occurs in a few heavy showers. The lee trough of the Andes, which affects this region in summer, is conducive to convection. At this time of year the western flank of the subtropical high steers low-level moisture into the region from the Atlantic. In winter, anticyclonic circulation, including the western flank of the South Atlantic High, acts to suppress rainfall. In the south, where winter rains prevail, rainfall is predominantly frontal and linked to the westerlies. The reason for the winter maximum is not well understood, but it is probably at least partly a result of weaker cyclonic activity in summer, when pressure and temperature gradients are weaker.

There is a pronounced east–west rainfall gradient in this region (Fig. 15.7). In the subtropical latitudes of the Monte Desert and the savanna grasslands to the east, rainfall is lowest just east of the Andes Mountains and increases toward the Atlantic coast, with semi-arid then subhumid climates nearer the coast. The reason for the reversal appears to relate to the ridge over the Andes and the trough further east. In these latitudes, the continent extends far enough eastward that in summer the coast often comes under the influence of the trough. That circulation regime is more conducive to rainfall and intensification of disturbances from either the subtropics or the polar regions. Interestingly, in these same latitudes west of the Andes, a winter rainfall regime prevails.

In the higher latitudes of Patagonia the gradient is reversed and rainfall increases from the Atlantic coast inland. The reasons for this gradient are not completely clear. One factor might be upslope winds associated with easterly flow during the relatively infrequent cyclone passages, although the centers of these systems are usually too far from the mountains to generate much upslope flow. Another possibility is a "spillover" of Pacific disturbances that are orographically lifted by the Andes (Trewartha 1970). The intensity of the Southern Hemisphere westerlies, which steer these systems, makes this a likely possibility.

Temperature patterns are determined primarily by latitude (Fig. 15.18), although the maritime and cordilleran influence on temperatures is apparent, especially in summer. Gradients are relatively weak in summer, especially in the subtropical and mid-latitudes to about 40° S. The gradients intensify in winter, especially in the lower latitudes. Compared with other dryland regions, relative humidity is fairly high throughout the Monte-Patagonian deserts, owing to the widespread maritime influence. Except for the northernmost sectors, relative humidity is highest in the winter.

THE MONTE

The Monte Desert is a floristically rich semi-desert shrub steppe with a large diversity of vegetation. A common species is "*jarillal*" or creosote. The Monte has numerous distinctive habitats and plant associations, including communities both of large cacti and of low shrubs. The cacti are usually on the higher slopes, the shrubs at lower elevations. The northern Monte,

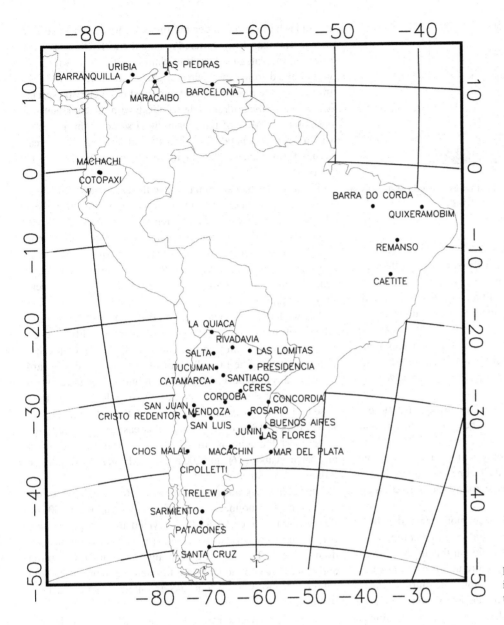

Fig. 15.17 Map of stations used to illustrate temperature and precipitation in the various biomes.

with a summer rainy season but occasional winter rains, has a more complex vegetation cover. However, it lacks both the summer and winter ephemerals found in other deserts with bimodal rainfall. The colder southern regions have mainly low shrubs. The region's soils range from clayey to rocky; clayey and sandy soils are common at lower elevations, especially in riparian gullies. There are only scattered and small dune systems, except for one area of large, moving dunes in the Salta Province.

Changes in the vegetation cover provide evidence of recent climatic change. In certain areas of the northern Monte Desert, wide bands of mostly dead cacti are located well below their current altitudinal limit. This change would have occurred within the last 50–100 years (Mares et al. 1985).

The Monte Desert consists of numerous enclosed basins surrounded by mountains; each is on the order of 30–50 km and has an arid nucleus of less than 100 mm rainfall annually, compared

with over 300 mm/year on the surrounding highlands. The rainfall gradient within such basins can be large. In the Andalgala Basin, for example, rainfall changes from 77 to 344 mm/year within a distance of 50 km and an elevation change of less than 600 m (Mares et al. 1985). To the north, roughly 60% of the annual rainfall occurs during the summer months and only 3% or less in winter. In central sectors, this changes to about 40% in summer and 10% in winter. More Mediterranean-type winter precipitation occurs further south and even exceeds summer precipitation at the southernmost stations, but even there the precipitation regime shows no clear winter peak and summer minimum. Thunderstorms occur on roughly 15–20 days per year in the north, but 5 or fewer as the summer rainfall regime fades toward the south. Fog occurs on only a few days per year.

The rainfall regimes of the northern, central, and southern Monte are illustrated in Fig. 15.18 with the stations at San Juan,

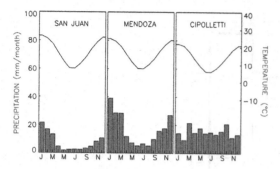

Fig. 15.18 Precipitation (mm) and temperature (°C) at three stations in the Monte Desert: San Juan, Mendoza, and Cipoletti.

Mendoza, and Cipoletti. At San Juan (31° S), the driest station with 87 mm/year, rainfall occurs mainly in summer, and the winter months are almost completely dry. At Mendoza (33° S), with mean annual rainfall of 197 mm, rainfall occurs in all months, but there is a clear minimum in winter. At both stations, January is the wettest month, June and July the driest. At Cipoletti (in the transition to the Patagonian Desert at 39° S) mean annual rainfall is 176 mm and there is neither a well-defined rainy season nor a dry season. However, May and October receive nearly twice as much rainfall as the other months.

The wind regime follows the general latitudinal pattern of global climate in most parts, except where mountains interrupt the terrain and impose regional-scale circulations. Thus, in the southern Monte, winds are generally westerly, but they become southerly in the far north. Hot dry foehn winds often descend from the cordillera in summer, desiccating the vegetation and drying the surface.

The temperature regime is relatively moderate and uniform (Fig. 15.18). Monthly mean temperatures in summer range from about 20° to 22°C in the south to 26°C in the north. Monthly means for the coldest months are more uniform, being generally about 6°–8°C throughout the region. The diurnal range is generally somewhat greater in summer than in winter and it is also greater in the higher latitudes. Typical values in summer are 11°–19°C, in winter 10°–17°C, but the temperature can rise to 25° or 26°C in the driest regions. The annual range is about 16°–18°C. Mean relative humidity is on the order of 50–60%, ranging from around 40–50% in the summer to about 60–75% in the winter. Monthly average cloud cover ranges around 30–50%, with greater cloudiness in winter. In the more arid regions, such as San Juan, it is as low as 20–30%.

PATAGONIA

In Patagonia there are desert and semi-desert biomes (Soriano *et al.* 1983). These are bounded to the south by a grass steppe where mean annual rainfall exceeds 300 mm, to the north by tropical forest, and to the east by savanna grasslands. The dryland biomes of Patagonia are generally grasslands or scrub. In the west, near the slopes of the high cordillera, the dominant plant form is tussock grasses, especially *Stipa* species. Low

perennial herbs grow under the tussocks, with occasional low shrubs (less than 1 m) scattered among them. To the east in the drier central plateau are cushion-plant growth forms, which are well adapted to a windy climate; grasses are rare. In the central region, the most dominant and most characteristic species is *Nassauvia glomerulosa*, a dwarf shrub with the characteristic cushion form. Most of this area is dwarf scrub or shrub steppe, with a cover of generally less than 40%, although in some areas shrubs form thickets. Halophytic vegetation is found in some parts.

The latitudes of Patagonia receive the full force of the westerlies. At the surface, these are the dominant winds throughout the year; their persistence and strength is the most characteristic feature of the Patagonian climate. The wind speed is relatively high, averaging 5–8 m/s. Potential evapotranspiration in the region is high as a result of the winds rather than warm temperatures. Frosts brought by incursions of polar air occur all year round.

The westerlies originate over the Pacific and are very moist when they reach the Andes, but lose their humidity in these mountains, which can receive as much as 6000 mm of precipitation annually. The rainfall gradient from the Andes toward the central semi-desert is therefore extreme: along an eastward transect at 41° S, annual rainfall drops from 4000 mm at their base to 1000 mm over a distance of less than 50 km. It drops to less than 300 mm only 75 km further east. From there to the coast, rainfall conditions are relatively uniform, varying generally between about 160 and 200 mm over the next 300 km to the Atlantic (Soriano *et al.* 1983).

The vegetation gradient from the Andes to the Atlantic is correspondingly pronounced. Several forest zones lie above 1000 m. Below this to the east begin the dryland biomes, progressing from scrub to a narrow steppe zone with tussock grasses, to dwarf shrubs. The soils are generally coarse-textured; gravel and stones are commonly found through the depth of the soil column. Regosols and solonchaks are interspersed throughout the region.

Patagonia is affected by transient cyclones of both polar and subtropical origin. Those from higher latitudes are relatively infrequent, occurring on average twice in summer and three times in winter. They tend to be steered from SW to NE by the ridge over the Andes and are weakened by the prevailing subsidence. Those from the subtropics are steered by the trough further east, moving southeastward and generally intensifying in the trough. The combined result of these systems is that Patagonia is a transition zone with a complex, nearly year-round (but meager) rainfall regime. The rainfall regime at several typical stations is shown in Fig. 15.19. Of the stations shown, only Chos Malal, on the eastern slopes of the Andes, has a definite winter maximum and summer minimum.

Other characteristics of the rainfall regime are relatively uniform throughout most of the Patagonian semi-desert. Annual means range from about 200 to 300 mm and the coefficient of variation is generally about 40%. Precipitation occurs on roughly 40 days per year. Some 70–80% of this is associated with frontal

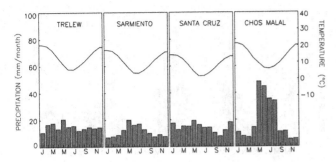

Fig. 15.19 Precipitation (mm) and temperature (°C) at four stations in Patagonia: Trelew, Sarmiento, Puerto Santa Cruz, and Chos Malal.

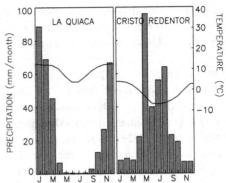

Fig. 15.20 Precipitation (mm) and temperature (°C) at two stations in the Puna: La Quiaca and Cristo Redentor.

activity, but thunderstorms and hail are nevertheless rare. Much of the winter precipitation is in the form of snow. Cloud cover is greatest during the winter rainy season. The annual average is about 50–60% cover; it is somewhat lower in the interior than near the cordillera or coast.

In the Patagonian region the continent narrows, resulting in a strong maritime influence throughout the region. This factor, combined with the consistent and intense westerly flow and frequent breaks in the high cordillera, permits humid Pacific air to readily penetrate the region. The result is a markedly maritime climate, with relatively uniform temperatures throughout Patagonia, except for higher elevations, and relatively small annual and diurnal variation.

Mean annual temperature is about 8°–9°C in central Patagonia, 13°C in the north, and 5°C in the south; the annual range is about 12°–15°C. Mean monthly temperatures range from 2° to 4°C in winter (but up to 7°C near the coast) to 10°C in summer in the south, 20°C in summer in central Patagonia. The diurnal range is about 7°–10°C in winter, 11°–15°C in summer. June or July is generally the coldest month, January the warmest.

Winters are kept fairly warm by the maritime influence and the strong foehn winds created when the westerlies descend the eastern slopes of the Andes. Even in the coldest months, the mean daily minimum is generally only −1° to −3°C, but no month is frost-free. The cold season lasts about 4 months in the semi-desert, 5 months elsewhere. Frost occurs on over 100 days per year.

Relative humidity is on the order of 70% in the south, 55–60% in the central portions of the Patagonian Desert. However, it undergoes large seasonal fluctuations, the annual range being 40% in northern Patagonia, 25% in central and southern Patagonia. Fog frequency is generally less than 10 days per year, and less than 2 days per year in much of the area. The annual number of days with thunderstorms is generally about 2–8.

THE PUNA

The Puna is a cold, windy montane grassland in the high plateaus and cliffs of the central Andes (Mares *et al.* 1985). It comprises the region between the treeline (generally 3200–3500 m ASL) and the permanent snowline (generally 4500–5000 m). Most of it lies between 3400 and 4500 m ASL. The Puna

extends from central Peru in the north, across the altiplano of Peru and Bolivia, and along the spine of the Andes into Argentina and Chile. The soils are those typical of desert and mountain regions. Volcanic and saline soils and regosols are common. The vegetation consists mainly of bunchgrasses interspersed with herbs, grasses, cushion plants, and occasional low shrubs. Some 80–90% of the precipitation falls in summer in the form of rain or hail. Occasional winter snows accompany cold air intrusions from the higher latitudes. Frost frequency is high. At La Quiaca, for example, it occurs on 176 days per year. One unusual characteristic for this latitude is that the diurnal temperature range can equal or exceed the annual range. Day-to-day changes are also severe.

Climatic conditions within the Puna can be quite diverse (Fig. 15.20). Both temperature and precipitation vary greatly with altitude and latitude; microclimatic effects of the steep terrain also create a mosaic of conditions. Consequently, several ecological regions can be distinguished in the region, primarily on the basis of precipitation. The most common distinction is between the wet *puna* and dry *puna*: northern and southern areas of the ecosystem, respectively.

Conditions in the wet *puna* depend on altitude. At high elevations, the region consists of bunchgrass, wetlands, small shrubs and trees, and herbaceous plants. The landscape is typically mountainous, with snow-capped peaks, high lakes, plateaus, and valleys. Nighttime freezes can occur anytime within the year. Annual precipitation is less than 700 mm and it occurs mainly as snow and hail. In the altiplano, at elevations between 3700 and 4200 m ASL, mean annual precipitation is on the order of 500–700 mm and the vegetation is predominantly grasses and shrubs. Precipitation decreases from north to south. In the north, near Lake Titicaca (16° S), rain occurs during eight months of the year, but in the Puna's southern extreme the season is reduced to one or two months. Frost occurs from March to October.

Within the dry *puna*, annual precipitation is generally less than 400 mm, but many areas receive less than 100 mm, and some receive less than 50 mm. Precipitation is highly seasonal; eight months are absolutely dry. The vegetation is primarily montane grassland, herbaceous plants, and dwarf shrubs. Salt flats, locally known as *salares*, are a characteristic feature of this region. Throughout most of the region there is a summer rainfall regime; winter rains prevail in the southernmost tip, south of about 25° S.

Stations typical of the Puna are La Quiaca in the north (22° S), with summer rainfall, and Cristo Redentor in the south (33° S), with winter rainfall (Fig. 15.20). Both receive comparable amounts of precipitation (300 mm/year and 357 mm/year, respectively). At Cristo Redentor (3829 m ASL) the mean temperature is below zero during eight months of the year. The diurnal range varies from 7°C in winter to 10°C in summer, compared with an annual range of about 10°C. Absolute minima are −30°C in winter, −9°C in summer. The mean temperature of the warmest month, January, is 4°C; in July, the coldest month, it is −7°C. Thunderstorms are rare but fog occurs on 26 days per year. Cloudiness is on the order of 20% in the dry summer, but nearly 50% in the winter. Snow occurs on 86 days per year. Strong winds blow steadily from the southwest. Wind speed exceeds 12 m/s on most days and on almost every day during the summer. At La Quiaca the mean monthly temperature ranges from 13°C in December to 4°C in June. The diurnal range is 14°–16°C in summer, 18°–24°C in winter. Mean relative humidity ranges from 35% to 65%.

15.5 THE SEMI-ARID XEROPHYLLOUS FORESTS AND WOODLANDS

The semi-arid plain to the east of the Monte is occupied by two semi-arid vegetation formations: the xerophyllous thorn forest in the more northern tropical–subtropical section and a xerophyllous woodland, regionally termed *espinal*, in the more southern section with more temperate climate. The boundary between the forest and woodland is roughly the 19°C annual isotherm. The *espinal* woodland consists of a grassland cover with occasional microphyllous (small-leafed) trees. The xerophyll forest characteristically has a continuous woody cover that includes abundant leafy shrubs, succulents, thorny tree species, and aphyllous plants. Some of the thorny plants possess bark photosynthesis. This region includes the thorn forest in the semi-arid Chaco region of Argentina, which is the warmest region in South America.

Mean annual rainfall ranges from about 200–300 mm to 700 mm in most of this region, but it approaches 1000 mm at the transition to more humid regions. The 700 mm isohyet is the approximate boundary between the *espinal* woodland and the grasslands to the northeast. The 1000 mm isohyet roughly delineates the xerophyllous and moist forests. The mean number of rain days generally varies from roughly 50 to 70 per year, but the number rises to 90 per year in the wetter Chaco region. The rainfall at typical stations in the xerophyllous forest and woodland regions is shown in Fig. 15.21. Those far to the north in the Chaco are shown in Fig. 15.22.

In the xerophyllous forest, rainfall shows a clear summer maximum. In the *espinal*, however, the seasonal distribution is

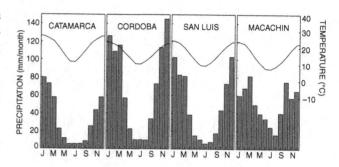

Fig. 15.21 Precipitation (mm) and temperature (°C) at three stations in the forest region (Catamarca, Cordoba, San Luis) and one in the woodland or *espinal* region (Macachin).

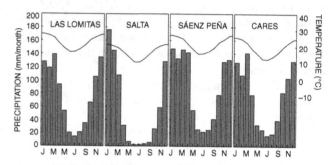

Fig. 15.22 Precipitation (mm) and temperature (°C) at four stations in the Chaco of Argentina: Los Lomitas, Salta, Sáenz Peña, and Ceres.

more complex, as with other areas along the eastern flank of the continent. At Macachin (Fig. 15.21), for example, there is a distinct minimum during the winter months, August being the driest with a monthly mean of 16 mm. The highest rainfall occurs during the transition season, the wettest months being March and October. The region is thus affected both by the tropical systems of summer and the winter cyclonic regime. Thunderstorms are common, the frequency being on the order of 20–50 days per year. Fog occurs on 10–15 days per year in most of the region.

In the xerophyllous forest, mean annual temperatures are roughly 19°–22°C. The coldest month is July throughout the region, with monthly means of 12°–18°C in the northern Chaco sector, 7°–12°C farther south. Summers are warm, with mean daily maxima exceeding 30°C in most places. In parts of the Chaco, where temperatures as high as 49° and 50°C have been recorded, mean daily maxima are as high as 35°–39°C. Nearly everywhere, January is the warmest month, with mean temperatures ranging from 22°C in the south to 29°C in the north. The diurnal range is roughly 11°–17°C, being greatest in winter or spring in the more northern parts of the Chaco, but in summer in remaining areas of the xerophyllous forest. In the *espinal* woodland farther south, temperatures are somewhat cooler than in the xerophyllous forest throughout the year and the diurnal range slightly higher. July temperatures are generally 6°–8°C, January temperatures, 21°–25°C.

Throughout most of the xerophyllous forest and woodland, the prevailing winds are from the N, NE or E during most or all of the year. This is a consequence of the relative stability of the South Atlantic High to the east and the region's position on its northwestern flank. In parts of the Chaco, however, the prevailing winds vary considerably because of complex topography.

In the Chaco, mean monthly relative humidity tends to be about 75–80% during the most humid months of autumn or winter. Means for the driest months are on the order of 50–60% throughout most of the Chaco. In the southern sector of xerophyllous forest, relative humidity is about 50% in the driest months and 70% in the most humid months of autumn. Further south in the Espinal, relative humidity is higher due to the coastal influence and lower temperatures.

The mean monthly cloudiness reaches a minimum of about 30% and a maximum of about 50% in the Chaco, with the minimum shifting from summer to autumn toward the south. In the remaining xerophyllous forest and woodland, the minimum is in autumn and cover is somewhat reduced, with monthly means reaching as low as 20% and means falling to 30–40% during the cloudiest months.

15.6 THE TEMPERATE GRASSLANDS

In northeastern Argentina, Uruguay, and southern Brazil lies a region of temperate subhumid climate occupied by the Río de la Plata grasslands (Fig. 15.23). Bunch and tuft grasses (i.e., hummock and tussock forms) are both common. The percent cover ranges from 60% to 80% in the drier areas to nearly 100% in more humid regions. The tallest species reach about 1 m in height. Locally, there are some areas of short grass steppe or shrubs, as well as occasional isolated stands of trees.

To the south, in Argentina, is the *pampas*; to the north is the *campos* of Uruguay and Brazil. The distinction between the two is geographical, but it generally coincides with a distinction based on water balance. The *pampas* tends to be a dry subhumid environment, with negative or near zero water balance, while the *campos* is a humid environment, with positive or near zero water balance (Soriano 1992). In the "flooding *pampas*" region, much of the land is seasonally or perennially inundated and halophytic vegetation is common.

Mean annual rainfall is on the order of 800–1200 mm. In most of the region there is no well-defined dry season. In the west, where annual rainfall is as low as 200 mm, winters are quite dry. Typical stations include Concordia, Buenos Aires, Rosario, Junin, Las Flores, and Mar del Plata (Fig. 15.23). Cloud cover is moderate, ranging from a minimum of about 30% to about 50% during the winter months. Both thunderstorms and fog are common, with about 50 thunderstorm days per year and fog frequency ranging from about 20 to 50 days.

The region's climate is mild, with mean annual temperatures ranging from 12° to 18°C, an annual range of 12°–14°C, and mean annual rainfall on the order of 400–1600 mm. Mean temperatures in January, the warmest month, range from 20° to 26°C; for July, the coldest month, the range is roughly 7°–12°C. Lowest temperatures are toward the coast in summer, but in

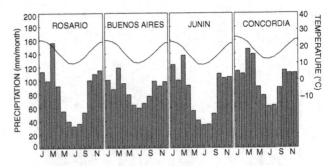

Fig. 15.23 Precipitation (mm) and temperature (°C) at four stations in the grasslands of Argentina: three in the *pampas* (Rosario, Buenos Aires, Junin) and one in the *campos* (Concordia).

the south in winter. Absolute maxima range from about 40° to 48°C. Absolute minima range from about −8° to −12°C inland, but no lower than −1°C closer to the coast. The diurnal range is about 8°–10°C in winter, 12°–16°C in summer. Frost occurs in much of the region, its frequency being lower/higher during El Niño/La Niña events (Muller *et al.* 2000).

Compared with most of the dryland regions of southern South America, the grasslands are relatively humid, due in part to their relative proximity to the Atlantic. Monthly means are generally 60–70% during the driest summer months, but 83–88% in winter. The humid conditions relate to the region's location on the western flank of the stable South Atlantic High. This creates a relative constancy of the wind regime throughout the year, with prevailing winds from a northerly or easterly direction.

15.7 THE DRY ZONE OF NORTHEAST BRAZIL

One of the most interesting dryland regions, from a climatic point of view, is the near-equatorial dry zone on the northeastern tip of Brazil, within the region locally referred to as *Nordeste*. This semi-arid region is unusual both in its tropical location and its distance inland, with the more humid climates lying directly on the coast. The region is an upland plain, generally below 500 m in elevation and extending from about 4° to 14° S. The vegetation in the region, referred to as *caatinga*, is a dry shrubland and thorn forest (Fig. 15.24). Annual rainfall is on the order of 500–750 mm in most places, but falls below 400 mm in some spots. An equally large surrounding subhumid area receives between 750 and 1200 mm annually.

Northeast Brazil includes two markedly different climatic regions. The seasonality of rainfall and the factors that control it are distinctly different, as are the degree to which the regions are impacted by drought. The northern part is the semi-arid region mentioned above and the overriding influence is the ITCZ. The rainfall maximum occurs in March and April, and winter and spring are very dry. A prominent feature of the region is the occurrence of lengthy and extreme droughts. The southern part of *Nordeste* is more humid and is influenced by several factors, an important one being the South Atlantic subtropical high

(a)

(b)

Fig. 15.24 (a) A typical *caatinga* landscape in semi-arid Nordeste region. (b) Cacti exist in many parts of the region, especially areas of degradation.

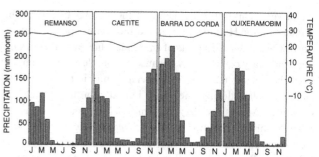

Fig. 15.25 Precipitation (mm) and temperature (°C) at four stations in Northeast Brazil: Remanso, Caetite, Barra do Corda, and Quixeramobim.

(Aragao *et al.* 2007). Rain is produced by westward propagating disturbances linked to the ITCZ, the SACZ, or frontal incursions (Chavez and Cavalcanti 2001). Here the rainfall maximum occurs in December and severe drought is not a regular feature of the region.

The contrast between the northern and southern regions is illustrated by the stations Quixeramobim (5° S, 39° W) with 763 mm of rainfall, and Remanso (10° S, 42° W) with 494 mm (Fig. 15.25). Remanso experiences maximum rainfall in December, while at Quixeramobim the maximum occurs in March, and December is nearly rainless.

The general aridity of the northern sector can be understood in terms of the annual cycle and the fact that the ITCZ essentially "follows the sun." During most of the year, the zone of maximum sea-surface temperature (SST) and minimum surface pressure lies in the Northern Hemisphere. These features control the location of the ITCZ and the convection associated with it (Hastenrath 1991). Rainfall occurs only when the ITCZ can penetrate far enough southward. This occurs mainly in March and April, when the maximum heating and minimum pressure

move toward the Southern Hemisphere. The unusually far equatorward and westward position of the South Atlantic High may play some role by making it difficult for the ITCZ to penetrate southward into the dry zone (Satyamurty *et al.* 1998).

Because of the dominance of a single control, this dryland region is prone to both droughts and floods. The droughts, locally called *secas*, are associated with an anomalous northward displacement of the ITCZ over the Atlantic (Hastenrath 2006). The ITCZ's displacement, and hence the character of the year's rainfall, is controlled by SST in the nearby tropical Atlantic (Andreoli and Kayano 2006). Dry conditions are favored by anomalously cold/warm conditions in the equatorial South/North Atlantic. Wet years are favored by the converse. The effect of the warm temperature in the equatorial North Atlantic is to enhance the annual cycle, producing abnormally large latitudinal excursions of the ITCZ. This displaces the ITCZ to the north of Northeast Brazil during the February–April rainy season. The region tends to suffer drought when the western coast of South America experiences the wet El Niño years (de Souza *et al.* 2005).

In Northeast Brazil as a whole, rainfall is mostly convective. However, thunderstorms are relatively infrequent, generally occurring on fewer than 10 days per year. Cloud cover is also generally convective, although stratus frequently occurs in winter. In much of the region, cloudiness ranges seasonally between 40% and 60%, reaching a maximum during the wettest months. It is greater to the far west, but less persistent in the interior of the region. At Remanso, for example, it ranges from 24% in August, a rainless month, to 37% in January and March, two of the wettest months. There, the development of convective clouds is hindered by the strong trade inversion.

In the warmest month, mean monthly temperatures are 26°–28°C, compared with 20°–26°C for the coldest month. Mean annual range is on the order of 3°C. The diurnal range is quite diverse within the region; at individual stations values averaged for the year as a whole vary between about 8° and 16°C. Seasonally the diurnal range varies by only about 2°C and is consistently greater during the driest months. In comparison, the mean annual range is on the order of 1°–3°C.

Relative humidity depends primarily on proximity to the ocean, with typical values ranging from 80% or higher in near-coastal locations to 40–50% in the interior of the northeast dry zone. Fog occurrence is therefore also variable, being rare in some areas but occurring up to 50 days per year in others. Dry haze or dust is common in much of the region.

15.8 THE NORTHERN CARIBBEAN COAST

Along the northern littoral of Venezuela and Columbia, at about 10° N, lies an extensive region of semi-arid steppe in which annual rainfall is less than 1000 mm (Fig. 15.7). The vegetation here is generally sparse grass, mesquite, or cacti. This dryness is unusual for these latitudes, especially since the low-level winds have an onshore component most of the year and the coastal highlands force the flow to ascend. Within this region, rainfall is quite variable. Numerous locations receiving less than 500 mm or even 300 mm/year are interspersed with humid highlands.

A combination of factors leads to the region's aridity. The main ones are related to the persistent easterly winds in the region. Their impact is twofold. Friction-induced divergence is one: because the low-level winds blow roughly parallel to the coast, which is particularly rugged, the frictional contrast between the land and water is strong. The easterly flow also produces coastal upwelling via Ekman transport, with the cold water stabilizing the atmosphere. The Caribbean low-level coastal jet, which parallels the Caribbean coast just north of Columbia and Venezuela, probably enhances these effects. This easterly jet stream has a core near 925 mb (at roughly 500 m) and reaches 16 m/s (Muñoz et al. 2008). Topography may also play some role.

Being near the equator, some aspects of the climate are relatively constant throughout the year. Temperatures for some typical stations (shown in Fig. 15.26) are moderate, with little seasonal variation. Cloudiness is relatively high, ranging from 40–50% in winter to 70–80% in summer. The wind regime is also reasonably constant, being northerly or northeasterly most of the year. Topographic effects dramatically change this pattern locally. Thunderstorms are common, occurring on roughly 40–80 days per year, but fog is relatively rare.

The rainfall patterns along the north Caribbean coast are complex (Fig. 15.26). The rainfall amounts are dependent to some extent on elevation but also tend to increase away from the coast. Throughout the region, winters are fairly dry. To the north and west there is a pronounced summer maximum.

In the south and east there are two maxima during the spring (generally May or June) and autumn (generally October), a consequence of the Caribbean easterly waves that bring most of the rainfall. In between lies a transition zone with three distinct maxima in spring, summer, and autumn (the summer maximum is often only one month). Somewhat unusual for this latitude, most of the rainfall is nocturnal, with an early morning maximum. Orography seems to play a role in this.

Except for higher elevations, such as Caracas at 1035 m, mean annual temperature is generally 28°–30°C, although it is as low as 26°C in some locations. The annual range is only 2°–3°C, compared with a diurnal range of 8°–10°C in most locations at lower elevations.

REFERENCES

Andreoli, R. V., and M. T. Kayano, 2006: Tropical Pacific and South Atlantic effects on rainfall variability over Northeast Brazil. *International Journal of Climatology*, **26**, 1895–1912.

Aragao, M. R. D., M. C. D. Mendes, I. F. A. Cavalcanti, and M. D. Correia, 2007: Observational study of a rainy January day in the Northeast Brazil semi-arid region: synoptic and mesoscale characteristics. *Quarterly Journal of the Royal Meteorological Society*, **123**, 1127–1141.

Carvalho, L. M. V., C. Jones, and B. Liebmann, 2004: The South Atlantic convergence zone: Intensity, form, persistence, and relationships with intraseasonal to interannual activity and extreme rainfall. *Journal of Climate*, **17**, 88–108.

Cerne, S. B., and Coauthors, 2007: The nature of a heat wave in eastern Argentina occurring during SALLJEX. *Monthly Weather Review*, **135**, 1165–1174.

Chavez, R. R., and I. F. A. Cavalcanti, 2001: Atmospheric circulation features associated with rainfall variability over southern Northeast Brazil. *Monthly Weather Review*, **129**, 2614–2626.

de Souza, E. B., M. T. Kayano, and T. Ambrizzi, 2005: Intraseasonal and submonthly variability over the Eastern Amazon and Northeast Brazil during the autumn rainy season. *Theoretical and Applied Climatology*, **81**, 177–191.

Durkee, J. D., T. L. Mote, and J. M. Shepherd, 2009: The contribution of mesoscale convective complexes to rainfall across subtropical South America. *Journal of Climate*, **22**, 4590–4605.

Fuenzalida, H. A., R. Sánchez, and R. D. Garreaud, 2005: A climatology of cutoff lows in the Southern Hemisphere. *Journal of Geophysical Research – Atmospheres*, **110**, D18101.

Garreaud, R. D., and R. C. Muñoz, 2005: The low-level jet off the west coast of subtropical South America: structure and variability. *Monthly Weather Review*, **133**, 2246–2261.

Garreaud, R. D., and J. Rutllant, 2003: Coastal lows along the subtropical west coast of South America: numerical simulation of a typical case. *Monthly Weather Review*, **131**, 891–908.

Garreaud, R. D., and J. M. Wallace, 1998: Summertime incursions of midlatitude air into subtropical and tropical South America. *Monthly Weather Review*, **126**, 2713–2733.

Garreaud, R., M. Vuille, and A. C. Clement, 2003: The climate of the Altiplano: observed current conditions and mechanisms of past changes. *Palaeogeography, Palaeoclimatology, Palaeoecology*, **194**, 5–22.

Gonzalez, M., and V. Barros, 1998: The relationship between tropical convection in South America and the end of the dry period in subtropical Argentina. *International Journal of Climatology*, **18**, 1669–1685.

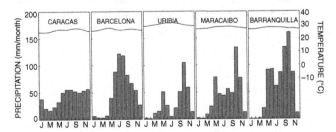

Fig. 15.26 Precipitation (mm) and temperature (°C) at five stations in the arid region along the northern Caribbean coast of Venezuela and Columbia: Caracas, Barcelona, Uribia, Maracaibo, and Barranquilla.

Grimm, A. M., B. Varros, and M. Doyle, 2000: Climate variability in southern South America associated with El Niño and La Niña events. *Journal of Climate*, **13**, 35–58.

Hastenrath, S., 1991: *Climate Dynamics of the Tropics*. Kluwer Academic Publishers, Dordrecht, 488 pp.

Hastenrath, S., 1999: Equatorial mid-tropospheric easterly jet over the eastern Pacific. *Journal of the Meteorological Society of Japan*, **77**, 701–709.

Hastenrath, S., 2006: Circulation and teleconnection mechanisms of Northeast Brazil droughts. *Progress in Oceanography*, **70**, 407–415.

Lettau, H. H., 1978: Explaining the world's driest climate. In *Exploring the World's Driest Climate* (H. H. Lettau and K. Lettau, eds.), Institute for Environmental Studies, University of Wisconsin, Madison, WI, pp. 182–248.

Liebmann, B., G. N. Kiladis, C. Vera, A. C. Saulo, and L. M. V. Carvalho, 2004: Subseasonal variations of rainfall in South America in the vicinity of the low-level jet east of the Andes and comparison to those in the South Atlantic convergence zone. *Journal of Climate*, **17**, 3829–3842.

Liebmann, B., and Coauthors, 2009: Origin of convectively coupled Kelvin waves over South America. *Journal of Climate*, **22**, 300–315.

Marengo, J. A., T. Ambrizzi, G. Kiladis, and B. Liebmann, 2002: Upper-air wave trains over the Pacific Ocean and wintertime cold surges in tropical-subtropical South America leading to freezes in southern and southeastern Brazil. *Theoretical and Applied Climatology*, **73**, 223–242.

Marengo, J. A., W. R. Soares, S. Saulo, and M. Nicolini, 2004: Climatology of the low-level jet east of the Andes as derived from the NCEP-NCAR Reanalyses: characteristics and temporal variability. *Journal of Climate*, **17**, 2261–2280.

Mares, M. A., J. Morello, and G. Goldstein, 1985: The Monte Desert and other subtropical semi-arid biomes of Argentina, with comments on their relation to North American arid areas. In *Hot Deserts and Arid Shrublands* (M. Evenari, I. Noy-Meir, and D. W. Goodall, eds.), Ecosystems of the World 12A, Elsevier, Amsterdam, pp. 203–238.

Muller, G. V., M. N. Nunez, and M. E. Seluchi, 2000: Relationship between ENSO cycles and frost events within the Pampa Humeda region. *International Journal of Climatology*, **20**, 1619–1637.

Muñoz, E., A. J. Busalacchi, S. Nigam, and A. Ruiz-Barradas, 2008: Winter and summer structure of the Caribbean low-level jet. *Journal of Climate*, **21**, 1260–1276.

Nesbitt, S. W., R. Cipelli, and S. A. Rutledge, 2006: Storm morphology and rainfall characteristics of TRMM precipitation features. *Monthly Weather Review*, **134**, 2702–2721.

Nicolini, M., and A. C. Saulo, 2006: Modeled Chaco low-level jets and related precipitation patterns during the 1997–1998 warm season. *Meteorology and Atmospheric Physics*, **94**, 129–143.

Nicolini, M., and Coauthors, 2004: South American low-level jet diurnal cycle and three-dimensional structure. *CLIVAR Exchanges*, **9**, 6–9.

Nogués-Paegle, J., and K. C. Mo, 1997: Alternating wet and dry conditions over South America during summer. *Monthly Weather Review*, **125**, 279–291.

Nogués-Paegle, J. N., and C. Vera, 2009: South American precipitation regimes. *Proceedings of the Ninth International Conference on Southern Hemisphere Meteorology and Oceanography*. American Meteorological Society, Boston, MA.

Salio, P., M. Nicolini, and E. J. Zipser, 2007: Mesoscale convective systems over southeastern South America and their relationship with the South American Low-Level Jet. *Monthly Weather Review*, **135**, 1290–1309.

Satyamurty, P., and M. E. Seluchi, 2007: Characteristics and structure of an upper air cold vortex in the subtropics of South America. *Meteorology and Atmospheric Physics*, **96**, 203–220.

Satyamurty, P., C. A. Nobre, and P. L. Silva Dias, 1998: South America. In *Meteorology of the Southern Hemisphere* (D. J. Karoly and D. G. Vincent, eds.), American Meteorological Society, Boston, MA, pp. 119–139.

Schwerdtfeger, W., 1976: *Climates of Central and South America*. World Survey of Climatology, Vol. 9, Elsevier, Amsterdam, 544 pp.

Seluchi, M. E., R. D. Garreaud, F. A. Norte, and A. C. Saulo, 2006: Influence of the subtropical Andes on baroclinic disturbances: a cold front case study. *Monthly Weather Review*, **134**, 3317–3335.

Siqueira, J. R., and L. A. T. Machado, 2004: Influence of the frontal systems on the day-to-day convection variability over South America. *Journal of Climate*, **17**, 1754–1766.

Siqueira, J. R., and V. D. Marques, 2008: Occurrence frequencies and trajectories of mesoscale convective systems over southeast Brazil related to cold frontal and non-frontal incursions. *Australian Meteorological Magazine*, **57**, 345–357.

Soriano, A., 1992: Río de Plata grasslands. In *Natural Grasslands* (R. T. Coupland, ed.), Ecosystems of the World 8B, Elsevier, Amsterdam, pp. 367–408.

Soriano, A., and Coauthors, 1983: Deserts and semi-deserts of Patagonia. In *Temperate Deserts and Semi-Deserts* (N. West, ed.), Ecosystems of the World 5, Elsevier, Amsterdam, pp. 423–460.

Trewartha, G. T., 1970: *The Earth's Problem Climates*. University of Wisconsin Press, Madison, WI, 334 pp.

Vera, C. S., and P. K. Vigliarolo, 2000: A diagnostic study of cold-air outbreaks over South America. *Monthly Weather Review*, **128**, 3–24.

Vera, C., P. K. Vigliarolo, and E. H. Berbery, 2002: Cold season synoptic-scale waves over subtropical South America. *Monthly Weather Review*, **130**, 684–699.

Vera, C., and Coauthors, 2006a: Toward a unified view of the American monsoon systems. *Journal of Climate*, **19**, 4977–5000.

Vera, C., and Coauthors, 2006b: The South American low-level jet experiment. *Bulletin of the American Meteorological Society*, **87**, 63–77.

Zhou, J. Y., and K. M. Lau, 1998: Does a monsoon climate exist over South America? *Journal of Climate*, **11**, 1020–1040.

Zipser, E., P. Salio, and M. Nicolini, 2004: Mesoscale convective systems activity during SALLJEX and the relationship with SALLJ events. *CLIVAR Exchanges*, **9**, 14–18.

Zipser, E. J., D. J. Cecil, C. Liu, S. W. Nesbitt, and D. P. Yorty, 2006: Where are the most intense thunderstorms on earth? *Bulletin of the American Meteorological Society*, **87**, 1057–1071.

16 Sub-Saharan Africa

16.1 CONTINENTAL OVERVIEW

The African continent includes some of the driest and some of the wettest locations on earth. Some locations receive several meters of rainfall each year; Mt. Cameroon receives over 10 m. Vast desert regions receive less than 5 mm on average. The major African deserts (Fig. 16.1) include the Sahara, the Somali-Chalbi and the Namib. Much of the rest of the continent is semi-arid, including such regions as the Sahel of West Africa, the Kalahari of southern Africa, and much of eastern Africa (Fig. 16.2).

A rather remarkable aspect of Africa's climate is the tendency for climatic fluctuations to impact nearly the entire continent. The most common pattern is one of aridity throughout most of Africa, on a yearly (e.g., Fig. 21.8) or decadal basis. During the 1980s, for example, average rainfall for the decade was below normal throughout all but a few isolated sectors, mainly in the highlands of East Africa. Quite often the equatorial sector is out of phase with the rest of the continent. The 1950s, one of the wettest periods in the Sahel in recent times, showed such a pattern. Collectively, these large-scale anomaly patterns, extending over nearly 80° of latitude, implicate very large-scale controls on African climate.

Africa has received much attention as a result of a devastating multidecadal period of drought in the semi-arid Sahel. The continent as a whole experienced a long drying trend (Fig. 11.28), with 30-year means decreasing by 10–30% in most of Africa's dryland regions. Some suggested this represented a long-term process of desertification at the hands of mankind, with the desert "marching" southward, overtaking the savanna lands. This idea has since been discounted, and at least a partial recovery of the rainfall has occurred in the Sahel (Nicholson 2005) and elsewhere.

This chapter presents an overview of African climate and a detailed look at some of the continent's arid and semi-arid regions. Some regions are discussed in other chapters because the dominant climatic controls are different from those prevailing over the continent as a whole. The northern Sahara is included in Chapter 17 on Mediterranean climates, and the coastal deserts of the western Sahara and Namib are included in Chapter 20. Rainfall is stressed more than other climatic elements because it is the most variable and is the limiting environmental factor over most of Africa. Also, our understanding of rainfall over Africa has changed markedly in recent decades, and an in-depth, updated look is required.

16.1.1 WIND AND PRESSURE SYSTEMS

The African continent spans over 70° of latitude, from 35° S to 37° N, and is thus affected by wind and pressure systems of equatorial, subtropical, and mid-latitude origin (Fig. 16.3). However, most of the continent lies in the tropics. Consequently, the prevailing winds are easterly and the dominant surface circulation features are the equatorial pressure trough and the Intertropical Convergence Zone (ITCZ). The subtropical highs are also influential surface features. Traditionally, the march of the seasons over Africa is described in terms of the north–south migration of these features, and rainfall is assumed to be associated with the ITCZ. This picture of the tropics is most valid over the oceans. Its validity requires that more local influences, such as large-scale topography and land/sea contrast, do not play a major role. This is not the case over most of Africa, so that this simplistic picture does not hold up to scrutiny (Nicholson 2009a). A more detailed picture is presented in the regional sections of this chapter.

In the boreal summer, a distinct surface ITCZ transects the continent, stretching across the Sahara (Fig. 16.3). The tropical rainbelt, i.e., the zone of maximum rainfall, lies well to the south of it. Both migrate north and south with the seasons, following the path of the sun. In the austral summer, the convergence zone over the continent becomes relatively indistinct over much of the continent because of the complex topography in equatorial and southern Africa.

Poleward of the ITCZ are subtropical high-pressure cells in both hemispheres. Africa is influenced by the North and South

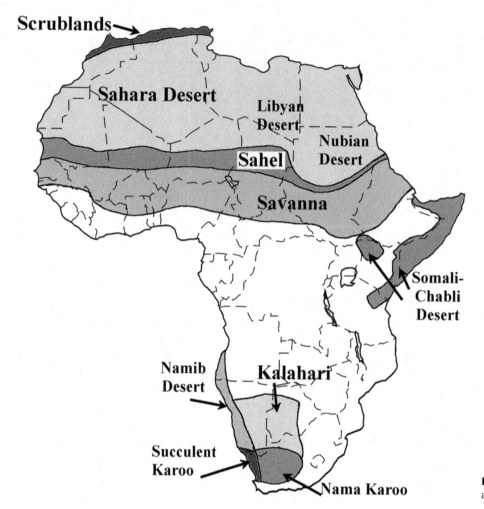

Fig. 16.1 Location of major deserts and semi-arid regions over Africa.

Atlantic Highs (the former is also called the Azores or Bermuda High, the latter the St. Helena High) and the South Indian Ocean High (also termed the Mascarene High). In winter these merge with high-pressure regions over the continent itself, especially in the Southern Hemisphere, where there is a quasi-continuous belt of high pressure. Because of the subsiding air and weak winds, the high pressure promotes aridity.

The seasonal switch between high and low pressure produces the basic pattern of climates over Africa, which tend to be organized according to latitude. This basic pattern is complicated by the seasonal temperature changes in higher latitudes of the continent, which influence the pressure and wind fields. During the boreal summer, an intense heat low forms over the Sahara (Fig. 16.3) and low pressure extends throughout most of Northern Hemisphere Africa (Lavaysse *et al.* 2009). This system is shallow and is overridden by the Saharan High (Chen 2005). Interactions between the heat low and orography appear to influence the onset of the rainy season (Drobinski *et al.* 2005). Over southern Africa a pronounced high-pressure center is the dominant influence in this season. However, the higher latitudes are often traversed by mid-latitude low-pressure systems that are displaced equatorward during the cool season.

A dramatic change takes place in boreal winter. A high-pressure cell extends over most of northern Africa as the desert cools and the rainbelt moves into the Southern Hemisphere. The northern extreme of the continent is under the influence of eastward moving mid-latitude low-pressure systems and westerly winds. Over southern Africa, in latitudes south of 20° S, weak high pressure still prevails except in eastern parts of the continent.

In July there is cyclonic flow around a heat low that builds up over the Sahara, and the ITCZ separates moist southwesterly flow over West Africa from the dry northeast trade winds, also called the *harmattan* (Fig. 16.3). A second convergence zone in central Africa separates air flow off the Indian Ocean from that originating over the Atlantic. Traditionally this was referred to as the Zaire or Congo Air Boundary, but some of the more recent literature calls this the Inter-Ocean Convergence Zone (IOCZ). The southeast trade winds (the SE monsoon) are dominant east of that zone. These arise within the anticyclonic flow around the high over southern Africa.

In January, when the ITCZ moves equatorward and into the Southern Hemisphere, there is anticyclonic flow around the high-pressure cell over the Sahara. The NE harmattan along

Fig. 16.2 Typical savanna landscapes of (a) the low tree and shrub savanna of the Sahel, (b) the Kalahari, and (c) an East African grassland, and (d) a baobab tree, typical of the savanna woodland of West Africa.

its southern margin is dominant throughout most of North and West Africa, while NE trades originating over the Indian Ocean (the NE monsoon) prevail over eastern Africa. An anticyclonic circulation exists around the high over southern Africa, producing the SE trades south of the ITCZ and westerly and north-westerly flow over the southernmost latitudes (Fig. 16.3).

The wind patterns at higher levels are less complex. Throughout the summer hemisphere the dominant flow aloft is easterly (Fig. 16.4). A shallow, low-level westerly flow is evident in the equatorial region throughout the year but this is more extensive in the Northern Hemisphere and in western sectors of Africa. In some years it becomes a well-developed jet stream, with the shallow flow being replaced by westerlies extending into the mid-troposphere (Nicholson and Webster 2007). In the winter hemisphere, easterlies prevail in equatorial and in sub-tropical latitudes, but westerlies are dominant in the higher lati-tudes. A protracted region of weak equatorial westerlies is also apparent around 10° S. Newell *et al.* (1972) show this feature as extending up to nearly 500 mb.

Within the easterly wind regime there are three major jet streams: the tropical easterly jet (TEJ) of the upper tropo-sphere at 200 mb (12,000 m) and two mid-level easterly jets at roughly 650 mb. Both are termed African easterly jets (AEJ),

but Nicholson and Grist (2003) used the designations AEJ-N and AEJ-S to distinguish them. The Northern Hemisphere AEJ occurs year-round, migrating latitudinally with the seasons. It is primarily a result of the temperature contrast between the Sahara and the cooler Atlantic to the south. The Southern Hemisphere AEJ is evident only in some seasons and is primarily a result of the temperature contrast between the equatorial rainforest region and the drier/warmer woodlands to the south.

Superimposed on this east–west pattern of winds are two vertical circulation systems: the north–south oriented Hadley circulation and the east–west oriented Walker circulation. The Walker circulation is confined to near-equatorial latitudes and develops over the Atlantic Ocean during the boreal winter and over the Indian Ocean in the boreal autumn (Hastenrath *et al.* 2002; also see Chapter 4). When present, this circulation produces a tendency for rising motion over the western parts of equatorial Africa, and subsidence on the eastern edge. A Walker-type circulation (east–west overturning) is also evident over the higher latitudes of southern Africa, where rainfall is significantly lower in southwestern regions than in the south-east, and over the western Sahara. The Hadley cell produces the general tendency for low rainfall in its descending branch in

(a)

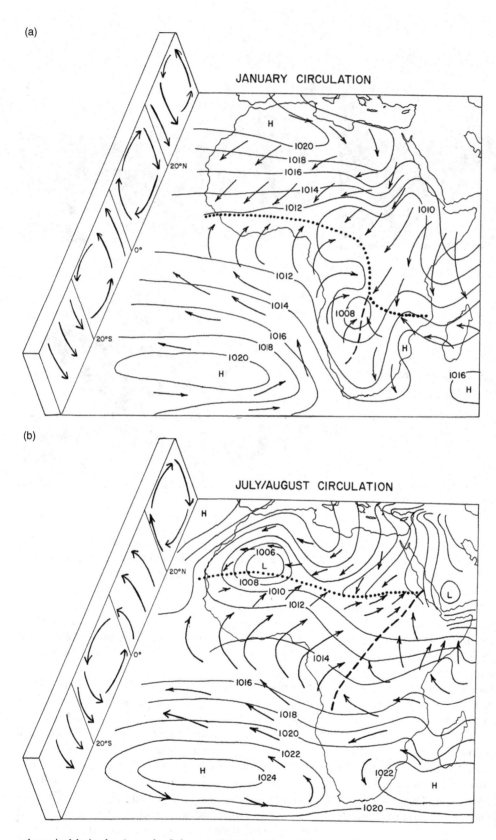

JANUARY CIRCULATION

(b)

JULY/AUGUST CIRCULATION

Fig. 16.3 Mean pressure and wind fields over Africa in (a) January and (b) July/August. The dotted and dashed lines represent the ITCZ and Zaire Air Boundary, respectively.

subtropical latitudes (e.g., the Sahara and Kalahari) and high rainfall in the equatorial regions of ascent. The intensity of the Walker and Hadley overturnings over North Africa varies greatly from year to year (Nicholson 2009b).

The rain-bearing disturbances that affect the continent are closely linked to the wind systems. Although localized thunderstorms can occur, most rainfall is associated with large-scale disturbances, which propagate across the continent. On the poleward extremes of both hemispheres, rainfall is produced by extra-tropical low-pressure cells embedded in the westerlies. Along the southern margin of the Sahara most precipitation accompanies easterly waves. In subtropical regions marking the transition from mid-latitude to tropical climate and circulation systems (20–30° N and S), the rain-bearing disturbances are often "hybrids," involving both tropical and extra-tropical systems (e.g., Todd and Washington 1999a; Knippertz et al. 2003). Waves in the extra-tropical westerlies interact with tropical, surface low-pressure systems to create a large-scale disturbance, which is often fairly stationary.

The rain-producing systems of the equatorial regions are the least well understood. In general, intense rainfall is coupled with regional-scale convergence in the wind field, although much localized rainfall occurs as a result of intense afternoon surface heating. There is some evidence that transient pressure disturbances, including easterly waves, also traverse equatorial latitudes, producing convective clouds and rainfall.

16.1.2 TEMPERATURE, RAINFALL AND OTHER CLIMATIC ELEMENTS

Since most of Africa lies in relatively low latitudes, temperatures are high throughout the year and vary more from daytime to nighttime than during the course of the year (Fig. 16.5). A true cold season occurs only in the extra-tropical regions (north of the Sahara and limited areas of southern Africa); elsewhere, small seasonal changes may result from changes in cloudiness or incoming solar radiation. Nearly everywhere in Africa, rainfall is the limiting factor in agriculture and is the most variable characteristic of climate in both time and space.

Both annual and diurnal temperature range are strongly dependent on both aridity and latitude. Consequently, the seasonal changes in minimum and maximum daily temperatures are quite different in the two hemispheres (Fig. 16.6). The annual range varies from roughly 24°C in the core of the northwestern Sahara to less than 2°C in central equatorial regions. In southern Africa the annual range reaches roughly 14°C in the Kalahari, much lower than in the corresponding latitudes of the Northern Hemisphere. The diurnal range is markedly greater in winter than in summer. In the driest locations (e.g., Sahara and Kalahari) it is on the order of 15°–18°C in winter and 12°–15°C in summer.

The pattern of precipitation over Africa is more complex; it includes some of the world's driest and wettest locations. Its precipitation regimes are also highly diverse in their nature and causes. Annual mean rainfall varies from less than 1 mm/

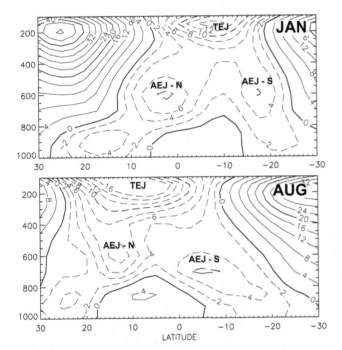

Fig. 16.4 Mean zonal winds (m s⁻¹) over Africa in January and August (from Nicholson and Grist 2003).

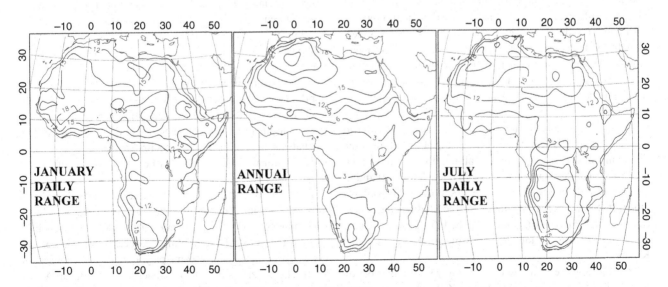

Fig. 16.5 Annual temperature range and diurnal range in January and July (°C).

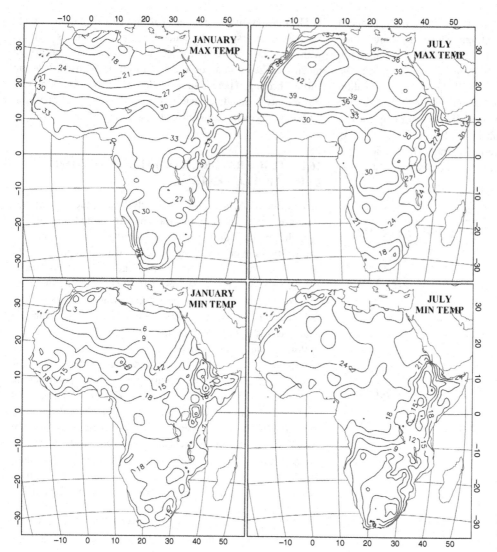

Fig. 16.6 Mean daily minimum and maximum temperatures in January and July (°C).

year in parts of the Sahara to over 5000 mm in some areas of tropical rainforest (Fig. 16.7). Most of the continent, however, is subhumid and experiences a prolonged dry season or seasons during the course of the year. Throughout the continent, rainfall is strongly seasonal (Fig. 16.8), with over 50% falling in the wettest quarter of the year. In parts of West Africa, the proportion increases to 70–80% or more. Rainfall is highly variable from year to year, with the coefficient of variation (CV) being generally on the order of 20–50% in semi-arid regions and 60–100% in desert regions (Fig. 16.9). Rain is the dominant form of precipitation in all months, except in high mountainous regions (which receive frequent snowfall) and in some coastal deserts, where fog provides most of the available moisture. Snow is generally limited to the Mediterranean sections of the Sahara (see Chapter 17), the Ethiopian highlands, the Ruwenzori range, and the peaks of Mt. Kenya and Mt. Kilimanjaro.

The traditional explanation for the rainy season in most of the continent is the presence of the ITCZ. The surface conver-gence associated with the ITCZ was assumed to provide sufficient ascent to induce rainfall. The classical picture of the relationship between rainfall and the ITCZ over West Africa is shown in Fig. 16.10. The ITCZ marks the surface intersection of the moist, southwest monsoon layer and the dry northeasterly harmattan air from the Sahara. In this picture, the depth of the moist layer is seen to increase equatorward of the ITCZ. It was assumed that the depth of the moist layer limits the development of convective clouds and associated rainfall, so that the maximum rainfall occurs several degrees of latitude equatorward of the ITCZ, where the layer is sufficiently deep.

This concept appeared adequate when it was believed that most of the rain over West Africa was associated with localized thunderstorms. Since the mid-1970s, however, it has been realized that most of the rainfall is associated with large-scale systems and that the jet streams play a major role (see Section 16.2). These systems (Fig. 16.11) affect most of tropical Africa, except for eastern equatorial regions (Zipser *et al.* 2006).

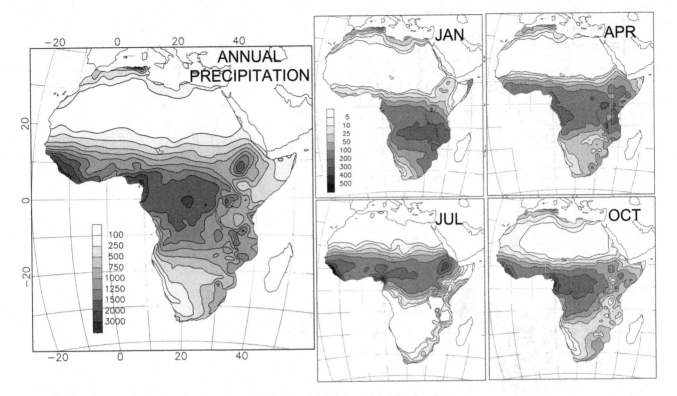

Fig. 16.7 Mean annual rainfall and monthly means for January, April, July, and October (mm).

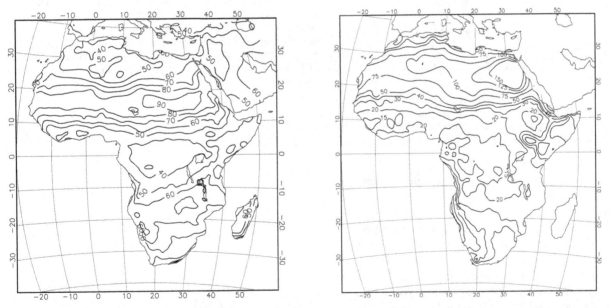

Fig. 16.8 Percentage concentration of precipitation in the wettest quarter of the year.

Fig. 16.9 Coefficient of variation of annual rainfall (%).

Several other aspects of the classical ITCZ/rainfall picture over Africa are also incorrect (Nicholson 2009a). For example, the latitudinal position of the zone of maximum rainfall, henceforth termed the "tropical rainbelt," is more closely linked to the latitudinal positions of the AEJ and TEJ than the ITCZ (Fig. 16.12). Also, the rainfall maximum lies at 10°–12° N, some 10° of latitude south of the surface ITCZ (Fig. 16.13), which is relatively fixed at the latitude of the core of the Saharan heat low during boreal summer. Moreover, analyses based on newer data sets indicate that the depth of the moist layer does not increase equatorward, but has a maximum between the cores of the AEJ and the TEJ (Nicholson and Grist 2003). Finally, the surface ITCZ becomes hard to track in Southern Hemisphere Africa because of the ubiquitous presence of major highlands

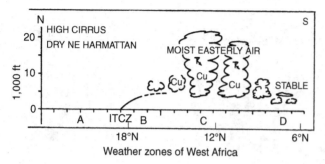

Fig. 16.10 Classical picture of the ITCZ over Africa during the boreal summer (from Griffiths 1972).

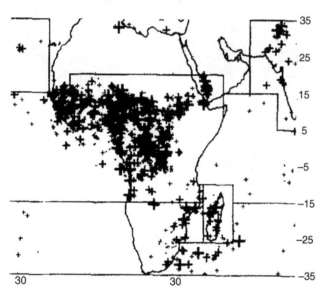

Fig. 16.11 The occurrence of intense mesoscale convective systems (MCSs) over Africa during one year (Toracinta and Zipser 2001). Each symbol indicates an occurrence of a system; the size of the symbol is related to the system's intensity.

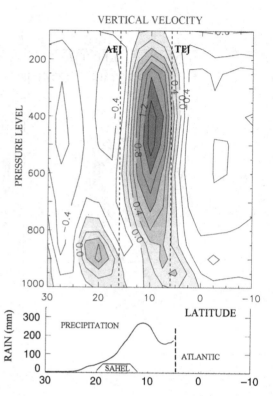

Fig. 16.12 Schematic of the rainbelt over West Africa (from Nicholson 2009a). Top diagram indicates the intensity of vertical motion (10^{-3} hPas^{-1}). The main region of ascent lies between the axes of the African easterly jet (AEJ) and the tropical easterly jet (TEJ). A shallow region of ascent corresponds to the surface position of the Intertropical Convergence Zone (ITCZ) and the center of the Saharan heat low. The bottom diagram gives rainfall as a function of latitude, with the location of the Sahel indicated on the latitudinal axis.

(van Heerden and Taljaard 1998), so that a highly idealized picture of convergence zones (Fig. 16.3) has been created to fit with the overall ITCZ concept.

The ITCZ concept is nonetheless useful in explaining the seasonality of rainfall over most of Africa. The basic pattern is aridity (i.e., the dry season) in association with the subtropical highs, and the rainy season in association with the ITCZ. The seasonal cycle is basically a shift of all systems toward the summer hemisphere and the pattern of wet and dry seasons is due to this displacement. The wet and dry seasons begin and end as the dominant pressure system in a given region shifts between the ITCZ and the subtropical high. The exception is the poleward extremes of the continent, where the rainfall is associated with the mid-latitude westerlies and low-pressure systems, which shift equatorward in winter.

The seasonal shift of atmospheric circulation features and rain-bearing systems is associated with a synchronous shift in the tropical rainbelt over Africa. This is clearly seen in Fig.

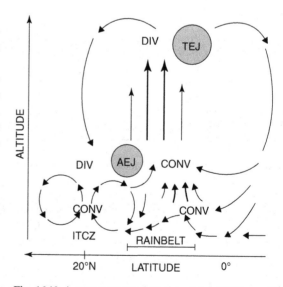

Fig. 16.13 A new conceptual model of the ITCZ over West Africa (from Nicholson 2009a).

16.7. The rainbelt lies deep in the Southern Hemisphere from December through February. It gradually shifts northward from March through June, at which time an abrupt shift or "jump" to roughly 10–12° N occurs in some years (Ramel 2006). The average position of the rainbelt then remains relatively stationary through August, when it moves progressively southward and eventually reaches well into the Southern Hemisphere.

A consequence of this seasonal displacement of the rainbelt is that the seasonal pattern of rainfall can be broadly generalized by latitude. In equatorial regions, which always lie within the rainbelt, year-round rains occur but with two maxima occurring in the transition seasons as the rainbelt crosses the equator. In somewhat higher latitudes, there are two wet seasons (also associated with the passage of the rainbelt) and two dry seasons. The dry seasons occur after the solstices (December and June), the larger and more intense one occurring near the winter solstice. In the outer tropical latitudes, there is generally one dry season and one wet season; the latter occurs during the summer season of the respective hemisphere and, with increasing latitude, it progressively becomes shorter and commences later. The most arid conditions are found in the subtropics (e.g., the Sahara in the Northern Hemisphere, the Kalahari and the Namib deserts in southern Africa). These mark the transition to extra-tropical climates along the poleward extremes with winter rainy seasons.

Superimposed upon this basic zonation are numerous regions with patterns of rainfall dictated by more localized factors. Some of the more important of these include topographic relief, shoreline and maritime effects, and local wind systems. Examples are numerous. In some southeastern areas of the continent, the combined influences of topography, coastline, tropical systems, and mid-latitude disturbances in the westerlies result in year-round rainfall. In East Africa, two monsoon wind systems, several convergence zones, and the Rift Valley lakes, mountains and highlands establish a highly diverse pattern rainfall.

The most difficult influences to generalize are those of topography (Fig. 16.14). In most cases, relief will tend to enhance rainfall. This is clearly evident over the Ethiopian, East African, and Rift Valley highlands, the Atlas Mountains, the Tibesti and Hoggar massifs in the central Sahara, the Guinean highlands, the Jos Plateau of Nigeria, and the mountainous terrain of Cameroon, Namibia, South Africa, Swaziland, and Lesotho. The degree of enhancement and the altitudinal distribution of rainfall vary greatly from one location to another and are determined both by the interaction of the topography with the large-scale wind and moisture fields and by the more local mountain–valley and slope wind systems (Oettli and Camberlin 2005). The highlands, such as Darfur, also appear to be important in triggering the development of the mesoscale convective complexes (MCCs) that bring rainfall to much of Africa (Mohr and Thorncroft 2006).

The resultant patterns of rainfall seasonality over the continent are considerably more complex than the latitudinal patterns described earlier (Fig. 16.15). There is a general latitudinal gradient in the length of the rainy season, but this is interrupted in eastern and far southern Africa. The length of the rainy season ranges from one month along the southern margin of the Sahara

Fig. 16.14 Major topographic features over Africa.

to 11 or 12 months in most equatorial regions (Fig. 16.16). In southern Africa, the wet season's duration decreases with latitude, but it generally exceeds five months everywhere but in the Namib Desert and other arid sectors of southwestern Africa. The pattern of a single rainy season in the subtropics and two wet and two dry seasons in the tropics is also apparent (Fig. 16.16).

The rainiest month tends to be December or January in Mediterranean Africa and January in most of southern Africa, but August throughout most of the northern subtropics (Fig. 16.16). The pattern is much more complex in areas with maxima during the transition seasons. In the Guinean zone of West Africa, it is generally June or September. In the western equatorial regions, it is October or November, but most commonly April or May in the eastern equatorial region. It is February, March, or April in western portions of southern Africa and in much of South Africa.

16.2 NORTHERN AFRICA

A vast area of continuous desert extends over 1000 km north to south and some 5000 km east to west across northern Africa and Saudi Arabia. It is termed the Sahara over Africa but the Arabian Desert further east. Along its northern bounds, this region gives way to the extra-tropical semi-arid Mediterranean lands. To the south of the desert lies the West African Sahel, a tropical semi-arid region. This section will concentrate on the desert core and regions to the south of it, all of which are under the influence of the same meteorological factors. This region is roughly bounded by 10° N and 25° N latitude, with annual rainfall ranging from near zero in the north to about 1200 mm in the south, a gradient of about 1 mm/km.

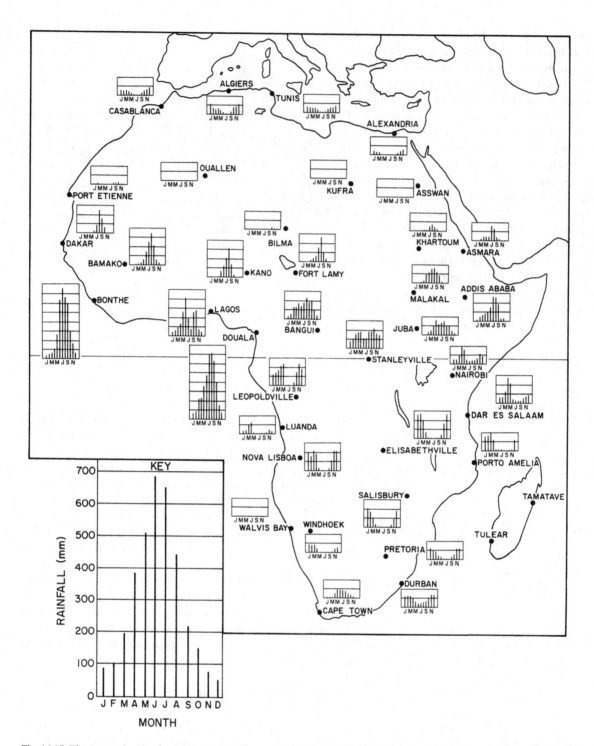

Fig. 16.15 The seasonal cycle of precipitation (mm) over Africa, illustrated via select stations.

Within this region flow six major rivers: the Senegal, Gambia, Niger, Ubangi, Chari, and the Nile, the only permanent river in the Sahara itself. The only major lake is Lake Chad, fed by the Ubangi-Chari system. The desert surface is primarily regs, stone pavements covering about 68% of the Sahara (LeHouerou 1986); another 22% is occupied by ergs, huge sand seas comprised of a hierarchy of dunes. Rocky mountain massifs and

hamadas, large stone plains, comprise the remaining 10%. The mountain massifs, Air in Niger, Tibesti in Chad, the Hoggar in Algeria, and Darfur in the Sudan generally reach 2000 m in elevation. In the Sahara most of the "soils" are pedocals, which consist mainly of parent material. Most common are saline solonchaks and calcisols. The former are found in depressions that are seasonally or permanently waterlogged and in ancient

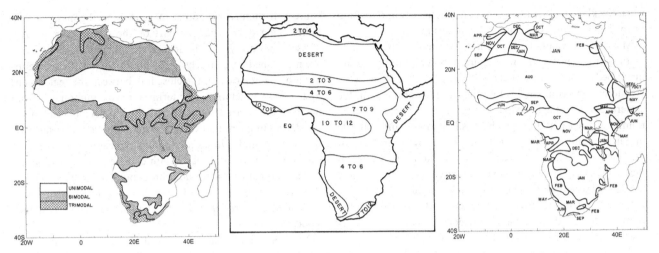

Fig. 16.16 (left) Rainfall seasonality, (center) length of the rainy season (months), and (right) month of maximum rainfall.

Fig. 16.17 The fertility of the Sahel is reduced by the high iron content of the soil; in this photo the large chunks of rock are composed largely of iron compounds.

lake beds. In some areas are weakly developed leptosols, soils limited in depth by an impervious hard layer such as laterite, or highly calcareous materials. True soils are generally confined to the semi-desert and semi-arid and subhumid regions further south. The most common are weakly developed arenosols or sand-rich soils, and well-developed lixisols – clay-rich soils that often contain large amounts of iron, aluminum, and titanium oxides (Fig. 16.17).

16.2.1 THE CENTRAL SAHARA

The Sahara is the largest desert on earth and one of the driest. The causes of this expansive arid zone and the reasons for its extreme aridity are not completely clear. The dominant high-pressure cells play a major role, but they are absent in the summer season, when the core of the Sahara remains dry. No other desert belt similarly dominated by a subtropical high is so extensive in east–west extent. The tropical easterly

jet may be a factor in its aridity (Webster and Fasullo 2003). Subsidence prevails in the left forward quadrant, which in the case of the TEJ coincides with the Sahara (see Chapter 4). Additional factors are likely, such as compensatory subsidence related to the broad ascent of air over the monsoon region to the east and/or the failure of either the mid-latitude cyclones or West African wave disturbances to penetrate into the desert core.

The transition seasons are the interesting ones in the Sahara. The heat low is still pronounced in the west and a second low and associated trough develops in the east over the Sudan. The trough (termed the Sudan trough) often spawns light rains, and occasional downpours. The eastern Sahara also experiences *khamsin* depressions. These generally occur three or four times per month during the March–May season. Prior to their onset, the air is warm, dry, dust-laden, and southerly. The depressions can be vigorous and produce severe dust- and sand storms. They infrequently bring light rain, but on occasion can produce thunderstorms, heavy rain, and hail.

In the western Sahara, the comparable system is the Soudano-Saharan depression, which develops in the lee of the Atlas Mountains, generally in association with a trough in the westerlies. It occurs from September through May, but is most common during the transition seasons. This system brings most of the rain to the central Sahara. The southern margin of the Sahara also receives occasional rainfall from the easterly waves and associated cloud clusters and squalls that traverse the semi-arid lands to the south.

The seasonality of rainfall in the Sahara is complex and is a direct consequence of the seasonal occurrences of these main rain-bearing systems. In the central Sahara there is little seasonal preference, but peak rainfall tends to occur in the transition seasons. The seasonal concentration is somewhat lower in the west than in the east (50–80%, compared with 70–90%). In the west, the meager rains tend to occur in September or October, when the Soudano-Saharan depressions extend far to the south. In the east, it tends to be the spring months, particularly May, when the khamsins are most common. In the far southern limit of the Sahara, August is usually the wettest month.

Mean annual rainfall is higher in the western Sahara because of the better-developed low-pressure system and the more frequent occurrence of disturbances. There it is generally between 20 and 50 mm, compared with a higher hyper-arid core in the eastern Sahara that is nearly rainless. In some years, annual rainfall can reach 100 mm in the western Sahara and 200–300 mm in the central Saharan highlands. In the hyper-arid east, it rarely exceeds 25 mm. The mean number of rain days is 1 or 2 in the east, 2–4 in the west. Relative humidity is not extremely low but rarely exceeds 40%. Hence, fog is uncommon.

Insolation tends to reach a maximum in May and July in the central Sahara and a minimum in December. In the annual average, cloudiness is less than 10% throughout most of the desert, and some regions are essentially cloudless. Cloudiness attains a minimum in July and August and a maximum in January–April, when it can be as high as 20–30% in some areas. In the core regions of the desert in the far west and east, cloudiness is below 10% throughout the year.

Throughout the central Sahara and its southern margins, temperatures remain relatively high throughout the year. Freezing temperatures do not occur south of 15°–20° N. The highest mean monthly temperatures are on the order of 32°–38°C and occur in June or July. The thermal extremes can reach 50°–55°C or −5° to –10°C. The mean for the coldest month is on the order of 15°–20°C, but it is as low as 12°C in the Saharan highlands and as high as 25°C in the eastern Sudan and near the western coast. The annual temperature range is 16°–20°C in most of the desert, but it reaches 24°C in the northwest. The diurnal range, generally on the order of 15°–20°C, exceeds the annual in much of this region.

16.2.2 THE SEMI-ARID ZONES: THE SAHEL AND SOUDAN AND THE DRY ZONE OF THE GUINEA COAST

GEOGRAPHICAL OVERVIEW

Over West Africa, the entire semi-arid region south of the Sahara is often referred to as the Sahel. However, most sources distinguish a number of geobotanical zones, the true Sahel being only a narrow part of the region (Fig. 16.18). These zones are variously defined by different authors. They are mentioned here largely to illustrate the geographical zonation of the region and the close link between the region's climate and its vegetation. The terms in most common use, from north to south, are the Sahelo-Saharan, Sahelian, Soudanian, Soudano-Guinean, and Guinean zones. Seasonal patterns of temperature and rainfall for stations typical of each zone are shown in Fig. 16.19. Clearly both climate and vegetation are largely a function of latitude throughout this region. Within these regions there is a gradual transition between semi-desert grassland, low tree and shrub savanna, savanna woodland, woodland, and forest.

No precise limits can be assigned to these zones either geographically or on the basis of rainfall, but some rough idea of the corresponding rainfall regime is given in Table 16.1. The southern desert boundary is generally given as 50 mm/year, with no preferred season of rainfall. Semi-arid climates with mean annual rainfall exceeding 50 mm begin at about 20°–22° N. Within the semi-arid zones, rainfall systematically shifts from

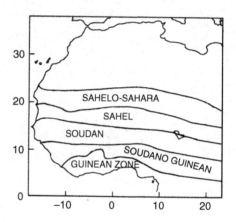

Fig. 16.18 Vegetation zonation of West Africa (five zones): the Sahelo-Sahara is desert steppe with widely spaced grass clusters, the Sahel is semi-desert grassland, the Soudan is a savanna grassland, the Soudano-Guinean zone is woodland, and the Guinean zone is primarily forest.

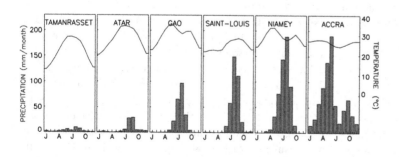

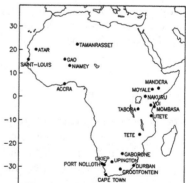

Fig. 16.19 Temperature (°C) and rainfall (mm) at typical stations over West Africa. The adjacent map gives the location of these and other stations utilized in climate diagrams.

Table 16.1. *Approximate climatic characteristics of four vegetation zones in West Africa: mean annual rainfall (mm), coefficient of variation (CV, %), length of the rainy season (months) (from Nicholson 1981).*

Vegetation zone	Rainfall (mm)	CV (%)	Season (months)
Sahelo-Sahara	50–100	50	1–2
Sahel	100–400	30–50	2–3
Soudan	400–1200	20–30	3–5
Soudano-Guinean	1200–1600	15–20	5–8

50 mm/year with a rainy season lasting 1–2 months, to 1000 mm/year with a rainy season lasting 4–5 months. In the north, about 80–90% of the rainfall occurs during the wettest quarter of the year; in the south, this figure falls to about 50%. The Guinean zone marks the beginning of the humid climates.

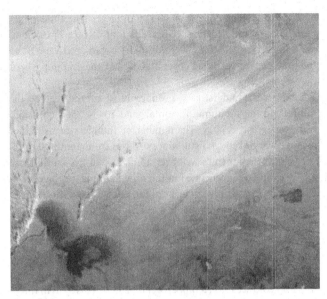

Fig. 16.20 Dust plumes emanating from the Bodélé Depression (photo from NASA).

CLIMATOLOGICAL OVERVIEW

The semi-arid nature of this region is a result of the dominance for most of the year of the climatological conditions producing the Sahara: a high-pressure regime, subsidence, and an absence of rain-bearing disturbances. The further south you go, the shorter the dominance of these conditions and the less semi-arid the character of the region. The exception is a relatively dry zone along the Guinea coast, centered near the southern Ivory Coast, Ghana, and Togo (Adejuwon and Odekunle 2006). The reasons for the aridity of this region, where rainfall is generally between 800 and 1200 mm, are not clear but they relate in part to the mid-summer dry season. They probably include high pressure, cold coastal waters, and winds parallel to much of the coast line.

Throughout almost the entire semi-arid region the seasonal cycle of rainfall is characterized by a single maximum, almost always occurring in August (Fig. 16.16). In the northwestern extreme of the region, a slight secondary maximum occurs in the coldest months, and in the far south the maximum occurs in September rather than August. The uniformity of this distribution lies in the common meteorological origin of the rainfall throughout the region: the squall lines and mesoscale convective systems which propagate across the east–west extent of this zone (see below). These systems are concentrated in the latitudinal zone lying between the cores of the African easterly jet and the upper-level tropical easterly jet – a configuration related to the role of these jet streams in the origins of the disturbances.

Two of the most prominent climatic features of this region are the pronounced seasonal alternation between intense rainfall and parching aridity and the pervasive presence of mineral dust in the atmosphere. West Africa produces some 50% of the global atmospheric mineral dust content (Prospero *et al.* 2002). Intense dust outbreaks are a frequent occurrence throughout the year, although their frequency is reduced when the rainy season

commences. The dust can rapidly reduce visibility to less than 5 km during almost any season of the year. The source of the dust is mainly the dry playas, remnants of Holocene lakes that dried up and left behind fine-grained sediments (Prospero *et al.* 2002). Dust storms occur most frequently near these source regions, especially in southern Mauritania, northern Mali, Niger, and Chad. The Bodélé Depression in central Chad (Fig. 16.20) is probably the world's largest source of dust (Warren *et al.* 2007). The intensity of the outbreaks in this depression are related to a low-level jet stream (the Bodélé jet) that is particularly effective at dust mobilization and transport (Washington *et al.* 2006). The best descriptions of dust storms come from the Sudan, where they are called *haboobs*. In the central Sudan, 20 per year occur on average, mostly between May and August. Their frequency declines when the rainy season begins.

THE RAINFALL REGIME

Rainfall is associated primarily with large-scale systems that propagate westward across the continent (see Chapter 5). These are collectively termed mesoscale convective systems (MCSs) and include line squalls, mesoscale convective complexes (MCCs), organized convective systems (OCSs), and cloud clusters (Fink *et al.* 2006). The latter three types differ primarily in terms of intensity and size, but definitions and limits vary between various authors. Within West Africa, the frequency of MCS occurrence is greatest in the Sahel, from *c*. 10° N to 15° N, but they are also common in the equatorial latitudes (Zipser *et al.* 2006; Jackson *et al.* 2009). As few as 2% of these features produce 60–90% of the rainfall over the Sahel (Nesbitt *et al.* 2006). These are probably the particularly intense form of MCS, the OCS described by Mathon *et al.* (2002). By definition, this includes at least one cloud cluster with temperatures of 213 K or colder, a size of 5000 km² or larger, a lifetime longer than 3 hours, and a mean speed greater than 10 m/s. These comprise some 12% of MCSs over West Africa, but produce 80% of the total convective cloud cover and 90% of the rain in parts of the Sahel.

Localized thunderstorms produce most of the rest of the rainfall. In the subhumid region around 10° N, their contribution is as great as 40%. The number of thunderstorms (either localized or linked to disturbances) varies from about 10 or 20 per year in the north to well over a hundred in the south; these are, of course, concentrated in the rainy months. In the central Sahel, they frequently occur at night, coinciding with the passage of disturbance lines. Hail is almost unknown in the hotter, northern regions, but in southern areas it occurs about once in 5 years.

The semi-arid zones of West Africa occasionally receive precipitation from diagonal cloud bands that occur in the transition and cold seasons (Knippertz et al. 2003). This system rarely produces more than 25 mm of rainfall, but nevertheless contributes significantly to the overall availability of moisture in the northwestern extremes of the region. These systems, also called tropical plumes, can even bring rainfall into subequatorial regions along the Guinea coast, especially in the boreal winter season (Knippertz and Fink 2008).

EASTERLY WAVES, CONVECTION, AND THE AFRICAN EASTERLY JET

Two of the better-known features of the atmospheric circulation over West Africa are the mid-tropospheric African easterly jet and African easterly waves (Burpee 1972). These systems are closely linked (e.g., Leroux and Hall 2009) and both also show a close association with the squall lines and mesoscale systems that bring rainfall. Easterly waves (see Chapter 5) propagate across West Africa along two tracks (Fig. 16.21). One is centered at roughly 20° N and is coincident with the surface position of the ITCZ. The latitude of the other is somewhat variable, as it tends to lie equatorward of the AEJ, which shifts north or south from year to year (Grist 2002). Generally constrained to lie between the cores of the AEJ and TEJ, this second track lies within the latitudinal zone of 5°–15° N (Nicholson 2009a). In a typical year some 60 or 70 waves traverse West Africa (e.g., Pasch and Avila 1994). Most originate as far east as the Sudan and traverse most of Africa's east–west extent, leading to a rather remarkable tendency for interannual variations of rainfall to affect most of the zone. The number, intensity, and frequency of the waves change from year to year. Important factors in their variability include the strength of the tropical easterly jet, its latitudinal displacement with respect to the AEJ, and the vertical and horizontal wind shear in the vicinity of these jets (Nicholson et al. 2008).

Convective activity tends to occur in specific geographic locations with respect to both the jet streams and the easterly waves. Mesoscale convective systems and line squalls, which bring most of the rainfall to West Africa, are limited to the equatorward

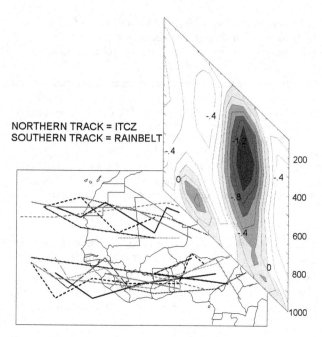

Fig. 16.21 Tracks of easterly waves over West Africa (from Baum 2006).

side of the African easterly jet, which steers them westward (Mohr and Thorncroft 2006). Thus, the seasonal contribution of these disturbances to rainfall in a given location is linked to the seasonal movements of the AEJ.

In the southernmost part of the semi-arid zone around 10° N, the contribution of MCSs peaks in the spring and in late summer or autumn. In mid-summer, when the AEJ is far to the north, thunderstorms provide most of the rainfall. In the Sahel, the contribution of MCSs peaks in mid-summer and the systems that bring rain early and late in the season are quite different (Lélé and Lamb 2009). The waves appear to organize the large-scale convection and probably enhance rainfall. Many reach the Atlantic and Caribbean, where they sometimes develop into hurricanes (Chen et al. 2008).

Most of the organized convection over West Africa consists of coherent episodes that span an average distance of some 1000 km and last about 25 hours, but some span nearly 1500 km and last up to 36 hours (Laing et al. 2008). Besides the waves, this is influenced by several factors, such as the Indian monsoon (Janicot et al. 2009), westerly wind anomalies, and intrusions of extremely dry polar air in the upper troposphere (Roca et al. 2005).

Much of the rain associated with these convective systems is stratiform, produced by stratiform clouds that develop during the night as the systems spread. In the Sahel, stratiform rain becomes increasingly important as the wet season evolves, with 35–50% of the rainfall in August and September being stratiform (Jackson et al. 2009). Consequently, in the zones with the greatest number of MCSs, there are two peaks in the diurnal cycle (Mohr 2004). A peak in the evening hours is associated with convective clouds, and a nocturnal peak is associated with the stratiform cloud that develops as the system matures.

THERMAL REGIME

In the drylands of West Africa, the thermal regime results from the combined influence of the seasonal cycles of solar insolation and cloudiness. In the Sahel and Soudan, cloudiness attains a minimum during December and January, when it is about 20–30%. However, the maximum insolation tends to be in April or May and October or November. Insolation is lower during the June–September rainy season, when cloudiness increases to about 40% in the north and 70% in the south.

In the semi-arid Sahel and Soudan, the highest mean monthly temperatures are on the order of 30°–35°C throughout the region. The lowest monthly means are generally at the most northern stations and are on the order of 20°C. The exception is highland regions, such as Tamanrasset, where the monthly mean can be as low as about 14°C. The seasonal pattern of temperatures follows that of insolation. Thus, the hottest months are April or May in most areas, before the rainy season commences.

SURFACE WINDS

The pattern of winds in this region, like rainfall, also shows a striking seasonality. Winds tend to be from the south or southwest during the summer season, when low pressure dominates, and from the east or northeast during the harmattan winter season. The southerly flow is often termed the SW monsoon, but this is not a true monsoon system. The duration of the harmattan is a function of latitude. Around 15° N it blows roughly half of the year (November through April), but at 18° N it prevails throughout the year. In the southern limits of this region near 10° S, it may prevail for as little as one month.

There are regional departures from this pattern, as a consequence of location, with respect to the pressure systems or proximity to the coast. In much of Senegal and western Mauritania, winds tend to be westerly for much of the year, as a result of sea breezes from the Atlantic and surface pressure patterns over the western Sahara.

16.2.3 INTERANNUAL VARIABILITY

Rainfall variability in West Africa has several striking characteristics. One is the frequent occurrence of multidecadal droughts. These include not only the recent dry period, which – at least temporarily – ended in the late 1990s (Nicholson 2005), but also several similarly extreme droughts in the historical past (Nicholson 2000). One such episode occurred in the 1820s and 1830s and was evident throughout the continent. A related characteristic is the exceedingly strong interdecadal signal, with Sahel rainfall in the 1950s and early 1960s being 50% greater than during the subsequent dry period.

Other striking characteristics are the vast spatial scale and the spatial coherence of the largest anomalies. The spatial coherence is so great that the time series shown in Fig. 16.22 approximately describes the variability throughout the region extending from

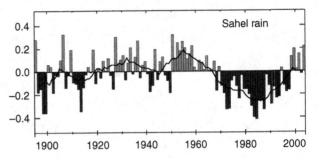

Fig. 16.22 Time series of rainfall in the Sahel, expressed as a regionally averaged standard deviation (%). The solid line shows 10-year running means.

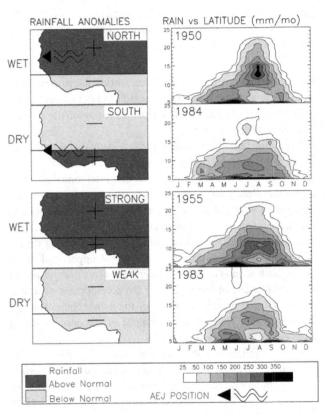

Fig. 16.23 The four principal anomaly types for West Africa. These are associated with either a change in the intensity or a latitudinal shift of the rainbelt. The dipole patterns correspond to the latter, as does a northward or southward shift in the axis of the African easterly jet (from Nicholson and Webster 2007).

10° N to 18° N and 15° W to 30° E. These implicate large-scale rather than local factors governing interannual variability.

Because of the strong spatial coherence, two basic spatial modes of variability describe most of the rainfall variability. These are (1) an out-of-phase relationship between the Sahel and the Guinea coast south of *c.* 10° N and (2) a pattern of increased or decreased rainfall throughout the region. Consequently, the annual rainfall anomalies over West Africa fall into one of the four patterns depicted in Fig. 16.23. The two cases with opposite anomalies over the Sahel and Guinea coast

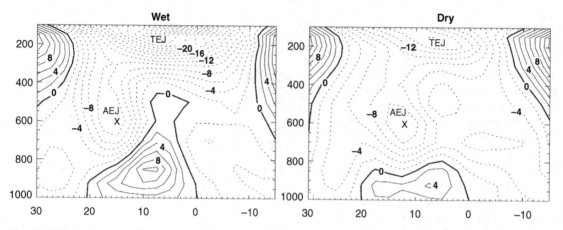

Fig. 16.24 Vertical cross-sections of zonal wind (m s⁻¹) in wet and dry years. Dashed lines/solid lines correspond to easterly/westerly winds. The tropical easterly jet (TEJ) and the two mid-tropospheric African easterly jets (AEJ) are indicated (from Grist and Nicholson 2001). Data are averaged for the years indicated in Fig. 16.23.

are linked to a north–south displacement of the tropical rainbelt. Those with increased or decreased rainfall throughout are associated with an intensification or weakening of the tropical rainbelt (Nicholson 2008).

For a long time it was assumed that year-to-year fluctuations in the Sahel were related primarily to shifts in the latitude of the ITCZ. It is now well established that interannual variability is more closely linked to factors in the circulation aloft (see, e.g., Zhang *et al.* 2006; Nicholson 2008, 2009b). The common denominator in wet years is a very strong tropical easterly jet over West Africa (Fig. 16.24), which promotes strong disturbances via its influence on wind shear and upper-level divergence (Nicholson *et al.* 2008). The critical factor in determining the displacement of the rainbelt is the development of a low-level equatorial westerly jet south of the Sahel (Nicholson and Webster 2007). This, in turn, develops in response to strong cross-equatorial pressure gradients.

Sea-surface temperature (SST) patterns play a critical role in interannual variability over the Sahel. A particularly important factor is an SST dipole, with SST anomalies of the opposite sign in the equatorial and subtropical Atlantic (e.g., Hastenrath 1990; Lamb and Peppler 1992), which appears in conjunction with the changes in the pressure gradient. High/low rainfall in the Sahel is associated with anomalously high/low SSTs in the subtropical latitudes and low/high SSTs in the equatorial latitudes.

The ocean's influence on Sahel rainfall goes beyond this Atlantic dipole. Associations have also been shown with global interhemispheric SST contrasts (Ward 1998) and SSTs in the Indian Ocean (Lu 2009), the Mediterranean (Rowell 2003), the Pacific (Rowell 2001), and the South Atlantic (Nicholson and Webster 2007). Atlantic warmings, termed the Atlantic Niño, have a strong influence on rainfall along the Guinea coast (Polo *et al.* 2008).

These seeming contradictions between studies variously implicating different regions of the global oceans can be readily explained by the interrelationships between these ocean regions and by the fact that SST–rainfall relationships are time- and space-scale dependent. The local importance of a given factor may be different from its importance for continental-scale variability; likewise, factors influencing intraseasonal, seasonal, and interannual variability can also be quite different.

16.3 EASTERN AFRICA

In this chapter, eastern Africa refers primarily to the countries of Ethiopia, Eritrea, Djibouti, Somalia, and the three countries traditionally referred to as East Africa: Kenya, Uganda, and Tanzania. The others are referred to as the "Horn of Africa." The region straddles the equator, extending from roughly 18° N to about 12° S. Except for coastal sectors (including most of Somalia), elevation throughout the region generally exceeds 1000 m, with many peaks attaining more than 4000 m. The major highlands include the Ethiopian highlands, the Ruwenzori to the west, and the areas surrounding Mts. Kenya and Kilimanjaro; most of the region is high plateau. The major lakes include Victoria (which provides the source of the White Nile), Turkana, Tanzania, and Malawi.

The central parts of Kenya and Tanzania are semi-arid, while deserts exist in northeastern Kenya, Somalia, and much of Eritrea. In Ethiopia, small desert basins are interspersed among the highlands. Rainfall is generally less than 600–800 mm/year outside of the highland regions (Fig. 16.25). The best-known deserts within eastern Africa are the Somali-Chalbi of northern Kenya, southern Ethiopia, and Somalia, and the Danakil Desert of Eritrea. In these deserts, mean annual rainfall is as low as 100–200 mm/year. Rainfall is extremely variable, even by comparison with other arid and semi-arid regions. The coefficient of variation of rainfall exceeds 50% throughout most of eastern Africa and exceeds 75% in many areas (Fig. 16.9).

In the semi-arid region of Kenya, vegetation tends to be *Acacia* bushland and thicket, with some areas of thicket mosaics that include grassland. Further south and west, in more humid regions, are woodlands and forests. In the driest desert

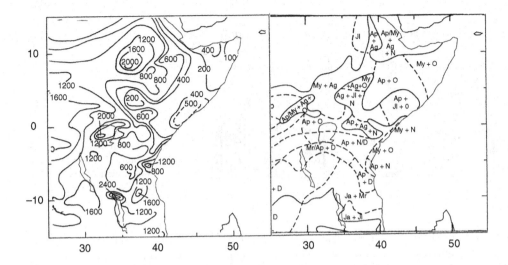

Fig. 16.25 Map of rainfall characteristics over eastern Africa (from Nicholson 1996b). (left) Mean annual rainfall (mm); (right) months of peak rainfall.

areas of Kenya, the dominant vegetation is shrub grassland, as it is along many of the desert coastal regions, including the Danakil Desert. The rest of the Somali-Chalbi is mainly *Acacia* bushland and thicket.

16.3.1 CLIMATIC CONTROLS

The wind and pressure patterns governing the region's climate are illustrated in Fig. 16.3. There are three major airstreams and three convergence zones. The airstreams are the westerly and southwesterly Congo airstream, the northeast monsoon, and the southeast monsoon, which takes on a southwesterly component locally in some seasons. Both monsoons are thermally stable, divergent, and relatively dry. The westerly flow from the Congo is – in contrast – humid, convergent, and thermally unstable. These airstreams are separated by two surface convergence zones, the ITCZ and the Congo Air Boundary; the former separates the two monsoons, the latter separates the easterlies and westerlies. Aloft, at about 2400–4000 meters, a third convergence zone is produced by the intersection of the dry, stable, northerly flow of Saharan origin and the more humid southerly flow.

There is a dramatic reversal of flow between the Northern Hemisphere (NH) summer and NH winter seasons, a situation that prompted the use of the term "monsoon." However, unlike true monsoons, the reversal is due to seasonal shifts of the trade winds and ITCZ, rather than to a reversal of the heating differential between land and water. The NE monsoon, which prevails during the NH winter, is made up of two quite different airstreams: a relatively humid one originating from and traversing the Atlantic Ocean, and a drier one that has traversed the eastern Sahara. The SE monsoon, which prevails during the NH summer season, often splits into two streams when it encounters the coast. One continues westward and the other turns abruptly northward to parallel the Somali coast. In the transition seasons there is a strong onshore component to the flow, and the elevation and friction of the shoreline promote rising motion

and rainfall. During both the winter and summer seasons, the monsoon flow tends to parallel the coast. The shoreline thus provides no uplift, and divergence and subsidence are produced by the differential friction of the land and water (see Chapter 5). Consequently, for the region as a whole, the transition seasons tend to be wet while the winter and summer seasons, with "monsoon" flow, tend to be dry.

16.3.2 THE RAINFALL REGIME

Meteorologically, eastern Africa is one of the most complex sectors of the African continent (Nicholson 1996b). The large-scale tropical controls, such as the convergence zones, are superimposed upon regional factors associated with the lakes, topography, and the maritime influence of the Indian Ocean and Red Sea. Northern and western sectors also mark the transition between the summer rainfall regime typical of semi-arid West and Central Africa, and the equatorial rainfall regime with a dry summer season. The boundaries of these regimes shift markedly from year to year, dramatically altering the seasonal cycle of rainfall.

This complexity manifests itself clearly in the patterns of rainfall, which are highly localized and strongly influenced by topography (Oettli and Camberlin 2005). Both the amount and seasonality of rainfall change markedly over relatively short distances (Fig. 16.25). Deserts are interspersed with tropical forests. Mean annual rainfall ranges from less than 200 mm/ year in the Somali-Chalbi Desert of northern Kenya, Somalia, Djibouti, and Ethiopia, to over 1600 mm/year, with humid climates that are confined mostly to the numerous highlands and a few coastal strips (Fig. 16.25). In some parts of the desert, rainfall is as low as 20 mm/year, while it exceeds 2000 mm/year in some highland regions.

Three primary seasonal patterns of rainfall prevail. These are illustrated for select stations in Fig. 16.26. Most areas have a largely bimodal distribution, with maxima during the transition seasons, March–May and October/November, and

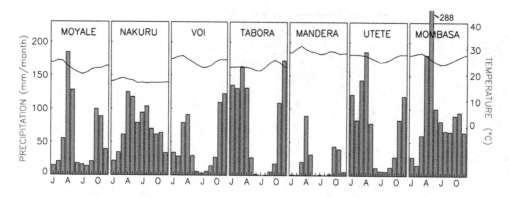

Fig. 16.26 Rainfall (mm) and temperature (°C) at typical stations in eastern Africa. See Fig. 16.19 for location of stations.

dry seasons in the summer and winter months of the higher latitudes: Moyale and Mandera, in the northeast, are good examples. The bimodal pattern is the equatorial regime and generally prevails from about 10° N to 10° S (Fig. 16.16). Aridity is most intense in the dry season of the boreal summer; very few areas experience rainfall from June to September. Further south, the rainfall regime is largely unimodal, with a peak in the austral summer. The month of maximum rainfall shifts progressively later, from November to January, with increasing latitude. The station Tabora (Fig. 16.26) is a good example, but it still has a minor peak in March, during the "long rains." Voi and Utete represent a transition between this regime and the bimodal equatorial regime, showing the "long rain" peak and a second peak in December. In western Ethiopia, northwestern Kenya, and parts of Uganda, rainfall is similarly unimodal, with a peak in the boreal summer (Nicholson and Selato 2000). The coastal sectors of Kenya and Somalia are exceptions to these three patterns (Camberlin 1996) and are discussed below.

In the areas with a strongly bimodal distribution, the earlier rainy season (roughly March–May) tends to have the highest and most widespread rainfall. This is termed the "long rains" in Kenya and surrounding areas, but the "small rains" in Ethiopia. The second rainy season (mostly October and November) is generally weaker and shorter. Hence this season is termed the "short rains" in Kenya and nearby areas of Tanzania and the Horn of Africa.

The two rainy seasons have very different characteristics (Nicholson 1996b; Camberlin et al. 2009). The "short rains" are much more variable from year to year than the "long rains" and rainfall occurs more persistently during the "short rains." The spatial coherence is much higher during the "short rains," with most stations having anomalies of the same sign during a given season. The "short rains" are also more temporally homogeneous, with October and November rainfall being highly correlated (Nicholson and Entekhabi 1986). In contrast, the correlation between March and April rainfall is relatively weak, and April and May are completely uncorrelated (Camberlin and Philippon 2002; Zorita and Tilya 2002). The origin of rainfall in these sub-seasons is different, as are the extra-regional teleconnections. Rainfall in March and April shows strong links to the ITCZ

latitude and ENSO (El Niño/Southern Oscillation), while May rainfall shows a strong link to the Asian monsoon.

The coastal belt, up to 50 km inland, has a very distinctive rainfall regime. The climate is largely maritime (Camberlin and Planchon 1997). It is characterized by an enhanced and delayed "long rains" season, and the dry season in the boreal summer is absent along the coast. Most rainfall occurs at night or in the early morning. Rainy spells occur along the whole coast when strong easterly or southeasterly winds weaken the sea breeze.

The coast along the Red Sea Trench also has a distinctive regime. One of the most abrupt changes of seasonality in the world occurs here (Flohn 1965). In the lowest elevations a winter rainfall regime, as typified by the station Massawa, occurs from at least 12° N to 26° N. Higher upslope in the trench a summer rainfall regime prevails, as typified by Asmara. These stations are separated by a distance of only 63 km but 2000 m in elevation. In between lies an area of year-round rainfall and lush, green vegetation.

The patterns of rainfall become very complex where the summer and transition-season rainfall regimes converge. Just to the southeast of the region with summer rainfall is an area with two rainfall maxima: the maximum occurs during the summer season (generally in August) and a second rainy season occurs in the April–May transition season (see Nakuru in Fig. 16.26). Further south and east, where the summer rainfall regime meets the classical bimodal regime of eastern Africa, there are three peaks in the annual cycle: a rainfall peak occurs in the NH summer season (generally with an August maximum) and in both transition seasons. Some semblance of this regime is evident at Nakuru. In southern sectors of the summer rainfall region, many locations have a small secondary maximum in March.

16.3.3 CAUSES OF ARIDITY IN EASTERN AFRICA

The classical literature, such as Griffiths (1972), considered the aridity of eastern Africa to be perplexing. The area between approximately 3° and 10° N is the most arid. Such dryness in equatorial latitudes is quite uncommon, particularly at the eastern extremity of a continent. The cause of the dry conditions is unclear, but it is likely that several factors play a role (Nicholson

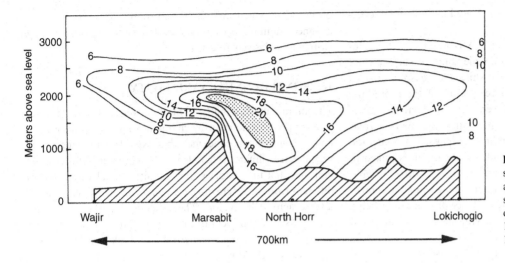

Fig. 16.27 Typical easterly wind speeds (m s^{-1}) of the Turkana jet as a function of altitude (meters above sea level); shading indicates speeds exceeding 20 m s^{-1} (from Nicholson 1996b, as modified from Kinuthia 1992).

1996b). One is the thermal stability of the two monsoons and the fact that they tend to be parallel to the coast during the extreme seasons. Another factor is that the moist airstreams in the region are relatively shallow; dry, stable air from anticyclones over the Sahara and Arabia prevails aloft. The highlands also play some role: the moist, unstable, westerly air from the Congo is blocked by the highlands, especially in Ethiopia and Somalia. It also subsides and dries out in these regions, which are leeward rain shadows. This leads to an extremely complex pattern of rainfall and aridity over the Ethiopian highlands, with pockets of humid climate alternating with arid ones within a few tens of kilometers.

The local low-level jet streams probably contribute to the aridity as well. The Turkana jet (Kinuthia 1992) has core speeds of over 20 m/s at an elevation of about 1500 m (Fig. 16.27). This jet blows southerly and southeasterly into the Turkana channel between the Ethiopian and East African highlands (Fig. 16.28). Patterns of divergence and descending flow associated with it enhance aridity throughout the Turkana basin. Moreover, the airstream entering the Turkana Channel is also divergent. The low-level Somali jet just offshore promotes aridity by way of the subsidence induced through the friction differential between land and water (see Chapter 4). It also produces the upwelling of cold water along the coast.

16.3.4 FACTORS CONTROLLING RAINFALL AND ITS SEASONALITY

The question of what produces rainfall in eastern Africa is somewhat obscure (Nicholson 1996b). Certainly the humid, unstable, westerly Congo airstream plays a role, and incursions of this airstream into eastern Africa generally bring high rainfall (van Heerden and Taljaard 1998). Easterly disturbances from the Indian Ocean also affect the region (Okoola 1989). These are most frequent in May–June and October–November in areas near the Arabian Sea, including eastern Sudan and Ethiopia. However, many of the phenomena associated with rainfall occurrences elsewhere in the tropics, such as easterly waves and

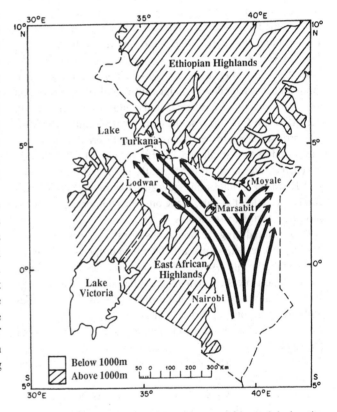

Fig. 16.28 Small-scale topography over eastern Africa and the location of the Turkana jet (from Nicholson 1996b, modified from Kinuthia 1992).

large mesoscale convective systems, appear to be lacking here (Toracinta and Zipser 2001a; Zipser et al. 2006).

Rainfall predominantly occurs as part of organized patterns of convection that propagate eastward, generally on time scales of 40–50 days (Pohl and Camberlin 2006a). They are associated with the Madden–Julian Oscillation. Clear linkages to large-scale disturbances or to the climatological convergence zones are not apparent, except in the more northern extremes, where upper-level troughs and cold fronts of mid-latitude origin occasionally bring rainfall.

Much of the literature on eastern Africa suggests that rainfall events are linked to synoptic-scale zones of surface wind discontinuities or convergence, while the seasonal march of precipitation is assumed to be a result of the latitudinal migration of the ITCZ. However, synoptic analysis generally fails to detect such convergence zones, and the rain areas do not show any particular relationship with the position of the ITCZ. Nevertheless, the idea that rainfall is linked to synoptic-scale pressure fields, which in turn perturb the wind field and trade inversion, is becoming increasingly accepted. However, there is also increasing evidence of rain events being linked to upper-level disturbances with few manifestations at the surface.

December–February is when the NE monsoon prevails, and extensions of the Saharan and Arabian highs affect the northern extremes of eastern Africa, including Somalia. High rainfall is limited to southwestern regions, a few highland regions in Ethiopia and elsewhere, and in a narrow strip in Eritrea along the Red Sea Trench.

In April and May, a low develops over the Ethiopian highlands and southerly flow dominates over southern Ethiopia and areas further south. Convergence and rainfall are associated with the Ethiopian Low and also the Congo Air Boundary, which is then situated along the western fringes of eastern Africa. This is the period of the "small rains" in Ethiopia and Eritrea (locally called the *belg* rains). Further south, particularly in Somalia and Kenya, the "long rains" prevail. Here rainfall in this season is due, at least in part, to the onshore component of the SE monsoon during these months.

Convection during the "long rains" is associated with westerlies at low levels, up to at least 700 mb, and upper-level easterlies (Camberlin and Okoola 2003). The westerlies are associated with two low-pressure systems on either side of the equator. During this season, occasional cold waves in the westerlies aloft and even some upper-level mid-latitude cold fronts extend into Ethiopia, also resulting in intense convective activity (van Heerden and Taljaard 1998).

The summer months of June–September are the driest in the region as a whole, even though low pressure develops over the Indian Ocean and Arabian Sea. However, this is the wettest season in roughly half of Ethiopia, with the percent concentration in this season increasing northward (Segele and Lamb 2005). This season, locally termed the *kiremt* rains in Ethiopia and the *gu* rains in Somalia, follows the *belg* rains of March–May, but the two seasons are separated by a distinct minimum in the seasonal cycle (Segele and Lamb 2005). Much of this rainfall is a result of local convective instability, which peaks in summer with the intense heating of the Ethiopian highlands. Consequently, spatial variability is high (Conway *et al.* 2004).

The transition season of October and November is the period of the "short rains" in the equatorial latitudes of eastern Africa (locally called the *der* or *deyr* rains). Rainfall occurs throughout most of the region during these months. The relatively strong

onshore component of the SE monsoon at this time probably enhances rainfall. During this season, northern Ethiopia and Eritrea are relatively free of rainfall.

16.3.5 INTERANNUAL VARIABILITY

The interannual variability of rainfall over East Africa is characterized by two noteworthy characteristics. One is the exceedingly strong spatial coherence that extends across regions of very diverse climates (Nicholson 1996a, 1996b). This strongly indicates that the factors governing interannual variability are different from those creating the mean climate, which is highly diverse across the region. The other is that the briefer and weaker "short rains" contribute most of the interannual variability, despite the much higher rainfall totals during the "long rains" of March–May.

The strong spatial coherence of rainfall variability is illustrated in Fig. 16.29. The indicated loading pattern roughly describes the first principal component for both annual totals and seasonal totals. It explains 36% of the annual variance, but over 50% of the variance during October–November.

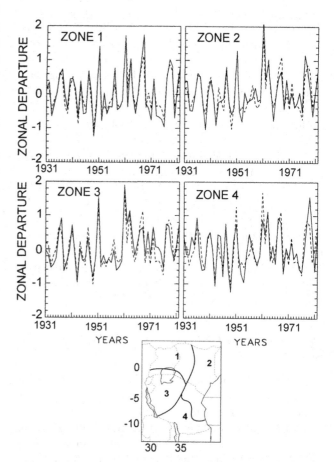

Fig. 16.29 The coherence of rainfall variability throughout eastern Africa. Graphs show the annual rainfall in each of four sectors shown in the inset, compared with variability averaged throughout the region as a whole. Rainfall is expressed as a regionally averaged standard departure (from Nicholson 1996b).

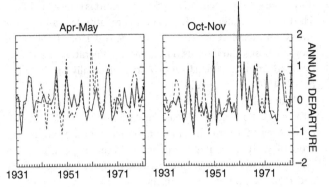

Fig. 16.30 Time series of April–May and October–November rainfall compared with annual rainfall over eastern Africa. Units are regionally averaged standard departures (from Nicholson 1996b).

Table 16.2. *November 1961 rainfall at select Kenyan stations, compared with mean November rainfall.*

Meteorological station	Mean November rainfall (mm)	November 1961 rainfall (mm)
Lodwar	16	190
Lamu	35	212
Lokitaung	39	302
Eldama	48	402
Makindu	48	478
Mandera	52	193
Malindi	54	238
Kitale	57	365
Wajir	58	612
Nakuru	60	280
Rumuruti	79	350
Garissa	81	412
Moyale	86	362
Mombasa	99	217
Marsaabit	143	612
Murango	187	610
Embu	189	605
Machakos	189	600

Consequently, the interannual variability in areas of predominantly summer rainfall is nearly identical to that of the regions with the bimodal seasonal cycle.

The importance of the "short rains" is seen clearly in the comparison of the time series of the seasonal rainfall with annual rainfall (Fig. 16.30). The correlation between October–November rainfall and the annual total is 0.71, compared with 0.53 for the correlation between April–May and annual rainfall. Consequently, considerable emphasis has been placed on predicting the "short rains," but with only moderate success (Hastenrath 2007).

The interannual variability of the "long rains" appears to be even less predictable, although there has been some success in predicting extremely wet or dry years (Diro *et al.* 2008). Interannual variability of the "long rains" is seen not only in the total amount of rainfall, but also in the frequency and length of dry and wet spells (Mapande and Reason 2005), the dates of onset and cessation, and the daily intensity of rainfall (Camberlin *et al.* 2009).

ENSO has a strong influence in the region, particularly during the short rains of October–December. The short rains tend to be anomalously high during El Niño events (Hastenrath *et al.* 2004; Bowden and Semazzi 2007) and anomalously low during La Niñas (Nicholson and Selato 2000). The influence during this season is probably tied to El Niño's impact on SSTs in the Atlantic and Indian Oceans (Nicholson *et al.* 2001). Strong positive SST anomalies appear in the equatorial Indian Ocean and southeastern Atlantic during October and November of El Niño years. El Niño also tends to decrease rainfall during the long rains of March–May but the impact is weak, particularly in May (Camberlin and Philippon 2002). ENSO also influences the summer rainfall that prevails in parts of eastern Africa (Eritrea, western Ethiopia, parts of Uganda and Kenya). La Niña/El Niño events are associated with anomalously high/low summer rainfall (Korecha and Barnston 2007).

There have been many studies examining the relationship between East African rainfall in various regions and seasons and sea-surface temperatures over various parts of the globe. These include studies of interannual variability (e.g., Ummenhofer *et al.* 2009; Segele *et al.* 2009a, 2009b; Nicholson and Entekhabi 1987), as well as case studies of droughts and floods (e.g., Hastenrath *et al.* 2007; Kijazi and Reason 2009a, 2009b). Links to both the Atlantic and Indian Oceans have been found.

A spectacular example of an extreme event linked to Indian Ocean SSTs occurred in 1961 (Kijazi and Reason 2009b). During this year Lake Victoria rose several meters and reached levels unattained since the nineteenth century. The heaviest rainfall occurred during October and November (Table 16.2). At stations in northern Kenya November rainfall was several times greater than the monthly mean and approached or exceeded annual means in many sectors. For example, Wajir, Eldama, and Lokitaung received 612, 402, and 302 mm, respectively, during November, compared with monthly means of 58, 48, and 39 mm.

16.3.6 THE THERMAL REGIME

The thermal regime over eastern Africa is at least as complex as the rainfall regime. It is a function not only of geographical factors, such as location and elevation, but also of the conditions of cloudiness and rainfall. Perhaps the overriding factor throughout the region is elevation; it correlates closely with mean annual maximum and minimum temperatures and with annual range. The distance from the coast and, to a lesser extent, from a lakeshore, also plays a role, with both diurnal and annual ranges increasing with distance inland. Latitude primarily affects the annual range, including whether or not distinct cool seasons are evident. The diurnal and annual ranges are reduced by rainfall, ground moisture, and clouds, so that – other factors being equal – the largest variability is evident in the driest regions. In Ethiopia and Eritrea, elevation and, to a lesser extent, latitude are major factors. In the Horn of Africa, aridity and proximity to a coastline produce opposite effects on the thermal regime. Elsewhere in eastern Africa, moderately high elevation (a high plateau) influences the thermal regime, with distance from the Indian Ocean playing a secondary role.

Throughout most of the region the diurnal temperature range exceeds the annual. The annual range is generally on the order of 2°–6°C (Figs. 16.5 and 16.26), being lowest near the equator, but in the winter rain region of the Red Sea Trench a temperature range of 8°–11°C is more common (Griffiths 1972). The diurnal range in the Horn of Africa is roughly 5°–15°C, being a function of the distance inland. In East Africa and Ethiopia, the diurnal range is less variable. In Kenya and Tanzania it is generally 10°–13°C, although 7°–9°C is more usual near the coast or lakeshores. In Ethiopia, the diurnal range depends greatly on the season. This region experiences two "cool" seasons, one a result of latitudinal effects during the winter season, the other a result of cloudiness during the rainy season. During the hot period, diurnal temperature range is generally 14°–17°C, but about 15°–20°C during the dry cool season. It is reduced to roughly 10°C during the wet cool season.

In eastern equatorial Africa, the warmest temperatures tend to occur during the summer season of the respective hemisphere. Thus, the hottest months are variable, being April, May, or June in Eritrea, northern Ethiopia, and Somalia, but March or April in southern Ethiopia and Somalia and most of Kenya and Uganda. The hottest month shifts to October or November in Tanzania. There are some peculiarities along the coast, where the warmest month shifts progressively later with increasing northern latitude. Thus, it is February on most of the Tanzanian coast, March in Kenya, April along the Indian Ocean coast of Somalia, but May along the Gulf of Aden and the Red Sea coast of Somalia, Ethiopia, and Eritrea.

In most of East Africa and southern Somalia, the coldest month is July or August, even though this is generally the dry season. In northern Somalia it tends to be December or January. There are two cool seasons in most of Ethiopia; as with the warm seasons, they are determined both by the timing of the rains and the position of the sun and are quite variable within the country.

The mean daily maximum temperature is generally on the order of 25°–30°C in East Africa, Eritrea, and Ethiopia, dropping to about 20°C at elevations exceeding 2000 m, but higher at low elevations and in the desert regions. Maximum temperatures are higher in the Horn of Africa, where arid conditions prevail. Here they are usually on the order of 30°–35°C. Near to the coast, 20°–30°C is more usual, but mean maximum temperatures of 40°C or higher are reached at a few stations. Absolute maxima are lowest in East Africa, at equatorial latitudes; there they are on the order of 30°–35°C in most locations and months. Temperatures in excess of 40°C have been reported at

only three stations, generally in desert locations (Griffiths 1972). The range is greater over Ethiopia, where latitude is an important factor; absolute maxima are on the order of 30°–40°C in most locations and months. The highest temperatures have been reported from the arid Horn of Africa, generally on the order of 35°–45°C at most stations.

Mean daily minimum temperatures are on the order of 15°–20°C in most of equatorial eastern Africa, except at high elevations. Above 2000 m, mean minimum temperatures are on the order of 8°–10°C. Ethiopia experiences lower temperatures than other countries in the region because of both higher latitude and higher elevation. Here, mean minimum temperatures are usually on the order of 10°–15°C, although they may be as high as 20°C in some areas. The Horn of Africa experiences higher minima because proximity to the coast prevents extreme temperature drops. There the mean minima are 20°–25°C in most areas, but they range from as low as 10°C to as high as 30°C at a few stations. The absolute minima are lowest for Ethiopia and vary widely within the Horn of Africa. At most Ethiopian stations, they are on the order of 0°–10°C, although subfreezing temperatures have been recorded at a number of stations. In Kenya, absolute minima are around 3°–5°C for highland locations, but 10°–15°C in the lowlands. The range over the Horn of Africa is about 5°–25°C, with the lowest values furthest inland.

16.3.7 OTHER CLIMATIC ELEMENTS

The relative humidity is generally constant over equatorial eastern Africa, showing little day-to-day or seasonal variation. In the early morning it is on the order of 85–95% at most stations, except in the desert, where it is generally 65–75%. Mid- or late afternoon minima are around 50% in semi-arid regions and 35% in arid regions. Over the Horn of Africa, it is around 70–85% in the early morning, but anywhere from 30% to 80% in the afternoon, with proximity to the coast playing a big role. Over Ethiopia, relative humidity varies seasonally. Many stations have early morning readings of 60–70% most of the year, but these values reach 85–95% during the rainy season. Afternoon readings are generally 20–40% during dry periods and 65–80% during wet periods. Fog is uncommon in the eastern African drylands except in some of the highlands and along some coastal sectors, particularly in Somalia.

Other elements of climate are difficult to generalize, due to the complexity of the causal factors. Winds and cloudiness, for example, are at least as dependent on local topography and coastal effects as on the general atmospheric circulation. There is a tendency for winds from northerly or easterly directions along the coast and over Somalia from November to April, with southwesterlies the rest of the year. Inland, winds are more variable.

Cloudiness tends to be greater inland than along the desert coasts. At Moyale it has a pronounced annual cycle, ranging from 38% in January and February to 75% in April through October. At Magadi and Garissa it is 50–60% year-round, but generally only 10–25% year-round over the desert regions of Somalia. In the Ethiopian deserts, interspersed among the highlands, winds and cloudiness are strongly controlled by local topography.

16.4 SOUTHERN AFRICA

Drylands cover nearly all of southern Africa from *c*. 22° S to the southernmost tip of the continent. The exceptional humid environments are limited to the highland regions, where rainfall is orographically enhanced, or to the temperate southeast margin, where year-round rains occur. The region can be subdivided into an arid or desert province and a semi-arid savanna. Southern portions of Zambia, Zimbabwe, and Malawi, and parts of Namibia and Angola fall into the latter. Much of Botswana and South Africa, and parts of Namibia fall into the former. The Kalahari "desert" of southern Africa spans both regions.

The vegetation in the drier regions furthest south is generally shrubland of various types, including the thornveld of Botswana (an *Acacia*-dominated shrub savanna) and the karoo of South Africa (Fig. 16.31). In the Mediterranean climates of the Cape, however, is a shrubland with a heath-like vegetation called "fynbos." In the higher elevations of northeastern South Africa is highveld grassland (Fig. 16.31) that gives way to woodland

further north. In the wetter savannas to the north (principally in Angola, Zambia, and Zimbabwe) there is a north to south progression from woodlands to scrub woodlands dominated by the mopane tree (*Colophospermum mopane*), then dry woodland termed "miombo." The soils are mainly weakly developed arensols (the Kalahari sands) in the west and better developed lixosols in the east, while calcisols prevail in the Namib Desert.

16.4.1 GENERAL ATMOSPHERIC CIRCULATION

The pressure field over southern Africa is dominated by the subtropical high-pressure belt of the Southern Hemisphere. This feature exerts the dominant control on the region's climate (van Heerden and Taljaard 1998). In summer there are two distinct cells over both oceans (the South Atlantic or St. Helena High and the South Indian High) and a shallow heat low over the land centered at about 15° S. Aloft, above 700 mb, is again high pressure and anticyclonic circulation. In winter, the high-pressure cells move several degrees equatorward and substantially westward. Over the southern African subcontinent, a ridge of high pressure prevails. Thus, the atmosphere over southern Africa is dynamically relatively stable throughout the year.

Stability is enhanced by a subsidence inversion that prevails throughout southern Africa for most of the year, in association

(a)

(b)

(c)

Fig. 16.31 Vegetation types (arid and semi-arid) of southern Africa: (a) karoo, (b) Kalahari thornveld, (c) grassland of the highveld.

Fig. 16.32 An ever-present inversion layer is made visible by the high dust content in the mixed layer beneath it. Often clouds form, but development is limited by the inversion, resulting in numerous small cumulus at the inversion base.

with the prevailing high pressure. Its height varies but is typically between about 650 and 750 mb and is more intense in winter. In summer the inversion is interrupted in the vicinity of the convergence zones, principally over Zambia, eastern Angola, southern Tanzania, and northern Mozambique. The inversion layer is often clearly visible either at the top of fields of shallow cumulus clouds or at the top of the dense dust layer (Fig. 16.32).

The influence of the high-pressure cells is fourfold. They buffer the core and equatorward extremes of the region against the travelling depressions and anticyclones of the mid-latitudes. The steadiness of the highs creates a day-to-day persistence of the pressure field at the level of the southern African plateau, despite considerable changes in the upper air flow and weather. The steadiness also results in virtually rainless dry seasons throughout most of southern Africa. The trade winds on their equatorward side make subtropical latitudes accessible to the tropical cyclones of the southwestern Indian Ocean.

Poleward of the highs lie the mid-latitude westerlies and associated transient disturbances, short-lived migrating cyclones and anticyclones (Fig. 16.4). Westerly waves and disturbances in the mid- to upper troposphere often override the anticyclonic circulation below. Thus, unlike in the equivalent latitudes of Northern Hemisphere Africa, the tropical and extra-tropical circulations are virtually superimposed upon each other. Consequently, many of the rain-bearing systems have both tropical and extra-tropical components and react to vagaries of the circulation in both regions.

Equatorward of the subtropical highs is a region where the primary influence is the various convergence zones of the outer tropics (Fig. 16.3). The primary air masses associated with these airstreams are the NE monsoon, which can be dry or moist, depending on its track; the dry SE trades; and the Congo airstream northwest of the Congo Air Boundary, which is very moist and unstable. The ITCZ is not strongly evident at the

surface but is marked on a day-to-day basis by regions of convection associated with propagating disturbances with lifetimes measured in days.

16.4.2 CONTROLS ON DAY-TO-DAY WEATHER

The disturbances that control day-to-day weather over southern Africa are linked to the tropical easterlies or mid-latitude westerlies or are hybrid features that combine characteristics of tropical and mid-latitude systems (van Heerden and Taljaard 1998). Pressure systems from higher latitudes are a major control on the day-to-day weather patterns in extra-tropical southern Africa.

The prevalence of the various systems is largely a function of latitude. In the summer season, precipitation in extra-tropical southern Africa is linked to frontal systems and to tropical cyclones from the Indian Ocean. In somewhat lower latitudes, easterly waves and tropical low-pressure systems also play a role. The disturbances of winter are primarily cold outbreaks and cut-off lows. The patterns of rainfall associated with these systems are frequently quite different in the coastal regions of high latitudes than in the interior and subtropical latitudes. This is particularly true in winter, when precipitation is generally limited to the south and southeast coasts and the winter rains region of the Cape Province in the southwest.

The most common synoptic situation, "normal undisturbed weather," is high-pressure dominant throughout the subcontinent both at plateau level and in the mid-troposphere (Tyson 1986). While bringing fair weather to most of the subcontinent, these conditions produce rain in the Cape region of winter rainfall. Similar situations occur 60–70% of the time in winter. When at the same time a coastal low appears along the west or south coast, this situation results in warm "berg" (mountain) winds, which subside and warm adiabatically in the lee of the highlands.

During winter and spring, occasional cold outbreaks, and even thunderstorms, occur in association with upper-level troughs in the westerlies and surface lows. Such situations bring a persistent cold snap to areas as far equatorward as Zimbabwe, but the precipitation associated with them generally affects only areas of the Cape province (the winter rainfall region) and the south and southeast coasts, where snowfall may result. The interior and other areas of central Africa remain dry.

The heaviest rains experienced in southern Africa are those caused by "cut-off lows," which bring warm, moist equatorial air from the northwest. These systems, also called subtropical lows, can occur at any time of year, but are most common from March to May (Singleton and Reason 2007). A second maximum occurs in September. On average about 11 occur per year. Roughly half of these systems bring copious rainfall to large portions of the interior. When unusually persistent, the result may be extreme flooding, particularly when followed by an anticyclone bringing in moist air from the south that is orographically forced to ascend. Generally two or three produce flooding each year. Winter occurrences are often accompanied by snowfall in the southeast. A cut-off low brought 600 mm of rain to the Kwazulu-Natal province during the period September 26–29, 1987, and over 900 mm to northern coastal areas of South Africa (van Heerden and Taljaard 1998). The ensuing flooding took more than 300 lives and caused more than one billion rands' worth of damage.

A typical summer situation that promotes rainfall is that of the diagonal cloud bands (see Fig. 5.13) associated with a trough in the upper-level westerlies (tropical/temperate troughs) (Todd and Washington 1999b). They generally occur when there is a surface- to mid-level low over central Africa, a high to the southeast, and the wave in the upper-level westerlies overrides this system. Once established, this situation can persist for up to 14 days, bringing protracted rainy weather over the northern and eastern portions of the interior.

In the more tropical latitudes of Botswana, Zambia, Zimbabwe, or Malawi, the rain-bearing systems also tend to be a result of tropical/extra-tropical interaction. Four types of westerly wave interactions with tropical systems affect rainfall in this region (Kumar 1978). Distinguished on the basis of their effect on the tropical system, these include intensification and displacement of the low over Angola, direct enhancement of rainfall, temporary disintegration of the ITCZ over the subcontinent, and intensification of the coastal low over southeastern Africa with an associated southward displacement of the convection and tropical low-pressure systems. These may overlap with the tropical/temperate troughs.

16.4.3 THE RAINFALL REGIME OF THE SOUTHERN AFRICAN DRYLANDS

Over southern Africa, rainfall generally decreases with increasing latitude until about 25° S (see Section 16.1.2). This pattern is dictated by the increasing influence of the zone of high pressure over southern Africa. Further south, isohyets are oriented much more north and south, with drier conditions prevailing in the west and wetter conditions in the east. The east–west contrast is linked to the contrasting meteorological conditions on the eastern versus western portions of the subtropical highs. Relief also augments the rainfall on the eastern side of the continent.

Figure 16.33 shows the seasonal cycle of rainfall at select stations in southern Africa. Rainfall occurs mainly in summer from about 10° S to 30° S, with the maximum generally occurring in

January. Tete and Gaborone are representative stations. Rainfall on the west and southwest coast and in parts of the interior of southern Africa follows a different pattern. This area has a late summer maximum; peak rainfall generally occurs in February or March, but may occur as late as April or May in some areas. This regime is illustrated by Upington and Grootfontein. In the extreme southwest, winter rainfall prevails, although summers are generally not entirely rainless. Okiep and Cape Town are representative stations.

Since much of the precipitation is convective, thunderstorms are fairly common throughout southern Africa. The average annual number of days with thunder is between 30 and 70 in arid and semi-arid sectors, except for the arid southwest, where it is rare. The number of days increases to over 100 in regions of Zambia and Zimbabwe where mean annual rainfall is on the order of 900–1000 mm. Hail is quite uncommon in the dryland sectors. It occurs in parts of Zimbabwe and Zambia, but not as frequently as might be expected from the high thunderstorm frequency in these areas. Snow rarely falls in the dryland regions, probably never occurring further equatorward than 23° S (Schulze 1965).

Cloudiness generally reaches a maximum around January, ranging from about 20% in the driest regions to over 50% in the wettest semi-arid areas. It falls to around 10–20% in July, which is the dry season over most of the subcontinent. Unlike in the Sahel, the cycle of insolation is roughly in phase with the cycle of cloudiness, reaching a maximum in July or August and a minimum in December–February. Fog is quite common in the winter rains region of the southwest coast (see Chapter 20).

Throughout most of southern Africa the rainfall regime has the characteristics of both extra-tropical frontal rainfall and tropical convective rainfall. Consequently, rainfall tends to persist for longer periods, with rainy spells often lasting several days. It also tends to be less intense (i.e., lower monthly totals) than in comparable semi-arid tropical regions of the Northern Hemisphere.

16.4.4 INTERANNUAL VARIABILITY OF PRECIPITATION

Two interrelated factors appear to control the lion's share of interannual variability in southern Africa: the occurrence of easterly versus westerly winds and the location of the cloud

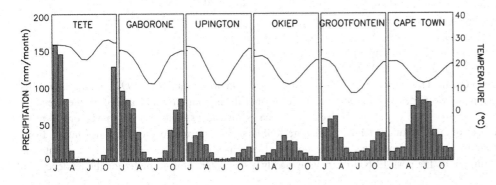

Fig. 16.33 Rainfall (mm) and temperature (°C) at six typical stations in southern Africa. See Fig. 16.19 for location of stations.

bands associated with the tropical-temperate troughs that develop over the region. The cloud bands have two preferred positions: one over South Africa and the other over the Indian Ocean (Harangozo and Harrison 1983; Fauchereau *et al.* 2009). Dry episodes and wet episodes over southern Africa are manifestations of changes in the frequency of cloud bands, which in turn are related to the upper-level winds. In the case of westerly flow aloft over central Africa, the bands tend to lie over the Indian Ocean, but are generally situated over the continent when tropical easterly flow prevails. Thus, easterly winds are linked to above-normal rainfall. The rainfall response to El Niño in this region appears to relate to this factor. In subtropical regions, such as Angola and Zambia, wet spells are characterized instead by low-level westerly anomalies (Hachigonta and Reason 2006).

Interannual variability is strongly coherent, with the most prominent spatial anomaly patterns indicating anomalies of like sign throughout most of southern Africa, sometimes up to equatorial latitudes. This is consistent with the large-scale systems governing the variability. Another notable feature is the predominantly out-of-phase relationship between equatorial and southern Africa (Nicholson 1986a, 1986b). This association can be readily explained by the impact of El Niño on these regions.

Rainfall in southern Africa is very strongly influenced by both the Atlantic and Indian Oceans. However, ENSO is a driver of much of the variability in the equatorial and southern portions of these oceans. It is difficult to distinguish between these three factors, as they are all interrelated (Nicholson *et al.* 2001).

Various studies have also shown many diverse SST patterns that are related to wet or dry conditions over southern Africa. The ocean sectors exerting the strongest influence appear to be the SW Indian Ocean, the Agulhas Current, the Benguela Current in the SE Atlantic, and the southwestern Atlantic off the coast of Brazil, in addition to the Indian Ocean dipole (e.g., Washington and Preston 2006; Williams *et al.* 2008). SSTs in large sectors off the east coast of southern Africa play a critical role in interannual variability, with warm SSTs promoting high rainfall (Reason and Mulenga 1999). South African rainfall also appears to be very sensitive to the Indian Ocean SST dipole (Reason 2002). Rainfall in the winter rainfall region of the Cape responds more strongly to Atlantic SSTs (Reason and Jagadheesha 2005). Extreme rainfall events can also be produced by either the Atlantic or the Indian Ocean.

The years 1974 and 1976 were some of the wettest on record, with rainfall several standard deviations above normal over most of the subcontinent (Fig. 16.34) (Nicholson *et al.* 1988). These rainfall anomalies were associated with warm anomalies in the subtropical SW Indian Ocean (Fig. 16.35) and cool anomalies in the northern SW Indian Ocean. This same configuration of anomalies has been seen in many other cases of very wet conditions.

The influence of ENSO is likewise complex. Drought tends to occur in southern Africa during most El Niño episodes, generally during the late summer or autumn of the post–El Niño

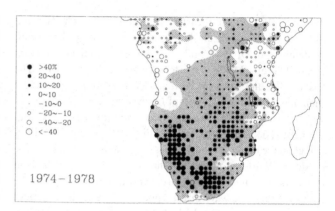

Fig. 16.34 Rainfall over southern Africa for the period 1974–1978. Rainfall is expressed as a regionally averaged percent departure from the long-term mean.

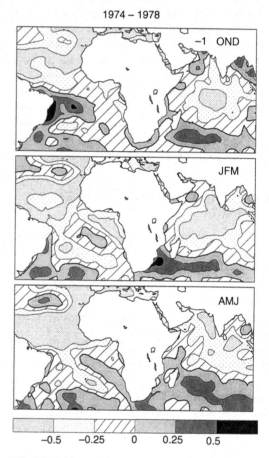

Fig. 16.35 SST anomalies (°C) for the period 1974–1978 (from Nicholson *et al.* 2001).

year (Fig. 16.36) (Nicholson and Kim 1997). Rainfall tends to be above normal during summer and spring of the previous year. La Niña is instead associated with above-normal rainfall. The ENSO signal appears to propagate northward from the poleward extremes of the continent, reaching equatorial regions nearly a year later. This accounts for the tendency for

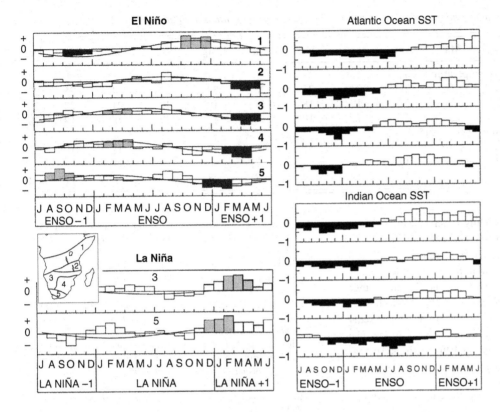

Fig. 16.36 El Niño-composite rainfall and SST anomalies and La Niña rainfall anomalies for select African sectors (see inset map for locations) (based on Nicholson 1997; Nicholson and Kim 1997; and Nicholson and Selato 2000). Time series commence in July in the year prior to the El Niño or La Niña episode and continue until the year following the episode. Light/dark shading indicates seasons of maximum positive/negative response. Amplitude is an index based on rank. The SST sectors roughly correspond latitudinally with the rainfall regions indicated in the map (see Nicholson 1997 for precise locations).

rainfall anomalies of the opposite sign in equatorial and southern Africa.

This rainfall response occurs in roughly three out of four episodes. Those episodes that do not result in drought over southern Africa are those that failed to produce a response in SSTs in the Atlantic and Indian Oceans, suggesting that the principal factor governing interannual variability is SST in the ocean regions surrounding the subcontinent (Nicholson *et al.* 2001).

16.4.5 THE THERMAL REGIME

Most of southern Africa experiences a relatively mild climate, although a distinct winter occurs in areas south of about 20° S (Fig. 16.6). The thermal regime is latitudinally dependent, but elevation, coastal location, and aridity modify it significantly. Mean annual temperatures throughout the region are on the order of 20°–22°C in the tropical latitudes equatorward of 23° S, 17°–20°C in the subtropical latitudes further south. Temperatures at select stations are shown in Fig. 16.33. The annual range is generally about 10°C in the tropical latitudes, 16°–18°C in the subtropics. The warmest months are generally October in the tropics and January in the subtropics; June and July are the coldest months throughout the region. Diurnal range varies with latitude and the degree of aridity; in general, the diurnal range is largest during the dry season months. In the tropical latitudes, it ranges from about 10°–12°C during the wet season up to 18°C during the dry season. In the subtropical latitudes it ranges seasonally from about 14°–20°C in areas with summer rainfall, but is reduced in areas with winter rainfall. In

the drier regions, with mean annual rainfall on the order of 200 mm or less, the diurnal range tends to exhibit less seasonal variation and varies locally from around 15° to 20°C.

In the tropical latitudes of the drylands of southern Africa, the mean daily maxima tend to be highest in October and are on the order of 29°–30°C. They are generally lowest during June and July, being on the order of 21°C. Mean daily minima tend to be highest during the summer season, on the order of 16°–17°C, and lowest during June and July, when they attain 7°–9°C. In the drier regions of the tropical and subtropical latitudes, temperatures tend to be higher, but with lower minima during the dry season. During October and November, mean daily maxima reach 32°–35°C in areas of the Kalahari in Botswana and South Africa, compared with mean daily minima of 17°–19°C. During the coldest months of June and July, mean daily maxima are on the order of 21°–25°C, minima, 4°–6°C. In the more temperate latitudes of South Africa, daily minima range seasonally from 1° to 3°C in winter, from 19° to 23°C in summer. Maxima range seasonally from 15°–20°C in winter to 27°–36°C in summer. At even moderate elevations in these latitudes, mean monthly minima can be below 0°C for one or more months of the year.

Examples of coastal influences are seen for Durban, on the east coast, Port Nolloth on the cold-water west coast, and Cape Town in the winter rains region of temperate latitudes (Table 16.3). The diurnal and annual range are very different from those at the interior locations described above, but each of those locations shows notable differences related to cold versus warm coastal waters and latitude. At Cape Town the annual range is about 8°C and the diurnal range is roughly 9°C, being

Table 16.3. *Mean daily maximum and minimum temperatures (°C) for three coastal locations in South Africa.*

Location	Jan	Feb	Mar	Apr	May	Jun	Jul	Aug	Sep	Oct	Nov	Dec
Daily maximum temperature (°C)												
Durban	28	28	28	26	24	23	22	23	23	24	25	27
Cape Town	26	26	25	22	19	18	17	18	19	21	23	25
Port Nolloth	20	20	20	19	19	19	18	17	17	18	19	20
Daily minimum temperature (°C)												
Durban	21	21	20	18	14	11	10	12	15	17	18	20
Cape Town	16	16	14	12	10	8	7	8	9	11	13	15
Port Nolloth	13	13	12	11	10	9	8	8	9	10	11	12

slightly higher in summer and slightly lower in winter. Mean daily maxima reach 25°–26°C in summer, 17 °–19°C in winter. Mean daily minima are around 15°–16°C in summer, 7°–9°C in summer. Durban, being at a somewhat higher latitude, is a few degrees warmer in summer and winter. The mean annual range is 7.5°C and the mean diurnal range is 9°C. Port Nolloth, bordered by the waters of the cold Benguela Current, is markedly colder than Durban, although at nearly the same latitude. The current dramatically moderates seasonal fluctuations, so that the annual range is less than 4°C. The mean diurnal range is similar to Durban, at 8°C.

As is typical of semi-arid climates, relative humidity exhibits strong seasonal variation. In the tropical north it is on the order of 40–50% during the dry season, 70–80% during the wet season. In the more arid regions, it reaches 70–80% in the wet season, but falls as low as 30–35% in the dry season. In the subtropical latitudes to the south, it ranges seasonally from about 30–40% to 50–60%, being somewhat more humid in coastal locations.

The wind regime is complex over the subcontinent. Several features of the general atmospheric circulation affect the region, and topographic effects are superimposed upon these. In the semi-arid region of the tropical latitudes of Botswana, Zambia, Zimbabwe, and Malawi, winds tend to be from an easterly sector throughout the year, most commonly northeasterly. In the central interior of South Africa, they tend to be from a northerly sector. In the subtropical latitudes of Namibia and South Africa, they exhibit more seasonal variation, with a general tendency for winds from a northerly sector in winter, a southerly sector in summer. The pattern is significantly modified by topography and coastal influences, although no generalizations about these can be made. In parts of the arid interior, the winds are occasionally strong enough to generate dust storms.

REFERENCES

Adejuwon, J. O., and T. O. Odekunle, 2006: Variability and the severity of the "little dry season" in southwestern Nigeria. *J. Climate*, **19**, 483–493.

Baum, J. D., 2006: African easterly waves and their relationship to rainfall on a daily timescale. MS Thesis, Department of Meteorology, Florida State University, 171 pp.

Bowden, J., and F. H. M. Semazzi, 2007: Empirical analysis of intraseasonal climate variability over the greater Horn of Africa. *Journal of Climate*, **20**, 5715–5731.

Burpee, R. W., 1972: The origin and structure of easterly waves in the lower troposphere of North Africa. *Journal of the Atmospheric Sciences*, **29**, 77–90.

Camberlin, P., 1996: Intraseasonal variations of June–September rainfall and upper-air circulation over Kenya. *Theoretical and Applied Climatology*, **54**, 107–115.

Camberlin, P., and R. E. Okoola, 2003: The onset and cessation of the long rains in eastern Africa and their annual variability. *Theoretical and Applied Climatology*, **75**, 43–54.

Camberlin, P., and O. Planchon, 1997: Coastal precipitation regimes in Kenya. *Geografiska Annaler – Series A: Physical Geography*, **79A**, 109–119.

Camberlin, P., and N. Philippon, 2002: The East African March–May rainy season: associated atmospheric dynamics and predictability over the 1968–97 period. *Journal of Climate*, **15**, 1002–1019.

Camberlin, P., V. Moron, R. Okoola, N. Philippon, and W. Gitau, 2009: Components of rainy seasons' variability in equatorial East Africa: onset, cessation, rainfall frequency and intensity. *Theoretical and Applied Climatology*, **98**, 237–249.

Chen, T. C., 2005: Maintenance of the midtropospheric North African summer circulation: Saharan high and African easterly jet. *Journal of Climate*, **18**, 2943–2962.

Chen, T. C., S. Y. Wang, and A. J. Clark, 2008: North Atlantic hurricanes contributed by African easterly waves north and south of the African Easterly Jet. *Journal of Climate*, **21**, 6767–6776.

Diro, G. T., E. Black, and D. I. F. Grimes, 2008: Seasonal forecasting of Ethiopian spring rains. *Meteorological Applications*, **15**, 73–83.

Drobinski, P., B. Sultan, and S. Janicot, 2005: Role of the Hoggar massif in the West African monsoon onset. *Geophysical Research Letters*, **32**, L01705.

Fauchereau, N., B. Pohl, C. J. C. Reason, M. Rouault, and Y. Richard, 2009: Recurrent daily OLR patterns in the Southern Africa/Southwest Indian Ocean region, implications for South African rainfall and teleconnections. *Climate Dynamics*, **32**, 575–591.

Fink, A. H., D. G. Vincent, and V. Ermert, 2006: Rainfall types in the West African Sudanian zone during the summer monsoon 2002. *Monthly Weather Review*, **134**, 2143–2164.

Flohn, H., 1965: Contributions to a synoptic climatology of the Red Sea Trench and adjacent territories. In *Studies on the Meteorology of Tropical Africa*. Bonner Meteorologische Abhandlungen 5, pp. 2–35.

Griffiths, J. F., 1972: *Climates of Africa*. World Survey of Climatology 10, Elsevier, Amsterdam, 604 pp.

Grist, J. P., 2002: Easterly waves over Africa. Part I. The seasonal cycle and contrasts between wet and dry years. *Monthly Weather Review*, **130**, 197–211.

Grist, J. P., and S. E. Nicholson, 2001: A study of the dynamic factors influencing the interannual variability of rainfall in the West African Sahel. *Journal of Climate*, **14**, 1337–1359.

Hachigonta, S., and C. J. C. Reason, 2006: Interannual variability in dry and wet spell characteristics over Zambia. *Climate Research*, **32**, 49–62.

Harangozo, S., and M. S. J. Harrison, 1983: On the use of synoptic data indicating the presence of cloud bands over southern Africa. *South African Journal of Science*, **79**, 413–414.

Hastenrath, S., 1990: Decadal-scale changes of the circulation in the tropical Atlantic sector associated with Sahel drought. *International Journal of Climatology*, **10**, 459–472.

Hastenrath, S., 2007: Circulation mechanisms of climate anomalies in East Africa and the equatorial Indian Ocean. *Dynamics of Atmospheres and Oceans*, **43**, 25–35.

Hastenrath, S., D. Polzin, and L. Greischar, 2002: Annual cycle of equatorial zonal circulations from the ECMWF reanalysis. *Journal of the Meteorological Society of Japan*, **80**, 755–766.

Hastenrath, S., D. Polzin, and P. Camberlin, 2004: Exploring the predictability of the 'short rains' at the coast of East Africa. *International Journal of Climatology*, **24**, 1333–1343.

Hastenrath, S., D. Polzin, and C. Mutai, 2007: Diagnosing the 2005 drought in equatorial East Africa. *Journal of Climate*, **20**, 4628–4637.

Jackson, B., S. E. Nicholson, and D. Klotter, 2009: Mesoscale convective systems over western equatorial Africa and their relationship to large-scale circulation. *Monthly Weather Review*, **137**, 1272–1294.

Janicot, S., F. Mounier, N. M. J. Hall, S. Leroux, B. Sultan, and G. N. Kiladis, 2009: Dynamics of the West African monsoon. Part IV. Analysis of 25–90 day variability of convection and the role of the Indian monsoon. *Journal of Climate*, **22**, 1541–1565.

Kijazi, A. L., and C. J. C. Reason, 2009a: Analysis of the 1998 to 2005 drought over the northeastern highlands of Tanzania. *Climate Research*, **38**, 209–223.

Kijazi, A. L., and C. J. C. Reason, 2009b: Analysis of the 2006 floods over northern Tanzania. *International Journal of Climatology*, **29**, 955–970.

Kinuthia, J. H., 1992: Horizontal and vertical structure of the Lake Turkana Jet. *Journal of Applied Meteorology*, **31**, 1248–1274.

Knippertz, P., A. H. Fink, A. Reiner, and P. Speth, 2003: Three late summer/early autumn cases of tropical–extratropical interactions causing precipitation in northwest Africa. *Monthly Weather Review*, **131**, 116–135.

Knippertz, P., and A. H. Fink, 2008: Dry-season precipitation in tropical West Africa and its relation to forcing from the extratropics. *Monthly Weather Review*, **136**, 3579–3596.

Korecha, D., and A. G. Barnston, 2007: Predictability of June-September rainfall in Ethiopia. *Monthly Weather Review*, **135**, 628–650.

Kumar, S., 1978: *Interaction of Upper Westerly Waves with Intertropical Convergence Zone and their Effect on the Weather over Zambia during the Rainy Season.* Government Printers, Lusaka, 36 pp.

Laing, A., R. Carbone, V. Levizzani, and J. Tuttle, 2008: The propagation and diurnal cycles of deep convection in northern tropical Africa. *Quarterly Journal of the Royal Meteorological Society*, **134**, 93–109.

Lamb, P. J., and R. Peppler, 1992: Further case studies of tropical Atlantic surface atmospheric and oceanic patterns associated with sub-Saharan drought. *Journal of Climate*, **5**, 476–488.

Lavaysse, C., and Coauthors, 2009: Seasonal evolution of the West African heat low: a climatological perspective. *Climate Dynamics*, **33**, 313–330.

Lélé, M. I., and P. J. Lamb, 2010: Variability of the Intertropical Front (ITF) and rainfall over the West African Sudan-Sahel zone. *Journal of Climate*, **23**, 3984–4004.

LeHouerou, H. N., 1986: The desert and arid zones of Northern Africa. In *Ecosystems of the World*, **12B**, *Hot Deserts and Arid Shrublands* (M. Evenari, I. Noy-Meir, and D. W. Goodall, eds.). Elsevier, Amsterdam, pp. 101–147.

Leroux, S., and N. M. J. Hall, 2009: On the relationship between African Easterly Waves and the African Easterly Jet. *Journal of the Atmospheric Sciences*, **66**, 2303–2316.

Lu, J., 2009: The dynamics of the Indian Ocean sea surface temperature forcing of Sahel drought. *Climate Dynamics*, **33**, 445–460.

Mapande, A., and C. J. C. Reason, 2005: Interannual rainfall variability over western Tanzania. *International Journal of Climatology*, **25**, 1355–1368.

Mathon, V., H. Laurent, and T. Lebel, 2002: Mesoscale convective system rainfall in the Sahel. *Journal of Applied Meteorology*, **441**, 1081–1092.

Mohr, K. I., 2004: Interannual, monthly, and regional variability in the wet season: diurnal cycle of precipitation in sub-Saharan Africa. *Journal of Climate*, **17**, 2441–2453.

Mohr, K. I., and C. D. Thorncroft, 2006: Intense convective systems in West Africa and their relationship to the African easterly jet. *Quarterly Journal of the Royal Meteorological Society*, **132**, 163–176.

Nesbitt, S. W., R. Cipelli, and S. A. Rutledge, 2006: Storm morphology and rainfall characteristics of TRMM precipitation features. *Monthly Weather Review*, **134**, 2702–2721.

Newell, R. E., J. W. Kidson, D. G. Vincent, and G. J. Boer, 1972: *The General Circulation of the Tropical Atmosphere and Interactions with Extratropical Latitudes.* Vol. I. MIT Press, Cambridge, 258 pp.

Nicholson, S. E., 1981: Rainfall and atmospheric circulation during drought periods and wetter years in West Africa. *Monthly Weather Review*, **109**, 2191–2208.

Nicholson, S. E., 1986a: The spatial coherence of African rainfall anomalies: interhemispheric teleconnections. *Journal of Climate and Applied Meteorology*, **25**, 1365–1381.

Nicholson, S. E., 1986b: The nature of rainfall variability in Africa south of the equator. *International Journal of Climatology*, **6**, 515–530.

Nicholson, S. E., 1997: An analysis of the ENSO signal in the tropical Atlantic and western Indian Oceans. *International Journal of Climatology.*, **17**, 345–375.

Nicholson, S. E., 1996a: Environmental change within the historical period. In *The Physical Geography of Africa* (A. S. Goudie, W. M. Adams, and A. Orme, eds.), Oxford University Press, Oxford, pp. 60–75.

Nicholson, S. E., 1996b: A review of climate dynamics and climate variability in eastern Africa. In *The Limnology, Climatology and Paleoclimatology of the East African Lakes* (T. C. Johnson and E. Odada, eds.), Gordon and Breach, Amsterdam, pp. 25–56.

Nicholson, S. E., 2000: The nature of rainfall variability over Africa on time scales of decades to millennia. *Global and Planetary Change Letters*, **26**, 137–158.

Nicholson, S. E., 2005: On the question of the "recovery" of the rains in the West African Sahel. *Journal of Arid Environments*, **63**, 615–641.

Nicholson, S. E., 2008: The intensity, location and structure of the tropical rainbelt over west Africa as factors in interannual variability. *International Journal of Climatology*, **28**, 1775–1785.

Nicholson, S. E., 2009a: A revised picture of the structure of the "monsoon" and land ITCZ over West Africa. *Climate Dynamics*, **32**, 1155–1171.

Nicholson, S. E., 2009b: On the factors modulating the intensity of the tropical rainbelt over West Africa. *International Journal of Climatology*, **29**, 673–689.

Nicholson, S. E., and D. Entekhabi, 1986: The quasi-periodic behavior of rainfall variability in Africa and its relationship to the Southern Oscillation. *Meteorology and Atmospheric Physics*, **34**, 311–348.

Nicholson, S. E., and D. Entekhabi, 1987: Rainfall variability in Equatorial and Southern Africa: relationships with sea-surface temperatures along the southwestern coast of Africa. *Journal of Climate and Applied Meteorology*, **26**, 561–578.

Nicholson, S. E., and J. P. Grist, 2003: On the seasonal evolution of atmospheric circulation over West Africa and Equatorial Africa. *Journal of Climate*, **16**, 1013–1030.

Nicholson, S. E., and J.-Y. Kim, 1997: The relationship of the El-Niño Southern Oscillation to African rainfall. *International Journal of Climatology*, **17**, 117–135.

Nicholson, S. E., and J. C. Selato, 2000: The influence of La Niña on African rainfall. *International Journal of Climatology*, **20**, 1761–1776.

Nicholson S. E., and P. J. Webster, 2007: A physical basis for the interannual variability of rainfall in the Sahel. *Quarterly Journal of the Royal Meteorological Society*, **133**, 2065–2084.

Nicholson, S. E., J. Kim, and J. Hoopingarner, 1988: Atlas of African rainfall and its interannual variability. Dept. of Meteorology, Florida State University, 237 pp.

Nicholson, S. E., D. Leposo, and J. Grist, 2001: On the relationship between El Niño and drought over Botswana. *Journal of Climate*, **14**, 323–335.

Nicholson, S. E., A. I. Barcilon, and M. Challa, 2008: An analysis of West African dynamics using a linearized GCM. *Journal of the Atmospheric Sciences*, **65**, 1182–1203.

Pasch, R. J., and L. A. Avila, 1994: Atlantic tropical systems of 1992. *Monthly Weather Review*, **122**, 539–548.

Oettli, P., and P. Camberlin, 2005: Influence of topography on monthly rainfall distribution over East Africa. *Climate Research*, **28**, 199–212.

Okoola, R. E., 1989: *Westward-moving Disturbances in the Southwest Indian Ocean*. Kenya Meteorological Department, Nairobi, 30 pp.

Pohl, B., and P. Camberlin, 2006: Influence of the Madden–Julian Oscillation on East African rainfall. I. Intraseasonal variability and regional dependency. *Quarterly Journal of the Royal Meteorological Society*, **132**, 2521–2539.

Polo, I., B. Rodriguez-Fonseca, T. Losada, and J. Garcia-Serrano, 2008: Tropical Atlantic variability modes (1979–2002). Part I: Time-evolving SST modes related to West African rainfall. *Journal of Climate*, **21**, 6457–6475.

Prospero, J. M., P. Ginoux, O. Torres, S. E. Nicholson, and T. E. Gill, 2002: Environmental characterization of global sources of atmospheric soil dust identified with the NIMBUS 7 Total Ozone Mapping Spectrometer (TOMS) absorbing aerosol product. *Reviews of Geophysics*, **40**, 1002, doi:10.1029/2000RG000095.

Ramel, R., 2006: On the northward shift of the West African monsoon. *Climate Dynamics*, **26**, 429–440.

Reason, C. J. C., 2002: Sensitivity of the southern African circulation to dipole sea-surface temperature patterns in the south Indian Ocean. *International Journal of Climatology*, **22**, 377–393.

Reason, C. J. C., and H. Mulenga, 1999: Relationships between South African rainfall and SST anomalies in the Southwest Indian Ocean. *International Journal of Climatology*, **19**, 1651–1673.

Reason, C. J. C., and D. Jagadheesha, 2005: Relationships between South Atlantic SST variability and atmospheric circulation over the South African region during austral winter. *Journal of Climate*, **18**, 3339–3355.

Roca, R., J. P. Lafore, C. Pirious, and J. L. Redelsperger, 2005: Extratropical dry-air intrusions into the West African monsoon midtroposphere: an important factor for the convective activity over the Sahel. *Journal of Atmospheric Science*, **62**, 390–407.

Rowell, D. P., 2001: Teleconnections between the tropical Pacific and the Sahel. *Quarterly Journal of the Royal Meteorological Society*, **127**, 1683–1706.

Rowell, D. P. 2003: The impact of Mediterranean SSTs on the Sahelian rainfall season. *Journal of Climate*, **16**, 849–862.

Schulze, B. R., 1965: *Climate of South Africa, Part 8. General Survey*. South African Weather Bureau, Pretoria, 330 pp.

Segele, Z. T., and P. J. Lamb, 2005: Characterization and variability of *Kiremt* rainy season over Ethiopia. *Meteorology and Atmospheric Physics*, **89**, 153–180.

Segele, Z. T., P. J. Lamb, and L. M. Leslie, 2009a: Season-to-interannual variability of Ethiopia/Horn of Africa monsoon. Part I. Associations of wavelet-filtered large-scale atmospheric circulation and global sea-surface temperature. *Journal of Climate*, **22**, 3396–3421.

Segele, Z. T., P. J. Lamb, and L. A. Leslie, 2009b: Large scale atmospheric circulation and global sea surface temperature associations with Horn of Africa June–September rainfall. *International Journal of Climatology*, **29**, 1075–1100.

Singleton, A. T., and C. J. C. Reason, 2007: Variability in the characteristics of cut-off low pressure systems over subtropical southern Africa. *International Journal of Climatology*, **27**, 295–310.

Todd, M., and R. Washington, 1999a: Extreme daily rainfall in southern African and southwest Indian Ocean tropical-temperate links. *South African Journal of Science*, **94**, 64–70.

Todd, M., and R. Washington, 1999b: Circulation anomalies associated with tropical-temperate troughs in southern Africa and the south west Indian Ocean. *Climate Dynamics*, **15**, 937–951.

Toracinta, E. R., and E. J. Zipser, 2001: Lightning and SSM/I-ice-scattering mesoscale convective systems in the global tropics. *Journal of Applied Meteorology*, **40**, 983–1002.

Tyson, P. D., 1986: *Climatic Change and Variability in Southern Africa*. Oxford University Press, Cape Town, 220 pp.

Ummenhofer, C. C., A. Sen Gupta, M. H. England, and C. J. C. Reason, 2009: Contributions of Indian Ocean sea surface temperatures to enhanced East African rainfall. *Journal of Climate*, **22**, 993–1013.

van Heerden, J., and J. J. Taljaard, 1998: Africa and the surrounding waters. In *Meteorology of the Southern Hemisphere*, American Meteorological Society, Boston, MA, pp. 141–174.

Ward, M. N., 1998: Diagnosis and short-lead time prediction of summer rainfall in tropical North America at interannual and multidecadal timescales. *Journal of Climate*, **11**, 3167–3191.

Warren, A., and Coauthors, 2007: Dust-raising in the dustiest place on earth. *Geomorphology*, **92**, 25–37.

Washington, R., and A. Preston, 2006: Extreme wet years over southern Africa: role of Indian Ocean sea surface temperatures. *Journal of Geophysical Research –Atmospheres*, **111**, D15104.

Washington, R., M. C. Todd, S. Engelstaedter, S. Mbainayel, and F. Mitchell, 2006: Dust and the low-level circulation over the Bodélé Depression, Chad: observations from BoDEx 2005. *Journal of Geophysical Research*, **111**, D03201, doi:10.1029/2005JD006502.

Webster, P. J., and J. Fasullo, 2003: Monsoon: dynamical theory. In *Encyclopedia of Earth Sciences* (J. R. Holton, J. A. Pyle, and J. A. Curry, eds.), Academic Press, pp. 1370–1366.

Williams, C. J. R., D. R. Kniveton, and R. Layberry, 2008: Influence of South Atlantic sea surface temperature on rainfall variability and extremes over southern Africa. *Journal of Climate*, **21**, 6498–6520.

Zhang, C. D., P. Woodward, and G. Gu, 2006: The seasonal cycle in the lower troposphere over West Africa from sounding observations. *Quarterly Journal of the Royal Meteorological Society*, **132**, 2559–2582.

Zipser, E. J., D. J. Cecil, C. Liu, S. W. Nesbitt, and D. P. Yorty, 2006: Where are the most intense thunderstorms on earth? *Bulletin of the American Meteorological Society*, **87**, 1057–1071.

Zorita, E., and F. F. Tilya, 2002: Rainfall variability in northern Tanzania in the March to May season (long rains) and its links to large-scale climate forcing. *Climate Research*, **20**, 31–40.

17 The Mediterranean lands

17.1 OVERVIEW AND GENERAL GEOGRAPHY

Strictly speaking, the Mediterranean lands are those bordering the Mediterranean Sea separating Africa and Europe (Fig. 17.1). This includes southern Europe, the Middle East, and Africa north of the Sahara. The essence of a "Mediterranean-type climate" is one with mild, wet winters and warm, dry summers. This description best fits regions near the Mediterranean Sea, semi-arid steppes with mean annual rainfall on the order of 100–350 mm. Further south and east are true deserts, where only slight to moderate rains occur during the cool season. The demarcation between desert and semi-arid steppe is roughly 70–100 mm, the lower threshold for rainfed vegetation. The largest are the Saharan, Arabian and the Negev deserts. In these deserts winters can become quite cold and summers are dry but can be unbearably hot.

This chapter considers all of the arid and semi-arid portions of the Mediterranean countries, as far south as the southern limit of the winter rainfall regime. For the sake of convenience, the summer rainfall region of the southern Arabian Peninsula is also included here. The Mediterranean of southern Europe is predominantly subhumid and is hence omitted. The information and data in this chapter come primarily from the Meteorological Office (1962), Dubief (1963), LeHouerou (1986), Orshan (1986), and Takahashi and Arakawa (1981). Littman and Berkowicz (2008) also provide an excellent review.

The dominant vegetation is strongly dependent on mean annual rainfall, but topography, elevation, and soils also play a role. In semi-arid regions near the coast the vegetation is mainly sclerophyll shrubland, except in the highlands. There are steppe grasslands in most of Turkey and Iraq and part of Israel (Fig. 17.2). In the desert zones of North Africa and the Arabian Peninsula, further south, is semi-desert scrub alternating with barren areas of dunes. In areas where rainfall ranges from less than 70 mm to 100 mm/year, vegetation is diffuse and restricted to small drainage channels, or runnels, and other microhabitats with favorable conditions of rainfall, topography, or soils.

Most of the terrain is low-lying and relatively flat. Exceptions are the Atlas Mountains of northwestern Africa, the plateaus of eastern Turkey and Arabia, small areas of mountains and hills in the Near East and Arabian Peninsula, and other isolated massifs. These relatively small topographic features markedly enhance precipitation locally.

17.2 GENERAL CLIMATOLOGY AND ATMOSPHERIC CIRCULATION

17.2.1 LARGE-SCALE CIRCULATION

The Mediterranean is a region of strongly seasonal climate and the major climatic controls exhibit a pronounced switch between winter, with dominant high pressure, and summer, with dominant low pressure. In the arid southern sector a similarly dramatic shift occurs aloft, from the dominance of the westerly subtropical jet stream in winter to the tropical easterly jet in summer. These changes are forced by the reversal of heating over the Himalayan plateau, and occur abruptly during the transition seasons, taking place within 2–3 week periods in May/June and October/November.

During winter, the prevailing influences are an anticyclonic ridge, the low-pressure systems of the Mediterranean, and the upper-level westerlies (depicted in Fig. 16.4). The ridge appears to be partly an eastward extension of the Bermuda High over the North Atlantic and partly linked to the Siberian High much further east.

In the summer, the western extremity of the monsoon trough exists over the Arabian Gulf and a stationary thermal low is established over the western Sahara. Consequentially, the prevailing surface flow tends to be northerly or northeasterly over most of the Mediterranean. This persistently brings warm, dry, subtropical air from the deserts of western Asia into the region, producing excessive summer heat waves.

An important feature of the transition seasons is the Red Sea Trough (RST), a surface low-pressure trough that extends from

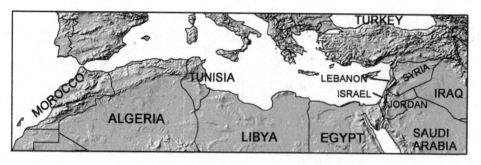

Fig. 17.1 Map of general topography of the Mediterranean drylands.

Fig. 17.2 (left) Sparse grasses of the Iraqi steppe (photo courtesy of Tracie Mitchell); (center) northwest Negev landscape of dwarf shrubs and sand dunes; (right) close up of Negev dune shrubs.

eastern Africa northward along the Red Sea, into the eastern Mediterranean. RSTs are generally accompanied by an upper-level cyclonic storm (Ziv *et al.* 2005). Tsvieli and Zangvil (2005) suggest that their origin is the "Sudan monsoon low" centered over eastern Africa or Arabia. They occur most frequently in October and November, but have a secondary maximum in April. They also occur during the winter but disappear entirely in summer.

As in the Sahara, the aridity appears to be related primarily to the absence of major storm tracks through the region, especially in summer. Then the Mediterranean drylands lie south of the usual limit of penetration of the extra-tropical cyclones but north of the limit to which the summer tropical rains penetrate. Summer aridity is also enhanced by mid- and upper-tropospheric subsidence associated with the tropical easterly jet (Ziv *et al.* 2004) and by a persistent inversion layer produced by the subsidence (Bitan and Saaroni 1992). Mesoscale influences also produce intense low-level inversions in summer, especially in coastal regions of the eastern Mediterranean. Thus, summers are generally completely rainless.

17.2.2 PRECIPITATION REGIME

Figure 17.3 shows mean annual precipitation for the Mediterranean region. In the drylands of North Africa and the Near East, isohyets tend to run parallel to the shores of the Mediterranean, and precipitation decreases with increasing distance eastward and southward. Mean annual precipitation is as high as 200–400 mm along the North African coast, decreasing rapidly southward to 50 mm or less throughout the northern

Sahara. In the hyper-arid core of the eastern Sahara, a vast region receives less than 5 mm/year on average, with many areas being virtually rainless. Along the eastern Mediterranean coast of Syria, Lebanon, and Israel, mean annual precipitation is 800–1000 mm, but falls off rapidly inland, with most of the region receiving less than 100 mm/year on average. Annual rainfall is generally as high as 600–800 mm in the Atlas Mountains of North Africa and other highlands in the Near East. Isolated sectors of the mountains have annual means well in excess of 1000 mm.

The depressions that bring precipitation are classified according to their origin. The five types that are recognized include Atlantic depressions (which originate outside the Mediterranean region), thermal lows, lee depressions, wave depressions, and depressions formed in association with troughs (Fig. 17.4). The westerlies aloft steer these systems across the full extent of the Mediterranean. Over 90% of the depressions affecting this region form as lee or wave depressions. Most (70%, or 50 per year) originate over the western Mediterranean Sea. About 18% (or 14 per year) form in the lee of the Atlas and are termed Saharan depressions. Their formation is illustrated in Fig. 5.14. Another 4% (or 3 per year) form in the eastern and central Mediterranean. Systems linked to the polar-front jet occasionally enter the region in winter. Cut-off lows also occur frequently and generally bring very heavy rainfall (Porcu *et al.* 2007).

Winter rains occur throughout the region, a consequence of the southward displacement of the mid-latitude westerlies and associated disturbances (Fig. 7.5). Over most of the Mediterranean region only 40–50% of the precipitation is concentrated in the wettest quarter of the year (Fig. 17.6). Light

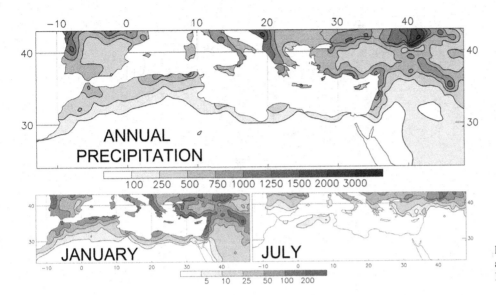

Fig. 17.3 Mean annual, January and July precipitation (mm) for the Mediterranean region.

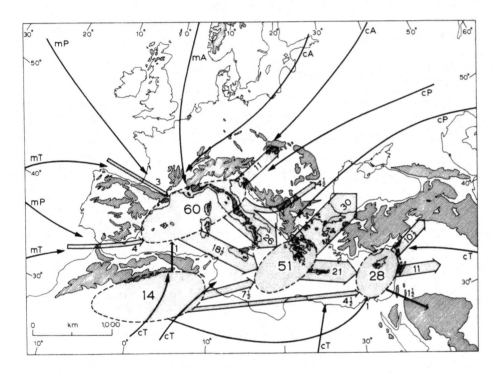

Fig. 17.4 Tracks of depressions over the Sahara and the Near East, showing annual frequencies and air-mass sources (Meteorological Office 1962).

showers commence in early autumn, September and October, but the heavier and more prolonged rains begin in November. The rainy season ceases by mid-May and the region is generally rainless until September or October. Thus, the rainy season is relatively long in much of the region. In January (Fig. 17.3), in the core of the rainy season, mean monthly rainfall is on the order of 10–30 mm, reaching 40 mm along some coastal areas and 50 mm in the interior of Turkey. In July (Fig. 17.3), the driest month for most of the region, only a small area in northern Turkey receives any rainfall.

Snow is unknown in most of the desert regions, being confined primarily to highlands and some coastal regions in the east. However, it occurs with some regularity in the central

Sahara (Fig. 17.7a) and has even reached the southern Arabian Peninsula.

Some parts of the region experience a rainfall maximum during the transition seasons. In these areas, most rainfall is associated with disturbances developing over the northern Sahara, especially in the lee of the Atlas Mountains. These propagate rapidly eastwards. The transition season maximum is strongly evident in the western Sahara and in the dry interior of Turkey. In autumn, cold outbreaks, which may produce a cut-off low in the eastern Mediterranean, can trigger precipitation and thunderstorms in the Red Sea and areas of the Arabian Peninsula, Iraq, and other parts of the Near East.

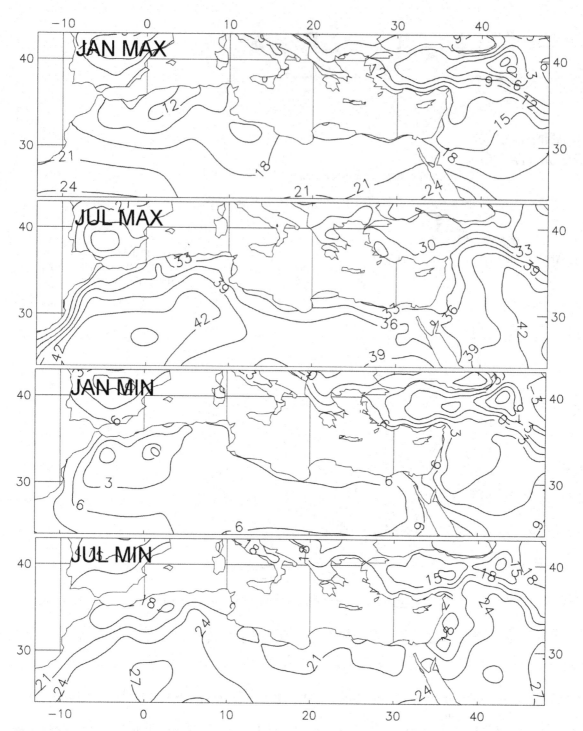

Fig. 17.5 The seasonal cycle of precipitation over North Africa and the Middle East.

Although precipitation is brought by extra-tropical cyclonic disturbances and fronts, it is generally convective in character. Consequently, it is intense, of short duration, and erratic in its onset and its time and space distribution. Its variability exceeds 70% in many areas (Fig. 17.8). Convective activity is more common in spring than in autumn, because the rapid heating produces instability, while the autumn cooling produces more stable conditions. The instability makes sand- and dust storms more frequent in spring.

Cloud cover is low in most of the region, so evaporation is high. Cloud cover is generally 10–20% on average in the desert and 20–30% in the semi-arid regions nearer the coast. Maximum cloud cover occurs during the rainy season, but even then it is generally only about 30% in the desert regions.

17.2.3 THERMAL REGIME

Being in relatively high latitudes, the annual temperature range in the Mediterranean drylands is large: 15°–20°C in most of the region, except for the western Sahara, where it is roughly 20°–25°C (Fig. 17.9). The annual range increases both with latitude and with increasing distance from the Mediterranean, so that it is greatest in the central desert. The diurnal range (Fig. 17.9) is comparable, being 15°–20°C in most of the desert, but 12°–15°C nearer the Mediterranean coast. It varies by only 2°–5°C from month to month.

Generally, January is the coldest month, while summer temperatures are relatively uniform from June to August. During the warmest months, mean daily maxima are 42°–45°C in the central and western Sahara and in the Arabian Desert, but 35°–40°C in much of the desert in Libya, Egypt and the Near East, and in the highlands of the central Sahara (Fig. 17.10). Absolute maxima exceed 50°C in most of the Sahara and Arabian deserts. Mean daily minima in the warmest months vary mainly with distance from the Mediterranean, ranging from about 18°–20°C along the North African coast (except in the Atlas Mountains and surrounding regions) and the eastern Mediterranean to 28°C in the southern Sahara and southern Arabian Peninsula (Fig. 17.10). Mean winter minima are on the order of 3°–9°C, while daily maxima are on the order of 20°–27°C inland, but 18°–20°C

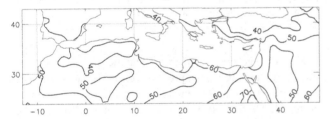

Fig. 17.6 Percent concentration of precipitation in the wettest quarter of the year.

along the coasts. The lowest minima are in mountainous regions, such as in Turkey, the eastern Mediterranean, and the Atlas.

The cold outbreaks of winter have different source regions. Many arrive from the north and northwest, brought by maritime polar air that is transformed over the Mediterranean Sea. Iraq experiences occasional cold waves due to the advection of continental air from Central Asia. These can bring heavy rain along the frontal zone where the continental air mass meets the Mediterranean air mass. These cold outbreaks can bring temperatures as low as −10°C into northern Iraq, eastern Syria, and western Iraq and −20°C into the highlands of eastern Iraq. In Israel they occur most frequently in December, with many cold spells lasting 4–5 days (Saaroni *et al.* 1998a).

17.2.4 REGIONAL WIND SYSTEMS

Many regional winds also bring cold air in winter. A bora blows into northern parts of the Near East from the Zagros Mountains of Iran. It is called the *n'ashi* in the Persian Gulf. Similar winds (the *belat*) blow from December to March on the coast of the Arabian Peninsula from Cape Sadjir to Masira and in the Gulf of Aden (the *kharif*). A strong and cold northwesterly wind called the cold *shamal* affects southern Iraq and the northern Persian Gulf.

Eastward-moving depressions coming off North Africa and Arabia are preceded by a regional wind known as the *sirocco* in Syria, Jordan, Lebanon, Israel, and much of Iraq. These winds, with regional names such as the *samoom*, *simoom*, or *chlouk*, blow from the south or southeast. Occurring in spring and autumn, they are warm, dry, and frequently dust-laden. These depressions cause heat waves, sand storms and dust outbreaks as far east as western Iran and the western coast of the Persian Gulf. Occasionally they pick up enough moisture over the Arabian Gulf to produce rainfall, often coupled with thunderstorms.

(a)

(b)

Fig. 17.7 (a) Snow in the Sahara (from Dubief 1963) and (b) over Jerusalem (photo courtesy of Simon Berkowicz, Hebrew University of Jerusalem).

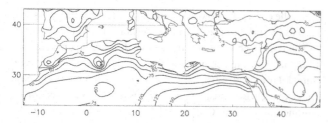

Fig. 17.8 Coefficient of variation (%) for annual precipitation in the Mediterranean region.

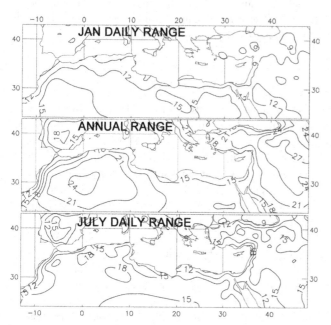

Fig. 17.9 Annual temperature range and diurnal temperature range in January and July (°C).

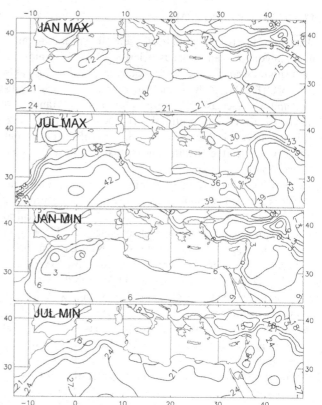

Fig. 17.10 Mean daily minimum temperature and mean daily maximum temperature in January and July (°C).

The *samoom* is stronger on the coast of Israel and Lebanon than inland, because the mountain barrier causes it to take on *foehn* characteristics. It can last a few hours or a few days. In the latter case, it is called the fire wind, the breath of death, or the *shobe*. Here it occurs on roughly 40 days per year.

17.3 INTERANNUAL VARIABILITY

Interannual variability in the dryland regions of the southern and eastern Mediterranean is tied closely to global-scale processes. The influence of both the North Atlantic Oscillation (NAO) and Arctic Oscillation (AO) is strong (e.g., Turkes and Erlat 2005, 2008). The AO appears to influence mainly temperature. Some tropical influence, including that of ENSO, is evident, although geographically limited (Alpert *et al.* 2005). El Niño often brings wetter than normal conditions to the Middle East in autumn and winter, but appears to influence warm-season precipitation in Turkey (Karabork *et al.* 2005). The Mediterranean Sea appears to have only a secondary influence on interannual variability (Xoplaki *et al.* 2004).

The NAO governs the path of Atlantic mid-latitude storm tracks, modulating winter precipitation in Turkey, Syria, and elsewhere. Strong NAO winters favor dry and cool conditions in the Middle East (Pagano *et al.* 2003) and are often linked to severe droughts in Turkey. Systems emerging from eastern Europe and Siberia also modulate winter precipitation in Turkey (Turkes *et al.* 2009).

17.4 NORTH AFRICA AND THE NORTHERN SAHARA

This section considers the regions of North Africa that are arid or semi-arid and that are either rainless or receive rainfall primarily during the winter and transition seasons. The region covers nearly 6 million km². The southern Sahara, where rainfall is concentrated in the high-sun season, is treated in Chapter 16. The western littoral of the Sahara is also treated in more detail in Chapter 20 on coastal deserts.

Within Mediterranean North Africa are various degrees of aridity, ranging from semi-arid to extremely arid. Arid and semi-arid regions are primarily steppes, in a continuous and diffuse pattern. Typical steppe vegetation in the region is low or half-shrubs 0.2–0.5 m high; *Artemisia* is a common genus. Extensive grass steppes and steppes with halophytic shrubs or succulents also occur. In isolated sectors there are desert savanna, i.e., scattered trees interspersed with tall perennial grasses and small shrubs, open woodlands (generally in areas of dunes), and even isolated stands of open forest (trees over 5 m high and over 100 single stems per hectare), usually of pine or juniper with an understory of shrubs. On hills where annual rainfall is between about 200 and 400 mm there is *mattoral* or *garrigue* vegetation cover (equivalent to the American chaparral), with evergreen tall shrubs or small trees.

In contrast, the extremely arid desert consists of large tracts of barren land where perennial vegetation is confined to run-off channels in a clustered or "contracted" pattern. Four main types of land form cover the surface: rocky or stony hills, like the mountain massifs of Tibesti, Air, Adrar, the Hoggar; reg; hamada; and ergs or sand seas. The regs cover about 68% of the desert, the ergs 22%, and the rocky mountains and hamadas 10%. Mature soils are exceptional anywhere in the region, even in the vegetated "arid" zone.

Except for the exogenous Nile, there are no permanent rivers in the region. The hydrology is characterized by *wadis* (also called *oueds* in North Africa). A rain event of 10–20 mm can trigger local flow. About 90–95% of the precipitation is lost to evaporation, so that only 5–10% of the rainfall provides the water that feeds the main wadi systems, such as the Wadi Dr'aa in Morocco, or the Wadi Cheliff in Algeria. The frequency of wadi flow is dependent on rainfall, so that the endogenous wadis in the arid zone flow 1–5 times per year, but those in the desert zone flow once in 2–100 years (or less often in the extreme desert).

The region is underlain by vast aquifers. They are commonly located at the border of large, sandy massifs (e.g., the Grand Erg Occidental and the alluvial basin of the Wadi Saoura in the Sahara). Some are partially or totally fossil. An aquifer in southern Algeria and Tunisia covers nearly 1 million km^2 and yields 8.5 m^3/s of water that is 4000–8000 years old. The Al Kufrah aquifer in the Libyan Desert, nearly 2 million km^2 in extent, contains water that is some 20,000–30,000 years old. Both arose from humid intervals in the Sahara's climatic past.

Most of the Sahara receives, on average, less than 20 mm rainfall per year, and in much of Libya and Egypt, annual rainfall ranges from less than 1 mm to 5 mm. Nearly all of the region experiences rainless years, but in parts of the eastern Sahara, periods of 5 years or more have passed without any fall of rain exceeding 0.1 mm (Fig. 17.11). In the desert the interannual variability of rainfall is generally 75–100% (Fig. 17.8).

There are three rainfall gradients in the Sahara: latitudinal, longitudinal, and altitudinal. Highlands such as Tibesti or the Hoggar generally average above 100 mm/year, compared with 10–25 mm on the surrounding plains. The north–south gradient is abrupt, so within 200–300 km rainfall is reduced from 400 to less than 100 mm/year. The east–west gradient is relatively slight, with mean annual rainfall in the desert core ranging from about 20 mm/year in the west to near zero in the virtually rainless area of the east.

In the northern Sahara, snow occurs regularly at high elevations, including many of the foothills; it becomes less frequent at lower elevations and further south (Fig. 17.12). At Laghouat, it has occurred about every second year during the twentieth century, but it fell every year between 1883 and 1896, a period of relatively cool climates over much of the globe (Fig. 17.13). The mean number of days with snow each year ranges from about 0.1 to nearly 15 in northern Algeria (Dubief 1963). Snow is generally limited to the months of November–March, but it has been recorded as late as May.

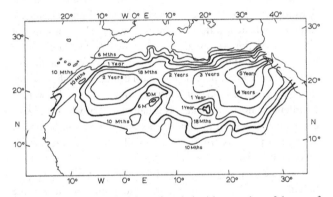

Fig. 17.11 Maximum duration of period without at least 0.1 mm of rainfall in 24 hours (from Dubief 1963).

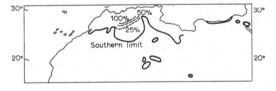

Fig. 17.12 Locations where snow can occur in the Sahara (from Dubief 1963).

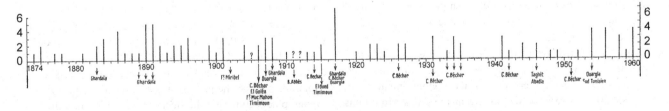

Fig. 17.13 The occurrence of snow in the central Sahara, by year, 1874 to 1960 (from Dubief 1963). The numbers refer to the frequency of occurrence per year for Laghouat; arrows indicate other low-lying locations where snow occurred.

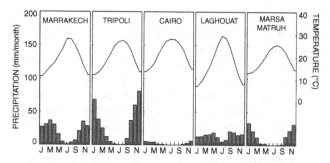

Fig. 17.14 Percent concentration of precipitation in the month of maximum precipitation.

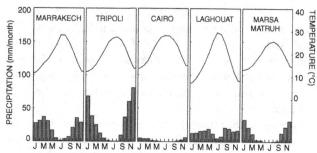

Fig. 17.15 Mean monthly rainfall (mm) and temperature (°C) at five representative stations of North Africa: Marrakech, Tripoli, Cairo, Laghouat, and Marsa Matruh.

Relative humidity depends on distance from the coast and decreases southward. It usually exceeds 70% on both coasts, but is 60–65% in the arid zone in winter and 35–40% in summer. In the desert it reaches 40–55% in winter but only 20–25% in summer. Frost and dew are known to occur in the interior, but quite rarely.

Rainfall occurs mainly during the winter months. The length of the rainy season progressively decreases and the onset of the season occurs progressively later moving southward along the rainfall gradient toward the central Sahara. The relationships between duration, onset, and amount are clearly related to the causality of rainfall: the Mediterranean cyclones traversing the region. These are linked to the mid-latitude westerlies, which move progressively further south in winter. Thus, the wettest month in the northern Sahara is usually January, when the westerlies and associated cyclone tracks reach their southernmost position. In the eastern Sahara, rainfall is so rare that there is no regular season.

The exception to this pattern is a small part of the western Sahara, where the precipitation maximum falls generally in the transition seasons (most frequently March) (Fig. 17.14). Here the Saharan depressions generated in the lee of the Atlas are the most important rain-bearing system and the conditions conducive to their formation are most likely to occur in spring or autumn, when the mid-latitude westerly regime lies close to the tropical regime. These conditions, sketched in Fig. 5.14, are an elongated wave trough in the upper-level westerlies overlying a low-level tropical depression. Because these systems are steered eastward by the westerlies, they produce a small secondary rainfall maximum in parts of the central and eastern Sahara, generally in May.

Mean conditions of precipitation and temperature are shown in Fig. 17.15 for select stations in North Africa. Tripoli, in Libya, and Marsa Matruh, in Egypt, clearly show the winter rainfall maximum and the Mediterranean's moderating influence on temperature. The seasonal cycle of temperature and rainfall is similar at Cairo, but the city is nearly rainless, as is much of the surrounding area. Laghouat, in Algeria, and Marrakech, in Morocco, are representative of the regions impacted by the Saharan depressions developing in the Atlas Mountains. Temperatures are also markedly higher, and the annual range greater, indicative of a more inland location.

Along the Mediterranean coast of North Africa, July is generally the warmest month, December and January the coolest. Mean daily minima in winter are generally 5°–7°C, with lower temperatures in the central desert than along the coast or the tropical margin (Fig. 17.10). In summer, mean daily maxima (Fig. 17.10) are on the order of 20°C near the coast; in the central desert they are 22°–25°C in the east and 25°–30°C in the west. Lower temperatures are experienced in the Atlas Mountains throughout the year. The absolute maximum is on the order of 45°–55°C, with a temperature of 58°C having been recorded once in Al Aziziah (Libya) in September 1922.

The annual range (Fig. 17.9) is strongly dependent on latitude and distance from the coast, ranging from about 12°–15°C along the coast to over 20°C a short distance inland. In the central Sahara, the range varies from about 15°C along the tropical margin of the desert to 20°–25°C in the desert core. The diurnal range is generally 15°–20°C, but less near the Mediterranean coast. It shows little annual variation, but has a weak maximum in April in the central desert and in July further north (Fig. 17.9).

The Sahara and surrounding regions experience a number of hot regional winds. A variety of local names have been used for these: *sirocco, ghibli, chergui, khamsin*. Depending on location, these occur about 20–90 days per year, most frequently in spring and autumn. Sand storms are also frequent in the Sahara, especially in the Sudan, where they are referred to as *haboobs*.

17.5 THE ARABIAN PENINSULA

The Arabian Peninsula extends from about 35°–60° E and 12°–30° N, comprising approximately 2.6 million km² of mostly arid land. Only some 45,000 km² of the peninsula is arable, of which about 10% is under cultivation. The Empty Quarter, which occupies 650,000 km², is the largest continuous body of sand in the world. It includes stabilized sand seas and mobile dunes which, in many cases, attain 150–200 m in height. The remaining area of the peninsula is also mostly sand. Principal topographic features are the steep escarpments along the western and southwestern edges of the peninsula and the coastal plain along the Red Sea.

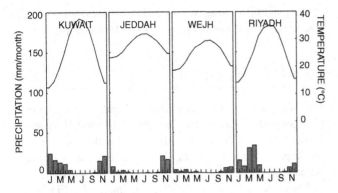

Fig. 17.16 Mean monthly rainfall (mm) and temperature (°C) at four representative stations of the Arabian Peninsula: Kuwait, Jeddah, Wejh, and Riyadh.

In central Arabia, the dominant vegetation in sandy areas is *Artemisia* and *Calligonum*, with some perennial and annual herbs. There are large areas with steppe vegetation. *Stipa* is the characteristic and most abundant cool-season annual, but drought may prohibit growth of herbs and grasses for several years in succession. Vegetation is quite different in the eastern region and along the Red Sea coast, where the terrain reaches higher elevations. In the northeast there are numerous dry wadis and shallow playas, and coastal salt bushes occupy small sand hills. Halophytes occupy saline areas there and along the Red Sea coast. Junipers are found in higher elevations along the Red Sea coast, but various *Acacia* species with tussock grasses and herbs grow below 1500 m elevation. Woodlands are found on the coastal plain and in some drainage channels; tussock-grass savanna occupies the lowest areas of the plain.

The climate within the peninsula depends on the proximity to the sea and topography. Climatic elements for typical stations are shown in Fig. 17.16. Hot desert climate prevails in two regions: a narrow coastal strip along the eastern shore of the Red Sea and the vast desert plain. Hot semi-arid steppe regions are found in the elongated plateau parallel to the Red Sea and the Oman plateau along the western coast of the Gulf of Aden.

Topography and proximity to the surrounding seas likewise determine the temperature regime. In the coastal zone, the annual temperature range (Fig. 17.9) is reduced to 10°–15°C and the diurnal range (Fig. 17.9) to 15°–19°C, depending on the season. In winter, nighttime temperatures tend to be 10°–15°C on the Red and Arabian Sea coasts and in the Rub' al Khali, and over 20°C at the southern end of the Red Sea and Gulf of Aden. Mean January maxima are higher along the Red and Arabian Sea coasts (20°–25°C) than along the Persian Gulf and Gulf of Oman (15°–20°C). In summer the mean daily minimum can be as high as 30°C along the coast of the Persian Gulf. It generally exceeds 25°C along the coasts of the Red Sea and the Gulfs of Aden and Oman. The summer maxima are on the order of 35°–40°C in the northern half of the peninsula. They are lower in the southern half of the peninsula, ranging from around 20°C in the Yemeni Mountains to nearly 35°C in the southeastern lowlands.

In the interior of the Arabian Peninsula, conditions are more extreme. The annual temperature range is 15°–20°C, while the diurnal range is 10°–15°C in winter, 20°–25°C in summer. Mean monthly temperature ranges from about 15°C in winter to 35°C in summer. In winter the mean daily minimum is generally on the order of 10°C (Fig. 17.10), but it can be below 5°C in northern Saudi Arabia and the Yemeni Mountains. Mean daily summer maxima are generally around 40°–45°C (Fig. 17.10).

Absolute minimum temperatures vary greatly, depending mainly on proximity to the coast. Temperatures as low as −6° or −7°C are occasionally reported in the interior of the peninsula and in the Yemeni Mountains. In the Persian Gulf, the northern Red Sea, and the Gulf of Aqaba the temperature occasionally drops to 0°–5°C; in the Gulf of Oman 5°–10°C has been recorded. The southern Red Sea coast and the Gulf of Aden are very warm, with absolute minima generally being higher than 15°–20°C. The absolute maximum temperature can reach 45°C in southern peninsula. The absolute annual range is generally 40°–45°C.

Over most of the peninsula, the climate has a Mediterranean seasonality with meager rainfall occurring in winter and dryness prevailing in summer. Cold-season rainfall is typical (Fig. 17.16), but some stations, especially in the south and west, have a maximum in spring, a result of the Saharan depressions that originate in the Atlas Mountains. Riyadh is a good example. In a small southwestern sector of the peninsula, a semi-arid climate with summer rainfall, similar to that in the Sahel, prevails.

Orography has a major influence, so that rainfall is lower in the central lowlands but much higher over the southwestern plateau, where orographic effects enhance the summer rainfall regime prevailing there. Some areas experience a peak in both winter and summer or spring (Marcella and Eltahir 2008). Along the southern Arabian Sea coast, the upwelling of cold water enhances the summer aridity. Annual means are generally 50–100 mm or less over most of the peninsula, and few stations receive more than 20–30 mm/month even during the wettest months. Some high-altitude stations, which experience nearly year-round rainfall, receive more precipitation, but generally not more than a few hundred millimeters per year. Except for the mountains, snow is unknown in most of the peninsula. In the far north, however, it falls once every few years.

The winter rains in the region are associated mainly with Mediterranean cyclones. However, cyclones that develop in the Zagros Mountains of Iran and Iraq also influence the region, as do air masses from parts of equatorial Africa, such as Sudan and Ethiopia (Barth and Steinkohl 2004). On interannual time scales, the winter rains are influenced by ENSO (Marcella and Eltahir 2008).

Spring is part of the rainy season in the south and southwest, where local land–sea and mountain–valley circulations are important factors in spring rainfall. These circulations are weak in autumn, when rainfall is largely confined to the northern edge of the peninsula, which is occasionally traversed by Mediterranean depressions during this season.

Over most of the peninsula, where a winter rainfall regime prevails thunderstorms are fairly uncommon. They occur most frequently in autumn and spring and are almost absent in summer. In the southwestern region with summer rainfall, thunderstorms can occur as often as 6 days per year, and peak in July and August.

The Arabian Peninsula is generally sunny and dry. Cloudiness in most locations is 20–30% in winter, but less than 10–20% in summer. Relative humidity is strongly affected by proximity to the coasts, where it is on the order of 65–70% in summer and over 70% in winter. In the interior, however, relative humidity is generally less than 20% in summer, 30–40% in winter. Fog occurs in the interior regions most frequently in late summer and early autumn, preceding the occasional winter rains. Along the coasts, fogs are rare. Those that do occur are usually light and persist for only a few morning hours. The coasts of the Red Sea and Persian Gulf commonly experience heavy dews, which can be so heavy as to give the appearance of rainfall on the surface. In Sana (Yemen) dew occurs on 40 nights per year. Visibility can also be reduced by a heavy haze along the Red Sea coast and the Persian Gulf. Unlike fog, this results from a combination of sea salt, dust, and moisture. In summer, this can reduce visibility to less than half a kilometer.

Dust storms can affect immense areas of the peninsula. Their frequency ranges from as few as 5 days per year to over 80 days per year at Riyadh. Dust devils are also common. In the Persian Gulf, dust storms are largely due to the northwest wind called the *shamal*, which blows in summer. In parts of the peninsula, the wind velocities are sufficiently high to produce sand storms in dune-covered regions.

Sand storms are of two main types: one resulting from turbulence produced by extreme surface heating, and the other associated with low-pressure centers passing through the region. The locally induced storms are most common during the daytime hours and in the hottest months, when the sand temperature can reach 85°C. At Al-Hofuf, the percentage of daytime hours in which sand drift occurs ranges from 76% in February to 91% in June. The sand storms associated with lows generally last longer, with sand blowing incessantly from 24 hours up to nearly 48 hours. In the Gulf of Aden, both dust storms and sand storms can be induced by hot, dry winds coming off the Somali tableland from June through August.

Surface winds are a response to a combination of surface pressure and local orography, including coastal effects. In the western coastal strip along the Red Sea, the prevailing wind in northern sectors is north to west for most of the year. In the southern sector, winds are west to northwest in summer, south to southwest in autumn and winter, and westerly in spring. On the southern coasts, winds are most frequently from an easterly quadrant in winter but are quite variable at other times. On the eastern coasts, winds are generally northwesterly in winter and from the northeasterly quadrant at other times. Elsewhere, winds are from the northwesterly quadrant in summer, and are otherwise easterly or southerly. In the northernmost sectors

of the peninsula, winter winds tend to be dry northeasterlies, emanating from the winter high over the Armenian plateau.

17.6 THE NEAR EAST

The Near East comprises large tracts of arid or semi-arid land with generally Mediterranean climates. The rainfall decreases from north to south and also inland, so that the desert regions are largely confined to Iraq, Jordan, and Israel. The most extreme deserts are in the Negev and Sinai, adjacent to the Red Sea. Climatic characteristics are summarized in Figs. 17.10 and 17.17, representing winter minimum temperatures, summer maximum temperatures, and precipitation and temperature at individual stations. Goldreich (2003) provides a detailed description of the weather systems affecting the region.

In the Mediterranean region, steppes are mainly widely scattered trees with a sparse understory of generally herbaceous perennial grasses; vegetation is sparse. In the more arid habitats, trees are replaced by shrubs and the understory is mostly composed of dwarf shrubs such as *Sarcopoterium*, *Artemisia*, and *Salsola*. The desert is marked by a more contracted growth of the same species and by sands and sand dunes. Further south the desert begins to take on some characteristics of a savanna, the upper story is deep-rooted Sudanian trees and shrubs, most commonly *Acacia* sp., but the understory consists of shallow-rooted desert elements.

17.6.1 TURKEY

The drylands of Turkey lie in its interior plateau, a region of continental climate with severe winters and warm but dry summers. Winter weather conditions are particularly severe when Black Sea depressions approach the plateau, bringing a rapid change from warm southerly air to cold fronts. These fronts often bring heavy snow in the mountains and moderate snowfall elsewhere.

Snow is confined to the mountain ranges in the plateau district. It falls mainly in winter, but occasionally occurs in spring or autumn. At Ankara, snow falls on 15 days per year, mostly in January and February, but it has been recorded as late as April and as early as November. At higher elevations, such as Erzurum (1829 m) and Sivas (1285 m), it occurs on as many as 30–60 days each year.

Typically continental, both the annual and diurnal temperature ranges are relatively large in Turkey's interior. In winter, mean daily minimum ranges from about −3°C to −13°C, with mean daily maxima between about 0° and 5°C. In summer, mean daily minima and maxima are on the order of 10°–15°C and 25°–30°C, respectively.

Cyclonic storms, originating primarily over the Mediterranean, bring most of the precipitation in autumn and winter (Kadioglu 2000). These tend to influence the entire country. Topographic influences prevail in the eastern half of the country. Maximum precipitation occurs during winter in the western portions of the country, but in May in eastern portions.

In the interior, mean annual rainfall is generally 350–500 mm, hence the region is semi-arid. Maximum rainfall occurs in June. Mean monthly rainfall is about 50–75 mm in the wettest months, 5–30 mm during the driest (usually summer). The number of days with rainfall exceeding 1 mm is about 65–80; rainfall exceeds 10 mm on about 10–15 days per year. The annual frequency of fog days is about 10–25, except at low elevations, where it occurs only infrequently. Fogs are most frequent during the coldest months and are rare in summer. Thunderstorms average about 10 per year, usually in spring and autumn. Relative humidity is generally 70–80% in winter, but can be less than 20–30% in summer in some areas.

Cloudiness is greater during the winter rainy season than during the summer. It is on the order of 75% in the north and 50% in the south. In summer it remains high near the coast, but is as low as 5% in the interior, where as many as 70–80% of the days are clear.

The prevailing winds shift seasonally. In the interior of Turkey, they tend to be from northerly directions in spring and summer, southwesterly in autumn and winter.

17.6.2 ISRAEL

PRECIPITATION REGIME

The distribution of rainfall Israel is complex and strongly affected by topographic and coastal influences (Sharon and Kutiel 1986). Along the Mediterranean annual rainfall is about 500 to 800 mm. In the Upper Galilee, in the far north, mean annual rainfall exceeds 900 mm in some places. Other areas of high rainfall are the highlands west of Jerusalem, where orography enhances rainfall, and the northern coast near Haifa. In both locations, mean annual rainfall exceeds 600 mm in many areas.

Most of Israel lies within the Negev desert, where mean annual rainfall ranges from roughly 25 to 300 mm. The number of rainfall occurrences is relatively small in the Negev, ranging from 16 days per year on average in the southern central highlands to 35 days per year at Beersheba in the north. Rainfall is also low in the rain shadow of the Jordan valley, a valley running from the Yam Kinneret (Sea of Galilee) to the Dead Sea.

All of Israel experiences a Mediterranean climate. Most rainfall is associated with Mediterranean depressions or cyclones, such as Cyprus Lows and Sharav Lows (Goldreich et al. 2004; Rubin et al. 2007). The latter originate over North Africa and are most common in the transition seasons. The Cyprus Lows are generally a winter feature. These various cyclones influence primarily northern portions of the country, as well as surrounding countries such as Syria, Lebanon, Jordan, and Iraq, and also bring cold fronts to the region. Maximum rainfall occurs in January, when these systems reach their southernmost tracks. Global-scale phenomena also influence precipitation. It tends to be above normal during El Niño years and below normal during La Niña years (Price et al. 1998). Thunderstorms are common in winter and the transition seasons, occurring on 5–10 days per year.

Snow occurs occasionally in the hills and ridges, where it falls a few days per year, and occasionally around Jerusalem. It occurs with somewhat higher frequency in the Golan Heights. In some years there are relatively spectacular occurrences following the passage of an intense depression. A storm in early January 1992 blanketed the entire city of Jerusalem with some 40 cm of snow (Fig. 17.7b). Snowstorms typically occur during El Niño years (Price et al. 1998). In Jerusalem, snow was a fairly common phenomenon in the late nineteenth century, occurring during roughly one in two years.

Winter cloud cover is generally 40–60%, but during the summer clear skies generally prevail, with mean cloud cover generally being less than 10%. Summer rainfall occurs on occasion, but is generally fairly light, only a few millimeters (Saaroni and Ziv 2000). It does not appear to be forced by synoptic systems, but by a local weakening of the prevailing thermodynamic conditions that suppress precipitation in summer.

Fog is fairly common, but shows strong geographical contrasts. In northern Israel on Mount Kenaan (alt. 936 m) it occurs about 45 nights per year and in the Jezreel Valley on about 54 nights. In the southern Beersheva Valley, fog occurs on 42 nights and in the central Negev hills it occurs on about 25 nights. The maximum frequency tends to be in the early spring in the north and mid-summer in the central to northern Negev. In the hills district the maximum frequency is in winter (26 days per year in Jerusalem). At Jericho (258 m below sea level), near the Dead Sea, it occurs less than one day per year, on average.

Dew is also an important form of precipitation in much of the Mediterranean. The average number of nights with dew formation reaches nearly 200 in the central Negev, where it brings roughly 33 mm of water per year, or about one third the average annual rainfall (Evenari et al. 1971, Zangvil 1996). In drought years, dew has been found to exceed rainfall. Dew can wet the first millimeters of soil and it is important for lower plants, biological sand crusts and some animals and soil fauna.

A remarkable aspect of Israel's climate is intense floods that dramatically interrupt the aridity of the Negev and Judean Desert. These are generally associated with Red Sea Trough (RST) or Mediterranean cyclones centered over Syria (Ziv et al. 2005). Of 52 flood events examined by Kahana et al. (2002), 37 were associated with these two synoptic situations. In the eastern Negev 84–100% of the runoff in ephemeral streams is related to these systems (Kidron and Pick 2000). Floods in the Negev are also triggered by "tropical plumes" (see Chapter 5) that emanate from Africa and other upper-level troughs over the Mediterranean (Rubin et al. 2007; Funatsu et al. 2008) and by features termed "potential vorticity streamers," parcels of energy that likewise emanate from the tropics and provide a dynamic situation conducive to convection (Krichak et al. 2007).

THERMAL REGIME

Throughout most of Israel, summers are warm and winters are cool or mild. In summer mean daily maxima vary from 32° to 36°C in the north and from 38° to 42°C elsewhere. Summer

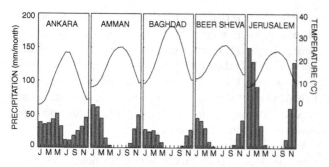

Fig. 17.17 Mean monthly rainfall (mm) and temperature (°C) at five representative stations of the Near East: Ankara, Amman, Baghdad, Beer Sheva, and Jerusalem.

nighttime temperatures are more variable, with mean daily minima ranging from 14° to 19°C in the north and from 22° to 26°C in the south of Israel. In winter there is a pronounced temperature gradient between the coast and the hilly district and from north to south. Near the coast, mean daily minima are on the order of 7°–11°C, compared with 4°–8°C in the hills. Inland mean daily minima range from below freezing in the north, but 10°–14°C elsewhere. Mean daily maxima in winter range from 14° to 19°C near the coast and from 9° to 14°C inland in the hills. In some locations, such as the Hulle Valley, subfreezing temperatures occur on occasion. The mean diurnal range is on the order of 10°–12°C in winter; in summer the diurnal range is much more spatially variable, but on the order of 12°–18°C.

The temperature regime of the Negev Desert itself is illustrated with data from Beer Sheva, in the northern Negev (Fig. 17.17), Mitzpeh Ramon (a highland location in the central Negev), and Eilat, in the southern-most tip of the Negev. The highest temperatures generally occur in August. Mean annual temperature ranges from 18°C at Mitzpeh Ramon to 26°C at Eilat, with minima ranging from 0.5°C at Mitzpeh Ramon to 5°C at Eilat. Frost rarely occurs in the Negev, even though at a few locations the absolute minimum is near zero. The annual temperature range exceeds 22°C at Eilat, but is only about 15°C at Beer Sheva and Mitzpeh Ramon.

Relative humidity is strongly dependent on elevation and distance from the coast. It decreases on the eastern slopes of the hills, but increases in the regions around the Dead Sea and the lakes of Kinneret (Sea of Galilee) and Hula. It is lowest in summer, when it is on the order of 70–75% in the coastal district and 50% in the hilly district. Inland, it averages 45–65% in the northern parts in summer and 25–45% in the middle and southern areas. In winter, the rainy season, it generally averages 70–80%.

WINDS

The wind regime is generally light to moderate, with prevailing directions showing strong geographical contrasts imposed by the coast and the inland topography. Over much of the country there is a tendency for southerly and westerly directions in winter, northerly and northwesterly directions in summer, but

directions change with the passage of depressions. Easterly windstorms occur from time to time during the cold season, often causing significant damage (Saaroni *et al.* 1998b). In northern Israel, southeasterlies are common in summer.

Israel also experiences the khamsin winds, which blow mostly from an ESE direction, after passing over the Arabian or Sinai deserts. This wind is so dry that the relative humidity generally falls below 10%. Temperatures associated with it are extremely high, reaching 42°C or more in Eilat. Temperatures exceeding 48°C have occurred during a khamsin in the Arava Valley.

17.6.3 SYRIA

There are five climate divisions in Syria: coastal, western inland, northeast district, the eastern district, and the desert district. The coastal climate in the west is typically Mediterranean. Further inland are dry steppes and then a desert interior. The western inland is colder and somewhat less rainy than the coast in winter. It is mild and dry the rest of the year. The northeast is somewhat rainier and colder than the western inland. The east is hot and dry in summer, cold and with light rain in winter. The desert district, which includes the great desert of the Syrian plateau (500 m ASL to above 1000 m ASL), is excessively hot and rainless in summer and the transition season, and dry and cool in winter, with warm days and extremely cold nights.

In most of the country, mean annual rainfall is between 100 and 200 mm. Winter is the rainiest season, but significant precipitation also occurs during the transition seasons. January is generally the wettest month, with about 25–35 mm in the dry zones. The amount falls to 10 mm by May, and summer is nearly rainless. Thunderstorms are infrequent in winter, but fairly common in April/May and October/November. Snow is rare in the dryland sectors, occurring on average less than one day per year. It is generally confined to the wetter regions of the coast and to the north and west. At Damascus, Aleppo, and Kamishli it occurs on 3–4 days per year, on average.

Mean cloudiness is generally on the order of 40–50% in winter, but 10–15% or less in the dry, summer months. Humidity is also lowest in summer, when it is on the order of 25–35% in these regions. Fog occurs on 5–12 days per year, mostly in winter. It is rare in summer.

In Syria, the lowest temperatures occur in January, the highest in July and August. In winter, mean temperature is generally 10°–11°C in coastal areas but 6°–8°C in the arid regions inland. Mean daily maximum temperatures in the arid regions are around 13°–18°C, mean minima 2°–4°C. In summer, the mean minima and maxima are 22°–27°C and 39°–42°C in the arid regions.

There are two surface wind regimes. A northeasterly regime prevails in summer, a westerly/southwesterly regime at other times. These are manifested differently at various locations within the country. In the desert, flow tends to be westerly to northwesterly for most of the year.

17.6.4 LEBANON

Lebanon is extremely dry in summer, but winter rainfall can be considerable. Mean annual rainfall can reach 1000–1400 mm in the mountains and almost that along the coast. Dry climates, with annual rainfall less than 400 mm, are found only in small inland regions; the driest areas, in the south, receive about 200 mm/year. Cloudiness is on the order of 40–60% in winter and 10–20% or less during the dry summer. Thunderstorm frequency is generally 20–30 days per year, except in the more arid south, where it is as low as 5–10 days. Fog occurs relatively infrequently.

Temperatures are more moderate than in nearby Syria, a consequence of its higher elevation. Winter mean temperatures range from 10°–15°C near the coast to 0°–5°C in the highest mountain elevations, which are generally snow-covered in winter. The diurnal range is only a few degrees. In summer, mean temperatures range from 25°–28°C along the coast to 10°–15°C in the mountains. The annual range is dependent on elevation and proximity to the coast, but is generally on the order of 12°–15°C at lower elevations and as great as 18°–20°C at higher elevations.

The prevailing winds shift markedly with the season and vary greatly with geographic location. The topography and coastal effects provide for a complex wind regime.

17.6.5 JORDAN

Jordan has a semi-arid climate in the west, which is nearest the coast, but most of the country is desert (Tarawneh and Hadadin 2009). The desert district is hot and dry in summer, cold with mostly clear skies in winter. Annual precipitation in the desert district is generally less than 100 mm/year, on average. It is concentrated in winter: January and February are the wettest months. In the lowlands of the Jordan (the Ghor district, where land is below sea level), annual rainfall is 150–250 mm. In the wettest region, rainfall is on the order of 500 mm/year. There cloudiness can be on the order of 50% during the winter rainy season but less than 10% during the summer months.

Snowfall is rare, falling not more than once or twice per year even in high terrain. Dew is an important source of precipitation in the desert regions. It is most common during the dry season under conditions of cloudless skies. In occurs on 160 nights in the Jordan Rift Valley and on 70 nights in Amman. Fog and thunderstorms are also most common in the hills. In Mafraq, fog occurs on 25 days per year, peaking in both winter and summer. Only a few thunderstorms or fogs per year occur in the desert and Ghor. In the desert, relative humidity ranges from 22% in summer to about 50% in winter. Elsewhere it can be relatively humid. Outside the desert, maximum relative humidity is generally 60–80%, but it can rise above 80% for most of the night and early morning. It reaches 95–100% for long periods on many days.

In winter, mean daily minimum temperatures range from about 9°C in the Ghor to about 2°–4°C in the desert; it

frequently falls below zero. Mean daily winter maxima in these regions are 18°–20°C and 12°–15°C, respectively. In summer, the mean maximum is about 39°C in the Ghor and 35°C in the desert; the mean summer minimum is about 22°–25°C in the Ghor and 18°–20°C in the desert.

Surface winds are generally northwesterly in summer but westerly or southwesterly during other seasons. The westerly winds often bring strong foehn winds into the Rift Valley of Jordan, as they descend off the mountains to the west. Southerly or southeasterly winds may blow ahead of fronts during the passage of depressions, causing dust storms. Dust outbreaks are frequent in the Ghor and the desert, especially in summer; they occur with a frequency of 25–60 days per year.

17.6.6 IRAQ

Most of Iraq is semi-arid, with dry, hot steppe in central and southern regions, and dry, hot desert in the west. Here the winter rainy season begins in October. Clear skies prevail most of the year, even during the rainy season. Summer clouds appear on only a few days. Winter cloudiness is on the order of 30% in some months. Even then, clouds appear on only 35–40 days.

Nearly all of the rainfall is produced by Mediterranean depressions. Mean annual rainfall is less than 200 mm in most of the country, but it falls to less than 100 mm in the extreme southwest. In spring, most of the rainfall is accompanied by thunderstorms. The mountains in northern Iraq receive heavy snowfall in winter, but in the dry central plains it is rare and usually melts immediately. It occurs on one day per year on average at Mosul and Kirkuk, in the north. It is extremely rare in the south and the arid southwest. Relative humidity drops to about 25% in summer, but it occasionally reaches zero in the desert. Fog is very rare in the desert, especially in spring and summer.

The temperature regime is extremely continental, with the annual range varying from 21° to 27°C. The diurnal range is 15°–20°C. In winter, mean daily temperatures range from 6°C inland to 12°C near the gulf. Temperature can fall below freezing in many locations. In summer the desert tends to be cooler than other locations, owing to its higher elevation. In some areas, temperatures in excess of 50°C have been recorded. The mean minimum is on the order of 2°–8°C in winter and 22°–28°C in summer. The mean maximum temperature in winter ranges from about 12° to 15°C in the north and from about 16° to 20°C in the south. In summer the mean maximum temperature is about 40°–43°C throughout most of the country.

In winter the prevailing winds are generally northwesterly, shifting with the passage of cyclones. Northwesterlies also prevail in summer but much more persistently. The sand- and dust storms are most frequent in central and southern areas. Their frequency is highest in July and lowest in December and January. The annual frequency varies geographically from 20 to 30 days. In spring and autumn, the sirocco occasionally blows, bringing hot and dry air in association with cyclone passages.

REFERENCES

Alpert, P., and Coauthors, 2005: Tropical tele-connections to the Mediterranean climate and weather. *Advances in Geosciences*, **2**, 157–160.

Barth, H. J., and F. Steinkohl, 2004: Origin of winter precipitation in the central coastal lowlands of Saudi Arabia. *Journal of Arid Environments*, **57**, 101–115.

Bitan, A., and H. Saaroni, 1992: The horizontal and vertical extension of the Persian Gulf pressure trough. *International Journal of Climatology*, **12**, 733–747.

Dubief, J., 1963: *Le Climat du Sahara*. Université d'Alger, Institut de Recherches Sahariennes, Algiers, 2 volumes.

Evenari, M., L. Shanan, and N. H. Tadmor, 1971: *The Negev: the Challenge of a Desert*. Harvard University Press, Cambridge, MA, 345 pp.

Funatsu, B. M., C. Claud, and J. P. Chaboureau, 2008: A 6-year AMSU-based climatology of upper-level troughs and associated precipitation distribution in the Mediterranean region. *Journal of Geophysical Research–Atmospheres*, **113**, doi:10.1029/2008JD009918.

Goldreich, Y., 2003: *The Climate of Israel: Observation, Research and Application*. Springer, Berlin, 298 pp.

Goldreich, Y., H. Mozes, and D. Rosenfeld, 2004: Radar analysis of cloud systems and their rainfall yield in Israel. *Israel Journal of Earth Sciences*, **53**, 63–76.

Kadioglu, M., 2000: Regional variability of seasonal precipitation over Turkey. *International Journal of Climatology*, **20**, 1743–1760.

Kahana, R., B. Ziv, Y. Enzel, and U. Dayan, 2002: Synoptic climatology of major floods in the Negev Desert, Israel. *International Journal of Climatology*, **22**, 867–882.

Karabork, M. C., E. Kahya, and M. Karaca, 2005: The influences of the Southern and North Atlantic Oscillations on climatic surface variables in Turkey. *Hydrological Processes*, **19**, 1185–1211.

Kidron, G. J., and K. Pick, 2000: The limited role of localized convective storms in runoff production in the western Negev Desert. *Journal of Hydrology*, **229**, 281–289.

Krichak, S. O., P. Alpert, and M. Dayan, 2007: A southeastern Mediterranean PV streamer and its role in December 2001 case with torrential rains in Israel. *Natural Hazards and Earth Science Systems*, **7**, 21–32.

LeHouerou, H. N., 1986: The desert and arid zones of northern Africa. In *Hot Deserts and Arid Shrublands* (M. Evenari, I. Noy-Meir, and D. W. Goodall, eds.), Ecosystems of the World 12B, Elsevier, Amsterdam, pp. 101–147.

Littman, T., and S. M. Berkowicz, 2008: The Regional climatic setting. In *Arid Dune Ecosystems. The Nizzana Sands in the Negev Desert* (S.-W. Breckle, A. Yair, and M. Veste, eds.), chap. 4, 49–63. Ecological Studies 200, Springer, Berlin.

Marcella, M. P., and E. A. B. Eltahir, 2008: The hydroclimatology of Kuwait: explaining the variability of rainfall at seasonal and inter-annual time scales. *Journal of Hydrometeorology*, **9**, 1095–1105.

Meteorological Office, 1962: *Weather in the Mediterranean. Volume 1: General Meteorology*. Her Majesty's Stationery Office, London, 362 pp.

Orshan, G., 1986: The deserts of the Middle East. In *Hot Deserts and Arid Shrublands* (M. Evenari, I. Noy-Meir, and D. W. Goodall, eds.), Ecosystems of the World 12B, Elsevier, Amsterdam, pp. 1–28.

Pagano, T. C., S. Mahani, M. J. Nazemosadat, and S. Sorooshian, 2003: Review of Middle Eastern hydroclimatology and seasonal teleconnections. *Iranian Journal of Science and Technology – Transaction B*, **27**, 1–16.

Porcu, F., A. Carrassi, C. M. Medaglia, F. Prodi, and A. Mugnai, 2007: A study on cut-off low vertical structure and precipitation in the Mediterranean region. *Meteorology and Atmospheric Physics*, **96**, 121–140.

Price, C., L. Stone, A. Huppert, B. Rajagopalan, and P. Alpert, 1998: A possible link between El Niño and precipitation in Israel. *Geophysical Research Letters*, **25**, 3963–3966.

Rubin, S., B. Ziv, and N. Paldor, 2007: Tropical plumes over eastern North Africa as a source of rain in the middle east. *Monthly Weather Review*, **135**, 4135–4148.

Saaroni, H., and B. Ziv, 2000: Summer rain episodes in a Mediterranean climate, the case of Israel: climatological-dynamical analysis. *International Journal of Climatology*, **20**, 191–209.

Saaroni, H., A. Bitan, P. Alpert, and B. Ziv, 1998a: Continental polar outbreaks into the Levant and eastern Mediterranean. *International Journal of Climatology*, **16**, 1–17.

Saaroni, H., B. Ziv, A. Bitan, and P. Alpert, 1998b: Easterly wind storms over Israel. *Theoretical and Applied Climatology*, **59**, 61–77.

Sharon, D., and H. Kutiel, 1986: The distribution of the rainfall intensity in Israel, its regional and seasonal variations and its climatological evaluation. *Journal of Climatology*, **6**, 277–291.

Takahashi, K. and H. Arakawa (eds.), 1981: *Climates of Southern and Western Asia*. World Survey of Climatology 14, Elsevier, Amsterdam, 348 pp.

Tarawneh, Z., and N. Hadadin, 2009: Reconstruction of the rainy season precipitation in central Jordan. *Hydrological Sciences Journal*, **54**, 189–198.

Tsvieli, Y., and A. Zangvil, 2005: Synoptic climatological analysis of 'wet' and 'dry' Red Sea troughs over Israel. *International Journal of Climatology*, **25**, 1997–2015.

Turkes, M., and E. Erlat, 2005: Climatological responses of winter precipitation in Turkey to variability of the North Atlantic Oscillation during the period 1930–2001. *Theoretical and Applied Climatology*, **81**, 45–69.

Turkes, M., and E. Erlat, 2008: Influence of the Arctic Oscillation on the variability of winter mean temperatures in Turkey. *Theoretical and Applied Climatology*, **92**, 75–85.

Turkes, M., T. Koc, and F. Saris, 2009: Spatiotemporal variability of precipitation total series over Turkey. *International Journal of Climatology*, **29**, 1056–1074.

Xoplaki, E., J. F. González-Rouco, J. Luterbacher, and H. Wanner, 2004: Wet season Mediterranean precipitation variability: influence of large-scale dynamics and trends. *Climate Dynamics*, **23**, 63–78.

Zangvil, A., 1996: Six years of dew observations in the Negev Desert, Israel. *Journal of Arid Environments*, **32**, 361–371.

Ziv, B., H. Saaroni, and P. Alpert, 2004: The factors governing the summer regime of the eastern Mediterranean. *International Journal of Climatology*, **24**, 1859–1871.

Ziv, B., U. Dayan, and D. Sharon, 2005: A mid-winter, tropical extreme flood-producing storm in southern Israel: synoptic scale analysis. *Meteorology and Atmospheric Physics*, **88**, 53–63.

18 Australia

18.1 THE AUSTRALIAN DESERT

The Australian desert is the Southern Hemisphere analog of the Sahara, in that it represents the dry transition between the tropical summer rains and extra-tropical winter rains. Most of the region, however, is more akin to the semi-arid Mediterranean steppes or Sahelian savanna than to the Sahara. Approximately 50% of the Australian continent is arid land and over a quarter is semi-arid. The most important desert regions include the Great Sandy Desert, Gibson Desert, the Great Victoria Desert, the Simpson (Arunta) Desert, and the Sturt Desert (Fig. 18.1).

The desert surface types include sand deserts, stone deserts, mountain and shield deserts, and riverine and clay plains (Fig. 18.2). The largest area, almost 2 million km^2, is the sand deserts; these areas are almost devoid of surface water but are generally vegetated and reasonably stable. The stone deserts are the second largest, nearly 1 million km^2. These often have integrated drainage basins terminating in large salt lakes, such as Lake Eyre. The shield deserts are relatively featureless, with drainage becoming disconnected in more arid regions and with a few salt lakes. There is relatively little high terrain in Australia; most of the land lies below 600 m. The few mountainous regions generally have elevations between 600 and 1200 m. Those in the central desert tend to enhance the meager rainfall. Many of the soils are deep red, porous sands. Calcareous and siliceous loams, shallow and gray or gray-brown, are associated with salinas; cracking clays of moderate depth occupy alluvial plains and uplands.

Despite a relatively strong uniformity of climate throughout much of the Australian desert, the continent contains a remarkable complexity of vegetation types (Fig. 18.3). The complexity is a result of the vegetation's strong sensitivity to topography, drainage, and soil type, as well as aridity gradients. The vegetation of most of Australia's drylands consists of low shrubland and perennial grassland, with scattered trees (Fig. 18.4). The areas classified as desert include those with typically arid vegetation formations: hummock and tussock grasslands (Fig. 18.5), *Acacia* shrublands (Fig. 18.6), and low chenopod shrublands.

Wetter regions are typically *Eucalyptus* shrublands, arid and semi-arid low woodlands (e.g, *Casuarina* sp.), and semi-arid shrub woodlands. Many of the *Eucalyptus* species grow with multiple stems from an underground tuber and are limited in height. This form, called "mallee," is a commonly occurring type of landscape in Australia (Fig. 18.7). It is a natural growth form for some eucalypts, but in other species the form develops during post-fire regrowth.

The most arid regions include the hummock and tussock grasslands, where the grasses take on clustered cushion and bunch or tuft forms, respectively (Fig. 18.8). The hummock grassland, also known as "spinifex" grassland (Fig. 18.9), is largely dominated by species of *Triodia* and *Plectrachne*, with scattered low trees and shrubs. The hummocks are widely spaced, separated by bare ground in dry years, ephemerals in wet years. They can reach a diameter and height of up to 6 m and 1.5 m, respectively, and tend to cover 5–30% of the ground. The tussock grasslands are predominantly *Astrebla* species, and consist of well-spaced tussocks 0.5–2 m apart and 0.1–0.5 m in diameter.

While the grasslands roughly correspond to the sand deserts, shrubland is more common over rockier surfaces. The shrublands are highly diverse in character and climatic conditions. The *Acacia* shrublands (Fig. 18.10) tend to occupy somewhat wetter sectors of the central and southern desert. The *Eucalyptus* (Fig. 18.11) and *Chenopodium* shrublands occupy mainly the southern margins of the desert. In the latter, also termed "shrub steppe," shrubs range from 30 cm to more than 1 m in both diameter and height, and density ranges up to 500 or more per hectare.

The eastern margin of the desert is occupied by semi-arid shrub woodlands that have a canopy of *Eucalyptus* sp., an underlying shrub layer, and a ground layer of *Aristida* grasses in regions of summer rainfall, or *Stipa* grasses and other species in regions of winter rainfall. Trees are 10–30 m tall, with the canopy covering 10–30% of the surface. The arid and semi-arid low woodlands that lie along the tropical margin of the

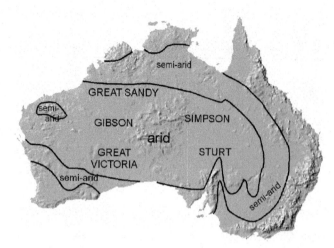

Fig. 18.1 Desert and semi-arid regions of Australia (from Williams and Calaby 1985).

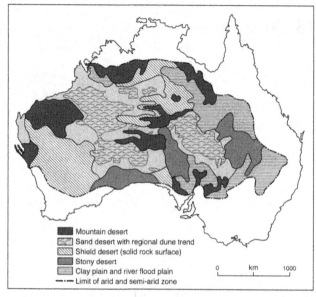

Fig. 18.2 Map of Australian surface types (from Goudie and Wilkinson 1980).

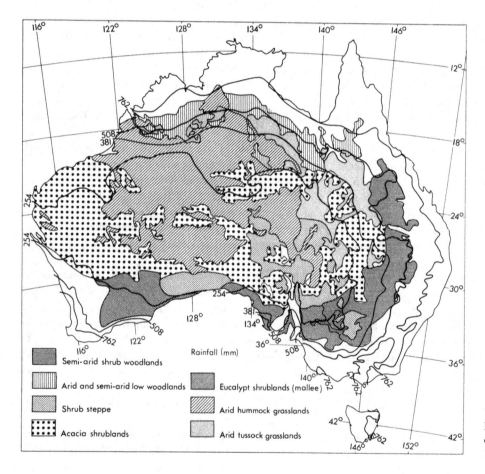

Fig. 18.3 Map of Australian vegetation (from Williams and Calaby 1985).

(a)

(b)

Fig. 18.4 Typical Australian landscapes consist of grasses, shrubs, and scattered trees: (a) spinifex grassland and mulga on sand plain and sand dune; (b) hummock grassland with *Acacia* overstory.

Fig. 18.5 Grassland with hummock grasses.

desert include trees such as *Casuarina* (Fig. 18.12) and dwarf *Eucalyptus*, a shrub layer, and a grass layer. Trees are generally 5–10 m in height, with 5–30% cover. A variety of species, including various *Acacia* species, occupy the shrub layer, while the composition of the grass layer shifts from *Aristida* dominance, to *Triodia* dominance, then *Triodia* hummock grassland, with increasing aridity.

18.2 CLIMATIC CONTROLS

The primary cause of Australia's aridity is its latitudinal position in the subtropical zone of high pressure, separating the tropical circulation prevailing to the north from the mid-latitude westerlies and associated disturbances to the south. The seasonal movement of this zone and the circulation features to the north and south of it control the march of the seasons over Australia.

In winter (Fig. 18.13) high pressure prevails in the mean over the continent and over the adjacent oceans. This pressure field produces dry southeasterlies over most of the continent, but a humid westerly flow from the Indian Ocean over southern Australia (Fig. 18.14). The winter rains that prevail in southern Australia are associated with transient depressions in the westerlies and frontal activity in these depressions. Northwest-oriented cloud bands aloft sometimes interact with these systems to produce prolonged rain episodes.

In summer (Fig. 18.13), the zone of high pressure is displaced southward, and northern Australia comes under the influence of the equatorial or monsoon trough. A heat low also develops over northwestern Australia. The northwest monsoon is the dominant wind system in the north (Fig. 18.14). This flow is mainly a result of Northern Hemisphere trades taking on a westerly component when they cross the equator. In the northwest, the prevailing northwesterly winds originate instead in the Southern Hemisphere: Indian Ocean air masses drawn in to the heat low. Gentili (1971) terms this the "pseudo-monsoon." In the northeast, the prevailing summer winds are the northeast trades. Northern Australia comes under the influence of summer disturbances associated with the warm, moist monsoon flow. Tropical depressions and tropical cyclones also produce summer rainfall; these can originate in either the Indian or the Pacific Ocean. Elsewhere over the continent, dry southeasterlies prevail in summer, producing generally dry weather.

Australia is one of the driest continents, with mean annual rainfall being less than 250 mm on about 37% of the land (Fig. 18.15). In total, 68% has an annual mean of less than 500 mm. However, the degree of aridity in Australia is relatively moderate. This is particularly true along the west coast, where extreme coastal deserts exist over the other Southern Hemisphere continents (Table 18.1). In Australia, mean annual rainfall along the coast exceeds 200 mm everywhere, while large stretches of other continental coasts receive less than 100 mm annually (in many cases between 0 and 25 mm). But even the core of the desert, centered around Lake Eyre in the south, experiences only moderate aridity, in part because small pockets of high relief enhance rainfall in the central desert. The area in which mean annual rainfall drops below 100 mm is relatively small.

The nature of the pressure field moderates the degree of aridity prevailing over the continent. In contrast to the Atlantic and Pacific, where semi-permanent subtropical highs persist throughout the summer, over the Indian Ocean in summer

Fig. 18.6 (left) *Acacia* shrubland dominated by mulga, with its typical fan shape. (right) Mistletoe clinging to mulga shrubs; slight desertification is indicated by bare and eroding soil.

Fig. 18.7 *Eucalyptus* shrubland (mallee) (left), close-up on right.

Fig. 18.8 *Triodia*, a typical grass of hummock or cushion form.

Fig. 18.9 Spinifex grassland.

there are instead a series of travelling anticyclones. In the mean, these produce a zone of relatively high surface pressure. This in itself tempers the aridity over Australia, because rain can occur between successive anticyclones, interrupting the prevailing dry conditions. More importantly, the absence of a semi-permanent high results in the absence of persistent upwelling along this coast, although weak upwelling is present during part of the summer season (Gersbach *et al.* 1999). This is a unique characteristic of Australia's climate. Without the high, there is also little advection of polar water equatorward; the southerly flow

Fig. 18.10 *Acacia* shrubland.

(a)

(b)

Fig. 18.11 Many species of *Eucalyptus* are found in Australia: (a) old *E. camaldulensis* in a drainage line; (b) a typical gum tree, with characteristic light-colored bark.

(a)

(b)

Fig. 18.12 (a) *Casuarina*, or "desert oak"; (b) foliage and capsule detail of this *Casuarina* tree.

that does exist, termed the Capes Current, draws in water from comparably low latitudes.

Therefore Australia's climate is moderate in its thermal regime, compared with other continents in subtropical latitudes. This is a result of the prevalence of air masses of maritime origin, with few areas of major relief to transform or deflect them, and the previously noted absence of a quasi-permanent high-pressure cell over the Indian Ocean. Winter temperatures, in particular,

are kept moderate by the lack of continental polar air masses, which create extreme cold conditions in the deserts of Asia.

The three primary air masses that affect the continent are the Australian continental tropical, Indian maritime tropical, and Pacific maritime tropical. The continental tropical air is dry and relatively cold in winter, warm in summer. Its origin is in subsiding air in the center of high-pressure cells. This air, which warms and dries out as it descends, can be traced upward to at least 10,000 m. The maritime tropical air masses are humid and of moderate temperature during both winter and summer; they often bring summer thunderstorms during the monsoon season. Subpolar maritime air occasionally reaches Australia, but it is relatively mild, seldom being colder than 2°C in winter or 15°C in summer.

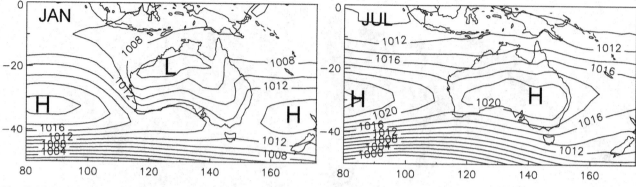

Fig. 18.13 Mean surface pressure (mb) in (left) January and (right) July.

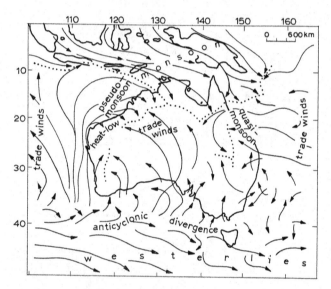

Fig. 18.14 Mean flow pattern in January, showing the areas of dominance of various wind regimes (from Gentili 1971).

18.3 ATMOSPHERIC CIRCULATION

In northern Australia, winds tend to be northerly or westerly in summer, a consequence of the prevailing northwest monsoon. Winds are more variable in the northern desert, which experiences the dry, weak winds of the anticyclone belt. Near Alice Springs, the prevailing direction switches to southeasterly. From there southward, the prevailing summer winds are from southerly quadrants, generally with a westerly component in the west and an easterly component in the far east. The mid-latitude westerlies are displaced far to the south, to Tasmania.

In winter, the direction reverses in the north, with southeasterly trades prevailing in northern and central Australia. The anticyclonic belt narrows and breaks up into individual cells, and the prevailing mid-latitude westerlies control the weather in southern Australia. In the southern desert, winds are extremely variable, marking the transition between the tropical and extra-tropical regimes. The surface circulation is strongly zonal in winter, but more meridional in summer, when monsoon

northwesterlies and southeasterly trades alternately prevail in northern Australia.

The southeasterly trades are probably the dominant wind system over Australia. They prevail for more than 11 months of the year in central Australia (Fig. 18.16). They also prevail for 9–11 months of the year in much of northern Australia, but for less than 6 months per year in the southern half of the continent, south of roughly 25° S.

The troposphere over Australia is dominated by westerly winds, with an upper-tropospheric jet stream having its core near 12 km (Fig. 18.17). In winter the jet is particularly strong, with a single core near 30° S and core speeds in excess of 60 m/s. In summer the jet is substantially weaker (maximum speeds generally about 35 m/s) with two cores, near 35° S and 50° S. The subtropical core is not a constant feature and its position varies considerably from year to year. Limited areas of easterlies occur in the lower-to-mid-troposphere, between about 0° and 25° S in winter, but between 15° and 35° S in summer. In summer their maximum speed is generally less than 35 m/s (Hendon and Liebmann 1990).

Imbedded within them is an easterly jet stream at 600 mb which attains core speeds of about 30 m/s. Like the African easterly jet, it is a response to the thermal gradient between the desert and the tropical ocean in summer, and it promotes easterly wave activity along its equatorward margin. Unlike those of West Africa, the waves are partially over land and partially over water. Another low-level jet exists over northern Australia (Riley 1989). Like that over the US Great Plains, it is a nocturnal boundary flow with a wind maximum between 600 and 1000 m. A transient winter feature, this jet is generally a southeasterly or easterly current with maximum speeds on the order of 10–12 m/s.

18.4 THE RAINFALL REGIME

The most arid part of the continent is the interior. To the north lies the region of tropical, warm-season rains, to the south is the region of mid-latitude, cold-season rains. The core of the arid zone is to the south and east of Alice Springs, in the Sturt

Table 18.1. *Mean annual rainfall (mm) on the western coasts of subtropical continents (from Gentili 1971).*

Latitude N or S	Australia	North America	South America	Southern Africa	North Africa
20°	300	875	0	25	100
22°	300	750	0	25	50
24°	225	100	0	25	25
26°	250	100	25	25	75
28°	400	100	50	50	175
30°	575	200	100	100	200
32°	700	250	275	175	375

Desert east of Lake Eyre. Mean annual rainfall is less than 120 mm in this core and is as low as 80 mm at some stations, such as Puttaburra. In general, Australia is drier in the west than in the east, where arid conditions are far distant from coastal areas. Rain occurs on 20–50 days per year in arid and semi-arid regions of Australia, but in many locations these days are spread out over a period of 10–12 months.

Cloud cover generally increases in all directions from the desert core. Stratiform cloud is common in winter in the south, which is influenced by frontal systems associated with the mid-latitude westerlies. In the north, cumuliform clouds are the most common in summer, but the winter dry season is generally cloud-free. In the central desert, at Alice Springs, cloudiness is

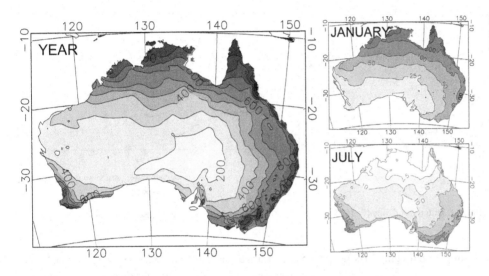

Fig. 18.15 Mean annual rainfall (mm) and mean rainfall (mm) for January and July.

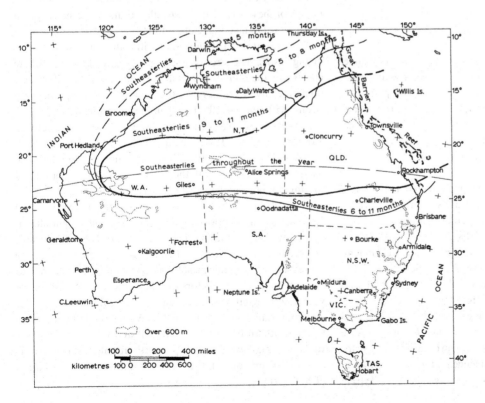

Fig. 18.16 Dominance of surface southeasterlies over Australia (from Gentili 1971). The area bounded by a thick line experiences SE trades throughout the year. Elsewhere the number of months of dominance is indicated.

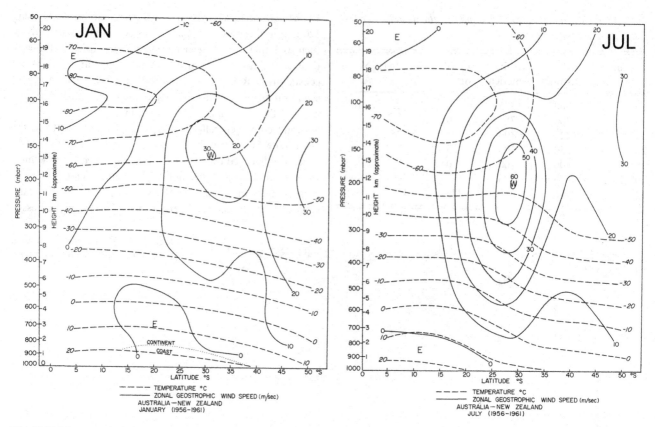

Fig. 18.17 Mean zonal winds over Australia in January and July (from Gentili 1971). Mean temperatures (°C) are also indicated.

very low, with daytime cloud cover ranging from roughly 10% in winter to 25–30% in summer.

18.4.1 SEASONALITY

The seasonal cycle of rainfall in the Australian drylands includes three dominant patterns: a summer maximum, a winter maximum, or – as in most arid regions – little seasonal preference for the meager rainfall (Fig. 18.18a). Over most of Australia there is either a single peak in the annual cycle or little seasonal preference (Fig. 18.18b). The few areas with bimodal regimes generally represent locations in the transition zones between winter and summer rainfall.

The annual march of both precipitation and temperature at typical arid and semi-arid stations are shown in Fig. 18.19. Mean annual precipitation at these stations ranges from 148 mm at Oodnadatta to 555 mm at Halls Creek.

All of these stations show a pronounced annual cycle in temperature. Winter monthly minima are on the order of 10°–15°C, except at Wittenoom, Carnarvon, and Halls Creek, where coastal location provides some protection from the winter cold.

Typical of the summer regime are Alice Springs, Giles, and Wittenoom (Fig. 18.19). At central desert stations, such as Alice Springs, a shift to a decisive winter maximum occurs in many years. In the southwestern desert, there is a tendency for an autumn or an autumn and winter maximum, and summer is generally dry. Typical

stations are Carnarvon and Wagin. Kalgoorlie, Meekatharra, and Carnarvon clearly show the double maximum. In the southern margin of the desert, rainfall generally occurs year-round. Typical stations are Oodnadatta, Cobar, and Cook.

The concentration of the rainfall within the wet season is shown in Fig. 18.20. The strongest seasonality is in the north, where 60–70% of the rainfall occurs in the wettest quarter. Elsewhere, even in relatively arid regions, the seasonality is not particularly strong, with only 30–50% falling in the wettest quarter.

In many cases, isolated, unseasonal rainfall events occur, even in monsoonal regions. These are important for ecological processes (Cook and Heerdegen 2001). The contribution of temporally isolated rainfall events to both the duration of the rainy season and the amount of rainfall tends to increase with latitude.

The wettest month is January in the north and east and February in the interior (Fig. 18.21). In January, rainfall can occur anywhere on the continent (Fig. 18.15). The summer rains are widespread and intense. In the winter rainfall region, the wettest month is quite variable, generally ranging from June to August, but as late as October in the extreme southeast. In the west desert region with an autumn maximum, tropical cyclone activity produces a peak in March. Winter rainfall is confined to the southern half of the continent and is not very intense. In July, rainfall exceeds 25 mm/month in only a small area. It exceeds 100 mm only in the extreme southwest and a few areas of high terrain.

(a)

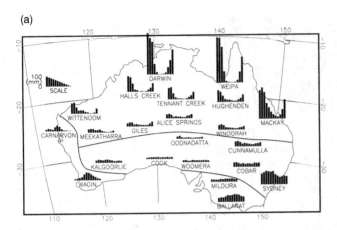

(b)

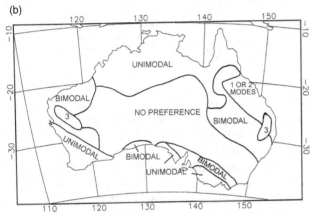

Fig. 18.18 (a) Rainfall seasonality map, showing the seasonal cycle at select stations. The solid lines delineate areas of summer rainfall (southernmost areas), winter rainfall (northern Australia), and little or no seasonal preference (central area). (b) Number of peaks in the annual cycle: indicated are areas of unimodal and bimodal seasonal cycles and with little or no seasonal preference. Two small areas with three peaks are indicated and are transition regions.

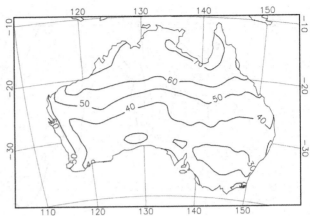

Fig. 18.20 Percentage concentration of precipitation in the wettest quarter of the year.

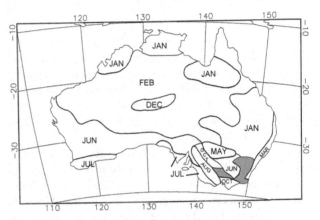

Fig. 18.21 The wettest month of the year. In the two shaded areas, there is no preferred month of peak rainfall.

For the continent as a whole, spring is the driest season. However, in the summer rainfall region, the driest month is August in the north and east, but September in the interior. In the winter rainfall region, the driest month is as diverse as the wettest. It is January or February in most locations, but varies from October to December in the west, where tropical cyclones often bring rain in January or February.

18.4.2 RAIN-PRODUCING SYSTEMS

Gentili (1971) identifies five independent surface systems that produce rainfall over Australia, three of which are of tropical origin. These include the monsoon rains, the trade winds, tropical cyclones, mid-latitude depressions, and inter-anticyclonic fronts. Each system occurs seasonally and the degree of dominance of each system locally produces the seasonal cycle of rainfall. Consequently, the rainfall maximum occurs in January in the monsoon belt in the north and interior, in June in the frontal belt of southern Australia, and in March in areas receiving rainfall mainly from tropical cyclones (Fig. 18.22). Any of these systems can bring rainfall to desert regions of the continent;

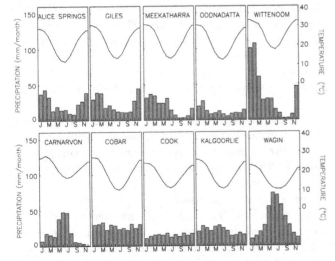

Fig. 18.19 Mean monthly precipitation and temperature at typical stations in the drylands of Australia. See Fig. 18.18a for station locations.

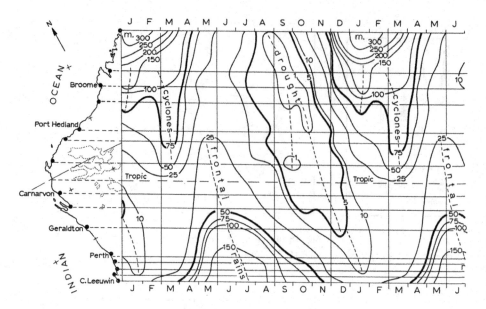

Fig. 18.22 Mean monthly rainfall (in mm) along a west coast transect. This illustrates the penetration of both mid-latitude frontal rains and rains from tropical cyclones toward the interior and the long period of droughts in between (from Gentili 1971).

aridity is related to their less frequent penetration into these regions. Any of these systems can even traverse the desert under exceptional circumstances. Pook *et al.* (2006) identify additional upper-level systems that produce rainfall in southern Australia. These include cut-off lows, waves in the easterlies, and deep upper-level troughs in the easterlies.

The heaviest rainfall tends to be monsoonal or associated with tropical cyclones. Frontal rains tend to be more persistent and widespread, but result in lower daily totals. Maximum daily intensities in the dryland regions are generally on the order of 100–150 mm/day. They can be higher in the western margin of the desert, which is subjected to passages of tropical cyclones. At Kalgoorlie, in the winter rainfall zone, February cyclones once brought 315 mm of rainfall, with 178 mm falling on one day. The mean annual rainfall there is 243 mm. At Port Hedland, where the mean annual rainfall is 327 mm, cyclones caused 500 mm to fall in one January. In some cases, the cyclonic rains have been extraordinary. A storm that hit in April 1898 brought 914 mm to Whim Creek, near Cossack, almost three times the annual mean.

The monsoon is a northerly or northwesterly flow of maritime tropical air (Hendon and Liebmann 1990). The low-level monsoon westerlies are most common in summer (November–March), but they are shallow except in January and February, when they can extend up to 10 km. Occasionally the monsoon flow penetrates into the desert, as far as 32° S, resulting in heavy rains and local flooding.

The onset of the Australian monsoon is very abrupt, with the transition occurring within 2–5 days (Troup 1961). The decay is equally abrupt. The onset is marked by a poleward shift of the subtropical jet and an expansion of the upper-level tropical easterlies (Hendon and Liebmann 1990). The onset date varies markedly. During the period 1957–1987 its onset at Darwin ranged from November 23 to January 23.

The monsoon rains are mostly intermittent and localized convective thunderstorms, due to the instability of the humid maritime tropical air. The occurrence of convection in the monsoon is closely linked to the Madden–Julian intraseasonal oscillation (Wheeler *et al.* 2009). The monsoon can bring heavy and prolonged wet spells if coupled with the presence of the ITCZ (Intertropical Convergence Zone). Monsoon activity is affected by the circulation in both hemispheres, especially by the presence of a strong upper-level jet. Peak rainfall occurs in January, and heavy rains continue through February. In March, the monsoon rains diminish as tropical cyclones become more frequent.

The tropical cyclones are active in summer or early autumn, commencing in December and peaking in March. They produce heavy and widespread rainfall when they are slow-moving, but heavy and localized rainfall when they move quickly. Cyclones occur infrequently, generally only 2–3 per year, but they can bring torrential rains. They can reach nearly every part of the mainland, but their passage over the desert is often blocked by the transient anticyclones. There are two tropical cyclone belts, one in the east and one in the west. Those in the latter affect the interior desert more frequently. The frequency of tropical cyclones over Australia is strongly negatively correlated with sea-surface temperatures in the central and western equatorial Pacific (i.e., regions typically linked to El Niño) (Ramsay *et al.* 2008).

The trade winds (also called tropical easterlies) are southeasterlies originating on the equatorward side of anticyclones. They are relatively constant and are strongest in the northeast of Australia. The trades produce orographic rainfall along a small part of the northeast coast.

The travelling anticyclones are a typical weather system over the continent. They are particularly common in central Australia, where they can remain almost stationary for several days. They also affect southern Australia during summer. The magnitude and latitude of these systems is related to the speed and magnitude of the jet stream. Between two consecutive

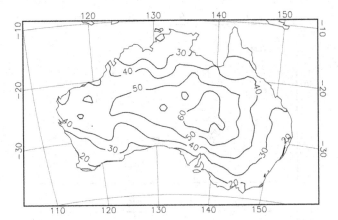

Fig. 18.23 The coefficient of variation (CV, %) of annual precipitation.

anticyclones is a brief change of winds to southerly and consequently a rapid onset of cool weather. This event, called a "cool change," is essentially a dry cold front. These inter-anticyclonic fronts frequently bring moderate, widespread rainfall because they are often associated with an upper-level trough and a cold air pool. When secondary waves develop aloft, they produce localized areas of heavy rainfall. Their influence is enhanced if the jet stream overrides them, because the jet enhances the low-level convergence and vertical motion.

The frontal rains associated with mid-latitude depressions affect primarily southern Australia. They move onto the continent mainly from the southwest. These depressions only skirt the southern border in summer, but move considerably further north in winter. This displacement commences in March, when there is a rapid northward spread and intensification of the associated rainbelt. These systems peak in June, so that areas where they are dominant have a June rainfall maximum. The depressions then retreat southward and diminish from August to October.

In southeast Australia, over half of the rainfall is associated with cold-core cut-off lows, described and catalogued by Pook *et al.* (2006). Most of the rest is associated with cold fronts. The cut-off lows are sometimes followed by subtropical warm fronts that can bring extreme rainfall and flooding (Griffiths *et al.* 1998).

18.4.3 INTERANNUAL VARIABILITY OF RAINFALL

As is typical of dryland regions, interannual variability is high. The coefficient of variation of annual rainfall is roughly inversely proportional to mean annual rainfall, being on the order of 50–60% in desert regions and 30–40% in semi-arid regions (Fig. 18.23). At Alice Springs, in the central desert, annual rainfall totals have varied between 903 mm in 1974 and 51 mm in 1985.

Much of Australia experienced significant changes in rainfall during the last century (Gallant *et al.* 2007; Smith *et al.* 2008). Particularly noteworthy are major declines in the southeast and the southwest and increases in the frequency and intensity of rainfall in arid central Australia. The decrease in rainfall in the drylands has been on the order of at least 20 mm per decade

since early in the twentieth century. During the period 1950–2006, autumn rainfall decreased by about 40% in southeastern Victoria (Cai and Cowan 2008).

The reasons for these changes are not completely clear, but the most extreme anomalies generally involve the simultaneous occurrence of several large-scale factors. One of the strongest influences is ENSO (El Niño/Southern Oscillation). Typically, drought occurs during El Niño, especially across the eastern two-thirds of the country (Wang and Hendon 2007). The impact tends to be greatest in the austral spring (September –November). La Niña tends to enhance rainfall.

Despite these generalities, there are considerable variations in the Australian rainfall/ENSO relationship, particularly on decadal time scales. The near–record strength El Niño of 1997 was associated with near-normal rainfall (Wang and Hendon 2007). In contrast, the moderate El Niño of 2002/03 produced near-record drought in eastern Australia. The 2007 La Niña failed to produce the anticipated increase in rainfall (Gallant and Karoly 2009).

The link between Australian rainfall and El Niño may be weak or strong, depending on regional factors and on variations in the characteristics of the individual El Niño episodes, particularly the distribution of sea-surface temperature (SST) anomalies in the equatorial Pacific. Factors modulating the link to El Niño include shifts in the Antarctic circumpolar vortex (White 2000), SSTs in the Indian Ocean (England *et al.* 2006; Ummenhofer *et al.* 2008; Cai *et al.* 2009), and shifts in the subpolar westerlies (Evans *et al.* 2009).

Many major Australian drought episodes, particularly multi-year episodes, are unrelated to El Niño, being impacted more by Indian Ocean conditions than by those in the Pacific. This is true of the so-called "Big Dry," a drought that has gripped a large region of southeast Australia from 1996 to 2007 (Ummenhofer *et al.* 2009; Daniell 2009). For the decade, the rainfall averaged over southeastern Australia was roughly 30% below the climatological mean.

18.5 THERMAL REGIME

The absolute maximum recorded air temperature in Australia, 53.1°C, occurred at Cloncurry (Queensland). There are unofficial reports of temperatures reaching 57°C. The hottest time of the year occurs progressively later from the north, where it is November, to the far south, where it is February. In January, which is the warmest month over nearly the entire continent, the mean daily minimum is 20°–25°C over the desert and as low as 15°–18°C in the semi-arid region poleward of the desert (Fig. 18.24a). The mean daily maximum in most areas is 35°–37°C, but exceeds 39°C in several lowland cores of the central desert, particularly in the west (Fig. 18.24b). It is about 32°–35°C in the semi-arid region south of the desert. Nearly every part of the continent is susceptible to bursts of desert air, with temperatures of 46°C or more. Many stations have recorded temperatures in excess of 49°C. Thus, heat waves are a common occurrence (Tryhorn and Risbey 2006).

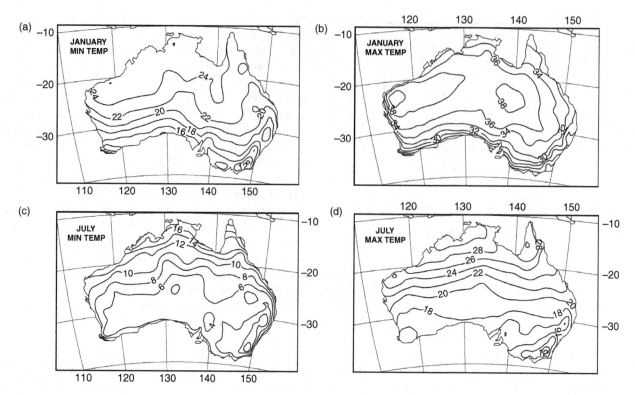

Fig. 18.24 Mean daily temperature extremes over Australia (°C): (a) January minimum, (b) January maximum, (c) July minimum, (d) July maximum.

Australia does not experience exceedingly cold conditions in winter because its low to moderate latitudinal location, low altitude, flat relief, and ocean surroundings are not conducive to extreme cooling. July is generally the coolest month. Throughout most of the desert and semi-arid regions, the mean daily minimum in July is generally between 5° and 10°C, although it can be as high as 15°C in northern, semi-arid regions (Fig. 18.24c). Mean daily maxima in winter are quite diverse, ranging from 15°–18°C in the south to 30°C in the north (Fig. 18.24d). In the dryland regions, the absolute minima are generally on the order of −5° to −10°C. The length of the frost period (time between first and last day with minima below 2°C) ranges from 0 in the northern desert to between 50 and 100 days in the southern desert and adjacent semi-arid regions. The number of days with frost is on the order of 20–30 in the central and southern desert.

Neither the annual nor the diurnal temperature ranges are particularly large. The annual temperature range depends on latitude and continentality. It ranges from 5°C in the north to 17°C in the interior (Fig. 18.25a). Due to the influence of the ocean, the annual range generally does not exceed 11°C in southern Australia. The diurnal temperature range (Fig. 18.25b, c) is on the order of 15°C at most stations in the desert (Gentili 1971).

18.6 OTHER CLIMATIC ELEMENTS

Maximum and minimum relative humidity in the drylands are on the order of 20–50% and 20–30% in summer. In winter they are generally 40–70% and 20–50%, respectively. Because the relative humidity is low, fog is neither common nor widespread in Australia. Fog occurs more frequently in the south, where Canberra experiences 47 days per year on average. It occurs fairly infrequently in the north. Darwin, for example, averages 2 days per year. At Alice Springs, in the central desert, there are 12 days of fog on average. Fog frequency is highest in winter, particularly between May and August.

Because of the moderate thermal regime, snow is unknown in northern Australia and occurs rarely in southern Australia. It has not been recorded equatorward of 20° S except in the Australian Alps in the southeast, where snow falls annually. It is essentially unknown in the desert regions.

Hail is rare, although it can accompany cold fronts in winter and spring in southern Australia. The number of thunderstorms generally increases from south to north and along the west and east coasts (Kuleshov *et al.* 2002). In the desert, thunderstorm frequency ranges from 10 days per year in the south to 30 days per year in the north; it is also more common in western desert areas, which are more frequently affected by tropical cyclones.

Dust storms can affect nearly any part of Australia, but they are most frequent in the southeastern desert and the adjacent semi-arid regions (Buckley 1987). They are most common in spring and summer, very unusual in winter. The frequency ranges from less than 1 per year in the western desert of Australia to over 30 per year in parts of the southeast.

Dust devils are a common phenomenon in the central desert (Fig. 13.14). They are readily visible because of the brilliant red

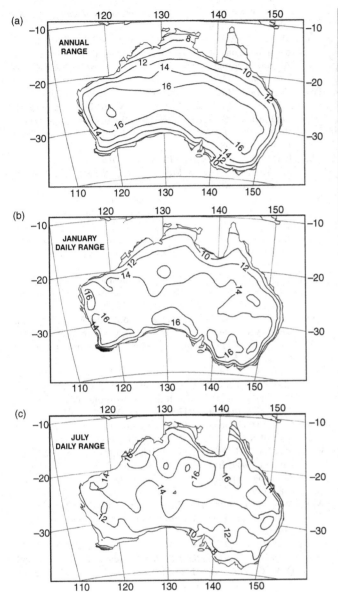

Fig. 18.25 Mean temperature range (°C): (a) annual, (b) diurnal in January, (c) diurnal in July.

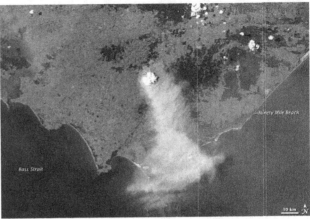

Fig. 18.27 The bushfire of February 2009 was one of Australia's most devastating (photograph courtesy of NASA).

color of the soil they entrain. Often, several vortices are visible at any one time, making a stark contrast with the brilliant blue skies behind them.

Another weather-related hazard plaguing the continent is bushfires (Fig. 18.26) (Powell 1983). Australia may be the most fire-prone country in the world. In most of the country, the fire season is 3–4 months. Wildfires are least common in the most arid regions, because of the low vegetation density. Over most of the country they occur less than once in 20 years; in the more humid southeast, the frequency can be as high as once in 3 years.

The primary condition for bushfires is essentially dryness. Fire risk is highest in the driest and hottest months and is enhanced by drought conditions. The highest risk in southwestern Australia is in December and January (McCaw *et al.* 2007). Synoptic conditions also play an important role in determining fire outbreaks. In Victoria, the most extreme fire events occur when the prevailing wind is from the north or northwest, i.e., out of the arid core of the continent (Long 2006). The bushfires that raged through Victoria in February 2009 (Fig. 18.27), leaving 160 people dead, were associated with a long-term drought, a record heat wave (with temperatures over 47°C), and winds exceeding 100 km/h.

A situation very conducive to fire is a rapid drop in nearsurface relative humidity, such as observed just prior to the severe bushfires of January 2003 and January 2005. This drying can take place on time scales of an hour, quickly desiccating the vegetation to produce tinder for fuel (Mills 2008).

REFERENCES

Buckley, B., 1987: *Duststorm Occurrence in Western Australia.* Meteorological Note 174, Bureau of Meteorology, Melbourne, 19 pp.

Cai, W., and T. Cowan, 2008: Dynamics of late autumn rainfall reduction over southeastern Australia. *Geophysical Research Letters*, **35**, L09708, doi:10.1029/2008GLO033727.

Cai, W., T. Cowan, and T. Sullivan, 2009: Recent unprecedented skewness towards positive Indian Ocean Dipole occurrences and its

Fig. 18.26 Fire scar from a recent bushfire.

impact on Australian rainfall. *Geophysical Research Letters*, **36**, L11705, doi:10.1029/2009GL037604.

Cook, G. D., and R. G. Heerdegen, 2001: Spatial variation in the duration of the rainy season in monsoonal Australia. *International Journal of Climatology*, **21**, 1723–1732.

Daniell, T., 2009: The implications of a decade of drought in Australia (1996–2007). *Secheresse*, **20**, 171–180.

England, M. H., C. C. Ummenhofer, and A. Santoso, 2006: Interannual rainfall extremes over southwest Western Australia linked to Indian Ocean climate variability. *Journal of Climate*, **19**, 1948–1969.

Evans, A. D., J. M. Bennett, and C. M. Ewenz, 2009: South Australian rainfall variability and climate extremes. *Climate Dynamics*, **33**, 477–493.

Gallant, A. J. E., and D. J. Karoly, 2009: Atypical influence of the 2007 La Niña on rainfall and temperature in southeastern Australia. *Geophysical Research Letters*, **36**, L14707, doi:10.1029/2009GL039026.

Gallant, A. J. E., Hennessy, K. J., and J. Risbey, 2007: Trends in rainfall indices for six Australian regions: 1910–2009. *Australian Meteorological Magazine*, **56**, 223–239.

Gentili, J., 1971: *Climates of Australia and New Zealand*. World Survey of Climatology 13, Elsevier, Amsterdam, 405 pp.

Gersbach, G. H., C. B. Pattiaratchi, G. N. Ivey, and G. R. Cresswell, 1999: Upwelling on the south-west coast of Australia: source of the Capes Current? *Continental Shelf Research*, **19**, 363–400.

Goudie, A., and J. Wilkinson, 1980: *The Warm Desert Environment*. Cambridge University Press, New York, 88 pp.

Griffiths, M., M. J. Reeder, D. J. Low, and R. A. Vincent, 1998: Observations of a cut-off low over southern Australia. *Quarterly Journal of the Royal Meteorological Society*, **124**, 1109–1132.

Hendon, H. H., and B. Liebmann, 1990: A composite study of onset of the Australian summer monsoon. *Journal of the Atmospheric Sciences*, **47**, 2227–2240.

Kuleshov, Y., G. de Hoedt, W. Wright, and A. Brewster, 2002: Thunderstorm distribution and frequency in Australia. *Australian Meteorological Magazine*, **51**, 145–154.

Long, M., 2006: A climatology of extreme fire weather days in Victoria. *Australian Meteorological Magazine*, **55**, 3–18.

McCaw, L., P. Marchetti, G. Elliott, and G. Reader, 2007: Bushfire weather climatology of the Haines index in Southwestern Australia. *Australian Meteorological Magazine*, **56**, 75–80.

Mills, G. A., 2008: Abrupt surface drying and fire weather. Part 1. Overview and case study of the South Australian fires of 11 January 2005. *Australian Meteorological Magazine*, **57**, 299–309.

Pook, M. J., P. C. McIntosh, and G. A. Meyers, 2006: The synoptic decomposition of cool-season rainfall in the southeastern Australian cropping region. *Journal of Applied Meteorology and Climatology*, **45**, 1156–1170.

Powell, F. A., 1983: Bushfire weather. *Weatherwise*, **36**, 130–131.

Ramsay, H. A., L. M. Leslie, P. J. Lamb, M. B. Richman, and M. Leplastrier, 2008: Interannual variability of tropical cyclones in the Australian region: role of large-scale environment. *Journal of Climate*, **21**, 1083–1103.

Riley, P. A., 1989: *The Diurnal Variation of the Low-Level Jet over the Northern Territory*. Meteorological Note 188, Bureau of Meteorology, Melbourne, 31 pp.

Smith, I. N., L. Wilson, and R. Suppiah, 2008: Characteristics of the northern Australian rainy season. *Journal of Climate*, **21**, 4298–4311.

Tryhorn, L., and J. Risbey, 2006: On the distribution of heat waves over the Australian region. *Australian Meteorological Magazine*, **55**, 169–182.

Troup, A. J., 1961: Variations in upper tropospheric flow associated with the onset of Australian summer monsoon. *Indian Journal of Meteorology and Geophysics*, **12**, 217–230.

Ummenhofer, C. C., A. Sen Gupta, M. J. Pook, and M. H. England, 2008: Anomalous rainfall over southwest western Australia forced by Indian Ocean sea surface temperatures. *Journal of Climate*, **21**, 5113–5134.

Ummenhofer, C. C., M. H. England, P. C. McIntosh, G. A. Meyers, M. J. Pook, J. S. Risbey, A. S. Gupta, and A. S. Taschetto, 2009: What causes southeast Australia's worst droughts? *Geophysical Research Letters*, **36**, L04706, doi:10.1029/2008GL036801.

Wang, G., and H. H. Hendon, 2007: Sensitivity of Australian rainfall to inter-El Niño variations. *Journal of Climate*, **20**, 4211–4226.

Wheeler, M. C., H. H. Hendon, S. Cleland, H. Meinke, and A. Donald, 2009: Impacts of the Madden–Julian Oscillation on Australian rainfall and circulation. *Journal of Climate*, **22**, 1482–1498.

White, W. B., 2000: Influence of the Antarctic circumpolar wave on Australian precipitation from 1958 to 1997. *Journal of Climate*, **13**, 2125–2141.

Williams, O. B., and J. H. Calaby, 1985: The hot deserts of Australia. In *Hot Deserts and Arid Shrublands* (M. Evenari, I. Noy-Meir, and D. W. Goodall, eds.), Ecosystems of the World 12A, Elsevier, Amsterdam, pp. 269–312.

19 Asia

19.1 INTRODUCTION

The drylands of Asia have several common denominators: their cold winters (in all but the extreme south), the prevailing influence of the seasonally shifting Asian monsoon system, the westerly disturbances affecting western and central areas, and the role of topography as a major climatic control. These drylands represent a gradual, east–west transition between summer and winter (or Mediterranean) precipitation regimes. The cold winters distinguish them from the subtropical deserts of most other continents, with extremely low temperatures caused by the persistent presence of the Asiatic High and the relatively high latitude of the region. Insolation is low in winter because the days are short and the sun is low on the horizon, while the high promotes radiative cooling in the clear skies and dry air, and also insulates the region from warmer air masses. Temperatures can plunge lower than −50°C.

Dust is another common denominator. As a result of the combined forces of high winds and dry surfaces, Asia is one of the world's dustiest regions. Even in relatively humid regions the frequency of dusty conditions can be quite high (Fig. 19.1). In most of the dryland regions of Asia the number of days with dust storms averages 5–20 per year. They are less common in more northern areas of the Former Soviet Union, but in parts of central and western Asia the average frequency is as high as 40–80 days per year.

The general aridity of the region is primarily due to the presence of the high that prevails in winter. However, topography (Fig. 19.2) and a deeply interior location also suppress the summer rains by producing numerous lee rain shadows and frequently preventing humid maritime air from penetrating the region. Topographic effects are particularly strong here since the summits of numerous mountain ranges (Tien Shan, Pamir, Himalayas, Hindu Kush, Tibet) rise above 6000–7000 m, and numerous basins lie between them. The topography also induces wind systems and mesoscale circulation systems that impact rainfall, temperature, and dust concentration in the atmosphere (e.g., Seino *et al.* 2005).

Asia includes both warm and cold deserts and steppes. The warm Iranian and Thar deserts lie to the south. The cold drylands of the north are concentrated in four locations (Fig. 19.2): the Central Asian deserts of China and Mongolia; Middle Asia (primarily the Turkestan Desert); eastern Siberia, where steppes prevail; and the predominantly semi-desert Caspian lowlands. In Middle Asia, deserts extend southward toward Afghanistan and Iran, while the southern border of Central Asia is the high plateau of Tibet and Pamir. To the north are the remaining Asian drylands, the semi-deserts of the Kazakhstan and Caspian lowlands and, to the east, the steppes of eastern Siberia.

The Middle Asian deserts provide an interesting climatic contrast with Central Asia (the Taklamakan and Gobi deserts), from which they are separated by the Dzungaria Basin and the highlands of Tien Shan. Middle Asia receives the remnants of eastward-moving disturbances from the Atlantic. Consequentially, precipitation falls predominantly in the winter and decreases from west to east. Central Asia is affected instead by frontal systems and cyclonic disturbances from Eastern Asia, which move westward in summer. Thus, summer rainfall prevails, and winter and spring are extremely dry. The transition between the winter and summer rainfall regimes is the Qaidam Desert, at an elevation of about 3000 m. The climatic contrast between the regions creates a floristic contrast as well (see Sections 19.3.2 and 19.3.3).

The contrast is similar between the Caspian lowlands and eastern Siberia, two somewhat less arid regions lying to the northwest of Middle and Central Asia, respectively. The Caspian lowlands receive primarily winter rainfall from eastward-moving systems from the Atlantic or Mediterranean, and the rainfall gradient is from north to south. This is a semi-desert transition between the steppes of Eastern Europe and the deserts of Asia. Much of eastern Siberia can receive rainfall in almost any month, but summer is generally the wettest season. The precipitation pattern is quite complex because of the rough and irregular terrain.

This chapter begins with an overview of the climatic elements for the region as a whole. Then the warm and cold deserts are

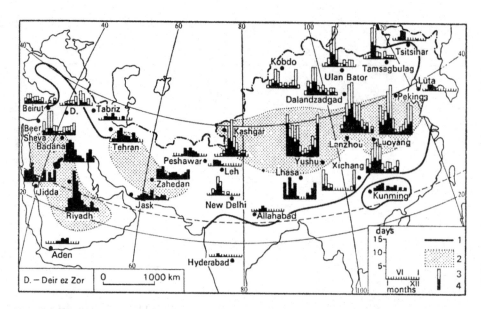

Fig. 19.1 Monthly frequency of dust storms at select locations (from Martyn 1992).

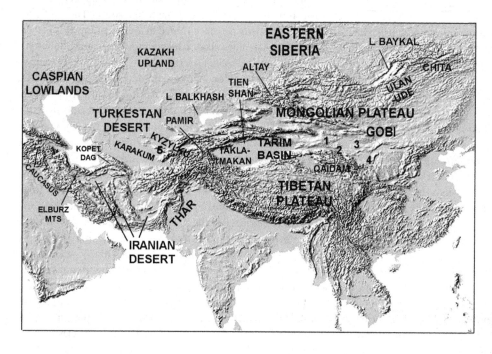

Fig. 19.2 Topography and general location of the drylands of Asia.

considered separately, with each section commencing with a discussion of the major climatic controls for the region. The complex topography of Asia creates a patchwork of climates too difficult to describe in great detail. Precipitation is generally higher in the mountains, but elevation, slope, and exposure produce large contrasts over relatively small distances. For this reason, the emphasis is on the major dryland areas in the lower elevations.

Unfortunately, the treatment of various regions in this chapter is not uniform, because research has been more intense in some regions. Most of the knowledge of the region came by way of research expeditions in the late nineteenth and early twentieth centuries. Relatively few recent studies have examined these regions in the context of global meteorological processes (Aizen

et al. 1997). Middle Asia has been well studied because research stations were established long ago in Karakum and Pamir, while comparatively little is known about Central Asia and Afghanistan. The picture is rapidly changing in Kazakhstan and Central Asia with the establishment of additional desert research stations. The Tarim Basin, for example, has become the focus of both observational programs (e.g., Chen et al. 1999; Sato et al. 2007) and modeling studies (e.g., Uno et al. 2005; Seino et al. 2005), generally in the context of dust storms (Kai et al. 2008; Kim et al. 2009). Reports of such studies have proliferated in the international meteorological literature during the last decade, as a consequence of the availability of global data sets (see Chapter 1).

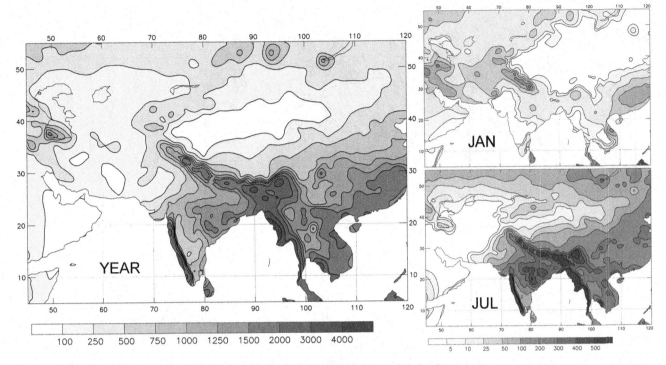

Fig. 19.3 Mean annual precipitation and mean precipitation in January and July over Asia (mm).

Excellent and detailed descriptions of the climate and environment of the Asian drylands, including details of the highland climates, are found in Lydolph (1977), Arakawa (1969), Arakawa and Takahashi (1981), Walter and Box (1983a, 1983b, 1983c, 1983d, 1983e, 1983f, 1983g), Walter *et al.* (1983), Breckle (1983), Domroes and Peng (1988), and Martyn (1992). Unless otherwise indicated, the information in this chapter derives from a synthesis of these sources.

19.2 THE CLIMATE OF ASIA

19.2.1 AN OVERVIEW OF PRECIPITATION, TEMPERATURE, AND OTHER CLIMATIC ELEMENTS

For a landmass the size of Asia, the patterns of climate are hard to generalize. Perhaps the most pronounced feature is the dominance of relatively dry climates. Over most of the continent, mean annual precipitation is below 500 mm (Fig. 19.3). More humid conditions are found in various mountain regions and in the southeast and east. The driest conditions are, in general, in the continental interior. In most of the region, precipitation is less variable from year to year than in other dryland regions with comparable annual means (Fig. 19.4). Over most of the region it is on the order of 25–30%. It is considerably higher, on the order of 40–60% or more, in the warm deserts and in the driest deserts of Central Asia.

Throughout most of the region, precipitation is highly seasonal, with some 40–70% falling in the wettest quarter of the year (Fig. 19.5). The seasonality of precipitation in the region is determined by location with respect to the major features of the

general atmospheric circulation. Across the continent there is a general transition from cold-season precipitation in the west, associated with Mediterranean and mid-latitude cyclone tracks, to summer precipitation in the east, in most cases associated with the Asian monsoon (Figs. 19.6 and 19.7).

The pattern is more complex, however, due to the shifting of the mid-latitude storm tracks and seasonal changes in their frequency (Schiemann *et al.* 2008). In Middle Asia, cyclone frequency is high from autumn through spring, with a maximum around spring. The tracks are located further south in winter. In summer they rarely occur in this region. Consequently, the wettest month in Middle Asia is generally April or May (Fig. 19.8) and summer is rainless. Further south, in the warm deserts of Iran and Afghanistan, the maximum shifts to March, then to the winter months in southern Iran.

To the east, the shift to summer rains is abrupt (Fig. 19.8). In northern areas such as the steppes of the Kazakh uplands, the summer maximum is a consequence of the shift in storm tracks as the polar front and jet streams move poleward. The wettest month is July. Eastward from there, many central sectors have year-round precipitation, but with a meager and distinct summer maximum. Throughout nearly all of Central Asia there is a summer precipitation maximum, with local variations induced by topography, and winters are rainless. The precipitation maximum also shifts from the cold season to summer in the transition from the Iranian desert to the Thar Desert. As in Central Asia, winters are nearly rainless, a result of the Asian monsoon being the main rain-producing system.

In southeastern Asia, where the wettest climates are found, cloudiness tends to be high year-round, ranging from about 60%

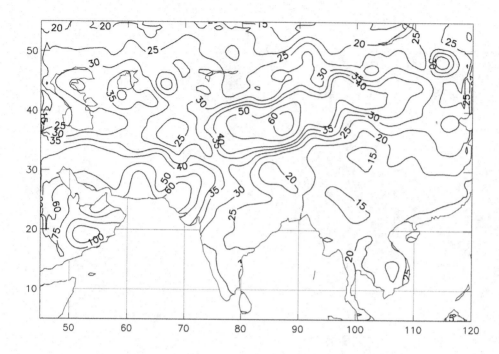

Fig. 19.4 Coefficient of variation (CV, %) of annual precipitation.

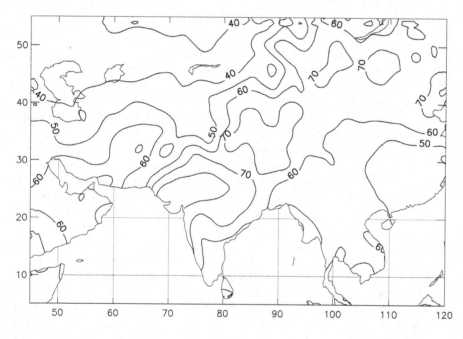

Fig. 19.5 Percentage contribution of precipitation falling in the wettest quarter of the year.

to 90% throughout most of the year. In the more arid southwest there is a pronounced annual cycle roughly in phase with the seasonal cycle in rainfall. Far to the west, the maximum cloud cover occurs in winter. Elsewhere there is a summer maximum, a consequence of the summer monsoon. Mean monthly cloud cover can exceed 80% or 90%. Further north, in the countries of the Former Soviet Union, data are sparse, so that only a rough picture can be derived. It appears that over most of the country afternoon cloudiness exceeds 60% or 70% during both winter and summer. It is somewhat lower over the Caspian lowlands and areas south and west of the Caspian during summer, with relatively clear skies much of the time.

The temperature regime across the drylands of Asia is extremely complex, owing to the combined influences of solar radiation, topography, and location with respect to maritime influence and to the Siberian High (Fig. 19.9). In winter, the daily maximum temperatures reflect the availability of solar radiation, increasing from north to south. Mean daily maxima in January are on the order of 0° to −12°C in the cold deserts. In the warm deserts the mean daily maxima increase rapidly with decreasing latitude, reaching 24°C in southern Iran. Daily minimum temperatures instead decrease eastward, reflecting the increasing influence of the Siberian High, below which an intense cold dome develops. In summer, topographic effects

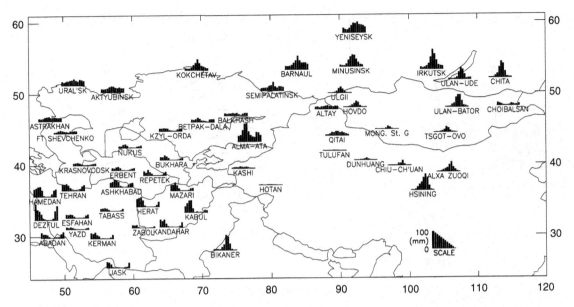

Fig. 19.6 The seasonality of precipitation shown for weather stations throughout Asia.

prevail. The diurnal range (Fig. 19.10) increases both inland and southward, being on the order of 9°–12°C in the mid-latitudes, but 12°–15°C in the lower latitudes.

In summer the impact of terrain on the thermal regime is very strong. As with daily winter minima, July temperatures very roughly decrease from east to west, but the lowest temperatures are clearly over the highest terrain. Mean daily maxima for July are on the order of 33°–42°C over the warm deserts and most of Middle Asia, but fall to 27°C over eastern Siberia and 18°C in the Gobi to the east of the Tibetan plateau. The spatial pattern of mean minimum temperatures is remarkably similar, reflecting a relative uniformity of the diurnal range throughout the region in summer. Both maximum and minimum temperatures in the low-lying deserts are some 10°–15°C higher than in the drylands in the higher terrain of Central Asia. The diurnal temperature range is more uniform throughout the region (Fig. 19.10). The impact of elevation is relatively small in summer, but in winter the diurnal range is some 3°C greater in the higher-elevation drylands.

In contrast, the annual range (Fig. 19.10) is largely a function of latitude, ranging from some 12°C in the tropics to about 30°C at 40° N. Further north, the annual range reflects the location with respect to distance from the Atlantic and with respect to the Siberian High. To the west, the annual range is on the order of 36°C. East of 90° E, in the higher terrain and beneath the Siberian High, it increases to roughly 40°–45°C.

19.2.2 CLIMATIC CONTROLS

Asia is climatically analogous to North America in many ways, such as the continentality and the dominant role played by mid-latitude cyclones. However, differences in the orientation of the landmass, its topography, and the dramatic seasonal shift of the dominant pressure patterns impart a different character to the Asian drylands. This shift between a semi-permanent high over the land in winter and a semi-permanent low in summer is accompanied by a seasonal, "monsoonal" shift of the prevailing wind regime that is absent in most of North America.

In contrast to the north–south mountain chains of North America, there are several highland regions of Asia primarily oriented east–west. Major topographic influences include the high mountains through Mongolia and western China, on the southern border of the Former Soviet Union, and the Tibetan plateau, and the Caucasus Mountains; the mountain peaks generally attain 2000–3000 m while basins sink to 400–1000 m. The Caucasus act as a barrier to the penetration of cyclones from the north and northwest into Middle Asia, but the vast plain to the north and east leaves the region open to cold Siberian air. Extensive mountains to the south and east prevent penetration of the Asian monsoon and tropical air into the region. In Central Asia, mountains effectively block both the southward penetration of cold polar air masses and the northward penetration of warm, humid monsoon flow. The Tibetan plateau, via both orographic and heating effects, regulates the surface Asian monsoon flow and probably intensifies the dry climates of Central Asia (Wu *et al.* 2007).

Winter circulation is dominated by the Siberian or Asiatic High (Fig. 19.11), thermally induced over the cold, dry interior and centered on Mongolia (Gong and Ho 2002). Year-to-year changes in its intensity are strongly correlated with rainfall variability. They appear to also exert a large degree of control on climatic variability in the whole Northern Hemisphere (Cohen *et al.* 2001). The high is sustained by anticyclones moving southward from polar latitudes, and is occasionally interrupted by cold fronts and cyclones from this regime. In the northwest, the high protrudes poleward as a northwest–southeast oriented ridge. This feature blocks Mediterranean air and cyclones and

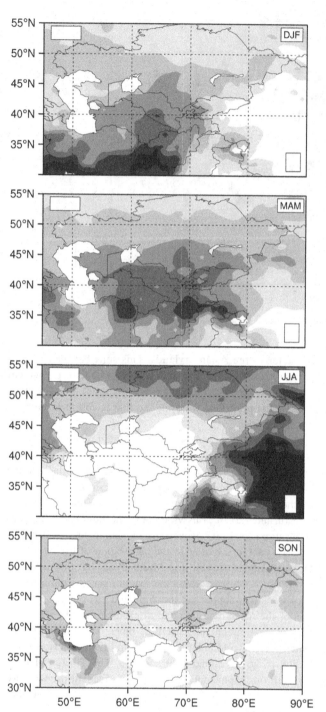

Fig. 19.7 The seasonality of precipitation (mm) over Asia (from Schiemann *et al.* 2008).

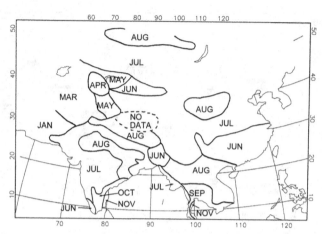

Fig. 19.8 Month of maximum precipitation in different parts of Asia.

increase through the layer on the order of 10°–20°C (Lydolph 1977). The inversion is more intense further west and further east; at Yakutsk, the average temperature increase is 18°C but up to 27°C has been observed (Flohn 1947). Another inversion layer exists aloft, produced by warm air overriding the cold dome built up in the low-lying basins.

The Siberian High is shallow; it is replaced at 850 mb by an elongated trough oriented northwest–southeast and arising from a merging of the Aleutian and Icelandic Lows. The trough intensifies aloft until, at 500 mb, a closed low exists. In winter the eastern portion of the continent lies primarily under the influence of the low pressure. Occasionally, however, the high extends up to 500 mb, creating a blocking situation.

The high produces a relative constancy of the surface wind regime in winter, with flow being primarily northwesterly from the interior of the continent to the periphery over the north and east of Russia (Fig. 19.12). These persistent offshore winds reduce the maritime influence of the Pacific, and cold polar continental air prevails. This produces extremely severe winters, with temperatures even lower than in the Arctic. In the deserts of China and Central Asia, northeasterlies along the southern flank of the high prevail. In the Siberian steppe, on the northern flank of the extended ridge, the flow is generally southerly. However, the simplicity of this pressure regime belies an extreme complexity of synoptic situations and weather patterns (Inagamova *et al.* 2002).

In summer a ridge extends eastward from the Azores High into Kazakhstan, but the dominant pressure feature over Asia is the monsoon trough, centered over northern India (Fig. 19.12). Consequently, the flow is primarily northerly or northwesterly over Middle Asia, and northeasterly over Mongolia and the Gobi. The Siberian steppe experiences a less persistent wind regime, a consequence of the transient cyclones and anticyclones affecting the region in summer. At the same time, the subtropical high prevails aloft in the lower troposphere over the western extremes of Middle Asia, inhibiting rainfall in regions such as Iran (Alijani *et al.* 2008).

acts to steer mid-latitude cyclones either to the north or to the south of it (Fig. 19.12). The ridge causes the wind to diverge and also divides the region's air masses.

The high is associated with deep and intense inversion layers. Surface cooling produces a low-level inversion 700–900 m thick. Observations at Ulan Bator in Mongolia indicate a temperature

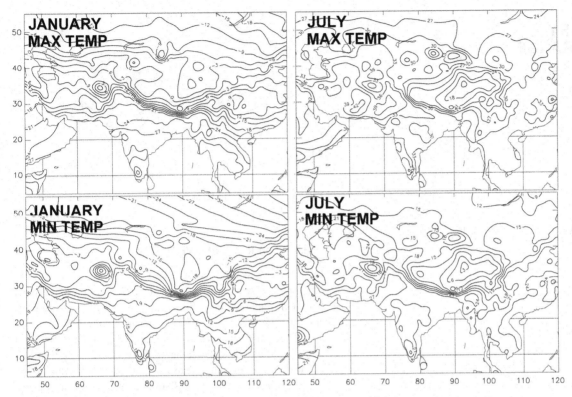

Fig. 19.9 Mean daily maximum and minimum temperatures (°C) for January and July.

In the winter, the polar front lies south of the Siberian High, somewhat south of the Former Soviet Union. Cyclones form in its western portions, often over the Mediterranean, and move east-to-northeastward. Cyclones originating in the Icelandic Low and the Mediterranean are shunted northward or southward by the westward protrusion of the Siberian High. Those shunted southward often intensify over the Black and Caspian seas, which are also areas of cyclogenesis in winter. High-pressure cells are most frequent in the south and east of the Former Soviet Union, especially near Lake Baikal, where sea-level pressures have reached 1075 mb.

During the summer, numerous regions within the interior of Asia become areas of cyclogenesis, in contrast to winter, where most systems enter the region from outside. In most of the region, cyclone frequency is greater in summer than in winter. Much of the cyclonic activity is associated with the polar front.

Throughout the drylands of northern Asia, interannual variability shows links to global-scale processes. In the semi-arid region of northern China, where rainfall is on the order of 200–500 mm, rainfall tends to be subnormal during El Niño years. The quasi-biennial oscillation (QBO) also influences precipitation in this region (Wang and Li 1990) and in the regions surrounding the Caspian and Aral seas (Ye 2001). Precipitation in Siberia varies on a 4–5 year time scale, the peak time scale of ENSO (El Niño/Southern Oscillation), and is closely linked to tropical Pacific sea-surface temperatures (SSTs). In northern Afghanistan, southern Kazakhstan, and eastern Turkestan, rainfall tends to be above normal during El Niño episodes (particularly the transition seasons) and in the positive phase of the North Atlantic Oscillation (Syed *et al.* 2006; Mariotti 2007).

19.3 THE COLD DESERTS OF CENTRAL ASIA

A major climatic influence in Central Asia is topography. Its rain shadow effects are especially important in the Taklamakan Desert, which lies deep in the Tarim Basin (Fig. 19.13). The terrain produces mesoscale wind patterns (Mikami *et al.* 1995a, 1995b) and a mosaic of rainfall patterns superimposed upon the prevailing dry character of the climate. In the north are two major mountain ranges with tops generally between 3000 and 5000 m ASL: the Altai Mountains, lying along the western fringe of Mongolia, and the Tien Shan range, which runs east–west across the middle of Sinkiang Province.

A common denominator throughout most of the region, one that distinguishes it from other dryland regions, is extensive snow cover throughout the winter (Fig. 19.14). Snow cover is well established by November or earlier and persists through at least April. From the latitude of roughly 45° N, mean snow cover depth ranges from 20 cm to over 80 cm throughout winter and spring (Kripalani and Kulkarni 1999). Deepest cover is in March, with the maximum cover being over Siberia. However, there are large spatial variations induced by vegetation and topography (Hirashima *et al.* 2004).

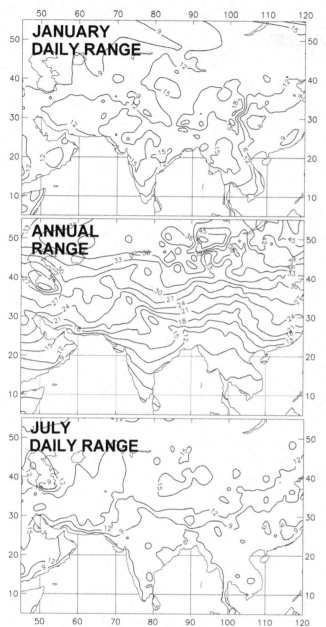

Fig. 19.10 Mean annual temperature range and diurnal range (°C) in January and July.

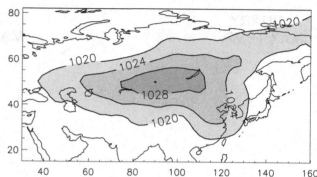

Fig. 19.11 The Siberian High in winter (sea-level pressure in mb, data from NCEP).

19.3.1 GENERAL CLIMATOLOGY

In winter, when the Siberian High intensifies and spreads to cover most of Asia, northeasterlies (the NE or winter monsoon) persist throughout the region. Cold, dry, stable polar continental air prevails. Commencing in October, the NE monsoon flow is gradually established in successive northeasterly bursts of this cold air mass. These bursts recur periodically throughout winter and are referred to as "cold waves" or "surges" of the NE monsoon. The Tien Shan and Altai mountains block all but the most vigorous outbursts, which may be channeled throughout the Dzungaria Basin to reach as far as southeastern Sinkiang. Cold waves from the Arctic produce winter rains in a few exposed parts of Sinkiang. During March and April the high gradually weakens, and the cold surges become indistinct and less frequent, with increasing incursions of warm, moist tropical air from south to east. March and April are the wettest months in much of the region.

In summer, when the semi-permanent thermal low develops over the southwestern provinces of China, flow is southwesterly below 600 m, except over the Mongolian plateau, where northwesterlies intrude. The southwesterlies (called the SW monsoon) are warm, moist and unstable, having passed over the Indian Ocean and South China Sea.

In most of the Central Asian desert region, mean annual rainfall is on the order of 100–200 mm, although the highlands are wetter (Fig. 19.3). In the driest regions it falls to about 40 mm. Maximum precipitation occurs during the summer monsoonal season (Yihui and Zunya 2008) and winter is comparatively dry (Figs. 19.6 and 19.7). The pattern of rainfall is a mosaic, with both annual means and the amount of cold-season rainfall being greatly influenced by topography. Some regions have a slight October maximum; in other parts of Dzungaria, north of the Tien Shan, cold fronts penetrating the mountains produce a winter maximum. To the south of Tien Shan in the Tarim Basin, the maximum occurs in spring, with 60–70% of the annual precipitation falling in March–May.

In the mountainous terrain of the Tien Shan and the Pamir, precipitation is largely a function of altitude (Schiemann *et al.* 2008), increasing steadily with elevation. These regions also receive significant winter precipitation. The precipitation

In the basins between these ranges elevation falls to 400–1000 m. Being in the lee of mountains on all sides and shielded from both the summer monsoon of the Pacific and China Sea and the moist southwesterly monsoon, these basins are exceedingly dry. The two major deserts in the region are the Gobi, south and east of the Tien Shan and Altai mountains, and the Taklamakan, in the core of the Tarim Basin between the Tien Shan and Tibetan plateau (Fig. 19.13). Most of the Tarim Basin lies within the Sinkiang region of northwestern China. Smaller desert basins are the Pei Shan, Hexi Corridor, Ala Shan, Ordos, and Qaidam, a transition region (Fig. 19.2). All are deserts with hot summers. Cool-summer deserts lie in the higher lands of Tibet and the Pamir.

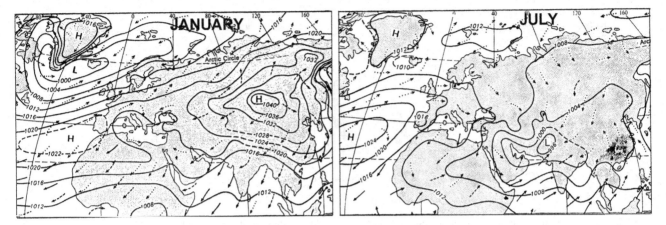

Fig. 19.12 Mean surface pressure and wind over Asia in winter and summer (from Martyn 1992).

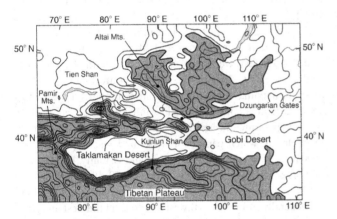

Fig. 19.13 Map of dryland locations and topography in Central Asia (from Warner 2004).

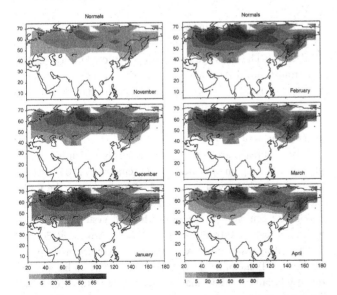

Fig. 19.14 Mean snow cover (cm) by season, with kind permission from Springer Science+Business Media: Kripalani and Kulkarni (1999: fig. 2).

maximum is generally around 3500 m ASL in the northern and western Tien Shan and the Pamir, but lies even higher in the central Tien Shan. The pattern of seasonality is complex in these regions. The maximum is generally in summer in higher elevations of the central and northern Tien Shan, but lower elevations have a bimodal distribution with maxima in spring and autumn. High in the northern Tien Shan, monthly rainfall exceeds 100 mm from May through July. In comparison, monthly rainfall seldom exceeds 30 mm in the lowland deserts (Schiemann *et al.* 2008). The bimodal distribution also prevails in the western Tien Shan and the Pamir, with a pronounced peak around March–May and a secondary peak in November. Aizen *et al.* (2001) describe the large-scale circulation corresponding to these regions.

Precipitation is very sparse in the north, but in the northeast and as far south as Shanghai, cold spells are ushered in by violent snowstorms. In the Taklamakan, around the edges of the Tarim Basin and in the Turfan, there are minor snowfalls on 4–8 days between mid-November and early March. In Dzungaria, just to the north of Tien Shan, snow falls on 30–40 days between mid-October and mid-April. Snow becomes more frequent in the mountains of Sinkiang and plays an important part in the economy of the arid Tarim Basin. About 80% of the water used for irrigation in Kashgar comes from mountain snow, the rest from springs.

19.3.2 THE GOBI

The Gobi is a region of desert, semi-desert, and desert steppe. Average elevation in the region is 1000–1300 m, but basins sink to 700 m. It is a vast region with little or no external drainage and few oases. Only in the south and east do any river systems reach the region. The steppe is characterized by short, perennial *Stipa* grasses, although some shrub steppe occurs. In the true desert, which occupies only a narrow strip, perennial grasses are largely absent and shrubs and semi-shrubs are dominant instead. Cover is generally less than 10% (and as little as 1–2% in some areas). *Haloxylon ammodendron*, a halophytic tree or shrub also called black saksa'ul, is the most widely distributed arboreal species. It can reach 14 m in height.

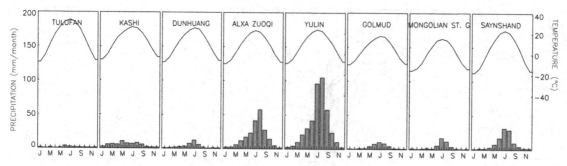

Fig. 19.15 Temperature (°C) and precipitation (mm) at eight Central Asian stations.

Mean annual precipitation is generally on the order of 100–150 mm in the steppe, but less than 40 mm in much of the desert. In the Transaltai-Gobi Desert, in the western Gobi, it averages 18 mm/year. Some years can be entirely rainless. In most of the Gobi, rainfall is limited to summer, with August being the wettest month. In the westernmost sector, the Dzungarian Gobi, precipitation is fairly evenly distributed throughout the year, but the maximum is in summer (see Qitai in Fig. 19.6). Because winter is dry, snow cover is thin and discontinuous.

In the desert region, the continentality of climate is extreme. Representative stations are Mongolian Station G for the desert and semi-desert and Saynshand for the desert steppe (Fig. 19.15). Monthly mean temperature is on the order of −15°C in January and 20°–25°C in July. Temperature can be above freezing in winter because of high insolation. The absolute diurnal temperature range can reach 60°C (Zhang *et al.* 2003) and the annual range can exceed 70°C. Surface temperatures as high as 43.6°C and as low as −28.5°C have been recorded. Cloud cover is rare, so that solar radiation is high enough to create a 178-day frost-free period.

Strong winds, generally westerly, and dust storms occur frequently. The number of dust storm days in the Gobi and western Mongolia ranges from 61 to 127. The storms generally last 3–6 hours and 61% occur in spring (Natsagdorj *et al.* 2003). They occur in connection with synoptic-scale cyclones, most commonly in their southwestern sector, where fronts and cold air churn up the dust (Takemi and Seino 2005).

19.3.3 THE TARIM BASIN AND TAKLAMAKAN DESERT

The Tarim Basin in far western China is surrounded on three sides by high mountain chains. Deep within lies the Taklamakan Desert, the most arid part of Central Asia. It includes vast areas of mobile sands, especially barchan dunes. The region is rich in groundwater but almost devoid of vegetation. Annuals grow after the summer rains fall.

The region's climate is extreme: hyper-arid, strongly continental, and usually dusty. Precipitation is very meager, ranging from about 50–80 mm in the northwest to 10 mm in the east. Its distribution is relatively even throughout the year, but with a distinct summer or spring maximum. Representative stations are Kashi, within the Tarim Basin, and Tulufan, within the Taklamakan itself (Fig. 19.15). Interannual variations of the region's summer rainfall are associated with east (west) shifts of the Tibetan High in wet (dry) years (Yatagai and Yasunari 1995).

Open to the east, the basin is readily penetrated in winter by cold intrusions from Siberia and Mongolia. Mean temperatures for January are about −5° to −10°C, with absolute minima below −27°C. In contrast, July means are about 25°C, with absolute maxima above 40°C. Relative humidity ranges from about 20% in summer to 70% in winter, a consequence of the cold temperatures rather than atmospheric moisture content. Dust clouds, which persist for weeks and reach up as high as 4500 m, reduce both solar insolation and visibility. Usually visibility is below 1–1.5 km and is often only 50–100 m.

The region has a persistent wind regime, with winds blowing generally from the northwest quadrant in the west and from the northeast quadrant in the east (Zu *et al.* 2008). Speeds are generally highest in April or May, except in central regions, where winds are strong from spring through summer. Complex mesoscale circulations also exist within the Tarim Basin, including strong downslope winds out of the Tien Shan (Uno *et al.* 2005).

Moisture transport during the summer rainy season is predominantly from the northwest (up to 90% of the time) (Sato *et al.* 2007). The moisture comes not from low-latitude regions but from Central Asia and western Siberia. Heavy precipitation is associated with the less frequent occurrence of moisture transport from the south, in association with the Asian summer monsoon. These cases are particularly frequent in very wet years (Yang *et al.* 2008).

Dust outbreaks are common in the basin, which is a major global source region of mineral dust (Seino *et al.* 2005). In contrast to the relatively short-lived dust outbreaks in the Gobi, the dust episodes in the Taklamakan are long-lasting. The outbreaks are generally associated with wind systems associated with mesoscale cold fronts (Uno *et al.* 2005; Aoki *et al.* 2005).

The outbreaks of dust are limited by strong nocturnal inversion layers (Tsunematsu *et al.* 2005). These form readily and frequently because cold air masses accumulate in the deep basin. They prevent the winds out of the highlands from reaching the basin floor. When the inversion breaks up in the morning, strong winds appear at the surface, kicking up the dust.

19.3.4 OTHER DESERT BASINS OF CENTRAL ASIA

East of the Taklamakan and south of the Gobi lie a series of deserts: the Ala Shan, Pei Shan, Kansu or Hexi Corridor, Ordos, and Qaidam. Climate changes considerably from east to west, but the region is overall remarkably dry, with most of the precipitation occurring in summer. The climate of several representative stations is shown in Fig. 19.15: Dunhuang (Kansu Corridor), Alxa Zuoqi (Ala Shan), Yulin (Ordos), and Golmud (Qaidam). Precipitation ranges from about 45 to 100 mm/year within the Kansu Corridor. In the Ala Shan, rainfall is about 40–70 mm in the west but climbs to over 200 mm toward the mountains in the east. About 65% of the surface is sand-covered, half by barchan dunes. The Pei Shan is a low mountain landscape; unlike other Central Asian deserts, its surface is largely rocky hamadas. Mean annual rainfall is on the order of 40–85 mm, except in the mountains, where it can reach 500 mm. The Ordos region is a sand desert that gives way to a loess plain toward the steppe region in the southeast. The west is quite dry but in the east intense monsoon rains bring as much as 400 mm/year.

The Qaidam Basin, at about 2700 m ASL, is one of the largest arid intermontane basins in Asia (Zeng and Yang 2008). It represents the transition from the Mongolian plateau to the Tibetan plateau. Rainfall is on the order of 15–25 mm, and July mean temperature does not exceed 18°C. Cloudless skies result in a relatively long growing season here (roughly September–May) in which temperature averages 12°C. The region contains many salt flats and much is vegetated, with cover reaching 70–80% in some areas (Walter and Box 1983f).

19.4 THE COLD DESERTS AND STEPPES OF MIDDLE ASIA

Middle Asia is a term most commonly used in the vast Russian literature on Asian deserts to describe a dryland region extending east and northeastward from the Caspian Sea. Lydolph (1977) uses the term Central Asia in this context. The region is delimited in the south and east by the foothills of Tien Shan and the Pamir-Alay Mountains. It includes primarily the Turkestan Desert and the deserts and semi-deserts of the central Karajasthan north and east of the Aral Sea (Fig. 19.16). The Turkestan Desert, deep within the Turanian lowland, is largely occupied by sand deserts, the major ones being the Karakum between the Caspian Sea and the Amu-Dar'ya River and the Kyzulkum Desert between the Amu-Dar'ya and Syr-Dar'ya rivers.

Climatically speaking, the northern boundary of Middle Asia is the westward protrusion of the Siberian High. In winter the region lies along the western periphery of the Siberian High, the dominant pressure feature of the season. Winter cold is intense. Absolute minima are on the order of −30°C in the south and −50°C in the north. In summer, the region lies along the southern periphery of the Azores High. Absolute maximum temperatures are on the order of 40°–50°C.

Middle Asia represents the transition between the dry, subtropical air of the Arabian–South Iranian region and the continental, temperate air masses to the north. It also represents the transition between winter and summer rains. In the central Karajasthan to the north, precipitation occurs mainly during the warmer months when the westerlies and associated cyclones are displaced northward. The climate is largely Mediterranean in character further south in Turkestan, with Mediterranean cyclones bringing precipitation in winter and spring (December–April) when the westerlies move toward the equator. El Niño has an impact on winter rainfall in much of the region (Ye 2001). The summer (May–September) is nearly rainless and very hot.

19.4.1 GENERAL CLIMATOLOGY

CYCLONIC DISTURBANCES

Middle Asia experiences on average about 32 cyclonic storms per year. Most originate in the southwesterly quadrant and traverse deserts in Iran, Iraq, and Afghanistan before reaching Middle Asia. Humid air from the Atlantic or Mediterranean seldom reaches the region. A mid-winter minimum occurs in February, when the Siberian High is most intense. A secondary maximum occurs in October, especially in northern regions.

Cyclones reaching the region originate in three areas: south of the Caspian, western Iran and Iraq, and the upper Amu-Dar'ya River basin in the southern part of the Tadzhik Republic. Those from the Caspian, which generally develop in the Mediterranean, occur most frequently in winter and bring most of the region's snowfall. They are the only system that affects the region in summer. The associated cold fronts bring much snow; frontal passage is followed by deep intrusions of cold air and a rapid temperature drop. Cyclones from the Murgab, associated instead with deep tropical air masses, bring rapidly rising temperatures and the disappearance of snow cover. Passage of the cold front brings much precipitation. The upper Amu-Dar'ya cyclones, in southern parts of the Tadzhik Republic, originate in southeastern Iran and western Pakistan and have the most eastern trajectory of the three types.

LOCAL WINDS

The deserts of Asia are known for a variety of local intense winds that produce extremes of heat, cold, and dust. Many of them are associated with the cold air intrusions from the west and northwest behind rapidly moving cold fronts. These are often accompanied by dust storms, with drifting sand and dust up to 3 km or higher, remaining for several days. Dust storms are so frequent and widespread that the atmosphere over Middle Asia is hazy with a whitish hue. In the mountains these same frontal passages produce intensely cold bora winds. The cold air rushes downslope, producing winds of 25–30 m/s in the basins. Foehn winds, which bring unusual warmth, tend to arrive from a southerly direction ahead of cyclonic storms and the associated frontal passages. These occur most frequently in winter and in association with the Caspian cyclones.

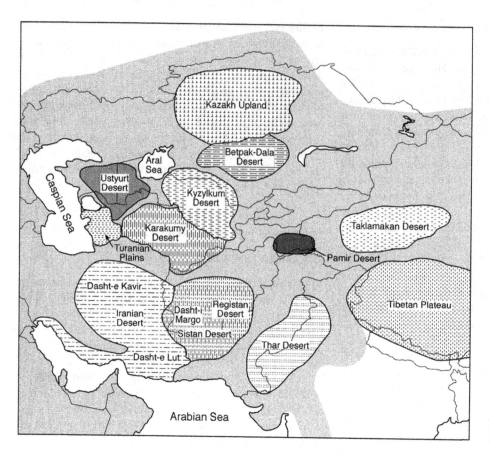

Fig. 19.16 Map of dryland locations in Middle Asia (from Warner 2004).

Dust storms are most frequent and severe in the Karakum. It has been estimated that storms have removed nearly 5 m of dust from the Karakum Desert over time. The maximum number of dust storm days is in western Turkmenistan, where the average is 146 per year (Orlovsky *et al.* 2005). Near the mouth of the Amu-Dar'ya River the number of days falls to 34, and as low as 4 at Tashkent, and 6 in the southwest of the Tadzhik Republic. Dust storms also occur frequently in the Pamir-Alay range, which has continually strong winds.

One characteristic post-frontal wind is the Afghanets (locally also called the *kara-buran*). Blowing from the west, it produces tremendous dust storms in Middle Asia, particularly in the upper Amu-Dar'ya River basin, which it reaches after crossing the Karakum Desert. This wind can reach up to 3500 m and last up to 5 days; the dust may take 10 days to settle. The wind can produce intense cold and unusual electrical phenomena, such as static electricity. It is strongest in summer, when the desert is most prone to turbulence and dust generation.

Foehns are characterized by high speeds on the order of 30 m/s, low relative humidities, high temperatures, and usually dust. Their aridifying influence is extreme. The relative humidity in a foehn can drop abruptly from 70–90% to 2–5%, while temperature increases almost instantaneously by as much as 25°C. The foehns produce a rapid thaw, even direct sublimation to vapor; if the surface refreezes, a layer of ice may form on the snow cover. Two days of foehn winds can melt as much snow

as 2 weeks of sunshine. Foehns in the Tien Shan keep pastures open year-round.

Although summer foehns are usually weak, they can be damaging. The Harmsil in Tadzhikistan has brought the temperature to 48°C and the relative humidity to 8% with speeds of only 6 m/s. Trees drop their leaves and fruit; a few hours of foehn conditions can decrease cotton yields by up to 50%. These winds are even more hazardous when followed by a cold front with a rapid temperature drop.

A particularly fierce foehn occurs in the Dzungarian Gate on the Russian–Chinese border east of Lake Balkhash. The 10 km wide passage through the gate produces winds of 60–80 m/s, strong enough to mobilize not only dust and sand but even gravel. Strong winds occur here over 100 days per year, quickly bringing temperatures from −30° or −20°C to thawing, causing clouds to dissipate, and rapidly lowering the visibility to less than 5 km.

MOISTURE REGIME

Annual rainfall varies from 200 mm along the northern fringes of Middle Asia and at the foot of the Kopet Dag along the Afghanistan border to less than 30 mm in the Hungry Steppe southwest of Tashkent. Precipitation is only about 80–100 mm/year in the delta of the Amu-Dar'ya River, one of the driest regions, and a little over 100 mm along the Caspian coast. In the southern mountains, however, mean rainfall may exceed 1000 mm/year, while areas above 2500 m ASL may be as dry as the driest desert core. In summer, even in the desert, the air contains much moisture because of proximity to the Caspian and Aral seas.

In southern parts of Middle Asia, precipitation is associated with cyclonic storms, which are most frequent in spring and winter. The wettest month is generally March or April (Schiemann et al. 2008). Interannual variations in winter precipitation in much of the region, particularly southern Kazakhstan and eastern Turkestan, are influenced by the North Atlantic Oscillation (Syed et al. 2006). Summers, from June to September, are dry despite a prevailing surface low; the shallow heat low does not override the stabilizing influence of dry, hot surface air and a subsidence inversion linked to the Azores High.

Along the northern border of Middle Asia, the month of maximum precipitation abruptly changes from April or May to June or July, as the primary control switches from the winter cyclones of the south to the summer cyclones of the northern steppes. Local convection also augments the summer rains in this region.

On the plains, snow falls infrequently and accumulation is light or absent. The amount of snowfall is lowest just east of the Caspian, where the mean for November–March is less than 50 cm. It increases rapidly to the north and slowly to the east, reaching 100–150 cm (Fig. 19.14). The number of days of snow ranges from 6 at Krasnovodsk on the eastern shore of the Caspian to 15 at Askhabad and 24 at Tashkent. It increases further north. In the desert plains southwest of Syr-Dar'ya and the northern Aral and Caspian sea regions, snow cover does not exceed 10 cm and is usually less. Along the northern fringe of the region it reaches, at most, 20–30 cm.

Thunderstorms and hail are uncommon, because summers are dry and winter precipitation is associated with stratiform clouds. In the plains, thunderstorms occur about 5–25 days per year, most frequently from April to June in the south and from May to August in the northern desert. They are seldom accompanied by rain. The plains generally experience hail once in 2 or 3 years, usually within the months from February to May.

Middle Asia is known for its frequent cirrus cover, the frequency of which increases toward higher elevations. Cloudiness is often stratiform, accompanying the cyclones of winter and early spring. Much of the area of the plains is overcast about 60% of the time in January, but only about 5–20% in July. Clear skies prevail in summer in the thermal low of the southern plains (i.e., the Murgab and upper Amu-Dar'ya valleys). Fog occurs infrequently in Middle Asia because of low relative humidity and high winds. For the same reason, dew is infrequent in the plains in summer. The plains experience 15–25 days of fog per year, mostly in winter.

19.4.2 THE CENTRAL KAZAKHSTAN

For the desert and semi-desert regions of the central Kazakhstan, relatively little climate information is available and

most comes from the semi-desert. There, mean annual precipitation is on the order of 150–300 mm; it can occur in any month but 60–70% is concentrated in the warm season in most regions (Fig. 19.5). Winter, with predominantly calm and clear weather, lasts about 5 months. Snow falls from October to April; it covers the ground from mid-November to spring. Its duration is on the order of 120–140 days; its thickness is generally 20–30 cm. In some years, snow cover is completely lacking. The January mean temperature is about −26°C, with absolute minima on the order of −40°C or colder. The July mean is about 22°–25°C, but absolute maxima of 40°–45°C are common. Frosts occur between late September and May or June.

More extreme conditions are found in the Betpak-Dala Desert, west of Lake Balkhash. Some parts of this desert are completely dry. In summer, temperatures typically rise to 42°–44°C, but ground temperatures readily reach 65°–70°C. In winter, temperatures sink to −40°C or below. A snow cover 15–20 cm thick lasts about 4½ months.

Climate is somewhat different in the lowlands north of the Aral Sea. Conditions at Saksaul'skiy (Fig. 19.17) are representative. Mean annual precipitation is about 120 mm and surface water is completely lacking, except for spring inundations of the salt pans and takyry (areas of clay soils that are prone to flooding, but crack into polygonal surfaces upon drying). Snow 5–8 cm thick covers the ground from mid-December to mid-March. In February the permafrost is about 1.5 m deep.

In the Sary-Ishikotrau Desert east of Lake Balkhash are vast sand seas. Here the precipitation regime changes as the winter becomes progressively drier. Representative climate data are available from a research station in the Taukum, to the southwest. The region is somewhat warmer and moister than the northern Aral and vegetation cover is about 50%. At Aydarly (Fig. 19.17) annual precipitation averaged about 225 mm for the period 1955 to 1972, but annual totals as high as 300 mm have been recorded. The precipitation maximum is in spring, and both winter and summer are dry.

19.4.3 THE TURKESTAN DESERT

The Turkestan Desert is a cold winter desert occupying part of the Aral-Caspian (Turan) lowland east of the Caspian. It joins up with the Kazakhstan deserts to the north and extends south to the Kopet Dag, rising over 2000 m along the border with Afghanistan (Fig. 19.2). Sand deserts, most notably the Karakum between the Caspian Sea and the Amu-Dar'ya River and the Kyzulkum Desert between the Amu-Dar'ya and Syr-Dar'ya rivers, occupy most of this region. Vegetation cover is relatively low except in favorable niches. The vegetation in the desert portions includes dwarf shrubs, halophytes, and ephemerals in the deserts.

The climate of the northern Turkestan Desert is represented by Nukus in the north, Ashkhabad in the south, Krasnovodsk and Chekishlyer in the west, and Kerki in the southeast (Fig. 19.17). Summers are generally dry; maximum precipitation occurs in spring, with a secondary maximum in winter. Mean annual

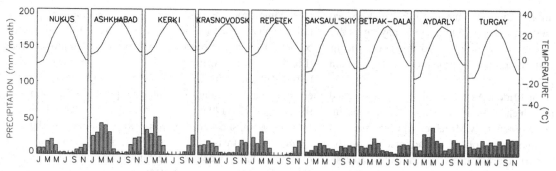

Fig. 19.17 Temperature (°C) and precipitation (mm) at nine stations in the cold deserts and steppes of Middle Asia.

precipitation is generally 100–200 mm, but is only 80–100 mm in the Amu-Dar'ya Delta, the driest section of the region.

Daily maxima in summer average about 30°C, with a diurnal range of 15–20°C. The absolute maximum is on the order of 50°C. Cool days begin in late September and become frequent in October–November. In general, the presence of the inland seas moderates the winter climate. Although the region is completely open to the north in winter, enabling cold Siberian air from the Siberian High to frequently penetrate the desert, the absolute minimum temperature is about −26°C. Mean January temperatures are well below freezing, except in the south, but thaws occur intermittently so that, unlike the Kazakhstan, there is no continuous snow cover. Sometimes heavy snowfall occurs in spring. Frost can occur until the end of April.

Climate is somewhat milder along the coast of the Caspian, particularly in the southwest, north of the Atrek Delta. Protected by the Kopet Dag from the cold northeasterly winds, temperatures here do not fall below −10°C. The climate of Kara-Kala is almost subtropical, with trees bearing olives, figs, almonds, dates, and citrus fruits. Spring is the most favorable season. The ground moisture from winter snow is augmented by spring rains. There spring ephemerals are striking.

The Karakum is the southern part of the Turkestan Desert. It has been extensively studied, as a research station was opened in Repetek in 1912. It is mostly sand desert (80%), with a few takyry and salt pans. The region includes a variety of dune forms, including nebkas, barchans, barchan fields or chains, and sand ridges. Numerous rivers flow through the region, the main one being the Amu-Dar'ya. Much of the region is well supplied with surface water and groundwater is at a depth of just 4–5 m in interdune valleys. Parts of the region have no vegetation cover at all, but in the southern Karakum vegetation cover is so dense that the region does not have a desert-like character. On its northwest border, numerous freshwater lakes are formed from standing groundwater.

The climate of the region is represented by the station Repetek in the southeast, 80 km west of the Amu-Dar'ya (Fig. 19.17). Repetek's climate is quite similar to the rest of the Turkestan Desert. Monthly mean temperature exceeds 20°C from the end of April to the end of October and exceeds 0°C in all months. The mean frost-free period is about 280 days (mid-May to October). The warm season is the second half of February to

early December. An absolute maximum of 50°C was recorded in July 1915; an absolute minimum of −31°C in January 1969.

Mean annual precipitation at Repetek is 114 mm, with 27% in winter, 49% in spring, 2% in summer, and 12% in autumn. Nearly 70% of the precipitation occurs during a 2–3 month period around March or April and is usually associated with heavy downpours. The months from July to September are nearly rainless. Annual totals at Repetek vary from about 24 to 230 mm. The average number of rainy days is 32. Snowfall is rare and a snow cover of 2–5 cm is present only about 8–10 days per year. Precipitation in the form of fog, dew or frost occurs about 40–50 days each winter. Relative humidity is highest (about 65–70%) in January, lowest (20–25%) in July and August; it can fall to a few percent.

19.5 EASTERN SIBERIA

Eastern Siberia is characterized by the highest degree of continentality on earth, with the absolute temperature range reaching or exceeding 100°C in some locations. In summer the daily maximum temperature commonly reaches 40°C in the south and 35°C in the north, but nocturnal cooling is high and frost can occur even in July. Extensive cloud cover characterizes much of the region, averaging around 70% during most of the year, but falling to about 60% in winter (Sun and Groisman 2000).

In summer there is an alternation between cyclones and anticyclones, with almost equal dominance. The cyclones are associated with the polar front and often form near Mongolia, moving east and northeastward. Shifts in cyclone tracks associated with stationary waves modulate the amount of summer precipitation received over eastern Siberia, creating a negative correlation between summer rainfall here and in western Siberia (Fukutomi et al. 2007).

Most of eastern Siberia can receive precipitation in any month of the year, but the maximum is generally in summer, especially July and August. Many locations experience a secondary maximum in autumn. In autumn, cyclones move into the region from Kazakhstan and from more northern parts of Siberia. Cyclone passage is followed by deep intrusions of cold Arctic air. Most of the region is relatively dry in winter and spring, but the transition to the more humid summer months

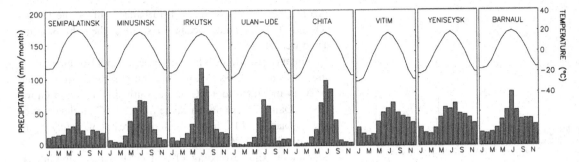

Fig. 19.18 Temperature (°C) and precipitation (mm) at eight stations in eastern Siberia.

is rapid. Snow occurs in winter, but snow cover is generally low in most of the region, ranging from about 20 cm (during the 10 days of deepest cover) just east of Lake Baikal to 50–60 cm just to the north in the cyclonic corridor.

Lydolph (1977) identifies several geographical provinces within eastern Siberia (Fig. 19.2), all of which receive relatively little rainfall. Many of these, described below, are near the Altai Mountains. The climate of representative stations is indicated in Fig. 19.18.

In the extreme west and southwest of the Altai Mountains lies a semi-arid steppe, a low plain where the elevation is generally 100–200 m ASL. The region contains many salt lakes. During the summer rainfall maximum, which generally occurs in July, the region receives heavy showers that rapidly run off the dry surface. These are due to frontal convection. A secondary rainfall maximum occurs in autumn and early winter. The driest conditions occur from January to April, during which time the western extension of the Siberian High prevails in the region. Winter is cold and clear, normally with a thin snow cover of 15–30 cm. The driest area is along the Irtysh River. Semipalatinsk is a typical location, with mean annual precipitation of 264 mm. Precipitation increases eastward toward the mountains, reaching 464 mm at Barnaul near the Ob' River. Summer is generally warm: surface temperature reaches 42°C at Semipalatinsk, compared with an absolute minimum of −49°C, and hence an absolute range of 91°C. Mean daily temperatures are about 20°–25°C in summer, but −15°C in winter.

The Altai Mountains themselves present a very complex climatology, with vast differences between basins and slopes or between slopes with varying degrees of exposure. Slopes may remain warm while the cold air sinking into the basins drops below −50°C. Moisture-bearing winds arrive from the southwest, so that rainfall is greatest on southerly and westerly slopes. Annual rainfall may typically be 1500 mm on a western slope, compared with 200–300 mm on eastern slopes. The precipitation maximum is in summer. The region experiences extreme foehn winds, with many areas averaging well over 100 days per year with foehns. Most frequent in October and least frequent in summer, these winds often precede cold fronts.

The region lying east and north of the Altai Mountains is known for dryness and extreme temperatures. Typical is the Minusinsk Basin, a semi-arid grassland where the absolute range

is from +39° to −53°C and mean annual rainfall is on the order of 250 mm. Little of the rain, less than 12%, falls during the winter period of November–March. Winter weather is stable, with light winds. Blizzards are rare and there is little snow cover; instead a deep permafrost layer forms. Foehns accentuate the dryness. To the north, in the Yenisey Valley (Yeniseysk), rainfall is somewhat higher and more evenly distributed through the year. There is a summer maximum, with the wettest month being August in the Yenisey Valley, July in the Minusinsk Basin.

Conditions are more extreme in the Tuva Basin to the south, another semi-arid grassland region. At elevations just over 500 m, cold air stagnates in the basin in winter, forming one core of the Siberian High. The region has a summer precipitation maximum, with 60% occurring in July and August. It usually falls in heavy showers. In winter, fronts reach the region from the west but are held aloft by the cold dome, so winter is dry. The absolute minimum air temperature in the region is −58°C, but a frontal passage can produce a temperature rise to −20°C. In spring, temperature rises rapidly and frequent foehns lift dust and fine sand from the surface and reduce relative humidity to 15–20%. July is the only month without frost.

Lake Baikal has a moderating influence on nearby locations. Its influence continues until the lake freezes in January. Frequently bora winds descend the river valleys to the lake in winter. The best known is the "sarma" in the Sarma River valley. It occurs 113 days per year with winds of 15–40 m/s and can extend as much as 20–30 km over the lake. This wind is dust-laden when it occurs in summer. Maximum precipitation and maximum cyclonic activity occur in July and August.

East of Lake Baikal, in the southern half of Transbaikalia, lies a range of NE–SW oriented mountains merging with a plateau to the north. The region includes two basins at about 600 m elevation: Chita in the east and Ulan-ude in the west. Similar to the Minusinsk and Tuva basins, these are semi-arid grasslands. The rest of the region is topographically complex, with alternating regions of forests and steppe grasslands. The latter can extend up to 60° N in the river valleys.

The climate of Chita is typical of these basins. In winter, the Siberian High is dominant, bringing clear skies, extreme cold and dry conditions. Since snow cover is thin, there is a deep permafrost layer. Dryness prevails until April, when cyclones start bringing rainfall to the region. Relative humidity remains

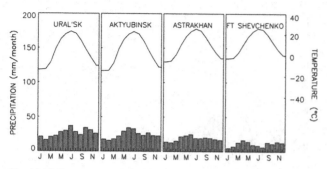

Fig. 19.19 Temperature (°C) and precipitation (mm) at ten stations in the Caspian lowlands.

low, on the order of 30%. Dust storms occur frequently early in summer, but late summer is warm and relatively rainy. July and August, with 183 mm of the annual mean of 343 mm, are the wettest months. Mean daily maxima and minima are on the order of −12°C and −40°C in winter, but around 35° and 0°C in summer. The diurnal range is about 20°C during the warm half of the year, but 13°–14°C in the cold seasons.

On the plateau further north, winter is colder, slightly wetter, and much cloudier. Daily winter minima are on the order of −50° to −60°C at Vitim. Precipitation is more evenly distributed during the year, with less occurring in summer and more in winter than in the more southern basins.

19.6 THE COLD SEMI-DESERT OF THE CASPIAN LOWLANDS

This region (Fig. 19.2) represents an ecotone both in terms of climate and vegetation, a transition between the more westerly steppes and the more easterly deserts. The region generally becomes increasingly arid both from north to south and from west to east. Just outside of the lowland in the dry steppe, stations such as Ural'sk receive about 300 mm of precipitation per year (Fig. 19.19). Typically, stations along the northeastern border of the semi-desert, such as Aktyubinsk, receive about 250 mm. Astrakhan and Fort Shevchenko, to the south, have a desert climate with about 150–180 mm of precipitation. While one or the other biome is dominant in the many microclimates of the region, in most areas the two biomes coexist.

In the north is a semi-desert biome with many large and small salt lakes (called limany), salt swamps, and marshes. Spring snowmelt moistens the steppe, which comprises about 25% of the region; after abnormally wet winters it also fills the saline lakes and limany. Here the typical cover is bunch grasses, such as *Stipa*. Salt-tolerant species thrive on the saline surfaces. The south is more desert-like, with extensive dunes; here mainly *Artemisia* and various grasses grow.

During winter, a westward protrusion of the Siberian High prevents Mediterranean cyclones from moving northward into the region and Atlantic cyclones from penetrating southward into the region. In summer, an eastward protrusion of the

Azores High similarly keeps disturbances out of the region. The rainfall maximum occurs in spring, when the Siberian High weakens.

Winters are cold, the NE winds ushering in Siberian air; there are few winter thaws. Mean temperatures range from about −5°C in the north to +5°C in the south (Elguindi and Giorgi 2006). Daily winter minima range from −30°C on the coast of the Caspian to −40°C inland. Summers are hot, with July means on the order of 25°C (though markedly cooler in the north) and absolute maxima exceeding 40°C. The SE winds bring intense summer dust storms.

The station Dzhanybek (roughly halfway between Ural'sk and Astrakhan) is typical of the region's climate. Precipitation averages 274 mm/year, split evenly between the warm and cold seasons. Relative humidity falls below 30% about 64 days per year. Sukhovey (hot dust storms) occur about 80 days per year. The temperature extremes are on the order of −36° to + 41°C. Snow cover, which usually persists from November to March or April, is less than 10 cm on the plains but 40–50 cm in depressions.

19.7 THE WARM DESERTS OF ASIA

The warm deserts of Asia are distinguished on the basis of absolute maximum temperatures exceeding 40°C (Evenari 1985), but in practical terms the distinction between warm and cold deserts is that the latter experience cold winters. Warm deserts occupy much of Pakistan and southern areas of the countries of Iran and Afghanistan (Fig. 19.20), but in this section the whole desert area of these countries is discussed.

The prevailing seasonal pressure fields, circulation patterns, and depression tracks are those described in Chapter 17, dealing with the Mediterranean and Near East. Thus, the reader is referred to that chapter and to the following sections for further detail.

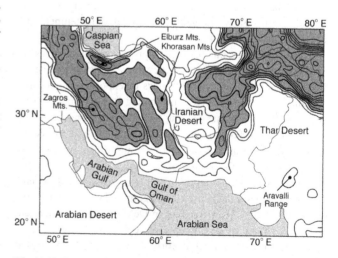

Fig. 19.20 Topography and location of the warm deserts of Asia (from Warner 2004).

19.7.1 CLIMATIC CONTROLS

The causes of aridity are probably a combination of those influencing North Africa and Asia. The Siberian High suppresses the winter rains by hindering the penetration of cyclones into the region. In western sectors with summer dryness, the cross-circulations associated with the tropical easterly jet, theorized to play a major role in creating the Sahara, probably also suppress rainfall. Additionally, the storm tracks of winter tend to miss the region. Over Iran and Afghanistan, an additional factor in the aridity is the split of the subtropical jet into two branches as it encounters the Hindu Kush mountains of Afghanistan and the Tibetan plateau. Where the branches depart to flow north and south of the jet, the divergent wind field produces large-scale subsidence (Pagano *et al.* 2003).

In this region, as in the cold deserts and steppes elsewhere in Asia, the major climatic controls are the seasonally shifting pressure systems over Asia and the mid-latitude cyclonic disturbances associated with the westerly jet streams. In winter, most of the region is still under the influence of the Siberian High, its importance increasingly diminishing toward the west. The western sector also falls under the influence of ridges of the Azores High and local anticyclones over the Armenian and Afghanistan plateaus and the Pamir-Alay Mountains to the east. The polar front frequently stretches across the region, ushering in maritime polar air and depressions. However, mountain barriers usually prevent the cold air north of the front from penetrating most of the region.

In summer, the main controls are the Azores High (Alijani *et al.* 2008) and the Asian Low. The former is the dominant influence in Iran and throughout most of Afghanistan. In summer, the anticyclonic regime associated with the high inhibits precipitation. On occasions when the high weakens, convective summer rains occur (Alijani 1997). The influence of the Asian Low and the associated monsoon increases eastward, so that the region as a whole represents a transition between winter Mediterranean rainfall and summer monsoon rainfall. In this sense, it is analogous to the Middle Asian dryland belt further north.

Thus, in the deserts of Iran and Afghanistan, rainfall occurs mainly in winter or spring and tends to be associated with cyclonic depressions traveling eastward along preferred tracks, mostly through the Mediterranean (Ghasemi and Khalili 2008). During particularly severe winters, the mountainous snowpack can bring torrential floods when it melts in spring (Davenport 2008). Depressions enter the region along southerly tracks from Iraq and the Persian Gulf and along a northerly track from the Caspian and Black seas. The Thar Desert in the east receives primarily summer monsoon rainfall that is subtropical in character.

19.7.2 IRAN AND AFGHANISTAN

The rugged mountains of Afghanistan and Iran have a dominant influence on the region's climate. Desert and semi-desert basins are scattered throughout the mountain and hill regions, as well as in interior locations in the mountains' rain shadows.

Both Iran and Afghanistan are primarily arid, with over half of Iran receiving less than 200 mm of rainfall, on average.

The true deserts occupy mainly two core basins in eastern Iran and southern Afghanistan; Meigs (1957) refers to this region's drylands collectively as the Iranian desert. The Afghan portion of the desert is bordered in the east by the Afghan Mountains and in the south by the higher terrain of western Pakistan. The desert core in Iran consists of the Kavir Plains or Desert (Dasht-e Kavir) and the Lut Desert (Dasht-e Lut), The vast Dasht-e-Kavir is a group of many playas covering 45,000 km^2. These deserts are surrounded by the Zagros Mountains in the west and southwest, the Alborz Mountains to the north and northwest, the Kopet Dag to the northeast, and the Khorasan Mountains and various hill regions to the east and southeast. The Alborz and Zagros mountains prevent moisture from the adjacent water bodies (Persian Gulf and Caspian Sea) from reaching the desert interior. The Alborz Mountains, along the southern shore of the Caspian, also prevent the moderating influence of this large sea from penetrating very far inland (Raziei *et al.* 2008).

The rivers of the region are primarily inland, dying in the deserts or steppes or flowing into inland lakes. However, northern Iran drains into the Caspian and a number of rivers link northern Afghanistan to the Amu-Dar'ya, which drains into the Aral Sea. The Zagros Mountains are the source of several large rivers: the Karkheh, the Dez, and the Karoon (Raziei *et al.* 2008). Several endoreic systems characterize central Iran (Breckle 1983). These are vast basins, called *kavirs*, which resemble the shotts of North Africa, with clay or salt swamps, and the playas of North America. Large areas are also gravel plains, but a variety of surface types are found, including sand ergs.

The drylands of Iran show much diversity. Vast areas of inner Iran are almost bare of vegetation. Saline plains where vegetation cover may be as low as 1% occupy most of the region. The Kavir is a vast system of plains that includes desert, semi-desert, dry steppe, and semi-desert mountain environments. These form a mosaic among rocky and saline kavir areas of the plains. Over half of the 283 species in this area are annuals. The best represented family are the chenopods (Chenopodioideae). These dominate the plant communities on non-sandy areas and have the peculiar characteristics of flowering and being active during the dry summer season. Elsewhere are semi-deserts dominated by *Artemisia*. In the far eastern areas under the influence of the monsoon are semi-desert open woodlands, dominated by *Pistacia atlantica*, with an understory of shrubs.

In the deserts and semi-deserts of northern and southern Afghanistan *Calligonum* and *Aristida* (*Stipagrostis*) cover large areas of mobile dunes. The cover is generally 1–10%, but exceeds 25% in some dune valleys. These areas include many annuals and perennials.

THERMAL REGIME

Iran and Afghanistan experience a dry and strongly continental climate except along the coasts. The annual temperature range exceeds 20°C nearly everywhere and averages 30°–40°C in many

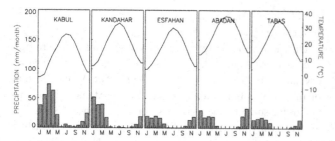

Fig. 19.21 Temperature (°C) and precipitation (mm) at five stations in the warm deserts.

locations. The absolute temperature range exceeds 70°C in some places. Absolute minima are commonly on the order of −25°C, but in some mountain basins (e.g., Shaharak in Hazarajat), temperatures below −50°C have been recorded. The extremes occur in January and July. The diurnal range is also large, often exceeding 20°C in the dry months.

Summer is hot and nearly everywhere dry. In most locations autumn is also hot and predominantly dry (Fig. 19.21). A typical July mean temperature is about 25°C, with an absolute maximum around 45°C. Mean July temperatures are higher along the Persian Gulf coast of Iran, reaching 33°–36°C. A temperature of 50°C has been recorded at Abadan (Martyn 1992). Except near the coast, where the summer is sultry, relative humidity is low, with annual means usually below 50%. Values lower than 5% have been recorded. In July mean daily minima are on the order of 24°–27°C and mean daily maxima are on the order of 39°–42°C, but daily temperatures in excess of 44°C commonly occur.

Winter is cold, with temperatures frequently below freezing, especially in the mountains. The region is characterized by the regular occurrence of frost, with most areas experiencing more than 30 frost days per year. Typical mean January temperatures are 0°–5°C, except along the gulf coast, where the moderating influence of the water keeps mean temperatures as high as 14°C. Severe winters are rare but occasionally occur; temperatures have fallen as low as −21°C in northern Iran and to −31°C in northern Afghanistan. The Siberian High has a large influence on the interannual variability of winter temperature in this region (Gong and Ho 2002). Winter temperatures in the region are largely a function of latitude (Fig. 19.19). Mean daily minima in January range from around 15°C near the coast to freezing in the northernmost desert latitudes. Maxima are on the order of 24°–12°C in these same latitudes.

PRECIPITATION REGIME

Roughly half of Afghanistan and Iran receives less than 200 mm/year, on average (Fig. 19.22). At some locations mean annual precipitation is only 50 mm or less (e.g., 50 mm at Seistan and Registan, 37 mm at Zarand). In the desert cores, nearly all months are dry, with none receiving more than 20 mm on average. Rainfall variability is high from year to year. It ranges from as low as 20% in the mountainous regions near the Caspian Sea

to over 75% in the far southeast. Most of this region receives primarily winter rainfall. During both winter and summer, cloudiness is relatively light, ranging from about 20–25% in winter over most of the region to as little as 5–20% in July, the clearest month. Greater amounts of rainfall and cloudiness are found in the mountains and along the Caspian Sea.

The precipitation regime is strongly influenced by local factors, including the complex topography of the Alborz Mountains and Zagros Plateau, the Caspian Sea, the desert basins, and coastal effects along the Persian Gulf and Arabian Sea (Modarres 2006; Raziei et al. 2009). The mountains have a profound effect, with the highest rainfall occurring over the highlands. However, there is no clear-cut altitudinal profile, as other aspects of the topography (e.g., slope and geographic location) play an important role (Alijani 2008). Precipitation is generally about 200–500 mm in the mountains and at their bases and exceeds 1000 mm only in higher elevations.

Maximum snowfall occurs in the mountains of the north and west and provides most of that region's precipitation. A considerable amount falls in winter in the highest elevations. The plateau experiences the most snowfall in the northeast, the minimum on the south and southeast margins. Snow falls occasionally in the lower regions but it does not stay for long.

The precipitation maximum generally falls in winter in the drier, lowland regions, but in spring at higher elevations, such as the Iranian plateau, and along the western and southern coasts. Most precipitation is associated with moisture transport from the west (Domroes et al. 1998), but southerly fluxes are disproportionately associated with large rainfall events. These peak in March and April. The spring maximum is particularly pronounced in the mountains of Afghanistan, where rainfall is relatively high. In the areas with a spring maximum, the wettest month tends to be April in Afghanistan but March in Iran. In contrast to the rest of the region, the Caspian littoral has an autumn maximum and a relatively high frequency of thunderstorms. Summer is dry because an anticyclonic regime prevails aloft, associated with the Azores High. During the May–October dry season, rain is rare throughout the region. On occasions when the high weakens, convective summer rains occur (Alijani 1997).

The patterns of climate and climatic variability contrast strongly between northern and southern Iran (Raziei et al. 2008). The northern plateau is open to dry, cold continental current from the northwest, and the mitigating influence of the Caspian is limited to the northern region of the Alborz Mountains. For stations with comparable annual rainfall totals, the more northern stations report more days of rain and cloud cover. In winter and spring, precipitation occurs on 10–16 days per month in northern Iran and Afghanistan, but only 4–6 days per month in central Iran. Heavy rainfall, with high surface runoff, is more common in the south.

WINDS

Mountains affect the wind regime, which is generally steady, especially in winter. Northerlies prevail in winter and calms

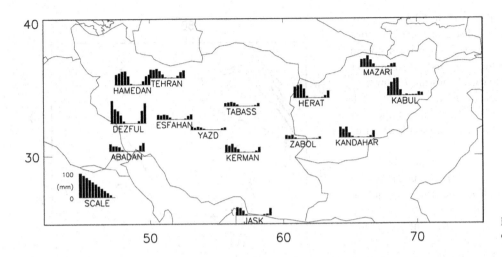

Fig. 19.22 Mean precipitation (mm) over Iran and Afghanistan.

are frequent. Several bora-type winds occur in winter in various parts of the region, blowing off the snowy mountains. The convergent topography of the Lut valley produces a northerly low-level stream downslope into the valley (Liu *et al.* 2000). It is primarily a channeled gap flow, but is influenced by diurnal forcing over the complex terrain and is strongest at night.

In summer, the prevailing direction depends on location with respect to the monsoon low over Asia: it tends to be north and northwesterly in the northern and central parts of Iran and Afghanistan, but west and southwesterly further south. In the Kabul Basin of Afghanistan, strong katabatic winds prevail. The summer winds tend to be persistent and strong. A good example is the Seistan winds that prevail from June to September in northwestern and western Afghanistan and eastern Iran. Called the "120-day wind," this system blows from a NW or NNW direction and has a constancy of 73–84%. These winds are hot, dry, and dusty and sufficiently strong and persistent to mobilize sand dunes. Dust storms occur in desert areas of Iran and Afghanistan about 20 days per year.

CLIMATIC VARIABILITY

The interannual variability of rainfall is high throughout this region, but especially in the arid and semi-arid portions of the country (e.g., Ghahraman 2008). Many global mechanisms of variability influence the region, and the patterns of variability are complex (Rahimzadeh *et al.* 2009). The global factors include ENSO (Nazsemosadat and Cordery 2000; Mariotti 2007; Dezfuli *et al.* 2009), the North Atlantic Oscillation (NAO) (Syed *et al.* 2006), the Madden–Julian Oscillation (MJO) (Bartow *et al.* 2005), and the Arctic Oscillation (AO) (Ghasemi and Khalili 2008).

ENSO has an influence on precipitation in Iran, but predominantly in autumn (Dezfuli *et al.* 2009). The ENSO response is relatively weak compared with other regions and appears to be both inconsistent and transient (Pagano *et al.* 2003). Over most of the country there is tendency for above-normal rainfall in autumn during El Niño episodes. The impact is strongest in the southern foothills of the Alborz Plateau, northern

Iran, and central desert regions. El Niño tends to reduce the intensity and frequency of autumn drought, especially in southern regions. La Niña tends to favor the development of drought. A weak influence appears in winter, particularly in northern Iran.

The NAO, MJO, and AO all appear to have a stronger influence than ENSO. The NAO and MJO both influence rainfall, while the Arctic Oscillation influences temperatures. Variations of the AO account for up to 46% of the variability of winter surface air temperature in Iran and roughly 30% of the summer variance (Ghasemi and Khalili 2006).

The NAO is negatively correlated with rainfall in southwestern Iran; autumn drought is extremely unlikely to occur when the NAO is negative in the preceding summer. Spring drought shows a similar correlation (Dezfuli *et al.* 2009). The response is different in northwestern Iran, where autumn (winter) rainfall has a positive (negative) correlation with the NAO (Pagano *et al.* 2003). This contrasting response in autumn is likely due to the tendency for a northward shift of the North Atlantic storm track during the positive phase of the NAO (Syed *et al.* 2006). The positive phase of the NAO is also associated with increased precipitation in Afghanistan. The influence of the MJO is spatially more uniform and very consistent from year to year. During its positive phase, rainfall in Iran is about 25% above normal.

19.7.3 THE THAR DESERT

In India, areas with roughly 250 mm of precipitation or less and a mean diurnal temperature range of 10°C or more are considered to be deserts; semi-deserts receive between 250 and 500 mm of rainfall and have a diurnal range of more than 10°C (Pramanik *et al.* 1952). This climatic definition agrees well with limits established on the basis of vegetation. Many other definitions appear in the literature, but these produce only minor changes in the boundaries of the main desert regions.

The Thar is a region of diverse environments: desert peneplains, dunes, sandy plains, floodplains, salt playas, lakes, and marshland. Four subdivisions of the Thar are recognized: the Thar (*sensu stricto*), the Rajasthan, the Pata, and the Ghaggar plains. The true Thar is a vast region of sand seas and sand hills oriented NE–SW by the monsoon. The Rajasthan is an upland steppe on a rocky plateau along the north and west fringes of the Thar Desert. The Pata is a variegated landscape of low sand hills, sandy flats, impervious clay surfaces, and numerous salt lakes. The Ghaggar alluvial plains, a relict of a gigantic river system that still floods occasionally, is the most fertile sector of the Thar.

Vegetation in the Thar is quite diverse. In the desert zone are mostly thorny shrubs and small trees. *Calligonum* and *Haloxylon* are most common, but there are also some *Acacia* species. In the dry savanna zone, the dominant species are *Acacia* and *Salvadora*, with *Prosopis* in the drier parts. Riverine and mountain environments sustain a variety of forest types.

The desert peneplains, extending from the Gujarat plains northeastwards to Delhi, are the areas referred to as the Rajasthan. The Luni River divides the landscape into two parts: undulating sand dunes in the north (called *thali*) and to the south, sand hills on a plateau enclosing small playas that produce enormous amounts of salt. Vegetation on these hills is mostly *Anogeissus pendula* and *Acacia senegal*, but *Euphorbia* grows in regions of human disturbance. Soils are generally shallow and alluvial.

The dunes region, or the Marusthali, includes longitudinal, transverse, and barchan types, covering over 100,000 km^2 and having a N–S orientation dictated by the SW monsoon. Typically, the vegetation consists of shrubs with clumps of grasses, although isolated trees occur. There are verdant oases in the interdunal areas, with sandy loam aridisols.

The sandy plains, the most distinctive feature of the Thar, are scattered through the desert and are studded with occasional sand hills called *bhits*. Under the older plains lies an impervious layer of limestone, so precipitation cannot penetrate downward to recharge groundwater and much of it is lost to evaporation. In more recently developed plains, without this layer, there is considerable groundwater recharge. The soils here are generally regisols.

The primary cause of aridity in the region is low-level divergence and associated subsidence up to 0.8 km, together with marked divergence between 2 and 6 km aloft over the region (Gupta 1986; Datta and George 1964). This is essentially a rain shadow effect, as the dry continental air overrides the moist monsoon flow and subsides in the lee of mountains to the north and west. The subsidence creates a temperature inversion that enhances the aridifying influence.

The region is one of predominantly summer rainfall, with hot summers and cool winters. However, it marks a very abrupt transition between regions of summer monsoon rainfall and those with a winter cyclonic regime lying just to the west (Fig. 19.23). Although classified as a "warm desert," temperature can fall below 0°C in some places and frost does occur. Rainfall is associated primarily with the summer SW monsoon, with 90% falling between June and September (Gupta 1986). Mid-latitude cyclones from the west, in association with disturbances

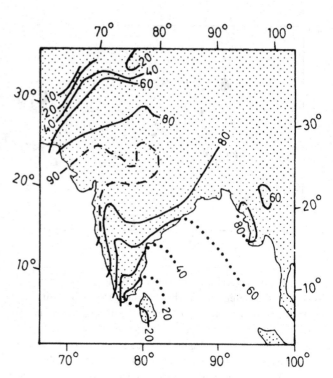

Fig. 19.23 Proportion (%) of mean annual precipitation linked to the Indian monsoon.

in the westerlies, occasionally penetrate into the region in winter. These occur mainly in January–May. Most of the region experiences low cloudiness; at Bikaner, it ranges from less than 10% in some months of spring and fall to about 30% during the summer monsoon season.

The dominance of the monsoon imparts a climatic character quite different from that of most deserts. The rains are more intense and more regular in seasonality and the air is relatively moist. The wind regime is fairly regular, being SW or W during the monsoon months and, during the rest of the year, mainly SE in the morning and NE in the afternoon.

The Thar is the last region reached by the monsoon, and the first from which it withdraws. The monsoon commences around July 2 in Jodhpur, around July 8 in Barmer, Bikaner, and Ganganagar, withdrawing in early September. Widespread rainfall generally occurs only in association with monsoon depressions. These are relatively rare, with only 24 moving through the region in the 90 years from 1871 to 1960 (Datta and George 1964). Mean annual rainfall ranges from about 100 to 400 mm. Isohyets runs roughly SW to NE, with the driest conditions lying in the northwestern parts of the region and the wettest being in the south and southeast (Goyal 2004). Rainfall variability is roughly 40–60%, but reaches 80% in the extreme west.

A typical intensity of the monsoon rains is 90 mm/day, but on one day nearly 500 mm fell (Ramaswamy 1975). The heaviest rains are associated with the movement of tropical depressions and can continue several days. Although rain occurs infrequently, humidity both at ground level and in the upper atmosphere can be as high as in nearby semi-arid and subhumid regions. Rainfall is so erratic that the entire year's rainfall may occur on one day (Goyal 2004).

The region's temperature regime is also affected by the monsoon, which reduces temperatures during the summer season. The cold season lasts from December to February, with January being the coldest month. Temperatures start rising in mid-March and reach a maximum in May; June is also hot. Temperatures are cooler during the following monsoon season, except in the extreme west where the monsoon only briefly intrudes. Temperatures fall rapidly between September or October and December or January. The average monthly minimum is generally 1°–8°C in the coldest months, but 23°–27°C in the warmest months. Average maximum temperature is about 24°–30°C in the coldest months, but 42°–46°C during the warmest, pre-monsoon months. Temperatures in excess of 50°C have been recorded. The diurnal range is on the order of 15°–18°C.

The Thar is frequently affected by dust, but fogs and hail are rare; snow is virtually unknown. Dust increases after February, reducing visibility during the hot season from April to June or July, when wind speeds are high. The monsoon rains then scavenge it from the air. Dust storms average about 5–15 per year in the Rajasthan plateau, but the number rises to about 20–30 for locations in the Thar Desert (Rao 1981). Fog is occasionally reported in the cold months, especially December–February. Hail was recorded only once in 20 years at Ganganagar, but once every 2–3 years elsewhere.

REFERENCES

Aizen, E. M., V. B. Aizen, J. M. Melack, and J. Dozier, 1997: Climatic and hydrologic changes in the Tien Shan, Central Asia. *Journal of Climate*, **10**, 1393–1404.

Aizen, E. M., V. B. Aizen, J. M. Melack, T. Nakamura, and T. Ohta, 2001: Precipitation and atmospheric circulation patterns at mid-latitudes of Asia. *International Journal of Climatology*, **21**, 535–556.

Alijani, B., 1997: *Climate of Iran*. Payam Nour University, Tehran, Iran (in Farsi).

Alijani, B., 2008: Effect of the Zagros Mountains on the spatial distribution of precipitation. *Journal of Mountain Science*, **5**, 218–231.

Alijani, B., J. O'Brien, and B. Yarnal, 2008: Spatial analysis of precipitation intensity and concentration in Iran. *Theoretical and Applied Climatology*, **94**, 107–124.

Aoki, I., Y. Kurosaki, R. Osada, T. Sato, and F. Kimura, 2005: Dust storms generated by mesoscale cold fronts in the Tarim Basin, Northwest China. *Geophysical Research Letters*, **32**, L06807.

Arakawa, H. (ed.), 1969: *Climates of Northern and Eastern Asia*. World Survey of Climatology 8, Elsevier, Amsterdam, 248 pp.

Arakawa, H., and K. Takahashi, 1981: *Climates of Southern and Western Asia*. World Survey of Climatology 9, Elsevier, Amsterdam, 334 pp.

Barlow, M., M. Wheeler, B. Lyon, and H. Cullen, 2005: Modulation of daily precipitation over Southwest Asia by the Madden-Julian Oscillation. *Monthly Weather Review*, **133**, 3579–3594.

Breckle, S.-W., 1983: Temperate deserts and semi-deserts of Afghanistan and Iran. In *Temperate Deserts and Arid Semi-Deserts* (N. E. West, ed.), Ecosystems of the World 5, Elsevier, Amsterdam, pp. 271–319.

Chen, W. N., D. W. Fryrear, and Z. T. Yang, 1999: Dust fall in the Takla Makan Desert of China. *Physical Geography*, **20**, 189–224.

Cohen, J., K. Saito, and D. Entekhabi, 2001: The role of the Siberian High in Northern Hemisphere climate variability. *Geophysical Research Letters*, **28**, 299–302.

Datta, R. K., and C. J. George, 1964: Some climatological and synoptic features of the arid zones of West Rajasthan. In *Proceedings of the Symposium on Problems of Indian Arid Zone, Jodphur*. UNESCO, Paris, pp. 626–630.

Davenport, S., 2008: Floods threaten Afghanistan. *Weather*, **63**, 86.

Dezfuli, A. K., M. Karamouz, and S. Araghinejad, 2009: On the relationship of regional meteorological drought with SOI and NAO over southwest Iran. *Theoretical and Applied Climatology*, **70**, doi:10.1007/s00704–009–0157–2.

Domroes, M., and G. Peng, 1988: *The Climate of China*. Springer, Berlin, 361 pp.

Domroes, M., M. Kaviani, and D. Schaefer, 1998: An analysis of regional and intra-annual precipitation variability over Iran using multivariate statistical methods. *Theoretical and Applied Climatology*, **61**, 151–159.

Elguindi, N., and F. Giorgi, 2006: Simulating multi-decadal variability of Caspian Sea level changes using regional climate model outputs. *Climate Dynamics*, **26**, 167–181.

Evenari, M. 1985: The desert environment. In *Hot Deserts and Arid Shrublands* (M. Evenari, I. Noy-Meir, and D. W. Goodall, eds.), Ecosystems of the World 12A, Elsevier, Amsterdam, pp. 1–22.

Flohn, H., 1947: Zum Klima des freien Atmosphäre über Siberien. II. Die regionale winterliche Inversion. *Meteorologische Rundschau*, **1**, 75–79.

Fukutomi, Y., K. Masuda, and T. Yasunari, 2007: Cyclone activity associated with the interannual seesaw oscillation of summer precipitation over northern Eurasia. *Global and Planetary Change*, **56**, 387–398.

Ghahraman, B., 2008: Investigation of annual rainfall trends in Iran. *Iranian Journal of Agricultural Science and Technology*, **10**, 93–97.

Ghasemi, A. R., and D. Khalili, 2006: The influence of the Arctic Oscillation on winter temperatures in Iran. *Theoretical and Applied Climatology*, **85**, 149–164.

Ghasemi, A. R., and D. Khalili, 2008: The association between regional and global atmospheric patterns and winter precipitation in Iran. *Atmospheric Research*, **88**, 116–133.

Gong, D.-Y., and C.-H. Ho, 2002: The Siberian High and climate change over middle to high latitude Asia. *Theoretical and Applied Climatology*, **72**, 1–9.

Goyal, R. K., 2004: Sensitivity of evapotranspiration to global warming: a case study of arid zone of Rajasthan (India). *Agricultural Water Management*, **69**, 1–11.

Gupta, R. K., 1986: The Thar Desert. In *Hot Deserts and Arid Shrublands* (M. Evenari, I. Noy-Meir, and D. W. Goodall, eds.), Ecosystems of the World 12A, Elsevier, Amsterdam, pp. 55–99.

Hirashima, H., T. Ohata, Y. Kodama, H. Yabuki, N. Sato, and A. Georgiadi, 2004: Nonuniform distribution of tundra snow cover in Eastern Siberia. *Journal of Hydrometeorology*, **5**, 373–389.

Inagamova, S. I., T. M. Mukhtarov, and T. S. Mukhtarov, 2002: *Characteristics of Synoptic Processes of Central Asia*. Central Asian Hydrometeorological Research Institute, Tashkent (in Russian).

Kai, K., and Coauthors, 2008: The structure of the dust layer over the Taklimakan desert during the dust storm in April 2002 as observed using a depolarization Lidar. *Journal of the Meteorological Society of Japan*, **86**, 1–16.

Kim, H. A., Y. Nagata, and K. Kai, 2009: Variation of dust layer height in the northern Taklimakan Desert in April 2002. *Atmospheric Environment*, **43**, 557–567.

Kripalani, R. H., and A. Kulkarni, 1999: Climatology and variability of historical Soviet snow depth data: some new perspectives in snow–Indian monsoon teleconnections. *Climate Dynamics*, **15**, 475–489.

Liu, M., D. L. Westphal, T. R. Holt, and Q. Xu, 2000: Numerical simulation of a low-level jet over complex terrain in southern Iran. *Monthly Weather Review*, **128**, 1309–1327.

Lydolph, P. E., 1977: *Climates of the Soviet Union*. World Survey of Climatology 7, Elsevier, Amsterdam.

Mariotti, A., 2007: How ENSO impacts precipitation in southwest central Asia. *Geophysical Research Letters*, **34**, L16706.

Martyn, D., 1992: *Climates of the World*. Developments in Atmospheric Science 18, Elsevier, Amsterdam, 435 pp.

Meigs, P., 1957: Arid and semiarid climate types of the world. *Proceedings, International Geographical Union, 17th Congress, 8th General Assembly*, Washington, DC, pp. 135–138.

Mikami, M., T. Fujitani, and X. M. Zhang, 1995a: Basic characteristics of meteorological elements and observed local wind circulation in Taklimakan Desert, China. *Journal of the Meteorological Society of Japan*, **733**, 899–908.

Mikami, M., T. Fujitani, and X. M. Zhang, 1995b: Long-term meteorological observation in Taklimakan Desert, China. *Journal of Arid Land Studies*, **4**, 103–117 (in Japanese).

Modarres, R., 2006: Regional precipitation climates of Iran. *Journal of Hydrology*, **45**, 13–27.

Natsagdorj, L., D. Jugder, and Y. S. Chung, 2003: Analysis of dust storms observed in Mongolia during 1937–1999. *Atmospheric Environment*, **37**, 1401–1411.

Nazemosadat, M. J., and I. Cordery, 2000: On the relationships between ENSO and autumn rainfall in Iran. *International Journal of Climatology*, **20**, 47–61.

Orlovsky, L., N. Orlovsky, and A. Durdyev, 2005: Dust storms in Turkmenistan. *Journal of Arid Environments*, **60**, 83–97.

Pagano, T. C., S. Mahani, M. J. Nazemosadat, and S. Sorooshian, 2003: Review of Middle Eastern hydroclimatology and seasonal teleconnections. *Iranian Journal of Science and Technology, Transaction B*, **27**, 1–16.

Pramanik, S. K., S. P. Hariharan, and S. K. Ghosh, 1952: Analysis of the climate of the Rajasthan desert and its extension. *Indian Journal of Meteorology and Geophysics*, **3**, 131–140.

Rahimzadeh, F., A. Asgari, and E. Fattahi, 2009: Variability of extreme temperature and precipitation in Iran during recent decades. *International Journal of Climatology*, **29**, 329–343.

Ramaswamy, C., 1975: Floods in the Desert. In: *Environmental Analysis of the Thar Desert* (R. K. Gupta and I, Prakash, ed.), English Book Depot, Dehradun, pp. 134-137.

Rao, Y. P., 1981: The climate of the Indian subcontinent. In *Climates of Southern and Western Asia* (H. Arakawa and K. Takahashi, eds.), World Survey of Climatology 9, Elsevier, Amsterdam, pp. 67–182.

Raziei, T., I. Bordi, and L. S. Pereira, 2008: A precipitation-based regionalization for Western Iran and regional drought variability. *Hydrology and Earth System Sciences*, **12**, 1309–1321.

Raziei, T., B. Saghafian, A. A. Paulo, L. S. Pereira, and I. Bordi, 2009: Spatial patterns and temporal variability of drought in Western Iran. *Water Resources Management*, **23**, 439–455.

Sato, T., and Coauthors, 2007: Water sources in semiarid northeast Asia as revealed by field observations and isotope transport model. *Journal of Geophysical Research*, **112**, D17112.

Schiemann, R., D. Lüthi, P. L. Vidale, and C. Schär, 2008: The precipitation climate of Central Asia: intercomparison of observational and numerical data sources in a remote semiarid region. *International Journal of Climatology*, **28**, 295–314.

Seino, N., and Coauthors, 2005: Numerical simulation of mesoscale circulations in the Tarim Basin associated with dust events. *Journal of the Meteorological Society of Japan*, **83A**, 205–218.

Sun, B., and P. Y. Groisman, 2000: Cloudiness variations over the former Soviet Union. *International Journal of Climatology*, **20**, 1097–1111.

Syed, F. S., F. Giorgi, J. S. Pal, and M. P. King, 2006: Effect of remote forcings on the winter precipitation of central southwest Asia. Part 1. Observations. *Theoretical and Applied Climatology*, **86**, 147–160.

Takemi, T., and N. Seino, 2005: Dust storms and cyclone tracks over the arid regions in east Asia in spring. *Journal of Geophysical Research–Atmospheres*, **110**, D18S11.

Tsunematsu, N., T. Sato, F. Kimura, K. Kai, Y. Kurosaki, T. Nagai, H. F. Zhou, and M. Mikami, 2005: Extensive dust outbreaks following the morning inversion breakup in the Taklimakan Desert. *Journal of Geophysical Research–Atmospheres*, **110**, D21207.

Uno, I., K. Harada, S. Sataske, Y. Hara, and Z. F. Wang, 2005: Meteorological characteristics and dust distribution of the Tarim Basin simulated by the nesting RAMS/CFORS dust model. *Journal of the Meteorological Society of Japan*, **83A**, 219–239.

Walter, H., and E. O. Box, 1983a: Overview of Eurasian continental deserts and semi-deserts. In *Temperate Deserts and Arid Semi-Deserts* (N. E. West, ed.), Ecosystems of the World 5, Elsevier, Amsterdam, pp. 3–7.

Walter, H., and E. O. Box, 1983b: The Caspian lowland biome. In *Temperate Deserts and Arid Semi-Deserts* (N. E. West, ed.), Ecosystems of the World 5, Elsevier, Amsterdam, pp. 9–42.

Walter, H., and E. O. Box, 1983c: Semi-deserts and deserts of central Kazakhstan. In *Temperate Deserts and Arid Semi-Deserts* (N. E. West, ed.), Ecosystems of the World 5, Elsevier, Amsterdam, pp. 43–78.

Walter, H., and E. O. Box, 1983d: Middle Asian deserts. In *Temperate Deserts and Arid Semi-Deserts* (N. E. West, ed.), Ecosystems of the World 5, Elsevier, Amsterdam, pp. 79–104.

Walter, H., and E. O. Box, 1983e: The Karakum Desert, an example of a well-studied EU-biome. In *Temperate Deserts and Arid Semi-Deserts* (N. E. West, ed.), Ecosystems of the World 5, Elsevier, Amsterdam, pp. 105–159.

Walter, H., and E. O. Box, 1983f: The orobiomes of Middle Asia. In *Temperate Deserts and Arid Semi-Deserts* (N. E. West, ed.), Ecosystems of the World 5, Elsevier, Amsterdam, pp. 161–191.

Walter, H., and E. O. Box, 1983g: The Pamir: an ecologically well-studied high-mountain desert biome. In *Temperate Deserts and Arid Semi-Deserts* (N. E. West, ed.), Ecosystems of the World 5, Elsevier, Amsterdam, pp. 237–269.

Walter, H., E. O. Box, and W. Hilbig, 1983: The deserts of Central Asia. In *Temperate Deserts and Arid Semi-Deserts* (N. E. West, ed.), Ecosystems of the World 5, Elsevier, Amsterdam, pp. 193–236.

Wang, W. C., and K. R. Li, 1990: Precipitation fluctuation over semiarid region in northern China and the relationship with El-Niño Southern Oscillation. *Journal of Climate*, **3**, 769–783.

Warner, T. T., 2004: *Desert Meteorology*. Cambridge University Press, Cambridge, 595 pp.

Wu, G. X., and Coauthors, 2007: The influence of mechanical and thermal forcing by the Tibetan plateau on Asian climate. *Journal of Hydrometeorology*, **8**, 770–789.

Yang, Q., W. S. Wei, and J. Li, 2008: Temporal and spatial variation of atmospheric water vapor in the Taklimakan Desert and its surrounding areas. *Chinese Science Bulletin*, **53**, 71–78.

Yatagai, A., and T. Yasunari, 1995: Interannual variations of summer precipitation in the arid/semi-arid regions in China and Mongolia: their regionality and relation to the Asian summer monsoon. *Journal of the Meteorological Society of Japan*, **73**, 909–923.

Ye, H., 2001: Characteristics of winter precipitation variations over northern central Eurasia and their connections to sea surface temperatures over the Atlantic and Pacific Oceans. *Journal of Climate*, **14**, 3140–3155.

Yihui, D., and W. Zunya, 2008: A study of rainy seasons in China. *Meteorology and Atmospheric Physics*, **100**, 121–138.

Zeng, B., and T.-B. Yang, 2008: Impacts of climate warming on vegetation in Qaidam Area from 1990 to 2003. *Environmental Monitoring and Assessment*, **144**, 403–417.

Zhang, Q., L. Song, R. Huang, G. Wei, S. Wang, and H. Tian, 2003: Characteristics of hydrologic transfer between soil and atmosphere over Gobi near Oasis at the end of summer. *Advances in Atmospheric Sciences*, **20**, 442–452.

Zu, R., X. Xue, M. Qiang, B. Yang, J. Qu, and K. Zhang, 2008: Characteristics of near-surface wind regimes in the Taklimakan Desert, China. *Geomorphology*, **96**, 39–47.

20 Coastal deserts

20.1 ORIGIN AND LOCATION

A common location of deserts is the western littoral of continents in the vicinity of the subtropical high. Coastal deserts are particularly well developed along the subtropical west coasts of South America and Africa, less so in the United States and Australia (Fig. 20.1). The desert along the Somali Coast of eastern Africa and the Danakil Desert to the north of it exhibit some of the same characteristics but these deserts are not linked to any of the subtropical high-pressure cells. The major coastal deserts, also called fog deserts, have similar causes of aridity. They also have certain characteristics that distinguish them from other desert regions. These include not only the high frequency of fogs, but also moderate temperatures, limited annual temperature range, moderate diurnal range, and high relative humidity, compared with inland deserts.

The fogs are an important part of the coastal ecosystem in these deserts (Gutiérrez *et al.* 2008) and vegetation patterns are very sensitive to their distribution. Dense fog "oases" are common in various niches in the hyper-arid Peruvian-Atacama Desert (Cereceda *et al.* 2008a), but they are found in many other deserts. In many locations, the fog actually sustains forests. Examples are the cloud forests in eastern Mexico (Vogelmann 1973), relict forests in Oman (Hildebrant and Eltahir 2006), and forest niches in north central Chile (del-Val *et al.* 2006). In some cases, fog is the dominant or sole moisture source. The fogs provide not only moisture, but also shading, facilitating vegetation survival through periods of drought (Fischer *et al.* 2009). The shading, a result of the low clouds producing the fog, may be more important than the accompanying moisture (Williams *et al.* 2008).

The classical explanation for the origin of the coastal deserts includes three factors:

1. the aridifying influence of the subtropical highs (subsidence, divergence and temperature inversions);
2. the cold, upwelled, coastal current produced by air circulation on the eastern flank of the high;
3. the frictionally induced divergence produced when winds blow parallel to the coast, due to the difference in friction between the land and the water (the stress-differential mechanism described in Section 5.5).

All three factors are present in the coastal deserts of Namibia and Mauritania (western coasts of Africa), Peru and Chile (South America), and Baja California (North America). Additional dynamical factors come into play when a north–south mountain barrier lies just inland of the coastal desert, a situation that exists in all of the cases except Mauritania. These include coastal jets (e.g., Lettau 1978a; Douglas 1995; Parish 2000; Muñoz and Garreaud 2005; Nicholson 2010; Monaghan *et al.* 2010), the blocking effect of the mountains, and rain shadow effects (Houston and Hartley 2003).

Figure 20.2 shows the rainfall gradient as a function of latitude along the five major desert coasts. The extent and degree of aridity are highly variable. The Peruvian-Atacama Desert is by far the most arid and the most extensive, the Australian the least so. In all but the Australian coastal desert, rainfall approaches zero at some locations and the greatest aridity is along the coast itself rather than inland. Within each of these deserts lies the transition between the tropical summer rains and extra-tropical winter rains, associated with the equatorward displacement of the westerlies and mid-latitude cyclones. The arid cores lie in between.

Curiously, the poleward extents of these deserts are relatively invariant, at about 24°–26° of latitude, while large differences are apparent in the position of their equatorward margins (Fig. 20.2). This suggests that either the aridifying influence of the subtropical high is more readily overridden by the mid-latitude cyclones bearing the winter rains than by summer convective systems, or that the poleward extent is more controlled by global dynamics (e.g., hemispheric temperature gradients, position of the Hadley cells) than is the tropical margin, with more localized convective systems.

The absence of a well-developed coastal desert in western Australia and the contrast with the Peruvian-Atacama Desert demonstrate the importance of the three "classical" factors in aridity. The Peruvian-Atacama Desert is the driest region in the world, with vast stretches being virtually rainless. It is bordered by the most intense upwelling in the world. Also, the coastline parallels the prevailing winds over some 30° of latitude. Over Australia there is no quasi-permanent subtropical high, but rather transient anticyclones, and consequently there is no well-developed cold coastal current and little upwelling. In fact, coastal water temperatures in the region are actually higher than further seaward. Likewise, the prevailing winds do not blow parallel to the coast, so that frictionally induced divergence is not present.

In this chapter the four major coastal deserts are treated in detail. Emphasis is on regionally specific characteristics; for the large-scale context in the framework of the general atmospheric circulation, the reader is referred to the chapters on the individual continents. Much of the information presented here is a synthesis of many sources or is based on an analysis of meteorological data at individual stations. Some of the most complete information on coastal deserts comes from older, classical sources such as Meigs (1966), Amiran and Wilson (1973), Johnson (1976), and Miller (1976). More recent studies provide

greater detail but are often focused on relatively small areas (e.g., the very comprehensive work of Garreaud and colleagues in the Fray Jorge relict forests of Chile), often in the context of studies of coastal fog. Unfortunately the classical sources occasionally provide information that is contradicted by newer studies. This is especially true of studies of coastal climates of South America, where there is tremendous spatial variability of meteorological phenomena. Here, an attempt is made to reconcile the various studies, while providing an appropriate overview of the coastal meteorology.

20.2 THE ROLE OF COASTAL JETS

Low-level jet streams parallel to the coast are an important factor in the development of several coastal deserts. Their influence results from two factors, the upwelling of cold water and the development of frictionally induced divergence along the coast. A coastal jet generally results from the thermal contrast between land and water, but a positive feedback exists such that the resultant upwelling enhances the jet.

The upwelling occurs because of Ekman transport of near-surface water to the left of the wind in the Southern Hemisphere and to the right of the wind in the Northern Hemisphere (see Chapter 4). When the wind blows such that the Ekman transport is away from the coast, the transported water is replaced with cold water from below. The divergence results because of lower friction over the water, and higher friction over the land (see Chapter 5) and it both produces subsidence and suppresses ascent. Both effects are enhanced when high terrain parallels the coast because the frictional contrast is enhanced, as is the temperature contrast between the land and water that produces the jet stream in the first place.

Lettau (1978a) pointed out that the Peruvian-Atacama Desert of South America clearly stands out from the others by virtue of its extreme aridity and latitudinal extent and by the abrupt transition to humid climates on its equatorward side. He put forth a complex theory, based on the development of a coastal jet, which appears to account for the unique aspects of this desert. In this theory, the north–south mountain barrier is of critical importance and, along with the land–water thermal contrast, plays a role in the development of the coastal jet. Accordingly,

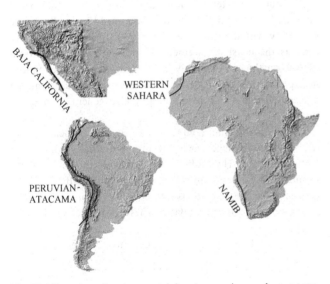

Fig. 20.1 Location of major coastal deserts, superimposed upon topographic features of the continents.

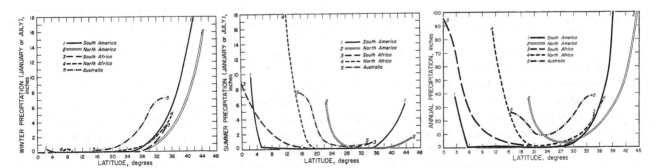

Fig. 20.2 Rainfall (mm) as a function of latitude in the major coastal deserts. Left to right: winter, summer, annual rainfall (from Lydolph 1973).

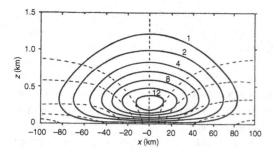

Fig. 20.3 Schematic of the low-level jet stream over the Pampa de la Joya, Peru (based on Lettau 1978a). Isotachs in m/s are indicated. The *x*-value of 0 indicates the coastline. The *y*-axis is height above ground.

there is a diurnal "tide" of air between the highlands and the coasts as the temperature gradients between them reverse from day to night, a combination of land–sea breezes and mountain–valley winds. Lettau calls this a thermo-tidal wind and suggests that the induced pressure gradients combined with Coriolis effects produce a low-level jet stream parallel to the coast. The Andes limit its inland extent.

The core of the Peruvian jet has a mean velocity on the order of 6 m/s and generally lies about 100–200 m above ground (Fig. 20.3). Coastal aridity is enhanced by a feedback between the jet and the coastal current: the stronger the jet, the stronger the winds parallel to the shore and the stronger (and colder) the upwelling, further intensifying the jet. The jet is associated with vertical circulations which, depending on latitude and location with respect to the jet core, produce subsidence or ascent, suppressing or enhancing rainfall, respectively.

The theory properly predicts the daytime jets observed along the Somali and South American coastal deserts and their aridifying influences. It accounts as well for the nighttime low-level jet in the Great Plains and the associated nocturnal rains (see Chapter 14). A similar low-level jet exists along the coast of Namibia (Nicholson 2010), where "thermo-tidal" winds like those in Peru also exist (Lindesay and Tyson 1990), and may serve to enhance the aridity of the Namib. Low-level coastal jets also exist in the Gulf of California near arid Baja California (Anderson *et al.* 2001), and along the Pacific coast of southern California (Parish 2000). In both locations, the flow is constrained by mountains extending roughly north–south near the coast.

20.3 GENERAL CLIMATIC CHARACTERISTICS

The most common characteristics of the coastal deserts are frequent fog and stratus cloud, a temperature inversion, cold water along the shore, and subsidence associated with the quasi-permanent presence of a subtropical high-pressure cell. The latter three features all contribute to the coastal aridity. These deserts are relatively humid but thermally moderate compared with interior deserts. Most lie westward of expansive highlands or mountain ranges. As a result, ephemeral highland streams that are prone to occasional flooding traverse the coastal deserts.

The contrast with interior deserts is striking. Morning relative humidity can be on the order of 80–90%, dropping to 60–70% in the afternoon. In the interior deserts, the humidity is strongly dependent on the season, with morning/afternoon relative humidity on the order of 40%/25% in the warmest months, but 65%/40% in the coolest months. The annual and diurnal temperature range in the coastal deserts is comparatively small, in stark contrast to the thermal extremes found in interior deserts. The annual range is typically about 5°–6°C very near the coast, compared with over 20°C in interior deserts. The diurnal range is quite variable and it increases rapidly inland, but is typically on the order of 5°–10°C very near the coast. By comparison, it is on the order of 20°–25°C in interior deserts, but the absolute range on an individual day can exceed 40°C.

In the coastal deserts, climatic characteristics are a function of distance from the coast and topography. Isotherms and isohyets tend to parallel the coast, with conditions being relatively uniform within the desert. Gradients are strong from the coast inland. The characteristics that distinguish coastal deserts from inland deserts are usually confined to relatively short distances from the coast, generally 50–150 km.

The coastal deserts represent the transition from tropical to mid-latitude climates, summer rainfall on their equatorward flank and winter rainfall on their poleward flank. The degree of aridity within them is quite variable. In less arid regions, such as the coast of southern California, the configuration of the coastline can strongly influence rainfall. It is enhanced near peninsulas, suppressed where the coast is concave and sea breezes diverge. Coastal topography also generally enhances rainfall.

The prevailing wind regime is generally that associated with the subtropical highs, having a strong southerly component in the Southern Hemisphere, but northerly in the Northern Hemisphere. Since the pressure gradients are usually weak, the land- and sea-breeze systems may be well developed, especially if topographic relief lies inland. These and other local winds can be very important.

The coastal deserts owe their moderate climate not only to the influence of the cold water, but also to the frequent occurrence of stratus cloud. The cloud deck further suppresses extreme fluctuations in temperature between day and night or during the course of the year. The annual and seasonal distributions of marine stratus (including fog) show three pronounced maxima along the western littoral of South America, North America, and southern Africa (Fig. 20.4). Stratus also characterizes the west coasts of North Africa and Australia, but the occurrence is much lower (Fig. 20.5). Maximum cloud cover occurs in July (i.e., the cold season) along the two Northern Hemisphere cold-water coasts and primarily in the cold season along the Southern Hemisphere coasts. The reasons for this are not clear. The contrasts between the regions are striking, with the cloudiness maximum approaching 75% along the Peruvian and Namibian coasts, 70% along the California coast, but only 45% near Australia and 35% near the Canary Islands along the North African coast.

Annual Stratus Cloud Amount

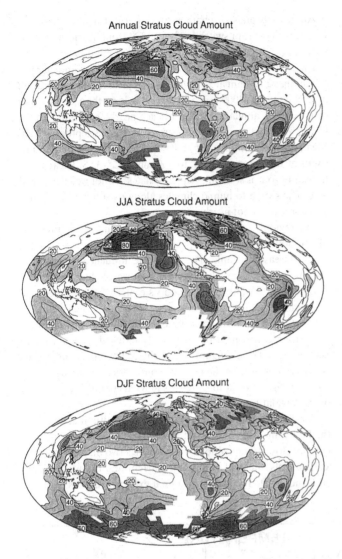

Fig. 20.4 Global distribution of (top) annual, (center) summer, and (bottom) winter frequency of marine stratus cloud (including fog) (from Klein and Hartmann 1993).

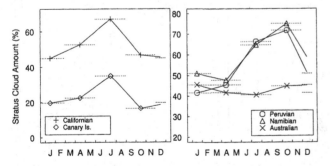

Fig. 20.5 Seasonal march of stratus cloud amount (%) for five coastal deserts (from Klein and Hartmann 1993).

Another notable characteristic is that the cloudiness maximum is generally some distance offshore. This is a result of the more stable conditions over the coldest water, i.e., adjacent to the coast. Stability increases the amount of marine stratus, probably

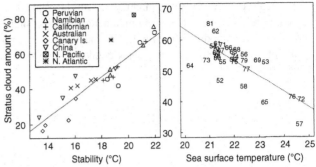

Fig. 20.6 Relationship of stratus cloud amount at coastal deserts to sea-surface temperature and low-level static stability (from Klein and Hartmann 1993).

because the inversion layers that produce the stable conditions trap moisture below the inversion (Klein and Hartmann 1993) (Fig. 20.6). Surface temperature is inversely proportional to stratus occurrence. The offshore maximum is probably a result of the higher temperatures at a distance from the coast, where the upwelling is strongest. The presence of the stratus possibly creates a radiative feedback that enhances the inversion layer.

In most of the coastal deserts, fog-water is a significant source of moisture and is critical to biological processes. There is some evidence that, like the marine stratus, the fog is most intense at some distance offshore. However, except for southern California, the frequency of fogs has not been well examined. Moreover, many authors fail to distinguish between fog and stratus (see Leipper 1994). The reason is, in part, that the distinction is a matter of perspective: fog is at ground level, stratus aloft, and the coastal stratus deck becomes fog at higher elevations. Here an attempt will be made to use fog to refer only to those occurrences in the coastal lowlands. Fog frequency is highly variable in space and seasonally; also, the inland and seaward penetration of fog is quite diverse in the various deserts.

20.4 THE PERUVIAN-ATACAMA DESERT

20.4.1 CLIMATIC OVERVIEW

The most extreme example of a coastal desert, the Peruvian-Atacama, lies along the western littoral of South America. The region has perhaps the longest history of arid conditions of any desert on earth (Clarke 2006), having existed since at least late Triassic times. The Peruvian-Atacama Desert extends from about ½° S to about 30° S, but only some 80–140 km inland, bordered on the east by the high Andean cordillera and on the west by the Pacific and the cold Humboldt or Peru Current (Fig. 20.7). Mean annual rainfall (Fig. 20.8) is less than 50 mm from about 5° to 30° S, with many stations recording no rain for one or two decades at a time. Within the belt from 18° S to 23° S, the coast is nearly rainless.

This desert possesses the typical features of the coastal deserts described in Section 20.3. It has one additional feature: the occasional occurrence of extreme precipitation events.

As much as 10 or 20 years might pass without rain, then torrential rains of several hundred millimeters during a single day will create a flooding situation. In Calama, Chile, where no rainfall had ever been recorded, a thick blanket of snow fell in June 1911.

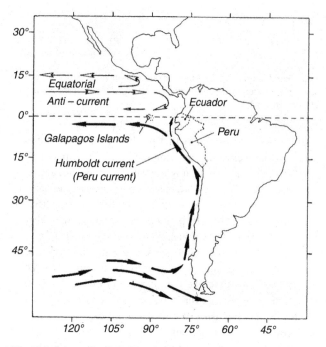

Fig. 20.7 Schematic of the Humboldt Current and coastal topography (partly based on Rauh 1985).

An equally unusual characteristic of the Peruvian-Atacama Desert is the abrupt switch at about 1° N from humid to dry climates and a dramatic change in vegetation (Lettau 1978a). To the north, rains and squalls occur all year round. Mean annual rainfall reaches 2000 mm at 2° N and lush tropical rainforests prevail; at 3° N it reaches 4000 mm. (At this latitude on the eastern slopes of the Andes, annual rainfall can reach nearly 9000 mm.) To the south, in a brief transition zone, rainfall decreases progressively. Mean annual rainfall is about 400 mm at the desert margin just south of the equator, falling to 200 mm at 2°–3° S. Because the transition zone is so small, steppe climates are not well developed here. The switch to humid climates along its southern margin is also abrupt, but not nearly as extreme.

The region's aridity is largely a result of the steady presence of the South Pacific subtropical high and the southerly winds along its eastern flank. The aridifying influence of the high also includes a generally divergent pattern of flow and a pronounced trade wind inversion on its eastern flank. The southerly winds produce the upwelling of cold water in the Humboldt Current. Blowing at the same time parallel to high terrain, there is a strong component of frictionally induced divergence (see Section 5.5). The configuration of the coast and the highlands is also important, as their north–south orientation over some 20 to 30 degrees of latitude results in winds parallel to the coast over a vast latitudinal extent.

The extreme aridity and unusually large latitudinal extent of the Peruvian-Atacama Desert are a result of these characteristics and close juxtaposition of the region's two major geographical controls, the cold Humboldt Current and the Andes (Figs. 20.1 and 20.7). The large latitudinal extent of the current, and

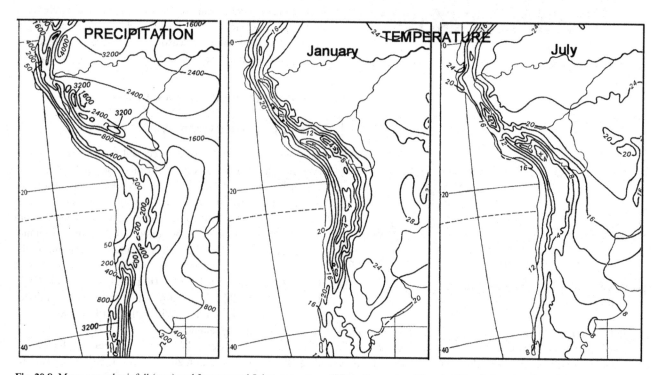

Fig. 20.8 Mean annual rainfall (mm) and January and July temperatures (°C) in the western littoral of South America (from Martyn 1991).

its continuation to near the equator, are unusual compared with the coastal currents of other continents. Offshore water temperatures in the area between 30° S and 2° S range from 14° to 20°C in winter, with temperatures increasing from south to north. In summer, water temperatures are 18°–25°C. These are perhaps the coldest water temperatures in these low latitudes anywhere. Just inland and parallel to the coast are the coastal cordilleras and, further east, the high Andes, with peaks reaching over 6000 m only some 100 km from the coast. This is an effective barrier to the humid air and weather systems to the east; it also dramatically limits the inland penetration of the maritime influence.

The climatic controls are thus the coastal maritime influence, the cordillera, and the large-scale general atmospheric circulation, with the last factor imposing the seasonality of precipitation. The coastal influence controls both rainfall and temperature, so that both rainfall isohyets and isotherms run parallel to the coast, i.e., roughly north to south (Fig. 20.8). Rainfall increases rapidly with distance from the coast. As with other coastal deserts, temperatures in most locations are relatively low all year round and the diurnal and annual ranges are relatively small compared with inland deserts at the same latitude. Because of the pervasive influence of the cold water, the north–south temperature gradient (Fig. 20.8) is relatively small, with mean monthly temperatures changing by only about 2°–4°C over coastal stretches spanning some 20° of latitude.

Temperatures are moderated not only by the cool water offshore, but also by the frequent presence of fog or low stratus. Within the desert, contrasts in temperature are related to insolation, so that the cloudy tropical and subtropical latitudes may be colder in winter than the extra-tropical latitudes on the southern margin of the desert. The occurrence of fog is strongly controlled by coastal topography and the configuration of the shoreline. These factors vary remarkably over short distances. As a consequence, over much of the coast, barren dunes and sands are interspersed with sporadic pockets of closed-cover vegetation where the fogs frequently occur. These are termed "fog oases" and life in these regions is adapted to fog as its main source of water.

The wind regime in the coastal desert is quite steady, owing to the presence of the subtropical high. Prevailing winds all along the coast are southerly or southwesterly throughout the year, although the winds become light and somewhat variable in the regions of lowest pressure gradient. In northern regions, occasional westerly wind events occur, generally bringing rainfall (Douglas et al. 2009). A noteworthy phenomenon in this desert is the strength and regularity of the diurnal land–sea breeze system. The latter frequently reaches almost gale force by mid-afternoon. This has important effects on local temperatures, cloud formation, fog, and even precipitation. In the interior deserts and highlands the wind regime is considerably more complex and dependent on local and regional circulation systems, such as sea breezes.

Two low-level coastal jets also impact the desert. That over Peru was described in Section 20.2. A more recently identified jet prevails along the Chilean coast, from ~26° S to 36° S, within the persistent southerly winds of the South Pacific high (Muñoz and Garreaud 2005). This is evident 60% of the time in spring. The jet's axis lies some 150 km offshore; its east–west extent is on the order of 500 km. Mean core speeds are roughly 10 m/s. The jet lies at the top of the marine boundary layer, within the large-scale subsidence of the inversion layers. Its climatological impact is not completely clear, but it likely enhances aridity in some sectors or may enhance it downstream of the core, where a mesoscale area of mean upward motion exists.

20.4.2 THE MOISTURE REGIME

The dry climates of the Peruvian-Atacama Desert are limited to the coast, the low levels of the western slopes of the cordillera, and the intermontane valleys. Everywhere rainfall increases rapidly inland, enhanced in part by the influence of the Andes. In the northern desert, within 500 km of the coast, rainfall increases to 2000 mm/year or more. In the southern desert, it increases inland to only about 200–300 mm/year. The coefficient of variation (CV), like the rainfall, is a function of distance from the coast. It ranges from more than 100% right at the coast to roughly 20–25% in the highlands.

The factors bringing rainfall include the Intertropical Convergence Zone (ITCZ) and westerly waves from the Pacific on the equatorward margin, excursions of humid air from the Amazon Basin to the east, and protrusions of the polar front and westerly waves on the poleward margin of the desert. The tropical ITCZ rains to the north of the coastal desert occur in the boreal summer, but their influence extends into the desert in the boreal spring/austral autumn, when the ITCZ is displaced equatorward. Along the desert's equatorward margin, maximum rainfall occurs in March–May. The polar-front rains of the poleward margin impact the region almost exclusively in winter. The southern limit of the desert, at roughly 30° S, represents the usual equatorward limit of the penetration of winter frontal systems (Houston and Hartley 2003). The arid core of the coastal desert is rarely influenced by any of these systems. That is the region of frequent fogs, which provide a greater moisture source than rainfall.

The trade inversion has is a major control on the aridity and on the occurrence of fog and stratus clouds (Fig. 20.9). It is present all year round, but is relatively weak in summer, when its base is somewhat lower and the temperature increase across the inversion layer is only a few degrees C. In winter, the inversion base is about 700–800 m in the north near Lima. Its depth is 400–1000 m and the temperature increases about 9°C across the inversion (Schwerdtfeger 1976; Rauh 1985; Rutllant 1994). Commensurate changes occur in the marine stratus deck and fog layer (McKay et al. 2003).

Further south, in southern Peru and northern Chile, both the base of the stratus and its upper limit rise progressively. Thus, the desert moves further upslope in the higher latitudes. In northern Chile, where the inversion base is 800–900 m in winter, the stratus base can be 500–800 m and the deck less than 200 m thick. Its base can be as low as 200 m in central Chile, around

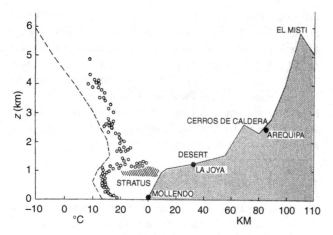

Fig. 20.9 Schematic of coastal topography, temperature inversions, and stratus cloud along the Peruvian coast (based on Lettau 1978b). Dashed line indicates radiosonde measurement; circles indicate daytime temperature measurements along highways.

the southern margin of the desert. There, coastal cloudiness is on the order of 60–65% year-round. The coastal cloudiness diminishes rapidly southward from there, with summer cloudiness being as low as 40% at ~34° S.

The fog occurs most frequently in the form of a low and uniform stratus deck with a variable base that averages between 350 and 550 m ASL (Trewartha 1970). Its formation is related to the combined effects of subsidence, linked to the subtropical high, and the cold waters of the Humboldt Current. The top of the fog layer commences at approximately the base of the inversion. Turbulence and a warm land surface usually keep it from penetrating to the ground or reaching the coast. Where the land is more elevated, the fog is commonly at ground level, especially when the exposure of the terrain is south or southwest. Thus, the slopes of the coastal range report more fog than the immediate coastal lowlands.

The stratus and fogs have a pronounced seasonality. In summer, the stratus deck is higher and less persistent, often breaking up during the day. Consequently, the fogs are also infrequent in summer. Thus, the whole coast tends to be sunny in summer and the vegetation disappears, giving the coastal desert a very dry appearance. In winter, the deck is thicker, lower, and more continuous, especially at night. In August and September, the stratus deck may persist for weeks on end.

The stratus is particularly low and extraordinarily persistent in central Peru. At Lima, 90% of all days were overcast in the winter of 1967. There the base is usually at 150–300 m, but fog is rather infrequent along the coast at levels below 100 m. It has a distinct upper boundary ranging from about 500–700 m, so that the slopes above 800–1000 m experience no fogs and are extraordinarily dry and nearly devoid of vegetation. Here the rock and debris deserts begin. The fogs creep inland some 30–40 km, not extending beyond the coastal range in Peru. The fog layer is more persistent but weaker during the summer months and somewhat more condensed and shallower in the winter months.

Rain rarely falls from the stratus fog, but where conditions are favorable, such as where the coastal terrain is elevated, a fine, penetrating drizzle, called *garua*, will precipitate from the fog deck. The base of the fog rises rapidly inland so that drizzle is rarely experienced more than 20 km from the coast. *Garua* is common in Peru, but it is localized and regionalized, being more common on slopes of the coastal range than immediately along the coast. It is rare or absent in coastal northern Chile; the few occurrences are lighter than in central Peru.

Garua, like the fogs, is most common in winter and is often initiated by the passage of a cold front from the Antarctic. An elevation of the stratus deck and development of small cumuliform clouds often accompany a frontal passage. The *garuas* disappear in November or early December; along with them, the vegetation cover fades and disappears. They commence again the following May. At Lima, Peru, *garuas* occur on about 25% of the days in winter, but bring only a few millimeters of moisture. Although they provide more water than the rare rains, the fine drizzle from the *garuas* seldom wets the soil below 1 or 2 cm. Where the fogs intercept the slopes, *garuas* can bring several hundred millimeters, particularly on those slopes oriented S or SW.

20.4.3 EL NIÑO

In the northern sector of the desert, El Niño largely controls the interannual variability of rainfall. As the upwelling ceases, coastal waters warm, and torrential rainfall rapidly replaces the hyper-arid conditions. Most El Niño events are associated with dramatically increased rainfall at both the northern and southern margins of the desert (Garreaud *et al.* 2008; Douglas *et al.* 2009), but it has little impact in the extreme arid core of the desert (McKay *et al.* 2003). Flow in the coastal streams is also affected, but in a complex manner (Houston 2006). ENSO also affects the marine stratus deck and associated fogs, with both increasing during La Niña events and decreasing during El Niños (Garreaud *et al.* 2008).

In the summer rainfall region of northern Peru and Ecuador, on the northern periphery of the coastal desert, El Niño tends to modulate rainfall in March and April. There it has minimal impact on coastal rainfall, and maximum impact on the coastal plain between the shoreline and the highlands (Garreaud *et al.* 2008). The impact of the El Niño event begins in the north, drifting slowly southward, so that the duration of warm water temperatures and high rainfall is a function of latitude (Fig. 20.10). Likewise the increase in rainfall is a function of latitude, and the most arid regions experience rainfall that is 6–10 times the normal. In the winter rainfall region on the southern periphery of the desert, El Niño's impact is more moderate. At La Serena (~30° S), with winter rainfall, annual totals are on the order of 200 mm during most El Niños, compared with 75 mm or less during La Niñas (Garreaud *et al.* 2008).

The El Niño rains of March 1925 were particularly spectacular. That month Puerto Chicama received 96 mm, compared

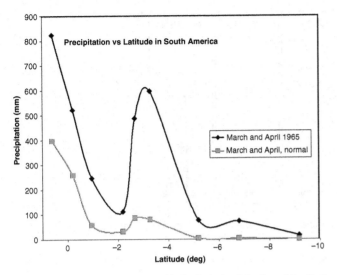

Fig. 20.10 Rainfall as a function of latitude during March and April of a typical El Niño event (1965), compared with the long-term mean for those two months (based on data from Schwerdtfeger 1976).

with an annual average of 4 mm (Cornejo 1970). At Trujillo, where the mean annual rainfall is about 30 mm, the El Niño brought over 390 mm, most of it during three days (Murphy 1926). At the same time Lima, Peru, where the mean annual rainfall is about 50 mm, received 1500 mm of rain (Meigs 1966). Similar rains in 1971 and 1983 left whole streets there and in Piura submerged for days (Rauh 1985).

Probably nowhere else in the world does the El Niño evoke such a dramatic effect. In fact, our knowledge of the phenomenon arose here, where its appearance has been known for centuries. An early El Niño that struck around the year 1100 AD destroyed the irrigation system that sustained the Chimu civilization in the Moche Valley of Peru (Moseley 1978). Only in recent decades have the global manifestations of El Niño been realized.

20.4.4 GEOGRAPHIC REGIONS OF THE DESERT

The precipitation regime can be used to distinguish three divisions of the coastal desert: a northern sector (northern Peru, southern Ecuador), with predominantly tropical summer rainfall; the central fog desert (central and southern Peru); and the southern sector (northern Chile), with mainly winter precipitation (Fig. 20.11). Climatic parameters for typical stations are shown in Fig. 20.12. These three sectors of the coastal desert correspond to vastly different vegetation provinces. These are described in Section 20.4.5. The distribution of drylands in the highlands is more complex, being strongly dependent on a combination of coastal characteristics and topography.

NORTHERN SECTOR

This sector of the coastal desert, extending from roughly ½° S to 8° S, is distinguished from the rest primarily by the absence of fog. To the north of this region rainfall occurs mainly in March

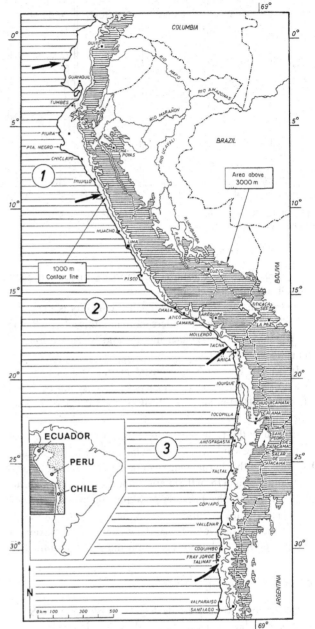

Fig. 20.11 Regionalization of the coastal desert in relationship to the nature and seasonality of precipitation: (1) northern "tree desert" of southern Ecuador and northern Peru, (2) the garua-loma or fog desert of central and southern Peru, (3) the southern desert of northern Chile (from Rauh 1985).

through May and it is associated with the ITCZ. Thunderstorms occur frequently in the northernmost latitudes, but they are not always accompanied by precipitation. Piura, at 5° S, roughly represents the southward limit of penetration of the ITCZ rains. Here mean annual rainfall is only about 25 mm and further south the coast becomes extremely arid. Rainfall is intense, episodic, and generally linked by westerly wind events replacing the usual southwesterly flow (Douglas *et al.* 2009). These westerlies might enhance rainfall by promoting orographic lifting,

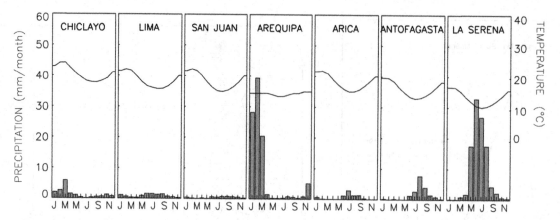

Fig. 20.12 Precipitation (mm) and temperature (°C) at seven typical arid and semi-arid stations: Chiclayo, Lima, San Juan, Arequipa, Arica, Antofagasta, and La Serena.

but tropospheric wave disturbances probably trigger convection as well (Liebmann *et al.* 2009).

Despite the absence of fogs, relative humidity is high year-round. Typical is Chiclayo, where mean monthly relative humidity is 73–75% in the summer months and 75–80% during the remainder of the year.

In winter, the persistent coastal marine stratus deck (see below) reaches its northernmost position, residing over Peru (Klein and Hartmann 1993). Broken cloud decks or overcast skies prevail.

In summer this region contrasts sharply with the central desert to the south. Clear skies are more common here, so temperatures are high and the diurnal range can be extreme. Sometimes dust and sand blown in from the interior produce dense haze. During the warmest months, February or March, daily minimum temperatures average about 20°C, maxima about 26°–28°C. The winter months have on average daily minima and maxima of about 1° and 25°C. August is generally the coldest month.

CENTRAL FOG DESERT

An abrupt transition to the fog desert occurs at about 8° S. The shift is probably a consequence of the cold Humboldt Current veering westward away from the shore north of 8° S. From there to 18° S, fog and low stratus cloud are common, particularly in winter and at specific altitudes; consequently temperatures are low and the diurnal and annual ranges are small. At Lima, the amount of cloud cover ranges from 40% in February–April to 90% in June–September. Rain is exceedingly rare, with summer showers occurring only every few years.

Mean annual temperature is on the order of 18°–21°C and is quite uniform throughout this sector of coastal desert. The annual range is 13°–14°C in the north, 11°C in the south. July or August is usually the coldest month, February the warmest month. Diurnal minima in winter decrease from about 15° to 13°C from the northern to southern extreme, while maxima decrease from about 2° to 18°C along this stretch. The diurnal range is only 4°–6°C. The fogs help to minimize this range. The diurnal range

is larger in summer, when the fogs become infrequent. Maximum temperatures are on the order of 26°–28°C, minima on the order of 18°C, so the diurnal range is about 8°–10°C. Further inland, away from the influence of the fogs, temperatures are more extreme and the annual and diurnal ranges are larger.

Relatively humidity is high year-round. It tends to be somewhat lower in summer, when monthly means range from about 83% at Lima in the north to 75% at San Juan in the south. In winter, highest monthly means range from about 79% to 87%.

SOUTHERN SECTOR

The southernmost sector of coastal desert extends from northern Chile southward to about 30° S. As elsewhere, stratus fogs are common. Here the fogs are called *camanchacas*. Their base is higher, they are thinner, and they occur less frequently than in regions further north. The fog zone extends from 650 to 1200 m ASL and yields some 1–7 liters of water per square meter each day (Cereceda *et al.* 2008b). Fogs also occur inland on the slopes, but they are rare at the coast, their frequency generally not exceeding one or two per year. The associated *garua* are rare or absent throughout this sector. This characteristic, together with a concentration of the meager precipitation in the winter season, distinguishes this sector from the two further north. Far to the south the frequency of fogs increases toward the mid-latitudes, reaching 26 per year at La Serena, near the southern boundary of the desert.

Cloudiness in the coastal desert is moderately high, with a minimum in summer (generally February or March) and a maximum in winter (generally August or September). In the north, at Arica (18° S) and Iquique (20° S), it is about 30–45% increasing to 65–75% in the winter months of June–September. At Antofagasta (23° S) it ranges from about 25% in summer to 55% in winter. Further south at La Serena (30° S) it is relatively persistent throughout the year at about 50–65%. Humidity is generally high, with relatively little annual variation. Unlike cloudiness, it generally increases from north to south. At Arica it ranges from 68% to 80%, at La Serena from 78% to 81%.

Dew occurs frequently, especially during the night, when relative humidity is greatest (McKay *et al.* 2003).

In this sector the marine stratocumulus deck extends from the coast to ~100° W. Further westward, towards Easter Island, the clouds shift to patchy trade cumulus. This cloud deck attains its maximum cover in the austral spring (September–November), when temperatures near the coast reach a minimum. During October the stratocumulus deck is present 60% of the time (Garreaud *et al.* 2001). The persistent subsidence inversion associated with the cloud deck is highly sensitive to transient synoptic-scale disturbances.

The development of fog in this sector is dependent not only on synoptic situation but also on local thermal and mechanical effects, so that the distribution is very complex (Cereceda and Schemenauer 1991; Cereceda *et al.* 2002). Coastal fog results when the coastal topography intercepts the marine stratus deck. Consequently, the fogs are prevalent at isolated locations (the "fog oases") (see Section 11.5.1). When coastal topography is low enough that it underlies the cloud base, fog and vegetation patterns are interrupted (Garreaud *et al.* 2008). Since the low-level flow is mostly parallel to the coast, orographic uplift plays, at most, a secondary role in their development.

Fogs occur least frequently in winter, when the stratus deck reaches its northernmost position (Klein and Hartmann 1993). The frequency increases in spring and reaches a maximum from October to December, when the stratus deck reaches its southward extreme. Observations in the Fray Jorge relict forests in semi-arid Chile (~30° S) show an average of some 15 days per month from October to December, compared with 4–6 fog days per month during summer (Garreaud *et al.* 2008). Daytime heating in summer also increases the possibility of fog dissipation on summer afternoons (Rutllant *et al.* 2003).

The desert is most arid in the latitudes 18°–27° S. This sector is beyond the reach of both the westerly disturbances to the south, which do not penetrate equatorward of 30° S, and the excursions of humid arid from the east in summer; subsidence associated with the subtropical high is marked in this sector and the rain shadow of the Andes and its blocking effects on the flow also play a role in the extreme aridity (Houston and Hartley 2003; Rutllant *et al.* 2003). Here three longitudinal divisions can be defined: the coast and coastal range, constantly under the influence of the Pacific maritime air; barren, arid plains east of the coastal range; and the highlands above 2700 m, where the influence of occasional, moist easterly currents is felt.

The Copiano River at 27° S marks the limit of extreme aridity. Rainfall increases from 25 mm/year at that latitude to 200 mm at 31° S, as troughs in the westerlies become more common in winter. Southern regions of this desert support some grass and shrubs. Here, the patterns of rainfall seasonality are extremely complex, with some areas receiving occasional rains in both summer and winter. In general, however, the winter precipitation regime begins here, comprising 90% of annual totals at La Serena (30° S) and areas further poleward. Throughout this sector of coastal desert, thunderstorms are quite rare, averaging from about two per year to less than one per year.

North of the Copiano River, the southerly flow along the eastern flank of the subtropical high persists almost continuously year-round. In winter, this wind regime extends to the southern limit of the desert at 31° S. At the coast itself, winds are relatively weak, in some sectors averaging only a few meters per second. However, foehn winds can exceed 12 m/s (McKay *et al.* 2003). In the sector from 24° S to 31° S, wind tends to be persistently southerly to westerly, but further north it becomes more variable as the prevailing pressure gradient weakens.

In this desert, temperatures are still strongly influenced by the cold coastal current. In the markedly uniform maritime air layer, the annual temperature range is about 7.5°C, the diurnal range about 6°C. Mean annual temperature is around 19°C in the far north. The north–south gradient is slight, being only 4°C between Arica and La Serena at 18° S and 30° S, respectively, at the northern and southern extremes. Fluctuations can nevertheless be extreme. During a 4-year observation period in the most arid part of this sector, the maximum temperature recorded was 39.7°C, while the minimum was −5.7°C (McKay *et al.* 2003). Temperature decreases with elevation at a rate of 0.7°C/100 m (Cereceda *et al.* 2008b). The warmest months are January and February, with mean temperatures of 22°C in the north and 18°C in the south and a diurnal range of 8°C. The coldest month is August in the north, with a mean temperature of 16°C and a diurnal range of 6°C. In the south, July is the coldest month, with a mean of 12°C and a diurnal range of 8°C.

WESTERN CORDILLERAN SLOPES AND INTERMONTANE VALLEYS

Throughout the desert region of western South America, the cordillera follows the coastline, generally lying some 100–150 km to the east. In most sectors, summer rainfall occurs above about 2500–3000 m, being either orographic in origin or associated with excursions of humid air from the Amazon Basin to the east. These areas receive occasional winter precipitation, with that in the southern regions being associated with equatorward excursions of the polar front. Above about 4000 m, sleet and snow occur regionally. The lowest slopes, which are part of the coastal range in most areas, generally receive moisture in winter from the *garua* or stratus decks. The slopes in between are generally exceedingly dry, as are many of the interior valleys within the cordillera.

The interior deserts include such areas as the Pampa de la Joya in Peru and the Tamarugal Pampa of Chile, both altiplanos situated between the coastal ranges and the western slopes of the Andes. There are also many dry valleys within the high cordillera. The Atacama Desert of Chile is one of these. These are less humid than the coastal strips and generally experience less cloudiness, hence more extreme variations in temperature.

The dry portions of the highlands present a stark contrast to the coastal desert, especially the central fog desert. Here skies are generally clear, insolation high, and diurnal temperature cycles pronounced. In northern Chile, for example, the typical diurnal range in the highlands is about 20°C in summer and

30°C in winter. They also experience a significant annual range, with temperatures below freezing in winter, but reaching 30°–32°C in summer.

The highlands further north experience much less variation during the year, but in many cases a more pronounced diurnal range. Typical is the Pampa de la Joya, an inland desert on a high altiplano only 20–40 km from the coast. This is essentially a continental desert. At about 1200 m above sea level, Pampa de la Joya lies above the coastal stratus and fog. Almost cloudless skies prevail during the daylight hours. During the summer there is frequent cloud cover at night, probably originating in the highlands and moving downslope with the valley breeze. The mean air temperature ranges from about 14°C in winter to 18°C in summer. The typical diurnal range is 20°C, but a diurnal variation of 64°C has been recorded on a dune surface (Lettau and Lettau 1978). The region is also much less humid than the coastal desert. The relative humidity is on the order of 50% in summer and 40% in winter.

The wind regimes in these deserts are quite diverse. In many, pronounced mountain–valley breezes prevail. In others, light and variable large-scale winds prevail. In extensive desert plains, the low surface friction can lead to a moderate and steady wind regime, strong enough to develop dune systems. In the Pampa de la Joya, for example, mean winds are on the order of 3–4 m/s in most months.

20.4.5 VEGETATION

The vegetation in the Peruvian-Atacama Desert is quite diverse (Arnesto *et al.* 2007; Rundel *et al.* 2007) because it is dependent on several factors, including the amount and seasonal distribution of the rainfall, the presence or absence of fogs, and the elevation. The region is a mosaic of microhabitats determined by moisture availability, with areas of lush vegetation thriving in the midst of arid/semi-arid environments.

The most northern stretch of dryland, from southern Ecuador to the fog belt beginning *c.* 8° S, is a "tree desert" or cactus savanna. To the south, in the fog desert, is *loma* vegetation: growth forms that utilize fog-water rather than rainwater. To the south, in the dry, fog-free desert beginning near the Chilean border at about 18° S, ephemeral herbaceous vegetation is found. Through the desert, streams coming off the cordillera create comparatively green valleys; in some years, these streams cut through the dry coastal desert to the sea.

The "tree desert" to the north is a mixture of savanna and deciduous rain-green forest. The savanna is a tree and xerophytic parkland with giant cacti. The forest includes species such as the giant *Bombacaceae*, with an enormous barrel-like trunk. Numerous cactus species dominate the undergrowth. To the south, near the Peru–Ecuador border at *c.* 5–6° S, the cactus savanna changes to true desert. There begins a rain-green grassland: generally dry and barren, the area is covered with dense grasses during the occasional years with copious rains.

The cacti differ from those of the rain-green cactus "forests" further north because the loma cactus species have special adaptations to the fog conditions. These include, for example, a thick wax layer and special root structures. In the Chilean section of fog desert, other unusual types of plants are found – the "soil cacti." These are partly buried in the soil; the volume of their succulent subterranean parts is greater than that of their above-ground organs. Buried deeply are thick, fleshy, succulent tubers, above which there are numerous, small, lateral roots.

South of about 8° S the fog loma desert begins. The region includes barchans and large sand seas, as well as a great diversity of life forms and plant communities. Their distribution is controlled by the intensity and frequency of fog, which in turn depend on coastal orography. Species-rich ephemeral or annual vegetation develops in winter when the fogs are present. Perennial communities also thrive, such as *Tillandsia*, an endemic plant that is the dominant loma vegetation in much of the region (Pinty *et al.* 2006). This unusual plant, which may be the only genuine angiosperm fog plant, thrives in the region by absorbing water from the air through its leaves. The region also includes cactus lomas; both cacti and *Tillandsia* are found in sandy areas. In areas where the coast is studded with cliffs there are herbaceous lomas at elevations around 300–500 m, where the fog condenses. The vegetation disappears with the fogs at the end of the winter, usually November or December.

Within the fog desert are pockets of particularly moist conditions referred to as "fog oases." Here cloud forests, with predominantly evergreen trees, develop from the fog-water. The trees usually start at about 500 m in elevation, inhabiting the zone that is humid year-round; most are densely covered with moss, lichens, and lianas. In this region, the amount of condensed fog-water is about four times the usual rainfall, which is on the order of 150 mm. The trees can condense up to 1500 mm of water per year; the water runs down the stems to the ground, providing sufficient moisture for a year-round ground cover.

The relatively fog-free northern Chilean coastal desert is one of the world's driest areas. It is an area of ephemeral herbaceous vegetation dependent on the irregular and episodic rains. About 90% of the land may be devoid of plant life for years at a time, but it becomes a flower garden if rainfall exceeds about 20 mm during June and July. In this region, most of the mean annual rainfall occurs in winter, commonly with half of it falling in one day. Along the desert margin to the south is a dwarf-shrub succulent community that likewise develops a rich cover of annuals after heavier rainfall.

20.5 THE NAMIB DESERT

The Namib is one of the world's smallest and most unusual deserts. It includes many rare plant and animal species that are adapted to its unique environments. Stretching north to south along the southwestern coast of Africa, the Namib occupies a narrow strip some 1500 km long but only about 80–130 km wide (Fig. 20.13). The desert is sandwiched between the cold southeastern Atlantic and the various highlands of Namibia.

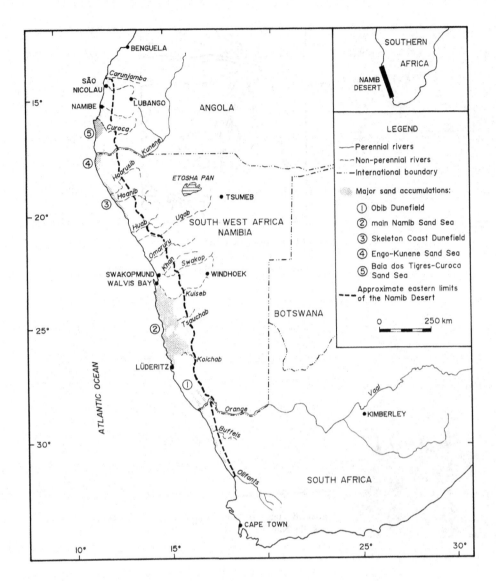

Fig. 20.13 Geographic sketch map of the Namib Desert (based on Ward and Corbett 1990, Lancaster 1982).

Its geographic situation is very similar to that of the Peruvian-Atacama Desert of South America and it rivals that desert in terms of the latitudinal extent of the desert and the degree of aridity. Some desolate areas of its northern sector, the Skeleton Coast, are virtually rainless. Extensive environmental and climatic information is available as a result of decades of research carried out at the Gobabeb Training and Research Centre in the central Namib (Fig. 20.14).

Despite the harsh environment, the Namib has a number of natural resources that help the national economy. These include minerals such as diamonds, uranium, and phosphate. Occasional ghost towns (Fig. 20.15), such as the town of Kolmanskop, are legacies of the former prosperity of the diamond industry. The cold Benguela Current, spanning the latitudinal extent of the desert, provides nutrients for rich fisheries resources. This attracts tremendous numbers of shorebirds (Fig. 20.16a), supporting an active guano industry. The cold waters also attract a colony of seals (Fig. 20.16b). In recent years, ecotourism has prospered.

20.5.1 GEOGRAPHIC ASPECTS

The overriding influence on the climate is the cold Benguela Current (Fig. 20.17), which is only 12°C in its coldest sectors in August and September, when the upwelling is most pronounced. The extent of the Namib's most arid region is clearly linked to the extent of the upwelling. The cold Benguela Current also produces the frequent fogs upon which its ecosystem is so dependent (Fig. 20.18).

It has long been known that El Niño–like coastal warmings occur here with some regularity (Covey and Hastenrath 1978; Nicholson and Entekhabi 1987; Florenchie *et al.* 2004). They occur more or less in sync with those in the Pacific (Nicholson 2010). The warm events are triggered by wind stress in the west-central equatorial Atlantic. Cold events of nearly equal magnitude also occur along this coastal region.

Termed Benguela Niños by Shannon *et al.* (1986), these warm events can bring copious rains both to coastal areas and further inland (Nicholson and Entekhabi 1987). Their impact on

Fig. 20.14 The Gobabeb Training and Research Centre, Namibia.

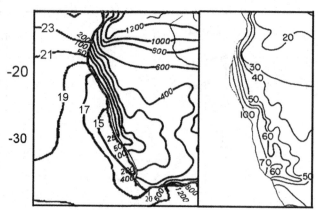

Fig. 20.17 Maps of (left) mean sea-surface temperature, mean annual rainfall and (right) coefficient of variation of annual rainfall (based on Nicholson *et al.* 1988; Reynolds 1982; Walker 1990).

Fig. 20.15 Cemetery in an abandoned settlement near the ghost town of Kolmanskop, near Lüderitz.

Fig. 20.18 Fogs form over the cold Benguela Current and move inland across the Namib.

Fig. 20.16 (a) In the cold waters of the Benguela Current, dense fish populations attract vast numbers of shore birds. (b) The water is particularly cold during the summer season, when a colony of up to 210,000 Cape fur seals is attracted to the cold water near Cape Cross, along the barren Skeleton Coast of the northern Namib.

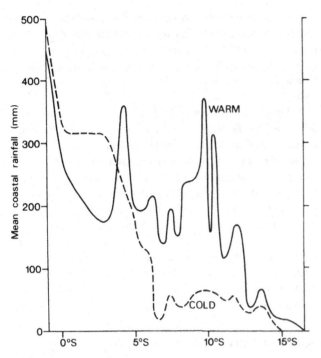

Fig. 20.19 March rainfall along the southwestern coast of Africa (Gabon and Angola) associated with warm and cold sea-surface conditions (from Nicholson and Entekhabi 1987).

Fig. 20.20 The Kuiseb River (strip of dense vegetation) sharply delineates the dune fields of the southern Namib and the gravel and rock plains to the north.

Fig. 20.21 This dune in Sossusvlei, 300 m in height, is reputedly the world's tallest.

rainfall is strongest in areas equatorward of 15° S (Fig. 20.19). In a zone extending from 7° S to 12° S, coastal rainfall in March during anomalously cold years is on the order of 50 mm, compared with roughly 200–400 mm during the warm events. In the most arid part of the Benguela coast, the major impact of these events is a coastal warming and reduction in coastal wind speed. Such extreme events can also be triggered by air–sea interaction in the region of the warm Agulhas Current (Rouault et al. 2002).

The resultant effects on the ecosystem can be disastrous. The warming events disrupt fish, bird, and mammal migrations and adversely affect fish populations (Binet *et al.* 2001). The Namibian stock of Cape anchovy virtually disappeared and the sardine stock was reduced to its lowest level on record during the 1995 Benguela Niño (Gammelsrød *et al.* 1998).

The Namib consists of dune fields in its southern and northern sectors, but the central Namib, running from the Kuiseb River in the south to the Huab River in the north, is predominantly a gravel plain cemented in places by gypsum or calcrete crusts (Fig. 20.20). Numerous east–west ephemeral streams, flowing from the highlands further east, dissect the desert at several spots. Some of the larger ones, such as the Kuiseb, bring torrential floods across the desert about once every 10 years. Their waters occasionally reach from the highlands to the Atlantic.

The Namib is the only desert in the world where extensive sand dunes occur in a cool, coastal climate. Just 1/30th the size of the Sahara, this desert contains virtually every type of sand dune known. It also boasts some of the world's tallest dunes (Fig. 20.21). The dunes are grouped at intervals into distinct fields with regular spacing. This is clearly illustrated in the Landsat photo in Fig. 20.22, which also shows how abruptly the southern dune field stops at the Kuiseb River. In the southern sand sea there is a coastal tract of transverse and barchan dunes, a central area of linear dunes, and network and star dunes along its eastern edge (Fig. 20.23). Photos of these dunes are shown in Chapter 2. The same sequence is observed in fossil dunes in the area. The northern area is very remote and little research has been done in the region, except for parts of the Skeleton Coast. There, the dunes are predominantly transverse and barchan types, with the barchans being further inland than the transverse dunes.

The environments of the Namib are quite diverse. In addition to the dune fields are extensive peneplains (Fig. 20.24a) with stone pavements, barren fields of rock, and river beds with dense vegetation when they are not in flood. Lichens (Fig. 20.24b) are the only permanent vegetation form in the peneplains; these thrive on the fog-water. Along the coast are nebkhas formed from hummock grasses that are generally

Fig. 20.22 Landsat photo illustrating the linear dune fields of the Namib Desert and other surface features bordering the Kuiseb River.

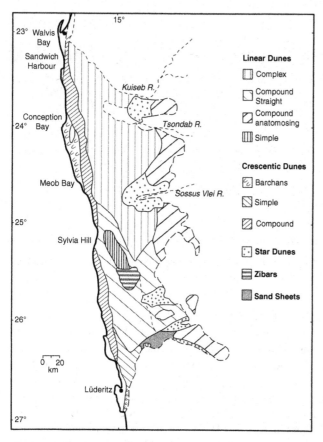

Fig. 20.23 Map of dominant dune types in the Namib Desert (based on Lancaster 1982, 1990).

halophytic (Fig. 20.24c, d). Succulents grow in wetter environments, such as at the foot of inselbergs and rocky ridges. In the river beds, arboreal species such as *Acacia* thrive. Several plants, such as the shrub *Acanthosicyos horridus* (Fig. 24.2 and 24.18) and the grass *Stipagrostis sabulicola* (Fig. 24.14), grow on the dunes; ephemeral vegetation develops in the interdune valleys, especially after heavy rains. Grassland (Fig. 20.24e)

lies inland, while some areas of the coast are nearly devoid of vegetation. In the higher terrain further inland, savannas prevail (Fig. 20.24f). The vegetation in the desert as a whole changes dramatically during years of high rainfall, with the biomass increasing several-fold and grasses thriving in many areas.

Although one of the world's driest deserts, the Namib contains some of the richest and most diverse desert flora and fauna, including several unusual endemic species, such as *Welwitschia* (Fig. 24.20). It probably has more endemic species (those found only in the region) than any other desert (Seely 1978). This diversity is due in part to the unique features of the Namib's ecosystem (Seely and Louw 1980), which include the absence of a significant microbiological decomposition loop, the fulfillment of this nutrient cycling function by tenebrionid beetles that feed on accumulated detritus, the dependence of the biota on fog, and their specific physiological and behavioral characteristics, partly as adaptations to the fog (see Chapter 24).

20.5.2 PRECIPITATION

Mean annual rainfall (Fig. 20.17) illustrates the aridifying influence of the cold Benguela Current. Near the coast it is less than 20 mm/year from roughly 15° S to 28° S. There is a strong gradient inland, so that rainfall increases to 100 mm within 100 km of the coast in most areas. The transition to humid climates is abrupt north of the desert, with mean annual rainfall reaching 400 mm at 10° S and 1200 mm at 5° S. South of the desert the gradient is less steep, and most of the coast is still relatively dry southward to the Cape region.

This region experiences extreme year-to-year variations in precipitation. The coefficient of variation reaches 100% in the driest northern part of the desert, the Skeleton Coast, but quickly drops to 50% around 12° S. In the southern Namib, where mid-latitude systems influence rainfall, the variability is comparatively low. In the desert it is on the order of 70%, but southward it rapidly falls to 30%. An example of this variability is the extreme rainfall event that occurred in April 2006. Lüderitz received 102 mm, which is six times its annual average (Muller *et al.* 2008). The rainfall during an 8-day period was twice the annual rainfall recorded in the previously wettest year. This unusual situation was related to anomalous positions of a tropical extra-tropical cloud band and a cut-off low (see Chapters 5 and 16) that occurred during La Niña conditions.

The patterns of rainfall and fog-water precipitation in the central Namib are complex. For most stations only a few years of data are available, so the patterns illustrated in Fig. 20.25 are valid only in their broadest details. At most locations, rain can occur in nearly every month, but no month has a mean in excess of 20 mm. The rainiest period usually falls within the months of January–April. At many stations a small secondary maximum occurs late in the year. The annual means do not vary much except at Ganab, the station furthest inland. The amount

Fig. 20.24 Typical landscapes of the Namib: (a) gravel peneplains, (b) lichens, (c) hummock shrub on coastal nebkha dunes, (d) halophytes on a salt crust, (e) small tussock grasses, and (f) savanna in the higher terrain of the eastern Namib.

of fog-water precipitation and the number of days of fog differ markedly from one station to the next and the annual cycle of the fogs is different at Swakopmund than it is inland.

At most stations the contribution of the fogs to total precipitation is considerably greater than that of rainfall. The cold, upwelled water offshore creates a high incidence of fog. A look at the interannual variability of rainfall and fog-water shows that the fogs are much more reliable sources of moisture (Fig. 20.26). This is confirmed by the data in Table 20.1, which show that the interval between rains can be much greater than the interval between fogs except at Ganab, 110 km inland. The variations in fog-water are strongly influenced by the degree of upwelling along the Benguela coast (Oliver and Stockton 1989).

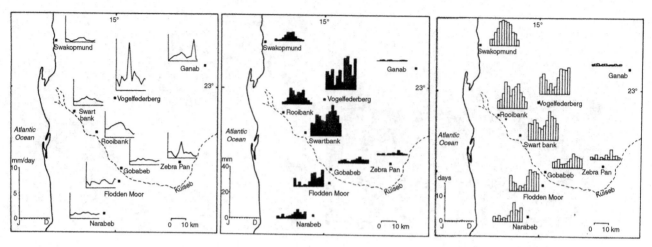

Fig. 20.25 Schematic map showing the annual distribution of mean monthly rainfall and fog-water precipitation (in mm) and mean number of days of fog-water precipitation (based on data in Lancaster *et al.* 1984).

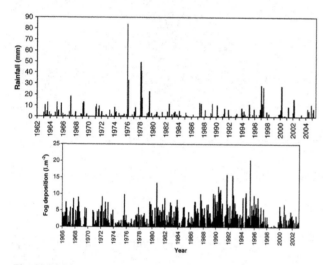

Fig. 20.26 Time series of total annual rainfall (top) and total annual fog-water precipitation (bottom) at Gobabeb in the central Namib (based on Henschel and Seely 2008). The fog-water shows much less variation from year to year.

20.5.3 CLIMATIC GRADIENTS

The Namib's physical and biological character can be attributed to a large measure to the climatic gradients that exist across its north–south and east–west extents. These gradients govern the nature of the dunes and the distribution of species. The seasonality and amount of rainfall varies from north to south. Wind, relative humidity, temperature, fog and rainfall vary with distance from the ocean, creating strong east–west gradients.

In most of the Namib there is a summer rainfall maximum, with the wettest month generally occurring between January and March (Fig. 20.27). Summer rainfall prevails in the north, but the central Namib is nearly rainless. In the southern extreme lies a transition zone with a slight amount of rainfall year-round, then a change to winter rainfall occurs. At about 32° S, the southern limit of the true desert, there is a shift to cold-season rainfall near the coast, but summer rainfall inland.

The variation of climate with distance from the coast is depicted in Fig. 20.28. Relative humidity decreases rapidly, especially within the first 30 km. Temperature, its annual range, and rainfall tend to increase inland. Mean annual rainfall is on the order of 20 mm for the first 50 km inland, then increases rapidly to over 100 mm 120 km from the coast. The pattern of fog is more complex, with the amount of fog-precipitation and the number of fog days increasing to a maximum about 30 km from the coast (Fig. 20.29). This distribution is due to the altitudinal distribution of the fog-water, which has a maximum from 300 to 600 m ASL. The terrain from about 30 to 60 km inland corresponds to these heights. At Swartbank (340 m ASL) fog-water reaches nearly 200 mm/year.

The north–south gradient in the Namib is illustrated with the stations Namibe at its northern extreme (15° S) and Alexander Bay near its southern extreme (29° S) (Fig. 20.30). At Namibe (mean annual rainfall = 50 mm), precipitation falls in summer and has a March maximum. The warmest months are March and April, with a mean monthly temperature of 29°C and a mean daily minimum of 23°C. The coldest month is July, with a mean temperature of 20°C and a mean daily minimum of 13°C. The annual range is 9°C and the diurnal range varies from 20°C in April to 13°C in July. In the Angolan portion of the coastal Namib the relative humidity seldom falls below 75% during the day and 80–90% at night. Cloudiness is relatively high year-round, ranging from about 50% in January–April to 90% in August.

At Alexander Bay (mean annual rainfall = 52 mm), a cold-season precipitation regime prevails (Fig. 20.30). A maximum occurs in both May and August. The average daily maximum temperature reaches 24°C during the warmest months, January–March, but 20°C during August, the coolest month. The daily minimum ranges from 9°C in July and August to 15°C in January and February. The daily range is fairly constant at about 10°C, being slightly higher in winter. Monthly average relative humidity ranges from 69% to 77%. Cloudiness ranges from 40% in winter to 60% in summer, when radiation is also most intense.

Table 20.1. *Maximum number of days between fogs and between rain events (based on Lancaster* et al. *1984).*

Location	Fog	Rain
Ganab	772	306
Gobabeb	63	278
Narabeb	77	301
Rooibank	46	351
Swakopmund	134	485
Swartbank	40	272
Vogelfederberg	33	–
Zebra Pan	118	–

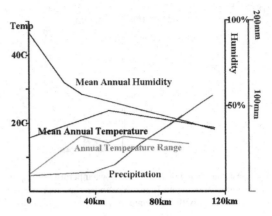

Fig. 20.28 The east–west climatic gradient of temperature, annual temperature range, humidity, and rainfall in the central Namib (based on data in Lancaster *et al.* 1984).

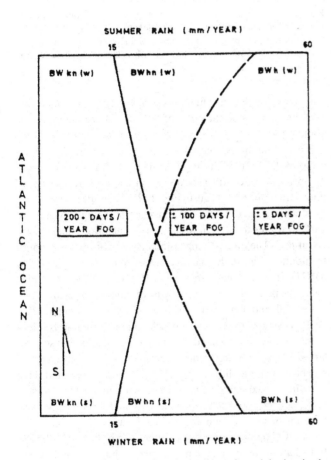

Fig. 20.27 The north–south climatic gradient of precipitation in the southern Namib dune area, indicating both summer and winter rainfall (mm/year) (from Lancaster *et al.* 1984).

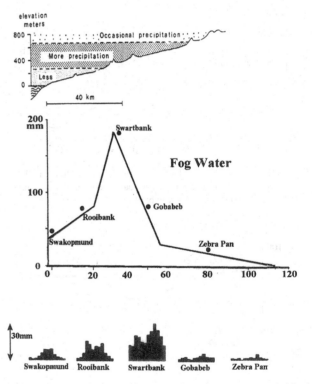

Fig. 20.29 Schematic of fog-water precipitation as a function of altitude and inland distance in the central Namib (based on Lancaster *et al.* 1984 and other sources). The bar graphs at the bottom indicate the seasonal cycle of fog-water precipitation at the five stations shown. Fog-water precipitation is greatest between about 300 and 600 m above sea level.

20.5.4 EAST–WEST GRADIENTS: COAST VERSUS INTERIOR

The annual march of climate in the Namib is contrasted for coastal and inland locations in Figs. 20.31–20.34. Gobabeb, about 50 km from the coast, represents the inland stations in the central Namib; Pelican Point (Walvis Bay) and Swakopmund represent the coast.

At Gobabeb, about 50 km inland, rainfall is greatest from January to March, averaging about 8 mm/mo, but a slight secondary maximum occurs in August (Fig. 20.31). Fog-water precipitation peaks in June to December at about 6 mm/mo, with a small peak again in February and March. Overall fog-water and rainfall are out of phase and the total precipitation therefore has two clear maxima in summer and winter. Relative humidity has a broad maximum from August to March (Fig. 20.32). Mean

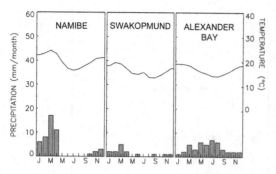

Fig. 20.30 The seasonal march of mean monthly temperature and precipitation at Namibe, Swakopmund, and Alexander Bay, representing the northern, central, and southern Namib, respectively.

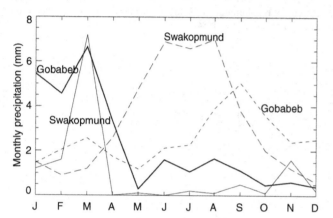

Fig. 20.31 Seasonal march of monthly rainfall (solid lines) and fog-water precipitation (dashed lines) at a coastal station (Swakopmund) and at a station 60 km inland (Gobabeb).

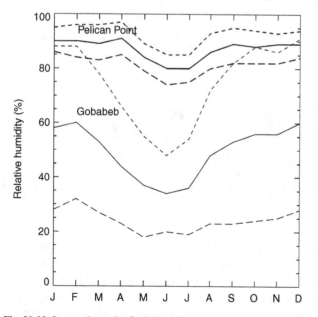

Fig. 20.32 Seasonal march of relative humidity (%) at a coastal station (Pelican Point) and at a station 60 km inland (Gobabeb). Mean daily relative humidity (solid line) and mean daily maximum and minimum relative humidity (dashed lines) are indicated.

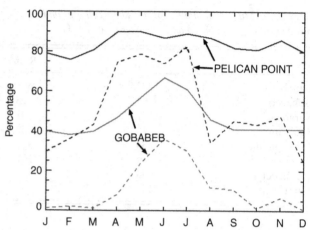

Fig. 20.33 Seasonal march of sunshine at a coastal station (Pelican Point) and at a station 60 km inland (Gobabeb). Sunshine is indicated by percent of possible sunshine hours (solid lines) and percent of days with at least 90% of sunshine hours (dashed lines) (based on data in Lancaster *et al.* 1984).

daily maximum relative humidity, which is generally reached in early morning, is on the order of 80%, the afternoon minimum about 30%.

Solar radiation is unimodal with a maximum in July and a prolonged minimum from September to March (Fig. 20.33). Temperature, in contrast, peaks in March and reaches its minimum from June to August (Fig. 20.34). The mean daily temperature (average of maximum and minimum) ranges from about 18°C in winter to 24°C in March; the annual range is therefore about 6°C. The diurnal range is relatively uniform from month to month; mean daily maxima range from 27°C in winter to 33°C in March, minima from 10° to 16°C (Fig. 20.34). The diurnal range is about 17°C, or nearly three times the annual range.

Rainfall and fog-water precipitation at Swakopmund (Fig. 20.31) is similar to that at Gobabeb, although rainfall is somewhat lower at Swakopmund. The march of radiation is also similar, with a single summer maximum, but the seasonal cycles of sunshine are quite different (Fig. 20.33). At Swakopmund there is a winter maximum, a consequence of the summer cloudiness that averages about 90% (compared with 50–60% in winter). In terms of relative humidity and temperature (Figs. 20.32 and 20.34) Swakopmund contrasts even more sharply with Gobabeb. Both daily maximum and minimum relative humidity at the coast (Pelican Point) are generally at least 80%. The mean daily temperature ranges from about 13°C in winter to 18°C in summer, so that the annual range is only about 5°C. Mean daily maximum and minimum temperatures are about 16°C and 9°C, respectively, in winter and 20° and 16°C, respectively, in summer (Fig. 20.34). The diurnal range is only about 7°C in winter and 5°C in summer, but it nevertheless equals or exceeds the annual range.

20.5.5 WIND REGIME

The wind regime of the central Namib is much more complex than the thermal and moisture regimes because the synoptic-scale winds are complemented by a low-level coastal jet and by meso-scale wind systems that include both mountain–plain winds and sea breezes (Tyson and Seely 1980; Lindesay and Tyson 1990). The mesoscale wind systems produce a marked change in the wind regime from the coast inland. The dominance of the sea- and land-breezes decreases and the topographically induced winds become progressively more important. The winds also have a strong seasonal and diurnal cycle. Daytime southwesterly sea breezes and northwesterly valley and plain–mountain winds dominate the air flow near the surface in summer. Nocturnal northeasterly land breezes and southeasterly mountain and mountain–plain winds, with substantial cool air drainage, occur preferentially in winter. Thus, at Gobabeb the prevailing winds shift from northwesterly in January to southeasterly in July.

The regime is somewhat simpler in the Namib's northern and southern extremes. In the south, at Alexander Bay, winds are southerly in January and relatively light in July with a slight dominance of easterly flow. In the northern Namib at Namibe (Moçamedes), winds tend to be westerly, with flow throughout the year from the southwest, west or northwest from 55% to over 70% of the time.

The winds have a dramatic effect on the Namib Desert. Day-to-day changes in the winds bring marked changes in the prevailing climate elements. They also produce the prevailing pattern of dunes in the central Namib, progressing landward from the coastal strip of transverse and barchan dunes; then linear dunes, and finally the more complex forms on the eastern edge (Lancaster 1990). The linear dunes in the central area are formed under the combined influence of moderate-energy southerly winds and high-energy but low-frequency easterlies, which produce the elongation.

Figure 20.35 illustrates the day-to-day weather changes at Gobabeb that accompany changes in wind direction. One case is a fog event, associated with a northwesterly wind, and the

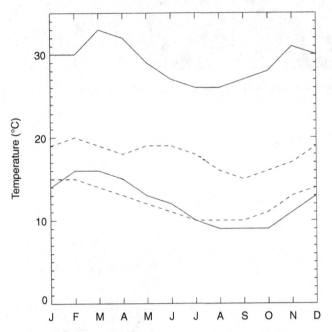

Fig. 20.34 Seasonal march of temperature at a coastal station (Pelican Point at Walvis Bay, dashed lines) and at a station 60 km inland (Gobabeb, solid lines). Mean daily maximum and minimum temperatures are indicated for each location (based on data in Lancaster *et al.* 1984; Nieman *et al.* 1978).

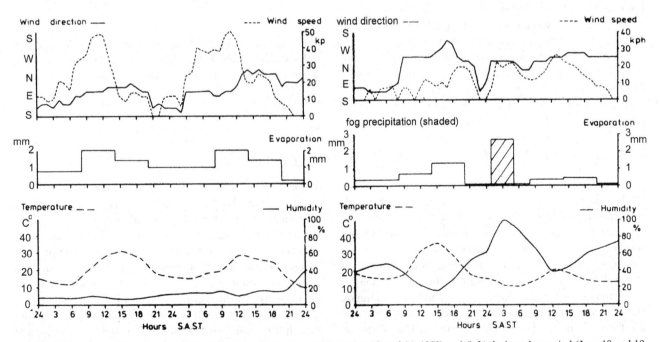

Fig. 20.35 Climate conditions at Gobabeb (right) during a fog event (September 13 and 14, 1979) and (left) during a berg wind (June 18 and 19, 1981) (from Lancaster *et al.* 1984).

Fig. 20.36 Sand plummeting over the crest of a dune and streaming across the Namib Desert near Swakopmund during a berg wind.

Fig. 20.37 A reversing barchan dune with a "lip" at the crest, indicating a change of wind direction during an easterly berg wind.

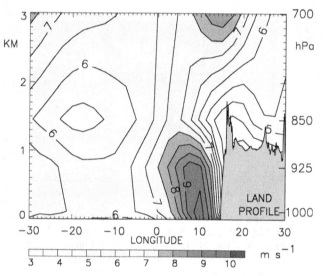

Fig. 20.38 Vertical cross-section of the Benguela jet (Nicholson 2010). Wind speeds are indicated in m/s.

other is a "berg" wind (a mountain wind from the east). The fog is limited to the hours of 01:00–05:00, coincident with the onset of relatively high wind speeds. During the fog, temperature decreases by 2°–5°C and relative humidity increases from 60% to 100%. At the same time, evaporation is reduced to zero. During the berg wind of the early morning hours, temperature increases by nearly 20°C and evaporation increases. Streams of sand emerge across the crest of the dunes and meander toward the coast (Fig. 20.36), reducing the visibility to near zero at times. The impact on dunes is readily apparent, with the directional change indicated by the formation of a "lip" at the top of the barchan dunes (Fig. 20.37).

The low-level coastal jet termed the Benguela jet (Nicholson 2010) appears to be a response to the temperature gradient between the cold, upwelled water and the warmer land. However, the presence of north–south highlands probably plays some role. This jet is imbedded in the southerly flow on the eastern flank of the South Atlantic High. The jet exhibits marked seasonal variations that parallel those of the high, reaching its southernmost latitude (26° S) in January and its northernmost (15° S) in May. The jet is well developed from July to November and from January to March. In October (Fig. 20.38), when sea-surface temperatures near the coast can fall below 15°C, the Benguela jet extends from 17° S to 32° S. Its maximum winds,

which average 10 m/s in October, are roughly 650 km offshore and some 200 m above the surface. The jet may play a role in enhancing the coastal aridity in this hyper-arid portion of the Namib.

20.6 THE WEST COAST OF NORTH AMERICA

20.6.1 OVERVIEW

Along the Pacific coast of North America, arid and semi-arid climates prevail from 23° N to 35° N. The northern region, California and northern Baja California, has a semi-arid Mediterranean climate with winter rainfall and summer aridity. The rest of the Baja peninsula has an arid climate with predominantly summer rainfall. The climate of this coastal dryland reflects the combined influence of the cold California Current and the North Pacific subtropical high (Fig. 20.39).

The California Current is a semi-permanent band of cold water just offshore, ranging from 25 to 50 miles in width and extending from 15° N to beyond 50° N. Temperatures in the current are some 5°C warmer than in the waters further west.

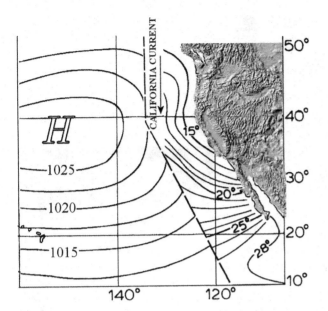

Fig. 20.39 Pacific subtropical high, sea-surface temperatures in the California Current, and coastal topography.

Relative to the respective latitudes, the temperatures of the Humboldt or Benguela currents are more extreme. For this reason, this coastal desert is not as dry as those of Peru and Chile or the Namib. Winter water temperatures in this current range from about 16° to 12°C off the southern California coast (30–40° N); further south near Baja California (22–30° N) the range is about 22°–16°C. Summer temperatures are 2°–4°C higher.

The trade wind inversion on the eastern flank of the subtropical high modulates both the climate and day-to-day weather. Over southern California in summer the inversion base ranges from about 300 to 500 m, its top (and the top of the marine layer) typically extends to 1500–2000 m (Neiburger *et al.* 1961). The temperature increase across the inversion layer is typically 8°C. In the latitudes where the inversion is strongest, around 34°–35° N, major urban areas such as Los Angeles are clustered. As a result, air pollution and smog are major features of the region's climate.

The subtropical high, and therefore the California Current, extend unusually far poleward, particularly in summer. Consequently, a maritime climate prevails as far as 55° N, although it is not as mild as the maritime climates of warmwater coasts, such as Europe. A semi-arid, Mediterranean climate with dry summers prevails along the coast from *c.* 31° to 42° N, but summers are relatively dry much further northward. Maximum subsidence on the eastern flank of the high occurs near 35° N. This reduces summer rainfall to near zero in the area from 30° to 38° N. In southern California the semi-arid coastal strip blends into the interior Mojave Desert. A true desert extends along the Baja peninsula, from its tip at 22° N to about 31° N. This is part of the Sonoran Desert.

As with the Namib and Peruvian-Chilean deserts, the coastal aridity is limited by a high cordillera to the east relatively close to the coast. In southern California, this includes a coastal range, with elevations of generally 500–1000 m, and the high Sierra Nevada further inland, generally reaching 2000–3000 m or higher. These are separated by the semi-arid Central Valley in central California, but merge to a single, near-coastal highland in southern California, with elevations typically 1000–2000 m. In Baja California, the coastal range is lower and relatively arid, and the coastal Sonoran Desert extends into the interior, blending with the Mojave and Chihuahuan deserts.

Two low-level jets also impact the climate of the region. A northerly jet exists off the coast of southern California extending between the latitudes of ~35° N to ~40° N. As with its Chilean counterpart, its core lies adjacent to the coast and just within the coastal inversion layer at an altitude of 300–700 m, and core speeds can exceed 30 ms^{-1} (Parish 2000). A low-level southerly jet also appears in the Gulf of California (Douglas *et al.* 1998). This jet, with a core near 450 m and speeds in excess of 7 ms^{-1}, is a summertime feature. It has a strong diurnal cycle, with a nocturnal maximum. Surges of this jet bring moisture into "monsoon" regions of the southwestern USA.

20.6.2 THERMAL REGIME

Temperatures in this region (Fig. 20.40) depend mostly on elevation and distance from the coast; the latitudinal gradient is comparatively small. Those in the Central Valley tend to be uniform except in its southern extreme. In winter, mean daily minima along the coast are around 4°–7°C in the northern and 4°–8°C in the southern limits of the Mediterranean region (central California and southern California, respectively). Temperatures in the coastal highlands are −3° to −1°C in the north and 0°–3°C in the south. In the Central Valley, minima are generally 2°–3°C. The coldest month is January throughout the region. Mean daily maxima are on the order of 15°–18°C in the coastal region of Mediterranean climate. They are on the order of 15°C in the coastal highlands of central California, but 11°–15°C in the higher coastal uplands of southern California. In the Central Valley of California, maxima are 12°–13°C.

In summer the effect of topography and distance from the coast plays an even greater role, particularly in determining mean monthly maxima. The temperature gradients become extremely pronounced inland. Mean monthly maxima along the coast are generally 15°–18°C in central California and 20°–22°C in southern California, but rapidly increase to 33°–36°C in the highlands. Maxima in the Central Valley are only slightly warmer than in the highlands, i.e., 36°–38°C. Mean monthly minima in summer are much more uniform, but have a stronger latitudinal gradient than the maxima. These are 9° to 10° or 11°C in central and 15°–17°C in southern California along the coast but 13°–15°C in the coastal highlands. Highest minima are in the Central Valley, 15°–19°C.

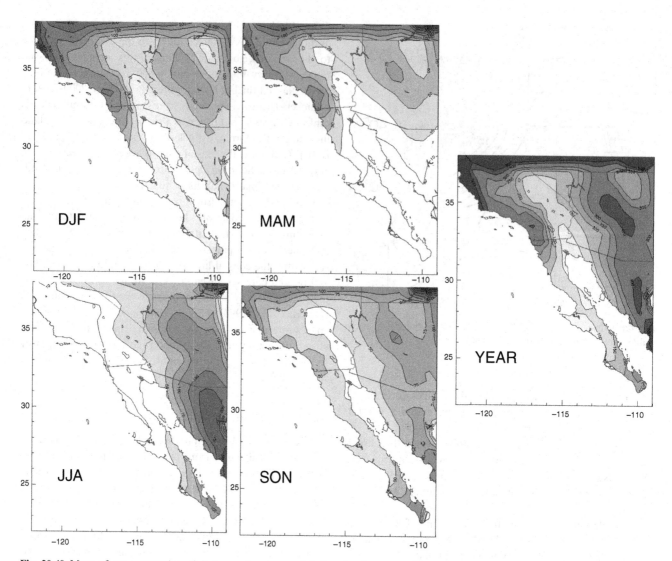

Fig. 20.40 Maps of mean annual precipitation and mean precipitation for the seasons December–February, March–May, June–August, and September–November.

For a maritime climate, both the diurnal and annual ranges of temperatures are relatively large (Fig. 20.41). In southern California the diurnal range is on the order of 8°–11°C, being greater in winter than in summer. Further inland it exceeds 14°C. The annual range is about 8°C.

In Baja California, the climate is somewhat more maritime because the peninsula is surrounded by water on three sides and no location is far from a coast. Monthly mean temperatures in the north range from about 12°C in winter to 22°C in summer. In the south they range from about 18°C in winter to 28°C in summer. Higher summer temperatures are experienced inland.

20.6.3 MOISTURE REGIME

As is typical of coastal deserts, the relative humidity is relatively high. In San Diego, for example, it varies from about 55% in summer to 65% in winter. Inland, humidity is usually high in winter and low in summer. In Baja California it is even higher, owing to the influence of the peninsula. At La Paz, in the center of the peninsula, relative humidity is on the order of 60–70% year-round. It decreases with increasing distance from the coast. In the interior valleys east of the coastal range, relative humidity is quite variable. There it is generally moderate to high in the winter, but can drop to below 10% during the warm season.

The California portion of this coastal desert experiences frequent fog, exceeding 60 days per year in most areas. Fogs occur much more often over the ocean; lighthouses record a five times greater frequency than coastal cities (Leipper 1994). These become more infrequent further south, decreasing to roughly 20 occurrences per year at San Diego, where instead there are persistent offshore low clouds. Fractional cloudiness is as high as 83% at 33° N (Albrecht *et al.* 1995). Some of the coastal ecosystems are dependent on the fogs, which help to reduce the stress of the occasional droughts (Fischer *et al.* 2009). Fog becomes less frequent in Baja California, where the coast is relatively far

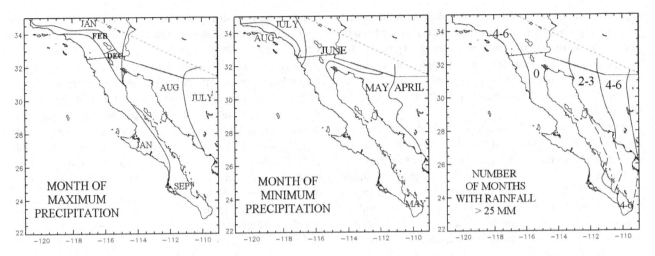

Fig. 20.41 Maps of months of maximum and minimum precipitation and number of months with mean precipitation exceeding 25 mm.

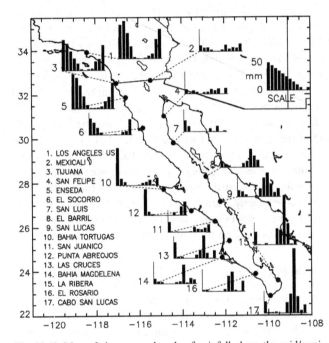

Fig. 20.42 Map of the seasonal cycle of rainfall along the arid/semi-arid West Coast of North America.

from the cold California Current. However, both fog and dew are common along the west coast of the peninsula and high humidity may reach 50 km inland.

Baja California is the most arid region of North America and the most geographically isolated peninsula in the world. In the driest sector of the coastal desert, from 24° N to 30° N, mean annual rainfall is less than 100 mm/year (Fig. 20.40). Annual means range from 100 to 150 mm/year over most of Baja California and up to 200 mm/year in southern California.

From there, mean annual rainfall increases progressively northward, reaching 500 mm/year near San Francisco at 38° N. Topography and shoreline configuration disrupt this pattern

(Fig. 20.40). Rainfall is higher in the mountains, lower in lee valleys. It is also higher where the coastline is convex, with a large land area protruding seaward. Rainfall is highly variable, with the coefficient of variation being on the order of 40% in southern California and increasing to over 50% in the southern half of the Baja California peninsula.

Rainfall is strongly seasonal and the seasonality varies with latitude (Pavia and Graef 2002). From California northward, rainfall occurs in winter (Fig. 20.42), when the high is displaced far to the south and its influence is replaced by that of the Pacific westerlies. However, even in relatively dry southern California and northern Baja California, rainfall can occur in any season except summer, when the cold California Current keeps this area nearly rainless (Fig. 20.39). Characteristically, the month of maximum rainfall along the Pacific coast is a function of latitude. This feature is a consequence of the progressive south-ward displacement of the winter jet streams and associated cyclones. Thus September is the rainiest month at Anchorage, Alaska (61° N), but further south the maximum occurs progres-sively later. It occurs in January in most of California; August is generally the driest month.

In Baja California the rainfall regime is more complex (Hastings and Turner 1965; Markham 1972). In the northern peninsula, January is likewise the wettest month. Further south on the western littoral of southern Baja California there is an abrupt change to summer rainfall (Figs. 20.40 and 20.42), although a secondary winter maximum is evident at most sta-tions. This change occurs at about 26° S on the western coast of the peninsula, but at 29° S on the east coast, along the Gulf of California. In this sector maximum rainfall occurs in September, shifting to August and July eastward in western Mexico and the southwestern USA (Fig. 20.41). The driest month is June in most of the region, shifting eastward in western Mexico to May, then April. Over most of the Baja peninsula, mean rainfall is below 25 mm in every month.

The climate of Baja California is analogous to that of the savannas of West Africa on the southern flank of the subtropical high. The summer rainfall season is linked to southwesterly flow resulting when the subtropical high shifts poleward and the moist trades cross the equator. The winter aridity is linked to the equatorward displacement of the subtropical high. Summer rainfall is further enhanced by the westward displacement of an upper-level trough into this region and by the hurricanes of the June–November season.

In most of southern California, where storms are linked to the subtropical jet, January is the rainiest month. In spring and summer the subtropical jet wanes rather than shifting back poleward, hence there is no spring or early summer precipitation maximum. A strong flood hazard is present because torrential rains occasionally occur, but most of the streams are small and intermittent. Flooding is usually a result of heavy precipitation over periods of 1–3 days. This situation tends to result when the subtropical high dissipates and the coast receives the full fury of the Pacific storm tracks.

Flooding in southern California and in northern Baja California is often coupled with the occurrence of El Niño (Andrews et al. 2004), which controls most of the interannual variability of rainfall in both regions (Schonher and Nicholson 1989; Reyes-Coca and Troncoso-Gaytan 2004). Its impact in northern Baja California is strongest during February and March, while La Niña reduces rainfall there, particularly during December and January (Minnich et al. 2000). While much of its impact relates to modulation of the subtropical jet stream, El Niño also affects the region via warming in the Gulf of California (Herrera-Cervantes et al. 2007). El Niño also reduces the frequency of Santa Ana winds (Finley and Raphael 2007).

A few other aspects of the moisture regime merit a mention. The desert of the western littoral of Baja California is perhaps the only one in the world that experiences hurricanes. During a 13-year period, at least 15 hurricanes hit the region, most originating over the waters to the west between 10° and 20° N (Trewartha 1970). Thunderstorms are relatively infrequent, occurring on average less than 20 times a year in Los Angeles and as few as 2–3 times per year further south. They are more frequent in wetter regions and extremely rare in Baja California. Snow has occurred at some time or another nearly everywhere in California, even as far south as Los Angeles. However, it falls infrequently west of the Sierra Nevada except at high elevations in the Coast Range. Frost is also rare outside of the higher elevations. In northern Baja California, snow is infrequent below elevations of 1500 m (Minnich et al. 2000). Neither snow nor frost occurs in the southern extreme of the coastal desert in Baja California.

20.6.4 WINDS

The wind is generally westerly most of the year in southern California, although some locations experience prevailing northeasterlies during part or most of the winter. Southern California also experiences regular occurrences of easterly or northeasterly Santa Ana winds, a type of chinook descending from the mountains. The prevailing meteorological situation is generally an intensified pressure ridge over central and northern California and a low-pressure center over the southwestern USA (Sommers 1981).

Santa Anas originate in the dry valleys east of the mountains; they bring desert air into coastal areas of southern California and the Baja peninsula (Castro et al. 2003). Already very dry, these winds lose much of their meager moisture in the highlands. Relative humidity is further decreased as they subside and warm on the leeward side. These winds can attain speeds of 150 km/hour. The extreme dryness and strong force of these winds creates a serious fire hazard. Tremendous blazes have been attributed to the Santa Ana.

The Santa Ana tends to develop in the driest and warmest months, the highest frequency of occurrence being in September and October, when Santa Ana weather types prevail on average 1 or 2 days per month (Monteverdi 1973). The fire risk is increased when the season is preceded by an abnormally warm summer or dry conditions in autumn (Keeley 2004). A particularly severe Santa Ana burned 80,000 acres (325 km^2) in southern California during November 1980 and caused more than $40 million dollars' worth of damage (Svejkovsky 1985).

Further south in Baja California, the wind regime becomes more complex. In winter (November through February or March) winds tend to be northeasterly throughout most of the peninsula. Some sources suggest that there is a monsoon-like shift in wind direction in summer, but this may be exaggerated. Over the Gulf of California, the summer shift to southeasterly flow is evident only in the far south (Parés-Sierra et al. 2003). In the northwestern extreme of the Baja peninsula, winds are generally from a northwesterly quadrant, but further east they tend to be easterly. Further south, on the Pacific side of the peninsula, the prevailing wind is generally southwesterly, but easterlies are common on the Gulf of California side.

20.6.5 VEGETATION

The vegetation distribution of southern and Baja California corresponds closely to the prevailing climate (Fig. 20.43). In the Mediterranean area, including the coastal highlands, is found predominantly sclerophyllous vegetation, mainly chaparral with scrub and dwarf forest. In the interior valley lies a prairie grassland. Further east in the Mojave is mainly short perennial shrubs and cacti. The Sonoran Desert further south includes these, plus trees and tall shrubs. The vegetation formations are described in more detail in Chapter 14.

Most of Baja California is considered to be a part of the Sonoran Desert. Typical of the Sonoran Desert, most of the peninsula has a bimodal rainfall regime but the percent concentration in summer generally increases southward. Six vegetation regions prevail. The Central Gulf Coast, with summer rainfall, has desert vegetation and roughly 15% plant cover. The Vizcaino region receives little rainfall in either winter or summer. This is mainly a "fog desert" with sparse vegetation and low plant forms. Plants that utilize the fogs, such as lichens and Tillandsia sp. (Spanish moss) are common, found clinging

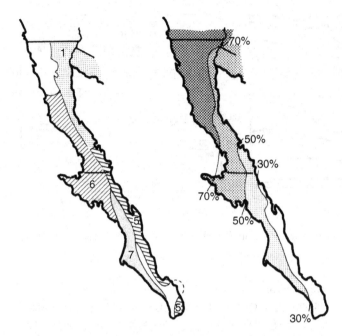

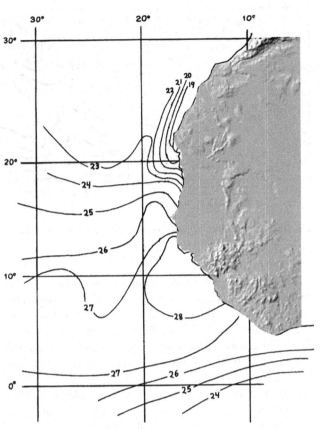

Fig. 20.43 (left) Map of vegetation and (right) percentage of total annual precipitation falling during winter (November–April) in the Sonoran Desert (from MacMahon and Wagner 1985). In the left-hand map, the numbers correspond to various provinces within the Sonoran Desert.

Fig. 20.44 Canaries Current: temperatures (°C) and topography in relation to the Azores High.

to stones. In the Magdalen region, with mainly summer rainfall, the vegetation is mainly dwarf trees and tall cacti, similar to the central Sonoran Desert. The region also experiences many morning fogs. The Lower Colorado (northernmost part) has mostly shrubs (creosote bush is dominant). This is the hottest and most arid region of North America. Rain is predominantly in winter. The "California" region has sclerophyllous coastal shrubland and chaparral ("dwarf forest"), and winter rainfall. Coniferous shrubs cover the highlands in the center of the peninsula.

20.7 THE WESTERN LITTORAL OF THE SAHARA

The Atlantic coastal desert of Morocco, the western Sahara, and Mauritania is merely the western extension of the Sahara Desert. However, along the shore flows the cold Canaries Current (Fig. 20.44), giving the region the typical thermal and moisture characteristics of a coastal desert until some 200–400 km inland. Unlike the rest of the Sahara, this region falls under the influence of the Azores High for most of the year. Thus, throughout most of this coastal desert, winds are northerly or northeasterly most of the year. Only in the area further south do the winds shift to southwesterly in summer.

The vegetation in this region is denser than in most parts of the Sahara and is comparatively rich in species. The vegetation in the coastal strip of true desert is mostly succulent shrubland (White 1983). However, only a few, such as *Euphorbia* and *Senecio*, extend south of Saguia el Hamra at 27° N. In northern

parts of this region, typically Mediterranean shrubs are also common, as are halophytes. In the south the vegetation is Sahelian, with *Acacia* and other shrubs and various grasses.

The Canaries Current is milder than those along other cold-water coasts (Marchesiello and Estrade 2009), but its temperature contrast with the land is sufficient to enhance aridity very close to the coast. Coastal waters are coldest in winter, being 16°–17°C in the area from approximately 21° to 26° N, the latitudes of permanent upwelling. Water temperature rises to 18°–19°C in summer in this area. The upwelling commences at roughly 30° N, extending to about 15° N in summer, 12° N in winter. Its extent and intensity vary markedly from year to year, especially in the area south of 20° N (Nicholson 1975; Santos *et al.* 2005). The upwelling enhances productivity in an area extending some 100–200 km offshore. By comparison, the California Current's impact on productivity extends only some 50 km offshore. Hence, the Canaries Current is a very important part of the coastal ecosystem.

The characteristics of a coastal desert are best developed where the water is coldest and the current flows roughly parallel to the shore. That is the stretch between 20° N and 30° N. The material in this chapter applies only to that sector. General characteristics will be described, but for extensive maps and discussion, the reader is referred to Chapter 16, on African deserts.

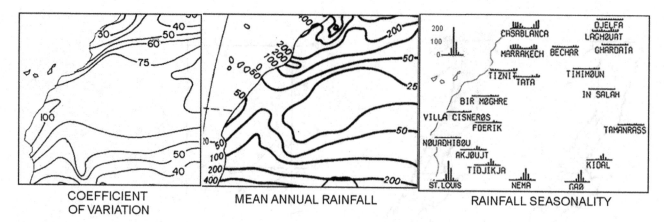

Fig. 20.45 Maps of mean annual rainfall in coastal northwest Africa and its coefficient of variation (from Nicholson *et al.* 1988). The seasonal cycle at typical stations is shown in the bar graphs.

Unlike most coastal deserts, rainfall does not increase continuously inland. Arid conditions prevail right at the coast, but only in the latitudes from roughly 20° N to 27° N. Outside of these latitudes mean annual rainfall exceeds 50 mm (Fig. 20.45). Some 150–200 km inland, rainfall also exceeds 50 mm through most of this latitudinal range. This is most likely a consequence both of local topography, as elevation increases inland, and of the tracks of the storms that form east of the Atlas Mountains. Further east, conditions again become arid, with many areas receiving less than 20 mm/year (Dubief 1963).

In the northern extreme of this sector, mean annual rainfall is about 200 mm and is confined to the winter and transition seasons. The summer months are completely dry. The southern sector can receive rainfall in summer or winter, the driest months being May and June. The wettest month is generally September in the south, November or January in the north. Mean annual rainfall is less than 50 mm in the entire sector from 20° to 28° N. Probably no sector receives less than 30 mm annually on average. The annual number of rain days ranges from about 5 in the south to 20 in the north. The coefficient of variation for annual rainfall is about 50% in the north, increasing to 100% in the south (Fig. 20.45).

The sources of rainfall are diverse: the Saharan depressions in the transition season, tropical disturbances in summer, and westerly waves in winter. Thunderstorms are rare or unknown. "Heug" weather is a common winter phenomenon occurring in association with persistent westerly troughs in the area. In this situation, light drizzle continues for days on end, in some cases lasting 10 days or more. Although only a few tens of millimeters fall overall, this amount is sufficient for a winter harvest in some regions.

The cold Canaries Current and the frequent cloud cover associated with it are major factors in the region's climate. As elsewhere, there is a coastal inversion, with a base typically at 500–750 m. Related to the inversion is a stratus deck and frequent coastal fog in the northern coastal desert. Consequently, solar radiation is significantly reduced along the coast, being about 20% less than further inland. Most sources suggest a fog

frequency of about 10 days per year along the coast, with a higher frequency over the ocean. The actual frequency is probably higher, since observations are scarce along the foggiest portion of the coast. The frequent dense fogs over the ocean led to the legend of the "dark sea" that dates back at least several centuries BC. Offshore, on Tenerife in the nearby Canary Islands, fog frequency in the early morning approaches 100% for July and averages 80% for the summer as a whole (Marzol 2008).

Minimum cloudiness occurs in a zone from about 16° to 23° N in which mean annual cloudiness does not exceed 20%. This stretch of coast is south of the most intense cold-water upwelling, where stratus is prevalent, but north of the sector with frequent convective cloudiness in summer in association with the advance of the ITCZ. The zone of maximum cloudiness lies over the ocean, but it moves toward the coast in summer.

Coastal cloudiness from May to August or September ranges from 30% in the south to 70% in the north. A few hundred kilometers inland, skies are nearly cloudless. Annually the number of clear days is greater than 100 in the summer, but less than 50 in the north. In winter total cloud cover is about 30–50% along most of the coast. At Nouadhibou (21° N) there is an October–December cloudiness maximum, reaching 40% in November. The minimum is in June, with less than 20% coverage. At Cap Juby (28° N) and at Nouakchott (18° N) there are summer maxima reaching 65% and 55%, respectively. Cloudiness falls to 40–45% at Cap Juby in winter and to less than 30% at Nouakchott in spring.

Since the cloud cover and cold current are most intense in summer along most of the coast, they dramatically reduce the annual range. On the coast it is about 7.5°C, but twice that some 200–300 km inland. Freezing temperatures have not been recorded except in the inland extreme of the coastal desert. Absolute maxima are about 40°–45°C on the coast, compared with 50°–55°C inland.

The coastal influence on temperatures is greatest in summer, when monthly means are on the order of 22°–23°C and isotherms are parallel to the shore. Some 200–300 km inland, they are on the order of 30°C. August is generally the warmest month; inland this shifts to June or July. Mean daily maxima are on the order of 25°C, increasing to 40°C inland; mean daily minima are 18°C at the coast and 22°–23°C inland.

In winter, temperatures increase with latitude, the coast having little influence. Monthly means are about 15°C in the north of the coastal desert, 20°C in the south. Mean daily maxima are about 20°C in the north and 25°C in the south; the coastal influence is manifested as a slight inland gradient in maximum temperatures, on the order of 2°C within about 400 km. Mean daily minima are on the order of 12°–14°C over most of the coast, and decrease slightly inland. January and February are the coldest months.

The diurnal range in winter is about 8°–11°C, depending on latitude. It is a few degrees higher inland. In summer, the diurnal range is a relatively uniform 7°C along the coast, compared with 17°–18°C inland. This again illustrates the stronger influence of the coastal current and cloud cover in summer. The annual range likewise increases inland, ranging from about 3° to 7°C along the coast but from 10° to 15°C inland.

Relative humidity is moderately high along the coast. At Nouakchott, for example, it ranges from about 30% in winter to 66% in summer during the afternoon hours. At Nouadhibou it is close to 50% nearly year-round. Humidity falls rapidly inland to 20–30%.

REFERENCES

Albrecht, B. A., M. P. Jensen, and W. J. Syrett, 1995: Marine boundary layer structure and fractional cloudiness. *Journal of Geophysical Research–Atmospheres*, **100**, 14209–14222.

Amiran, D., and A. Wilson (eds.), 1973: *Coastal Deserts: Their Natural and Human Environments.* University of Arizona Press, Tucson, AZ.

Anderson, B. T., J. O. Roads, S. C. Chen, and H. M. H. Juang, 2001: Model dynamics of summertime low-level jets over northwestern Mexico. *Journal of Geophysical Research*, **106**, 3401–3413.

Andrews, E. D., R. C. Antweiler, P. J. Neiman, and F. M. Ralph, 2004: Influence of ENSO on flood frequency along the California coast. *Journal of Climate*, **17**, 337–348.

Arnesto, J. J., M. K. Arroyo, and L. Hinojosa, 2007: The Mediterranean environment of central Chile. In *The Physical Geography of South America* (T. T. Veblen, K. R. Young, and A. R. Orme, eds.), Oxford University Press, New York, pp. 184–199.

Binet, D., B. Gobert, and L. Maloueki, 2001: El Niño-like warm events in the Eastern Atlantic (6° N, 20° S) and fish availability from Congo to Angola (1964–1999). *Aquatic Living Resources*, **14**, 99–113.

Castro, R., A. Parés-Sierra, and S. G. Marinone, 2003: Evolution and extension of the Santa Ana winds of February 2002 over the ocean, off California and the Baja California peninsula. *Ciencias Marinas*, **29**, 275–281.

Cereceda, P., and R. S. Schemenauer, 1991: The occurrence of fog in Chile. *Journal of Applied Meteorology*, **30**, 1097–1155.

Cereceda, P., and Coauthors, 2002: Advective, orographic and radiation fog in the Tarapacá region, Chile. *Atmospheric Research*, **64**, 261–271.

Cereceda, P., H. Larrain, P. Osses, M. Farías, and I. Egaña, 2008a: The spatial and temporal variability of fog and its relation to fog oases in the Atacama Desert, Chile. *Atmospheric Research*, **87**, 312–323.

Cereceda, P., H. Larrain, P. Osses, M. Farías, and I. Egaña, 2008b: The climate of the coast and fog zone in the Tarapaca region, Atacama Desert, Chile. *Atmospheric Research*, **87**, 301–311.

Clarke, J. D. A., 2006: Antiquity of aridity in the Chilean Atacama Desert. *Geomorphology*, **73**, 101–114.

Cornejo, A. T., 1970: Resources of arid South America. In *Arid Lands in Transition* (H. E. Dregne, ed.), American Association for the Advancement of Science Publication 90, Washington, DC, pp. 345–380.

Covey, D. L., and S. Hastenrath, 1978: The Pacific El Niño phenomenon and the Atlantic circulation. *Monthly Weather Review*, **106**, 1280–1286.

del-Val, E., and Coauthors, 2006: Rain forest islands in the Chilean semiarid region: fog-dependency, ecosystem persistence and tree regeneration. *Ecosystems*, **9**, 1–13.

Douglas, M. W., 1995: The summertime low-level jet over the Gulf of California. *Monthly Weather Review*, **123**, 2334–2347.

Douglas, M. W., A. Valdez-Manzanilla, and R. Garcia Cueto, 1998: Diurnal variation and horizontal extent of the low-level jet over the northern Gulf of California. *Monthly Weather Review*, **126**, 2017–2025.

Douglas, M. W., Mejia, J., N. Ordinola, and J. Boustead, 2009: Synoptic variability of rainfall and cloudiness along the coasts of northern Peru and Ecuador during the 1997/98 El Niño event. *Monthly Weather Review*, **137**, 116–136.

Dubief, J., 1963: *Le Climat du Sahara.* Université d'Alger, Institut de Recherches Sahariennes, Algiers, 2 volumes.

Finley, J., and M. Raphael, 2007: The relationship between El Niño and the duration and frequency of Santa Ana winds of southern California. *Professional Geographer*, **59**, 184–192.

Fischer, D. T., C. J. Still, and A. P. Williams, 2009: Significance of summer fog and overcast for drought stress and ecological functioning of coastal California endemic plant species. *Journal of Biogeography*, **36**, 783–789.

Florenchie, P., C. J. C. Reason, J. R. E. Lutjeharms, M. Rouault, C. Roy, and S. Masson, 2004: Evolution of interannual warm and cold events in the Southeast Atlantic Ocean. *Journal of Climate*, **17**, 2318–2334.

Gammelsrød, T., C. H. Bartholomae, D. C. Boyer, V. L. L. Filipe, and M. J. O'Toole, 1998: Intrusion of warm surface water along the Angolan-Namibian coast in February–March 1995: the 1995 Benguela Niño, Benguela dynamics. *South African Journal of Marine Science*, **19**, 41–56.

Garreaud, R., J. Barichivich, D. A. Christie, and A. Maldonado, 2008: Interannual variability of the coastal fog at Fray Jorge relict forests in semiarid Chile. *Journal of Geophysicas Research*, **113**, doi:10.1029/2008JG000709.

Garreaud, R., J. Rutllant, J. Quintana, J. Carrasco, and P. Minnis, 2001: CIMAR-5: A snapshot of the lower troposphere over the Southeast subtropical Pacific. *Bulletin of the American Meteorological Society*, **82**, 2193–2207.

Gutiérrez, A. G., and Coauthors, 2008: Regeneration patterns and persistence of the fog-dependent Fray Jorge forest in semiarid Chile during the past two centuries. *Global Change Biology*, **14**, 161–176.

Hastings, J. R., and R. M. Turner, 1965: Seasonal precipitation regimes in Baja California. *Geografiska Annaler*, **47**, 204–223.

Henschel, J. R., and M. K. Seely, 2008: Ecophysiology of atmospheric moisture in the Namib Desert. *Atmospheric Research*, **87**, 362–368.

Herrera-Cervantes, H., D. B. Lluch-Cota, S. E. Lluch-Cota, and G. D. V. S. Guillermo, 2007: The ENSO signature in sea-surface temperature in the Gulf of California. *Journal of Marine Research*, **65**, 589–605.

Hildebrant, A., and A. B. Eltahir, 2006: Forest on the edge: Seasonal cloud forest in Oman creates its own ecological niche. *Geophysical Research Letters*, **33**, doi:10.1029/2006GL026022.

Houston, J., 2006: Variability of precipitation in the Atacama desert: its causes and hydrological impact. *International Journal of Climatology*, **26**, 2181–2198.

Houston, J., and A. Hartley, 2003: The central Andean west-slope rainshadow and its potential contribution to the origin of hyper-aridity in the Atacama Desert. *International Journal of Climatology*, **23**, 1453–1464.

Johnson, A. M., 1976: The climate of Peru, Bolivia and Ecuador. In *Climates of South America* (W. Schwerdtfeger, ed.), World Survey of Climatology 12, Elsevier, Amsterdam, pp. 147–188.

Keeley, J. E., 2004: Impact of antecedent climate on fire regimes in coastal California. *International Journal of Wildland Fire*, **13**, 173–182.

Klein, S. A., and D. L. Hartmann, 1993: The seasonal cycle of low stratiform clouds. *Journal of Climate*, **6**, 1587–1606.

Lancaster, J., N. Lancaster, and M. K. Seely, 1984: Climate of the Central Namib Desert. *Madoqua*, **14**, 5–61.

Lancaster, N., 1982: Dunes on the Skeleton Coast, Namibia (South West Africa): geomorphology and grain size relationships. *Earth Surface Processes and Landforms*, **7**, 575–587.

Lancaster, N., 1990: Regional aeolian dynamics in the Namib. *Namib Ecology* (M. K. Seely, ed.), Transvaal Museum, Pretoria, pp. 39–46.

Leipper, D. F., 1994: Fog on the U. S. West Coast: a review. *Bulletin of the American Meteorological Society*, **75**, 229–240.

Lettau, H. H., 1978a: Explaining the world's driest climate. In *Exploring the World's Driest Climate* (H. H. Lettau and K. Lettau, eds.), Institute for Environmental Studies, University of Wisconsin, Madison, WI, pp. 182–248.

Lettau, H. H., 1978b: Introduction. In *Exploring the World's Driest Climate* (H. H. Lettau and K. Lettau, eds.), Institute for Environmental Studies, University of Wisconsin, Madison, WI, pp. 12–29.

Lettau, H. H., and K. Lettau, 1978: *Exploring the World's Driest Climate*. Institute for Environmental Studies, University of Wisconsin, Madison, WI, 264 pp.

Liebmann, B., and Coauthors, 2009: Origin of convectively coupled Kelvin waves over South America. *Journal of Climate*, **22**, 300–315.

Lindesay, J. A., and P. D. Tyson, 1990: Climate and near-surface airflow over the central Namib. In *Namib Ecology* (M. K. Seely, ed.), Transvaal Museum, Pretoria, pp. 27–37.

Lydolph, P. E., 1973: On the causes of aridity along a selected group of coasts. In *Coastal Deserts: Their Natural and Human Environments* (D. H. K. Amiran and A. W. Wilson, eds.), University of Arizona Press, Madison, WI, pp. 67–72.

MacMahon, J. A., and F. H. Wagner, 1985: The Mojave, Sonoran and Chihuahuan Deserts of North America. In *Hot Deserts and Arid Shrublands* (M. Evenari, I. Noy-Meir, and D. W. Goodall, eds.), Ecosystems of the World 12A, Elsevier, Amsterdam, pp. 105–202.

Marchesiello, P., and P. Estrade, 2009: Eddy activity and mixing in upwelling systems: a comparative study of Northwest Africa and California regions. *International Journal of Earth Sciences*, **98**, 299–308.

Markham, C. G., 1972: Baja California's climate. *Weatherwise*, **25**, 64–76.

Martyn, D., 1991: *Climates of the World*. Elsevier, 425 pp.

Marzol, M. Y., 2008: Temporal characteristics and fog water collection during summer in Tenerife (Canary Islands, Spain). *Atmospheric Research*, **87**, 352–361.

McKay, C. P., E. I. Friedmann, B. Gomez-Silva, L. Caceres-Villanueva, D. T. Andersen, and R. Landheim, 2003: Temperature and moisture conditions for life in the extreme arid region of the Atacama Desert: four years of observations including the El Niño of 1997–1998. *Astrobiology*, **3**, 393–406.

Meigs, P., 1966: *Geography of Coastal Deserts*. Arid Zone Research 28, UNESCO, Paris, 140 pp.

Miller, A., 1976: The climate of Chile. In *Climates of South America* (W. Schwerdtfeger, ed.), World Survey of Climatology 12, Elsevier, Amsterdam, pp. 113–146.

Minnich, R. A., E. F. Vizcaino, and R. J. Dezzani, 2000: The El Niño/Southern Oscillation and precipitation variability in Baja California, Mexico. *Atmósfera*, **13**, 1–20.

Monaghan, A. J., D. L. Rife, J. O. Pinto, C. A. Davis, and J. R. Hannan, 2010: Global precipitation extremes associated with diurnally varying low-level jets. *Journal of Climate*, **23**, 5065–5084.

Monteverdi, J. P., 1973: The 'Santa-Ana' weather type and extreme fire hazard in the Oakland–Berkeley Hills. *Weatherwise*, **26**, 118–121.

Moseley, M. A., 1978: An empirical approach to prehistorical agrarian collapse: the case of the Moche Valley, Peru. In *Social and Technological Management in Dry Lands* (N. Gonzalez, ed.), Westview Press, Boulder, CO, pp. 9–43.

Muller, A., C. J. C. Reason, and N. Fauchereau, 2008: Extreme rainfall in the Namib Desert during late summer 2006 and influences of regional ocean variability. *International Journal of Climatology*, **28**, 1061–1070.

Muñoz, R. C., and R. D. Garreaud, 2005: Dynamics of the low-level jet off the west coast of subtropical South America. *Monthly Weather Review*, **133**, 3661–3677.

Murphy, R. C., 1926: Oceanic and climatic phenomena along the coast of South America during 1925. *Geographical Review*, **16**, 26–54.

Neiburger, M., D. S. Johnson, and C.-W. Chien, 1961: *Studies of the Structure of the Atmosphere over the Eastern Pacific Ocean in Summer*. University of California Press, Los Angeles, 94 pp.

Nicholson, S. E., 1975: Sea-surface temperature variation off the west coast of Africa during Phase I of GATE. *Preliminary Scientific Results of the GARP Atlantic Tropical Experiment*, Gate Report 14, vol. 1, pp. 129–136.

Nicholson, S. E., 2010: A low-level jet along the Benguela coast, an integral part of the Benguela Current ecosystem. *Climatic Change*, **33**, 313–330.

Nicholson, S. E., and E. Entekhabi, 1987: Rainfall variability in Equatorial and Southern Africa: relationships with sea-surface temperatures along the southwestern coast of Africa. *Journal of Climate and Applied Meteorology*, **26**, 561–578.

Nicholson, S. E., J. Y. Kim, and J. Hoopingarner, 1988: *Atlas of African Rainfall and its Interannual Variability*. Florida State University, Tallahassee, FL.

Nieman, W. A., C. Heyns, and M. K. Seely, 1978: A note on precipitation at Swakopmund. *Madoqua*, **11**, 69–73.

Oliver, J., and P. L. Stockton, 1989: The influence of upwelling extent upon fog incidence at Lüderitz, Southern Africa. *International Journal of Climatology*, **9**, 69–75.

Parés-Sierra, A., A. Mascarenhas, S. G. Marinone, and R. Castro, 2003: Temporal and spatial variation of the surface winds in the Gulf of California. *Geophysical Research Letters*, **30**, doi:10.1029/2002GL016716.

Parish, T., 2000: Forcing of the summer low-level jet along the California coast. *Journal of Applied Meteorology*, **39**, 2421–2433.

Pavia, E. G., and F. Graef, 2002: The recent rainfall climatology of the Mediterranean Californias. *Journal of Climate*, **15**, 2697–2701.

Pinty, R., I. Barria, and P. A. Marquet, 2006: Geographical distribution of Tillandsia lomas in the Atacama Desert, northern Chile. *Journal of Arid Environment*, **65**, 543–552.

Rauh, W., 1985: The Peruvian-Chilean deserts. In *Hot Deserts and Arid Shrublands* (M. Evenari, I. Noy-Meir, and D.W. Goodall, eds.), Ecosystems of the World 12A, Elsevier, Amsterdam, pp. 239–268.

Reyes-Coca, S., and R. Troncoso-Gaytan, 2004: Multidecadal variation of winter rainfall in northwestern Baja California. *Ciencias Marinas*, **30**, 99–108.

Reynolds, R. W., 1982: A monthly averaged climatology of sea surface temperature. *NOAA Tech. Rep.*, NWS 31, 35 pp.

Rouault, M., S. A. White, C. J. C. Reason, J. R. E. Lutjeharms, and I. Jobard, 2002: Ocean–atmosphere interaction in the Agulhas Current region and a South African extreme weather event. *Weather and Forecasting*, **17**, 655–669.

Rutllant, J., 1994: *On the Generation of Coastal Lows in Central Chile*. IAEA/UNESCO Internal Report IC/94/167, International Centre for Theoretical Physics, Trieste, Italy, 15 pp.

Santos, A. M. P., A. S. Kazmin, and A. Peliz, 2005: Decadal changes in the Canary upwelling system as revealed by satellite observations: their impact on productivity. *Journal of Marine Research*, **63**, 359–379.

Schonher, T., and Nicholson, S. E., 1989: The relationship between California rainfall and ENSO events. *Journal of Climate*, **2**, 1258–1269.

Schwerdtfeger, W., ed., 1976: *Climates of South America*. World Survey of Climatology 12, Elsevier, Amsterdam, 381 pp.

Seely, M. K., 1978: The Namib Dune Desert: an unusual ecosystem. *Journal of Arid Environments*, **1**, 117–128.

Seely, M. K., and G. N. Louw, 1980: First approximation of the effects of rainfall on the ecology and energetics of a Namib Desert dune ecosystem. *Journal of Arid Environments*, **3**, 2–34.

Sommers, W. T., 1981: Waves on a marine inversion undergoing mountain leeside wind shear. *Journal of Applied Meteorology*, **20**, 626–636.

Svejkovsky, J., 1985: Santa Ana airflow observed from wildfire smoke patterns in satellite imagery. *Monthly Weather Review*, **113**, 902–906.

Rundel, P. W., P. Villagra, M. Dillon, S. Roig-Juñent, and G. Debandi, 2007: Arid and semi-arid ecosystems. In *The Physical Geography of South America* (T. T. Veblen, K. R. Young, and A. R. Orme, eds.), Oxford University Press, New York, pp. 158–183.

Rutllant, J. A., H. Fuenzalida, and P. Aceituno, 2003: Climate dynamics along the arid northern coast of Chile: the 1997–1998 Dinamica del Clima de la Region de Antofagasta (DICLIMA) experiment. *Journal of Geophysical Research–Atmospheres*, **108**, 4538.

Shannon, L. V., A. J. Boyd, G. B. Brundrit and J. Taunton-Clark, 1986: On the existence of an El Niño-type phenomenon in the Benguela system. *Journal of Marine Research*, **44**, 495–520.

Trewartha, G. T., 1970: *The Earth's Problem Climates*. University of Wisconsin Press, Madison, WI, 334 pp.

Tyson, P. D., and M. K. Seely, 1980: Local winds over the central Namib. *South African Geographical Journal*, **62**, 135–150.

Vogelmann, H. W., 1973: Fog precipitation in the cloud forests of eastern Mexico. *Biogeoscience*, **23**, 96–100.

Walker, N. D., 1990: Links between South African summer rainfall and temperature variability of the Agulhas and Benguela current systems. *Journal of Geophysical Research*, **95**, 3297–3319.

Ward, J. D., and I. Corbett, 1990: Towards an age for the Namib. *Namib Ecology: 25 Years of Namib Research* (M. K. Seely, ed.), Transvaal Museum Monograph 7, Pretoria, pp. 17–26.

White, F., 1983: The *Vegetation of Africa*. UNESCO, Paris, 356 pp.

Williams, A. P., C. J. Still, D. T. Fischer, and S. W. Leavitt, 2008: The influence of summertime fog and overcast clouds on the growth of a coastal Californian pine: a tree ring study. *Oecologia*, **156**, 601–611.

Part V Life and change in the dryland regions

21 Drought and other hazards

21.1 INTRODUCTION

Meteorological phenomena that pose hazards in dryland regions include droughts, floods, snow, wildfires, and dust storms. Drought is a protracted phenomenon; abnormally dry conditions must generally persist for a month or more before a drought situation arises. It results from changes in large-scale circulation patterns, usually quasi-continental or greater in spatial extent. Floods and dust storms are usually coupled to individual weather events, although floods may also result from long periods of excessive rainfall. Flood events are generally linked to smaller, mesoscale meteorological patterns, although their occurrence may be made more likely by conducive, large-scale atmospheric conditions. Drought is often a prelude to major dust storms in semi-arid regions, since the aridity promotes the mobilization of dust. Wildfires are also associated with prolonged drought. Snow is a relatively rare phenomenon in the warm drylands and is associated with very unusual weather conditions. Because it occurs infrequently, the inhabitants of these regions are generally not equipped to cope with it.

All of these hazards have major physical and economic consequences. Some of the impacts are qualitatively the same for several of these hazards. Drought reduces the surface vegetation cover and leaves the soil prone to erosion by both wind and water; crops fail and herds die off. Floods, fires, and dust storms similarly destroy the vegetation and remove or destroy fertile soil; agricultural and economic consequences can be comparable to those due to drought. The impact of floods can be exacerbated by their frequent occurrence during what is normally the dry season, a time when vegetation cover is reduced and the ground is more prone to runoff and erosion.

21.2 PHYSICAL ASPECTS OF DROUGHT

21.2.1 DEFINING DROUGHT

Drought and aridity are closely allied, but distinctly different, phenomena. In the older literature, drought is often used synonymously with general climatic aridity or with the dry season, but the term should be reserved for protracted departures from "normal" conditions of water availability. That is, drought is a relative condition that temporarily prevails rather than an absolute measure of aridity. Furthermore, unlike a flood, drought is not a distinct event. It often has neither a distinct beginning nor a recognizable end, since intermittent wet spells occur within long droughts.

For these reasons there is no straightforward or universal definition of drought. Its occurrence depends on "supply" and "demand"; hence drought is linked both to water balance and to societal needs and expectations (Wilhite 2000). The best definition is perhaps the most general: drought is a situation in which the water supply is insufficient to meet the habitual needs of people. How large the deficit must be and how long it must persist to be declared a drought thus depends on both expectations and needs. The expectations and requirements for Sahelian nomadic pastoralists are likely to be lower than for the residents of Palm Springs, where population density is high and abundant irrigation turns the desert into scores of grass-covered golf courses. The rainfall deficits producing drought may be relatively small in low latitudes or in regions where abnormally low rainfall is coupled with reduced cloudiness, because high evaporative loss enhances the impact of decreased rainfall. Table 21.1 gives some examples of criteria and definitions currently in use.

Recognizing this complexity, the American Meteorological Society (AMS), the World Meteorological Organization (WMO), and many international organizations have adopted several generic definitions of drought. These include, for example, "prolonged absence or marked deficiency of precipitation," a "deficiency of precipitation that results in water shortage for some activity or for some group," and "a period of abnormally dry weather sufficiently prolonged for the lack of precipitation to cause a serious hydrological imbalance" (Heim 2002). These generic definitions are advantageous because they have a universal range of applicability and work in a large number of situations (Redmond 2002).

Table 21.1. *List of drought criteria/definitions used in various countries (Heim 2002).*

Britain	15 consecutive days with less than 0.25 mm (0.01 in) rainfall
Britain	15 consecutive days with less than 1.0 mm (0.04 in) rainfall
India	Rainfall half of normal or less for a week, or actual seasonal rainfall deficient by more than twice the mean deviation
Russia	10 days with total rainfall not exceeding 5 mm (0.20 in)
Bali	A period of 6 days without rain
Libya	Annual rainfall that is less than 180 mm (7 in)

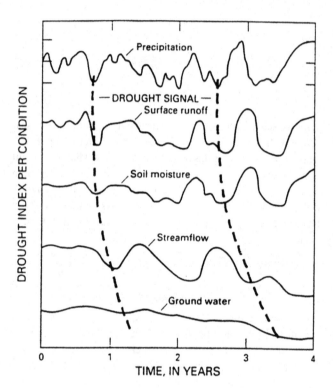

Fig. 21.1 Propagation of precipitation anomalies through the surface branches of the hydrological cycle (Changnon 1987).

21.2.2 TYPES OF DROUGHT

Traditionally three types of drought are distinguished: meteorological, agricultural, and hydrological. The first refers primarily to subnormal rainfall, the second to soil moisture deficits, and the third primarily to water storage in reservoirs such as lakes, rivers, and groundwater. More recently, a fourth type, socioeconomic drought, was proposed by the American Meteorological Society (AMS 2004). For each of these types of drought, the thresholds for occurrence are generally arbitrary and usually locally defined. A survey of the relevant literature shows that the thresholds and definitions vary substantially from author to author.

Meteorological drought depends only on precipitation statistics for a given place and time of year. Because of the large spatial variability of rainfall, especially in dryland regions, in any given year or season some location is likely to receive abnormally low rainfall. For this reason, some authors suggest using

the term drought only for periods of deficient water availability that are extensive in time and space.

Precipitation also plays a role in the other types of drought. For that reason, the remaining types are also commonly evaluated with meteorological statistics. However, they cannot be defined solely on the basis of precipitation. Agricultural drought, for example, depends largely on the type of crop and on the precipitation in the growing season, but evapotranspiration and runoff play a role. A severe meteorological drought might not be an agricultural one if there is a substantial reservoir of soil moisture. Conversely, small rainfall deficits can produce agricultural drought in marginal regions with little soil moisture reserve. Hydrological droughts necessarily relate to temporally and spatially extensive precipitation deficits, since hydrologic reservoirs integrate precipitation in time and space. These reservoirs will generally not be significantly affected by short-duration, localized dry conditions.

Figure 21.1 (from Changnon 1987) illustrates the relationships between these three types of drought. The most noticeable feature is the time lag between various elements of the water cycle. The lag between precipitation and soil moisture is relatively small, although it depends greatly on the soil type and soil conditions. The lag to streamflow and groundwater can be considerable, on the order of months or more. For this reason, a hydrological drought might still be occurring while precipitation or soil moisture has returned to normal or above-normal levels.

21.2.3 DROUGHT INDICES

A number of indices have been derived to describe drought intensity, but these are at least as varied as the definitions of drought. In fact, the lack of a standard definition for drought makes it impossible to design a universally applicable standard. Moreover the sampling periods, record lengths, and homogeneity of records differ tremendously from one location to another. This makes it difficult to produce comparable statistics and indices for diverse locations.

Drought indices are nonetheless valuable, even indispensable, in order to anticipate its occurrence and mitigate the resultant impacts. The importance of developing indices, as well as the difficulties in doing so, is shown by a series of articles on indices (Redmond 2002; Heim 2002; Keyantash and Dracup 2002) and monitoring (Svoboda *et al.* 2002; Lawrimore *et al.* 2002)

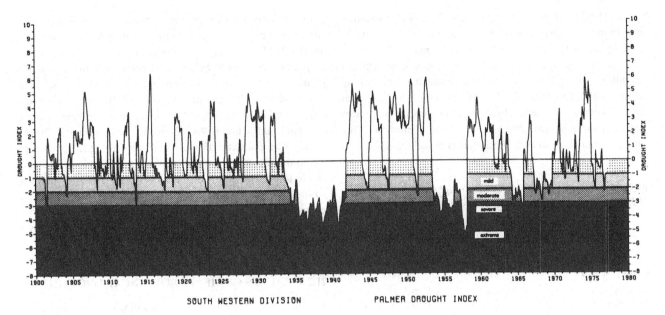

Fig. 21.2 Time series of the Palmer drought index (PDI) for southwestern Kansas, from 1900 to 1980 (Felch 1978). The long droughts in the 1930s and 1950s stand out. At the heart of the drought-stricken area, precipitation was roughly 30% below the long-term mean.

that appeared in the *Bulletin of the American Meteorological Society*. These provide excellent reviews and extensive detail on many of the recently developed indices and monitoring efforts. Much of the following discussion is based on these excellent sources.

The perplexing problem of developing drought indices has been considered for a long time. Friedman (1957) put forth several criteria that should be met:

1. the index should be applicable to the problem at hand and its time scale;
2. the index should be a quantitative measure of large-scale and persistent conditions;
3. a long, accurate past record of the index should be realizable.

Heim (2002) adds an additional criterion: indices for operational drought monitoring should be computable on a near-real-time basis.

Redmond (2002) suggests that, as a starting point in developing a drought index, one needs to ask three questions: what is its purpose, what is its audience, and what do they want to know? Unfortunately, when an index is developed for operational use, it is often implemented with little evaluation of its statistical properties, dependency on geographical or seasonal factors, response to input data, or relationship to climate impact information. In other words, there is a serious failure to determine whether the index has a reasonable relationship with the aspect of the drought or drought impact it is designed to monitor.

Probably the most widely used is the Palmer drought index (PDI), which is based on precipitation and cumulative moisture availability, or antecedent precipitation, plus moisture demand (Palmer 1965). An example of the PDI is shown for southwestern Kansas in Fig. 21.2. The droughts of the 1930s and 1950s clearly stand out.

What is termed the "Palmer drought index" is actually three indices that have come to be known as the PDSI, PHDI, and the Z index. The Palmer drought severity index (PDSI) is a measure of meteorological drought. The values of the PDSI corresponding to various categories of drought and wet conditions are given in Table 21.2. Except for the most extreme cases, the values range from -4 to $+4$ and mild droughts and wet spells begin at -1 and $+1$, respectively. A comparison of these three indices with several others is presented in Fig. 21.3. Considering the lag time between the cessation of a meteorological drought and the environmental recovery from the drought, Palmer also developed the PHDI, or Palmer hydrologic drought index. This assesses long-term hydrologic moisture conditions. The third index, the Z index, is essentially an accounting of a "moisture anomaly," based on the work of Thornthwaite (Thornthwaite and Mather 1957).

In the 1990s, the National Weather Service modified the PHDI to develop the PMDI, or Palmer modified drought index. A modified calculation was introduced that facilitates the operational use of this index. An example of its use to describe the drought of the 1950s is shown in Fig. 21.4. During September 1956, nearly 40% of the contiguous United States experienced severe to extreme drought.

Another widely used index is the standardized precipitation index (SPI) developed by McKee *et al.* (1995). It was designed primarily to recognize the importance of time scale in defining drought. It also recognizes that accumulated precipitation can be both in excess and in deficit, depending on the time scale. The SPI describes precipitation in absolute terms as well as absolute and percentage departures, and departures in frequency space (i.e., deciles). The SPI is extremely well correlated with the Palmer drought index at time scales of 6–12 months; the 9-month SPI generally correlates best with the PDI (Redmond 2002).

The previously described indices relate to meteorological and agricultural drought. Other indices are available for describing hydrological drought (e.g., Huff and Changnon 1964; Yevjevich 1967; Dracup *et al.* 1980). These are analogous to the indices developed for other drought types, but utilize streamflow as the basic parameter, rather than precipitation or soil moisture.

Table 21.2. *The values of the Palmer drought severity index (PDSI) corresponding to various categories of drought and wet conditions (from Heim 2002; Palmer 1965).*

Moisture category	PDSI
Extremely wet	≥4.00
Very wet	3.00 to 3.99
Moderately wet	2.00 to 2.99
Slightly wet	1.00 to 1.99
Incipient wet spell	0.50 to 0.99
Near normal	0.49 to −0.49
Incipient drought	−0.50 to −0.99
Mild drought	−2.00 to −2.99
Severe drought	−3.00 to −3.99
Extreme drought	≤−4.00

Most drought indices are not universally applicable to all situations or climate types. Some useful general conclusions are that the Palmer Z index is preferable to the commonly used crop moisture index (CMI) for evaluating agricultural drought (Karl 1986), that the Palmer drought index may not be useful in arid regions (Bhalme and Mooley 1979), and that it is not spatially comparable across broad regions or between months.

A particularly useful study is that of Keyantash and Dracup (2002), who evaluated 18 indices of meteorological, hydrological, and agricultural drought according to a series of criteria. Interestingly, the highest total scores were given to the most straightforward indices: rainfall deciles, total water deficit, and computed soil moisture. For all three drought types, the Palmer indices ranked low. Morid *et al.* (2006) also evaluated seven types of drought indices for Iran.

21.2.4 CAUSES OF METEOROLOGICAL DROUGHT

Two major characteristics of meteorological drought are (1) they result from a persistent deficit in rainfall, and (2) they are generally large in scale – at least regional and sometimes continental. The large spatial scale means that an anomalous feature in the general atmospheric circulation is likely involved. The persistence of the drought implicates slowly varying boundary forcings as important factors in its development. Thus, drought is essentially a regional manifestation of a persistent anomaly

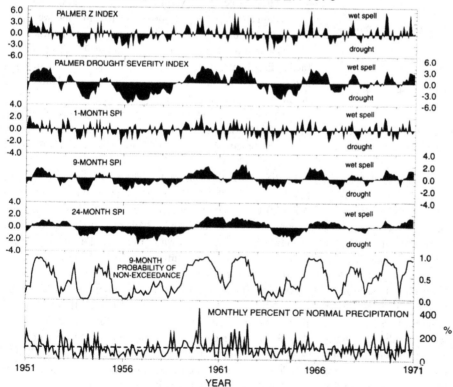

Fig. 21.3 Comparison of select drought indices for east-central Iowa for the period January 1, 1951 to December 31, 1970 (from Heim 2002).

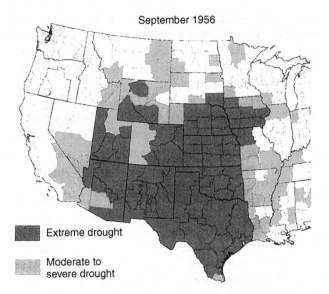

September 1956

Fig. 21.4 The Palmer modified drought index (PMDI) showing the cumulative long-term drought conditions for September 1956 (from Heim 2002).

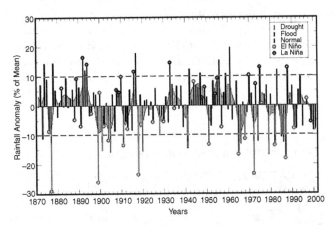

Fig. 21.5 All-India rainfall series, with La Niña and El Niño years indicated (courtesy of the Indian Institute of Tropical Meteorology). Rainfall is expressed as a regionally averaged percent departure from the long-term mean.

in the general atmospheric circulation, but the underlying cause is often found at the atmosphere's lower boundary. A common trigger is anomalous atmospheric heating associated with shifts in sea-surface temperature (SST) patterns.

Many factors produce meteorological drought and normally several play a role in any particular event. The causes of drought are also both regionally and seasonally specific. For this reason, it is difficult to make a concise statement about the causes of drought. However, a few generalizations can be made. One is that the generic, immediate cause of a meteorological drought is a change in the intensity of the rain-producing mechanism, a displacement of the rain-bearing circulation system, or a reduction in the spatial extent of that system.

In the higher latitudes, the immediate causes of drought are relatively easy to identify; usually stationary high pressure replaces

low pressure and accompanying disturbances. This may occur when the extra-tropical jet streams are displaced or unusually weak, thereby modifying storm development and storm tracks. Those changes, in turn, are generally associated with various surface or boundary forcings, such as SSTs or snow cover. In the US Great Plains, for example, the cause of drought appears to be an anomalous storm track, related to changes in the location of the jet streams (Trenberth and Guillemot 1996). This, in turn, is linked to Pacific SSTs. In the low latitudes, where atmospheric dynamics are quite different and rainfall is more localized, the causes are not as well understood. Quite often it can be shown that drought in the tropical or subtropical latitudes is initiated by abnormal SSTs either in nearby oceans or globally, yet the direct cause in the regional meteorology (e.g., pressure, winds, and disturbances) may still be unknown.

In many low-latitude regions, drought is often coupled with the occurrence of ENSO events. In India, for example, the most severe droughts tend to occur during El Niño years (Fig. 21.5). ENSO events are disruptions of the usual pattern of SSTs in the Pacific and they produce weather changes over much of the globe. There is generally more widespread and severe drought associated with El Niño (Fig. 21.6). However, some regions instead suffer drought during La Niña years.

The globally pervasive influence of ENSO accounts for the recurrent spatial patterns associated with drought. Notable droughts with a global "footprint" include several episodes in the late nineteenth century, the 1930s, and the 1950s (Herweijer and Seager 2008). Such links between spatially distant locations are referred to as climatic "teleconnections." Table 21.3 gives an example of such teleconnections between droughts in Australia, India, and the Sahel. In every case but one, drought occurred synchronously in two (and often all three) of these locations.

It should be mentioned that a large body of literature makes a connection between solar activity, notably sunspots, and drought. This issue is extremely controversial. Few studies of the link between sunspots and drought have any degree of statistical rigor. One possible exception is that of the link between sunspots and the extent of drought in the midwestern USA (Mitchell *et al.* 1979). Although that study is controversial, the association is reasonably strong. Recent work also suggests that solar activity plays a role in historical fluctuations of climate (Pang and Yau 2002) and appears to modulate the intensity of solar radiation reaching the earth (Lean and Rind 1998).

21.2.5 BIOGEOPHYSICAL FEEDBACK IN DROUGHT

Some years ago, in a widely publicized study, Charney (1975) proposed that meteorological droughts can have a human origin (see Chapter 6). His theory was specifically formulated for the West African Sahel, which was undergoing severe drought in the late 1960s and 1970s. This region had the prerequisites for the realization of his idea: a negative net radiation balance at the top of the atmosphere and a highly reflective soil (i.e., a light-colored soil with relatively high albedo). Charney

Table 21.3. *Historical occurrence of drought in Australia, India, and the West African Sahel (based on Foley 1957; Nicholls 1992; Nicholson 1976).*

Australia	India	Sahel
1789–1791	1790–1792	1790s
1793		1790s
1797		1790s
1798–1800		1790s
1802–1804	1802–1804	
1808–1815	1812–1813	1809–1813
1818–1821		1820–1822
1824	1824–1825	
1827–1829	1828	1828–1838
1833	1832–1833	1828–1838
1837–1839	1837–1839	1828–1838
1842–1843		1843–1844
1846–1847		1846–1853
1849–1852		1846–1853
1855		1855
1857–1859	1856–1858	
1861–1862		
1865–1869	1865–1866	1863–1866
1872		1871–1872
1875–1877	1875–1877	
1880–1881		1880
1884–1886	1884	

suggested that overgrazing in the Sahel had denuded the land, baring its soil and sending more radiation back to space. He hypothesized that this radiative loss would be offset by warming via subsidence of the air, a situation that would reduce rainfall, thus expanding the desert and producing drought in the Sahel. Unfortunately, Charney's idea has been taken as gospel in some disciplines and it is applied even to areas where the prerequisites are not present. Overall the physical relationships he describes are sound, but insufficient to produce drought.

Charney's ideas have essentially been reformulated to create the concept of biogeophysical feedback in droughts: various interactions between the land surface and atmosphere. Extensive evidence suggests that various physical feedback processes can modify an existing drought, causing it to intensify, spread, or persist. Some indicate that the feedback is even essential. "Development and growth of feedback mechanisms are what transform an anomalous perturbation into a persistent and prolonged drought" (Entekhabi *et al.* 1992). In other words, although the meteorological cause of drought is an anomalous condition of the general atmospheric circulation, this may be a necessary – but insufficient – condition for the development of severe and sustained drought.

A number of studies have evaluated feedbacks via physically based statistical models with interactive climatic and hydrologic processes (e.g., Entekhabi *et al.* 1992; Entekhabi and Rodriguez-Iturbe 1994). Several characteristics of the feedback emerged.

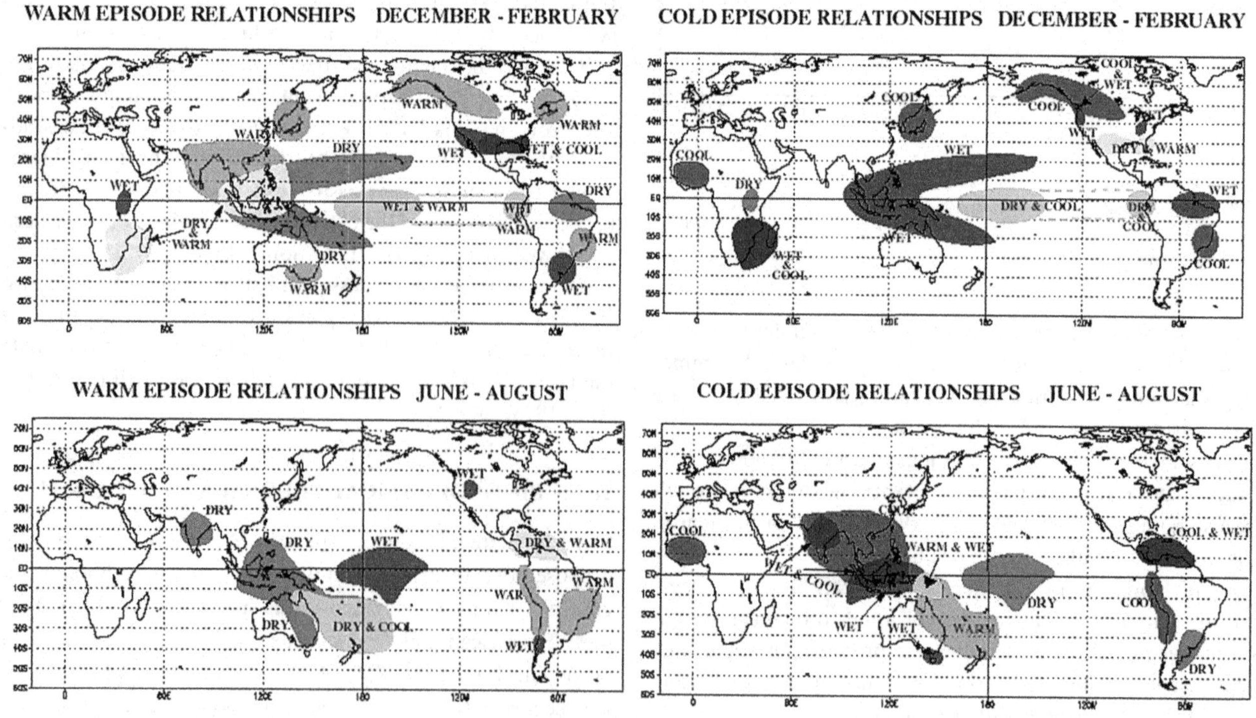

Fig. 21.6 Locations that tend to experience drought or wetter conditions in association with El Niño (warm episodes) and La Niña (cold episodes) (from NOAA website).

One is that the heat and moisture fluxes are governed by factors acting on different time scales. Also, the direction of some of the feedback loops is different for dry states that are "soil controlled" and wet states that are "atmosphere controlled." Perhaps most importantly, the feedbacks are stronger in the dry case than the wet case, strongly tending to amplify drought.

The evidence also suggests that changes in soil moisture, soil characteristics, or vegetation type are more likely to induce such feedback than albedo. In theory the feedback processes can be of natural or human origin. An example of the former is soil moisture reduction due to reduced rainfall. In general, human surface modification has not reached such a level that it is comparable with the natural feedback mechanism.

A handful of observations support the concept of feedback amplifying drought. Once a drought is initiated, the probability of a transition to a normal state is less in the interior (Karl *et al.* 1987), where precipitation recycling, a rough measure of the importance of land surface hydrology, is greatest (Brubaker *et al.* 1993). Also, the degree of recycling is greatest in West Africa, where the magnitude, duration, and extent of drought is particularly severe (Fig. 21.7). Finally, continental interiors tend to be more drought-prone, and droughts persist longer in the interior (Diaz 1983; Karl 1983).

The gist of the foregoing discussion is that severe, large-scale drought is not purely of meteorological origin. At least in the dryland regions, interactions between the land and atmosphere play a role, transforming an initial rainfall deficit into a persistent condition of dryness. The interaction goes in two directions. The rainfall deficit affects the conditions of the land and its fluxes to the atmosphere. These fluxes potentially interact with the atmosphere in ways that enhance the initial conditions producing the rainfall deficit.

This suggests that anthropogenic changes to the earth's surface can modulate drought, but they cannot produce it independently of the initial meteorological anomaly. In other words, people (i.e., the land surface changes caused by them) do not cause meteorological drought. However, they may exacerbate its effects by altering the capacity of the surface to store water. This influences whether or not a meteorological drought becomes an agricultural or hydrologic drought. Thus, land conservation is an important aspect of drought management (see Section 21.3).

21.2.6 CHARACTERISTICS OF METEOROLOGICAL DROUGHT

Meteorological drought is generally characterized by statistics of duration, frequency or return time, magnitude, and severity. The magnitude is the average rainfall deficit during the drought event, and the severity is the total deficit during the event. Thus severity is a combination of magnitude and duration. The data from which these statistics can be calculated are routinely compiled by the weather services of most countries. This essentially involves a hydrological analysis of rainfall data, as described in Chapter 11. However, the derivation of these drought statistics requires a working definition of meteorological drought.

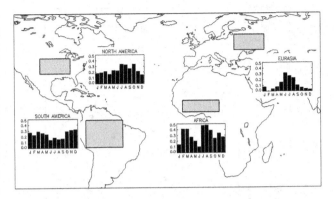

Fig. 21.7 Moisture recycling (in %) for January–December for each of the continental regions shown (based on Brubaker *et al.* 1993).

Definitions for the drought's onset and end are also required and these are difficult to delineate. Comparable statistical quantities can be derived also for agricultural and hydrological droughts, but these are generally more subjective since the question of impact is so critical in these cases.

A number of generalizations can be made about the spatial characteristics of drought. For one, because of the coupling of local and large-scale causal factors, severe droughts are unlikely to affect small areas. In general, the more severe a drought, the larger the area affected. In Africa and Australia, droughts can be nearly continental in scale. In 1983, rainfall was below normal throughout virtually all of tropical Africa (Fig. 21.8); this was part of a multi-year dry period. In 1982/83 a major drought affected the entire Australian continent. Record low rainfall occurred over most of the southeast and deficits were severe throughout most of the eastern half of Australia. During the US Great Plains droughts of the 1930s and 1950s, nearly 100% of the region suffered severe or extreme drought in some years, such as 1934 and 1956 (Fig. 21.9).

Another typical characteristic is the tendency for drought to exhibit recurrent spatial patterns. Oladipo (1986) shows, for example, that four patterns typify most of the variability in the central USA and that the greatest severity is usually in the Great Plains. The whole of central USA is seldom affected. In general, there is an opposition between the eastern and western sectors of the country, so that when drought prevails in the west, good rainfall is likely in the east (Diaz and Quayle 1980).

The spatial patterns can often provide some clue as to the cause of a drought. A quasi-continental-scale drought, as typically occurs in the Sahel, cannot result from purely local factors, such as the albedo mechanism proposed by Charney. A drought this extensive probably results from reduced intensity of the rain-producing mechanism. In contrast, the pattern of opposition between two regions suggests that the likely cause may be a displacement of the main rain-bearing systems. Nkemdirim and Weber (1999) conclude that the spatial character of drought in Canada in the 1930s indicated a southward displacement of the Alberta cyclone track from its normal path. The previously mentioned inverse relationship between drought in the eastern

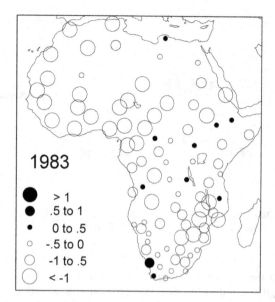

Fig. 21.8 Rainfall over Africa during 1983, expressed in units of standard departure (or deviation) from the long-term mean (from Nicholson 1993). The circles represent spatially averaged rainfall for some 90 regions of the continent, with the analysis based on some 1200 stations.

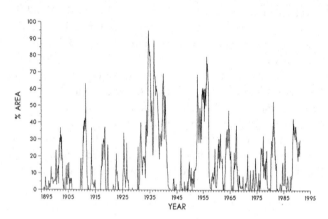

Fig. 21.9 Percentage of the US Great Plains experiencing severe or extreme values of the Palmer drought index from 1895 to 1995 (from Rasmusson and Arkin 1993).

and western United States indicates a shift in trough positions in the upper-level circulation.

21.2.7 DROUGHT PREDICTION

There are three distinct approaches to the long-range forecasting of seasonal weather anomalies such as drought (Hastenrath 1995, 2002):

1. empirical methods that are purely statistical (i.e., lacking any dynamical relationships in the forecast scheme);
2. empirical methods based on a combination of general circulation diagnostics (atmospheric dynamics) and statistical methods;
3. numerical models.

Early drought forecasting has traditionally relied on purely statistical models, but that picture is changing.

The empirical models that combine general circulation diagnostics and statistics are based on teleconnections and other relationships between meteorological variables. In these models, parameterizations are used to incorporate dynamical cause/effect relationships. The first step is evaluating the general atmospheric circulation to determine dynamic relationships between variables and to identify predictor candidates. At present, empirical models have a modest degree of forecasting skill for some seasons and some locations (Barnston *et al.* 2005). Operationally, a 3-month lead time is generally used, but in some cases skill has also been shown 6 months in advance.

Most methods based purely on the statistical properties of time series, in contrast, have little credibility. These rely primarily on concepts of persistence, trends, or cycles; all of these assume that future conditions of precipitation can be extrapolated from past events. Unfortunately, that assumption is faulty, as climate and weather conditions tend to change abruptly, so that the success of these methods is limited. Nevertheless, numerous authors still advocate them, although official forecast centers do not. More complex statistical models (e.g., Barnston and Ropelewski 1992; Penland and Magorian 1993) have proven to have good forecast potential (Hastenrath 1995).

"Cycles" in precipitation have been used by some to predict the occurrence of drought (e.g., Faure and Gac 1981; Dyer and Tyson 1977). However, like trends, they are extremely unreliable forecast tools because they are generally weak, irregular, and transient. Consider, as an example, a cycle of 15 years that has been shown to be *statistically* significant in the rainfall record at some stations during the period 1940–1990. This means only that its occurrence is more pronounced than would be expected from totally random events. Thus, a "significant" cycle might only explain a small fraction, perhaps a few percent, of the total variation in rainfall from year to year. Also, a cycle of 15 years signifies only an approximate frequency of recurrence; the actual interval could easily range from 12 to 18 years or more. A cycle might be apparent during some time interval then abruptly disappear. Cycles detected in relatively short records, such as the above example, are particularly unreliable.

Drought predictions based on teleconnections and other empirically determined linkages within the climate system have some potential, but only if (1) a predictor can be identified that is strongly linked to the occurrence of drought, (2) the predictor is slowly varying (as in the case of boundary forcing), and/or (3) a reliable time sequence of events can be established. These sorts of linkages have been commonly demonstrated, but much work is still required to convert such concepts into reliable forecasts of drought. Notable attempts have been made for select locations such as the arid region of Northeast Brazil, the Indian monsoon region, the Sahel, and Australia.

Reviews and evaluations of prediction and predictability are found in Hastenrath (1995), Kumar *et al.* (1996), Shukla (1998), Shukla *et al.* (2000), and Nicholls (1984). An excellent overview of the state-of-the art approach to long-term (i.e., seasonal)

Observed precipitation   Model precipitation

Fig. 21.10 Predicted versus modeled precipitation for the period 1998–2002 (modified from Hoerling and Kumar 2000).

forecasting is given by Barnston *et al.* (2005). Most forecasting schemes are two-tiered. The first tier is a prediction of slow variations in the earth's boundary conditions, such as sea-surface temperatures (SSTs). These conditions provide the "forcing" of seasonal climate; their long time scale of variation provides the forecast skill (Shukla 1981). The second tier involves determining the atmospheric response to this forcing. This response can be evaluated via numerical models, such as general circulation models (GCMs), or by empirical models. Figure 21.10 gives an example of a GCM precipitation forecast for the period 1998–2002. The skill is reasonable, but is lower for individual years.

An extensive direct comparison of the numerical and empirical/dynamical approaches has only been carried out for ENSO (Barnston *et al.* 1994) and for Northeast Brazil (Moura and Hastenrath 2004). These and other studies have shown that the performance of the empirical models generally equals or exceeds that of the GCMs (Anderson *et al.* 1999). Skill in predicting ENSO has much practical value because ENSO, or the Southern Oscillation index (SOI) (its atmospheric equivalent), is used directly in many empirical forecasting methods (e.g., Farmer 1988; Shukla and Paolino 1983; Nicholls 1985).

Any forecast scheme will be more reliable if the causes of drought are understood. In many arid regions of the world this is not the case. Progress is often impeded by a lack of meteorological data; in some deserts, the rainfall record is so poor that even the typical seasonality of rainfall cannot be reliably established. In regions where the physical causes of drought are understood, a combination of empirical and numerical models will provide the best prospect for long-range (e.g., seasonal) forecasting.

21.3 LIVING WITH DROUGHT

As the population of dryland regions increases, the management of drought and its accompanying hazards becomes increasingly important. The steadily increasing risk is associated not only with the number of people affected, but by higher standards of living, the reduced environmental knowledge of the population as people migrate into these regions, and the impact of people on the environment. A drought has relatively little impact in a desert with a small population compared with one in a highly populous desert city. A natural landscape will be less susceptible to the impact of drought than a landscape that has been degraded by badly managed agricultural practices. As population and development increase in drylands, strategies for management, reduction of vulnerability and impact, and early warning become increasingly important.

21.3.1 THE IMPACT OF DROUGHT

The impact of drought depends on its severity and duration, on societal expectation, and upon the ability of the affected people to manage the situation. A full tally of the impact is not possible except on a local level. Its economic consequences, however, can rival those of natural hazards such as tornadoes and hurricanes. Riebsame *et al.* (1991) estimated that the cost of the US drought in 1987/89 was $20 billion in agricultural and forest production losses and a $10 billion loss associated with increased food costs.

Severe drought episodes have environmental, economic, and societal or political consequences. The primary impact is a shortage of water, which can manifest itself in many ways. A drought may result in inadequate rainfall for crops or inadequate water in reservoirs for irrigation, domestic use, power generation, or industry. Secondary effects of drought include fires, dust storms, soil erosion, and heat waves.

The 1982/83 drought in Australia provides one example of impact. Nearly half of the continent suffered from severe drought and much of the southeast had record low rainfall. On the national scale, wheat production was 63% of the average for the previous five years and rural production fell from A$4600 million to A$2300 million. Conditions were worse in New South Wales and Victoria, where production fell to 29% and 16% of average, respectively. Bushfires associated

Fig. 21.11 Dust storm approaching Melbourne, Australia, on February 8, during the severe drought of 1983, a year of a strong El Niño (photograph courtesy of Bureau of Meteorology, www.bom.gov.au).

with the drought killed a quarter of a million sheep and at least 70 people; economic losses included at least $A130 million in insurance claims and $A145 million in forest damage (Powell 1983). Business failures and decreased revenue followed. Water supply was inadequate for electricity generation and domestic and industrial needs. Dust storms (Fig. 21.11) associated with that drought eroded 250,000 tons of topsoil (Gibbs 1984).

The historical record is replete with examples of the societal impacts of major droughts. In fact, historical records in dryland regions attest to their pervasive influence. References to drought are consistent and reliable enough that documentary reports and oral histories provide a reasonably accurate record of climate during historical times (Nicholson 1979, 2001). Figure 21.12, comparing historical indicators from newspaper accounts and settlers' diaries with tree-ring records for parts of the western USA, illustrates this. The drought periods that stand out are the modern ones in the 1930s and 1950s, the 1860s, and two periods around 1810 and 1820. Other more localized but severe droughts occurred around the 1880s or 1890s, the 1780s, and the early eighteenth century. A notable localized drought that occurred in the western Great Plains in 1845–1856 may have contributed to the decline of bison herds (Woodhouse *et al.* 2002).

Drought may have accounted for the disappearance of the colonists from the settlement at Roanoke Island, Virginia, just south of Jamestown. Termed the "Lost Colony of Roanoke," not a single colonist was found in 1591, just four years after settlement. Tree rings show that the disappearance coincided with the driest growing season and the driest 3-year period within 800 years (Stahle *et al.* 1998). A similar drought some 20 years later (1606–1612) probably caused a large part of the population of the Jamestown settlement to perish.

Some of the most tragic droughts occurred in Mexico and West Africa. A sixteenth-century megadrought killed 80% of the native population of Mexico. Most deaths were associated with hemorrhagic fever epidemics that accompanied the drought (Acuna-Soto *et al.* 2002). The Sahel drought of the late

1960s and early 1970s killed scores of people and decimated cattle herds throughout the region.

21.3.2 SOCIETAL RESPONSE TO DROUGHT

The management of drought involves both reduction of vulnerability and timely warning of a potentially disastrous event. This requires recognizing drought as an erratic but inherent part of the physical environment. This recognition is built into the traditions of longtime residents of dryland regions. Unfortunately, the lessons are often lost on the newcomers and urban dwellers in these regions. Development is often based on a false sense of environmental security, and the need for adaptation first becomes clear as a result of a major drought.

The Sahel drought of 1968–1974 is a case in point. The region may have been more vulnerable as a result of the relatively wet, drought-free interval that had prevailed for decades. Colonial development had prospered during those decades and the resultant agricultural and economic systems had been adjusted to that norm. Most of West Africa gained independence in 1960 and was struggling to deal with the new *status quo*. When rainfall suddenly decreased around 1968, nations were ill-equipped to deal with the consequences, and traditional coping strategies were no longer feasible. Millions fled to refugee camps and cities. Famine was rampant. Estimates of the death toll range from 200,000 to 500,000. Some 3.5 million head of cattle perished.

When a much more severe event occurred in the early 1980s, the impact was considerably less. The reasons are not completely clear. Reduced environmental pressure, following population and herd reductions in the drought-stricken regions, certainly played a role. However, the earlier drought also led to the development of adaptive strategies, international relief, and early warning systems. Some evidence suggests that adaptive strategies similarly reduced the impact of droughts in the US Great Plains after the disastrous "Dust Bowl" days of the 1930s (Warrick and Bowden 1981). Subsequent droughts in the region in the 1950s and 1970s were milder, so that a lesser impact would have been anticipated. However, in the 1970s, soil loss was actually as great as during the 1930s Dust Bowl (Gillette and Hanson (1989). Environmental factors, such as more intense winds, could account for the stronger erosion, as could greater susceptibility as a result of more intense agriculture. The modern center-pivot irrigation systems (Fig. 21.13) allow marginal lands to be plowed and these become extremely susceptible to deflation during droughts (Goudie 1983).

The response to drought and strategies to mitigate it depend on many factors, such as the degree to which a society is accustomed to the occurrence of drought, the proportion of rural versus urban population, and the economic situation of the population. Strategies are often different in the developing versus the developed world. In the former, water use may be very direct: a drought reduces grazing animals and food supply. In the developed world, demand can often be reduced before nutrition is threatened. For example, demand might be reduced by

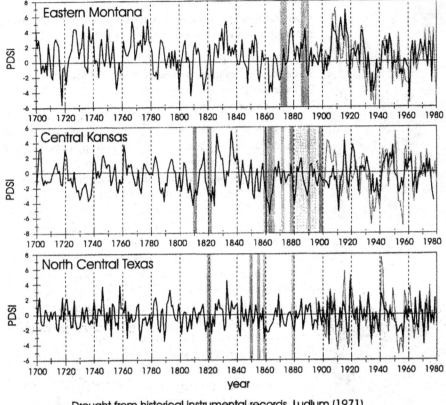

Drought from historical instrumental records, Ludlum (1971)

Pentads of below average precipitation from historical instrumental records, Rocky Mountain region, Bradley (1976)

Newspaper accounts of drought, Bark (1978)

Seasons or years of drought in >1 Great Plains regions, Mock (1991)

Travelers' accounts of eolian activity compiled by Muhs and Holliday (1995)

Fig. 21.12 Multi-source reconstruction of the Palmer drought severity index (PDSI) for eastern Montana, central Kansas, and north-central Texas (from Woodhouse and Overpeck 1998).

accepting lower levels of comfort (e.g., reducing demand for hydrologic power by reducing home heating or air conditioning) or limiting non-critical water use (e.g., by not watering lawns and golf courses).

One of the common traditional strategies for dealing with drought is migration. In some cases this produces serious social disruption. The westward migration from the Great Plains to California of millions fleeing the Dust Bowl during the 1930s drought is a classic image of American history. In that case, the ongoing economic depression exacerbated the problem. A long drought from 1276 to 1299 was probably a factor in the cliff-dwelling Anasazi people's abandonment of their Mesa Verde settlement in the southwestern USA (deMenocal 2001) (Fig. 21.14).

In other cases, the migration is purely adaptive. The nomadic lifestyle of desert dwellers is the best example. The pastoralists migrate to areas where the meager rains fell or remain near wells, if these have not been desiccated by the drought. Many years ago the pastoralists of Sahelian Africa typically took refuge with the sedentary farmers to the south during times of

drought. They provided labor on the farms, and their animals provided fertilizer and fed on the stubble of crops. This strategy was no longer possible when European colonists imposed national boundaries that inhibited migration.

21.3.3 TECHNOLOGICAL MANAGEMENT OF DROUGHT

Very little can be done to ameliorate a meteorological drought. In some cases, cloud seeding has been attempted, but with little success. This technique can in some circumstances enhance rainfall by 10–15% (Grant 1996), but natural variability of rainfall is on the order of 400% in the drier regions. Thus, even if successful, cloud seeding could do little to alleviate a typical drought with rainfall deficits of 30–80%. The seeding can work only in meteorological situations that are conducive to rainfall; this is generally not the case during drought. More can be done to reduce the impact of meteorological drought on agriculture, hydrology, and society.

Fig. 21.13 Soil erosion during the drought of 1977 over a field irrigated by a pivot system in the Portales Valley of New Mexico (from McCauley *et al.* 1981).

Rosenberg (1978) summarizes four generic strategies for ameliorating the impact of drought on agriculture. Any management strategy must do at least one of four things:

1. increase the capture of rainfall;
2. decrease the evaporative loss of stored water;
3. increase the amount of available water;
4. improve the efficiency of water use.

The first three augment the water supply. The fourth reduces demand. Variants of these strategies are also applicable to other drought impacts, including those in urban areas.

The supply of water during drought can be modulated in many ways. Contour and terrace farming are good ways to enhance rainfall capture. So is direct water harvesting, via run-off capture or rainwater collection. Irrigation, desalinization, recycling of water/wastewater use, and weather modification are used to increase the availability of water. Reducing tillage also helps to conserve water. Strip cropping and windbreaks reduce erosion, which is a secondary impact of drought. Efficiency of water use can be controlled via the crops or strains of crops that are planted, or the timing of planting.

Evaporative loss can be reduced via changes in the micro-climate around fields or storage reservoirs. Examples include windbreaks and stubble mulch. The stubble, which also helps to capture rainfall, reduces evaporative loss by reducing both wind and ground temperature and by transferring moisture to the subsurface layers. Attempts have also been made to reduce evaporation via thin films on the water surface, but wind generally breaks these up (Frenkiel 1965).

A common strategy that has been utilized in modern times has been to control the storage of water. The classic example has been dam building and active regulation of streamflow, and reservoir storage. Although this allows adjustment to the water supply in times of drought, the creation of dams has many negative impacts. Many of the largest projects have been controversial and numerous rivers and streams have been deregulated after many decades of control by dams. One such controversial case is the Aswan Dam, which has disrupted the natural irrigation and fertilization supplied by the Nile River. This and other examples are described in Chapter 23.

21.3.4 DROUGHT MONITORING AND EARLY WARNING

The management of drought is facilitated by reliable information on its status and early warning of its occurrence. This allows for adaptive responses to be put in place, in order to reduce vulnerability to drought and its impact. The goal of drought monitoring is assessing its status or likelihood on a near-real-time basis. Drought monitoring centers or programs have been set up in a number of countries, including the United States, Australia, Kenya, and Zimbabwe. One commonality of the centers is centralized and consolidated monitoring activities. Another is an assemblage of a variety of data types that provide information

Fig. 21.14 Cliff houses in Mesa Verde, Colorado. The Anasazi, or Ancestral Pueblo, left the region during an extended period of drought in the second half of the thirteenth century.

on meteorological, agricultural, and hydrological conditions (e.g., Kininmonth *et al.* 2000; Kunkel *et al.* 1998).

An example of a comprehensive monitoring program is the Drought Monitor in the USA. Because drought is more "nebulous" than other natural disasters, traditional methods of assessment and prediction are not applicable (Svoboda *et al.* 2002). In this case, much of the assessment is qualitative and subjective – the consensus of a varying group of experts. The Drought Monitor is not an index nor is it developed from a single index. Rather it is a composite consensus based on a large amount of information, including climate indices, numerical models, and the input of experts.

The Drought Monitor uses five severity categories (D0 to D5) that loosely correspond to the probability of an event of that magnitude occurring within the course of a year. Six key objective drought indicators, such as soil moisture and precipitation deficit, are considered, and an initial report is drafted by a single author. Authorship rotates between a small group of individuals. The initial report is reviewed by a group of experts selected from a relatively large roster of participants. Categories are mapped, and the associated impacts of each category for agriculture, fire, and water are indicated.

Other monitoring systems that merit mention include the North American Drought Monitor, the Australian "Drought Watch" system, the Drought Monitoring Centers (DMC) in Harare (Zimbabwe) and Nairobi (Kenya), and the AGRHYMET Regional Center in Niamey (Niger). The North American Drought Monitor is a follow-up to that for the USA. It includes cooperative efforts between the USA, Canada, and Mexico (Lawrimore *et al.* 2002). The Australian Drought Watch is a project of the Bureau of Meteorology. Its purpose is primarily to provide climatological statistics to the community.

The DMC in Harare is sponsored by the SADC (Southern African Development Community) and represents a collaborative effort of 14 southern African countries. A major product is the regularly published Africa Regional Climate Outlook Forum (SARCOF). The Drought Monitoring Center for the Greater Horn of Africa, based in Nairobi, is a collaborative effort of 10 countries in eastern equatorial Africa. Sponsored by several agencies, including the UNDP (United Nations Development Programme) and the WMO (World Meteorological Organization), the Center also assembles massive data sets to produce a regularly disseminated Climate Outlook. In partnership with the two African centers is the FEWS NET (Famine Early Warning System). This network relies primarily on remote sensing data, such as the normalized difference vegetation index (NDVI), to monitor droughts and floods in Africa, in order help secure the food supply. Seventeen countries in West, eastern, and southern Africa participate in this endeavor.

21.4 DUST STORMS

Dust storms are specific meteorological events in which fine material is mobilized from the surface, entrained into the atmosphere, and suspended therein for periods ranging from minutes to days. The essentials required for their formation include loose surface material of fine texture, strong winds for mobilization of the material, and vertical motion for entrainment and suspension. The need for vertical motion implies that turbulence and instability are also required. Although instability and turbulence are common in deserts, the deflatable materials and adequately strong winds are not. The requirements are more common in the semi-arid regions along the desert margins. Moisture and vegetation bind materials and hinder the development of dust storms. Therefore the frequency, geographical distribution, and seasonality of dust storms are closely linked to weather and climate.

Dust storms are associated with a number of meteorological phenomena on various time and space scales. These range from the dust devils, with a scale of meters and lasting from seconds to hours, to global wind systems that transport dust over thousands of kilometers. The commonality of these phenomena is wind that is intense enough to mobilize and transport dust (see Section 13.4). Pye (1987) provides a comprehensive review and identifies four main meteorological situations that are associated with large-scale dust outbreaks. These include the downdrafts of individual storm cells, areas of steep pressure gradients, easterly waves of North Africa, and depressions and cold fronts.

Dust storms are also commonly referred to as sand storms. The particulates mobilized by the storm include the larger sand particles as well as dust. The sand grains are too large to be entrained within the air mass for a long period of time, but they are a particularly destructive component of the storms.

21.4.1 THE DOWNDRAFT *HABOOB*

Several types of dust storms are distinguished on the basis of the synoptic situations that give rise to them (Goudie 1983; Pye 1987; Bhalotra 1963). Some of the most spectacular storms are associated with instability and downdrafts from thunderstorms. These are termed *haboobs*, an Arabic term for violent wind. The downdraft *haboob* is associated with towering cumulonimbus clouds. These storms are driven by horizontal density gradients often imposed by a cold air incursion and maintained by a system of warm updrafts and cold downdrafts. Downdraft *haboobs* generally occur in a region of instability (Lawson 1971; Bhalotra 1963) where cooler, often moisture-laden air converges with warm, dry air, forcing it aloft (Fig. 21.15).

The downdraft is a density current: a column of cold, dense, moisture-laden air that falls under the influence of gravity. It generally precedes an area of hail and rain and produces a sudden temperature drop of up to 13°C. The air is cold because it originates high in the atmosphere and because water droplets evaporate during its downward path through drier air. Density is high because the air is cold and moisture-rich. The downdraft slams into the ground at great speed and spreads in all directions, moving forward at speeds up to 50 km/hour. The warmer air is pushed aloft, its ascent enhanced by the instability that produced the storm. This ascending air gives further life to the storm.

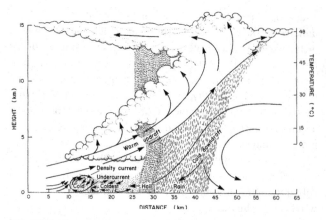

Fig. 21.15 The structure of a downdraft-type *haboob* dust storm (from Idso 1976).

Fig. 21.16 Dust front associated with a *haboob* approaching Phoenix, Arizona (from the website of Stan Celestian, Glendale Community College; photo from Arizona Department of Transportation).

The impact of the downdraft creates squalls at its forward edge. These turbulent vortices form as the cold air encounters the warm surface air, churning up tremendous quantities of dust in the unstable air. The dust raised by individual thunderstorm cells coalesces to form a wall of dust extending vertically thousands of meters (Idso *et al.* 1972). Dust concentrations can be as high as 40 μg/m³ (Lawson 1971).

In the Sudan, where the term *haboob* originated, downdraft *haboob*s occur on average 24 times per year. In the Khartoum area, they travel forward with a speed of roughly 50 km/hour and last between 30 minutes and an hour (Freeman 1952). They are most common around 12°–13° N latitude and are relatively rare in the more arid region north of latitude 15° N.

Similar storms in the American Southwest (Fig. 21.16) are termed "American *haboob*s" (Idso *et al.* 1972). They occur much less frequently than in the Sudan, on the order of once or twice per year. A common origin is the convergence of moist air from the Gulf of Mexico with local canyon winds. At Phoenix the *haboob* arrives most frequently in the afternoon or early evening, accompanied by a rapid drop in visibility to below 400 m and a temperature drop of 3°–15°C. The *haboob* is characterized by a solid wall of dust that extends as high as 3000 m above the surface. Roughly half of the region's dust storms are downdraft *haboob*s. Other situations giving rise to dust storms in this region include frontal passages, tropical disturbances, and upper-level cut-off lows (Brazel and Nickling 1986). Downdraft *haboob*s also occur in Australia.

21.4.2 LOCAL WINDS ASSOCIATED WITH STEEP PRESSURE GRADIENTS AND DUST OUTBREAKS

A second type of dust storm is associated with steep pressure gradients within persistent surface synoptic systems. Generally, the prevailing synoptic situation interacts with regional-scale effects, such as topography, to enhance the low-level pressure gradient. Examples of these systems are the anticyclone that gives rise to the Santa Ana winds (see Section 13.2) and the low-level trough that gives rise to the northwesterly, Middle

Eastern *shamal* (Fig. 13.7). The dust storms brought by easterly winds over Israel are a result of strong pressure gradients within a trough that extends from the Red Sea to the eastern Mediterranean.

The shamal occurs in response to the development of an intense summer monsoonal heat low over Pakistan and Afghanistan (Fig. 13.8). This produces northwesterly winds that are enhanced by the Zagros Mountains (Pye 1987) and by high pressure over North Africa. The intense winds are generally restricted to low levels, so that most dust storms during the shamal do not rise above 1000 m.

21.4.3 DUST STORMS ASSOCIATED WITH DEPRESSIONS, COLD FRONTS, AND EASTERLY WAVES

A third type of dust storm is associated with cold fronts and synoptic-scale depressions. Although less turbulent, the cold fronts are comparable to the density currents in the downdrafts of the *haboob*s. The cold and turbulent air abruptly lifts the warm surface air, which entrains dust as it overrides the cold front. The dust can readily be lifted to 3 km or higher.

The passage of a cold front linked to a surface cyclone is probably the most frequent cause of dust storms (Pye 1987). Dust storms in Israel are linked to cold fronts associated with low-pressure systems centered over Cyprus or with Saharan depressions initiated over the Atlas Mountains. These often create heat waves in Israel as well. Intense cold fronts associated with cyclones sweeping across Mongolia are the most favorable synoptic pattern for the development of Asian dust storms (Qian *et al.* 2002). A cold front initiated a dust storm (Fig. 21.11) that shrouded Melbourne, Australia, in dust during the 1983 El Niño year, one of extreme drought in that country. The height of the dust cloud extended to 3650 m at Mildura, 500 km from Melbourne, but was only 320 m at Melbourne.

The hot, dry dusty *khamsin* wind of Egypt is associated with such frontal disturbances. The dusty khamsin of Egypt is also initiated by the Saharan depressions and by cyclones traversing the Mediterranean or North African coast. Dust storms in the USA are typically associated with a cyclone centered over Utah and Colorado (Fig. 21.17). Southwesterly winds in the warm center generate most of the dust, so that maximum dust occurrence is over Texas and the Southern High Plains (Fig. 21.18).

Dust storms are also associated with the squalls and easterly waves that traverse West Africa near the mid-level African easterly jet. The squall lines are typically 300–500 km in length and oriented north–south (Sommeria and Testud 1984). At the leading edge, easterly winds with speeds up to 30 m/s produce a gust front similar to that of the *haboob*. A wall of dust arises at the front and frequently precedes heavy rain and thunderstorms. The entrained dust gets moved northward and westward, where it is merged into the dust-laden Saharan air layer.

21.4.4 RELATIONSHIP TO RAINFALL AND DROUGHT

Numerous studies demonstrate that dust storm frequency is highest in semi-arid regions. A global analysis by Goudie (1983) shows a sharp peak between 100 and 200 mm/year, and a marked drop above about 500 mm/year (Fig. 21.19). Regions with less rainfall tend to have less available source material. In regions with more rainfall, surface materials are more likely to be confined by the binding forces of water and vegetation. They are quite rare in regions with more than 1200 mm/year on average, except during periods of drought.

A regional analysis for Australia confirms the global relationship shown in Fig. 21.19, but there dust storms frequently occur even when annual rainfall is as high as 300 mm (McTainsh *et al.* 1989). The role of other environmental factors creates regional and local differences in the rainfall–dust storm relationship. These differences could include the nature of the source material, soil moisture, surface vegetation, surface temperatures, and typical wind speeds and synoptic situations. For example, the Australian cities of Kalgoorlie and Alice Springs have similar conditions of rainfall, but far fewer dust storms develop at Kalgoorlie, where there is more soil moisture.

The close relationship between rainfall and dust storm frequency is further demonstrated by the relationship between drought and dust storm frequency over Australia in the two wet years 1964 and 1965 and the two dry years 1974 and 1975 (Fig. 21.20). The overall frequency decreases in the drier years and the spatial pattern of dust storms corresponds reasonably well to the spatial pattern of drought.

Goudie and Middleton (1992) examined the interannual variability of dust storms at 30 locations worldwide. Their results further indicate the strong link to rainfall. There is a clear upward trend in dust storm frequency in regions such as West Africa, where a downward trend in rainfall prevailed in the Sahel zone since the late 1960s. Areas where rainfall gradually increased in more recent decades, such as Mexico City or Mongolia, showed a decreasing frequency of dust storms.

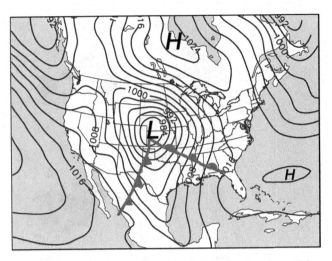

Fig. 21.17 Synoptic situation typical of Great Plains dust storms (after Warn and Cox 1951, Pye 1987, reprinted by permission of the *American Journal of Science*).

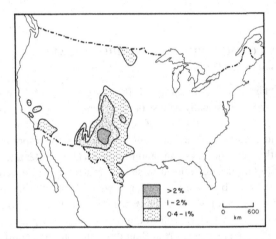

Fig. 21.18 Annual frequency of dust hours with visibility < 11 km in the United States (reprinted from Orgill and Sehmel 1976 with permission from Elsevier).

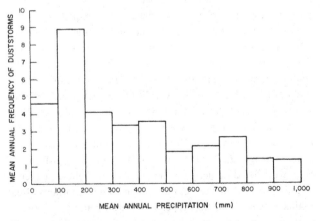

Fig. 21.19 Global dust storm frequency as a function of mean annual rainfall (from Goudie and Middleton 1992).

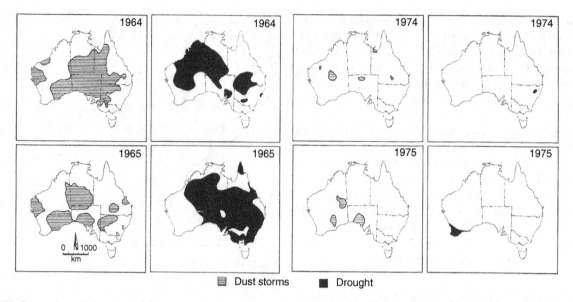

Fig. 21.20 Comparison of drought and dust storm occurrence over Australia during four years of contrasting conditions (from McTainsh *et al.* 1989).

21.4.5 GEOGRAPHICAL DISTRIBUTION AND SEASONALITY

A reliable global inventory of the frequency of dust storm occurrence is virtually impossible to generate because of inconsistent definitions, the lack of readily available data, and dramatic changes in frequency from year to year in response to weather conditions. Goudie and Middleton (1992) attempted to clarify terminology related to storm occurrence, so that comparisons can be made on a regional basis. Their definition of a *dust storm* requires that visibility be reduced below 1000 m and that dust be actively entrained during the event. Despite strong international agreement regarding this criterion, many authors have failed to adhere to it in their analyses of dust storms. This makes regional comparisons difficult.

Other categories of dust events include dust haze, blowing dust, and dust devils. *Dust haze* involves dust suspension rather than entrainment and generally visibility is above 1000 km (Fig. 21.21). Global inventories of haze frequency are feasible, since they can be assessed via satellite (see Chapter 13). *Blowing dust* applies to active but local entrainment with visibility above 1000 km. *Dust devils* or *dust whirls* are local and spatially limited columns of dust with brief life spans.

The Middle East has the world's highest frequency of dust storm occurrence (Middleton 1986; Goudie 1983). Figure 21.22 shows the frequency for Afghanistan. Many areas experience more than 20 per year; at Kandahar the number exceeds 79 per year. The highest reported frequency is at the Seistan Basin of Iran, with 80.7 days per year. In contrast, the highest frequency in the USA, at Lubbock in the drought-prone Southern Great Plains, is only 20–30 storms per year (Stout 2001).

Perhaps the only generalization that can be made about the seasonality of dust storms is that in most regions there are preferred seasons (Fig. 21.23). Dust storms are generally concentrated in a 4–6 month period of the year. In many, but not all, locations this coincides with the dry season. In regions such as Arizona, where most storms are of the downdraft type and associated with thunderstorms and squalls, the peak occurs during the wet season. In Jerusalem, Egypt, and the Sudan, dust storm frequency is similarly greatest during the wet season. In Texas, the maximum dust storm frequency occurs just prior to the summer rainy season. In India, there is a peak early in the rainy season. This is the case also for parts of Sahelian Africa. The cause of the diverse patterns of seasonality is the large number of factors controlling dust storm generation. Most important are a relatively dry and barren surface, but adequate winds. The latter are often coupled with the wet season disturbances.

21.4.6 IMPACT OF DUST STORMS

The impacts of dust storms include soil erosion, morphological changes of the land surface, reduced soil productivity, removal and deposition of soil and plant nutrients, disease transmission and other health consequences, property damage, visibility reduction, and accidents. Soil erosion is perhaps the most pervasive. Millions of tons of topsoil can be removed by a single dust storm (McCauley *et al.* 1981). A dust storm in New Mexico in 1977 eroded agricultural land to a depth of 7 m over a period of 7 hours. It removed all the wheat, plowed furrows, and loose soil. The dust plume began on a single farm near the Texas–New Mexico border (Fig. 21.24). Within two days the dust plume covered much of the southeastern USA and a huge region of the western Atlantic. The erosion removes mainly fine-grained topsoil, which contains essential nutrients. This loss also impairs the ability of the soil to hold water, reducing

Fig. 21.21 Dust haze at a camel market near Cairo. The numerous animals stir up the dust and the morning temperature inversion keeps the dust near the surface.

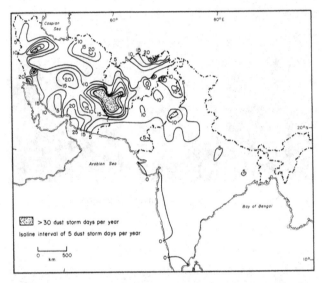

Fig. 21.22 The frequency of dust storms over Afghanistan (based on Middleton 1986).

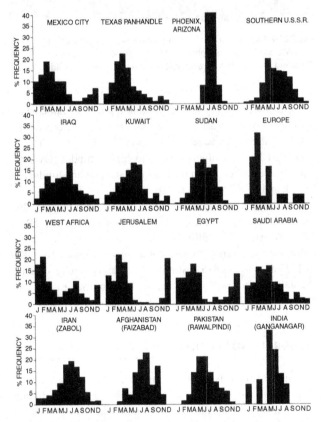

Fig. 21.23 Percentage frequency of dust storms by month for world-wide sites (from Pye 1987, copyright Elsevier).

water available in the root zone of plants. The result is reduced production (Fryrear 1981).

The soil particles carry with them numerous pathogens that are harmful to people, plants, and animals. The result is disease, respiratory ailments, and even suffocation. The dust produces lesions in plants and animals that allows pathogens to enter. In 1895 a dust storm caused the loss of 20% of the cattle in eastern Colorado (Idso 1976). Particles less than 2 microns in diameter are retained in the lungs and may affect health. The Kalahari of Botswana has the highest rate of death from lung diseases in the world (Lundholm 1979; Péwé 1981). The dust-transported pathogens include numerous fungal and yeast species that cause disease. One example is "valley fever" or "desert rheumatism" that occurs in the southwestern United States, parts of Mexico, and Argentina (Leathers 1981).

The vast amount of dust churned up and transported by dust storms suddenly reduces visibility, sometimes to near zero. Figure 21.25 shows a dust storm in Niamey, Niger, in August 1992. Moments earlier the sky had been cloudless and deep blue. The visibility reduction produced twilight in mid-afternoon. This storm occurred early in the rainy season, so that the rain that followed the arrival of the dust very effectively scavenged

Fig. 21.24 Development of a dust plume from sources along the Texas/New Mexico border on February 23, 1977 (from McCauley *et al.* 1981). A second plume can be seen developing over eastern Colorado.

the atmosphere. A crystal-clear swimming pool became nearly opaque 30 minutes after the rain and took on a deep orange color, similar to the water in the muddy Niger River. The haze generated by dust often extends to high altitudes. Over the Sahara, the dust can be entrained into tall cumulus clouds (Fig. 13.33), giving them a deep red appearance.

The abrupt visibility reduction associated with blowing dust can lead to dangerous and often fatal highway hazards (Péwé 1981). In Arizona, multi-vehicle accidents triggered by a rapid reduction to zero visibility are a common occurrence. From 1968 to 1975 as many as 886 dust-related accidents were recorded in the state. The toll was 36 fatalities and 720 injuries. In one area, dust is responsible for up to 15% of all fatalities.

21.5 DRYLAND FLOODS

Although drought is the most common natural hazard in drylands, floods are an additional inescapable risk in most dryland environments. Small amounts of rainfall in the uplands can almost instantaneously produce torrents in dry wadis or rapid sheets of overland flow. Some of these commonplace events can be quite destructive, though short-lived. In Niamey, Niger, one brief (1 hour) localized rain event that produced between about 50 and 150 mm of rainfall around the countryside was intense enough to destroy houses and bridges and wash away segments of road (Fig. 21.26). The more unusual flood-producing weather situations, which might persist for days, can ravage arid lands as savagely as drought. Three case studies are presented later in this section to illustrate the nature and impact of such events.

Fig. 21.25 A mid-afternoon dust storm in Niamey, Niger.

Fig. 21.26 Niger flood damage near Niamey, July 1992, from a rapidly arising torrent; within moments a highway bridge was washed away.

In humid regions, floods generally result when water overflows the banks of a river or reservoir. In some cases, saturated soil can produce or enhance flooding. These flood situations generally require persistent wet conditions that may not be particularly extreme. Soil saturation and overbank flooding seldom occur in dryland regions. There the most common mechanism of flood generation is torrential rainfall; flooding results when rainfall intensity exceeds the infiltration capacity of the soil. Because there is little vegetation to reduce the intensity and speed of flow, the powerful current often captures and carries along vast quantities of sediment and debris, including coarse materials and boulders. This material is left behind as the flood-waters recede.

Dryland floods and droughts differ in several ways. The most notable are the spatial extent and duration of the events. A flood is generally a localized phenomenon, while drought is at least regional and may even occur on a continental or subcontinental scale. Also, a drought is a cumulative effect of a relatively persistent deficit in expected rainfall. In arid regions, a dry spell is probably not felt unless it persists for at least a month. A drought requires several months or more of persistently dry conditions. In contrast, a flood can result from one brief event. The upper time scale of occurrence – generally a few days – is shorter than the shortest time scale for drought.

The different time and space scales underscore the contrasts in the meteorological conditions responsible for floods and droughts. Droughts are generally associated with persistent anomalies in the large-scale atmospheric circulation. Floods tend to be associated with much smaller, mesoscale anomalies in the regional meteorology. These are sometimes persistent, but are on scales of days rather than months.

In general, dryland floods are associated with brief torrential rains. One explanation for this intensity involves the atmospheric stability associated with the meteorological situations prevailing in many drylands. An example is aridity associated with a persistent and intense temperature inversion. In order for convective processes to overcome the stability and produce rainfall, they must be unusually strong, and hence result in torrential rains. As a result, the daily intensity of rainfall may actually increase with increasing aridity. This has been the case in parts of Israel (Alpert et al. 2002).

Another factor may explain the occasional occurrence of torrents when arid and humid climates occur in close proximity. A relatively small shift in circulation patterns can bring the rain-bearing systems into normally dry regions. In other situations, the development of unusual meteorological situations lead to flooding. The diversity of meteorological situations producing floods in dry regions is illustrated using examples from several extraordinary floods described below.

21.5.1 CASE STUDIES

1969 NORTH AFRICAN FLOOD

In September 1969, floods ravaged parts of Tunisia and Algeria, the result of a cyclonic storm in the Mediterranean that persisted for 7 days. Its intensity is illustrated by rainfall at Biskra, Algeria, where the mean September rainfall is 17 mm (Winstanley 1970). On two days alone, September 27 and 28, 299 mm fell at Biskra. This is roughly twice the mean annual rainfall. Just to the east of El Djem, Tunisia, over 400 mm fell. The flood was linked

to an infrequently occurring system described many years ago by Flohn (1971), a diagonal trough in the upper-level westerlies that extends from the Mediterranean across the Sahara southward to the Sahel or Guinea coast (see Chapters 4 and 16). The initial system was a Saharan depression that formed in the Atlas Mountains, but that later intensified and was held in place by the elongated tropical-temperate trough. Many of the region's floods occur in such situations.

The seven days of flood-producing rains resulted in 600 deaths and left a quarter of a million people homeless. The destruction of phosphate mines in Gafsa, Tunisia, deprived 25,000 miners of work and cost the Tunisian government about $5 million per week in revenues. Destruction included Roman bridges that had survived floods since their construction some 2000 years before.

A similarly devastating flood occurred in this same region in November 2001. Over two days, Algiers received about 285 mm of precipitation. This is roughly three times the November mean. The storm, which resulted from a mesoscale cyclone, caused 740 deaths (Tripoli et al. 2005).

NATAL FLOOD DISASTER OF 1987

In South Africa, a comparable tropical-temperature trough situation is frequently responsible for floods during the second half of the summer (Lindesay and Jury 1991). However, a different situation, a persistent cut-off low, was responsible for South Africa's greatest flood disaster, which occurred in September 1987 (Triegaardt et al. 1988; Mason and Jury 1997). It affected mainly Natal and the eastern Cape Province. Its toll included 300 deaths, 60,000 people displaced by flood waters, 15 bridge collapses, and 223 slope failures that produced torrents of mud. An international relief effort was set up to help deal with the consequences. At Durban, monthly rainfall totaled 402 mm, with most of the rain falling in the 7-day period from September 26 to October 2. Mean September rainfall at Durban is only 67 mm.

The gravity of the situation was compounded by a second period of flooding further westward and northward in February and March 1988. In a 22-day period in February, floods took 24 lives and displaced 10,000 in the Orange Free State, Natal, the Cape area, and the karoo plains. The economic toll was some $400 million. In March, a much larger area was affected, including parts of Zimbabwe, Botswana, and Mozambique; 18 died and around 24,000 people were displaced.

BIG THOMPSON CANYON, 1976

On July 31, 1976, a foot of rain (c. 30 cm) fell in the Big Thompson Canyon region of Colorado within three hours. This was the height of the tourism season and thousands of people were camped along the Big Thompson Creek. The volume of rainfall was related to the development of a stationary storm system over the Rocky Mountains. The explosive combination was persistent low-level easterly winds forced upward in the Rockies, an inversion layer, and upper-level westerlies to move the system out of the region. During a 1-hour period, 8 inches (20 cm) of rain fell, swelling the Big Thompson Creek from a lazy 2-foot (60 cm) stream to a 19-foot (580 cm) torrent.

The devastation caused by the ensuing flood was related in part to the layout of the terrain. The canyon was lined with sheer rock walls. When the floodwaters wiped out the highway into the canyon, the only means of escape was by climbing the canyon walls. Within two hours the storm killed 145 people, destroyed 418 houses, and resulted in damage costing about $40 million.

21.5.2 FLOODS LINKED TO EL NIÑO

Although El Niño is most often associated with droughts, its typical signature in some arid regions is flooding. Examples are Peru, southern California, and Israel. These are usually not the flash floods described earlier, but rather a result of a persistent rainfall-inducing situation and many rainfall events.

In Peru, El Niño's influence is a direct consequence of the persistently warm coastal waters. During 1925, El Niño produced 1524 mm in Lima (mean annual rainfall 46 mm) and 394 mm at Chicama (mean annual rainfall 4 mm) (Goudie and Wilkinson 1980). The 1982/83 El Niño transformed coastal desert areas into grasslands and lakes. Whole streets were submerged for days in Chiclayo and Piura; rivulets that had been completely dry for years became torrents (Rauh 1985). Many stations registered daily totals well in excess of 100 mm. Areas with mean annual rainfall on the order of 200–300 mm received instead several meters of rain during that event (Goldberg et al. 1987). In parts of Ecuador and northern Peru, 2500 mm fell within a 6-month period.

El Niño has taken quite a toll on the region. A flood in 2004 left 70 dead and 22,000 homeless. A multidecadal period of strong El Niños and floods may have contributed to the decline of the Moche civilization that thrived in northern Peru from about 200 to 700 AD. Some speculate that similar conditions led to the demise of the Chimu in that region some five centuries later (Caviedes 1973; Moseley 1978; Wells 1987; DeVries 1987).

In southern California, El Niño's meteorological impact generally involves a disruption of the prevailing subtropical high-pressure cell and a strengthening of the subtropical jet stream over the Pacific. This brings a steady progression of disturbances into California, where the mountains enhance their impact. These storms produce floods and landslides on the western slopes of the Sierra Nevada. Nearly everywhere south of 35° N, El Niño increases the frequency of flooding along the California coast (Andrews et al. 2004). During the 1997/98 El Niño a series of powerful winter storms created a disaster area in this region, with hundreds of millions of dollars in property losses from floods. Houses literally slid down hillsides in the San Francisco Bay area, and landslides cut across Route 1, the coastal highway, isolating Big Sur for months.

El Niño produces floods in many other areas of the western USA as well. They occur in the winter months over much of the Pacific coast, including areas north of 35° N. They commonly occur in the spring or summer months in the western plains east of the Rocky Mountains. Areas from North Dakota and Montana southward to Texas and New Mexico frequently experience flooding in association with El Niño (Pizarro and Lall 2002).

In Israel, the occasional floods linked to El Niño similarly result from its effects on the jet stream (Price et al. 1998). It is displaced southward in winter, bringing frequent, intense storms into the eastern Mediterranean. During the 1991/92 El Niño, flooding occurred when Jerusalem experienced the longest sustained rainfall since 1949. Egypt was also severely hit, and its port of Alexandria was swamped by rivers and canals breaking through their banks. The 2004 El Niño brought 30–40 mm of rainfall (25% of the annual average) to the Negev in less than one hour. Roads and bridges in the Dead Sea and Arava regions were destroyed.

21.6 SNOW IN THE LOW-LATITUDE DESERTS

Snow is a common feature in the world's cold deserts, where temperatures can plummet in winter. The idea of snow seems incongruous with the hot deserts of Africa and the Middle East, where the general perception is one of extreme heat. However, snow does occur on occasion in the Sahara and the Negev and other subtropical deserts.

Snow can occur as far south as the tropic of Cancer in the central Sahara (see Fig. 17.12). While most of the snow occurs in the highlands or near the coast, its occurrence on the inland plains is not a rarity. A heavy snowfall creates havoc, since it cannot slide off the flat-roofed adobe dwellings and their roofs are not constructed to support a layer of snow. As a result, roofs and buildings collapse. Dubief (1963) chronicled the occurrence of snow at 24 locations in the Sahara during the late nineteenth and early twentieth centuries. It was a more or less regular feature of climate at many of these places. At Laghouat (see Fig. 17.13), for example, in the lee of the Atlas, snow fell in 53 of the 86 years between 1874 and 1960. It fell every year from 1885 to 1896, generally on 2–5 days each year.

Snow also makes a spectacular appearance when it blankets Jerusalem (see Fig. 17.7b). As over the Sahara, snow appears to have occurred more frequently during the late nineteenth century (Chaplin 1891). Its occurrence seems to be linked to El Niño (Alpert and Reisin 1986; Price et al. 1998). During the period 1975–1993 there was above-average snowfall during six of eight El Niños. Only one non-El Niño year saw above-average snowfall during that period. In recent decades four years had remarkable snow events. In March 1998, 20 cm of snow fell on Jerusalem just a few days after a record-breaking dust storm. Two large snow events occurred that year. Heavy snow fell also in the Golan and Palestine areas, and Jordan had its first snowfall in 50 years, with 10 cm falling in Amman. A record number of snow days, roughly 15, occurred in Jerusalem in 1991/92 and, before that, in 1982/83. All of the aforementioned years were El Niños. However, Jerusalem received a record 40 cm of snow in January 2000, a La Niña year. In the hyperarid Atacama Desert snow is rare, but it fell on several days in July 2011. A cold front brought some 80 cm near Arica in northern Chile.

21.7 WILDFIRES

Wildfires are a frequent consequence of drought. The three elements necessary to maintain a fire are fuel, heat, and oxygen (Powell 1983). Weather and climate influence all three. A situation that is conducive to fire is a wet spring that promotes vegetation growth, followed by a summer drought. The drought converts the vegetation to dry, flammable fuel and produces protracted heat. Oxygen is provided both by winds and by the fire itself, once it begins and draws in air from the surroundings.

Australia is the driest and most drought-prone continent. The fire season persists for 4–7 months during the dry season in each region. During every month some part of the continent is experiencing fires. The settlement of the continent by the Aboriginal people increased the incidence of fire, as did the European settlement in the eighteenth century. Fire incidence is now some 16 times greater than when only natural fires occurred.

Two of the most disastrous fires in Australia occurred in Victoria in January 1939 and in Tasmania in February 1967. The first is known as the "Black Friday" fire, and much of southeastern Australia was burning. Some 1.5 million hectares of land and 1000 homes were destroyed; at least 71 deaths occurred, 50 on one day alone. The Tasmanian fire claimed 62 lives, 1400 homes, and 263,000 hectares of land. Both fires were in the southeast, where conditions are right for disastrous fires. Intense westerlies exceeding 100 km/hour traverse the dry regions of the continent, rapidly turning vegetation into tinder and promoting rapid burning and spread. Entire houses were lost within 15–20 minutes. An equally devastating event occurred in 1983, another El Niño year (see Section 21.3.1).

California is equally well known for its disastrous wildfires (Fig. 21.27), usually associated with Santa Ana winds. The synoptic situation that produces the Santa Ana wind is a high-pressure center over northern California and southern Oregon and a low centered off the northwestern Sonora state of Mexico (Sergius *et al.* 1962) (Fig. 13.9). The resultant northeasterly

winds out of the desert regions of Nevada and Utah cross the Sierra Nevada chain of mountains into California. The air loses much of its moisture at higher elevations. It descends on the lee side, the subsidence making it both warmer and drier. A strong Santa Ana with winds in excess of 130 km/hour in January 2003 caused a 2200-acre (890 ha) wildfire that destroyed dozens of lavish homes in Malibu and blew over trees, trucks, and power lines. Over 42,000 people were left without power.

REFERENCES

Acuna-Soto, R., D. W. Stahle, M. K. Cleaveland, and M. D. Therrell, 2002: Megadrought and megadeath in 16th century Mexico. *Emerging Infectious Diseases*, **8**, 360–362.

Alpert, P., and T. Reisin, 1986: An early winter polar airmass penetration to the eastern Mediterranean. *Monthly Weather Review*, **114**, 1411–1418.

Alpert, P., and Coauthors, 2002: The paradoxical increase of Mediterranean extreme daily rainfall in spite of decrease in total values. *Geophysical Research Letters*, **29**(11), 1536, doi:10.1029/2001GL013554.

AMS (American Meteorological Society), 2004: Meteorological drought: policy statement. *Bulletin of the American Meteorological Society*, **85**, 771–773.

Anderson, J., and Coauthors, 1999: Present-day capabilities of numerical and statistical models for atmospheric extratropical seasonal simulation and prediction. *Bulletin of the American Meteorological Society*, **80**, 1349–1361.

Andrews, E. D., R. C. Antweiler, P. J. Neiman, and F. M. Ralph, 2004: Influence of ENSO on flood frequency along the California coast. *Journal of Climate*, **17**, 337–348.

Barnston, A. G., and C. F. Ropelewski, 1992: Prediction of ENSO episodes using canonical correlation analysis. *Journal of Climate*, **5**, 1316–1345.

Barnston, A. G., and Coauthors, 1994: Long-lead seasonal forecasts-where do we stand? *Bulletin of the American Meteorological Society*, **75**, 2097–2114.

Barnston, A. G., A. Kumar, L. Goddard, and M. P. Hoerling, 2005: Improving seasonal prediction practices through attribution of climate variability. *Bulletin of the American Meteorological Society*, **86**, 59–72.

Bhalme, H. N., and D. A. Mooley, 1979: On the performance of modified Palmer index. *Archiv für Meteorologie, Geophysik und Bioklimatologie Serie B*, **27**, 281–295.

Bhalotra, Y. P. R., 1963: *Meteorology of the Sudan.* Sudan Meteorological Service, Technical Memoir No. 6, Khartoum, 113 pp.

Brazel, A. J., and W. G. Nickling, 1986: The relationship of weather types to dust storm generation in Arizona (1965–1980). *Journal of Climatology*, **6**, 255–275.

Brubaker, K., D. Entekhabi, and P. Eagleson, 1993: Estimation of continental precipitation recycling. *Journal of Climate*, **6**, 1077–1089.

Caviedes, C. N., 1973: Secas and El Niño: two simultaneous climatological hazards in South America. *Association of American Geographers Proceedings*, **5**, 44–49.

Changnon, S. A., 1987: *Detecting Drought Conditions in Illinois.* ISWS/CIR-169-87, Illinois State Water Survey, Champaign, IL, 36 pp.

Chaplin, T., 1891: Das Klima von Jerusalem. *Zeitschrift des Deutschen Palästina-Verein*, **XIV**, 93–112.

Charney, J., 1975: Dynamics of deserts and drought in the Sahel. *Quarterly Journal of the Royal Meteorological Society*, **101**, 193–202.

deMenocal, P. B., 2001: Cultural responses to climate change during the late Holocene. *Science*, **292**, 667–673.

Fig. 21.27 A wildfire in the San Bernadino National Forest, August 2005, as seen from the International Space Station (courtesy of NASA).

DeVries 1987: A review of the geological evidence for ancient El Niño activity in Peru. *Journal of Geophysical Research*, **92**, 14471–14479.

Diaz, H., 1983: Some aspects of major dry and wet periods in the contiguous United States, 1895–1981. *Journal of Climate and Applied Meteorology*, **22**, 3–16.

Diaz, H., and R. G. Quayle, 1980: The climate of the United States since 1895: spatial and temporal changes. *Monthly Weather Review*, **108**, 249–266.

Dracup, J. A., K. S. Lee, and E. G. Paulson Jr., 1980: On the statistical characteristics of drought events. *Water Resources Research*, **16**, 289–296.

Dubief, J., 1963: *Le Climat du Sahara*, Vol. II, Institut de Recherches Sahariennes, Algiers, 275 pp.

Dyer, T. G. J., and P. D. Tyson, 1977: Estimating above and below normal rainfall periods over South Africa 1972–2000. *Journal of Applied Meteorology*, **16**, 145–147.

Entekhabi, D., and I. Rodriguez-Iturbe, 1994: An analytic framework for the characterization of space-time variability of soil moisture. *Advances in Water Resources*, **17**, 35–45.

Entekhabi, D., I. Rodriguez-Iturbe, and R. Bras, 1992: Variability in large-scale water balance with a land surface–atmosphere interaction. *Journal of Climate*, **5**, 798–813.

Farmer, G., 1988: Seasonal forecasting of the Kenya coast short rains, 1901–1984. *Journal of Climatology*, **8**, 489–497.

Faure, H., and J. Y. Gac, 1981: Will the Sahelian drought end in 1985? *Nature*, **291**, 475–478.

Felch, R. E., 1978: Drought: characteristics and assessment. In *North American Droughts* (N. J. Rosenberg, ed.), Westview Press, Boulder, CO, pp. 25–42.

Flohn, H., 1971: *Tropical Circulation Patterns*. Bonner Meteorologische Abhandlungen 15, Meteorologischen Institut der Universität Bonn, Bonn, 55 pp.

Foley, J. C., 1957: *Droughts in Australia: Review of Records from Earliest Years of Settlement to 1955*. Australian Bureau of Meteorology, Bull. 43, 281 pp.

Freeman, M. H., 1952: *Dust Storms of the Anglo-Egyptian Sudan*. Meteorological Report 11, HMSO, London.

Frenkiel, J., 1965: *Evaporation Reduction*. Arid Zone Research XXVII, UNESCO, Paris, 79 pp.

Friedman, D. G., 1957: *The Prediction of Long-continuing Drought in South and Southwest Texas*. Occasional Papers in Meteorology, No. 1, The Travelers Weather Research Center, Hartford, CT, 182 pp.

Fryrear, D. W., 1981: Long-term effect of erosion and cropping on soil productivity. In *Desert Dust: Origin, Characteristics, and Effect on Man* (T. L. Péwé, ed.), GSA Special Paper 186, pp. 253–260.

Gibbs, W. J., 1984: The great Australian drought: 1982–83. *Disasters*, **8**, 89–104.

Gillette, D. A., and K. J. Hanson, 1989: Spatial and temporal variability of dust production caused by wind erosion in the United States. *Journal of Geophysical Research*, **94D**, 2197–2206.

Goldberg, R. A., G. Tisnado, and R. A. Scofield, 1987: Characteristics of extreme rainfall events in northwestern Peru during the 1982–1983 El Niño period. *Journal of Geophysical Research*, **92**, 14225–14241.

Goudie, A. S., 1983: Dust storms in space and time. *Progress in Physical Geography*, **7**, 502–530.

Goudie, A. S., and N. J. Middleton, 1992: The changing frequency of dust storms through time. *Climatic Change*, **20**, 197–226.

Goudie, A. S., and J. Wilkinson, 1980: *The Warm Desert Environment*. Cambridge University Press, Cambridge, 88 pp.

Grant, L. O., 1996: Weather modification. In *Encyclopedia of Weather and Climate* (S. H. Schneider, ed.), Oxford University Press, New York, pp. 839–841.

Hastenrath, S., 1995: Recent advances in tropical climate prediction. *Journal of Climate*, **8**, 1519–1532.

Hastenrath, S., 2002: Climate prediction (empirical and numerical). In *Encyclopedia of the Atmospheric Sciences* (J. R. Holton, J. A. Curry, and J. A. Pyle, eds.), Academic Press, pp. 411–417.

Heim, R. R., Jr., 2002: A review of twentieth-century drought indices used in the United States. *Bulletin of the American Meteorological Society*, **83**, 1149–1165.

Herweijer, C., and R. Seager, 2008: The global footprint of persistent extra-tropical drought in the instrumental era. *International Journal of Climatology*, **28**, 1761–1774.

Hoerling, M., and A. Kumar, 2003: The perfect ocean for drought. *Science*, **299**, 691–694.

Huff, F. A., and S. A. Changnon Jr., 1964: *Relation between Precipitation, Drought and Low Streamflow*. Surface Water Publication 63, IAHS, Gentbrugge, Belgium, pp. 167–180.

Idso, S. B., 1976: Dust storms. *Scientific American*, **235**, 108–114.

Idso, S. B., R. S. Ingram, and J. M. Pritchard, 1972: An American haboob. *Bulletin of the American Meteorological Society*, **53**, 930–955.

Karl, T. R., 1983: Some spatial characteristics of drought duration in the United States. *Journal of Climate and Applied Meteorology*, **22**, 1356–1366.

Karl, T. R., 1986: The sensitivity of the Palmer Drought Severity Index and Palmer's Z-Index to their calibration coefficients including potential evapotranspiraiton. *Journal of Climate and Applied Meteorology*, **25**, 77–86.

Karl, T. R., F. Quinlan, and D. S. Ezell, 1987: Drought termination and amelioration: its climatological probability. *Journal of Climate and Applied Meteorology*, **26**, 1198–1209.

Keyantash, J., and J. A. Dracup, 2002: The quantification of drought: an evaluation of drought indices. *Bulletin of the American Meteorological Society*, **83**, 1167–1180.

Kininmonth, W. R., M. E. Voice, G. S. Beard, G. C. de Hoedt, and C. E. Mullen, 2000: Australian climate services for drought management. In *Drought: A Global Assessment* (D. A. Wilhite, ed.), Routledge, New York, pp. 210–222.

Kumar, A., M. Hoerling, M. Ji, A. Leetmaa, and P. Sardeshmukh, 1996: Assessing a GCM's suitability for making seasonal predictions. *Journal of Climate*, **9**, 115–129.

Kunkel, K. E., and Coauthors, 1998: An expanded digital daily data base for the climatic resources applications in the midwestern United States. *Bulletin of the American Meteorological Society*, **79**, 1357–1366.

Lawrimore, J., and Coauthors, 2002: Beginning of a new era of drought monitoring across North America. *Bulletin of the American Meteorological Society*, **83**, 1191–1192.

Lawson, T. J., 1971: Haboob structure in Khartoum. *Weather*, **26**, 105–112.

Lean, J., and D. Rind, 1998: Climate forcing by changing solar radiation. *Journal of Climate*, **11**, 3069–3094.

Leathers, C. R., 1981: Plant components of desert dust in Arizona and their significance for man. In *Desert Dust: Origin, Characteristics, and Effect on Man* (T. L. Péwé, ed.), GSA Special Paper 186, pp. 191–206.

Lindesay, J. A., and M. R. Jury, 1991: Atmospheric circulation controls and characteristics of a flood event in central South Africa. *International Journal of Climatology*, **11**, 609–627.

Lundholm, B., 1979: Ecology and dust transport. In *Saharan Dust* (C. Morales, ed.), John Wiley and Sons, Chichester, pp. 60–68.

Mason, S. J., and M. R. Jury, 1997: Climatic variability and change over southern Africa: a reflection on underlying processes. *Progress in Physical Geography*, **21**, 23–50.

McCauley, J. F., C. S. Breed, M. J. Grolier, and D. J. McKinnon, 1981: The U. S. dust storm of February 1977. In *Desert Dust: Origin,*

Characteristics, and Effect on Man (T. L. Péwé, ed.), GSA Special Paper 186, pp. 123–147.

McKee, T. B., N. J. Doesken, and J. Kleist, 1995: Drought monitoring with multiple time scales. In *Ninth Conference on Applied Climatology, Dallas, TX*, American Meteorological Society, pp. 233–236.

McTainsh, G. H., R. Burgess, and J. R. Pitblado, 1989: Aridity, drought and dust storms in Australia (1960–1984). *Journal of Arid Environments*, **16**, 11–22.

Middleton, N. J., 1986: A geography of dust storms in southwest Asia. *Journal of Climatology*, **6**, 183–196.

Mitchell, J. M., C. W. Stockton, and D. M. Meko, 1979: Evidence of a 22-year rhythm of drought in the western United States related to the Hale solar cycle since the 17th century. In: *Solar–Terrestrial Influences on Weather and Climate* (B. M. McCormack and T. A. Seliga, eds.), Reidel, Dordrecht, pp. 125–144.

Morid, S., V. Smakhtin, and M. Moghaddasi, 2006: Comparison of seven meteorological indices for drought monitoring in Iran. *International Journal of Climatology*, **26**, 971–985.

Moseley, M. A., 1978: An empirical approach to prehistorical agrarian collapse: the case of the Moche Valley, Peru. In *Social and Technological Management in Dry Lands* (N. Gonzalez, ed.), Westview Press, Boulder, CO, pp. 9–43.

Moura, A. D., and S. Hastenrath, 2004: Climate prediction for Brazil's Nordeste: performance of empirical and numerical modeling methods. *Journal of Climate*, **17**, 2667–2672.

Nicholls, N., 1984: The stability of empirical long-range forecast techniques: a case study. *Journal of Climate and Applied Meteorology*, **23**, 143–147.

Nicholls, N., 1985: Towards the prediction of major Australian droughts. *Australian Meteorological Magazine*, **33**, 161–166.

Nicholls, N., 1992: Historical El Niño/southern Oscillation variability in the Australasian region. In *El Niño: Historical and paleoclimatic aspects of the Southern Oscillation* (H. F. Diaz, and V. Markgraf, eds.), Cambridge University Press, pp. 151–173.

Nicholson, S. E., 1976: A climatic chronology for Africa: synthesis of geological, historical and meteorological information and data. Ph.D. Thesis, University of Wisconsin, Dept. of Meteorology, 324 pp.

Nicholson, S E., 1979: The methodology of historical climate reconstruction and its application to Africa. *Journal of African History*, **20**, 31–49.

Nicholson, S. E., 1993: An overview of African rainfall fluctuations of the last decade. *Journal of Climate*, **6**, 1463–1466.

Nicholson, S. E., 2001: A semi-quantitative, regional precipitation data set for studying African climates of the nineteenth century. Part I. Overview of the data set. *Climatic Change*, **50**, 317–353.

Nkemdirim, L., and L. Weber, 1999: Comparison between the droughts of the 1930s and the 1980s in the southern prairies of Canada. *Journal of Climate*, **12**, 2434–2450.

Oladipo, E. O., 1986: Spatial patterns of drought in the interior plains of North America. *Journal of Climatology*, **6**, 495–513.

Orgill, M. M., and G. A. Sehmel, 1976: Frequency and diurnal variations of dust storms in the contiguous USA. *Atmospheric Environment*, **10**, 813–825.

Palmer, W. C., 1965: *Meteorological Drought*. U. S. Weather Bureau Research Paper 45, 58 pp. [Available from NOAA Library and Information Services Division, Washington, DC, 20852].

Pang, K. D., and K. K. Yau, 2002: Ancient observations link changes in sun's brightness and earth's climate. *EOS*, **83**, 480–490.

Penland, C., and T. Magorian, 1993: Prediction of Niño 3 sea surface temperatures using linear inverse modeling. *Journal of Climate*, **6**, 1067–1076.

Péwé, T. L., 1981: Desert dust: an overview. In *Desert Dust: Origin, Characteristics, and Effect on Man* (T. L. Péwé, ed.), GSA Special Paper 186, pp. 1–10.

Pizarro, G., and U. Lall, 2002: El Niño-induced flooding in the U.S. West: what can we expect? *EOS*, **83**, 349 and 352.

Powell, F. A., 1983: Bushfire weather. *Weatherwise*, **36**, 126–131.

Price, C., L. Stone, A. Huppert, B. Rajagopalan, and P. Alpert, 1998: A possible link between El Niño and precipitation in Israel. *Geophysical Research Letters*, **25**, 3963–3966.

Pye, K., 1987: *Aeolian Dust and Dust Deposits*. Academic Press, London, 334 pp.

Qian, W., L. Quan, and S. Shi, 2002: Variability of the dust storm in China and climatic control. *Journal of Climate*, **15**, 1216–1229.

Rasmusson E. M., and P. A. Arkin, 1993: A global view of large-scale precipitation variability. *Journal of Climate*, **6**, 1495–1522.

Rauh, W., 1985: The Peruvian-Chilean deserts. In *Hot Deserts and Arid Shrublands* (M. Evenari, I. Noy-Meir, and D.W. Goodall, eds.), Ecosystems of the World 12A, Elsevier, Amsterdam, pp. 239–268.

Redmond, K. T., 2002: The depiction of drought: A commentary. *Bulletin of the American Meteorological Society*, **83**, 1143–1147.

Riebsame, W. E., S. A. Changnon, and T. R. Karl, 1991: *Drought and Natural Resources Management in the United States*. Westview Special Studies in Natural Resources and Energy Management, Boulder, CO, 174 pp.

Rosenberg, N. J., 1978: Technological options for crop production in drought. In *North American Droughts* (N. J. Rosenberg, ed.), Westview Press, Boulder, CO, pp. 123–142.

Sergius, L. A., G. R. Ellis, and R. M. Ogden, 1962: The Santa Ana winds of southern California. *Weatherwise*, **15**, 102–105.

Shukla, J., 1981: Dynamical predictability of monthly means. *Journal of the Atmospheric Sciences*, **38**, 2547–2572.

Shukla, J., 1998: Predictability in the midst of chaos: a scientific basis for climate forecasting. *Science*, **50**, 728–731.

Shukla, J., and D. A. Paolino, 1985: The Southern Oscillation and long-range forecasting of summer monsoon rainfall over India. *Monthly Weather Review*, **111**, 1830–1837.

Shukla, J., and Coauthors, 2000: Dynamical seasonal prediction. *Bulletin of the American Meteorological Society*, **81**, 2593–2606.

Sommeria, G., and J. Testud, 1984: COPT 81: a field experiment designed for the study of dynamics and electrical conductivity of deep convection in continental tropical regions. *Bulletin of the American Meteorological Society*, **65**, 4–10.

Stahle, D. W., M. K. Cleaveland, D. B. Blanton, M. D. Therrell, and D. A. Gay, 1998: The lost colony and Jamestown droughts. *Science*, **280**, 564–567.

Stout, J. E., 2001: Dust and environment in the Southern High Plains of North America. *Journal of Arid Environments*, **47**, 425–441.

Svoboda, M., and Coauthors, 2002: The drought monitor. *Bulletin of the American Meteorological Society*, **83**, 1181–1190.

Thornthwaite, C. W., and J. R. Mather, 1957: *Instructions and Tables for Computing Potential Evapotranspiration and the Water Balance*. Publications in Climatology 10, Drexel Institute of Technology, Laboratory of Climatology, Centerton, NJ, pp. 181–311.

Trenberth, K. E., and C. J. Guillemot, 1996: Physical processes involved in the 1988 drought and 1993 floods in North America. *Journal of Climate*, **9**, 1288–1298.

Triegaardt, D. O., D. E. Terblanche, J. van Heerden, and M. V. Laing, 1988: *The Natal flood of September 1987*. South African Weather Bureau Technical Paper 19, Southern African Weather Bureau, Pretoria.

Tripoli, G. J., and Coauthors, 2005: The 9–10 November 2001 Algerian flood. *Bulletin of the American Meteorological Society*, **86**, 1229–1235.

Warn, G. F., and W. H. Cox, 1951: A sedimentary study of dust storms in the vicinity of Lubbock, Texas. *American Journal of Science*, **249**, 552–568.

Warrick, R. A., and M. J. Bowden, 1981: The changing impacts of droughts in the Great Plains. In *The Great Plains: Perspective and Prospects* (M. P. Lawson and M. E. Baker, eds.), Center for Great Plains Studies, University of Nebraska, Lincoln.

Wells, L. E.., 1987: An alluvial record of El Niño events from northern coastal Peru. *Journal of Geophysical Research*, **92**, 14463–14470.

Wilhite, D. A., 2000: Drought as a natural hazard: Concepts and definitions. In (Wilhite, ed.) *Drought: A Global Assessment*, Routledge, pp. 3–18.

Winstanley, D., 1970: The North Africa flood disaster, September 1969. *Weather*, **25**, 390–403.

Woodhouse, C. A., and J. T. Overpeck, 1998: 2000 years of drought variability in the central United States. *Bulletin of the American Meteorological Society*, **9**, 2693–2714.

Woodhouse, C. A., J. J. Lukas, and P. M. Brown, 2002: Drought in the western Great Plains, 1845–56. *Bulletin of the American Meteorology Society*, **83**, 1485–1493.

Yevjevich, V. M., 1967: *An Objective Approach to Definitions and Investigations of Continental Hydrologic Drought*. Colorado State University Hydrology Paper No. 23, Fort Collins, CO.

22 Desertification

22.1 OVERVIEW OF THE PROBLEM

The term desertification is a relatively new one that has readily worked its way into both the scientific literature and the popular press. The term evokes an image of the "advancing desert," a living environment becoming sterile and barren, arable lands turning into deserts. In fact, this idea was at the heart of most early definitions of the phenomenon (Graetz 1991). The concept of desertification is inextricably linked to arid lands, and many sources state that virtually all drylands are at "risk" and that over 9 million km² are already desertified (Fig. 22.1). The United Nations (UN 1994) even suggested that 25% of the earth's land surface had been affected.

Such alarmist claims have greatly exaggerated the extent of the problem. As the summary of headlines in Table 22.1 shows, the press avidly jumped on these estimates. Consequently, the relevant issues have been clouded by controversy, misinformation, and a severe dearth of data and rigorous scientific study.

The picture of desertification that emerged in the 1970s and 1980s was that it took the form of vegetated land becoming denuded, often in pockets of unproductive land surrounding pressure points of intense activity (e.g., villages or watering holes). Associated with these changes were growing populations and livestock numbers, overcultivation, intensive irrigation, and deforestation (Kassas 1995). The scenario was that animals cluster around the few available wells, overgrazing the surrounding land; trees and shrubs are removed to produce fuelwood, charcoal, lumber, or land for agriculture. Consequently the barren land becomes increasingly impacted by wind and water erosion.

Many studies now take issue with much of the literature produced on desertification in the 1970s and early 1980s. It is now acknowledged that desertification is not nearly as widespread as once suggested (e.g., Prince et al. 1998; Prince 2002), that much misunderstanding exists concerning its causes (e.g., Mortimore and Turner 2005), and that, in the "desertified" Sahel, where recovery was thought unlikely, more verdant conditions have

re-emerged (Herrmann et al. 2005), along with better conditions of rainfall (Nicholson 2005).

During the 1990s much new research emerged to underscore the oversimplification provided by the above picture (see review in Nicholson et al. 1998). The complexity of the problem, the inappropriateness of applying such generalizations to diverse geographic regions and ecosystems, and the necessity of considering various time and space scales have all been recognized (Reynolds and Stafford Smith 2002a).

Nevertheless, desertification is a real problem and an environmental risk with which the 20% of the world's population living on arid, semi-arid, and dry subhumid land must contend. Appropriate information is a key to managing the problem. This chapter reviews the nature and history of desertification and summarizes the controversies and unresolved issues. The most important of these issues (Stafford Smith and Reynolds 2002) are:

- the causes and consequences of land degradation
- the extent to which land changes are natural versus anthropogenic
- whether or not observed changes are reversible
- how to determine the amount of land affected or at risk
- the role and success of various abatement efforts

This last issue is particularly concerned with the effectiveness of social versus scientific or technological approaches to abatement or recovery. Some of the most contentious issues are the extent and severity of desertification (is it a local or global phenomenon?), the role played by grazing animals, and the impact of supposedly desertified land on climate.

22.2 DEVELOPMENT OF THE CONCEPT

The concept of productive lands changing to deserts is an old one and there are numerous examples of cultural declines associated with desert advances in arid and semi-arid environments. Some are clearly linked to changing conditions of

Table 22.1. *Droughts, deserts. and death: headlines in the popular press illustrate the high-profile (and often melodramatic) interest in desertification (from Reynolds and Stafford Smith 2002b).*

Headline	Source	Date
Droughts, deserts and death	Nassau Guardian	5/13/1985
Threat of encroaching deserts may be more myth than fact	New York Times	1/18/1994
Orchard of Spain crumbles into dust: Drought, tourism and intensive cultivation all helped to transform lush farmland into a desert … in just 20 years	Guardian (London)	5/25/2000
Desertification threatens half of Tanzania's land	Africa Newswire	7/12/2000
Beijing's desert storm	Asiaweek	10/13/2000
Farming-pasturing area faces rapid desertification	Xinhua News	8/22/2000
Sahara jumps Mediterranean into Europe	Guardian (London)	12/20/2000
The arid expansion	Guardian (London)	1/11/2001
Expanding desert – an urgent problem	China Daily	6/30/2001
A harvest of bounty and woe: Experts estimate that 20 percent of Spain is turning into a desert	Christian Science Monitor	8/22/2001
30 per cent of the land surface is threatened by desertification	Narodnoye Slovo (Uzbekistan)	6/16/2001
Boom in hothouse farming yields more desert in Spain	International Herald Tribune	4/4/2002
Human activity turns China into desert	Australian Broadcasting Corporation	1/29/2002
Dust storms may add to US pollution: Increasing desertification that began in Africa in the early 1970s, and more recently in parts of China, is intensifying the giant dust storms	Herald-Sun (Durham, NC)	4/6/2002

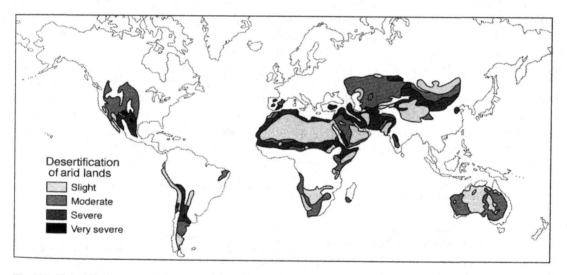

Fig. 22.1 United Nations map of the status of desertification over Africa (UNEP 1977).

climate, such as the decline of Saharan civilizations when the region changed from savanna to desert some 3000–4000 years ago. Other examples from China (Hou 1985) and the Middle East (Evenari *et al.* 1985) appear to be linked to human-induced degradation of the arid ecosystems. A case in point is the Mesopotamian civilization, the decline of which is frequently attributed to land degradation associated with irrigation in the "fertile crescent" between the Tigris and Euphrates rivers. In these cases of cultural decline, climate probably played some role, as they occurred at times of quasi-global climate change that produced a reduction in rainfall in many arid regions (Nicholson *et al.* 1998).

The concept of productive lands becoming deserts has re-emerged from time to time since the beginning of the twentieth century, but the most intense interest in the topic came on the heels of the 1970s drought in Sahelian West Africa. Numerous scientists, such as the late meteorologist Jule Charney (1975), claimed that the drought had been a *result* of desertification in the region. In 1977 the United Nations announced that globally some 35 km^2 of land had been affected by desertification and that perhaps another 35% of the earth's surface was at risk of undergoing similar changes (UNEP 1977). Tolba (1977), then director of UNEP, stated that "very severe desertification affects between 50 and 78 million" people, or roughly 10% of those living in dry regions. With the sounding of that alarm, desertification became the focus of much scientific, political, and institutional attention.

Throughout the twentieth century there have been numerous examples of dryland regions becoming increasingly desiccated (Nicholson 2004). As with recent claims of desertification, these incidents were surrounded by controversy concerning human vs. climatic causes. One of these was Schwarz's (1920) paper on the drying up of the Kalahari; a claim that led to the creation of a drought investigation committee to evaluate the situation. A proposed solution was a grandiose flooding scheme, the creation of Lake Kalahari, to bring back the waning rains.

The concept of the encroaching Sahara goes back to a paper by Bovill (1921). He observed the desiccation of the Senegal River and nearby wells, events that were roughly contemporaneous with the drying up of the Kalahari. Later Stebbing (1935) observed degradation of the forests in northern Nigeria, eastern Mali, and southern Niger. He attributed this to the advance of the Sahara, supposedly proceeding at a rate of 1 km/year for 300 years, and claimed that the advance was a result of human activities. Notably, the observations of Bovill followed a two-decade decline in rainfall throughout the western Sahel and a major drought episode in the 1910s, an episode that was evidenced throughout most of Africa, including the Kalahari.

Landscape degradation has also affected huge regions of Australia and the United States. Severe degradation occurred in Australia at the end of the nineteenth century, at about the same time as the desiccation of the Sahel and the Kalahari. The number of sheep in New South Wales declined from 13 million in 1890 to 4–5 million in 1900 (Graetz 1991). As in the case of the Sahel, this trend paralleled a climatic desiccation that affected nearly all the global tropics and was particularly apparent in Australia and Africa (see Chapter 25). Likewise, farmland was ruined and soil was eroded during the "Dust Bowl" days of the 1930s in the Great Plains of the USA. Here again, these changes could definitely be associated with some of the worst drought conditions on record in the region. In some cases, the supposed anthropogenic expansion of the desert was associated with reductions in rainfall of 25–50% (Nicholson 2004).

Clearly, there is an association between desertification and drought. This relationship is evident because the end-product of both processes is similar, despite their different origins. Vegetation cover is reduced, exposed soil is eroded, the water-holding capacity of the soil decreases, and vast amounts of dust are generated. Moreover, drought makes the environment more sensitive to disturbances created by other factors, such as grazing pressure or human mismanagement. Hence, the changes attributed to desertification occur more visibly in times of drought.

The trigger for the upsurge of interest in desertification was the drought that ravaged Sahelian West Africa in the late 1960s and early 1970s. It is virtually impossible to separate the impact of drought from that of desertification, but during the two periods of most severe conditions, c. 1969–1975 and c.

Fig. 22.2 Example of "exclosure" in Tunisia. Grazing is controlled inside the fenced in area, so that vegetation density is greater and the surface appears darker (from National Science Foundation 1977).

1982 to the late twentieth century, reportedly a million people starved, 40–50% of the populations of domestic stock perished, and millions of people took refuge in camps and urban areas and became dependent on external food aid (Graetz 1991). The case for desertification was enhanced by aerial and satellite photos showing evidence of large-scale human impact on the land. International borders separating grazed and ungrazed land were clearly seen from space: Afghanistan/Soviet Union, Namibia/Botswana, Sinai/Negev, southwestern USA/Mexico. Fenced "exclosures," i.e., huge fenced ranches where the extent of grazing was rigidly controlled, were likewise visible as dark patches in aerial photos (Fig. 22.2).

In view of such evidence, Charney (1975) wrote a classic but controversial paper suggesting that such desertification at the hands of humans had in fact caused the Sahelian drought to occur. His hypothesis was based on the bright appearance of overgrazed land on satellite and aerial photos: the degradation had exposed highly reflective soil which increased the albedo of the land, thereby returning more of the solar radiation back to space. His model showed that this would, overall, intensify the negative balance of radiation over the Sahara and adjacent Sahel. To compensate for the increased cooling, subsidence, which heats the air, would be enhanced over the region. The subsidence would further suppress rainfall, providing a positive feedback mechanism by which droughts could be self-accelerating, or that could even produce drought. Unfortunately, this mechanism is inconsistent with the factors that produce rainfall in the region and with the fact that the desiccation was not limited to the Sahel, but encompassed most of the continent (Nicholson 1996). That fact clearly speaks for large-scale, natural processes.

Charney's theory was followed up by a series of numerical models confirming that land-surface changes have the potential to cause drought in the region (see review by Nicholson 2000).

A number of scientists echoed Charney's claims that the Sahel drought was human-induced. While few hold such extreme views today, the studies of the 1970s clearly established that human degradation of the environment was in some instances severe and that the problem was widespread.

22.3 THE CONTROVERSY SURROUNDING DESERTIFICATION

From the onset of interest in desertification in the 1970s, the issue has caused considerable controversy. Some of the points of debate include whether the process is restricted to dryland regions; its causes; current extent, severity, and progression; remedies; and the reversibility of the process. Ellis *et al.* (2002) underscore three questions that are at the very heart of the controversy:

- Are deserts expanding?
- Is this a global phenomenon or restricted to local occurrences?
- Are the causes climatic or anthropogenic?

Much of the controversy results from the lack of scientific rigor in the studies of the 1970s and the information disseminated by the United Nations. The issue was highly politicized, as was the first United Nations Desertification Conference that took place in Nairobi in 1976.

An abundance of reports were produced, documenting case studies from around the world. In them was a virtual absence of ecological detail. Attention was paid to higher levels of ecosystem function, e.g., numbers of livestock and harvest quality, rather than manifestations of desertification on soils and vegetation (Graetz 1991). Sweeping conclusions were drawn from disparate observations, with no measurements or systematic assessments of the actual changes. There was no objective analysis of causes and consequences (Ellis *et al.* 2002). Much of the information was even anecdotal. UNEP (1977) produced, for example, a map of desertification severity (Fig. 22.1) that was largely extrapolated from risk factors in the environment; *de facto*, it was a map of vegetation and climate and contained virtually no information on desertification status. Consequentially, the extent and severity of desertification was greatly exaggerated.

An extreme example of the politicization relates to the Turkana pastoralists of northern Kenya (Ellis *et al.* 2002). An article published by UNEP stated that the pastoralists had overstocked, overgrazed and degraded their range to the point that it would no longer support cattle. The article cited the Turkana's use of the camel as evidence of this last point. At the same time, a long-term project termed STEP (the South Turkana Ecosystem Project) was under way to study these people and their ecosystem. It concluded that neither degradation nor desertification was detectable in the southern Turkana district and that it was unlikely to occur under the existing low-density of livestock (Little and Leslie 1999). Further the Turkana were found to keep cattle and camels in roughly equal numbers because the camels could maintain production during droughts or the dry season.

A major controversial point was the intentional elimination, in most works on desertification, of climate as one of its causes (e.g., Eckholm and Brown 1977). This may, in fact, be appropriate, but unfortunately the few assessments that were made extended over periods of drought or significant reduction in rainfall. An example of this problem is the work of Lamprey (1975), who attempted to quantify desert encroachment in the Sahel using maps from the 1950s and aerial surveys from 1975. He concluded that during the 17 years from 1958 to 1975 the desert had advanced southward in the western Sudan by 90–100 km. During this same period, rainfall in the Sahel as a whole declined by nearly 50%, although the reduction in the western Sudan was somewhat smaller. The widely disseminated UN figures on desert advance of 7 km/year stemmed from his measurements, so that changes that were partially or largely due to drought were claimed to be "anthropogenic" desertification.

Other aspects of the desertification issue have also been challenged. Several scientists at Lund University in Sweden did extensive studies in the Sudan to test some of the tenets of the desertification (Helldén 1984; Olsson 1985; Ahlcrona 1988). They showed through a combination of field work and analysis of satellite photos that there was neither a systematic advance of the desert or other vegetation zones, nor a reduction in vegetation cover, although degradation and replacement of forage with woody species was apparent. On the other hand, they clearly demonstrated that changes took place in response to drought, with full recovery of the land productivity at the end of the drought. Other studies (e.g., Tucker *et al.* 1994; Tucker and Nicholson 1999; Nicholson *et al.* 1998; Prince *et al.* 1998; Wessels *et al.* 2004) came to a similar conclusion for the Sahel as a whole. The Sahara was not, in fact, expanding into the Sahel, but was instead retreating, following the severe droughts of the early 1980s (Fig. 22.3).

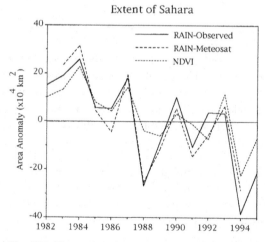

Fig. 22.3 The extent of the Sahara Desert, calculated as the area between the 200-mm isohyet and 25° N: solid line – as assessed from rainfall station data; dashed line – as assessed from Meteosat data; dotted line – as assessed from the normalized difference vegetation index (NDVI). Area, in 10^4 km², is calculated with respect to the mean area during the period 1980–1995 (from Nicholson *et al.* 1998).

None of the studies challenging earlier ideas about desertification indicate that it is not a problem. Land degradation has clearly affected many regions, including the African Sahel, but much of the problem had been exaggerated and the process is still insufficiently well understood. On the other hand, studies of desertification have established many useful things about the arid and semi-arid environments. One is that people, climate, and the environment are intricately linked. Over time a relatively stable equilibrium has been established, with feedbacks between these three components maintaining a healthy and functioning ecosystem. Stability has occasionally been disrupted by major climatic fluctuations, such as the end of the Ice Age (when a savanna dried up to produce the present Sahara). In recent times, human pressure on the land has increased, the impact of climatic fluctuations has become more severe, and the stability of the environment is threatened by more moderate climatic fluctuations. Studies of desertification have also demonstrated that the earth's dryland environments – the arid, semi-arid, and subhumid lands that support some 20% of its population – are truly at risk of severe degradation. At present, prudent strategies of land management are needed to prevent or reverse the process.

22.4 DEFINITION OF DESERTIFICATION

Part of the reason for the dearth of knowledge and assessment of desertification is the lack of any agreed-upon definition of the term. Some 100 definitions exist in the literature (Verstraete 1986; Mainguet 1991) and they differ vastly. The emphasis ranges from human impacts on the land to economic impacts, landforms and vegetation, and even phenomena on geologic time scales. Unfortunately, most definitions do not distinguish between *causes*, *mechanisms*, *manifestations*, and *impact*: all points of controversy surrounding the desertification issue. This failure has further added to the confusion surrounding the term.

Reynolds and Stafford Smith (2002b) provide an excellent summary of the better-known definitions and the extent to which they include ecological, meteorological, and human aspects of the problem. These authors wisely refrain from proposing yet another definition and instead attempt to deal with misunderstandings and differing interpretations of what is meant by land degradation (the core of the biophysical aspects) and why people or organizations do or do not consider it important. Ironically, the authors of most of the other chapters in their book provide their own definitions, necessitated by the lack of a commonly accepted one. This in itself underscores the problem. It is, in fact, far easier to say what desertification is not: the "advancing desert" or "marching sands" idea promoted in much of the literature.

A common aspect of most definitions is the emphasis on diminished land productivity. Most also restrict the use of the term to cases wherein the cause is human and not climatic. Dregne (1983), for example, states that it is the "impoverishment

of terrestrial ecosystems under the impact of man." The United Nations (UNEP 1977) definition is "the diminution or destruction of the biological potential of land [which] can lead to desert-like conditions." Some years later the United Nations (UN) Convention to Combat Desertification (CCD) revised their definition to read "land degradation in arid, semi-arid and dry subhumid areas resulting from various factors, including climatic variations and human activities" (UN 1994). The key change was to include climate as a potential causal factor – something that was missing from most previous definitions.

One controversial point is whether or not the occurrence of human (i.e., economic) consequences are a necessary part of the definition. In other words, if the physical processes do not impact humans, desertification does not take place (Stafford Smith and Reynolds 2002). It might be argued, however, that the addition of this caveat further clouds the issue, so that definitions restricting the process of "desertification" to its physical manifestations are preferable for several reasons. Perhaps the most important reason, as Stafford Smith and Reynolds admit, is that a definition requiring socioeconomic impacts does not readily lend itself to monitoring either. This creates a "moving target" that precludes quantitative assessment or global comparisons. Moreover, because the origins of the term relate to the physical environment, anyone unfamiliar with the myriad debates is likely to assume that the term refers to the biophysical processes of environmental change. Finally, the inclusion of "impact upon humans" as a prerequisite is largely a political decision, one that is a legacy of the highly politicized first UN Conference on Desertification.

Prince (2002) points out that sharpening and simplifying the definition by restricting the term to *biophysical changes* in the environment that are a *direct product of human activity* makes desertification more amenable to large-scale monitoring. He proposes as a definition that " desertification ... refers to the process by which changed biogeophysical conditions emerge owing to human actions that cannot be supported by the resource base (mainly rainfall) and that will not quickly return to their former, non-desertified conditions, either naturally or by application of minor management practices." Accordingly, the term "land degradation" would be reserved for cases in which the changes are readily reversible when the external pressure producing the degradation is removed. This definition serves to distinguish drought, in which vegetation and edaphic factors fully recover from a temporary reduction in rainfall, and desertification, in which over time there is incomplete recovery.

22.5 SUSCEPTIBILITY OF DRYLANDS TO DESERTIFICATION

Desertification has long been associated with land degradation in drylands. These regions are characterized by ecological and environmental traits that make them particularly susceptible to disturbance and hence to the processes of desertification. These include the patchy nature of vegetation, the meager amount of rainfall and its erratic distribution in time and space, and the characteristics of typical dryland soils. The soils in these regions are sensitive to disturbance because they contain only small amounts of organic matter, are often shallow, and have low aggregate strength.

Typical of the dryland climate is extreme year-to-year fluctuations; in some cases, long-term changes are also a feature of the environment. In the Sahel, for example, two or three decades of wet conditions alternate with dry periods of the same length. This pattern has been prevalent during at least the last few hundred years (Nicholson 1996) and is probably an inherent aspect of the Sahelian environment. The short-term variability alters the frequency and intensity of environmental "shocks" that must be weathered; the long-term change alters the overall resource base and may push the system beyond some critical threshold (Reynolds and Stafford Smith 2002b).

Typical dryland soils are sensitive to disturbance because they contain only small amounts of organic matter and have low aggregate strength. Both agricultural tillage and grazing can have profound effects in a relatively short time (Reynolds and Stafford Smith 2002a). Soil texture also plays a role. Sands are prevalent in many of the drylands. In northern China, sandy soils were found to be particularly prone to desertification (Jiang 2002). Disturbance resulted in shifting dunes, dune reactivation, sands spreading into grasslands, and wind erosion in dry farmland (Zhu and Wang 1993). In many locations, grazing and/or wood-cutting activated sand dunes, while adjacent clay-based soils supported the natural steppe vegetation, despite similar external pressures (Wu and Loucks 1992).

It was long assumed that grazing (Fig. 22.4) is a main factor in desertification of the drylands, some 88% of which sustain pastoral systems (UNEP 1997). The factor was especially stressed in the early literature on desertification. More recent research has shown this to be yet another fallacy. Grazing can create disturbances that can ultimately lead to desertification. However, at the drier edge of the Sahel, a region claimed to be under intense desertification, pastoral systems prevail and the risk of

environmental degradation is only moderate and mainly climate-driven (Hiernaux and Turner 2002). On the other hand, the risk of degradation is much higher in the crop–livestock systems of the southern Sahel and this is mainly management-driven. In many cases, the intensity of grazing pressure has been exaggerated (Tiffen *et al.* 1994) and the ability of the pastoralists to cope with environmental crises in a way that minimizes degradation has been underestimated (Mortimore and Turner 2005).

Overall, land use is less likely to degrade the inherent productivity of the land when (1) losses in primary productivity are tightly coupled to secondary production (e.g., herds), (2) when critical changes in productivity (whether primary or secondary) are quickly and easily observed and measured, and (3) when the initial detectable levels of change are reversible (i.e., the system is resilient) (Walker *et al.* 2002).

22.6 CAUSES OF DESERTIFICATION

In formulating a systematic approach to understanding and monitoring desertification, a critical point is that the causal factors have different levels of influence in different regions of the world and at different times (Stafford Smith and Pickup 1993). Failure to recognize this has contributed to many of the controversies surrounding desertification, including status, causes, processes, vulnerability, and appropriate abatement strategies and monitoring efforts.

For the developing world, many general causes have been identified, with the roots of desertification lying in societal changes such as increasing population, sedentarization of the indigenous nomadic peoples, breakdown of traditional market and livelihood systems, innovation of new and inappropriate technology in the affected regions, and poor strategies of land management (Warren 1996). Associated with these changes are growing livestock numbers, cultivation in marginal areas, intensive irrigation, slash/burn agriculture, and deforestation.

In developed regions like the southwestern USA, many other factors can be at play. In the Mojave the greatest disturbance

Fig. 22.4 In dryland regions, with relatively little surface cover, cattle and sheep grazing drylands can reduce the surface vegetation to stubble. This has often been assumed, incorrectly, to be the major cause of desertification.

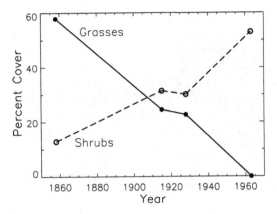

Fig. 22.5 Changes in proportion of grasses and shrubs in the Jornada Experimental Range of the Chihuahuan Desert (based on data from Buffington and Herbel 1965).

is from military maneuvers destroying stable soil crusts (Webb *et al.* 1986), off-road vehicle usage (Webb and Newman 1982), infrastructure and town building (Vasek *et al.* 1975), as well as agriculture.

In most cases, several causes of desertification operate simultaneously. These invariably include both socioeconomic (i.e., human) and biophysical (meteorological and ecological) factors. In many instances, the factors involved or their respective importance cannot be fully identified. The case of shrub encroachment in the southwestern USA illustrates this (Fig. 22.5). There is strong observational evidence that shrubs have invaded large areas that were previously grassland (Kurc and Small 2004). Various studies have implicated grazing, fire suppression, climatic change, and increasing atmospheric levels of carbon dioxide, but the cause is still unclear (Okin 2002).

22.7 THE MANIFESTATIONS OF DESERTIFICATION

The physical processes of desertification include removal of vegetation, excessive water consumption, irrigation, inappropriate use of agricultural technology, and other examples of poor land management practices. Natural vegetation is cleared for agriculture; savanna lands, fields and pastures are burned at the end of the dry season; wood is gathered for firewood, charcoal and building supplies (Fig. 22.6); animals overgraze grasslands. This generally leaves the surface more barren and more susceptible to wind and water erosion (Fig. 22.7). In many arid regions, the use of tractors and other technologies further enhances the erodibility. Animals trample the soil surface, compacting and aggregating materials in ways that hinder drainage and infiltration. When irrigated land is improperly drained, salt and other chemical residues are left behind.

Desertification involves reduction or loss of the biological and economic productivity and complexity of terrestrial ecosystems (including soils, vegetation, and other biota). It changes the vegetation cover and its spatial complexity, soil, and surface topography in ways that alter the ecological, biogeochemical,

Fig. 22.6 Implicated causes of desertification. (top) During the droughts of the 1970s and 1980s, animals clustered around the few available wells, degrading the surrounding vegetation. (center) Individual families gather wood and other materials and leave them by the roadside for collection, getting paid by weight. (bottom) Trees become lumber sold at the marketplace.

and hydrological processes operating within the ecosystem (Schlesinger *et al.* 1990; Graetz, 1991). There may also be shifts in natural fire cycles.

Desertification generally decreases primary production, leaf area, transpiration, and biodiversity (Pickup 1996). Vegetation composition may change, with rich grasses, forbs, and herbs

Fig. 22.7 Desertified landscapes: (a) healthy savanna of Niger, West Africa, (b) nearby degraded savanna, (c) onset of degradation in Australian savanna; sand is accumulating around the tree trunks, (d) dead mulga shrubs and drying soil in a desertified Australian landscape, (e) windswept and eroding dune, damaged through grazing by rabbits and other animals.

being replaced by less palatable shrubs (Todd and Hoffman 1999). Bush encroachment is a common problem in many degraded areas. The biomass and ground cover of native perennial plants and many associated microbial and animal populations may be reduced, while exotic and usually less desirable plant species may increase in dominance. The encroachment by woody shrubs also has a significant impact on hydrological processes (Huxman *et al.* 2005).

In drylands, soil erosion and sedimentation by both water and wind often result in a redistribution of topsoil, soil compaction and/or loss of soil aggregation, and dune and gully formation (Reynolds and Stafford Smith 2002a). Wind and water erosion remove vast amounts of soil; gullies and badlands form. Soil structure and texture are altered because the

erosion preferentially removes the organic topsoil and fine materials (Okin *et al.* 2001). This finer material contains much of the soil's resources, such as carbon, nitrogen, phosphorus, and other trace materials. Impervious horizons, such as lateritic crusts, may form. In many areas, the fertility of soils is reduced by salinization, alkalinization, or leaching. Water supplies also become enriched in salt. These changes all reduce soil fertility. Ultimately, bare ground is exposed and the vertical structure of vegetation changes. Sand dunes are built or mobilized, often encroaching upon vegetation.

Changes in surface hydrology and drainage patterns result from the altered conditions of vegetation and soil. With the removal of the fine materials, the infiltration capacity of the topsoil is generally reduced, as is the soil's capacity to store moisture. The threshold for the initiation of runoff is altered. There is a redistribution of surface water and the water profile in the soil.

Disruption of the biogeochemical cycle results in the redistribution of essential nutrients, decreased efficiency of nutrient cycling, and increased nutrient losses from the system (Reynolds and Stafford Smith 2002b). Nutrient-rich and diverse species are replaced by vegetation of poorer quality. Conversion from grassland to shrubland reduces and redistributes plant-available nitrogen, for example (Schlesinger et al. 1990). The spatial distribution of vegetation and water supply is unfavorably altered, with the landscape becoming more heterogeneous and resources shrinking. The carrying capacity of the land is dramatically reduced.

Graetz (1991) argues that desertification can be viewed as resulting from disturbance in an ecosystem that is otherwise stabilized by a series of feedbacks between society, people, herbivores, vegetation, soil, and climate. When a vegetated surface is grazed, the result is an alteration of soil water and nutrient flow in the vicinity of plants. The removal of litter by grazing and trampling of the soil surface changes the partitioning of rainwater on the surface. The removal of litter and exposure of soil has two positive feedbacks: the bare soil is more susceptible to crusting, to rainfall impact (rainsplash), and to wind and water erosion. A second feedback is the alteration of the microclimate over the disturbed land: the extremes of temperature and evaporation are amplified when plants are removed, thus making regrowth more difficult (Nicholson 1999).

22.8 DESERTIFICATION AS A PROCESS OF DISTURBANCE

Prince (2002) stresses that desertification is a *process* of disturbance. A land use disturbance decreases the herbaceous vegetation, soils become compacted in the intercanopy regions, and water runs off from the intercanopy toward the canopy (Breshears and Barnes 1999). As the desertification process proceeds, several transition states may exist, rather than a direct conversion to "desert" (Jeltsch et al. 1997).

In drylands the "state" of the vegetation includes "a range of conditions that typically vary from year to year but remain within a range of interconvertible states" (Prince 2002). The vegetation is not in a stable equilibrium but varies within a "domain," or transition state (Westoby et al. 1989). When the environment is disturbed by human factors or by episodic, rare events (such as flood or protracted drought), there can be a transition to a new domain. Because of the spatial complexity of the terrain, the vegetation, and the natural events that produce disturbance, some areas of the landscape are more affected than others. As a result, some parts of the landscape enter a new domain that persists even when the environment returns to "normal." Thus, two or more domains, i.e., vegetation states, exist within the same landscape under the same set of physical conditions.

The net effect is to alter the "patchiness" of the landscape, increasing the size of patches or eliminating them altogether (Meron et al. 2007). The disappearance of the patches dismantles the nutrient-rich islands. The exposed soil is eroded and the particles are often captured by vegetation, creating nebkha dunes. The landscape takes on an increasingly arid character (Schlesinger and Pilmanis 1998). As this happens, the soil becomes more vulnerable to the abiotic transport processes that characterize aridity, such as wind and water erosion. These become increasingly dominant at the expense of biotic transport processes (uptake by plants, redistribution by animals) that occur in semi-arid environments (Okin 2002).

Thus, desertification can be seen as a process in which human forces (such as cultivation or livestock grazing) or climatic forces (such as drought) trigger patch disruption. The consequences of this disruption are described in Chapter 3. As the "aridification" proceeds, the size and structure of the patches changes and increasingly more bare ground is exposed. The land becomes increasingly susceptible to wind and water erosion. Theoretical models, and some field observations, suggest that critical thresholds exist in the patch dynamics, beyond which recovery is not possible (e.g., Scheffer and Carpenter 2003; Rietkerk et al. 2004; see Section 3.5).

The consequences of this degradation are similar to those of short-term drought. With drought, the transition to a new state may be relieved within a few years of good rainfall, although in some cases the vegetation response may last 10–20 years (Wiegand and Milton 1996). When the drought is persistent, resilience is reduced and previously acceptable land uses may become unsustainable, rapidly shifting the vegetation into a new, more degraded domain. This transition may be irreversible or very slowly reversible (Ridolfi et al. 2008).

22.9 REVERSIBILITY: ISSUES OF RESILIENCE AND RESISTANCE

One of the most common myths resulting from the early, highly political literature on desertification is that grazing is an almost ubiquitous cause. Images were shown of browsing goats causing the Sahara Desert to advance southward. Although grazing (termed herbivory in the ecological literature) certainly has a significant impact on the landscape of a vast number of dryland regions, a body of knowledge has emerged to show that it does not necessarily have detrimental and long-lasting impacts and that many biophysical factors determine the nature of its impact.

The drylands globally are extremely diverse in both the biogeophysical conditions of the environment and the socioeconomic conditions governing human impact and response. Thus the relevant factors and feedbacks, and even the direction in which the feedback operates, are vastly different among the various dryland regions. Ecologically sound management in these diverse regions has been hindered by the inability to generalize about degradation from limited and widely separated studies – a problem that also plagues the desertification debate. To facilitate management, a framework is necessary that allows some generalization, so that results from one area can be extrapolated to another. This requires being able to determine the commonalities among various rangelands and to place any particular savanna in a context compared with others, so that conclusions about cause, effect, and solution can be drawn from the existing knowledge base. The concepts of resilience and resistance serve as a useful framework.

22.9.1 MODELS OF RESILIENCE AND RESISTANCE

Some landscape types can tolerate a disturbance more than others without such loss of function, or can completely recover (vegetation becomes re-established, soil organic matter re-accumulates), although recovery is generally slower than degradation (Ash *et al.* 2002; Walker *et al.* 2002). The ability of an ecosystem to persist in a well-functioning state is determined by the properties of resilience and resistance. Resilience is essentially a measure of how much disturbance can be tolerated with full recovery still being possible. Loss of resilience is a form of degradation. Resistance is a measure of how much disturbance can be tolerated before a marked change in the system occurs.

The relationship between rangeland conditions (e.g., productivity, palatable vs. unpalatable species) and external pressure reflects the ecological resistance. Resilience is more difficult to estimate, because a threshold must be established, beyond which recovery is no longer possible. The threshold can vary with season and time, and is also a function of ecosystem and region, so many observations are needed to detect the threshold. Once established, a manager can determine how much pressure (e.g., grazing) a system can withstand.

Carpenter *et al.* (2001) present three generic models of ecosystems varying in the degree to which they possess resilience and resistance (Fig. 22.8). The *x* and *y* axes, respectively, are external pressure (disturbance of the natural equilibrium) and rangeland condition, varying from good to poor. In all cases, as pressure is applied there is initially little change in condition, but at some point there is a very rapid and accelerating degradation. If the degradation continues beyond some threshold, complete recovery is no longer possible. The amount of external

pressure needed to produce the rapid decline is a measure of resistance; the location of the threshold on the scale of poor to good is a measure of resilience. Thus, model (a) is both resilient and resistant, model (b) is resistant but not resilient, and model (c) is not resistant but is resilient.

22.9.2 FACTORS INFLUENCING RESILIENCE

Walker *et al.* (2002) provide an excellent summary of some general rules governing resilience in rangelands. They synthesize diverse aspects of the issue by considering whether or not the landscape is what they term "self-organizing." By this is meant "the capacity to restructure under stress or disturbance such that function is approximately maintained." Self-organization is evident in cracking clays but not soil with a strong texture contrast. Deep sands are very resilient, but mainly because they are resistant to disturbance; they are not self-organizing. Redistribution processes are self-organizing where controlled by biology but not where soils and topography play the dominant role. Species diversity plays a role where it is large enough to sustain diverse animal responses. The factors that determine resilience are dependent on spatial scale, but they are strongly interactive both within and across scales (Peterson *et al.* 1998).

The climatic attributes affecting resilience at the regional scale are wind and various characteristics of the rainfall regime. Rainfall amount, intensity, seasonality, and variability all play a role (Ellis *et al.* 2002). Wind and rainfall intensity both affect the partitioning of the rainfall into soil moisture infiltration/ storage and runoff versus evaporation. They also control soil loss through erosion. By influencing the soil moisture regime, both factors also determine rain-use efficiency and therefore ground cover. Rainfall variability operates at seasonal, interannual, and interdecadal scales.

The effect at each scale is hard to generalize. A predictable wet season offers the greatest resilience, in that it facilitates good management. A warm growing season may be more resilient than a cool one, other factors being equal, but generally cool-season rainfall is more predictable. Interannual variability tends to decrease mean productivity, since a highly variable season is less reliable, and conservative land use precludes tuning productivity to the best years. High interdecadal variability can increase or decrease resilience. A wet decade provides time to recover, but a dry one promotes extended degradation.

At the patch scale, the processes leading to degradation include soil loss; changes in soil structure and texture; removal of fine materials, organic matter, and nutrients; and changes in vegetation structure and species (Walker *et al.* 2002). Soils without marked changes of texture within the soil profiles tend to tolerate grazing pressure and readily accept moisture. Sandy soils may resist grazing pressure because grazing does not greatly influence structure and hence infiltration. Cracking clay soils, which absorb water into their lattice with every rainfall, tend to be resilient because they reform their structure every

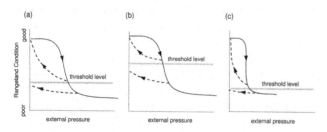

Fig. 22.8 Three models of resilience and resistance in ecosystems (Carpenter *et al.* 2001): (a) resilient and resistant, (b) resistant but not resilient, and (c) not resistant but resilient.

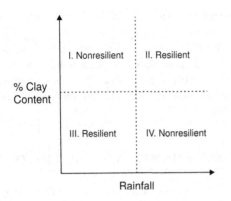

Fig. 22.9 Resilience in terms of fodder production as a function of amount of rain and percentage clay content (from Walker *et al.* 2002).

rainy season. However, these generalizations do not hold everywhere, as the case of northern China shows (Section 22.4). Degradation is generally limited to the sandy soils (Ellis *et al.* 2002).

In general, clays are resilient under conditions of high rainfall, but are easily degraded under conditions of low rainfall (Fig. 22.9). The opposite is true of sandy soils. Evolutionary history of grazing plays a role because it promotes adaptations in the vegetation that discourage foraging or browsing. Examples are spininess or chemical deterrents. Species diversity in terms of growth phenology and function also tend to enhance resistance or resilience. In some cases structural diversity is maintained by minor species of a structural class replacing the dominant species if the latter is lost as a result of grazing or other disturbance.

22.10 MONITORING DESERTIFICATION

Monitoring desertification has long proven to be a difficult task. Verón *et al.* (2006) provide an excellent review of various methods that have been utilized and conclude that assessment is still an unsolved issue. Assessment has shifted from broad-brush surveys of desert boundaries to complex, multivariate field surveys, with an emphasis on ecosystem function. However, progress is still hindered by conflicting definitions of desertification, a situation that has led both to politicization and misuse of the concept.

Most attempts at detecting desertification have been local case studies. Good examples include Pickup (1996) in Australia, Dean *et al.* (1995) in South Africa, Ringrose *et al.* (1997) in Botswana, Wu and Loucks (1992) in China, and Okin *et al.* (2001) in the southwestern USA, to name just a few. Unlike many "case studies" of desertification, these are careful scientific undertakings that include considerable ecological detail.

A handful of studies have attempted to produce global inventories of the process, the most notable being the map disseminated by UNEP (1977) and shown in Fig. 22.1. Some notable features of the map are that areas of "severe" desertification correspond to those of intense field study, such as the small

area so labeled in southern Tunisia, and that areas of extreme desert nearly devoid of humans, such as much of the Sahara, are nonetheless labeled as having "slight" desertification. Other attempts to portray desertification on a global scale have been made but, like the UNEP map, all lack readily measured, objective indicators.

A qualitative map worth noting is that of GLASOD (Global Assessment of Soil Degradation), undertaken by ISRIC (International Soil Reference Information Center) for UNEP. The goal was not specifically a desertification inventory, but a map of human-induced soil degradation. The approach, adapted from Dregne (1989), was termed "structured informed opinion analysis." In other words, the map was based on expert opinion rather than hard data. This includes a large degree of subjectivity. However, the resultant GLASOD map was based on more rigorous and consistent guidelines than previous assessments, and estimates were based on the input of a larger pool of regional soil degradation experts (Thomas and Middleton 1994).

Leemans and Kleidon (2002) conclude that global databases, including the GLASOD map, are of sufficient quality to allow vulnerability to land degradation to be assessed globally from information on climate, soil, land cover, and land use. Batjes (1996) used this approach to produce a global assessment of vulnerability to water erosion on a $0.5° \times 0.5°$ grid.

On the other hand, no techniques exist that can actually measure desertification on a global scale directly and objectively (Lambin *et al.* 2002). This is at least partly because of the ambiguities and complexity of the phenomenon. The variables used to measure the process must be closely related to human–biogeophysical interaction. Examples include erosion and net primary production (NPP) (Prince 2002). The absence of a clear definition of desertification, the inclusion of both biogeophysical and human dimensions, and possible reversibility owing to rehabilitation efforts all hinder approaches to measuring the phenomenon globally.

Prince (2002) strongly makes the case for quantitative global assessment through remote sensing. A number of local case studies have utilized various methods of remote sensing to assess desertification (e.g., Pickup 1996; Dahlberg 2000; Wessels *et al.* 2004, 2007). The major difficulties in such large-scale monitoring include finding the appropriate indicators of desertification, establishing a remote-sensing methodology for detecting them, and establishing a proper baseline. It is important that the indicators selected can be measured reliably, have appropriate temporal and spatial frequency, and have an explicit relationship with the process of interest (Prince *et al.* 1990; Rhodes 1991). Batterbury *et al.* (2002) summarize indicators that have been suggested by various authors (Table 22.2).

Nicholson *et al.* (1998) and Prince *et al.* (1998) have suggested using various aspects of production efficiency as an indicator. Okin *et al.* (2001) and others have suggested vegetation cover or structure and soil crusting or a combined approach using several of these indicators (Symeonakis and Drake 2004). All can be monitored using satellites. Prince (2002) indicates that

Table 22.2. *Indicators of desertification proposed by various authors, as summarized by Batterbury et al. (2002).*

Rain-use efficiency (ratio of net primary productivity (NPP) to rainfall)
Difference between potential and actual NPP
Fraction of vegetation cover (or patch distances)
Fraction of grass, shrubs, forests (or species change)
Crusting
Nutrient concentration in soil
Soil salinity
Water erosion rate
Wind erosion rate

---

net primary production or NPP and rain-use efficiency are particularly appropriate, since they can be reliably measured from space and because many definitions of desertification emphasize the carbon cycle. NPP itself may be useful, since low productivity is at the heart of many definitions.

Unfortunately, it is difficult to assess these indicators on a long enough temporal scale to determine declines and irreversibility. An alternative approach is examining NPP as a departure from potential NPP (or DNPP) or from a local average of NPP (LNPP) (Pickup 1996). DNPP is a departure from the productivity that biogeochemical models predict, based on available rainfall. Hence, it is an indicator that some unspecified factor has reduced productivity. LNPP assumes that, for a large area, there is little degradation and that small pockets of degradation show up as anomalously low productivity. Rain-use efficiency (RUE) is the ratio of NPP to precipitation, a measure that correlates well with desertification. As with NPP, it needs to be examined in the context of some reference frame, such as potential RUE or locally averaged RUE.

All three indicators (RUE, DNPP, LNPP) share a common problem: the non-equilibrium behavior of arid lands. Trends are masked by short-term temporal and spatial variations in rainfall, making it difficult to distinguish between natural and anthropogenic change. Herrmann *et al.* (2005) and Evans and Geerken (2005) have attempted to overcome this problem by looking at "residuals" from the NDVI (normalized difference vegetation index) predicted from rainfall relationships. It is important to stress that even when NPP or NDVI falls below the potential set by the biogeophysical conditions, the cause is not necessarily desertification. These and other remote sensing indicators are surrogate indices that can give false positive or negative signals because of the influence of other factors, including natural ecosystem dynamics. Thus, all trends suggested by such indicators must be confirmed at ground level (Prince 2002).

These concepts have been utilized in numerous studies assessing desertification over the Sahel and in southern Africa (e.g., Prince *et al.* 1998, 2007; Tucker and Nicholson 1999; Wessels *et al.* 2004, 2007; Anyamba and Tucker 2005). These have collectively demonstrated that desertification in the Sahel is not nearly as widespread as once believed, that the interannual variability of Sahelian vegetation mimics rainfall, and that the recent recovery has been commensurate with an increase in rainfall. Small areas of possible desertification have been identified (Prince *et al.* 2007; Wessels *et al.* 2007), but productivity has also appeared to increase in some areas once deemed desertified (Fig. 22.10).

22.11 IMPACTS ON WEATHER AND CLIMATE

A change in surface vegetation cover can alter many aspects of climate, such as maximum and minimum temperature, surface wind, and relative humidity. To examine the question of modification of large-scale climate and weather patterns, the main variables of interest are surface temperature and albedo, roughness, and evapotranspiration. A number of studies have assessed differences in climate-related variables between "desertified" and undisturbed land, or have utilized numerical models to evaluate their impact.

Three particular case studies including field measurements are worth noting. One took place in Tunisia and two evaluated the grazed and ungrazed sides of international borders in the Sonoran Desert (United States/Mexico) and in the Sinai and Negev (Egypt/Israel). Albedo contrasts were the first to be recognized (see photo in Fig. 6.10). Satellite photos (Otterman 1977, 1981) showed that the soil in the overgrazed Sinai had albedos of 0.4 in the visible and 0.53 in the infrared, for an average albedo of 0.46; in the protected area of the Negev, the visible and infrared albedos were 0.12 and 0.24, respectively, but in most of the region the albedo was about 0.25 on average. In Tunisia the albedo of protected vs. unprotected sites was 0.35 vs. 0.39 in one case and 0.26 vs. 0.36 in another, while oases had albedos on the order of 0.10–0.23 (Wendler and Eaton 1983).

Other differences between grazed and ungrazed land are more controversial. In the Sinai, where the soil surface is more reflective, radiometric ground temperatures on the order of 0.4°C were measured, compared with 45°C on the darker Negev side (Otterman and Tucker 1985). This apparent contradiction might reflect the differences of emissivity on the two sides and not real surface temperatures, but geometric arguments for a warmer surface where the vegetation cover is denser can be made. In contrast, in the Sonoran case surface temperatures were generally 2°–4°C higher on the supposedly brighter, more heavily grazed Mexican side of the border than on the US side (Balling 1988; Bryant *et al.* 1990). However, a more recent study (Michalek *et al.* 2001) found markedly smaller and inconsistent differences between the two sides (Table 22.3).

Numerous mathematical models have tested the effects of surface changes that might be evoked by desertification (e.g., higher surface temperatures, lower surface roughness, reduced evapotranspiration, increased albedo). Numerical models of the effect of vegetation patterns on mesoscale

Table 22.3. *Average albedo and radiant temperature from 25 km (east–west) transects near the Arizona/Sonora border for May 26 and August 30, 1998 Landsat TM data (from Michalek* et al. *2001).*

Transect location	Transect width (north–south) (m)	May 26, 1998		August 30, 1998	
		Albedo (%)	Temp. (°C)	Albedo (%)	Temp. (°C)
Arizona	100	19.62	37.44	18.29	36.21
Mexico	100	21.43	37.43	19.99	36.63
Arizona	2000	20.67	37.53	19.25	35.83
Mexico	2000	20.68	37.19	19.41	36.34

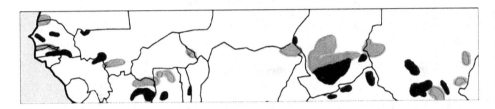

Fig. 22.10 Areas of the Sahel in which rain-use efficiency increased (light shading) or decreased (dark shading) over the period 1981–1999 (adapted from Prince *et al.* 2007). A decrease suggests possible areas of desertification.

circulation further support the idea that desertification can, at least potentially, have an influence on weather and climate. Most have concluded that a positive feedback exists, with such changes reducing rainfall and thereby further altering the vegetation and soil (see reviews in Pielke *et al.* 1998; Nicholson 2000). This implies that the processes associated with desertification could be self-reinforcing. However, the model simulations generally impose much greater differences in the surface parameters than are actually observed in the case of drought or desertification.

Despite conclusions from models on the importance of surface albedo, the albedo theory of Charney is losing support because it is contradicted by observational studies (see also Chapter 6). Field observations, aerial or satellite photos, and vegetation information all show that the albedo changes that occur are much smaller than Charney's theory requires and that they are inconsistent with the changes in rainfall and vegetation that occur (e.g., Courel *et al.* 1984). Some changes are temporary, and thus inconsistent with the central tenet of desertification: irreversibility. Regrowth of vegetation when an airfield was built in the Sinai is a case in point (Fig. 22.11). Moreover, a decrease in vegetation cover can actually decrease albedo, and in critical regions such as the Sahel, albedo is more a function of soil moisture than of vegetation (Ba *et al.* 2001). The effects of changes in evapotranspiration and soil moisture, as well as the increased dust generated by desertification or drought, appear to be more important than albedo.

Observational evidence of the climatic impact of desertification is more difficult to obtain. In the case of the Sonora, the temperature differences on the grazed and protected sides of the border were shown to have an impact on soil moisture and cloudiness, although no changes in precipitation could be identified (Balling 1988; Bryant *et al.* 1990). On the other hand,

Fig. 22.11 Satellite photo showing the greenness across the Sinai/Negev boundary also shown in Fig. 6.9. When an airfield was built on the Sinai side, the vegetation rapidly regenerated. (photograph courtesy of NASA).

there is evidence that irrigation can modify the local and meso-scale weather patterns, even to the extent of enhancing rainfall by 15–90% in some months and increasing thunderstorms, hail, and severe weather (Barnston and Schickedanz 1984). In the Sahel, a well-controlled observational study of Taylor and Lebel (1998) showed that a location receiving particularly high rainfall from an early-season storm is more likely to receive rainfall from subsequent storms (Fig. 22.12). Hence, at least regional-scale impacts on climate seem possible.

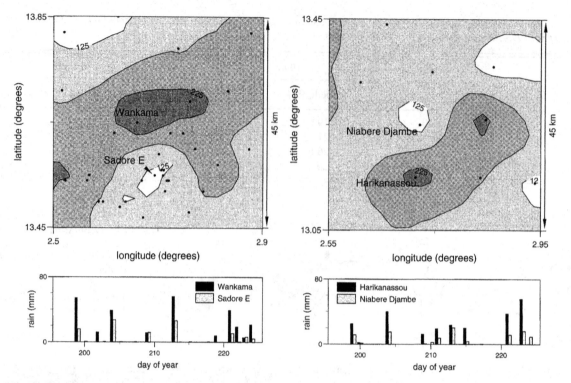

Fig. 22.12 (top) Accumulated 30-day rainfall totals in the Sahel region, and (bottom) daily rainfall at two pairs of gauges during the HAPEX-Sahel experiment, July 14 to August 12, 1992. Darkest shading indicates monthly rainfall in excess of 225 mm; lightest indicates monthly rainfall less than 135 mm. In both cases, the gauges are ~10 km apart, and the climatological gradient is small and in the opposite direction of that shown (from Taylor and Lebel 1998).

22.12 PREVENTION AND REMEDIES

Many grandiose schemes have been proposed to stop the "march" of the desert. Most have been based on little background research or understanding of the problem, but nevertheless many have been implemented. The most common example is the planting of trees to create "green belts" to stop the moving sands or to stem water erosion, even though trees are poorly suited to the task (Romero-Diaz *et al.* 1999). UNEP actually sanctioned such a scheme, a trans-Saharan green belt to stop the desert's advance. Little thought was given to whether trees could accomplish the task or could even subsist in the Sahara's arid climate. Another proposed scheme in northern China was flattening sandy lands by covering them with clay plus forests or grass.

Mainguet (1991) sketches a number of more realistic ways by which desertification can be prevented or reversed. Many of these have been long-established methods of soil, vegetation, and water conservation. Some are indigenous strategies going back at least hundreds of years in arid and semi-arid regions. Active management techniques can also be useful. Examples are terracing, plowing, improved water distribution techniques (e.g., drip irrigation), water harvesting, and fertilization. Controlled grazing helps as well. The introduction of plant species that increase forage quality, replace species lost

to desertification, or provide resistance to drought is another general strategy. Vegetation is also introduced to create windbreaks or to prevent soil loss.

Many of these methods have been used in arid lands to increase productivity (see Chapter 24). When carefully managed they can allow the desert to bloom or the man-made desert to recover its full potential. Most must be created for and adapted to specific regions; if they are not, they may instead worsen the land degradation.

REFERENCES

Ahlcrona, E., 1988: *The Impact of Climate and Man of Land Transformation in Central Sudan.* Lund University Press, Lund, Sweden.

Anyamba, A., and C. J. Tucker, 2005: Analysis of Sahelian vegetation dynamics using NOAA-AVHRR NDVI data from 1981–2003. *Journal of Arid Environments,* **63**, 596–614.

Ash, A. J., D. M. Stafford Smith, and N. Abel, 2002: Land degradation and secondary production in semi-arid and arid grazing systems: what is the evidence? In *Global Desertification: Do Humans Cause Deserts?* (J. F. Reynolds and D. M. Stafford Smith, eds.), Dahlem University Press, Berlin, pp. 75–94.

Ba, M. B., S. E. Nicholson, and R. Frouin, 2001: Temporal and spatial variability of surface radiation budget over the African continent as derived from METEOSAT. Part II. Temporal and spatial variability of surface global solar irradiance, albedo and net radiation. *Journal of Climate*, **14**, 60–76.

Batjes, N. H., 1996: Global assessment of land vulnerability to water erosion on a ½ by ½ degree grid. *Land Degradation & Development*, **7**, 353–365.

Balling, R. C., Jr., 1988: The climatic impact of Sonoran vegetation discontinuity. *Climatic Change*, **13**, 99–109.

Barnston, A. G., and P. T. Schickedanz, 1984: The effect of irrigation on warm season precipitation in the southern Great Plains. *Journal of Climate and Applied Meteorology*, **23**, 865–888.

Batterbury, S. P. J., and Coauthors, 2002: Responding to desertification at the national scale: detection, explanation, and responses. In *Global Desertification: Do Humans Cause Deserts?* (J. F. Reynolds and D. M. Stafford Smith, eds.), Dahlem University Press, Berlin, pp. 357–386.

Bovill, E. W., 1921: The encroachment of the Sahara on the Sudan. *African Affairs (London)*, **XX**, 174–185.

Breshears, D. D., and F. J. Barnes, 1999: Interrelationships between plant functional types and soil moisture heterogeneity for semiarid landscapes within the grassland/forest continuum: a unified conceptual model. *Landscape Ecology*, **14**, 465–478.

Bryant, N. A., L. F. Johnson, A. J. Brazel, R. C. Balling, C. F. Hutchinson, and L. R. Beck, 1990: Measuring the effect of overgrazing in the Sonoran Desert. *Climatic Change*, **17**, 243–264.

Buffington, L. C., and C. H. Herbel, 1965: Vegetational changes on a semidesert grassland range. *Ecological Monographs*, **35**, 139–164.

Carpenter, S., B. Walker, J. M. Anderies, and N. Abel, 2001: From metaphor to measurement: resilience of what to what? *Ecosystems*, **4**, 765–781.

Charney, J. G., 1975: The dynamics of deserts and droughts. *Quarterly Journal of the Royal Meteorological Society*, **101**, 193–202.

Courel, M., R. Kandel, and S. Rasool, 1984: Surface albedo and the Sahel drought. *Nature*, **307**, 528–538.

Dahlberg, A. C., 2000: Interpretations of environmental change and diversity: a critical approach to indications of degradation – the case of Kalakamate, north-east Botswana. *Land Degradation and Development*,**11**, 549–562.

Dean, W. R. J., M. T. Hoffman, M. E. Meadows, and S. J. Milton, 1995: Desertification in the semi-arid Karoo, South Africa: review and reassessment. *Journal of Arid Environments*, **30**, 247–264.

Dregne, H. E., 1983: *Desertification of Arid Lands*. Hardwood Academic, New York.

Dregne, H. E., 1989: Informed opinion: filling the soil erosion data gap. *Journal of Soil and Water Conservation*, **44**, 303–305.

Eckholm, E., and L. R. Brown, 1977: *Spreading Deserts: The Hand of Man*. Worldwatch Paper 13, Worldwatch Institute, Washington, DC, 40 pp.

Ellis, J. E., K. Price, F. Yu, L. Christensen, and M. Yu, 2002: Dimensions of desertification in the drylands of northern China. In *Global Desertification: Do Humans Cause Deserts?* (J. F. Reynolds and D. M. Stafford Smith, eds.), Dahlem University Press, Berlin, pp. 167–180.

Evans, J., and R. Geerken, 2005: Discrimination between climate and human-induced dryland degradation. *Journal of Arid Environments*, **57**, 535–554.

Evenari, M., I. Noy-Meir, and D. W. Goodall, 1985: *Hot Deserts and Arid Shrublands*. Ecosystems of the World 12A, Elsevier, Amsterdam, 458 pp.

Graetz, R. D., 1991: Desertification: a tale of two feedbacks. In *Ecosystem Experiments* (H. A. Mooney, E. Medina, D. W. Schindler, E.-D. Schulze, and B. H. Walker, eds.), Wiley, New York, pp. 59–87.

Helldén, U., 1984: Drought impact monitoring: a remote sensing study of desertification in Kordofan, Sudan. *Lunds Universitets Naturgeografiska Institution*, **61**.

Herrmann, S. M., A. Anyamba, and C. J. Tucker, 2005: Recent trends in vegetation dynamics in the African Sahel and their relationship to climate. *Global Environmental Change*, **15**, 394–404.

Hiernaux, P., and M. D. Turner, 2002: The influence of farmer and pastoralist manage practices on desertification processes in the Sahel. In *Global Desertification: Do Humans Cause Deserts?* (J. F. Reynolds and D. M. Stafford Smith, eds.), Dahlem University Press, Berlin, pp. 135–148.

Hou, R.-Z., 1985: Ancient city ruins in the deserts of the Inner Mongolian Autonomous Region of China. *Journal of Historical Geography*, **11**, 241–252.

Huxman, T., B. Wilcox, D. Breshears, R. Scott, K. Snyder, E. Small, K. Hultine, W. Pockman, and R. Jackson, 2005: Ecohydrological implications of woody plant encroachment. *Ecology*, **86**, 308–319.

Jeltsch, F., S. J. Milton, W. R. J. Dean and N. Van Rooyen, 1997: Analysing shrub encroachment in the southern Kalahari: a grid-based modelling approach. *Journal of Applied Ecology*, **34**, 1497–1508.

Jiang, H., 2002: Culture, ecology, and nature's changing balance: sandification on Mu Us Sandy Land, Inner Mongolia, China. In *Global Desertification: Do Humans Cause Deserts?* (J. F. Reynolds and D. M. Stafford Smith, eds.), Dahlem University Press, Berlin, pp. 181–196.

Kassas, M., 1995: Desertification: a general review. *Journal of Arid Environments*, **30**, 115–128.

Kurc, S., and E. Small, 2004: Dynamics of evapotranspiration in semiarid grassland and shrubland ecosystems during the summer monsoon season, central New Mexico. *Water Resources Research*, **40**, WO9305.

Lambin, E. F., P. S. Chasek, T. E. Downing, C. Kerven, A. Kleidon, R. Leemans, M. Lüdeke, S. D. Prince, and Y. Xue, 2002: The interplay between international and local processes affecting desertification. In *Global Desertification: Do Humans Cause Deserts?* (J. F. Reynolds and D. M. Stafford Smith, eds.), Dahlem University Press, Berlin, pp. 387–401.

Lamprey, H., 1975. *Report on the Desert Encroachment Reconnaissance in Northern Sudan*. National Council for Research/Ministry of Agriculture, Food and Natural Resources, Khartoum.

Leemans, R., and A. Kleidon, 2002: Regional and global assessment of the dimensions of desertification. In *Global Desertification: Do Humans Cause Deserts?* (J. F. Reynolds and D. M. Stafford Smith, eds.), Dahlem University Press, Berlin, pp. 215–232.

Little, M., and P. Leslie (eds.), 1999: *Turkana Herders of the Dry Savanna: Ecology and Biobehavioral Response of Nomads to an Uncertain Environment*. Oxford University Press, Oxford.

Mainguet, M., 1991: *Desertification, Natural Background and Human Mismanagement*. Springer Verlag, Springer.

Meron, E., H. Yizhaq, and E. Gilad, 2007: Localized structures in dryland vegetation: forms and functions. *Chaos*, **17**, 037109.

Michalek, J. L., J. E. Colwell, N. E. G. Roller, N. A. Miller, E. S. Kasischke, and W. H. Schlesinger, 2001: Satellite measurements of albedo and radiant temperature from semi-desert grassland along the Arizona/Sonora border. *Climatic Change*, **48**, 417–425.

Mortimore, M., and B. Turner, 2005: Does the Sahelian smallholder's management of woodland, farm trees, rangeland support the hypothesis of human-induced desertification? *Journal of Arid Environments*, **63**, 567–595.

National Science Foundation, 1977: Life at the desert's edge. *Mosaic*, Jan/Feb, 20–27.

Nicholson, S. E., 1996: Desertification. *Encyclopedia of Climate and Weather* (S.H. Schneider, ed.), Simon Schuster, New York, pp. 239–242.

Nicholson, S. E., 1999: The physical–biotic interface in arid and semiarid systems. *Arid Lands Management: Towards Ecological*

Sustainability (T. W. Hoekstra and M. Shachak, eds.), University of Illinois Press, Urbana, IL, pp. 31–47.

Nicholson, S. E., 2000: Land surface processes and Sahel climate. *Reviews of Geophysics*, **38**, 117–139.

Nicholson, S. E., 2004: Desertification. In *The Encyclopedia of World Environmental History* (S. Krech III, J. R. McNeill, and C. Merchant, eds.), Berkshire Publishing Group/Routledge, UK, pp. 297–303.

Nicholson, S. E., 2005: On the question of the "recovery" of the rains in the West African Sahel. *Journal of Arid Environments*, **63**, 615–641.

Nicholson, S. E., C. J. Tucker, and M. B. Ba, 1998: Desertification, drought and surface vegetation: an example from the West African Sahel. *Bulletin of the American Meteorological Society*, **79**, 815–829.

Okin, G. S., 2002: Toward a unified view of biophysical land degradation processes in arid and semi-arid lands. In *Global Desertification: Do Humans Cause Deserts?* (J. F. Reynolds and D. M. Stafford Smith, eds.), Dahlem University Press, Berlin, pp. 95–110.

Okin, G. S., B. Murray, and W. H. Schlesinger, 2001: Degradation of sandy arid shrubland environments: observations, process modelling and management implications. *Journal of Arid Environments*, **47**, 123–144.

Olsson, L., 1985: *An Integrated Study of Desertification*. Studies in Geography No. 13, Lund University, Sweden.

Otterman, J., 1977: Anthropogenic impact on the albedo of the earth. *Climatic Change*, **1**, 137–157.

Otterman, J., 1981: Satellite and field studies of man's impact on the surface in arid regions. *Tellus*, **33**, 68–77.

Otterman, J., and C. J. Tucker, 1985: Satellite measurements of surface albedo and temperatures in semi-desert. *Journal of Climate and Applied Meteorology*, **24**, 228–234.

Peterson, G., C. R. Allen, and C. S. Holling, 1998: Ecological resilience, biodiversity, and scale. *Ecosystems*, **1**, 6–18.

Pickup, G., 1996: Estimating the effects of land degradation and rainfall variation on productivity in rangelands: an approach using remote sensing and models of grazing and herbage dynamics. *Journal of Applied Ecology*, **33**, 819–832.

Pielke, R. A., Sr., R. Avissar, M. Raupach, A. J. Dolman, X. Zeng, and A. S. Denning, 1998: Interactions between the atmosphere and terrestrial ecosystems: influence on weather and climate. *Global Change Biology*, **4**, 461–475.

Prince, S. D., 2002: Spatial and temporal scales for detection of desertification. In *Global Desertification: Do Humans Cause Deserts?* (J. F. Reynolds and D. M. Stafford Smith, eds.), Dahlem University Press, Berlin, pp. 23–40.

Prince, S. D., C. O. Justice, and S. O. Los, 1990: *Remote Sensing of the Sahelian Environment: A Review of the Current Status and Future Prospects*. Technical Centre for Agricultural and Rural Cooperation and the Commission of the European Communities, Wageningen, The Netherlands.

Prince, S. D., E. Brown de Colstoun, and L. Kravitz, 1998: Evidence from rain use efficiencies does not support extensive Sahelian desertification. *Global Change Biology*, **4**, 359–373.

Prince S. D, K. J. Wessels, C. J. Tucker, and S. E. Nicholson, 2007: Desertification in the Sahel: a reinterpretation of a reinterpretation. *Global Change Biology*, **13**, 1308–1313.

Reynolds, J. F., and D. M. Stafford Smith (eds.), 2002a: *Global Desertification: Do Humans Cause Deserts?* Dahlem University Press, Berlin, 437 pp.

Reynolds, J. F., and D. M. Stafford Smith, 2002b: Do humans cause deserts? In *Global Desertification: Do Humans Cause Deserts?* (J. F. Reynolds and D. M. Stafford Smith, eds.), Dahlem University Press, Berlin, pp. 1–22.

Rhodes, S. L., 1991: Rethinking desertification: what do we know and what have we learned? *World Development*, **19**, 1137–1143.

Ridolfi, L., F. Laio, and P. D'Odorico, 2008: Fertility island formation and evolution in dryland ecosystems. *Ecology and Society*, **13**, 5.

Rietkerk, M., S. C. Dekker, P. C. de Ruiter, and J. van de Koppel, 2004: Self-organized patchiness and catastrophic shifts in ecosystems. *Science*, **305**, 1926–1929.

Ringrose, S., C. Vanderpost, and W. Matheson, 1997: Use of image processing and GIS techniques to determine the extent and possible causes of land management/fenceline induced degradation problems in the Okavango area, northern Botswana. *International Journal of Remote Sensing*, **18**, 2337–2364.

Romero-Diaz, A., L H. Cammeraat, A. Vacca, and C. Kosmas, 1999: Soil erosion at three experimental sites in the Mediterranean. *Earth Surface Processes and Landforms*, **24**, 1243–1256.

Scheffer, M. M., and S. R. Carpenter, 2003: Catastrophic regime shifts in ecosystems: linking theory to observation. *Trends in Ecology and Evolution*, **18**, 648–656.

Schlesinger, W. H., and A. M. Pilmanis, 1998: Plant–soil interactions in deserts. *Biogeochemistry*, **42**, 169–187.

Schlesinger, W. H., J. F. Reynolds, G. L. Cunningham, L. F. Huenneke, W. M. Jerrell, R. A. Virginia, and W. G. Whitford, 1990: Biological feedbacks in global desertification. *Science*, **247**, 1043–1048.

Schwarz, E. H. L., 1920. *The Kalahari or Thirstland Redemption*. T. Maskew Miller, Capetown.

Stafford Smith, D. M., and Pickup, G., 1993: Out of Africa, looking in: understanding vegetation change. In: *Range Ecology at Disequilibrium: New Models of Natural Variability and Pastoral Adaptation in African Savannas* (R. H. Behnke, Jr, I. Scoones, and C. Kerven, eds.), London: Overseas Development Institute and Intl. Institute for Environment and Development, pp. 196–244

Stafford Smith, D. M., and J. F. Reynolds, 2002: The Dahlem desertification paradigm: a new approach to an old problem. In *Global Desertification: Do Humans Cause Deserts?* (J. F. Reynolds and D. M. Stafford Smith, eds.), Dahlem University Press, Berlin, pp. 403–424.

Stebbing, E. P., 1935: The encroaching Sahara. *Geographical Journal*, **86**, 509–510.

Symeonakis, E., and N. Drake, 2004: Monitoring desertification and land degradation over sub-Saharan Africa. *International Journal of Remote Sensing*, **25**, 573–592.

Taylor, C. M., and T. Lebel, 1998: Observational evidence of persistent convective scale rainfall patterns. *Monthly Weather Review*, **126**, 1597–1607.

Thomas, D. S. G., and N. J. Middleton, 1994: *Desertification: Exploding the Myth*. John Wiley and Sons, Chichester.

Tiffen, M., M. Mortimore, and F. Gichuki, 1994: *More People, Less Erosion, Environmental Recovery in Kenya*. Wiley, Chichester.

Todd, S. W., and M. T. Hoffman, 1999: A fence-line contrast reveals effects of heavy grazing on plant diversity and community composition in Namaqualand, South Africa. *Plant Ecology*, **142**, 169–178.

Tolba, M. K., 1977: Desertification: a man-made process. *World Health*, July, 2–3.

Tucker, C. J., and S. E. Nicholson, 1999: Variations in the size of the Sahara desert from 1980 to 1997. *Ambio*, **28**, 587–591.

Tucker, C. J., W. W. Newcomb, and H. C. Dregne, 1994: AVHRR data sets for determination of desert spatial extent. *International Journal of Remote Sensing*, **15**, 3547–3566.

UN (United Nations), 1994: *UN Earth Summit: Convention on Desertification*. UN Conference on Environment and Development, Rio de Janeiro, Brazil, June 3–14, 1992. UN, New York.

UNEP (United Nations Environmental Programme), 1977: *Draft Plan of Action to Combat Desertification*. UN Conference on Desertification, Background Document. Nairobi.

UNEP (United Nations Environment Programme), 1997: *World Atlas of Desertification* (editorial commentary by N. Middleton and D.S.G. Thomas), Edward Arnold, New York.

Vasek, F. C., H. B. Johnson, and G. D. Brum, 1975: Effects of power transmission lines on vegetation of the Mojave Desert. *Modroño*, **23**, 114–130.

Verón, S. R., J. M. Paruelo, and M. Oesterheld, 2006: Assessing desertification. *Journal of Arid Environments*, **66**, 751–763.

Verstraete, M. M., 1986: Defining desertification: a review. *Climatic Change*, **9**, 5–18.

Walker, B. H., N. Abel, D. M. Stafford Smith, and J. Langridge, 2002: A framework for the determinants of degradation in arid ecosystems. In *Global Desertification: Do Humans Cause Deserts?* (J. F. Reynolds and D. M. Stafford Smith, eds.), Dahlem University Press, Berlin, pp. 75–94.

Warren, A., 1996: Desertification. In *The Physical Geography of Africa* (W.M. Adams, A. S. Goudie, and A. Orme, eds.), Oxford University Press, Oxford, pp. 342–355.

Webb, R. H., and E. B. Newman, 1982: Recovery of soil and vegetation in ghost towns in the Mojave Desert, southwestern United States. *Environmental Conservation*, **9**, 245–248.

Webb, R. H., J. W. Steiger, and H. G. Wilshire, 1986: Recovery of compacted soils in Mojave Desert ghost towns. *Soil Science Society of America Journal*, **50**, 1341–1344.

Wendler, G., and F. Eaton, 1983: On the desertification in the Sahel zone. Part I. Ground observations. *Climatic Change*, **5**, 365–380.

Wessels, K. J., S. D. Prince, P. E. Frost, and D. van Zyl, 2004: Assessing the effects of human-induced land degradation in the former homelands of northern South Africa with a 1 km AVHRR NDVI time-series. *Remote Sensing of Environment*, **91**, 47–67.

Wessels, K. J., S. D. Prince, J. Malherbe, J. Small, P. E. Frost, and D. van Zyl, 2007: Can human-induced land degradation be distinguished from the effects of rainfall variability? A case study in South Africa. *Journal of Arid Environments*, **68**, 271–297.

Westoby, M., B. Walker, and I. Noy-Meir, 1989: Opportunistic management for rangelands not at equilibrium. *Journal of Range Management*, **42**, 265–273.

Wiegand, T., and S. J. Milton, 1996: Vegetation changes in semiarid communities: simulating probabilities and time scales. *Vegetatio*, **125**, 169–183.

Wu, J., and Loucks, O., 1992: Xilingele. In *Grasslands and Grassland Science in Northern China: A Report of the Committee on Scholarly Communication with the People's Republic of China* (National Research Council, ed.), National Academy Press, pp. 67–84.

Zhu, Z., and T. Wang, 1993: Trends of desertification and its rehabilitation in China. *Desertification Control Bulletin*, **22**, 27–30.

23 People in the dryland environments

23.1 PEOPLE IN THE DRYLAND ENVIRONMENTS

As with other life forms, the human inhabitants of dryland regions need to cope with excessive thermal stress and maintain sufficient water supply. A critical need is for the temperature of the body to remain within tolerable limits, but comfort is also a factor. Water plays a major role in thermal regulation, and it is also required for cellular functioning. The ways in which humans avoid thermal stress are comparable to those of animals, especially in terms of physiological response to stress. Unlike plants and animals, humans can actively manipulate the thermal response of their body, through choice of clothing for example, and can actively manipulate their environment, through construction of their dwellings. They can also conceive of strategies to minimize water demand and optimize its availability. Humans also require energy. Modern technology provides additional options for coping with the dryland environment, but this chapter will focus primarily on traditional adaptations to dryland environments. The involuntary physiological responses of the body, as well as clothing, architecture, livelihood/lifestyle and ways to meet energy and water demands are examined.

23.2 HEAT BALANCE AND COMFORT OF THE HUMAN BODY

The surface heat balance concept presented in Chapter 6 can readily be applied to the human body (Fig. 23.1). The body receives heat through direct and diffuse solar radiation and solar radiation reflected from various surfaces in the environment. It also absorbs longwave radiation given off by the surface and atmosphere. The body cools via longwave radiation and sensible heat transfer (conduction and convection) from the skin and via latent heat transfer (evaporative cooling via sweating). Additional cooling mechanisms for the human body include warming of inhaled air and evaporating moisture into inhaled air. Heat is generated via metabolic activity, but much of this is lost to mechanical work (i.e., movement/exercise) (Young 1979).

The heat balance of the human body can be expressed using the surface heat balance equation (Eqs. 6.1 and 6.2) but adding a term for metabolic heat production M. Thus, any change in the body's heat storage (Δ $storage$) is the net result of the heat acquisitions and expenditures:

$$\Delta \, storage = R_{sw}{\downarrow}\,(1 - a_s) + R_{lw}{\downarrow} - R_{lw}{\uparrow} - LE - S + M. \quad (23.1)$$

The meaning of the terms must be modified from those of Eqs. (6.1) and (6.2). In this case, the "albedo" a refers to the reflectivity of the human body or clothing, $R_{lw}{\downarrow}$ includes heat from both the sky and ground surface, $R_{lw}{\uparrow}$ is radiation emitted from the body, S is sensible heat exchange (conductive and convective) between the person, ground surface and atmosphere, and LE is a latent heat exchange involving deposition, evaporation, and transpiration of moisture on and by the body.

In order to maintain thermal equilibrium, the change in heat storage must be zero. Although excess heat is stored, internal storage is limited because a body core temperature of 40°C (only 3°C above normal temperatures) is considered to be the maximum temperature that can safely be maintained (Young 1979). At 10°C the radiative and convective exchanges are nearly an order of magnitude larger than evaporative cooling, but at 30°C evaporation exceeds the total of the remaining processes. Clothing serves to shield the body from direct radiation and insulates the body (Morgan and de Dear 2003). Flowing, light-colored garments (Fig. 23.2) are common in hot desert climates because they reflect a large proportion of the incoming solar rays and they maximize ventilation, thus enhancing evaporative cooling. Metabolic heat gain is reduced by minimizing activity during the hottest time of the day, the traditional midday "siesta" or at least a period of rest being a common ritual in dryland regions of the low latitudes. Bright sunshine is avoided, shade is sought.

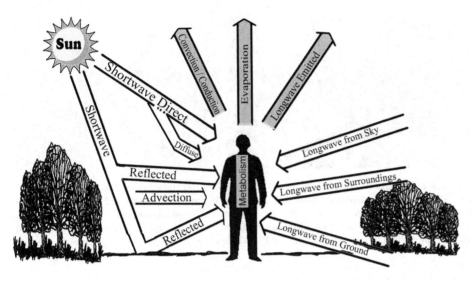

Fig. 23.1 Heat balance of the human body.

Fig. 23.2 Flowing white garments moderate temperatures by reflecting solar radiation and promoting air flow around the body. The photo on the left was taken in the Dallol Bosso of Niger, a dry arm of the Niger River. Soil moisture in the dallol supports a thriving agriculture, as seen in the background. The photo on the right was taken in a village of Niger.

When a balance is not achieved, a number of physiological responses act to restore the equilibrium. Thermoregulatory responses to heat include dilation of skin blood vessels, dilution of blood, extension to increase exposed body surface, decreased muscle tone, sweating, and inclination to reduced activity (Oliver 1981). Consequentially, urine volume is decreased, appetite is reduced, and the body has difficulty in maintaining blood supply to the brain and salt balance, leading to dizziness, heat exhaustion, and heat cramps.

Thermal stress or comfort is a function of six factors: four ambient environmental variables and two behavioral variables. The latter include metabolic activity (e.g., exercise) and clothing. The environmental or meteorological factors include ambient air temperature, humidity, wind, and solar radiation (Epstein and Moran 2006). In a cold climate the primary factors are temperature and wind, which affect the rate at which heat is lost from the body to the cold air. In hot climates, however, the level of thermal stress is largely controlled by temperature and humidity, with a lesser effect from wind and solar radiation.

A number of indices have been developed to quantify the level of comfort or discomfort in hot environments. Epstein and Moran (2006) give a list of 46 of the most common. A widely used index of heat stress or "sultriness," based only on temperature and relative humidity, is that utilized by the US National Weather Service. This index gives, for any combination of ambient air temperature and relative humidity, an "apparent temperature." This represents the comfort/discomfort level the body would feel if that were the ambient temperature but the relative humidity were at a standard level (Steadman 1979a, 1984). Nicholls *et al.* (2008) have developed a simple heat alert system, based on minimum and maximum temperatures, that can predict periods of severe thermal stress.

The impact of humidity on bodily comfort is largely imposed by its influence on evaporative heat loss. The effect of wind is less pronounced in humid regions than in arid regions. Whether it is a cooling or warming effect depends on relative humidity. At very low humidity, if the apparent temperature is below ambient, heat is transferred to the body by the air and a hot dry

Fig. 23.3 Adobe architecture can be traced back to ancient cultures, often with little change over the centuries. (left) The Sankore Mosque in Timbuktu, Mali, part of the fabled city's ancient university. The university dates back to at least the fourteenth century and the city was a major intellectual and spiritual center in the fifteenth and sixteenth centuries. (center) the Grand Mosque of Bobo Dioulasso, Burkina Faso. Although it looks very old, it was built in the late nineteenth century. (right) The Great Mosque of Djenné, Mali, the largest adobe building in the world. Djenné is the oldest city in sub-Saharan Africa. The modern city was founded around 800 AD but a settlement just outside the present city dates back to 250 BC.

wind increases the heat flow, but when perspiration becomes a factor in cooling at high humidity, wind has a cooling effect (Steadman 1979b).

23.3 ARCHITECTURE

The most basic function of dwellings is to provide an environment in which the atmospheric elements of temperature, moisture, and wind are within the range of human comfort. Traditional housing in dryland regions includes many design aspects that efficiently reduce thermal stress, protect from the occasional – sometimes torrential – rains, and shelter the inhabitants from hot winds, often laden with dust and sand. The priority, however, is maintaining a moderate range of temperature: keeping out the extreme heat of day but preventing excessive cooling at night or during the cold season. This is accomplished by shading, insulating construction materials, building orientation, and ventilation.

The degree of insulation is related both to the thickness of walls and the nature of the building materials. The goal is to prevent heat exchange between the inside of the dwelling and the air outside, thereby reducing both the diurnal and seasonal temperature ranges. The heat exchange through walls is analogous to the penetration of heat through a depth of soil (Chapter 6) and is described by the same equations (e.g., Eq. 6.9). The thicker the wall, the greater the dampening of the diurnal or annual waves, the lower the temperature range inside, and the later the temperature maxima and minima occur. If the walls are so thick that the temperature maximum inside corresponds to the temperature minimum outside, conduction of heat is reduced even further.

In these terms, adobe structures (Fig. 23.3), which are ubiquitous in arid regions, are ideally suited to the climate. Adobe is a natural material composed of sand, clay, water, and some kind of fibrous, organic material, such as sticks or straw (Fig. 23.4), and it is much more effective than wood or stone as a thermal insulator. The adobe is made into bricks and sundried. Generally, the same material, minus

Fig. 23.4 Close-up of adobe building material, with the imbedded straw clearly visible.

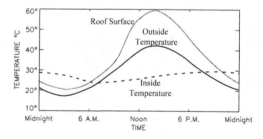

Fig. 23.5 Temperatures inside and outside a typical adobe dwelling in a hot, dry climate (modified from Fitch and Branch 1960).

the straw, is used as a mortar and as a plaster on the exterior walls. Figure 23.5 shows a typical diurnal temperature cycle inside and outside an adobe dwelling. The roof heats up most, attaining a temperature on the order of 60°C, over 15°C higher than the ambient air temperature, and cools by about 40°C at night. In contrast, inside temperatures range between about 24° and 30°C, reaching a maximum toward midnight.

Fig. 23.6 Adobe dwellings in northern Nigeria (left), New Mexico (center), and Timbuktu, Mali (right) are markedly similar. Note that the ovens (center and right photos) are outside, reducing heat load inside.

Fig. 23.7 Houses and granaries of the cliff-dwelling Dogon of Mali.

In hotter and more arid regions, the dwellings tend to have thicker walls and flat roofs. A wall thickness of 22 cm can reduce the temperature range in the dwelling's interior to 1/25 of that outside. In the arid regions of the higher latitudes, the solar rays are at an oblique angle; thus, a flat roof absorbs less radiation than a slanted one, which is more perpendicular to the solar beam. Such dwellings are remarkably similar in such diverse regions as the US Southwest and the African Sahara (Fig. 23.6). They contain few windows and small entrances, a form that both reduces heat exchange between the interior and exterior of the dwelling and minimizes exposure to wind-blown sand or dust. Ovens are constructed outside the building, to reduce heat load. In some areas, dwellings may be built into hillsides or may be partially subterranean, taking advantage of the more moderate temperature regime inside the earth or in the hills (Fig. 23.7).

In desert villages, additional protection is often provided by shading via surrounding walls or by closely packed dwellings, separated by narrow streets or passageways, reducing the amount of surface area exposed to solar radiation (Fig. 23.8). An elongated design, with the narrower sides facing the sun, further reduces this exposure. Orientation with respect to prevailing winds can increase ventilation. This is useful where the winds have a cooling effect. The streets of ancient Mesopotamia were so designed.

In wetter areas, adobe walls are less satisfactory because they can succumb to steady, intense rainfall (Fig. 23.9). Therefore, in the grasslands and savanna, mud brick gives way to thatching (Fig. 23.10). In drier parts of the savanna, dwellings may be constructed with adobe walls, but a thatched roof. The thatched roofs of grass huts are typically thick and steeply slanted (Fig. 23.11), making them highly impermeable and allowing them to readily shed the intense rainfall. An interesting construction is found in the wet/dry tropics of Nigeria: a double-shelled dome adapted for both the wet and dry seasons (Fitch and Branch 1960). The inner dome is mud and provides insulation; the outer one is thatched and protects the inner dome from the harsh rains. The air space between the two layers provides additional insulation. Another adaptation to the wet and dry seasons is seen in numerous parts of Africa: huts sheathed in woven mats that may be detachable (Fig. 23.12). The weave of the mats expands and contracts with atmospheric humidity, being more open in the dry season to allow ventilation but closed and nearly impermeable during the wet season.

The American equivalent of the thatched hut is the sod house, or "soddy," commonly found across the grasslands of the Great Plains in the nineteenth and early twentieth centuries (Fig. 23.13). Instead of clay, the building material was the thick, tough sod of the thick-rooted prairie grass (Stratton 1981). The sod was cut in large blocks, often 1 or 2 feet (30–60 cm) on a side and 6 inches (15 cm) thick. Like adobe, the sod provided excellent thermal insulation but was vulnerable to rain damage.

Other architectural designs that are well suited to dryland climates include the pit house and subterranean housing. The mass of ground surrounding the dwelling provides thermal insulation to moderate both summer and winter temperatures. The pit house, also known as a dugout, is a shelter that is partially sunk into a hole or depression in the ground. A pit house can be fully or partially recessed or dug into a hillside (Gilman 1987). These were very common structures in the American Southwest and elsewhere in the western USA during the first millennium AD (Farwell 1981; Smith 2003). Often the foundation of the dwelling consisted of poles covered with brush and adobe. The sunken floor (Fig. 23.14) was below the frost line, typically 6–18 inches (15–45 cm) deep but occasionally as deep as 4 feet (120 cm). These pit houses were the forerunners of the *kivas* of the later Pueblo cultures (Rohn and Ferguson 2006). Areas

Fig. 23.8 High-density building construction reduces the heat load on individual structures and is hence a common feature of settlements in arid lands.: (left) northern Nigeria, (right) New Mexico.

Fig. 23.9 The town of Shali, in the present-day Siwa Oasis, dates back to the thirteenth century, but was destroyed in 1926 by three days of heavy rain. The buildings were constructed of *kershef*, a material closely related to adobe that contains a mixture of salt and clay.

Fig. 23.11 Thick slanted roof on hut in the Sahel of Niger.

Fig. 23.10 Typical thatched hut in Sahelian West Africa.

Fig. 23.12 Hut with thatched woven mats attached behind the hut and on its roof.

Fig. 23.13 "Soddies" were common dwellings in the Great Plains in earlier times. Like adobe, they provide excellent insulation. Additional protection is provided by building them into a hillside (reprinted by permission of the Kansas State Historical Society).

Fig. 23.14 Sunken floor of pit houses of the ancient Anasazi culture of the American Southwest.

of traditional underground housing include Berber regions of Tunisia and the loess plateau of northern China, where an estimated 40 million people live in artificial caves called *yaodongs* (Golany 1992). In Coober Pedy, an opal-mining town in the Australian desert, much of the population lives in caves built into the hillside.

In the Mediterranean regions, where the dry season coincides with oppressive summer heat, houses are built to take advantage of the winds. As in many hot deserts, thick walls will provide insulation from the heat, but the house is often built around a central courtyard exposed to cool breezes. Fountains or pools may be used to enhance the cooling through evaporation as the breeze passes over the water surface. The passive cooling effects of the courtyard are used also in areas of the Middle East (Safarzadeh and Bahadori 2005).

There are numerous other examples of traditional ventilation schemes. In ancient Rome, heat was circulated and houses ventilated through fires built in the airspace beneath tiled floors. In Pakistan and China, wind catches that can be turned seasonally into the prevailing winds take advantage of regular monsoon

winds. A most ingenious system of ventilation and cooling is the wind tower (Fig. 23.15), a common architectural feature of Iran and other parts of the Middle East (Bahadori 1978; Bahadori *et al.* 2008).

The wind tower takes advantage of the large vertical temperature gradient, the extreme diurnal temperature range, and the strong thermal contrast between its walls and the outside air. The wind tower's operation depends on wind conditions and time of day. On a calm night it acts as a chimney, the hot tower walls heating the air inside to a temperature higher than the air outside. The hotter, less dense air in the tower rises and escapes through a top vent. At the same time, the cooler night air near the surface is drawn in through vents near the ground. If there is wind at night, the circulation is reversed, with cold night air sinking in the tower and being forced through the buildings. During the early part of the day, the tower is cooler than the air outside; it cools the air within, which then sinks because it is heavier than the outside air. The cold drafts then circulate through the building. A series of windows can be adjusted, depending on the ambient conditions, in order to promote the airflow and cooling.

A simpler system utilizes a domed roof with an air vent at the top. As air flows over the curved roof, its pressure is reduced (the Bernoulli effect) compared with the ambient air pressure. Hot air within the building rises to the dome and exits through the dome's vent. During hot periods, air rises internally to the dome and is then sucked out into the region of lower pressure. The curved surface of the dome also maximizes the surface area, thus enhancing cooling via radiation. Like many adaptations, the dome roof is cross-cultural. Figure 23.16 shows the domed roofs of granaries in Niger, probably constructed in this way so to minimize the heat load on the stored grain.

23.4 AGRICULTURE

Civilizations have thrived in many of the world's drylands for thousands of years. Livelihood systems were developed that efficiently utilized the available resources and helped contend

Fig. 23.15 Wind towers in Iran at Yazd (left) and Kermak (right) (photos courtesy of Dara Entekhabi, MIT).

Fig. 23.16 Domed roofs of granaries in Niger reduce the heat load on the stored grain by enhancing radiative cooling.

Fig. 23.17 Fishing along the Niger River in Mali.

with the environments' limitations. Some of the earliest inhabitants of dryland regions were simple hunters and gatherers. Since these practices had little impact on the land, they were perfectly adapted to the dryland environment. Traditionally, three forms of agriculture were also practiced: pastoralism (either sedentary or nomadic), rainfed cropping, and irrigation agriculture (Beaumont 1989; Heathcote 1983). Salt mining and trading were also practiced in some regions such as the Sahara, and fishing (Fig. 23.17) was practiced in the coastal deserts and around some inland seas and rivers.

The African Bushmen of the Kalahari, even today a subsistence-based society, are a prime example of hunter-gatherers. Other early subsistence-based societies in the drylands include the Shoshonean tribes of the Great Basin in the USA and the aborigines of Australia. Typically, each family might need as much as 300–400 square kilometers for subsistence. Therefore, in most regions this system was replaced by agriculture as population density grew. The pressure to make the change was

particularly strong in the drylands, since the area of territory needed per individual increases with aridity (Heathcote 1983).

Dryland pastoral systems use grazing or browsing animals (Fig. 23.18) to harvest a sparse crop of natural vegetation. In semi-arid regions this form of agriculture is often combined with crop production, but in arid regions pastoralism is dominant or exclusive, unless irrigation is possible. The herdsmen employ many strategies to cope with the inadequate moisture supply and sparse vegetation cover. Stock is spread thinly over large areas to reduce grazing pressure at any point and take advantage of the patchwork nature of the ecosystem resulting from the erratic nature of rainfall and the complex matrix of topography, soil quality, and vegetation that are characteristic of dryland regions. Pastures are rotated, and some rangelands are held ungrazed in some years in order to let moisture and vegetation accumulate. Some pastures are burned to facilitate the growth of more desirable plants. The herdsmen limit the size of herds, selling off animals during years of drought, and utilize a multitude of water points to dilute pressure.

Fig. 23.18 (left) A cattle herd in central Botswana. (right) A typical cattle kraal.

Fig. 23.19 Maize growing around a West African village.

In arid and semi-arid regions, rainfed cropping (Fig. 23.19) generally requires special techniques of water collection, storage, and conservation. Rain harvesting (Section 23.4.3), which has been practiced for millennia, is one example. Terracing and contouring to catch water, constructing wind breaks and minimizing tillage to reduce evaporation, and allowing fields to lie fallow in alternate years are all strategies that serve to retain moisture and hence increase the efficiency of water use (Rosenberg 1980). In modern times, drought-resistant varieties of crops have been utilized. These techniques and technologies have pushed cultivation to the limit, increasing the vulnerability of rainfed crops to droughts or climatic change. Rainfed agriculture has other disadvantages in dryland regions. It transforms the landscape more than pastoralism does and it makes land more prone to erosion. Also, the areas of cultivation tend to encroach into the pastoral regions, limiting that form of agriculture.

The constant to which life in the drylands must adapt is the inconstancy of the environment. The environment, especially moisture supply, is sporadic in time and space, and at any given location, adequate moisture supply will not persist for a long period of time. Other resources are equally scarce and rational exploitation requires controlled use of these resources. This can be accomplished through maintaining a low level of need; putting activities and demands in phase with resource availability; designing strategies for recycling energy and resources and minimizing depletion of non-renewables such as fossil aquifers; and isolating factors, such as herd size, that increase environmental impact and risk.

The nomadic pastoralists rationally and efficiently exploit the dryland environment, using mobility as protection from its erratic nature. They utilize a multiplicity of plants and animals: species with different needs, tolerances, and susceptibilities to disease. Diversifying resources provides more options for survival and also reduces pressure on the environment. Sheep utilize grass, goats utilize scrub, and camels feed on saltbush. Nomads also exploit a spatial and, where feasible, altitudinal range, wandering to where water and pasture are available. During the dry season they settle around wells, streams, and farmed areas. During the wet season they migrate to more distant pastures. Their migrations allow some pastures to lie "fallow," reducing grazing pressure on the land. The nomadic pastoralists also maintain a close relationship with sedentary farmers: goods are traded, stubble provides feed while manure provides fertilizer, and the nomads become farm laborers for part of the year. The farmer provides "insurance" against the hardships of the dry season or a long drought. This symbiotic relationship characterized human interactions of the central Soudan for centuries in pre-colonial times, but the system broke down when national borders prevented free movement of the population (Lovejoy and Baier 1975). This may have exacerbated the impacts of the twentieth-century droughts in the region.

23.5 WATER RESOURCES

23.5.1 WATER SUPPLY IN DRYLANDS

Dryland cultures have traditionally coped with water deficiency through active irrigation, rainwater catchment, and high-efficiency water management. Irrigation and water conservation date back to at least the third millennium BC (Beckinsale 1971). Floodplain irrigation allowed tremendous cities to flourish along the rivers of the Fertile Crescent of the Middle East, cities which were abandoned when the land became so salty that crops could no longer be grown on it (Pillsbury 1981). The seasonal inundation of the Nile's floodplain allowed Egyptian civilization to prosper. The ancient inhabitants of the Negev devised ingenious methods to enhance water supply. Runoff collected from uncultivated slopes was conveyed to areas of cultivation, using intricate constructions of diversions, drops, guide walls, and terraces to minimize the erosive effect on the soil (Wiener 1977). Cisterns for storing runoff allowed habitation and cultivation in areas with as little as 100 mm of rain per year, since 2000 BC or earlier (Evenari et al. 1961).

In other cases, exogenous streams in arid regions provide a steady supply of water. The Nile supports some 40 million people in the surrounding barren desert. The Colorado River supplies water to 2.1 million people, 85% of whom receive water diverted from the direct course of the river. Streams emanating from the highlands of deserts likewise provide an abundant supply of water. Hundreds of small streams from the Andes dissect the coastal deserts, carrying water for irrigation and replenishment of groundwater (Peterson 1970). Irrigation has existed in the region since Inca times. More recently, technology has improved the capacity to exploit such streams by damming them to create reservoirs.

Groundwater has also long been exploited, with perennial water supplies provided in deserts from underground sources through springs, wells, and tunnels. The Kharga Depression of Egypt, an oasis inhabited for thousands of years, is home to nearly 10,000 people in numerous small villages. The Siwa Oasis, in the hyper-arid Western Desert of Egypt, supports 23,000 people (Fig. 23.20). In Iran and Afghanistan, the Gobi, and the American deserts, large towns are supported through oasis agriculture (Peterson 1970). In many cases alluvial aquifers, relatively close to the surface, provide water for irrigation. The dallols of Niger (ancient arms of the Niger River) support rice cultivation and other crops (see Fig. 23.2).

Today there are technological options to increase the amount and reliability of the water supply. These generally provide at least one of four benefits: capturing and storing rainfall; decreasing evaporative loss of water stored in soil or reservoirs; increasing water availability through irrigation or precipitation enhancement; and improving water-use efficiency (Rosenberg 1980). More sophisticated designs and applications of traditional

techniques are utilized for irrigation, rainwater catchment, and groundwater exploitation. Recycled "gray" water is also being used (Juanico and Friedler 1999; Madungwe and Sakuringwa 2007). Flood control and irrigation is practiced through the construction of dams and the diversion of rivers. Techniques have been developed to desalinize water (Glueckstern et al. 2008) and, in many countries, attempts have been made to artificially enhance rainfall. Farming methods have also been modified to reduce water demand. In some countries, passive radiative cooling systems are used to promote dew condensation for water supplies (Beysens et al. 2006).

23.5.2 IRRIGATION

Irrigation agriculture is practiced in arid, semi-arid, and subhumid regions all over the world. Globally some 2,788,980 km^2 of land has the infrastructure to support irrigation (Siebert et al. 2005). Of this some 700,000 km^2 are in the arid lands in the Middle East, North Africa, and North America. In the drier regions of India and China, the figure exceeds one million. Irrigation can increase yield four- to sixfold, allows humans to occupy regions of insufficient rainwater supply, and can remove the drought risk and uncertainty of crop yield, providing for a more stable resource and economic system. On the other hand, modern irrigation systems are costly and, in many cases, technologically complex, requiring skilled and experienced management. If prudent principles of water management are not applied, irrigation can – and does – lead to numerous environmental problems.

There are two major techniques of irrigation: diversion of existing surface water flows and exploitation of underground water sources. The first approach utilizes gravity to capture water from lakes and rivers or from surface runoff, using the hydraulic gradient to allow water to flow naturally to lower areas. When rainwater is captured, this technique is often termed "water harvesting" or "runoff farming"; when water is diverted from lakes or rivers, the term "furrow irrigation" is used. Water is transported or captured via aqueducts, hillside conduits and terracing. The second technique requires mechanization to lift water from wells that tap alluvial aquifers or groundwater. Another set of techniques exploits the natural flooding of rivers; floodwaters are captured in low-lying regions along the banks. The Nile is a prime example, but similar small depressions along the Niger River in Mali, termed douanas, retain the flood waters and are sown as the water recedes. Throughout sub-Saharan Africa, small-scale irrigation is practiced at the local farm level using small depressions in valleys and floodplains (Adams and Carter 1987). This has met with considerable success in many regions, with few harmful consequences, in contrast to large-scale irrigation projects.

Most natural systems of water diversion have the advantage of also carrying and depositing silts containing natural nutrients. A disadvantage is that flow is sporadic in time and space, because these systems operate only where and when rainfall or streamflow is available. A second disadvantage is that they cannot be regulated. In some cases, the diverted water is stored

Fig. 23.20 The Siwa Oasis, in the hyper-arid Western Desert of Egypt, supports a population of 23,000.

Fig. 23.21 The town of Shifta, in the Negev, thrived in the Negev Desert over two thousand years ago.

Fig. 23.22 Remnants of a cistern in the ancient Israeli city of Shifta.

Fig. 23.23 A conduit transports rainfall runoff in the hills down to the plains near Avdat, Israel

Systems of irrigation and water diversion have been used since ancient times in the drylands of North Africa, the Middle East, Pakistan, India, and China. In ancient Mesopotamia, irrigation was used as far back as the sixth millennium BC. In Peru it dates back to the fourth millennium BC. The Nabataean culture of Israel and Jordan used systems of irrigation and water harvesting extensively in the sixth century AD. This allowed major towns such as Shifta, in Israel (Fig. 23.21), or Petra, in Jordan, to prosper in the desert. Huge cisterns were hewn into rock to channel rainfall (Fig. 23.22). Terraces and conduits (Fig. 23.23) were utilized to carefully guide runoff from higher elevations in a "catchment" area to a small "runon" area that was sown with crops. Stone mounds were also used to reduce runoff and also promote the condensation of dew (Lavee *et al.* 1997). Towns frequently had cisterns to store such water for general use. Variants of gravitational diversion are used even today in many areas of the world.

In several early societies, water was supplied by using a series of tunnels and tapping groundwater relatively high up a slope. The water would run downslope through the tunnel, being tapped by wells at several points and then emerging through the tunnel at some location downslope. Such an irrigation system is termed *galeria* in Mexico and *qanats* in Iran (Kirby 1969). Qanats were first used in ancient Persia *c.* 800 BC. The technology was later exported to other parts of the Arab world, including North Africa (Beaumont *et al.* 1989). Excellent examples were found in the Kharga Oasis of Egypt (Wuttmann *et al.* 2000). This is one of the oldest methods still in use today (Motiee *et al.* 2006).

Unfortunately, many irrigation systems also cause serious environmental problems that ultimately threaten resources in arid and semi-arid regions. Several cases of environmental problems associated with irrigation are described in Section 23.6.3. The most serious problems include salinization, alkalinization, and groundwater depletion. The first two situations arise under conditions of improper drainage: the evaporating water leaves behind saline or alkaline residues of formerly dissolved materials. Problems such as waterlogging and

and utilized at will, but the nutrient-rich silts are generally left behind in the storage vessels. Also, after the water is stored, gravity is no longer effective in allocating it and other energy sources must be utilized for distribution.

Fig. 23.24 Pivot irrigation systems in Namibia. In times of drought, or when plots are left fallow, the bare soil is highly susceptible to erosion.

water-borne diseases arise when too much water is applied. Irrigation's detrimental effects also include soil erosion and degradation (Fig. 23.24), overgrazing around the water source, and siltation of reservoirs.

Much of the irrigation practiced in dryland regions utilizes water from underground aquifers. These are often remnants of humid periods of past climate, dating back thousands or even millions of years. In most dryland regions, water is being drawn down faster than it is being replenished. Perhaps nowhere is the problem as severe as in the American West, where extensive irrigation systems tap mostly the vast Ogallala aquifer. In the 1930s only about 600 wells tapped in, but the number rapidly grew to over 150,000. In some areas, notably the dry western plains of Texas, the Ogallala is being drawn down as much as 18 times faster than it is being replenished (Boslough 1981). The rate exceeds 5 feet (1.5 m) per year in some places.

The ensuing problems include salt-water intrusions, sinking and cracking of the surface, and large drops in the water table. Some cracks are up to 25 feet (7.5 m) wide in the Southwest. Near Tucson, irrigation produced a 120-foot (38 m) drop in the water table. In the San Joaquin Valley of California (Fig. 23.25), the land had dropped nearly 30 feet (9 m) by 1980 (Adler 1981).

23.5.3 RAIN HARVESTING

Rain harvesting, also called rainwater catchment, has been utilized for centuries. Systems capture rainwater from rooftops or at ground-level, or trap surface runoff. Techniques as simple as construction of small dirt mounds to concentrate runoff of slopes (Fig. 23.26) can allow trees or crops to thrive in extremely arid regions. Roof catchment systems date back to early Roman times; they served as the principal water source for drinking and domestic purposes for whole cities (Gould 1992). The practice was widespread among the Phoenicians and Carthaginians. In Israel and Jordan, since 2000 BC or earlier, cisterns for storing runoff allowed habitation and cultivation in areas with as little as 100 mm of rain per year (Fig. 23.27) (Evenari *et al.* 1961).

Fig. 23.25 Markings indicating the subsidence of the land surface in the San Joaquin valley, southwest of Mendota, California, between 1925 and 1977. In the photo is USGS scientist Joe Poland (photo credit: Dick Ireland, USGS, 1977).

In the ancient city of Jawa (Jordan), macro- and microcatchments supplied water some 5000 years ago and the water was retained by deflection dams (AbdelKhaleq and Alhaj Ahmed 2007). The volume of water so-acquired may have been as high as 2 million cubic meters annually (Helms 1981). Nomadic peoples in the Turkestan Desert used the *takyry* surfaces as water catchments (Fleskens *et al.* 2007). Small-scale water collection in earthen pots and jars has been practiced in Africa, Asia, and Latin America for millennia (Gould 1992). At Um el-Jimal, in today's Jordan, dams, channels, and underground reservoirs were constructed during the Late Roman Period to divert water to irrigated fields (de Vries 1993; Alkhaddar *et al.* 2005).

In recent times there has been a resurgence of interest in rainwater harvesting, particularly in rural parts of the developing world. Modern variants of ancient methods are utilized in such areas as Africa (Tabor 1995; McPherson and Gould 1985),

Fig. 23.26 (left) Microcatchments in Morocco and Israel sustain more dense vegetation than the surrounding desert by capturing runoff on hillsides. (center) Trees grow upslope of each catchment mound in Morocco. (right) Microcatchments allowed for extensive afforestation of hillsides in the arid Negev of Israel.

Fig. 23.27 Avdat farm in Israel was created by the botanist Michael Evenari to test ancient methods of water harvesting first detected via aerial photos. The farm produces fruit and vegetables, despite its desert location.

Fig. 23.28 Rainwater catchment barrel on a home in Botswana. Rainfall is captured on the roof and funneled into the barrel.

Israel (Tenbergen *et al.* 1995), and Southeast Asia (Gould 1992). In Botswana, where the average distance to village water sources is on the order of a kilometer, nearly 50% of the households in a recent village survey collected some rainwater for domestic use. Some systems are as simple as buckets or oil drums placed under the eaves of roofs. Though used on traditional huts, these systems become much more effective on the corrugated iron roofs that have gradually begun to replace thatching (Fig. 23.28).

23.5.4 WEATHER MODIFICATION

In arid lands, the primary interest in weather modification is in precipitation enhancement, one of its least successful aspects. In general, it is impossible to artificially evoke rainfall from cloud-free skies, so that the approach involves increasing the amount of rain falling from already developed cloud systems. The basic technique is "cloud seeding," in which small particles such as silver iodide are introduced into clouds in order to trigger condensation and/or freezing of existing droplets (Mather *et al.* 1996). In some cases, the latent heat released by condensation produces additional instability and rising motion in clouds, making the precipitation process more effective by increasing

the size of droplets and by creating a more turbulent cloud environment, which is more conducive to further production of precipitation. In other cases, the added nuclei induce the freezing of supercooled water droplets. This also acts to increase droplet size, because ice crystals grow more rapidly than droplets in moist air. Rainfall occurs only when droplets reach sufficient size to overcome the effects of friction, updrafts, and evaporation during their fall to the surface.

A few studies have claimed moderate success with rainfall enhancement. The most accepted example is the Israel experiment (Levi and Rosenfeld 1996; Rosenfeld and Nirel 1996). However, even these results are controversial (Rangno and Hobbs 1995). A National Research Council (2003) panel was convened in the USA to assess the effectiveness of cloud seeding. The conclusion was that there has been little evidence of success. In contrast, Woodley and Rosenfeld (2004) evaluated operational cloud seeding in Texas and concluded that the efforts enhance rainfall on the order of 50%.

The American Meteorological Society (1998) has issued a statement that, at most, a 10% augmentation in rainfall can be achieved under the best combination of circumstances. This suggests that there is too little benefit to offset the costs of the technology. Furthermore, the techniques are of little use in droughts, when cloudless skies often prevail. Even if modification were possible in such years, the added 10% in rainfall hardly compensates for the deficits of 25–50% that are typical of drought years in arid and semi-arid regions.

For at least the last century, flooding of miscellaneous depressions has been proposed as a strategy to augment rainfall. After the droughts in southern Africa early in the century, Schwarz (1920) suggested forming Lake Kalahari by flooding the Kalahari Depression, in order to increase atmospheric humidity and, in turn, rainfall. Many other such schemes have been proposed. In response to such suggestions, MacDonald (1962) conducted a classic study in which he examined the impact of creating the hypothetical "Lake Fallacy." He concluded that the only significant impact would arise from the ascent of air triggered by mounds of dirt surrounding the lakeshore after excavation of the lake basin. MacDonald's conclusion makes sense because a lack of available atmospheric moisture is not a common aridifying factor in dryland regions. Segal et al. (1982), using a sophisticated mathematical model, similarly concluded that the proposed flooding of the Qattara Depression in Egypt (see Section 23.6.3) would have little impact on rainfall. This is likewise true of the effects of planting "green belts," although these at least have the effect of reducing soil erosion.

A related issue is the effect of irrigation on rainfall. The most comprehensive study is that of Barnston and Schickedanz (1984). They presented statistical evidence of rainfall enhancement as a result of irrigation in two regions of the Great Plains: the Texas panhandle and an area including parts of Kansas, Colorado, and Nebraska. Other apparent effects include changes in temperatures, cloudiness, thunderstorms, and severe weather. Points that underscore a possible relationship are that the links are evident during the seasons of irrigation and in the counties irrigated, but not during other times or in other counties. However, the meteorological situation in the Great Plains is somewhat unusual and the results cannot be extrapolated to other locations.

23.6 ENERGY RESOURCES

The dryland inhabitants, like those elsewhere, require energy for cooling, for fuel, and, in higher latitudes, for heating. The energy demand is reduced by thermally insulating housing designs, as described in Section 23.2. Nevertheless energy requirements are substantial and traditional energy sources, like wood, are relatively scarce in the drylands. This chapter does not attempt to provide a comprehensive review of energy resources in the dryland regions. Instead it seeks to demonstrate various means of reducing the need for non-renewable supplies by examining select traditional and modern strategies for cooling and energy supply. These include traditional passive cooling systems and solar and wind energy.

Two truly innovative passive cooling systems were used in Iran for centuries (Bahadori 1978). The cistern keeps water cold and the ice-maker provides a year-round supply of ice in the hot desert. Their designs illustrate a sophisticated understanding of atmospheric and thermal processes. The cistern is a reservoir sunk 10–20 m deep into the ground, filled with cold water in winter, and topped by a domed roof. In summer the roof, the air and the top layer of water are warmed. But before the summer heat can reach the deeper layers of water it is utilized to evaporate surface water. Wind towers maintain a circulation that removes vapor and promotes continued evaporative cooling. The ice-maker involves huge ponds used to freeze ice during the cold season, and deep storage pits (10–15 m in depth) to retain the ice throughout the year. Adobe walls that shield the ponds and ice from solar radiation are used to promote freezing and retard melting.

When such traditional techniques are combined with modern technology, the results can be impressive. One simple technique is to allow water to flow over the roof and walls of a building, absorbing the solar heat by day but releasing it at night. This is used in greenhouses in Israel. The process is extremely effective because of the high heat capacity of water.

The more moderate thermal environment deep in the ground can be exploited by using "earth cooling/heating tubes" (also known as ground-coupled heat exchangers) (Elifrits and Gillies 1983). Such tubes circulate the ambient air to a depth of some 1.5–3 m, where it is cooled in summer or heated in winter. In the desert outside Abu Dhabi, a new city named Masdar is being constructed with the goal of creating zero carbon demand. Energy requirements will be reduced through many of the traditional strategies described in Section 23.3. These include natural cooling through wind towers and fountains in courtyards, and reducing solar heating – and hence energy demand – by constructing buildings close together and enclosing them with a surrounding high wall.

Solar energy (Fig. 23.29) is the one commodity that is readily available in dryland regions; it can be exploited cheaply and effectively. The most common uses are for heating water and for generating electricity, but solar power can also be used for refrigeration and cooking. It is particularly useful for refrigeration, since the greatest availability of solar energy coincides with the greatest demand for refrigeration. Since 10,000 times more energy is received daily from the sun than is consumed worldwide, the potential for use is staggering. A number of countries are promoting its use. In semi-arid Botswana, for example, solar cells are a common sight on the roofs of houses, schools, and government buildings.

Wind power is also traditional, with the windmill (Fig. 23.30) being used worldwide in very diverse environments. Its use declined in industrial countries as other forms of energy came into use, but windmills are a major source of power in many developing countries. In drylands, however, wind power is of more limited use than solar energy because many dryland regions are in the subtropics, where winds are generally weak. But even in such locations, wind turbines can be used to generate power if constructed in proper microhabitats, such as on coastlines or atop rolling hills, where the winds are steadier.

In essence, the kinetic energy of the wind is converted to mechanical energy to turn the blades of a windmill. This can

Fig. 23.29 A solar cell in the Australian desert near Alice Springs.

Fig. 23.30 Windmills are still operative at many dryland farms.

be used to pump water, run motors, or generate electricity. The power generated is proportional to the cube of the wind speed and the cross-sectional area through which the blades rotate (Griffiths 1976). The wind velocity, however, must exceed a certain threshold to overcome the inertia of the system. Local wind energy potential is readily assessed from wind data. In the central Great Plains of the USA, mean annual wind power has been estimated to exceed 500 W/m² (Linscott *et al.* 1981). Many experimental wind energy farms have been set up in the United States, Australia, and many other countries. As of 2004, three wind energy farms in California (Fig. 23.31) provided 4258 million kilowatt hours, or 1.5% of the state's energy demands.

23.7 MODERN EXPLOITATION OF THE DRYLAND ENVIRONMENT: OASES, PLAYAS, RIVERS, AND LAKES

With expanding populations and demands for higher standards of living, the inhabitants of drylands have found increasingly sophisticated technological ways to exploit the natural resources of these environments. Most often exploitation is for water supply, but additional uses are for mining or hydroelectric power. Unlike traditional methods, many of the modern approaches have detrimental environmental consequences. In this section, a handful of examples of exploitation are described, some of

Fig. 23.31 Turbines at a wind energy farm near Palm Springs, California.

which have been economically beneficial and some of which have had disastrous consequences. In most cases, the consequences have been foreseeable. In some cases, they could have been avoided through better management, or better incorporation of traditional exploitation strategies. Often there is a trade-off between the beneficial and harmful effects.

Fig. 23.32 The Siwa Oasis played an important role in Egypt's history. The Temple of Amon was the site of Alexander the Great's visit to the Oracle of Amon in the fourth century BC and was one of the world's six most important oracles.

Fig. 23.33 The springs of the Siwa Oasis support some 70,000 olive trees and 300,000 date palms.

Fig. 23.34 Palm trees in a Californian oasis near Palm Desert.

23.7.1 OASES

Desert oases have been inhabited for thousands of years. Many sustain a prosperous agriculture and entire towns. The Siwa Oasis of Egypt's Libyan Desert has been inhabited since at least the tenth millennium BC (Fig. 23.32). The region receives less than 10 mm of rainfall annual, but olives (Fig. 23.33) and dates grow vigorously in the oasis, a consequence of the 200 springs in the region. In many of the Californian oases, palm trees thrive (Fig. 23.34).

The government of Niger chose to exploit desert oases to help the northern Sahelian population most devastated by the drought of the 1970s. At the Iférouane Oasis, in the north of the Air Massif, there are fertile soils 3–4 m thick. Before the drought, wells tapped a water table only 4 m below the surface at the end of the rain season and 8 m below the surface at the end of the dry season. By 1974, the water table had fallen to 20 m as a consequence of the dry years commencing in 1968 (GTZ 1977). The government took special labor-intensive measures to restore the oases to prior conditions of prosperity, in order to help the population recover from the long sequence of drought years.

23.7.2 PLAYAS, MINING, AND OTHER NATURAL RESOURCES

The microclimatic conditions of playas are markedly different from those of other locations in dryland regions. Some of the most notable differences are increased receipt of solar radiation and higher near-surface wind speeds (Malek *et al.* 2002). Consequently they have much potential for exploiting solar and wind energy. Currently, the most common uses are for extraction of minerals.

Minerals accumulate in deserts through two main processes: evaporation and leaching of minerals by groundwater. Some of the valuable minerals mined in arid regions include copper (USA, Chile, Peru, Iran), uranium (Niger, Namibia, Australia, USA), iron (Australia), chromite (Turkey), diamonds (Botswana, Namibia), and sodium nitrate (Chile). The evaporite deposits that are commonly mined include various sodium compounds (table salt, sodium nitrate, sodium carbonite, trona) and other saline minerals such as borate, gypsum, phosphate,

Fig. 23.35 Salt deposits on the edge of a playa in western Australia.

Fig. 23.36 Salt crust deposited along the edge of a salt lake.

and calcium. Petroleum is extracted from the dry deserts of the Middle East.

The evaporite deposits are typically found in playas (Fig. 23.35). Examples of exploitation are the borax mines in Searles Lake and other areas of the western USA, natron production in Niger and Chad, salt production in the Great Salt Lake of Utah, and soda ash and salt mining in Botswana. Borax production in the USA commenced in the pioneer days, with mule teams hauling it out of Death Valley being a memorable image of the Old West. Salt companies extract 2.5 tons of sodium chloride annually from the Great Salt Lake (Fig. 23.36). In Niger, natron is extracted from evaporation pans in the dallols, the dry former beds of the Niger River. Natron, which is a mixture of soda ash, sodium bicarbonate, and other sodium compounds, is locally processed and sold in large cylinders (Fig. 23.37). Like borax, it is used to make soaps and other cleaning compounds, as well as antiseptics, digestives, and preservatives. In Botswana, a salt and soda ash plant lies along the edge of the Sua Pan (Fig. 23.38), a segment of the Makgadikgadi pans. This plant is the leading producer of salt and soda ash in southern Africa and some 96% of its products are exported.

23.7.3 RIVERS AND LAKES

Various grandiose schemes have been put forward to manipulate hydrologic resources in arid regions. At least three schemes have been proposed to divert watercourses to bring water into the Sahara. One of these involved flooding some 10% of the continent. Flooding of the Qattara Depression of Egypt by building a canal to the Mediterranean was suggested as a way of producing hydroelectric power and extracting minerals from the saline water (Segal et al. 1982). The Jonglei Canal was conceived as a way to drain the *sudd* of the southern Sudan, a vast swamp formed by the White Nile (Petersen et al. 2007). The goal was to reduce the evaporation of Nile water in the sudd, conserving as much as 4.8×10^9 m^3 of water per year. Construction was started but never completed.

Numerous hydrologic projects have been put in place in dryland regions. Unfortunately, many of these have had unfortunate consequences. The Hoover Dam (Fig. 23.39), in the US Great Basin, has devastated the estuarine ecosystem of the Colorado River delta because of salinization. The Aswan High Dam of Egypt, a major engineering feat, has damaged much of the natural ecosystem upon which Egypt had thrived. Construction of the Karakum Canal and other canals in the Former Soviet Union made the desert "bloom" (Voskresenski 1977) but had disastrous consequences for the Aral Sea.

Often these large-scale projects are based on inadequate climatological information. A case in point is a dam in Morocco that, for at least the first 10 years of its existence, held no water (Fig. 23.40). Similarly, a major dredging project was planned for the Okavango Delta, based on climate information for one year, which happened to be exceedingly wet.

THE ARAL SEA

In the 1960s the Soviet government conceived a plan to enhance agricultural production in the drylands of Turkmenistan and Uzbekistan by diverting the rivers that fed into the Aral Sea, the Amu-Dar'ya and Syr-Dar'ya (Kasperson et al. 1995). Construction began in the 1940s and the desert prospered. Annual production was some 1 million tons. Plans were put in place to create pastures and provide areas to grow fruits and vegetables.

Unfortunately, the construction and management was poor and tremendous amounts of water leaked or evaporated out of the irrigation canals. In the largest, the Karakum Canal, the loss was 30–75%. The Aral Sea, once the world's fourth largest lake, began to shrink (Micklin 2007). By 1998 the sea's surface area had shrunk by approximately 60% and its volume by 80%. By 2007, it had declined to 10% of its original surface area, splitting into three separate lakes. Vast areas of the lake bottom became exposed, resulting in intense dust storms. The former ports of Aral'sk and Munyak soon lay tens of kilometers from the sea and boats were stranded in the "desert" (Ellis 1990).

(a)

(b)

Fig. 23.37 (a) Natron is prepared in a drying vat in the dallols of Niger. It is used as a digestive for animals and for many other purposes. (b) Natron is shaped into cylinders that are baked in ovens and sold in the marketplace.

Fig. 23.38 Soda ash and salt are mined in the Sua Pan of eastern Botswana.

Fig. 23.40 The Mansour Eddabhi Dam in Morocco, on the Draa near Ouarzazate. It was constructed in 1972, but by 1983, it had never contained water. In the photo is the power generating plant associated with the dam.

Fig. 23.39 The Hoover Dam, on the Colorado River near Las Vegas, provides hydroelectric power to most of Nevada and to some parts of California over 400 km away (photo courtesy of D. Klotter, Florida State University).

An environmental disaster accompanied these changes in the Aral Sea. Salinity climbed to 92 ppt, roughly three times the salinity of the ocean. The lake's water became too salty to support fish, destroying the local fishing industry and creating unemployment and economic hardship (Bissell 2004). The water table sank, local water supply was threatened, and health problems such as asthma increased dramatically. Efforts have been made to revitalize the Aral Sea but have been slow to proceed (Pala 2006).

THE SALTON SEA

The Salton Sea (Fig. 23.41) in the Imperial Valley of California has a similar, sad history (Micklin 2007). California's largest water body, the sea is surrounded by mountains on three sides and has no outlet except evaporation. Its surface lies approximately 227 feet (69 m) below sea level. The Salton Sea

Fig. 23.41 The Salton Sea in the Imperial Valley of California.

was formed by accident when the Colorado River breached its banks during a 1905 flood and overwhelmed irrigation canals. River water poured into the low-lying regions.

At present, the main inflow into the Salton Sea is irrigation drainage from the Imperial Valley. This fills the lake's water with fertilizers, salts, and pesticides. The consequences include eutrophication, algal blooms, and high salinity. The salinity of the Salton Sea is 44 ppt, compared with ocean salinity of 34 ppt.

The lake's pollution resulted in a tremendous loss of fish and bird life (Kaiser 1999). An estimated 7.6 million fish died on one day in 1999. Thereafter about 35 pelicans perished each day. Die-offs affected roughly one-fifth of the 380 species of birds that frequent the area around the lake.

THE COLORADO RIVER

The Colorado River is 2330 km long, commencing in the Rocky Mountains of Colorado and reaching into the Sonoran Desert of Mexico. It originally flowed into the Gulf of Mexico. However, most of its water is currently diverted for irrigation, so that it no longer reaches the sea in many years. Below Hoover Dam, in southern Nevada, 70% of the river's water is diverted to California via an aqueduct. The Colorado River Compact of 1922 allotted 7.5 million acre-feet annually to upper-basin states and the same amount to lower-basin states. A 1944 agreement gave another 1.5 million acre-feet to Mexico, making a total of 16.5 million acre-feet. Unfortunately, the rationing of its water was based on its flow during an abnormally wet period prior to 1922 and the river rarely carries more than 14.8 million acre-feet (Boslough 1981).

The irrigated fields near the river return to it runoff water with a high salt content. This has markedly increased the salinity of the river. At its source, the salt content is about 50 ppm. Prior to the extensive diversion of water for irrigation and urban water supply, its salinity was roughly 400 ppm by the time the river reached Mexico. During the 1960s, the salinity at the border reached 1200–1500 ppm (Lohman 2003). The USA agreed to build a desalinization plant at the border near Yuma, Arizona, to bring the salinity down to 800 ppm. Throughout much of the United States the salinity is still on the order of 700 ppm.

THE ASWAN HIGH DAM

In 1983 the Aswan High Dam was built across the White Nile in the barren desert 1000 km north of Khartoum. The Egyptian government cited several reasons for its construction, the most important being to increase water supply (Latif 1984). Construction of the dam would ensure adequate and dependable irrigation water for agriculture, even during low flood years, and convert agriculture from seasonal to permanent irrigation. The dam would also impound all the flood water of the wet years, without losses to the sea; protect lives, crops, and property against flood disaster; reduce the impact of droughts; and generate cheap hydroelectric power.

Unfortunately, the dam upset the delicate equilibrium of the Nile system (Ibrahim 1983). The most important impact was the loss of the annual Nile flood. The flood produced alluvial silt that provided a natural fertilizer and also replaced materials eroded during the previous agricultural season. Without the annual flood, the balance of erosion and deposition was upset. The fertile silt was replaced with chemical fertilizers that are both costly and environmentally detrimental. The lack of silt deposition also meant a loss of material for the manufacture of bricks used for housing construction. Other impacts included waterlogging, salinization, changes in water quality, diseases such as bilharzia, an increase of the rat population, and the growth of destructive water hyacinths. The effects were so severe that some environmentalists campaigned for the destruction of the dam.

REFERENCES

AbdelKhaleq, R. A., and I. Alhaj Ahmed, 2007: Rainwater harvesting in ancient civilizations in Jordan. *Water Science and Technology: Water Supply*, **7**, 85–93.

Adams, W. M., and R. C. Carter, 1987: Small-scale irrigation in sub-Saharan Africa. *Progress in Physical Geography*, **2**, 1–27.

Adler, J., 1981: The browning of America. *Newsweek*, Feb. 23, 26–37.

Alkhaddar, R., G. Papadopoulos, and N. Al-Ansari, 2005: *Water Harvesting in Jordan*. Amman, Jordan.

American Meteorological Society, 1998: Scientific background for the AMS policy statement on planned and inadvertent weather modification. *Bulletin of the American Meteorological Society*, **79**, 2773–2778.

Bahadori, M., 1978: Passive cooling systems in Iranian architecture. *Scientific American*, **238**, 144–154.

Bahadori, M. N., M. Mazidi, and A. R. Dehghani, 2008: Experimental investigation of new designs of wind towers. *Renewable Energy Journal*, **33**, 2273–2281.

Barnston, A. G., and P. T. Schickedanz, 1984: The effect of irrigation on warm season precipitation in the southern Great Plains. *Journal of Climate and Applied Meteorology*, **23**, 865–888.

Beaumont, P., M. E. Bonine, and K. McLachlan, 1989: *Qanat, Kariz and Khattara: Traditional Water Systems in the Middle East and North Africa*. Menas Press, Cambridge.

Bissell, T., 2004: *Chasing the Sea: Lost among the Ghosts of Empire in Central Asia*. Vintage Books, New York.

Boslough, J., 1981: Rationing a river. *Science* **81**, 26–34.

de Vries, B., 1993: The Umm El-Jimal Project, 1981–1992. In *Annual of the Department of Antiquities of Jordan*, 37th Kenneth Wayne Russell Memorial Volume, Amman, Jordan, pp. 433–460.

Elifrits, C., and A. Gillies, 1983: Earth pipes. *Earth Shelter Living*, **29**, 6–7.

Ellis, W. S., 1990: A Soviet sea lies dying. *National Geographic*, February, 73–93.

Epstein, Y., and D. S. Moran, 2006: Thermal comfort and the heat stress indices. *Industrial Health*, **44**, 388–398.

Evenari, M., L. Shannan, N. Tadmor, and Y. Aharoni, 1961: Ancient agriculture in the Negev. *Science*, **133**, 979–996.

Farwell, R. Y., 1981: Pit houses: prehistoric energy conservation? *El Palacio*, **87**, 43–47.

Fitch, J. M., and D. P. Branch, 1960: Primitive architecture and climate. *Scientific American*, **203**, 134–144.

Fleskens, L., A. Ataev, B. Mamedov, and W. P. Spaan, 2007: Desert water harvesting from Takyr surfaces: assessing the potential of traditional and experimental technologies in the Karakum. *Land Degradation and Development*, **18**, 17–39.

Gilman, P., 1987: Architecture as artifact: pit structures and pueblos in the American Southwest. *American Antiquity*, **52**, 538–564.

Glueckstern, P., M. Prietl, E. Gelman, and N. Perlov, 2008: Wastewater desalination in Israel. *Desalination*, **222**, 151–164.

Golany, G. S., 1992: *Chinese Earth-Sheltered Dwellings*. University of Hawaii Press, Honolulu, HI.

Gould, J. E., 1992: *Rainwater Catchment Systems for Household Water Supply*. Environmental Sanitation Reviews 32, ENSIC, Asian Institute of Technology, Bangkok.

Griffiths, J. F., 1976: *Applied Climatology*. Oxford University Press, Oxford, 136 pp.

GTZ, 1977: *Life to the Desert*. Deutsche Gesellschaft für Technische Zusammenarbeit (GTZ), Eschborn, Germany, 27 pp.

Heathcote, R. L., 1983: *The Arid Lands: Their Use and Abuse*. Longman, New York, 323 pp.

Helms, S. W., 1981: *Jawa, Lost City of the Black Desert*. Cornell University Press, New York.

Ibrahim, F., 1983: Der Assuan Staudamm. *Bild der Wissenschaft*, **4**, 76–83.

Juanico, M., and E. Friedler, 1999: Wastewater reuse for river recovery in semi-arid Israel. *Water Science and Technology*, **40**, 43–50.

Kaiser, J., 1999: Battle over a dying sea. *Science*, **284**, 28–30.

Kasperson, J., R. Kasperson, and B. L. Turner, 1995: *The Aral Sea Basin: A Man-Made Environmental Catastrophe*. Kluwer Academic Publishers, Dordrecht.

Kirby, A. V., 1969: Primitive irrigation. In *Introduction to Geographical Hydrology* (R. J. Chorley, ed.), Methuen, London, pp. 52–55.

Latif, A. F. A., 1984: Lake Nasser: the new man-made lake in Egypt (with reference to Lake Nubia). In *Lakes and Reservoirs* (F. B. Taub, ed.), Ecosystems of the World 23, Elsevier, Amsterdam, pp. 385–410.

Lavee, H., J. Poesen, and A. Yair, 1997: Evidence of high efficiency water-harvesting by ancient farmers in the Negev Desert, Israel. *Journal of Arid Environments*, **35**, 341–348.

Levi, Y., and D. Rosenfeld, 1996: Ice nuclei, rainwater chemical composition, and static cloud seeding effects in Israel. *Journal of Applied Meteorology*, **35**, 1494–1501.

Linscott, B. S., J. T. Dennett, and L. H. Gordon, 1981: *The Mod-2 Wind Turbine Development Project*. DOE/NASA/20305-5, NASA Technical Memorandum 82681, 21 pp.

Lohman, E., 2003: Yuma desalting plant. *Southwest Hydrology*, May/June, 20–22.

Lovejoy, P. E., and S. Baier, 1975: The desert-side economy of the Central Sudan. *International Journal of African Historical Studies*, **8**, 551–581.

Macdonald, J., 1962: The evaporation–precipitation fallacy. *Weather*, **17**, 168–177.

Madungwe, E., and S. Sakuringwa, 2007: Greywater reuse: a strategy for water demand management in Harare? *Physics and Chemistry of the Earth*, **32**, 1231–1236.

Malek, E., C. Biltoft, J. Klewicki, and B. Giles, 2002: Evaluation of annual radiation and windiness over a playa: possibility of harvesting the solar and wind energies. *Journal of Arid Environments*, **52**, 555–564.

Mather, G. K., M. J. Dixol, and J. M. de Jager, 1996: Assessing the potential for rain augmentation: the Nelspruit randomized convective cloud seeding experiment. *Journal of Applied Meteorology*, **35**, 1465–1482.

McPherson, H. J., and J. Gould, 1985: Experience with rainwater catchment systems in Kenya and Botswana. *Natural Resources Forum*, **9**, 253–263.

Micklin, P. P., 2007: The Aral Sea disaster. *Annual Review of Earth and Planetary Sciences*, **35**, 47–72.

Morgan, C., and R. de Dear, 2003: Weather, clothing and thermal adaptation to indoor climate. *Climate Research*, **24**, 267–284.

Motiee, H., E. McBean, A. Semsar, B. Gharabaghi, and V. Ghomashchi, 2006: Assessment of the contributions of traditional qanats in sustainable water resources management. *International Journal of Water Resources Development*, **22**, 575–588.

National Research Council, 2003: *Critical Issues in Weather Modification Research*. National Academies Press, Washington, DC, 144 pp.

Nicholls, N., C. Skinner, M. Loughnan, and N. Tapper, 2008: A simple heat alert system for Melbourne, Australia. *International Journal of Biometeorology*, **52**, 375–384.

Oliver, J. E., 1981: *Climatology: Selected Applications*. Edward Arnold, London, 260 pp.

Pala, C., 2006: Once a terminal case, the North Aral Sea shows new signs of life. *Science*, **312**, 183.

Petersen, G., J. A. Abya, and N. Fohrer, 2007: Spatio-temporal water body and vegetation changes in the Nile swamps of southern Sudan. *Advances in Geosciences*, **11**, 113–116.

Peterson, D. F., 1970: Water in the deserts. In *Arid Lands in Transition* (H. E. Dregne, ed.), AAAS, Washington, DC, pp. 15–30.

Pillsbury, A. F., 1981: The salinity of rivers. *Scientific American*, **245**, 54–65.

Rangno, A. L., and P. V. Hobbs, 1995: A new look at the Israeli cloud seeding experiments. *Journal of Applied Meteorology*, **34**, 1169–1193.

Rohn, A. H., and W. M. Ferguson, 2006: *Puebloan Ruins of the Southwest*. University of New Mexico Press, Albuquerque, NM.

Rosenberg, N. J., 1980: *Drought in the Great Plains: Research on Impacts and Strategies*. Water Resources Publications, Littleton, CO, 225 pp.

Rosenfeld, D., and R. Nirel, 1996: Seeding effectiveness: the interaction of desert dust and the southern margins of rain cloud systems in Israel. *Journal of Applied Meteorology*, **35**, 1502–1510.

Safarzadeh, H., and M. N. Bahadori, 2005: Passive cooling effects of courtyards. *Building and Environment*, **40**, 89–104.

Schwarz, E. H. L., 1920: *The Kalahari or Thirstland Redemption*. Oxford University Press, London.

Segal, M., Y. Mahrer, and R. A. Pielke, 1982: On climatic changes due to a flooding of the Qattara Depression (Egypt). *Climatic Change*, **5**, 73–83.

Siebert, S., P. Döll, J. Hoogeveen, J.-M. Faur ès, K. Frenken, and S. Feick, 2005: Development and validation of the global map of irrigation areas. *Hydrology and Earth System Sciences*, **9**, 535–547.

Smith, C., 2003: Hunter-gatherer mobility, storage, and houses in a marginal environment: an example from the mid-Holocene of Wyoming. *Journal of Anthropological Archaeology*, **22**, 162–189.

Steadman, R. G., 1979a: The assessment of sultriness. Part I. A temperature-humidity index based on human physiology and clothing science. *Journal of Applied Meteorology*, **18**, 861–873.

Steadman, R. G., 1979b: The assessment of sultriness. Part II. Effects of wind, extra radiation and barometric pressure on apparent temperature. *Journal of Applied Meteorology*, **18**, 874–885.

Steadman, R. G., 1984: A universal scale of apparent temperature. *Journal of Climate and Applied Meteorology*, **223**, 1674–1687.

Stratton, J. L., 1981: *Pioneer Women: Voices from the Kansas Frontier*. Simon and Schuster, New York, 319 pp.

Tabor, J. A., 1995: Improving crop yields in the Sahel by means of water-harvesting. *Journal of Arid Environments*, **30**, 83–106.

Tenbergen, B., A. Günster, and K.-F. Schreiber, 1995: Harvesting run-off: the minicatchment technique – an alternative to irrigated tree plantations in semiarid regions. *Ambio*, **24**, 72–76.

Voskresenski, L., 1977: Making the desert bloom. *World Health*, July, 10–15.

Wiener, A., 1977: Coping with water deficiency in arid and semi-arid countries through high-efficiency water management. *Ambio*, **1**, 77–82.

Woodley, W. L., and D. Rosenfeld, 2004: The development and testing of a new method to evaluate the operational cloud-seeding programs in Texas. *Journal of Climate and Applied Meteorology*, **43**, 249–263.

Wuttmann, M., T. Gonon, and C. Thiers, 2000: The qanats of 'Ayn-Manâwîr (Kharga Oasis, Egypt). *Journal of Achaemenid Studies and Researches*.

Young, K. C., 1979: The influence of environmental parameters on heat stress during exercise. *Journal of Applied Meteorology*, **18**, 886–897.

24 Plant and animal life in the desert

24.1 INTRODUCTION

As with humans, water plays an important role in the heat balance of plants and animals, and it is also required for other aspects of an organism's physiology. For plants, water serves four main functions (Goudie and Wilkinson 1980): cellular material functions only in its presence; it dissolves and transports nutrients; it provides raw material for photosynthesis; and, by absorbing heat and losing it through evapotranspiration, water provides a thermoregulatory mechanism so critical in the harsh thermal environments of deserts and semi-arid lands. It is particularly effective in this last role because of its high specific heat and because of the heat utilized or released upon evaporation or condensation (approximately 600 kcal/g). For animals, water is required for cellular function and it also plays a thermoregulatory role. The ultimate stress is cellular dehydration, and dryland species of plants and animals vary greatly with respect to their tolerance for water loss.

Survival in a desert thus requires that plants and animals maintain both body moisture and temperature within established ranges of tolerance. This is achieved through a variety of mechanisms designed to evade periods of stress or reduce stress by reducing heat gain, maximizing heat loss, optimizing water uptake, and reducing water expenditure (Evenari 1985). These mechanisms are strikingly similar in plants and animals and include adaptations involving morphology, physiology, biological rhythms, habitats and, in the case of animals, behavior.

Heat gain by radiation and conduction is minimized by reducing surface area exposed to radiation, coats of thermally insulating or reflective material, and a preference for sheltered habitats (e.g., cooler microclimates such as within the ground or under plant cover) or maintaining a distance from the hot ground surface (by burrowing or elevating the body). Convective loss of heat is enhanced through morphology of plants, rapid panting and thin coats in animals, and location at sites (e.g., dune crests) exposed to high winds. The efficiency of acquiring atmospheric moisture is enhanced through protrusions and surfaces that promote condensation or deposition, extraction of water from non-saturated air, and absorption of water through leaves or body surfaces. Specialized root systems also enhance the capture of surface and subsurface moisture. Moisture is conserved in plants via reduced leaf or stomatal area and reduced transpiration and metabolic rates. Animals conserve moisture by exhaling unsaturated air, reduced metabolic activity, and reduced urine volume. Ephemerals and annuals thrive only during periods of moisture availability; perennials inhabit wetter niches such as riverbeds. Many utilize fog-water. Animals migrate to wetter areas, retreat to shelter during periods of thermal stress, or confine their activities to more moderate times of the day or year.

24.2 PLANT ADAPTATIONS TO THE DESERT ENVIRONMENT

The major environmental characteristics that plants must tolerate are low and irregular water supply and thermal extremes. Thermoregulatory adaptations are primarily morphological, but some plants instead tolerate high internal or surficial temperature. The response to water deficiency distinguishes two basic classes of desert plants: those that *evade* moisture stress and those that *avoid* it. The former, the annuals and ephemerals, use biological strategies related to their life cycle. The latter, termed xerophytes, are perennials which *resist* or *tolerate* drought. They resist drought primarily through morphological and physiological mechanisms to optimize water uptake or reduce water loss. They tolerate drought by reducing water demand, primarily through biological adaptations involving the rhythm of growth. An example of such a biological adaptation is the CAM photosynthetic pathway (see Chapter 3). By closing the stomata during the day, water loss is reduced.

In some cases, plants must adapt also to environmental hazards such as high salinity or predators, e.g., grazing animals and insects. Salt-tolerant plants, termed halophytes, utilize some of the mechanisms that reduce moisture stress. Many halophytes,

468

Fig. 24.1 A variety of saltbush on growing mounds of salt.

Fig. 24.3 Spines on a cactus provide protection from predators and also enhance radiative cooling.

Fig. 24.2 Thorns of *Acanthosicyos horridus*.

(a)

(b)

Fig. 24.4 Succulent shrubs of the Namib take on unusual forms: (a) *Arthraerua leubnitziae* or pencil plant, (b) *Zygophyllum clavatum*, a leaf succulent.

such as a wide variety of saltbushes (*Atriplex* sp., Fig. 24.1), tolerate high salt concentrations in cell sap (Goudie and Wilkinson 1980). Like the halophytes, most xerophytes are salt-tolerant. Other halophytes avoid the hazard by excluding from their intake ions that are potentially toxic or by excreting toxic salts from cells when high concentrations are reached. Some, especially ephemerals, evade potential damage by regulating their growth to coincide with periods of moisture, when salinity is reduced. Water retention, characteristic of the succulents, also reduces the risk of salt damage.

Protection against predators includes bitter or poisonous tissues, sharp and prickly thorns (Fig. 24.2) and spines (Fig. 24.3), pungent odors, and camouflage. Most *Acacia* species are thorny; thorns are larger and more concentrated in the small, young trees, which are more accessible to predators. The Sodom apple of the Sahara contains a poisonous and irritant latex. The low, round form of the stone-plants (lithops) and some ice-plants (mesembs) of southern Africa provide camouflage. The succulent shrub *Zygophyllum clavatum* of the Namib takes on a similar form (Fig. 24.4).

The avoidance of thermal stress generally requires protection from both heat and cold. Heating is reduced by such adaptations as reduced surface area exposed to radiation, insulating coatings, highly reflective (e.g., waxy) surfaces, or selection of more moderate habitats. Sagebrush (*Artemisia tridentata*), a shrub common in the North American deserts (Fig. 24.5), is covered with fine, silvery hairs that promote cooling and minimize moisture loss. In grasslands, the formation of hummocks or tussocks serves to moderate the environment in the vicinity of the plant, as well as to capture water and nutrients (Fig. 24.6). The diminutive size and formless structure of species such as the stone-plant of the Namib serve to keep the plants cool, as well as reducing water loss by minimizing the area where evaporation can take place. A thick waxy coating protects against both extreme heat and extreme cold. Temperatures in the central water tissue of Peruvian cacti can be as low as 45°C, while the surrounding soil temperature is 60°–70°C (Rauh 1985). Convective heat loss can be promoted through the exposure of surface area to winds or through spines or protrusions which increase the turbulence of the airflow over the organism's surface.

Drought-resistant species optimize water uptake in a variety of ways. These include, for example, protrusions and surfaces that promote condensation and deposition, water absorption through leaves and cuticles, specialized root systems, spacing of individual plants, and selection of wetter habitats such as river beds. Some resist drought but maintain growth through a "decoupling" of the rates of transpiration and photosynthesis. Others can survive drought periods via absorption of fog-water in favorable environments (Corbin *et al.* 2005).

Plants that utilize fog-water are particularly common in the Namib and Peruvian-Atacama coastal deserts. The perennial shrub *Trianthema hereroensis* (Fig. 24.7), found in the Namib Desert, absorbs water through its leaves. The cactus *Discocactus horstii* of the Peruvian desert takes up fog-water through its thorns (Rauh 1985). Species of *Tillandsia*, in the same desert, absorb fog through leaves that are densely covered with specialized absorbing trichomes. One species is so efficient at absorbing moisture that there may be a 20% difference in relative humidity between the air around the shoots and the ambient air. The diminutive roots of *Tillandsia* have no function as far as water uptake is concerned.

Water loss can be reduced through reduced leaf or stomatal area, small ratios of above- to below-ground biomass, and reduced rates of transpiration and photosynthesis, especially during the dry period. Water need is also reduced through plant size, as smaller plants require less water. Many desert species are low or dwarf shrubs or lie in small clusters close to the ground (Fig. 24.8). Succulents, a large class of plants that includes cacti, yucca, and the Old World aloe and euphorbia, impound water in their leaves, roots, or stems (Fig. 24.9). The barrel cactus (Fig. 24.10), for example, stores enough water in its stem to provide an emergency supply to travelers. Mechanisms to retard transpiration include closure of the stomata or shedding or rolling leaves during the dry season, and the development of small fleshy leaves, thick and waxy coatings, or mats of downy hairs. The sclerophyllous trees of the Mediterranean-type forest (Fig. 24.11) possess a thick waxy coating, as do many succulent and non-succulent perennials of the desert, such as *Trianthema hereroensis* (Fig. 24.7), various aloes (Figs. 24.9d and 24.12a), euphorbia (24.12b), and the pencil plant (*Arthraerua leubnitziae*) (Fig. 24.4).

Widely spaced plants with extensive, horizontally spread, near-surface root systems, such as those of the Saguaro cactus,

Fig. 24.5 Sagebrush in Death Valley in the western USA.

Fig. 24.6 A hummock grass in Australia (*Triodia* sp.) (left) and a tussock grass in New Mexico (right).

Fig. 24.7 *Trianthema hereroensis* has diminutive, waxy leaves that minimize evaporative losses and heat gain.

Fig. 24.8 *Aloe asperifolia* in the Namib. The clusters radiating from a single growth point are often found at the base of a rock because of the rain and fog-water that collects there. The rock can also provide shading at certain times of the day.

(a)

(b)

(c)

(d)

Fig. 24.9 Old and New World succulents: (a) a Saguaro cactus attracts birds and other animals; (b) yuccas of the Mojave have a characteristic twisted shape and long, waxy leaves; (c) *Moringa ovalifolia* (phantom tree or sprokiesboom), an African stem succulent; (d) *Aloe brevifolia* grows to huge sizes in Botswana.

maximize their capacity to capture water while reducing competition for soil moisture. Such a root system also minimizes surface erosion by binding surface materials. In such cases, even minimal above-ground cover can effectively prevent erosional loss. *Stipagrostis sabulicola*, a perennial grass, also has an extensive and horizontally spread root system that lies just below the sand surface and extends up to 20 m from the main plant. This facilitates the absorption of fog-water deposited on sand

Fig. 24.10 Barrel cactus in the Sonoran Desert.

Fig. 24.11 Sclerophyllous trees in northeastern Australia.

(a)

(b)

Fig. 24.12 (a) The bright waxy coating of this *Aloe asperifolia* readily reflects solar radiation and resists heat transfer into the plant, reducing thermal extremes. (b) The form of this *Trichocaulon clavatum* also helps in minimizing heat gain by the plant.

surfaces. Phreatophytes possess instead very long taproots that penetrate tens of meters downward to the water table. Mesquite (*Prosopis* sp.) (Fig. 24.13) commonly has roots 50 m or more in length. *Trianthema hereroensis*, a leaf succulent perennial of the Namib Desert (see Section 24.3), has both horizontal roots, which are surface localized, and vertical roots.

Small and formless plants, often close to the ground, possess a reduced evaporative surface. A round shape also minimizes the surface area to volume ratio, thus reducing heat load. Typical of these are the stone-plants common in the southern African deserts. In some such species this form results from leaves being diminutive or absent; photosynthesis is carried out instead in the stems. In general, evaporative loss is reduced by minimizing the amount of plant biomass above ground. A high ratio of roots (i.e., below-ground material) to above-ground biomass is advantageous because it reduces the ratio of water gain to both water need and water loss. Thus, for many dryland species the amount of subsurface material far exceeds that above ground.

The drought-tolerant xerophytes survive dry periods primarily through their biological rhythm, i.e., the rhythm and rate of growth. This is typical of the creosote bush and of cacti. Cactus species do most of their growth in a brief moist period and can survive two years or more without water. The creosote bush of North America is one of the few drought-tolerant plants that survive dry periods in the vegetative stage. The water content of its leaves may drop to 50% of its dry weight, compared with forest species containing 100–300% of their dry weight in water. The bush recovers when the next rain falls. The creosote reduces growth through a controlled rate of photosynthesis; the stomata close during the day (stopping CO_2 intake) but open during the cool night. The presence of woody and tough material, which prevents the collapse of plant tissue during wilting, promotes drought survival in desert plants like the aloe.

The ephemerals and annuals respond to the dryland environment by evading drought, perishing between periods of moisture (Figs. 24.14 and 24.15). These are herbaceous plants and grasses, such as the California primrose, that germinate and bloom rapidly after a rain, suddenly transforming the desert into

Fig. 24.13 Mesquite (the large shrub behind the tall Saguaro cactus) is common in the Sonoran Desert and in other North American drylands.

Fig. 24.14 *Stipagrostis sabulicola* on a Namib dune. In a wet year, perennial grasses can rapidly expand on the dunes, in this case even on the slipface, which is usually barren.

Fig. 24.15 These grasses are growing in an area of the hyper-arid eastern Sahara that averages only a few millimeters of rain each year. They developed from seeds produced during a rare rain event that had occurred three years earlier.

Fig. 24.16 Large seed pods of *Acacia hebeclada*; their shape is aerodynamic and facilitates the spreading of seeds.

a landscape of green grasses and colorful flowering plants. In a favorable year, these grow prodigiously, producing abundant flowers, fruit and seeds, and a fairly dense ground cover. Unlike xerophytes, which are sparse, often spiny, and separated by patches of bare ground, ephemerals provide a full surface cover and are a good grazing crop. Characteristically, their growth cycle is compressed into a period of 6–8 weeks, but in some the cycle is much shorter and more rapid. *Boerhavia repens*, found in the African Sahel, has been observed to flower, produce seed, and die within 8 days following a rain shower (Walton 1969).

The key to the regeneration of ephemerals is the preservation and dispersal of seeds (Fig. 24.16). When the plants die, the seeds must lie dormant until favorable conditions return, sometimes surviving extreme heat and drought and high salinity for several years. Their profuse growth and abundant seed production helps to assure regeneration during the next moist period. In some cases, dispersal mechanisms scatter seeds over large areas to increase the likelihood that a few reach sites suitable for germination.

Wind transport is one important mechanism, but dispersal is also accomplished when barbs, burrs, and bristles attach the seeds to the legs and coats of grazing animals. In one case, a new species of chenopod was introduced to the Libyan Desert during World War II when seeds were carried there on the boots of Australian soldiers. The seeds of some desert grasses collect to form a dense ball that is easily blown across the windy landscape. Seeds of the tumbleweed and rose of Jericho (resurrection plant) roll across the desert, as do the seed-bearing gourd-like fruits of the Saharan colocynth. The seeds drop off into the soil upon contact and the seed ball gradually disintegrates. Other species retain the seeds within withered plant material, which uncurls when it becomes damp. This ensures that seeds are scattered only during moist periods. A hard seed coat protects from the heat and drought. The moisture requirements of ephemerals are generally greater than those of the xerophytes, thus ephemeral herbs and grasses are generally found in wetter regions, in stream channels, in wetter microclimates of arid regions, and in the steppes bordering desert regions.

24.3 PERENNIALS OF THE NAMIB DESERT

The Namib Desert contains some of the world's most unusual plants and some of the best examples of plant adaptation to peculiar environmental characteristics. Although the Namib is one of the driest deserts in the world, the vegetation is abundant in places and species are quite diverse. The source of moisture for many of them is fog-water, which in much of the Namib is considerably more abundant than rainfall. Three unusual perennial species inhabit the Namib dunes: *Stipagrostis sabulicola*, a grass (Fig. 24.17); *Acanthosicyos horridus*, or nara, a melon or cucurbit of the cucumber family (Fig. 24.18); and *Trianthema hereroensis*, a leaf succulent (Fig. 24.7). A fourth perennial, *Welwitschia mirabilis*, occupies mostly shallow dry stream beds of floodplains (Figs. 24.19). With the exception of *Welwitschia*, these perennials rely heavily on fog precipitation for growth and survival. In some cases fog-water is absorbed directly by leaves, but in other plants, such as *Stipagrostis sabulicola*, the roots exploit fog droplets condensed on the sand surface or concentrated by runoff from the leaves and stems. In some deserts this is a critical component of the plant's water balance. It is generally a more reliable source of moisture than rainfall, which is much more temporally variable (Shmida 1985).

Fig. 24.17 The extensive root system of the grass *Stipagrostis sabulicola* allows it to grow on the dunes of the central Namib.

Fig. 24.19 Stream beds are a common habitat for *Welwitschia mirabilis*, allowing it to survive in the most barren parts of the Namib.

Fig. 24.18 Dune perennials of the Namib include (left) the shrub *Acanthosicyos horridus* and (right) the vine *Citrullus ecirrhosus*.

Fig. 24.20 Male (left, smaller and narrower cones) and female (right, larger and rounder cones) *Welwitschia* plants.

Welwitschia, a plant that defies classification, is perhaps the most unusual desert plant in the world. Showing traits of both cone-bearing gymnosperms, such as pine trees, and flowering angiosperms, *Welwitschia* probably represents a parallel evolutionary line that reached a dead end. There are both male and female forms and the plant exhibits none of the usual formative characteristics of xerophytes or desert perennials (Fig. 24.20). On the contrary, it has leaves with an immense surface area, only a thin waxy coat, and a relatively shallow root system generally no deeper than 3 m (Bornman 1972). Each plant consists of only two leaves, but they are gigantic. One leaf was measured to be 1.8 m wide and 6.2 m in length (3.7 m of which was living tissue), making a total surface area of 21 m². Such a size is achieved partly because of the tremendous age of these plants, estimated to be over 2000 years in some cases. The leaves are folded and contorted, so that an individual plant is generally not more than about 3 m in diameter.

Welwitschia is endemic to (uniquely found in) the Namib. The plant's waxy coat, although thinner than that found in most desert species, is highly reflective and hence helps to keep it cool. It is so effective that the leaf surface temperature is generally only a few degrees above the air temperature (Marsh 1990), whereas the leaf–air temperature difference may be as much as 10° or 15°C in some succulents and other desert plants (Petrov 1976). It was long believed that this plant survives the harsh desert environment mainly by absorbing fog-water through stomata on the leaf. Each leaf contains millions of stomata (up to 22,000/cm²) on both its upper and lower surfaces, in contrast to most xerophytes, which have no stomata at all on the leaves. The stomata open as the fog arrives inland and close shortly after it lifts. The size and flat structure of the leaf provide a large surface exposed to the atmosphere, promoting both the deposition of fog droplets and the condensation of dew. However, recent research suggests that these plants do not utilize fog-water, but instead respond to the availability of rainwater (K. Soderberg, personal communication). A leaf may grow 10–20 cm in a dry year, but up to 10 cm/month in a wet year (Seely 1987). Though its roots are shallow, they are nevertheless effective at procuring moisture, since *Welwitschia* is often found in dry streambeds where the water table is close to the surface.

Fig. 24.21 The succulent shrub *Trianthema hereroensis* commonly grows on the dunes of the Namib.

Acanthosicyos horridus, an inhabitant of the desert dunes, instead utilizes subsurface water, which it takes in through its long roots (Fig. 24.18). To conserve water and reduce heat gain, the leaves of *A. horridus* are reduced to minute scales; photosynthesis occurs in stems and long thorns, which contain chlorophyll and have taken over the role of the defunct leaves. The plant forms tangled thickets, some of which cover about 3000 m² and are estimated to be over 100 years old. The root of one plant was measured to be 40 cm in circumference (Berry 1991). The nara can survive without rain for many years. It produces a pale orange fruit, which grows so prolifically that the Topnaar Khoikhoi people, Kuiseb River nomads, extract and sell about 30 tons of seed annually. The seed is used in baking as a substitute for almond flavoring.

Trianthema hereroensis (Fig. 24.21), which also occupies the dunes, provides an interesting example of root specialization. This species is restricted to the western half of the Namib, where rainfall rarely occurs but fogs are frequent. The fact that it survives by imbibing fog-water through its leaves indicates the importance of this moisture source. Most individual plants have exclusively horizontal or vertical roots and the distribution of root types appears to be affected by climate. A study by Nott and Savage (1985) of 19 specimens at Rooibank, a near-coastal location, found only three to have vertical roots. The plants tended to inhabit areas taking longer to dry out after fog has lifted. Generally each plant had two or three horizontal roots within the top 20 cm of soil. One was measured at 6 m in length. At Flodden Moor, a more inland location, all 20 specimens studied had vertical roots; four also had horizontal roots. Thus the roots tended to be horizontal at the foggier coastal locations, but vertical inland where fog is less frequent and water may lie within a few meters of the surface.

Lichen soil crusts also depend on the fog-water for photosynthesis. A study of three lichen species showed nocturnal uptake of water from fog or dew, with uptake ranging from 0.49 to 0.73 mm precipitation equivalent (Lange *et al.* 1994). Annual net primary production from this uptake was assessed as 16 g of carbon per square meter. Some species can also utilize water vapor in very humid air (Lange *et al.* 1991).

24.4 ANIMAL SURVIVAL IN THE DESERT

The mechanisms by which animals adapt to the desert environment are comparable to those of plants. Like plants, they have to avoid thermal stress and dehydration. The strategies involve the heat balance, the water balance, activity pattern, life cycle, and metabolism (Evenari 1985; Petrov 1976). Most strategies have their analog in the plant communities previously described.

Biological adaptations involve life cycle and activity pattern. Animals migrate to wetter ground during dry periods, retreat to shelter during periods of thermal stress, or confine their activities to moderate times of the day or year. In some cases the organism is completely dormant during periods of stress. In states of dormancy, basal metabolism is dramatically lowered. Many desert animals are either nocturnal or nocturnal during the dry season and diurnal during the rainy season. Others will seek habitats with a more favorable microclimate (Krasnov *et al.* 1996), by occupying shady niches, such as within or below plant cover, or by maintaining a distance from the hot desert surface by burrowing or elevating the body. Some 70% of desert mammal species are subterranean. Since subsurface temperature gradients are extreme in deserts and the diurnal and annual range decrease rapidly with depth, this is a particularly effective strategy.

A number of characteristics of reproduction help ensure species survival. Breeding may be synchronized with the rainy season or triggered by rain. Reproductive organisms may be specially insulated from thermal fluctuations and extremes. In some organisms, such as mollusks, longevity counterbalances the high mortality of the young. In some birds, clutch size is proportional to rainfall. Some animals will partake of large quantities of food and drink within a short, favorable period and store them in the body for long periods. A related adaptation is a low metabolic rate and the ability to survive without food for long periods.

Regulation of the heat balance involves insulating layers of air, tissue cuticle, or fleece; reflective colors; exposure to wind; and various internal temperature controls. Some animals also

tolerate heat and can maintain abnormally high body temperatures for lengthy periods. Body temperatures up to 52.6°C have been reported for lizard species in the Karakum Desert (Petrov 1976). Wind cools the body by convection; it can be enhanced by habitat selection (e.g., dune crests or flight) and body posture. Common internal thermostats include dilation and contraction of circulatory vessels and evaporative cooling by panting or sweating. Panting is common in mammals, but this is also an effective cooling mechanism for many desert lizards. Less commonly, cooling can be achieved through what Evenari (1985) terms "countercurrent heat exchange," a mechanism utilized by the oryx and described in Section 24.5.

The water balance is maintained through adaptations that optimize water uptake, minimize water loss, or allow an organism to withstand a high degree of dehydration. Moisture can be derived from unusual sources (Henschel and Seely 2008), such as water in food, unsaturated air, dew or fog, saline water, metabolically formed water, and in some cases recycling of water from urine. Animals utilize these sources by seeking moist microclimates, drinking from wet surfaces, consuming moist food, collecting water on the body, and absorbing water vapor in the air.

Examples are numerous. The desert cockroach (*Arenivaga investigata*) (O'Donnell 1977) and the silverfish (*Ctenolepisma tenebrans*) can extract water vapor from unsaturated air, in the case of the latter at relative humidities as low as 47.5% (Edney 1971). The cockroach utilizes a balloon-shaped organ protruding from its mouth, covered with numerous fine hairs and a saline solution that is hygroscopic. Collectively, the hairs and the solution are very effective in triggering condensation at relatively low humidities. The sidewinding adder *Bitis peringueyi* drinks droplets that have precipitated on its own body (Louw 1972). Some species can tolerate a high degree of dehydration; some frogs can withstand a water loss of more than 40% of their total body weight (Petrov 1976). Others have the capacity to rapidly replace water loss by excessive drinking (Evenari 1985). The kangaroo rat, which hardly drinks at all, exhales air with negligible humidity (Petrov 1976).

The most common mechanism by which desert animals economize on water is through excretion of relatively dry feces and urine. This appears to be an adaptation, at least to some degree, of almost all desert-dwelling animals. Water is reabsorbed in the kidneys or large intestine. In some animals, such as lizards, the process is so effective that essentially solid and non-soluble "urine pellets" are excreted.

24.5 SOME TYPICAL DESERT INHABITANTS

Lizards such as the *Meroles anchietae* (the sand-diving lizard) regulate their temperature by burrowing in the sand during extreme heat (Fig. 24.22) and emerging during cooler periods. This particular lizard also performs what is referred to as a "thermal dance," alternately lifting two of its limbs, so that at one time no more than two limbs touch the scorching sand

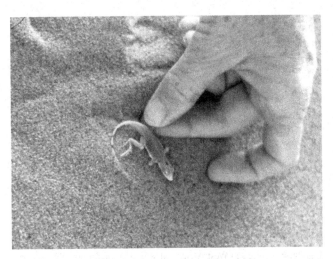

Fig. 24.22 Geckos burrow and other lizards (e.g., *Meroles anchietae*) "swim" through sand on dunes to escape the high temperatures.

Fig. 24.23 The land snail *Sphincterochila boissieri* is an important component of the Negev's ecosystem.

Fig. 24.24 Flocks of ostriches are common in the Namib; their feathers provide a cooling mechanism.

surface. *M. anchietae* drinks fog droplets deposited on vegetation in the dunes and stores this water for long periods.

Frogs and toads use a variety of adaptations to survive the harsh conditions of deserts (McClanahan *et al.* 1994). Many live in permanent water sources, such as small water holes. The toad *Scaphiopus couchii*, of the Colorado Desert in California, survives the dry season by burrowing into the ground, sometimes as deep as one meter. Some individuals have survived even two-year droughts. Another toad of the Colorado Desert, *Bufo punctatus*, recycles water from urine and can tolerate losing 40% of its body water. In comparison, camels can survive a loss of only 20%. A toad in the Gran Chaco of South America, *Lepidobatrachus laevis*, survives by burrowing into mud and constructing a multilayered cocoon to resist water loss. A tree frog of this region, *Phyllomedusa sauvagei*, coats itself with a waxy substance that similarly prevents water loss.

In the Negev Desert, land snails (*Sphincterochila boissieri*) are important components of the desert ecosystem. They fertilize and form soil by eating rocks (Jones and Shachak 1990). These snails (Fig. 24.23) survive the long summer drought through estivation. In one case, the body temperature of a snail estivating in the sun was observed to be 50.3°C (Schmidt-Nielsen *et al.* 1971). Temperatures inside the shell as high as 56°C have been measured. More commonly this species burrows in the sand and is generally active for only a few days after rainfall. Other adaptations include the white color of the shell, the thick shell, and the reduced aperture.

The ostrich (Fig. 24.24), like some of the beetles, regulates its temperature through its posture. It will raise its wings to expose the featherless thorax, which rapidly loses heat in the shade of the wings. It further enhances cooling by raising the feathers on its back. This effectively shields the body from solar radiation but, because the feathers are thin, the breeze penetrates through them and promotes convective heat loss. The ostrich conserves water by exhaling unsaturated air, with relative humidity in the range of 80–85%. This reduces the daily water requirement by about 25%.

The sociable weaver (*Philetairus socius*), a bird commonly found in arid regions of Namibia and Botswana, gets all of its water from insects. It also builds gigantic nests that house dozens of bird families and sometimes several generations (Fig. 24.25). The size and structure of the nest provide shade and serve to moderate the harsh thermal environment, greatly reducing the diurnal temperature cycle relative to the air outside.

The gerbil enters a state of torpor or metabolic sleep, also called estivation. In this state, water loss through transpiration and energy expenditure is minimized. This is accomplished partly through adaptive heterothermy, the ability of a mammal to allow its body temperature to fluctuate in response to environmental stress. The gerbil's "thermostat" is set at 22°C instead of the usual 37°C, reducing oxygen consumption and conserving between 50% and 70% of the usual energy expenditure. The gerbil's kidneys also extract water from its own urine and recycle it through its body, thereby reducing water loss by 95%.

Fig. 24.25 A sociable weaver's nest in Namibia; this bird builds some of the largest nests in the world.

Fig. 24.28 Camels in West Africa; fat stored in the hump helps to regulate both temperature and moisture.

Fig. 24.26 The bush baby or galago, found in the drier regions of southern Africa, avoids thermal stress by way of a nocturnal lifestyle.

The galago (Fig. 24.26), or bush baby, is a small primate that survives in the arid savanna regions of Africa by way of its pattern of activity. It is nocturnal, taking advantage of the intense nightly cooling in the drylands. It also seeks the moderate conditions in the branches of trees. The galago's strong hind limbs and long tail promote good balance, so that it can readily function within the tree canopy.

The oryx (Fig. 24.27) normally maintains a body temperature of about 39° or 40°C, but its temperature may reach 45°C when it is deprived of water (Louw 1972). For most mammals 42°C is lethal. The ability to withstand a higher internal temperature reduces the need for evaporative cooling. This temperature can be maintained because of a special network of blood vessels that keep the brain several degrees cooler than the rest of the body. In this network the relatively warm arterial blood flows very close to veins containing cooler blood and is cooled by conduction. Further cooling takes place when the animal pants and blood in the nasal sinuses is cooled. The oryx also reduces its body temperature by standing on rises or crests of dunes to catch the wind.

The springbok can also maintain body temperature above the usual 41°C. It also dissipates heat effectively through rapid panting, up to 274 times per minute. Its relatively thin coat also allows further rapid cooling and, being light-colored and reflective, it reduces the absorption of solar radiation.

The camel (Fig. 24.28) is one of the animals best suited to the desert environment. It has an unusual mechanism, body fat stored in the hump, for regulating temperature and moisture, even though it is almost constantly exposed to the hot sun. This is an energy reserve that can also be oxidized to produce water. Its body temperature fluctuates from about 34°C by night to 41°C by day. As a result of its thick coat, the excess heat of the day is given off at night without water loss. A camel can withstand dehydration of more than 20% of its body weight, but can drink more than 100 liters of water within 10 minutes (Petrov 1976). Also, most of its urine is absorbed in the blood and converted to usable protein, a mechanism that greatly reduces water loss.

Fig. 24.27 Oryxes are well adapted to survive the heat and aridity of the Namib as they can withstand body temperatures up to 45°C.

Because it is so well suited to desert environments, the camel was introduced into the Australian desert in the 1840s to aid in the exploration of the continent's interior. Having no natural enemies, the camel population has continually expanded since that time. Now over one million feral camels (Fig. 24.29) roam through central Australia, doing tremendous damage to cattle ranches and communities. In remote towns they commonly destroy bathrooms and pipelines in their search for water.

24.6 TENEBRIONID BEETLES OF THE NAMIB AND OTHER INSECTS

Many species of the tenebrionid beetle family (Fig. 24.30) inhabit the Namib Desert, using a variety of adaptations to the environment. They are important components of the Namib ecosystem, acting to recycle nutrients to the soil (Seely 1983). They do this by feeding on detritus (also called debris or litter). Detritus (Fig. 24.31) is derived mainly from plants and is composed of dry, broken off leaves, stems, roots, and seeds produced during the normal growth cycle or when plants die. It can also contain animal remains. The detritus is distributed by the wind and becomes trapped under stones, in rocky crevices, around plants, and on dune slipfaces. This provides a source of water for the beetles because the detritus can absorb moisture from non-saturated air, in some cases even when the relative humidity is as low as 70–80% (Tschinkel 1973). Because the Namib is dry, the detritus does not decompose readily and nutrients are tied up in it rather than being returned to the soil. However, the digestive tract of the beetles and other detrivores contains bacteria and fungi, which transform the detritus into mineral matter that is returned to the soil.

Two species of beetle, the sand-runner *Stenocara phalangium* and *Cauricara phalangium*, have exceptionally long legs. This allows them to take advantage of the steep temperature gradient above the sand and keep their body well above the hot surface of the dunes. *S. phalangium* has the longest legs of any beetle in the world, up to 6 cm in length. *C. phalangium* further avoids the extreme heat by standing atop small quartz stones and "stilting," or extending its legs. This stance also helps it to catch the desert breeze. Stilting keeps its body as much as 10°C cooler than it would be on the surface of the dune.

The nocturnal saucer beetle *Lepidochara* sp. is covered by a dense layer of moisture-absorbing scales and has one of the most unusual desert survival mechanisms. It erects trenches in the wet sand surface and extracts condensed fog-water from the raised sand ridges along their edges. The water content of the ridges is generally 2–5 times higher than in the surrounding sand. In one fog, the increase in water content in a population of beetles was measured to be 13.9% (Seely and Hamilton 1976). The trenches are up to a meter or more in length and generally a series of them are constructed in parallel. These become especially conspicuous when they trap the fog-water because their color darkens with the increasing moisture content. The trenches are laid out only during or just prior to the onset of

Fig. 24.29 Feral camels in Australia; they are not native, but were brought into the country in the nineteenth century to aid in exploration.

Fig. 24.30 A common diurnal Namib tenebrionid beetle, *Physodesmia globosa*.

Fig. 24.31 Detritus, comprising broken bits of organic matter, is an important component of the Namib's ecosystem.

Fig. 24.32 *Stenocara eburnea*, one of the few species of white beetles in the Namib; its coloration is thought to have an insulating effect.

Fig. 24.33 Beetles, bugs, and other organisms seek shelter in the comparatively cool microclimate of the *Welwitschia* plant.

fog, which generally occurs at night. The beetles return during the early morning stages of the fog to extract the water along the ridges. The trenches are oriented perpendicular to the fog-bearing winds during heavy fog, but during light fog they are erected on the highest part of existing ripples with no regard to direction. As the fog lifts, the beetles retreat below the surface of the sand, emerging only in the late afternoon or evening.

Two black beetles, *Onymacris unguicularis* and *O. laeviceps*, have evolved other mechanisms of water extraction. Both are generally active only during the day, but during the nocturnal fogs they emerge from the dunes at night. *O. unguicularis* utilizes a peculiar stance termed "fog-basking" to absorb water. It surfaces on the dune during a fog, generally on the slipface, and very deliberately climbs to the crest. Standing with its head pointed downward, the beetle orients its body to the wind to facilitate the condensation of fog on its body; the fog droplets then roll downward into its mouth. The water uptake from a fog increases a beetle's body weight on average by about 12%, but

an individual can increase its body weight by up to 40% during a single morning of fog. This beetle is found only in the western Namib, where fogs are a regular occurrence. *O. laeviceps* also takes advantage of the fog, but in contrast to its relative, it descends to the dune base, where it imbibes droplets that have precipitated on plants and plant material there.

Other adaptations of the dune beetles include low transpiration rates, waxy coats, light coloration (Fig. 24.32), and habitats with favorable microclimatic conditions (Fig. 24.33). Some tenebrionid species have the lowest transpiration rates of any arthropod, including the scorpion, which is known to be nearly "waterproof." The rate was recorded to be 0.11 mg/hour in one species, the lowest ever reported for any animal. Many of the beetles are covered by a waxy coat that reduces evaporation. When exposed to low humidity, some secrete lipids and proteins that form a second waxy extra-cuticular layer. This coating is called a wax bloom and consists of fine filaments that create a meshwork covering part or all of the beetle's surface. The rate of desiccation in beetles without this layer is markedly greater than in those with it. The bloom also plays a thermoregulatory role because it is highly reflective and an effective insulator. The incidence of beetle species with the bloom and the area of their bodies covered by the bloom increase from the cool foggy coast to the drier inland regions of the Namib. Hence, it is clearly an adaptation to the desert environment.

REFERENCES

Berry, C., 1991: Nara: unique melon of the desert. *Veld and Flora*, **77**, 22–23.

Bornman, C. H., 1972: *Welwitschia mirabilis*: paradox of the Namib desert. *Endeavour*, **31**, 95–99.

Corbin, J. F., M. A. Thomsen, T. E. Dawson, and C. M. D'Antonio, 2005: Summer water use by California coastal prairie grasses: fog, drought, and community composition. *Oecologia*, **145**, 511–521.

Edney, E. B., 1971: Some aspects of water balance in tenebrionid beetles and a thysanuran from the Namib desert of southern Africa. *Physiological Zoology*, **44**, 61–76.

Evenari, M., 1985: Adaptations of plants and animals to the desert environment. In *Hot Deserts and Arid Shrublands* (M. Evenari, I. Noy-Meir, and D. W. Goodall, eds.), Ecosystems of the World 12A, Elsevier, Amsterdam, pp. 79–92.

Goudie, A., and J. Wilkinson, 1980: *The Warm Desert Environment*. Cambridge University Press, Cambridge, 88 pp.

Henschel, J. R., and M. K. Seely, 2008: Ecophysiology of atmospheric moisture in the Namib Desert. *Atmospheric Research*, **87**, 362–368.

Jones, C. G., and M. Shachak, 1990: Fertilization of the desert soil by rock-eating snails. *Nature*, **346**, 839–841.

Krasnov, B., G. Shenbrot, I. Khokhlova, and E. Ivanitskaya, 1996: Spatial patterns of rodent communities in the Ramon erosion cirque, Negev Highlands, Israel. *Journal of Arid Environments*, **32**, 319–327.

Lange, O. L., A. Meyer, I. Ullmann, and H. Zellner, 1991: Microclimate conditions, water-content and photosynthesis of lichens in the coastal fog zone of the Namib Desert: measurements in the Fall. *Flora*, **185**, 233–266.

Lange, O. L., A. Meyer, H. Zellner, and U. Heber, 1994: Photosynthesis and water relations of lichen soil crusts: field measurements in the coastal fog zone of the Namib Desert. *Functional Ecology*, **8**, 253–264.

Louw, G., 1972: The role of advective fog in the water economy of certain Namib Desert animals. In *Proceedings: Comparative Physiology of Desert Animals*. Symposium of the Zoological Society, London, pp. 297–314.

Marsh, B. A., 1990: The microenvironment associated with *Welwitschia mirabilis* in the Namib Desert. In *Namib Ecology* (M. K. Seely, ed.), Transvaal Museum Monograph, Pretoria, pp. 149–154.

McClanahan, L. L., R. Ruibal, and V. H. Shoemaker, 1994: Frogs and toads in deserts. *Scientific American*, March, 82–88.

Nott, K., and M. J. Savage, 1985: Root distributions of *Trianthema hereroensis* in the Namib dunes. *Madoqua*, **14**, 181–183.

O'Donnell, M. J., 1977: Site of water vapor absorption in the desert cockroach, *Arenivaga investigata*. *Proceedings of the National Academy of Sciences USA*, **74**, 1757–1760.

Petrov, M. P., 1976: *Deserts of the World*. Halsted (Wiley and Sons), New York, 447 pp.

Rauh, W., 1985: The Peruvian-Chilean deserts. In *Hot Deserts and Arid Shrublands* (M. Evenari, I. Noy-Meir, and D. W. Goodall, eds.), Ecosystems of the World 12A, Elsevier, Amsterdam, pp. 239–268.

Schmidt-Nielsen, K., C. R. Taylor, and A. Shkolnik, 1971: Desert snails: problems of heat, water and food. *Journal of Experimental Biology*, **55**, 385–398.

Seely, M. K., 1983: Effective use of the desert dune environment as illustrated by the Namib tenebrionids. In *Proceedings of the VII. International Colloquium of Soil Zoology*, Louvain-la-Neuve, Belgium, pp. 357–368.

Seely, M. K., 1987: *The Namib*. Shell Oil SWA Ltd., Windhoek, Namibia, 104 pp.

Seely, M. K., and W. J. Hamilton III, 1976: Fog catchment sand trenches constructed by tenebrionid beetles, *Lepidochora*, from the Namib Desert. *Science*, **193**, 484–486.

Shmida, A., 1985: Biogeography of the desert flora. In *Hot Deserts and Arid Shrublands* (M. Evenari, I. Noy-Meir, and D. W. Goodall, eds.), Ecosystems of the World 12A, Elsevier, Amsterdam, pp. 23–77.

Tschinkel, W. R., 1973: The sorption of water vapor by windborne plant debris in the Namib Desert. *Madoqua Series II*, **2**, 21–24.

Walton, K., 1969: *The Arid Zone*. Aldine Publishing, Chicago, 175 pp.

25 Climatic variability and climatic change

25.1 INTRODUCTION

Drylands are thought to be the regions that global change will impact most severely. As described later in this chapter, our predictive capabilities are not up to the task of offering a view of these impacts. By evaluating the general nature of climatic variability, its causes, and its long-term trends in the drylands, this chapter provides a view of the broadest range of conditions that would likely prevail, as well as more realistic ones. The extremes are represented by conditions of the last 20,000 years, the late Pleistocene and Holocene. The historical period, roughly the last 2000 years, represents the more realistic scenarios. The majority of the examples in this chapter come from Africa because of the very visible and controversial changes over that continent, the large amount of available information, and the author's extensive experience in the region. However, these examples are used to illustrate broadly applicable concepts of climate variability. A discussion of El Niño and its impact is also included because of its almost ubiquitous influence on climatic variability in the drylands.

This chapter presents little information on recent climatic variability. The reason is that the available data and literature with analyses thereof are so voluminous that a brief synthesis is not feasible. Also, most of the literature on recent change is readily available in the meteorological literature and is therefore easy for the reader to access. A few notable papers looking at long-term variability during the instrumental period examine northern Eurasia (Wang and Cho 1997), the Tien Shan (Aizen and Aizen 1997), the US Pacific Coast (Chen *et al.* 1996), India (Krishnamurthy and Shukla 2000), the Mojave Desert (Hereford *et al.* 2006), the central United States (Hu *et al.*1998), China (Ye *et al.* 2003), Iran (Modarres and Sarhadi 2009), South America (Aguillar *et al.* 2005), and Africa (Nicholson 2000a, 2001a).

25.2 NATURE OF CLIMATIC VARIABILITY AND CLIMATE CHANGE

Climate is all too often described in terms of numbers representing a set of mean conditions. Consequently, climate is generally treated as an environmental constant, when in reality it is a dynamic environmental variable. The "mean" conditions can be significantly different from one century to the next or even one decade to the next. The climatic "mean" or "normal" also belies complex temporal and spatial patterns within the temporal or spatial span over which an average is taken. In this chapter, this complexity is described by considering the forms and causes of climatic variations.

Climatic variability can take on many forms. Figure 25.1 illustrates four common ones: a periodic variation, an abrupt change of mean conditions, a gradual trend, or a change in the year-to-year variability without a change in the mean (Hare 1979). In most cases the changes are not quite so distinct. Figure 11.27 shows a more realistic time series, that of rainfall in the West African Sahel; a decrease in the interannual variability is apparent after 1950, as well as a major change to drier conditions after 1967. The former might be viewed as a relatively abrupt change in the mean, or a downward trend between 1950 and 1973.

Drylands are particularly sensitive to climatic variability and change for two basic reasons. One is the relative scarcity of precipitation; small changes can have a large impact. A related reason is that most drylands exist along what might be termed "climatic ecotones." These are the transition zones between various prevailing climatic regimes, and the gradients in these transition zones are often quite sharp. The West Africa Sahel is a prime example: mean annual rainfall increases southward at a rate of 1 mm/km. When a climatic change occurs, it often involves the shifting of these boundaries. In the case of the Sahel, a 30–40% decline in rainfall between 1931–1960 and 1968–1997 was accompanied by only a 1° latitudinal shift in climatic boundaries (Nicholson 2001a).

Variability in precipitation can involve the total amount received at a given location or a change in its temporal rhythm. Changes in such characteristics as the timing and magnitude of the seasonal cycle, the season length, the year-to-year variability, the number and length of dry spells within the season, or the number and intensity of individual events can be as meaningful

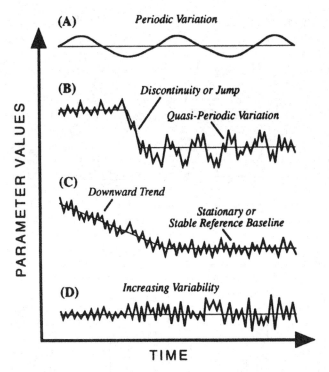

Fig. 25.1 Forms of climatic variability and change (from Marcus and Brazel 1984, reprinted with permission of the Office of the State Climatologist of Arizona).

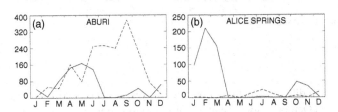

Fig. 25.2 (a) The seasonal cycle of rainfall (mm) at Aburi, Ghana (6° N) and (b) Alice Springs, Australia (24° S). For Aburi, a wet year (1968, dashed line) and a dry year (1958, solid line) are shown. For Alice Springs, a year with a winter maximum (1996, dashed line) and a year with a summer maximum (1976, solid line) are shown.

as a change in the overall amount. Such changes have dramatic implications for overall water balance (the partitioning into evapotranspiration and runoff, storage as soil moisture) and for agriculture, even if the total amount of precipitation remains invariant.

Several examples illustrate changes in seasonality. The most pronounced changes occur in individual years. Aburi, Ghana (Fig. 25.2a), lies in a dry zone along the Guinea coast, where a dry season normally occurs in summer. This is the case in 1958 and in the mean, but the 1968 rainy season had a single peak in June through August. In fact, in most really wet years, summer ceases to be the dry season and becomes the peak rainy season. At Alice Springs, Australia (Fig. 25.2b), the maximum may even shift from winter to summer. Such a shift occurs most commonly in drylands because these regions often lie at the

transition between winter and summer rainfall regimes. Over decades or centuries the seasonal maximum may exhibit persistent shifts. At Casablanca, Morocco, peak rainfall occurred in December or January early in the twentieth century but changed to a bimodal cycle, with one peak in March and the other in November (Fig. 25.3). At Santiago, Chile, the seasonal cycle shifted from a broad winter maximum to a sharp June peak, then back to a broad winter maximum (Fig. 25.3).

Variability can also be expressed as occasional unseasonal precipitation. In the western Sahel of Senegal and Mauritania, the rainy season is generally confined to mid-summer and the 6–9 month dry season is generally completely dry. In roughly one in two or three years, significant rainfall occurs during the core of the dry season, in December, January, or February. In December, 1905, rain fell on 5 days at Oualata and 12 days at Tichitt, Mauritania (Marty 1927). Winter rainfall was also abundant during 1906 in northern Senegal and southern Mauritania. Saint Louis, Senegal, received 67 mm in December, while Nouadhibou, Mauritania, received 132 mm (ORSTOM 1977). In the past this "unseasonal" rainfall occurred with enough regularity in Mauritania that a second harvest could be planned for.

It is often assumed that a wetter/drier year would be associated with either a longer/shorter rainy season and more/fewer rainfall events. However, this is seldom the case. A case in point is Bandiagara, Mali, where rain fell on 26 days during the wet year 1965 (annual total = 1009 mm) but on 42 days during the dry year 1941 (annual total = 439 mm). Furthermore, wetter or drier conditions tend to be associated with changes in particular months or seasons and these are not necessarily the peak rainfall months. The contribution of some months to the interannual variability of rainfall is disproportionate to their contribution to the seasonal mean. Examples from three dryland regions of Africa illustrate these concepts. Two of these, the Sahel and the Kalahari, are subtropical regions of summer rainfall; the third, East Africa, is an equatorial region with two rainy seasons and two dry seasons during the course of the year.

Figure 25.4 shows the mean monthly rainfall at Serowe (Botswana), Kidal (Mali), Moyale (Kenya), and Djibouti (Djibouti) and and contrasts it with mean monthly rainfall during the 10 wettest years and 10 driest years. In all cases, the contrast between the wettest and the driest 10 years is seen only during the rainy season, implying that wet/dry years are not associated with a significantly longer/shorter rainy season. At Kidal, in the Sahel, August is clearly the most variable month. Throughout the Sahel, August tends to contribute more to the interannual variability than do the early months of the rainy season. An example is shown for Ke-Macina, Mali (Fig. 25.5). Mean monthly rainfall in August is 190 mm, versus 145 mm for July, but the correlation (R) between annual rainfall and monthly rainfall is 0.68 for August and only 0.50 for July. At Serowe, in the Kalahari, November, December, and January show little variability, but the months of February–April produce most of the interannual fluctuations. An explanation (Bhalotra 1984) in this case is that the first half of the rainy season is dominated

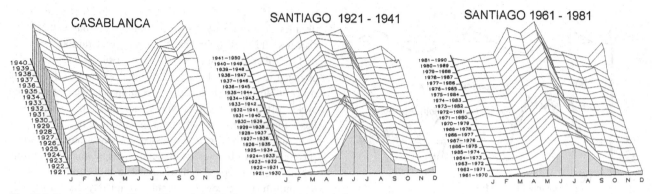

Fig. 25.3 Decadal means of monthly precipitation (mm) at Casablanca and Santiago, plotted every five years.

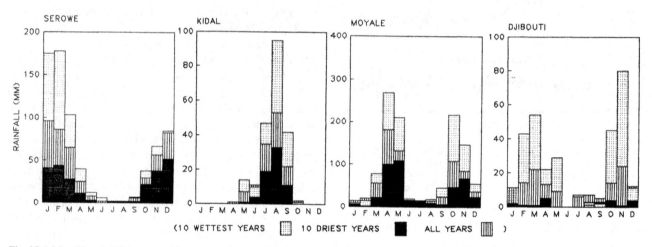

(10 WETTEST YEARS ☐　10 DRIEST YEARS ■　ALL YEARS ▤)

Fig. 25.4 Monthly rainfall in wet and dry years at Serowe, Kidal, Moyale, and Djibouti.

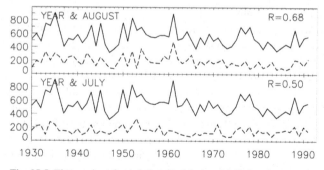

Fig. 25.5 Time series of rainfall at Ke-Macina, Mali, for the year, for August, and for July.

by temperate latitude influences, which are much more reliable than the tropical rainfall regime that prevails from mid- or late January.

A prominent case in point concerning seasonal contributions to interannual variability is East Africa. The main rainy season (the "long rains" of March/April/May) shows little relationship to the interannual rainfall variability, which instead is almost perfectly correlated with the "short rains" of October/ November (see Chapter 16). Thus, at both Djibouti and Moyale,

the difference between wet and dry years is particularly strong in October and November. These months contribute up to 50% of the variability, but only 10–35% of the mean rainfall. In this case, the explanation is that the latter season is more variable and much more strongly affected by El Niño.

It is also interesting to note that the number of rain days is fairly constant in the central Sahel, but in the Kalahari there is a clear tendency for the number of rainy days to increase during wet and decrease during dry years (Fig. 25.6). For the central Sahel, these statistics and many other studies demonstrate that an increase in annual precipitation is produced by a small increase in the number of really intense rain events, rather than by an increase in the total number of events. Other examples of this are presented in Chapter 11. Surprisingly, in the western Sahel (illustrated by Dakar, on the Atlantic coast), there is an association between the number of rain events and amount of annual rainfall, suggesting somewhat different causes of variability in this region of strong maritime influence.

Another important aspect of variability is its overall temporal character. Some time series show completely *random* variability. That is, each observation is completely independent of all others, with no regular pattern of occurrence. This is essentially "white noise." In contrast, some time series contain

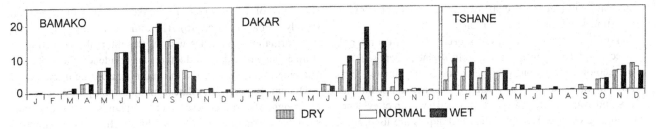

Fig. 25.6 Daily rainfall in wet and dry years at Bamako and Dakar, in the Sahel, and at Tshane in the Kalahari.

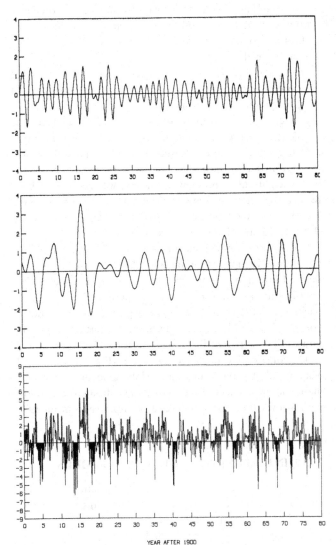

YEAR AFTER 1900

Fig. 25.7 Time series of the first principal component of global precipitation (from Lau and Sheu 1988). The series at the top has been filtered to retain variations on time scales of 18–36 months. The series in the middle has been filtered to retain variations on time scales of 36–120 months. The bottom series is the sum total of the top two.

persistence: that is, each observation bears some relationship to those immediately before it, but the relationship decreases with the time elapsed between the observations. This is termed "red noise." A time series can also be cyclic or periodic, although few such series exist in nature except for diurnal and annual cycles.

Figure 25.7 illustrates periodic time series derived by applying statistical filters to a time series representing the year-to-year fluctuations of the most common spatial mode of global rainfall variability. The filters isolate cycles on time scales of 18–36 months (termed the biennial time scale) and 36–120 months (termed the ENSO time scale). Together these account for most of the year-to-year variability in rainfall (Lau and Sheu 1988). However, the periodicities are irregular, with their magnitude clearly changing from year to year. Such a time series, more typical of climatic elements, is "quasi-periodic": there is a tendency for maxima and minima to occur at specific time intervals, but imperfectly so, and the periodicities account for only part of the variance. For climatic time series, the periodicities are generally not significant enough to provide any forecasting skill.

Climatic variability not only tends to show some degree of temporal organization, but also preferred spatial configurations of occurrence, i.e., spatial modes of variability. As an example, over the USA, warm/cold winters in the west are accompanied by cold/warm winters in the east. The statistical links between these regions are termed "teleconnections." A further example is that over Africa, there is an inverse relationship between rainfall in equatorial and subtropical latitudes that occurs in roughly one-third of all years (Nicholson 1986a). The "teleconnection" is particularly strong between East Africa and southern Africa, so that dry years in one area are generally wet years in the other (see Chapter 16). Such teleconnections can provide indications of the principal causes of variability, such as El Niño or shifts in the tropical rainbelt.

25.3 RECONSTRUCTING PAST CLIMATE

25.3.1 INDICATORS OF PAST CLIMATE

The most precise records of climate are instrumental observations. However, few of these are available prior to about 1600, and in most dryland regions the instrumental record barely extends beyond one century. As a result, other types of indicators, termed proxy records, are utilized to reconstruct the climate of the past. Proxy records are of two types: natural (physical or biological) and documentary (written archives or oral histories). The further back in time, the more imprecise the proxies, and different types of information are available during historical times (roughly the last two millennia) and geological times (time scales of millennia or longer).

On the longer time scale, ice cores, ocean sediment cores, pollen records, lake sediments, beach ridges, glacial deposits, and other geologic formations provide the basis for climate reconstruction. In dryland regions, sand dunes, aeolian sediments, lakes, and evaporite deposits have provided a wealth of evidence for climatic change. Various dating techniques are used, most notably radiocarbon dating.

Some highly visible remnants of wetter periods of the past are shown in Fig. 25.8. These include petrified wood from the hyper-arid core of the Egyptian Sahara and a virtual blanket of fossil sand dollars covering parts of the eastern Sahara. Though from millions of years ago, these sand dollars are nearly identical to modern ones. Paintings of savanna landscapes in Saharan caves trace back to a wetter period some 5000 years ago. Fossil dunes over Australia bear witness to older periods of increased aridity. Pottery scattered on the ground of the Djenné (Mali) archaeological site stem from a settlement period roughly 2000 years ago during a wetter episode. Sediments of former lake beds exist throughout the Sahara. Remains of a Roman camp site in the hyper-arid Qattara Depression of Egypt (Fig. 25.9) also bear witness to this wetter episode. Such archaeological evidence provides compelling proof of a period of human occupation in the Sahara 9000 years ago (Fig. 25.10).

For the last two millennia, tree-ring records, corals, varves (differential layers of sediment in lakes), archaeological findings, and documentary records add to the pool of information. On very short time scales, radiocarbon dating becomes less reliable, but for many indicators (notably trees, varves, and corals) dating can be accomplished via annual rings. Tree-ring records typically extend 500–700 years, although a few cover several millennia with annual resolution. Fossil wood can be used to extend the record. Corals provide annual or higher resolution and often provide continuous records covering up to 400 years. Varved lake sediments and ice cores likewise provide annual resolution. Varves generally cover centuries but ice cores can provide reliable information over at least the past 40,000 years. Pollen records can cover several millennia, but generally with time resolution of about 50 years at best.

Figure 25.11 shows some physical remnants of fairly recent wetter periods in various dryland regions. These include high water marks on trees in the Okavango Delta of Botswana, the dry bed of Lake Ngami in Botswana, dead trees along the edges of dry pans in the Namib, a dry river bed in the Kalahari, savanna animals in current grassland and desert regions, the cliff houses of the Anasazi people in the western USA, and camel tracks in the hyper-arid Egyptian desert, indicative of a period of greater water availability during the late nineteenth century.

During historical times, Lake Ngami alternated between being completely desiccated and being a large lake, deep enough to support waves that could push a hippopotamus to shore (Nicholson 1996b). In the 1950s it was deep enough that speedboat races were regularly held on the lake. Three former

shorelines of Lake Ngami are marked by lines of vegetation (barely visible in the photo). The estimated age of the vegetation formations coincides with wetter intervals reconstructed from either gauge data or documentary evidence: the 1950s, the late nineteenth century, and the mid-nineteenth century.

Documentary reports relate to many, diverse phenomena such as droughts, floods, harvest quality, and landscape observations such as the condition of rivers and lakes. In many cases actual weather diaries cover decades of time. Economic indicators such as wine quality, tax records, cereal prices, or the opening or closing of harbors and river navigation are often recorded for centuries. Phenological observations such as the dates of harvests or blossoming, first frost, and freezing or thawing of rivers and lakes are also useful.

In dryland regions, lake cores or remnants of now-dry lakes have been particularly useful in establishing the climate record on geologic time scales, while landscape descriptions – including vegetation – and reports of famines and drought provide excellent historical information. All too often, documentary evidence covering earlier dry periods is discounted. This was the case with an oral tradition of changes in Lake Ngami in the late eighteenth century, when the lake was said to have been replaced by a flowing river with trees along its banks. When the lake dried up, the dead trees and riverbed again became visible (Fig. 25.11d) and the story was confirmed.

For Africa, historical information is detailed enough to produce reconstructions of rainfall and lakes over roughly two centuries. Little information provides quantitative evidence, but semi-quantitative information can be derived (Fig. 25.12; see Nicholson 2001b). For lakes, some degree of quantification is possible (Fig. 25.13). In many cases, the documentary information can be supported by sediment records (Verschuren *et al.* 2000). For select locations, such as Angola (Fig. 25.14), Senegal, Mali, Chad, Algeria, Tunisia, and Malawi, documentary information allows for identification of at least droughts and wetter periods over the course of several centuries (Nicholson 1978, 1998).

25.3.2 INTERPRETATION OF PROXY RECORDS

The proxy records usually represent a combination of climatic and non-climatic factors. In many cases (tree rings, for example), a multitude of climatic factors may be involved: temperature, precipitation and its seasonality, and relative humidity over several years. In general, the use of proxy records in a quantitative sense requires the use of a statistical relationship, termed a transfer function, to establish a connection between climate and the variability of the proxy over some period of overlap.

What essentially needs to be established is the relationship between the climatic factor "forcing" the proxy indicator and the indicator's "response." This relationship often involves time lags, and the character of the response might differ markedly from the character of the forcing. Furthermore, some indicators are more sensitive than others. This problem is illustrated using examples of a glacier and a lake.

Fig. 25.8 Physical evidence of long-term climatic change in Africa: (a) sand-dollar field in Egypt and close-up of fossil sand dollar; (b) petrified wood and pottery from the Sahara; (c) seashells spread across the surface of the Sahara; (d) pottery littering the ground in Djenné; (e) cave painting in the Sahara; (f) stabilized dunes in Australia (dark spots are cloud shadows); (g) sediment layers of former Lake Taoudenni, Mali (photo courtesy of N. Petit-Maire).

Fig. 25.9 The vast Qattara Depression of Egypt (19,500 m²) is uninhabited, but evidence of former encampments during Roman times indicates a formerly more moderate climate. The cave on the left (dark holes are entrances) provided shelter, attested by the presence of a human skeleton near a former camp fire (right).

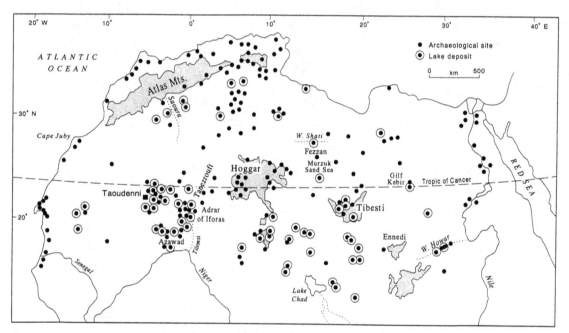

Fig. 25.10 Archaeological sites and pluvial lakes over North Africa *c.* 9000 BP (Goudie 1996, modified from Petit-Maire 1991).

In the case of a glacier, the net mass balance reflects both precipitation and temperature. If the net balance is simplified by prescribing merely glacial and non-glacial times (Fig. 25.15), the glacier grows steadily during the glacial period and retreats during the non-glacial. The response is a smooth curve, despite the abrupt climate discontinuity. Moreover, the times of maxima and minima of the glacier actually correspond to the times of the change from glacial to non-glacial. The dynamics of glacier movement further complicate the picture: the pressure of the ice can produce an advance even without a change in climate.

Analogous to the glacier's response to temperature change, a lake responds to the net water balance, so that the change in its depth reflects the net difference between input (precipitation plus inflow) and output (evaporation plus discharge). As with a glacier, its maxima and minima reflect the times of discontinuities, and a lake may maintain a high stand even when precipitation is not particularly high. Thus, Lake Victoria in equatorial East Africa rose rapidly in response to tremendous precipitation in the early 1960s and maintained a relatively high stand well into the 1990s (Nicholson *et al.* 2000). However, the precipitation over the lake was actually high for only a few years and the following drier conditions were reflected as a slow decline in lake levels (Fig. 25.16).

In the case of a lake, the response can include both its depth and surface area. The nature of the response depends on the

Fig. 25.11 Physical evidence of short-term climatic change in Africa: (a) water marks on trees in Okavango; (b) camel tracks in the eastern Sahara, dating to the late nineteenth century; (c) dry river through the Kalahari; (d) dried bed of Lake Ngami; (e) dead trees in the Namib; (f) cliff houses of the ancient Anasazi culture, near Mesa Verde, Colorado.

relative importance of inflow and outflow versus precipitation and evaporation over the lake, as well as on lake geometry. Closed basin lakes, those without an outlet, are particularly sensitive to climatic variations, as are those in which precipitation and evaporation are the prime factors in the water balance. Lake Victoria is an example of the latter.

The effect of geometry is illustrated in Fig. 25.17, which shows a lake with steep sides at depth, but a very shallow slope in the

upper levels. When the lake is relatively low, its depth will change rapidly but its surface area will not. When the lake reaches a high stand, its surface area will change rapidly in response to variations in the water balance but not its depth. Lake Chad, which has often been used as a proxy for West African climate, is an example of such a lake. The lake is relatively shallow and changes in precipitation are dramatically reflected as changes in surface area (see Fig. 1.17) until a dry stage is reached in which

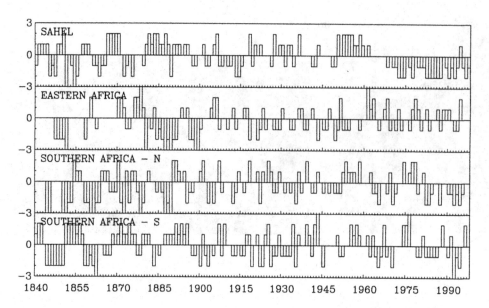

Fig. 25.12 Rainfall variability in four African regions, 1840–1997. The values range from −3, indicating extremely dry conditions, to +3, indicating extremely wet conditions. Instrumental data since 1900 have been converted to this same scale (from Nicholson 2001b).

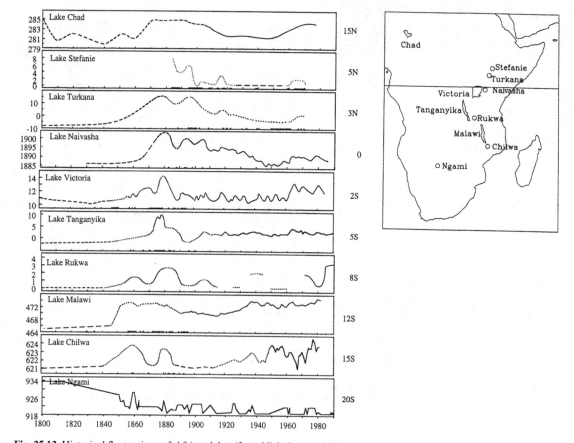

Fig. 25.13 Historical fluctuations of African lakes (from Nicholson and Yin 2001). Except for Lake Ngami, solid lines indicate modern measurements, short dashed lines indicate historical information, and long dashed lines indicate general trends. The dots on the x-axis represent years with actual historical references. The lakes are shown on the location map; the horizontal line represents the equator.

only a deep basin in the southeast retains water. At that point, the lake appears to remain relatively constant. Thus, knowledge of basin geometry is critical in establishing the response function of a lake to variations in climatic forcing.

Further complications arise when the lake or its basin is large enough to extend over several different climatic zones. Lake Chad lies in the semi-arid Sahel, but rivers providing inflow arise in the more humid equatorial region. Precipitation

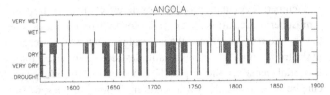

Fig. 25.14 Semi-quantitative time series of climatic conditions in coastal Angola since the late sixteenth century, based on numerous documentary sources. Several long drought intervals and a change to wetter conditions in the nineteenth century are clearly evident.

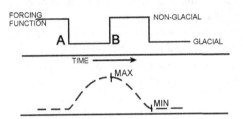

Fig. 25.15 Schematic of the change of glacier extent versus change in temperature, illustrating force versus response.

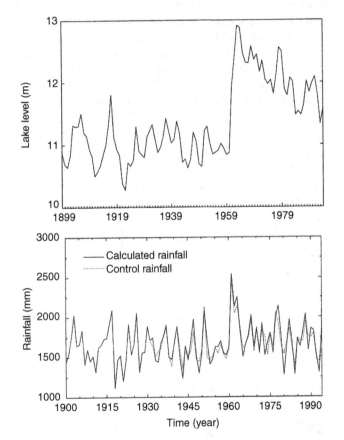

Fig. 25.16 Lake Victoria: levels versus precipitation (from Yin and Nicholson 2002).

fluctuations are often out of phase in these two regions, so that a rise in the lake can reflect either increased rainfall over the lake or changes in streamflow into it. Lake Bosumtwi is similarly complex. It lies south of the Sahel, in a region where

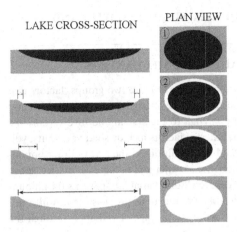

Fig. 25.17 Changes in surface area versus depth for a typical lake, as the lake dries down.

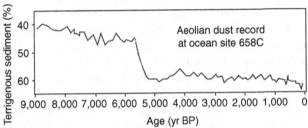

Fig. 25.18 Terrigenous material and estimated flux of material in North Atlantic sediment cores (from deMenocal *et al.* 2000).

rainfall fluctuations sometimes mimic those of the Sahel but equally often are in the opposite direction. Hence it is not an ideal indicator of Sahelian climate. A similar situation exists for Lake Malawi. The northern half of the lake is in the equatorial region and the southern half is in the subtropics, regions that also show precipitation fluctuations of the opposite sign. In such cases, the net change in the lake reflects a combination of influences and a statistical model is needed to interpret the lake in climatic terms. Also, the geochemistry of Lake Malawi, evident in the cores, is a response to wind not rainfall (Johnson *et al.* 2001).

A final example of a proxy indicator with complex interpretation is sediment cores, such as those in the eastern Atlantic that reflect dust coming from North Africa. The core in Fig. 25.18 shows a sharp discontinuity around 5500 BP that has been interpreted as an abrupt onset of arid conditions in the Sahara (deMenocal *et al.* 2000). However, the occurrence of this dust in the core also requires a prevailing easterly wind regime and a source of dust. The former requires a shift from the monsoonal wind regime to the northeast harmattan, a shift that is abrupt, and the desiccation of the Pleistocene lakes that ultimately became the source of most of the dust (Prospero *et al.* 2002). Hence, the sediment core simultaneously reflects the influence of several interrelated factors.

25.4 CAUSES OF CLIMATIC VARIABILITY AND CHANGE

25.4.1 GENERAL CAUSAL MECHANISMS

The factors governing climate fall into two groups: factors that are *external* to the system (also referred to as boundary forcing) and those that are *internal* to the system and involve only the atmosphere itself. External factors include solar variability, volcanic eruptions, and the conditions at the earth's lower boundary, such as the presence of water and ocean currents, ice, snow, bare land, and vegetation. Solar variability governs the amount of energy available to drive the atmospheric system, while volcanic eruptions influence the passage of radiation through the atmosphere. Because the atmosphere is relatively transparent to solar radiation and is heated most directly by the earth's surface, the boundary conditions govern the spatial and temporal partitioning and degree of energy transfer to the atmosphere. In this way boundary forcing influences climate and weather. There are also some aspects of randomness involved in climatic variability. Under the same set of boundary conditions, more than one atmospheric state can be stable; minor triggers can push the climate from one state to another.

The boundary conditions change relatively slowly in time, while internal factors within the atmosphere vary on comparatively short time scales (Shukla 1981). These internal factors are such complex facets of the atmosphere as the interactions between waves and jet streams, interactions between tropical and temperate circulation systems, or the effect of convection on atmospheric circulation. They tend to operate on time scales of days to months. In general, the internal interactions are weaker in the tropics than in the higher latitudes because the circulation is overall weaker in the tropics. For this reason, boundary forcing becomes particularly important in the low latitudes, where many of the drylands are situated.

The full spectrum of factors governing climatic variability is illustrated in Fig. 25.19. These act on quite diverse time scales from 10^0 to 10^9 years (Kutzbach 1976). Some, such as continental drift, act so slowly that they have little impact on modern or even historical climate variability. In most cases, however, global climate at a given instant is the net result of a full complex of factors. For this reason, it is exceedingly difficult to detect specific causes of variability or to answer such questions as whether current climatic trends are primarily a product of human activity.

GLACIAL TIME SCALES

It is well established that the main factors governing long-term variability (on the order of millennia) are the earth–sun relationships known as Milankovitch cycles (Berger 1977). Three aspects of earth–sun geometry vary in a regular fashion: the obliquity or tilt of the earth's axis, with a period of ~41,000 years; the eccentricity of the earth's orbit, with a period of about 100,000 years; and the seasonal timing (or precession)

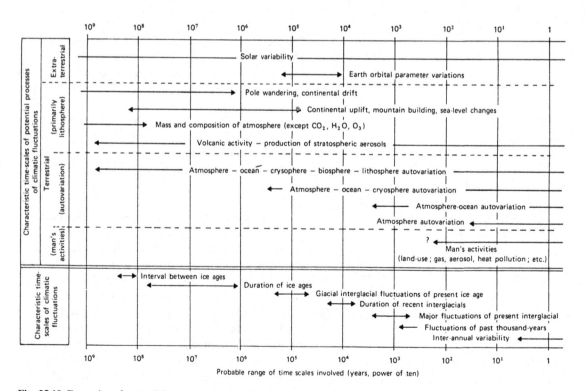

Fig. 25.19 Examples of potential processes involved in climatic fluctuations (top) and characteristic time scales (bottom) of observed climatic fluctuations (from Kutzbach 1976).

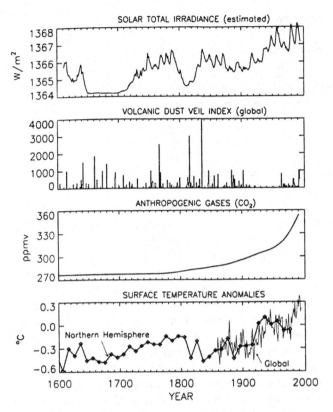

Fig. 25.20 Long-term variations in Northern Hemisphere surface temperature anomalies and various forcing mechanisms (from Lean and Rind 1998).

of the equinoxes, with a period of ~21,000 years. These modulate the spatial distribution of solar radiation, the magnitude and timing of its annual cycle, and the asymmetry between the hemispheres. Associated changes in solar insolation were clearly the prime drivers of the last Ice Age maximum some 18,000 years ago, the end of the Ice Age about 10,000 years ago, and the period of optimum warming about 5000 years ago in the middle of the post–Ice Age period called the Holocene.

Superimposed on these solar variations are boundary forcings that are related to the presence of the ice in the high-latitude oceans and on land, sea-surface temperature and ocean circulation changes, changing vegetation patterns over the continents, and dust mobilized from the surface or erupted from volcanoes.

TIME SCALES OF CENTURIES TO MILLENNIA

Climatic variability on the scale of centuries to millennia is the most difficult to explain. Current theories suggest that the main factors include volcanic activity and changes in solar output (Fig. 25.20) (Lean and Rind 1998; Crowley 2000). However, during the twentieth century, anthropogenic forcing via greenhouse gases appears to have overtaken these factors and become the dominant forcing of global temperatures (Crowley 2000).

The solar–climate connection has been discussed since at least the nineteenth century, when it was believed that sunspots

played a major role in regulating weather and climate. The theory fell out of favor until recently because most of the studies were poorly designed and because no physical basis for the relationship could be established. Once satellites provided evidence that solar output and sunspot activity are correlated, there was renewed interest (Kerr 1996) and numerous studies demonstrated apparent relationships between solar output and various meteorological phenomena, such as North Atlantic temperatures (Bond et al. 2001).

The relationship with volcanic dust has a stronger physical basis. Volcanic dust reflects solar radiation and absorbs both solar and terrestrial radiation. Its precise impact is dependent on intensity, season, location of the outbreak (especially latitude), length of residence time in the atmosphere, and whether or not the dust reaches the stratosphere. In general it will tend to cool the surface but heat the stratosphere (Robock 2000). Its impact is manifested mainly via its radiative effects, but it also plays a role in atmospheric chemistry (especially stratospheric ozone reactions), and dust also influences evaporation by blocking shortwave radiation. The cooling also influences vegetation productivity at higher latitudes (Lucht et al. 2002).

INTERANNUAL AND DECADAL VARIABILITY

Variability on interannual and decadal time scales is controlled largely by boundary forcing and internal dynamics of the atmosphere. Because internal factors tend to operate on shorter time scales, boundary forcing is the dominant factor. Sea-surface temperatures are the most important factor, particularly in the lower latitudes, but persistent snow and ice cover or soil moisture anomalies on a continental scale can influence year-to-year changes in weather. Volcanic eruptions also have some influence on these shorter time scales.

An interesting example of boundary forcing comes from Namias' (1978) study of the winter of 1976/77, when extreme cold prevailed in the eastern USA and extreme drought occurred in the west. Snow cover reinforced the prevailing circulation pattern, enhancing the abnormal conditions in both the east and the west.

Much of the interannual variability of rainfall is linked to global-scale phenomena, "modes" of variability that have distinct spatial patterns of variability and are associated with specific time scales of variations and often particular seasons. Most involve complex feedbacks between the atmosphere and oceans. El Niño is perhaps the best-known example. This and other important modes are discussed in greater detail in Chapter 5 (Section 5.6).

25.4.2 THE EL NIÑO/SOUTHERN OSCILLATION (ENSO) PHENOMENON

One of the most important factors in interannual variability is the El Niño/Southern Oscillation phenomenon, also known as ENSO. This tends to account for variations on time scales of roughly 2–7 years and plays a particularly important role in climate and vegetation in dryland regions. It not only forces the

Table 25.1. *Impact of the 1925 El Niño on coastal water temperatures at four locations along the Peruvian coast (from Schwerdtfeger 1976).*

				March temperature (°C)	
Location	Latitude	Period of El Niño	Duration (days)	Normal	1925
Lobitos	4°20′ S	Jan 20–Apr 6	76	22	27
Puerto Chicama	7°40′ S	Jan 30–Apr 2	63	21	27
Callao	12°00′ S	Mar 12–Mar 27	15	19	25
Pisco	13°45′ S	Mar 16–Mar 24	8	19	22

year-to-year variations in precipitation in many dryland regions, but it also imparts several characteristic aspects of the rainfall regime. Nicholls (1991) summarizes these for Australia, but his list is applicable for most dryland regions, with the exception of some continental interiors (see the section on droughts). These characteristics include:

- large interannual variability (enhancing an inherent characteristic of dry climates);
- droughts and wet periods with time scales of about one year;
- very large (often continental) spatial scales;
- anomalous rainfall that is phase-locked with the annual cycle;
- rainfall anomalies followed/preceded by anomalies of the opposite sign in the year following or prior.

The regular occurrence of El Niño in Australia is probably also responsible for numerous vegetation characteristics, because species will thrive that can tolerate the extreme fluctuations between wet and dry years. Nicholls suggests that El Niño plays a role in Australia's lack of succulents but large number of ephemeral species; in the extreme drought-tolerance and fire-resistance of its native vegetation; and in producing the relatively high proportion of trees (compared with the degree of aridity). Mulga, for example, can survive more than 50 years because of its ability to withstand even severe droughts. ENSO, particularly the extreme El Niño of 1877/78 and the subsequent La Niña, may have triggered a major vegetation change linked to the land management practices of the arriving Europeans. This was the transformation of a large area of grazing land into the Pilliga Scrub, an area with vegetation so dense that it has sustained timber harvesting (Austin and Williams 1988; Nicholls 1991).

El Niño's influence on the ecosystem is not confined to Australia. Holmgren *et al.* (2006) have shown El Niño's impact on arid and semi-arid ecosystems on three continents. Some South American ecosystems are particularly sensitive. In north-central Chile and northwest Peru, El Niño facilitates tree establishment and in some cases has triggered forest regeneration. Its impacts on climate can cause open dryland ecosystems to shift to permanent woodlands, hence El Niño events might provide a window of opportunity for the restoration of degraded dryland ecosystems (Holmgren and Scheffer 2001).

El Niño was first known from the catastrophic rainfall events it produced along the desert west coast of South America. The disappearance of cold upwelling and the appearance of warm currents along the coast mark the occurrence of El Niño, in the classical sense. Table 25.1 shows the timing and the rainfall anomalies during a typical episode, that of 1925. The replacement of the cold current with warm water begins in the north and progresses southward; both the duration and degree of warming decrease southward in the latitudes from 4° S to 14° S. The associated enhancement of precipitation also tends to increase southward but not uniformly so. In the 1965 episode, March/April rainfall was roughly twice the normal near the equator, but south of 1° S it was generally at least five times the normal for the two months (see Chapter 5). Significant rainfall occurred several degrees further south than normal.

It is now known that the South American El Niño is part of a global coupled ocean-atmosphere phenomenon that is related to pressure changes in the low latitudes. The core feature of El Niño is warming in the equatorial Pacific. It is usually associated with changes in the vertical Walker circulation in the equatorial plane, as described in Chapter 4: the tendency for rising motion on the east sides of low-latitude continents and subsidence on the west sides. Periodically the Walker circulation weakens and shifts longitudinally (see Fig. 4.6), evoking changes in the tropical pressure field referred to as the Southern Oscillation. Because the weakening of the Walker cell is linked to a weakening of the subtropical highs, El Niño is a particularly strong factor in dryland regions under the influence of the high-pressure centers.

The weakening of the Walker circulation reduces both the ascent and the subsidence of the component cells and at the same time it reduces the strength of the trade winds. The result of the latter is the development of an equatorial warm pool that moves eastward then southward along the coast of South America. A similar phenomenon appears to occur in the equatorial Atlantic and along the Benguela coast of Africa.

Because of the association between El Niño and the Southern Oscillation, the term ENSO is generally used. El Niño is its warm phase. These episodes, also termed "low index" or "warm" events, recur approximately every 3–5 years, although their frequency changes over decades and centuries.

The converse also occurs with pressure changes in the opposite direction and cold events over the Pacific; the anomalies of precipitation and temperature tend to be the opposite of those during the warm phase. This is variously termed La Niña, the cold-phase of ENSO, or a high index/cold event. These are generally weaker than El Niño and there tends to be less temporal and spatial consistency in the associated climate response.

The coastal deserts of Peru, southern California, Baja California, and southwestern Africa all tend to experience some degree of rainfall enhancement during El Niño episodes, although the timing of the anomaly differs between these regions. In at least some of these cases, reduced upwelling plays a role. In contrast, rainfall is reduced during El Niño episodes along some semi-arid eastern coasts, particularly in Australia and Northeast Brazil. An influence is further apparent in numerous interior dryland locations in China, the Great Plains, East Africa, southern Africa, India, Australia, and parts of South America. The influence on dryland precipitation is discussed in greater detail in Section 25.7.

25.4.3 LAND–ATMOSPHERE FEEDBACK AS A FACTOR IN CLIMATE VARIABILITY

Although weaker than the ocean's influence, feedback between the land and atmosphere appears to be a significant factor in climatic variability. The boundary forcing from the surface is manifested via the exchange of energy and mass, including particulates, water vapor, and other gaseous molecules. The latter include trace gases that are active in heating the atmosphere. As weather patterns change, so do the characteristics of the land surface, modifying the fluxes of both energy and mass.

A good example is a drought in a semi-arid region. When rainfall is reduced, so is soil moisture and vegetation cover. This results in decreased flux of latent heat and water vapor to the atmosphere, but it generally increases surface temperature, thus increasing the flux of sensible heat to the atmosphere. The reduced vegetation cover also reduces the carbon dioxide exchange and makes the surface prone to erosion, generating dust that may reside in the atmosphere for long periods of time and influence patterns of heating and cooling.

The land surface changes induced by drought in arid, semi-arid, and subhumid regions appear to evoke a positive feedback that reinforces the droughts, prolonging and intensifying them. Both numerical simulations and observations tend to support this view (Nicholson 2000a). Positive land–atmosphere feedback is likely a factor in the drought that has plagued Sahelian West Africa for over two decades. The degree of impact can be partially understood by examining the extent to which water vapor is recycled over various continents (Brubaker et al. 1993). The water vapor that goes to precipitation originates to a large extent over the land itself, rather than the ocean. Areas with strong recycling tend to be those where soil moisture and precipitation are closely coupled (Koster et al. 2004). Over Africa this degree of recycling is as high as 48% in August. However, dust generation during droughts might also have considerable impact (see Chapter 13).

There is substantial literature on the effects of land–atmosphere feedback, but most of it deals with theoretical models that are not sufficiently reliable to describe realistic impacts. The models also tend to exaggerate the changes that actually take place in the land surface. Clearly, surface properties such as soil moisture and snow cover have a significant impact on temperature fields in the lower atmosphere, and possibly regional precipitation (e.g., Pielke 2001; Notaro et al. 2006). A considerable body of evidence from general circulation models also suggests that large-scale changes in land surface characteristics can actually influence at least regional scale climate (e.g., Weaver and Avissar 2001; Xue et al. 2006). This conclusion, still controversial, has implications for human-induced changes of the land surface.

An interesting issue related to land–atmosphere feedback is the development of the Sahara Desert. Most of the region was a comparatively green savanna landscape in the early to mid-Holocene (Holmes 2008). Some numerical modeling experiments (e.g., Claussen et al. 1999; Liu et al. 2007) suggest that the change to an arid landscape around 5500 BP occurred very abruptly, citing land–atmosphere feedback as a controlling factor. This idea remains highly controversial, as field studies in the eastern Sahara show a much more gradual change in the vegetation cover (Kröpelin et al. 2008). However, land surface feedbacks still probably played a role in the development of the Sahara (Otto et al. 2009).

25.4.4 THE INFLUENCE OF MANKIND

Human beings have dramatically altered the global land surface and the composition of the atmosphere through such activities as deforestation, agriculture, and burning of fossil fuels. Two major areas of concern are the effects of global warming, due to increased levels of greenhouse gases in the atmosphere, and the direct effects of surface modification itself. Within the last century, the concentrations of the major greenhouse gases have dramatically increased at the hands of humans. Most climatologists agree that this will eventually lead to global warming and perhaps already has.

The general impact of global warming in dryland regions can be guessed at both from various modeling studies and by analogy with natural warm episodes of the past. Unfortunately, there is little consensus among the models as to what changes might occur (see Section 25.8). The climatic optimum of c. 5000 BP, perhaps the best analogy for greenhouse warming, saw a general increase in precipitation in most of the earth's drylands, but a decrease in some of the mid-latitude semi-arid regions, such as the US Great Plains. In contrast, the last glacial saw drier than present conditions in the tropics and wetter than present conditions in the mid-latitude deserts (Quade and Broecker 2009).

As for the direct effects of surface modification, that of greatest importance in dryland regions is desertification. As described in Chapter 22, there is observational evidence that desertified land can alter surface characteristics, notably temperature and perhaps albedo. Charney (1975) actually suggested that desertification via overgrazing caused drought in the Sahel. His mechanism was based on assumed changes in the surface; changes that did not actually occur (Nicholson et al. 1998).

A model using a more realistic representation of surface changes concluded that the impact is insufficient to produce drought in the Sahel (Taylor *et al.* 2002).

Another change of interest in dryland regions is irrigation. This both decreases surface temperature and increases the flux of water vapor and latent heat to the atmosphere. The discontinuities of these properties and fluxes when irrigated land stands adjacent to non-irrigated land may be even more important than the properties and fluxes themselves in modifying patterns of weather and perhaps climate (Moore and Rojstaczer 2001). A study of the Texas Panhandle found that during the months with the most irrigation, rainfall was up to 91% greater over irrigated locations than over non-irrigated control locations (Barnston and Schickedanz 1984). Irrigation also appears to enhance severe weather in this region (see review in Nicholson 1988).

25.5 GLOBAL CHRONOLOGY OF CLIMATE

25.5.1 THE LATE PLEISTOCENE AND HOLOCENE

The most extreme changes in global climate that have occurred are the Ice Ages and the interglacials. Not long ago, it was taught that there were four major ice advances (glacials) and that at the same time "pluvial" (wet) periods occurred in the tropics. Former strand lines of huge lakes provided witness to these pluvials. The development of advanced dating methods and isotopic techniques for studying past climate changed this picture in the 1970s. The new picture was one with manifold cycles of glacials and interglacials occurring on quasi-regular time scales. Their occurrence and timing is now known to be forced by changes in solar insolation associated with the Milankovitch cycles of earth–sun geometry (see Section 25.4.1). Evidence further suggested that many of the mega-lakes occurred after the Ice Age ended.

The periods that have been studied in greatest detail are the last major advance of ice (the late Pleistocene), and the post-glacial Holocene period that covers approximately the last 10,000–12,000 years (Fig. 25.21). The maximum development of glacial conditions is dated as roughly 18,000 BP. Within the Holocene a period of optimum temperatures occurred around 5000 BP, although warming in the Southern Hemisphere may have peaked around 9000 BP. The switch between the Pleistocene and Holocene occurred very quickly; in Greenland, temperature increased by as much as 10°C within a few years (Alley 2000).

25.5.2 THE LAST TWO MILLENNIA

The established scenarios of global climate fluctuations on the historical time scale stemmed largely from the compilations of Lamb (1977) and from pollen work in the Alps. With a pool of information largely restricted to northern Europe, Greenland, Iceland, and the North Atlantic, Lamb suggested that two major climatic episodes had occurred within the past two millennia: a Medieval Warm Period (MWP) followed by the Little Ice Age

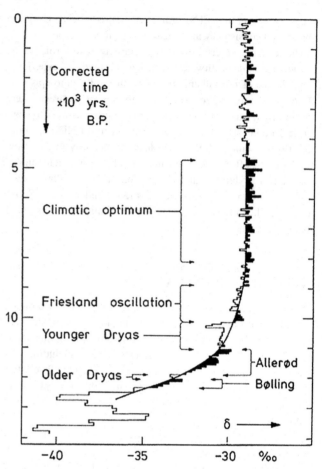

Fig. 25.21 Record of δO_{18} in an ice core from Camp Century, Greenland (from Dansgaard *et al.* 1969). Positive departures (shaded) are indicative of relatively warm conditions.

(LIA). Documentary evidence supporting the existence of the MWP included open water in the Canadian archipelago, whale hunting in now ice-covered areas, a northward extension of the Canadian forest, ice-free conditions in the North Atlantic, vineyards in England, and burials by the Norse colonists in parts of Greenland where there is now permafrost.

Published dates for these periods are quite diverse, but most generally encompass the period 900–1200 for the MWP and 1550–1900 for the LIA (Jones *et al.* 2001). However, many have suggested that the LIA spanned mainly the sixteenth through eighteenth centuries, with a brief but intense recurrence in the mid-nineteenth century.

Many recent works have examined questions concerning the temporal continuity of these periods and their global nature, but these questions are far from resolved (Bradley and Jones 1995; Hughes and Diaz 1994). Notably, both the LIA and the MWP are apparent in ice core records of the northern and southern polar regions. Bradley and Jones concluded, however, that there was no uniform cold period in the LIA, but rather numerous cold intervals that were not coincident in all regions affected, nor were all areas of the globe affected. Similarly, it can be concluded that a relatively warm period occurred in the global average around the tenth through thirteenth centuries, but it was not evidenced everywhere and there was considerable variability within the period and large diversity as to its timing.

25.5.3 GLOBAL TRENDS IN TEMPERATURE AND PRECIPITATION DURING HISTORICAL TIMES

The period during which human influence on global temperature is said to be greatest is roughly the last 150 years, commencing with the agricultural revolution in the late nineteenth century. Hemispheric and global temperature averages since 1855 all show a strong and nearly continual rise since around 1905. The rise has been particularly steep since around 1970. Similar trends are apparent in all seasons, although the increase has been somewhat larger in winter (Jones *et al.* 1999). The year-to-year variability of temperature has been greatest by far during the winter season of the Northern Hemisphere, where the large expanses of land stretching between the tropics and the polar latitudes create large fluctuations in temperature.

Bradley *et al.* (1987) and Diaz *et al.* (1989) have described the precipitation fluctuations that have occurred over the global land areas since the late 1800s. Several interesting features are apparent, although the changes are not as pronounced as the temperature changes. Overall precipitation has increased in both hemispheres since around 1900. In the Southern Hemisphere the trend continued into at least the 1980s, but in the Northern Hemisphere it peaked some time in the 1950s. The increase has been markedly greater in the Southern Hemisphere. There has been a similar rising trend in the tropics (25° N to 25° S), but only until around 1975, when droughts became common in many areas.

25.6 PALEOCLIMATES AND HISTORICAL FLUCTUATIONS IN DRYLAND REGIONS

In dryland regions, evidence of climatic change on geological time scales is plentiful because of the vast changes in surface hydrology and vegetation. The most important indicators in the drylands are lakes. However, pollen, dust, and aeolian landforms are also important indicators.

The picture is quite different for the historical period. The indicators that are useful elsewhere (e.g., tree rings, ice cores, pollen, and corals) are scarce in most dryland regions. Some of the notable exceptions are tree rings in the southwestern USA, the West Coast, and the Great Plains; glacial records in Peru and the Himalayas; tree rings in certain areas of South America; lake cores in East Africa; tree-ring records from extra-tropical northwestern Africa; and documentary reports from many areas.

25.6.1 THE LATE PLEISTOCENE AND HOLOCENE

The general picture of the late Pleistocene and Holocene was presented in Section 25.4.1. During the Last Glacial Maximum

(LGM), which peaked around 18,000 BP, aridity generally prevailed throughout the low latitudes. The tropical forests had retreated into a few small pockets of highland refuges, and deserts such as the Sahara expanded greatly. At the abrupt end of the Pleistocene around 10,000 BP, a pluvial period prevailed in the tropics, where lakes deepened and expanded. During the climatic optimum in the mid-Holocene a second phase of lakes was evident in many tropical regions, particularly in the Northern Hemisphere. In contrast, the world's deserts were reduced to a few hyper-arid cores, such as in the Western Desert of Egypt, or possibly some areas of the Namib Desert (Deacon and Lancaster 1988). In this chapter the regional manifestations of the global changes are described. Emphasis is on Africa, where some of the most extensive studies of past dryland climates have been carried out. Summaries are also presented for South America, Australia, parts of Asia, and the American West.

These fluctuations between more humid and more arid climates were accompanied by major ecological changes in the drylands. Over Africa, the Sahara pushed equatorward to as far as 10° N (Nicholson and Flohn 1980). The westerlies, displaced southward, kept the northern fringe of the desert moderately wet. The tropical rainforest all but disappeared, with cores remaining in mountain refuges and a few other niches. In the Holocene the forest again expanded and the savanna pushed into the deserts. Where Saharan dunes had before existed, vast lakes formed and tropical fauna roamed in the vast savanna. The Sahara existed in a hyper-arid core, primarily in the Western Desert of Egypt and Libya. Over North America the boreal forest and the deciduous woodlands were displaced into the southern USA during the late Pleistocene. During the Holocene, grasslands and prairie developed in the central Great Plains, while the woodland and boreal forest pushed northward into Canada.

AFRICA

In northern and equatorial Africa the general pattern of late Pleistocene aridity and abrupt change to more humid conditions early in the Holocene is almost ubiquitous (Garcin *et al.* 2007). The maximum of the early Holocene pluvial period occurred about 9000 BP. At this time (Fig. 25.10) archaeological sites and now-dry lakes existed throughout subtropical North Africa. The humid conditions extended across the Sahara and into the current desert areas of Arabia and the Middle East (Arz *et al.* 2003). A second pluvial period occurred in the mid-Holocene.

Figure 25.22b presents a histogram of lake levels over Africa during the last 30,000 years, showing the continuation of low stands throughout the period 21,000 BP to 13,000 BP. High stands commencing after 10,000 BP are evident throughout virtually the entire continent (Fig. 25.23). Both of these figures were prepared from data published in the 1970s or earlier. A wealth of more recent studies confirming these trends have been published for numerous other African lakes (see www.cambridge.org/9780521516495).

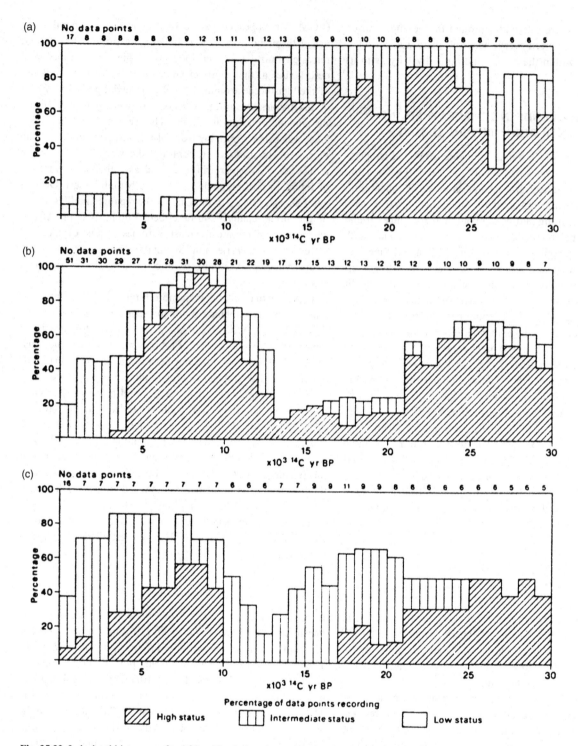

Fig. 25.22 Lake level histograms for Africa, North America and Australia (from Street and Grove 1979).

Relatively humid conditions prevailed until roughly 5000 BP, but this multi-millennium wetter period was interrupted in numerous locations by an arid interval that was strongly apparent in the now-dry lakes of the central Sahel and many other locations (Fig. 25.24). Various dates for this arid interval are given in the literature, but most center around 7000–8000 BP. After this interval, a second pluvial period produced high lake stands in North Africa around 6000 to 5000 BP, but more moderate stands in the equatorial regions. Throughout most of Africa, a trend toward drier conditions commenced some 5000 years ago, but the current state of aridity began about 3500 years ago.

Over southern Africa there are fewer indicators of long-term climate variability and those available, such as dunes

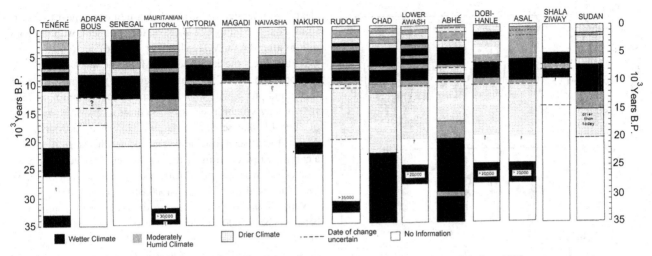

Fig. 25.23 Chronologies indicating the stands of select African lakes over the last 20,000 years (from Nicholson 1976).

and cave deposits, are more complex in their climatic interpretation. Thomas and Shaw (1991) provide a useful synthesis of available information. The trends in climate were quite different from those apparent elsewhere in Africa. For example, most indicators from southern Africa show that the abrupt shift from Pleistocene to Holocene corresponded to a drying up of the lakes, except for those in the southeastern Kalahari. Humid conditions corresponded to the pre-Holocene dry period evident in lakes from equatorial Africa (Fig. 25.23). Such an opposition between southern and equatorial Africa is a frequently occurring pattern in the modern record (Nicholson 1986).

Recently, a vast amount of information has become available on hydrologic changes in the southeastern Kalahari. This has been facilitated by satellite images that allow for excellent delineation of former beach ridges and new dating techniques. Vast paleo-lakes covered now-dry portions of the region (Burrough et al. 2009).

ASIA AND THE MIDDLE EAST

Pollen and lake studies indicate that the Holocene chronology is similar for northwestern India and probably much of the Arabian Peninsula, southern Iraq and Afghanistan, but that opposite trends prevailed in parts of the Near East (Roberts and Wright 1993; Enzel et al. 1999). Lakes Konya and Beysehir in Turkey were relatively high in the late Pleistocene, but lower than their present levels throughout much of the Holocene. In northwestern India, monsoon rains increased around 10,000 BP, probably following a dry late Pleistocene. Maximum stands of Indian lakes were maintained from about 6300 to 5500 BP, when an abrupt drying trend began that led to complete desiccation around 4800 BP. The last 5000 years have been similar to the present.

There are many parallels in the dryland regions of western China (Winkler and Wang 1993). The late Pleistocene, particularly the Last Glacial Maximum around 18,000 BP was

considerably drier than at present. The aridity gave way to conditions similar to present-day by the end of the Pleistocene around 12,000 BP. During the Holocene, climate was relatively unstable, but an arid episode prevailed at some point in the mid-Holocene. This is evidenced by pollen assemblages from lakes in Mongolia and the Tarim Basin (An et al. 2005; Chen et al. 2003).

SOUTH AMERICA

The late Pleistocene record over South America has provided some notable surprises. Two very important features are that the glacial maximum occurred between 22,000 and 19,500 BP and that Pleistocene lakes in tropical South America persisted well into the period of deglaciation (Seltzer et al. 2002). Lakes Junin and Titicaca, in the altiplano of Bolivia and Peru, and a lake in the Salar de Uyuni, further south on the altiplano, persisted until about 15,000 years ago. Precipitation here was out of phase with that over the Atacama Desert, where increased summer precipitation from about 16,000 to 10,500 BP (Latorre et al. 2002) recharged groundwater and produced a grass cover in the now-hyper-arid desert.

During the core of the Holocene, temperatures were higher and lake levels lower. In Peru the maximum warming was probably from 8400 to 5200 BP (Andrus et al. 2002). Lake levels in the altiplano were low at this time; low stands continued until about 3500 years ago but the lakes began to fall about 5000 years ago, marking the change to more arid conditions (Baker et al. 2001). At about the same time, coastal upwelling became more strongly established and El Niño became more frequent (Conroy et al. 2008). Elsewhere in South America, a shift to wetter conditions generally occurred around 4000 years ago (Marchant and Hooghiemstra 2004). Again, precipitation was out of phase over the altiplano and in the coastal desert, as an episode of higher groundwater levels prevailed in the central Atacama from 8000 to 3000 years ago (Betancourt et al. 2000).

clear in southeastern Australia, but conditions appear to depart from those prevailing elsewhere on the continent, with dry conditions after the LGM (Hollands *et al.* 2006).

As in Africa, the early and mid-Holocene was a time of wetter conditions (McCarthy and Head 2001). The maximum of the humid phase, with high lake stands, occurred around 7500 BP, roughly synchronous with the drier interval around 7000 or 8000 BP that separated the two Holocene "pluvials" over Africa. In the semi-arid rangelands of southeastern Australia, a complete restructuring of the vegetation occurred in the early Holocene. Trees and tall shrubs became conspicuous in stabilized dune fields from 9000 to 4000 BP (Cupper 2005).

Many signs of increasing aridity variously commenced between about 5000 and 4000 BP, synchronous with the decline of wet conditions over North Africa. Dunes became active in the deserts of south central Australia (Cupper 2005), lake levels declined, and the deposition of dust from the Lake Eyre region became evident as far away as New Zealand (Marx *et al.* 2009).

NORTH AMERICA

The regions described above all suggest a general pattern of aridity in the late Pleistocene and a rapid build-up of lakes early in the Holocene. A quite different pattern prevailed over North America, possibly because of the looming presence of the ice sheets over the continent. During the Last Glacial Maximum at 18,000 BP the southern edge of the ice extended to about 45° N over the western USA and 40° N over the eastern USA. In the West, large mountain glaciers extended down to about 35° N. Vast lakes such as Lakes Lahontan, Bonneville, Russell, and Jakes Lake formed in the areas south of the ice. As the glaciers melted and retreated in the early Holocene, the lakes rapidly dried up (Shuman *et al.* 2009). By 8000 BP most were dry, but a second period with intermediate lake levels occurred from about BP 5000 to 1000 (Street and Grove 1976).

Lakes in the Sonoran and Chihuahuan deserts of the southwestern USA were also high during the late Pleistocene and fell at the onset of the Holocene (Pigati *et al.* 2009). Increasing aridity also prevailed in northwestern Mexico toward the beginning of the Holocene (Ortega-Rosas *et al.* 2008). After a distinctly drier mid-Holocene, wetter conditions similar to the present day prevailed from about 4000 to 3000 years ago. Many of these same trends were apparent at pluvial Lake Mojave in California's Mojave Desert (Koehler *et al.* 2005).

Early to mid-Holocene aridity also prevailed in the Great Plains. The transition from prairie to forest was further east and north during the Holocene until roughly 6000 years ago (Williams *et al.* 2009). Sustained aeolian activity occurred from about 9600 to 6500 years ago and loess deposition commenced some 10,000 years ago (Miao *et al.* 2007).

25.6.2 HISTORICAL FLUCTUATIONS OF CLIMATE

Here "historical fluctuations" are loosely understood to be those prior to the availability of instrumental records, but within the

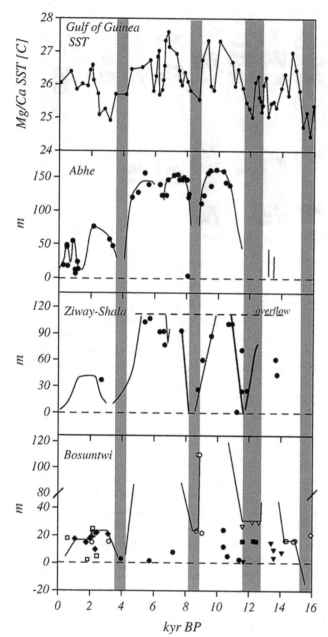

Fig. 25.24 African chronologies for equatorial lakes (Adhe and Ziway-Shala, Ethiopia, and Bosumtwi, Ghana) compared with sea-surface temperatures in the Gulf of Guinea: dots represent radiocarbon dates; the top curve is sea-surface temperature determined from magnesium/calcium ratios; and shading indicates periods of extensive drought (from Shanahan *et al.* 2006).

AUSTRALIA

Over most of Australia, but especially the northern and central regions, aridity tended to prevail during glacial periods. As in Africa, the Last Glacial Maximum (LGM) was a period of widespread aridity, particularly intense from *c.* 22,000 to 11,000 BP. The period was marked by deflation of lake sediments, low stands of lakes, and dune building (e.g., Cook 2009; Wyrwoll and Miller 2001; Fitzsimmons *et al.* 2007). The picture is less

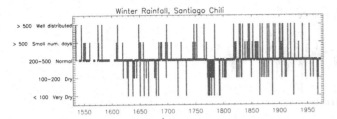

Fig. 25.25 Semi-quantitative time series of climatic conditions near Santiago, Chile, since the late sixteenth century, based on numerous documentary sources. A change to wetter conditions in the nineteenth century is clearly evident.

time in which written historical records exist. Thus, the time period covered varies from region to region. In West Africa, it extends roughly back in time from about 1900, but over Europe the last two or three centuries are well covered by instrumental records. In this section, records are summarized for the drylands of South America, Africa, Asia, and North America. Australia is not included because little information is available covering the historical period in its dryland regions.

SOUTH AMERICA

The regions with the greatest amount of historical material include the west coast (Peru and Ecuador), Chile, Argentina, and Brazil. The best-documented fluctuations are those related to El Niño, as described in Section 25.7. The arid coastal strip of the Atacama Desert, extending through Peru, is generally inundated during El Niño events, but the adjacent highlands generally experience drought. For the more southern areas, Chile and Argentina, the available data are primarily from tree-ring records. In Brazil, historical reports of drought provide the most important information.

Historical references to rainfall and floods at Piura (Pejml 1966), along the coast, and at Trujillo, in northern Peru (Garcia-Herrera *et al.* 2008), compare with historical El Niño reconstructions (e.g., Quinn and Neal 1992).

Based on general descriptions, ships' logs, meteorological observations, and cultural indicators, Pejml (1966) and others concluded that the desert west coast of South America and the Galapagos Islands experienced relatively dry conditions from about 600 to 1000 AD and about 1400–1650, with wetter intervals from 1000 to 1400 and from 1650 to 1850. The dry conditions intensified around 1500.

The ice and archaeological records compare reasonably well with these trends, indicating the anticipated out-of-phase relationship between precipitation along the coast and over the highlands. Wetter periods are described in the highlands from about 760–1040 and 1500–1720, and again after 1870. Drier periods occurred from about 540 to 730 (but with a brief wetter episode about 610–650), 1250–1310 and from roughly 1720 to 1860.

The historical fluctuations of climate in Chile have been reconstructed primarily from tree-ring chronologies (e.g., Boninsegna 1995; Le Quesne *et al.* 2009) and documentary

evidence (Vicuña-Mackenna 1877; Taulis 1934; Pejml 1966; Ortlieb 1994). An analysis of the documentary sources is shown in Fig. 25.25. It is based primarily on information in Taulis and Pejml and is assumed to be a proxy for precipitation in Santiago, a region of winter rains. The documentary records are in substantial agreement with the tree-ring records from the Santiago region (Boninsegna 1995). One obvious trend is relatively good rainfall from the mid-sixteenth century to the mid-eighteenth century, i.e., the core of the Little Ice Age. A prominent exception is intense drought in the 1630s. In contrast, there were continuous droughts in the 1770s and early 1780s, and numerous droughts or dry years until about 1804. A prolonged wet period occurred early in the nineteenth century, confirmed also by glacial evidence.

In Northeast Brazil the record of droughts only extends back about 200 years (Caviedes 1973). Notable droughts occurred around 1825 and in 1844 and 1845. These droughts were followed by a period of some 30 years of good rainfall and abundant harvests. However, in 1877 another drought occurred during what was one of the most intense El Niños on record, and another drought occurred in 1888 and 1889.

Tree-ring chronologies from southern South America provide information covering well over a millennium. In the central Patagonian Andes (Boninsegna 1992), a chronology from Rio Alerce (41° S, 72° W) indicates cold/moist conditions from about 900 to 1070, followed by a warm/dry period from 1080 to 1250, then cold/moist from 1280 to 1670. Tree-ring records have been used to reconstruct flow in the Nequen River and the Limary River, a drainage area from roughly 36° S to 44° S that lies in the Argentinean rain shadow east of the Andes. Similar trends are apparent in both rivers, with important low flow periods between 1630 and 1750 and a gradual rise to relatively high flow persisting from about 1760 to 1820. Relatively dry conditions occurred around the 1820s and 1830s, but the rest of the century was relatively wet. An abrupt downward trend occurred around 1890 that was particularly felt in the Neuquen River area; the dry period lasted from 1890 to about 1925. A tree-ring reconstruction of precipitation at Santiago (Fig. 25.26), west of the Andes, also shows relatively low rainfall during the seventeenth and eighteenth centuries and much wetter conditions commencing in the first few decades of the nineteenth century.

AFRICA

For Africa, proxy information covering the historical period is too plentiful to describe in great detail. The most useful indicators of climatic fluctuations on the historical time scale are lakes and documentary evidence. In some cases, high-resolution proxy records can be derived for several centuries. Adequate information from documentary sources and lakes is plentiful enough to provide a very detailed picture of the nineteenth century (Nicholson 2001b), for which few instrumental records are available except in the temperate extremes of the continent. The detailed bibliography of sources of historical climate information for Africa can be found at www.cambridge.org/9780521516495.

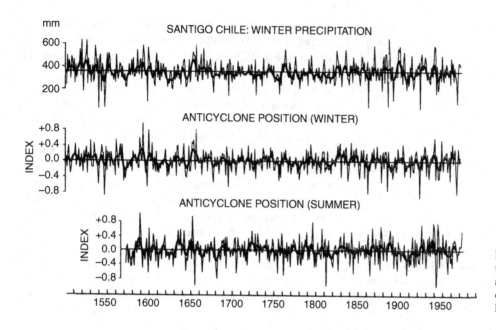

Fig. 25.26 Tree-ring estimates of winter precipitation at Santiago, Chile and the summer and winter positions of the subtropical highs over the Pacific (from Boninsegna 1995).

Tree-ring analyses are of limited use in Africa because few African tree species have annual rings. The few available tree-ring chronologies are from Morocco, Algeria, Tunisia, South Africa, Namibia, and Zimbabwe. These are complemented by a handful of records from ice cores and cave deposits, which are useful for both temperature and precipitation reconstruction.

One of the most important historical records is that of the Nile. Records of the summer minimum, a general indicator of equatorial rainfall, and the level of equatorial Lake Victoria (the origin of the Nile) are available nearly continuously from 622 AD (Toussoun 1925). Those of the flood stage, largely an indicator of summer precipitation over the Ethiopian highlands, are also available with less continuity; long gaps occur in the sixteenth and seventeenth centuries.

In addition to the Nile record, historical information has been used to produce long-term proxy records of precipitation and drought in many locations. Some of the longest available are those for Algeria, Tunisia, the Sahel (especially Nigeria, Chad, Mali, Mauritania, and Senegal), Angola, Mozambique, and South Africa (Nicholson 1996b, 2001b). For many cases, relevant and relatively well-dated material is available back through the sixteenth century. Figure 25.14 shows such a chronology for Angola.

Cores from a large number of lakes, especially in the equatorial region, provide very detailed information for the last few centuries. In some cases, varves are present in the cores, giving an annual resolution. The list of well-studied African lakes is long: Naivasha, Malawi, Tanganyika, Edward, Victoria, Baringo, Chad, Turkana, Kibengo, Kitagata, Bosumtwi, Abiyata, Chilwa, Rukwa, Stefanie, Oloidien, and Sonachi in the Lake Naivasha basin, plus Chibwera and Kanyamukali. Some of these are depicted in Fig. 25.13. Clearly, the lakes have exhibited major high and low stands during the last two thousand

years, but few generalizations can be made concerning their trends. This is not surprising in view of the complexity of equatorial climates over Africa, especially in East Africa (Nicholson 1996a) and the occurrence of several spatial modes of variability (see Chapter 16). This complexity makes it difficult to synthesize the records of the various lakes.

Extensive detail is available for the nineteenth century. In addition to the lake records, documentary evidence is plentiful. The combination of gauge data, documentary evidence, and lake level information suffices to produce time series for most of the continent with annual resolution since the early nineteenth century (Nicholson 2001b). The time series for several large sectors of the continent are shown in Fig. 25.12. Notable climatic episodes of that century include a period of nearly continent-wide aridity in the 1820s and 1830s (evident also in the lake records, Fig. 25.13), a period of wetter conditions throughout much of the continent late in the century, and a rapid trend toward increasing aridity commencing in the last few years of that century (Nicholson 1996b).

NORTH AMERICA

Climatic reconstructions for North America during historical times are based primarily on tree rings. Sufficient analysis has been carried out that spatial maps covering large portions of the continent have been reconstructed for several centuries (Fritts and Shao 1995). The tree-ring records have been utilized to produce multi-century time series of temperature, precipitation, and drought (e.g., Haston and Michaelsen 1997; Díaz et al. 2001; Woodhouse and Overpeck 1998; Meko et al. 2007), particularly for the dryland regions of the central and western USA. Most records commence in the sixteenth or seventeenth century, but the reconstruction of Meko et al. (2007) for the Colorado River Basin extends back to 800 AD and that of

Woodhouse and Overpeck (1998) for the central USA extends back 2000 years.

Fritts and Shao (1995) utilized tree rings to derive precipitation chronologies for five dryland regions: California valleys, the intermontane basins, the southwestern deserts, and the northern and southern high plains of the central USA (Fritts and Shao 1995). The trends are very similar in all regions. The similarity is particularly strong between the two southern regions (southwest and southern high plains), despite significant climatic contrasts, and between the two more northern regions (the intermontane basins and the northern high plains). Notable in all regions is a relatively wet period early in the seventeenth century and in the 1820s and 1830s. Exceedingly wet conditions prevailed in the southern regions at the beginning of the eighteenth century.

Tree rings also indicate that droughts were particularly widespread in the mid-eighteenth century, with a respite around 1750 (Stockton and Meko 1975). More prolonged but less widespread drought conditions occurred late in that century, early in the 1820s, and in the late nineteenth century. Briefer but pervasive drought episodes occurred in the 1840s and 1860s.

ASIA AND THE MIDDLE EAST

In comparison with Africa and the Americas, relatively little historical climate information is available for the vast Asian continent. A handful of tree-ring chronologies have been obtained, primarily for parts of China, Mongolia, Tibet, India, and Pakistan (e.g., Davi et al. 2006; Treydte et al. 2006). Drought chronologies extend back to the early sixteenth century (Mooley and Pant 1981; Zhang and Crowley 1989). Ice core records also cover several centuries (Thompson et al. 2000; Kaspari et al. 2007). Sediment cores (Anderson et al. 2002) provide an indication of monsoon strength, in terms of winds.

Tree-ring records from northern Pakistan suggest that precipitation was relatively low from the beginning of the last millennium through the early nineteenth century, with the twentieth century having been relatively wet (Treydte et al. 2006). Ice cores from the Himalayas (Thompson et al. 2000) similarly indicate that the twentieth century was relatively wet. An ice core from Mt. Everest indicates reduced penetration of the monsoon since ~1400 (Kaspari et al. 2007), consistent with the tree-ring evidence of centuries of drier conditions.

A few tree-ring chronologies have been derived for the Middle East (Waisel and Liphschitz 1968; Tarawneh and Hadadin 2009). More useful in the Middle East and western Asia are long-term records of the Caspian and Dead seas. The Dead Sea was particularly high from the late twelfth century to the early fourteenth century and again through the sixteenth and early seventeenth centuries (Klein and Flohn 1987). It has been very low since that time. References to the level of the Caspian Sea go back to the sixth century BC, but with little precision or temporal resolution. Its levels were particularly high from roughly 1730 to 1820 and very low until the early twentieth century, when it rose several meters (Issar et al. 1989).

25.7 IMPACT OF ENSO ON THE DRYLANDS

ENSO is one of the major influences on precipitation variability in the global drylands. Numerous regional-scale studies indicate that drylands on all continents are affected. However, the timing and consistency of the ENSO signal is quite diverse because the relationship is indirect in many locations, especially the drylands, and it is tied closely to the annual cycle of rainfall.

Examples of such indirect relationships are numerous. In eastern and southern Africa, ENSO generally influences precipitation only when it modifies temperatures in the Atlantic and Indian oceans (Nicholson et al. 2001), which in turn are the direct factors governing rainfall in those parts of Africa. Similarly, the influence on the South American altiplano appears to be limited to those episodes that influence the easterly/westerly wind regime aloft (Garreaud and Aceituno 2001). In southern California, precipitation is generally influenced only when the Pacific warming extends well into the winter at the end of the ENSO year, because southern California has a brief winter rainy season (Schonher and Nicholson 1989).

25.7.1 INFLUENCE OF EL NIÑO

AUSTRALIA AND SOUTH AMERICA

The strongest effects of El Niño on precipitation are probably in the drylands of South America and Australia, the continents lying at the eastern and western ends of the Pacific sector that is the core of El Niño. It produces torrential rainfall in the summer rainfall regions of coastal Peru/southern Ecuador at the northern extremity of the Peru Current (Caviedes 1973). El Niño also enhances winter rainfall in central Chile, but reduces rainfall in northwest South America at this same time (Aceituno 1988; Rutllant and Fuenzalida 1991). There is also a strong tendency for higher than average summer rainfall (November–January) at the end of an El Niño year in the grasslands and savannas of southern Brazil, northern Argentina, Uruguay, and Paraguay (Rogers 1988; Grimm et al. 2000; Boulanger et al. 2005).

In contrast, droughts tend to occur in Northeast Brazil during February–March–April of the El Niño year (Hastenrath and Heller 1977), in the altiplano of Bolivia and southern Peru, and along the dry Caribbean coast of Venezuela. In some cases, the anomalies extend through the central and equatorial Andes, across northern South America into the Atlantic. El Niño tends to reduce rainfall in the savanna grasslands of southern Brazil, at the same time increasing it in the grasslands to the south. This opposition is termed a "see-saw" by some authors (e.g., Nogués-Paegle and Mo 2002) and the node of the pattern lies at roughly 25° S.

El Niño's signal over Australia is much less complex. It produces intense droughts in eastern and northern Australia with striking consistency (Ropelewski and Halpert 1987) and also in the most arid central core of the continent. The season of influence varies by region, but for the continent as a whole its effects are particularly strong in September–October–November of the El Niño year. El Niño events reduce both the number and intensity of rain events (Nicholls and Kariko 1993) and delay the onset of the Australian summer monsoon (Joseph *et al.* 1991).

ASIA AND THE MIDDLE EAST

Less direct are the effects in India and northern China, both of which receive almost all of the year's rainfall during the summer monsoon. Monsoon rainfall in India is generally below normal during ENSO years and many droughts there are linked to ENSO. However, the relationship is inconsistent, being influenced by the timing of the warming in the Pacific and Indian oceans (Ihara *et al.* 2008). In China the relationship is equally complex, but also consistent. As in India, there is a tendency for drought or at least below-average rainfall in the semi-arid north, where mean annual precipitation lies between 200 and 500 mm (Wang and Li 1990).

El Niño also impacts precipitation in the Middle East, particularly the prevailing winter precipitation. It tends to increase winter rainfall over the Arabian Peninsula (Marcella and Eltahir 2008). In Israel, winter rainfall tends to be above normal during El Niño years and below normal during La Niña years (Price *et al.* 1998). The rare snowstorms also tend to occur during El Niño years.

NORTH AMERICA

ENSO influences most of the North American drylands of the USA and Mexico. In general, the El Niño phase tends to enhance rainfall, but the signal is strongly seasonally dependent. For example, in the southern Great Plains, precipitation is generally below normal during late autumn and winter of the El Niño year (Stahle and Cleaveland 1993), but El Niño generally enhances warm-season rainfall in this region (Ropelewski and Halpert 1986). El Niño also tends to produce summer drought in northwest and southwest Mexico.

El Niño tends to produce anomalously high rainfall along the California coast, parts of the American Southwest, and northwest Mexico (Schonher and Nicholson 1989; Swetnam and Betancourt 1990; Minnich *et al.* 2000). The signal, as indicated both by rainfall and river flow, is particularly strong in early winter at the end of the El Niño year, and in southern California, Arizona, and New Mexico. Wet summer conditions tend to occur during El Niño years in the central and much of the northern Great Plains and intermontane Great Basin (Brown *et al.* 2008).

AFRICA

The areas of Africa most strongly influenced by El Niño are equatorial East Africa and coastal sectors of the tropical Atlantic (the Guinea and Angolan coasts), where precipitation is enhanced, and southern Africa, where El Niño is associated with droughts (Nicholson and Kim 1997). Throughout most of East Africa its influence is strongest during the "short rains" of October–November. In the north, where precipitation falls mainly in the summer months of July–September (the Ethiopian highlands, Eritrea, Djibouti, Uganda, and western Kenya), ENSO is also associated with droughts during the main rainy season (Camberlin 1997). In southern Africa it is generally associated with drought late in the El Niño year or early in the year following, but wet conditions typically occur in some sectors early in the episode.

El Niño's influence on the semi-arid Sahel/Soudan zones is still hotly debated. However, Ward (1998), who argued that a signal is apparent in that region, has reconciled the inconsistencies reported in previous studies by showing that El Niño's influence in individual years is often masked by the decadal-scale trends produced by other factors.

25.7.2 INFLUENCE OF LA NIÑA

In general, the effect of La Niña years/cold events on precipitation tends to be the opposite of El Niño's effect. However, this symmetry is far from perfect (Hoerling *et al.* 1997). Further complicating the issue is the fact that there may also be a tendency for a reversal in the anomalies between the first and second halves of the El Niño or La Niña episode.

La Niña is also associated with drought in the Southwest, Great Plains and Great Basin regions of the USA (Cole *et al.* 2002; Brown *et al.* 2008), Baja California (Minnich *et al.* 2000), and much of Mexico; eastern equatorial Africa (Nicholson and Selato 2000); Israel (Price *et al.* 1998); and northern Argentina/southern Paraguay (Rogers 1988). It is associated with wet conditions over the Bolivian altiplano (Garreaud and Aceituno 2001); the northern Caribbean coast of South America and in Northeast Brazil (Rogers 1988); and in southern Africa (Nicholson and Selato 2000).

25.8 PREDICTING CLIMATE CHANGE AND ITS IMPACT

The marked increases in greenhouse gases and the well-documented warming are undoubtedly influencing global patterns of climate. A major issue is the conditions that will result as warming continues. Scores of studies have utilized general circulation models to predict these changes, and numerous scenarios for future temperature and precipitation regimes have followed. Unfortunately, these projections are marked with tremendous uncertainty, especially for the drylands, where the biggest concern is water availability.

Forecasting relies on statistical/empirical models, numerical/dynamical models, and hybrid approaches (Anderson *et al.* 1999). The numerical models are complex representations of the global atmosphere and oceans, such as general circulation models (GCMs), and include explicit physical-dynamical processes. Though much more sophisticated and process-oriented, the numerical models do not outperform the simpler statistical approaches (Anderson *et al.* 1999; Barnston *et al.* 1999). The hybrid approach involves using numerical models to predict changes in global sea-surface temperatures and statistical models to link these to climatic change at specific locations. This approach generally improves forecast skill.

Currently GCMs do not produce reliable forecasts of future climates. This is well illustrated using simulations of climate change in the West African Sahel. Druyan (2010) reviews the various simulations of twenty-first century rainfall in the Sahel and finds that they disagree even with respect to the sign of the projected trend. The reason is that the current generation of GCMs does a poor job in simulating West African climate (Lau *et al.* 2006). Even global patterns of precipitation are poorly reproduced (Dai 2006). The error is compounded when attempts are made at projecting changes on a regional basis, such as through downscaling of model results. In other words, at the moment, no reliable projections of climatic change in the drylands exist. Prudent management is perhaps the only reliable strategy for facing the future of the drylands.

REFERENCES

Aceituno, P., 1988: On the functioning of the Southern Oscillation in the South-American sector. 1. Surface climate. *Monthly Weather Review*, **116**, 505–524.

Aguillar, E., and Coauthors, 2005: Changes in precipitation and temperature extremes in central and northern South America, 1961–2003. *Journal of Geophysical Research–Atmospheres*, **110**, D23107.

Aizen, V. B., and E. M. Aizen, 1997: Climatic and hydrologic changes in the Tien Shan, Central Asia. *Journal of Climate*, **10**, 1393–1404.

Alley, R. B., 2000: The Younger Dryas cold interval as viewed from central Greenland. *Quaternary Science Reviews*, **19**, 213–226.

An, C. B., L. Y. Tang, L. Barton, and F. H. Chen, 2005: Climate change and cultural response around 4000 cal yr BP in the western part of Chinese loess plateau. *Quaternary Research*, **63**, 347–352.

Anderson, D. M., J. T. Overpeck, and A. K. Gupta, 2002: Increase in the Asian southwest monsoon during the past four centuries. *Science*, **297**, 596–599.

Anderson, J., H. van den Dool, A. Barnston, W. Chen, W. Stern, and J. Ploshay, 1999: Present-day capabilities of numerical and statistical models for atmospheric extratropical seasonal simulation and prediction. *Bulletin of the American Meteorological Society*, **80**, 1349–1361.

Andrus, C. F. T., D. E. Crowe, D. H. Sandweiss, E. J. Reitz, and C. S. Romanek, 2002: Otolith δ18O record of mid-Holocene sea surface temperatures in Peru. *Science*, **295**, 1508–1511.

Arz, H. W., F. Lamy, J. Pätzold, P. J. Müller, and M. Prins, 2003: Mediterranean moisture source for an early-Holocene humid period in the northern Red Sea. *Science*, **300**, 118–121.

Austin, M. P., and O. B. Williams, 1988: Influence of climate and community composition on the population demography of pasture species in semi-arid Australia. *Vegetatio*, **77**, 43–49.

Baker, P. A., and Coauthors, 2001: The history of South American tropical precipitation for the past 25,000 years. *Science*, **291**, 640–645.

Barnston, A. G., and P. T. Schickedanz, 1984: The effect of irrigation on warm season precipitation in the southern Great Plains. *Journal of Climate and Applied Meteorology*, **23**, 865–888.

Barnston, A. G., M. H. Glantz, and Y. He, 1999: Predictive skill of statistical and dynamical climate models in SST forecasts during the 1997–98 El Niño episode and the 1998 La Niña onset. *Bulletin of the American Meteorological Society*, **80**, 217–243.

Berger, A., 1977: Support for the astronomical theory of climatic change. *Nature*, **269**, 44–45.

Betancourt, J. L., C. Latorre, J. A. Rech, J. Quade, and K. A. Rylander, 2000: A 22,000-year record of monsoonal precipitation from northern Chile's Atacama Desert. *Science*, **289**, 1542–1546.

Bhalotra, Y. P. R., 1984: *Climate of Botswana. Part I. Climatic Controls*. Department of Meteorological Services, Gaborone, 68 pp.

Bond, G., and Coauthors, 2001: Persistent solar influence on North Atlantic climate during the Holocene. *Science*, **294**, 2130–2136.

Boninsegna, J. A., 1995: South American dendroclimatological records. In *Climate since A.D. 1500* (R. S. Bradley and P. D. Jones, eds.), Routledge, London, pp. 446–462.

Boulanger, J. P., J. Leloup, O. Penalba, M. Rusticucci, F. Lafon, and W. Vargas, 2005: Observed precipitation in the Parana-Plata hydrological basin: long-term trends, extreme conditions and ENSO teleconnections. *Climate Dynamics*, **24**, 393–413.

Bradley, R. S., and P. D. Jones, 1995: *Climate since A. D. 1500*. Routledge, London, 706 pp.

Bradley, R. S.., H. F. Diaz, J. K. Eischeid, P. D. Jones, P. M. Kelly, and C. M. Goodess, 1987: Precipitation fluctuations over Northern-Hemisphere land areas since the mid-19th century. *Science*, **237**, 171–175.

Brown, P. M., E. K. Heyerdahl, S. G. Kitchen, and M. H. Weber, 2008: Climate effects on historical fires (1630–1900) in Utah. *International Journal of Wildland Fire*, **17**, 28–39.

Brubaker, K. L., D. Entekhabi, and P. S. Eagleson, 1993: Estimation of continental precipitation recycling. *Journal of Climate*, **6**, 1077–1089.

Burrough, S. L., D. S. G. Thomas, and R. M. Bailey, 2009: Mega-Lake in the Kalahari: a late Pleistocene record of the palaeo-lake Makgadikgadi system. *Quaternary Science Reviews*, **28**, 1392–1411.

Camberlin, P., 1997: Rainfall anomalies in the source region of the Nile and their connection with the Indian summer monsoon. *Journal of Climate*, **10**, 1380–1392.

Caviedes, C., 1973: Secas and El Niño: two simultaneous climatological hazards in South America. *Association of American Geographers Proceedings*, **5**, 44–49.

Charney, J. G. 1975. The dynamics of deserts and droughts. *Quarterly Journal of the Royal Meteorological Society*. **101**, 193–202.

Chen, T.-C., J.-M. Chen, and C. K. Wikle, 1996: Interdecadal variation in U.S. Pacific coast precipitation over the past four decades. *Bulletin of the American Meteorological Society*, **77**, 1197–1205.

Chen, C.-T. A., H.-C. Lan, J.-Y. Lou, and Y.-C. Chen, 2003: The dry Holocene Megathermal in Inner Mongolia. *Palaeogeography, Palaeoclimatology, Palaeoecology*, **193**, 181–200.

Claussen, M., C. Kubatzki, V. Brovkin, A. Ganopolski, P. Hoelzmann, and H. J. Pachur, 1999: Simulation of an abrupt change in Saharan vegetation in the mid-Holocene. *Geophysical Research Letters*, **26**, 2037–2040.

Cole, J. E., J. T. Overpeck, and E. R. Cook, 2002: Multiyear La Nina events and persistent drought in the contiguous United States. *Geophysical Research Letters*, **29**, No. 1647.

Conroy, J. L., and Coauthors, 2008: Holocene changes in eastern tropical Pacific climate inferred from a Galapagos lake sediment record. *Quaternary Science Reviews*, **27**, 1166–1180.

Cook, E. J., 2009: A record of late Quaternary environments at lunette-lakes Bolac and Turangmoroke, Western Victoria, Australia, based on pollen and a range of non-pollen palynomorphs. *Review of Palaeobotany and Palynology*, **153**, 185–224.

Crowley, T. J., 2000: Causes of climate change over the past 1000 years. *Science*, **289**, 270–277.

Cupper, M. L., 2005: Last glacial to Holocene evolution of semi-arid rangelands in southeastern Australia. *The Holocene*, **15**, 541–553.

Dai, A., 2006: Precipitation characteristics in eighteen coupled climate models. *Journal of Climate*, **19**, 4605–4630.

Dansgaard, W., Johnsen, S. J., Moller, J., and Langway, C. C. 1969: One thousand centuries of climatic record from Camp Century on Greenland ice sheet. *Science*, **166**, 377–381.

Davi, N. K., G. C. Jacoby, A. E. Curtis, and N. Baatarbileg, 2006: Extension of drought records for central Asia using tree rings: west-central Mongolia. *Journal of Climate*, **19**, 288–299.

Deacon, J., and N. Lancaster 1988: *Late Quaternary Palaeoenvironments of Southern Africa*. Clarendon Press, Oxford, 225 pp.

deMenocal, P., and Coauthors, 2000: Abrupt onset and termination of the African Humid Period: rapid climate responses to gradual insolation forcing. *Quaternary Science Reviews*, **19**, 347–361.

Diaz, H. F., R. S. Bradley, and J. K. Eischeid, 1989: Precipitation fluctuations over global land areas since the late 1800s. *Journal of Geophysical Research–Atmospheres*, **94**, 1195–1210.

Díaz, S. C., R. Touchan, and T. W. Swetnam, 2001: A tree-ring reconstruction of past precipitation for Baja California Sur, Mexico. *International Journal of Climatology*, **21**, 1007–1019.

Druyan, L. M., 2010: Review studies of 21st century precipitation trends over West Africa. *International Journal of Climatology*, doi:10.1002/joc.2180.

Enzel, Y., and Coauthors, 1999: High-resolution Holocene environmental changes in the Thar Desert, northwestern India. *Science*, **284**, 125–128.

Fitzsimmons, K. E., J. M. Bowler, E. J. Rhodes, and J. M. Magee, 2007: Relationships between desert dunes during the late Quaternary in the Lake Frome region, Strzelecki Desert, Australia. *Journal of Quaternary Science*, **22**, 549–558.

Fritts, H. C., and X. M. Shao, 1995: Mapping climate using tree-rings from western North America. In *Climate since A. D. 1500* (R. S. Bradley and P. D. Jones, eds.), Routledge, London, pp. 269–295.

Garcia-Herrera, R., and Coauthors, 2008: A chronology of El Niño events from primary documentary sources in northern Peru. *Journal of Climate*, **21**, 1948–1962.

Garcin, Y., A. Vincens, D. Williamson, G. Buchet, and J. Guiot, 2007: Abrupt resumption of the African monsoon at the Younger Dryas-Holocene transition. *Quaternary Science Reviews*, **26**, 690–704.

Garreaud, R. D., and P. Aceituno, 2001: Interannual rainfall variability over the South American altiplano. *Journal of Climate*, **14**, 2779–2789.

Goudie, A. S., 1996: Climate: past and present. In *The Physical Geography of Africa* (W. M. Adams, A. S. Goudie, and A. R. Orme, eds.). Oxford University Press, Oxford, pp. 34–59.

Grimm, A. M., B. Varros, and M. Doyle, 2000: Climate variability in southern South America associated with El Niño and La Niña events. *Journal of Climate*, **13**, 35–58.

Hare, F. K., 1979: Climatic variation and variability: empirical evidence from meteorological and other sources. In *Proceedings of the World Climate Conference*. Publication 537, WMO, Geneva, pp. 51–87.

Hastenrath, S., and L. Heller, 1977: Dynamics of climatic hazards in Northeast Brazil. *Quarterly Journal of the Royal Meteorological Society*, **103**, 77–92.

Haston, L. L., and J. Michaelsen, 1997: Spatial and temporal variability of southern California precipitation over the last 400 yr and relationships to atmospheric circulation patterns. *Journal of Climate*, **10**, 1836–1852.

Hereford, R., R. H. Webb, and C. I. Longpré, 2006: Precipitation history and ecosystem response to multidecadal precipitation variability in the Mojave Desert region, 1893–2001. *Journal of Arid Environments*, **67**, 13–34.

Hoerling, M. P., A. Kumar, and M. Zhong, 1997: El Niño, La Niña, and the nonlinearity of their teleconnections. *Journal of Climate*, **10**, 1769–1786.

Hollands, C. B., G. C. Nanson, B. G. Jones, C. S. Bristow, D. M. Price, and T. J. Pietsch, 2006: Aeolian-fluvial interaction: evidence for Late Quaternary channel change and wind-rift linear dune formation in the northwestern Simpson Desert, Australia. *Quaternary Science Reviews*, **25**, 142–162.

Holmes, J. A., 2008: How the Sahara became dry. *Science*, **320**, 752–753.

Holmgren, M., and M. Scheffer, 2001: El Niño as a window of opportunity for the restoration of degraded arid ecosystems. *Ecosystems*, **4**, 151–159.

Holmgren, M., B. C. Lopez, J. R. Gutierrez, and F. A. Squeo, 2006: Herbivory and plant growth rate determine the success of El Niño Southern Oscillation-driven tree establishment in semiarid South America. *Global Change Biology*, **12**, 2263–2271.

Hu, Q., C. M. Woodruff, and S. E. Mudrick, 1998: Interdecadal variations of annual precipitation in the central United States. *Bulletin of the American Meteorological Society*, **79**, 221–229.

Hughes, M. K., and H. F. Diaz, 1994: Was there a 'Medieval Warm Period', and if so, where and when? *Climatic Change*, **26**, 109–142.

Ihara, C., Y. Kushnir, M. A. Cane, and A. Kaplan, 2008: Timing of El Niño-related warming and Indian summer monsoon rainfall. *Journal of Climate*, **21**, 2711–2719.

Issar, A., H. Tsoar, and D. Levin, 1989: Climatic changes in Israel during historical times and their impact on hydrological, pedological and socio-economic systems. In *Paleoclimatology and Paleometeorology* (M. Leinen and M. Sarnthein, eds.), NATO ASI Series, Kluwer, Dordrecht, pp. 525–542.

Johnson, T. C., S. L. Barry, Y. Chan, and P. Wilkinson, 2001: Decadal record of climate variability spanning the past 700 yr in the Southern Tropics of East Africa. *Geology*, **29**, 83–86.

Jones, P. D., M. New, D. E. Parker, S. Martin, and I. G. Rigor, 1999: Surface air temperature and its changes over the past 150 years. *Reviews of Geophysics*, **37**, 1730199.

Jones, P. D., T. J. Osborn, and K. R. Briffa, 2001: The evolution of climate over the last millennium. *Science*, **292**, 662–667.

Joseph, P. V., B. Liebmann, and H. H. Hendon, 1991: Interannual variability of the Australian summer monsoon onset: possible influence of Indian summer monsoon and El Niño. *Journal of Climate*, **4**, 529–538.

Kaspari, S., and Coauthors, 2007: Reduction in northwest incursion of the South Asian Monsoon since ~1400 AD inferred from Mt. Everest ice core. *Geophysical Research Letters*, **34**, L16701.

Kerr, R., 1996: A new dawn for sun–climate links. *Science*, **271**, 1360–1361.

Klein, C., and H. Flohn, 1987: Contributions to the knowledge of the fluctuations of the Dead Sea level. *Theoretical and Applied Climatology*, **38**, 151–156.

Koehler, P. A., R. S. Anderson, and W. G. Spaulding, 2005: Development of vegetation in the Central Mojave Desert of California during the late Quaternary. *Palaeogeography, Palaeoclimatology, Palaeoecology*, **215**, 297–311.

Koster, R. D., and Coauthors, 2004: Regions of strong coupling between soil moisture and precipitation. *Science*, **305**, 1138–1140.

Krishnamurthy, V., and J. Shukla, 2000: Intraseasonal and interannual variability of rainfall over India. *Journal of Climate*, **13**, 4366–4377.

Kröpelin, S., and Coauthors, 2008: Climate-driven ecosystem succession in the Sahara: the past 6000 years. *Science*, **320**, 765–768.

Kutzbach, J. E., 1976: The nature of climate and climatic variations. *Quaternary Research*, **6**, 471–480.

Lamb, H. H., 1977: *Climate: Present, Past and Future*, Vol. 2. Methuen, London, 835 pp.

Latorre, C., J. L. Betancourt, K. A. Rylander, and J. Quade, 2002: Vegetation invasions into absolute desert: a 45,000-year rodent midden record from the Calama-Salar de Atacama basins, northern Chile (Lat 22 degrees–24 degrees S). *Geological Society of America Bulletin*, **114**, 349–366.

Lau, K.-M., and P. J. Sheu, 1988: *Journal of Geophysical Research–Atmospheres*, **93**, 10975–10988.

Lau, K. M., S. S. P. Shen, K.-M. Kim, and H. Wang, 2006: A multimodel study of the twentieth-century simulations of Sahel drought from the 1970s to 1990s. *Journal of Geophysical Research*, **111**, D0711.

Lean, J., and D. Rind, 1998: Climate forcing by changing solar radiation. *Journal of Climate*, **11**, 3069–3094.

Le Quesne, C., C. Acuna, J. A. Boninsegna, A. Rivera, and J. Barichivich, 2009: Long-term glacier variations in the Central Andes of Argentina and Chile, inferred from historical records and tree-ring reconstructed precipitation. *Palaeogeography, Palaeoclimatology, Palaeoecology*, **281**, 334–344.

Liu, Z., and Coauthors, 2007: Simulating the transient evolution and abrupt change of Northern Africa atmosphere-ocean-terrestrial ecosystem in the Holocene. *Quaternary Science Reviews*, **26**, 1818–1837.

Lucht, W., and Coauthors, 2002: Climatic control of the high-latitude vegetation greening trend and Pinatubo effect. *Science*, **296**, 1687–1690.

Marcella, M. P., and E. A. B. Eltahir, 2008: The hydroclimatology of Kuwait: explaining the variability of rainfall at seasonal and interannual time scales. *Journal of Hydrometeorology*, **9**, 1095–1105.

Marchant, R., and H. Hooghiemstra, 2004: Rapid environmental change in African and South American tropics around 4000 years before present: a review. *Earth-Science Reviews*, **66**, 217–260.

Marcus, M. G., and S. W. Brazel, 1984: *Climate Change in Arizona's Future*. Arizona State Climate Publication No. 1. Office of the State Climatologist, Arizona State University, Tempe.

Marty, P., 1927: Chroniques de Oualata et de Nema. *Revue des Etudes Islamiques*, 355–426, 531–575.

Marx, S. K., H. A. McGowan, and B. S. Kamber, 2009: Long-range dust transport from eastern Australia: a proxy for Holocene aridity and ENSO-type climate variability. *Earth and Planetary Science Letters*, **282**, 167–177.

McCarthy, L., and L. Head, 2001: Holocene variability in semi-arid vegetation: new evidence from *Leporillus* middens from the Flinders Ranges, South Australia. *The Holocene*, **11**, 681–689.

Meko, D. M., and Coauthors, 2007: Medieval drought in the upper Colorado River Basin. *Geophysical Research Letters*, **34**, L10705.

Miao, X. D., and Coauthors, 2007: A 10,000-year record of dune activity, dust storms, and severe drought in the central Great Plains. *Geology*, **35**, 119–122.

Minnich, R. A., E. F. Vizcaino, and R. J. Dezzani, 2000: The El Niño/Southern Oscillation and precipitation variability in Baja California, Mexico. *Atmosfera*, **13**, 1–20.

Modarres, R., and A. Sarhadi, 2009: Rainfall trends analysis of Iran in the last half of the twentieth century. *Journal of Geophysical Research*, **114**, D03101.

Mooley, D. A., and G. B. Pant, 1981: Drought in India over the last 200 years. In *Climate and History* (Wigley, T. M. L., M. J. Ingrams, and G. Farmer, eds.), Cambridge University Press, Cambridge, pp. 465–478.

Moore, N., and S. Rojstaczer, 2001: Irrigation-induced rainfall and the Great Plains. *Journal of Applied Meteorology and Climatology*, **40**, 1297–1309.

Namias, J., 1978: Multiple causes of the North America abnormal winter 1976–77. *Monthly Weather Review*, **106**, 279–295.

Nicholls, N., 1991: The El Niño/Southern Oscillation and Australian vegetation. *Plant Ecology*, **91**, 23–36.

Nicholls, N., and A. Kariko, 1993: East Australian rainfall events: interannual variations, trends, and relationships with the Southern Oscillation. *Journal of Climate*, **6**, 1141–1152.

Nicholson, S. E., 1976: A climatic chronology for Africa: synthesis of geological, historical, and meteorological information and data. PhD dissertation, University of Wisconsin, Madison, WI, 324 pp.

Nicholson, S. E., 1978: Climatic variations in the Sahel and other African regions during the past five centuries. *Journal of Arid Environments*, **1**, 3–24.

Nicholson, S. E., 1986a: The spatial coherence of African rainfall anomalies: Interhemispheric teleconnections. *Journal of Climatology and Applied Meteorology*, **25**, 1365–1381.

Nicholson, S. E., 1988: Land surface atmosphere interaction: physical processes and surface changes and their impact. *Progress in Physical Geography*, **12**, 36–65.

Nicholson, S. E., 1996a: A review of climate dynamics and climate variability in eastern Africa. In *The Limnology, Climatology and Paleoclimatology of the East African Lakes* (T.C. Johnson and E. Odada, eds.), Gordon and Breach, Amsterdam, pp. 25–56.

Nicholson, S. E., 1996b: Environmental change within the historical period. In *The Physical Geography of Africa* (W.M. Adams, A. S. Goudie, and A. Orme, eds.), Oxford University Press, Oxford, pp. 60–75.

Nicholson, S. E., 1998: Historical fluctuations of Lake Victoria and other lakes in the northern Rift Valley of East Africa. In *Environmental Change and Response in East African Lakes* (J. T. Lehman, ed.). Kluwer, Dordrecht, pp. 7–35.

Nicholson, S. E., 2000: The nature of rainfall variability over Africa on time scales of decades to millennia. *Global and Planetary Change Letters*, **26**, 137–158.

Nicholson, S. E., 2001a: Climatic and environmental change in Africa during the last two centuries. *Climatic Research*, **17**, 123–144.

Nicholson, S. E., 2001b: A semi-quantitative, regional precipitation data set for studying African climates of the nineteenth century. Part I. Overview of the data set. *Climatic Change*, **50**, 317–353.

Nicholson, S. E., and H. Flohn, 1980: African environmental and climatic changes and the general atmospheric circulation in late Pleistocene and Holocene. *Climatic Change*, **2**, 313–348.

Nicholson, S. E., and J.-Y. Kim, 1997: The relationship of the El Niño–Southern Oscillation to the African rainfall. *International Journal of Climatology*, **17**, 117–135.

Nicholson, S. E., and J. C. Selato, 2000: The influence of La Nina on African rainfall. *International Journal of Climatology*, **20**, 1761–1776.

Nicholson, S. E., and X. Yin, 2001: Rainfall conditions in equatorial East Africa during the nineteenth century as inferred from the record of Lake Victoria. *Climatic Change*, **48**, 387–398.

Nicholson, S. E., C. J. Tucker, and M. B. Ba, 1998: Desertification, drought, and surface vegetation: an example from the West African Sahel. *Bulletin of the American Meteorological Society*, **79**, 815–829.

Nicholson, S. E., D. Leposo, and J. Grist, 2001: On the relationship between El Niño and drought over Botswana. *Journal of Climate*, **14**, 323–335.

Nogués-Paegle, J., and K. C. Mo, 2002: Linkages between summer rainfall variability over South America and sea surface temperature anomalies. *Journal of Climate*, **125**, 279–291.

Notaro, M., Z. Liu, and J. W. Williams, 2006: Observed vegetation-climate feedbacks in the United States. *Journal of Climate*, **19**, 763–786.

ORSTOM (Office de la recherché scientifique et technique outré-mer), 1976: *Précipitations journalières de l'origine des stations à 1965.* République du Mauritanie. Etienne Julienne, Paris, 314 pp.

Ortega-Rosas, C. K., M. C. Penalba, and J. Guiot, 2008: Holocene altitudinal shifts in vegetation belts and environmental changes in the Sierra Madre Occidental, Northwestern Mexico, based on modern and fossil pollen data. *Review of Palaeobotany and Palynology*, **151**, 1–20.

Ortlieb, L., 1994: Major historical rainfalls in central Chile and the chronology of ENSO events during the 15th–19th centuries. *Revista Chilena de Historia Natural*, **67**, 463–485.

Otto, J., T. Raddatz, M. Claussen, V. Brovkin, and V. Gayler, 2009: Separation of atmosphere–ocean–vegetation feedbacks and synergies for mid-Holocene climate. *Global Biogeochemical Cycles*, **23**, L09701.

Pejml, K., 1966: *Studie o kolísání klimatu v historické dobe na západním pobrezí jizní ameriky* (In Czech). Hydrometeorological Institute, Prague, 82 pp.

Petit-Maire, N. (ed.), 1991: *Paléoenvironments du Sahara: lacs holocènes à Taoudenni (Mali)*. Paris.

Pielke, R. A., Sr., 2001: Influence of the spatial distribution of vegetation and soils on the prediction of cumulus convective rainfall. *Reviews of Geophysics*, **39**, 151–178.

Pigati, J. S., J. E. Bright, T. M. Shanahan, and S. A. Mahan, 2009: Late Pleistocene paleohydrology nears the boundary of the Sonoran and Chihuahuan Deserts, southeastern Arizona, USA. *Quaternary Science Reviews*, **28**, 286–300.

Price, C., L. Stone, A. Huppert, B. Rajagopalan, and P. Albert, 1998: A possible link between El Niño and precipitation in Israel. *Geophysical Research Letters*, **25**, 3963–3966.

Prospero, J. M., P. Ginoux, O. Torres, S. E. Nicholson, and T. E. Gill, 2002: Environmental characterization of global sources of atmospheric soil dust identified with the Nimbus 7 Total Ozone Mapping Spectrometer (TOMS) absorbing aerosol product. *Reviews of Geophysics*, **40**(1), 1002, doi:10.1029/2000RG000095.

Quade, J., and W. S. Broecker, 2009: Dryland hydrology in a warmer world: lessons from the Last Glacial period. *European Physical Journal – Special Topics*, **176**, 21–36.

Quinn, W. H., and Neal, V. T., 1992: The historical record of El Niño events. In *Climate Since A. D. 1500* (R. S. Bradley and P. D. Jones, eds.). Routledge, London, pp. 623–648.

Roberts, N., and H. E. Wright Jr., 1993: Vegetational, lake-level, and climatic history of the Near East and Southwest Asia. In *Global Climates since the Last Glacial Maximum* (H. E. Wright Jr., J. E. Kutzbach, T. Webb III, W. F. Ruddiman, F. A. Street-Perrott, and P. J. Bartlein, eds.), University of Minnesota Press, Minneapolis, MN, pp. 194–220.

Robock, A., 2000: Volcanic eruptions and climate. *Reviews of Geophysics*, **38**, 191–219.

Rogers, J. C., 1988: Precipitation variability over the Caribbean and tropical Americas associated with the Southern Oscillation. *Journal of Climate*, **1**, 172–182.

Ropelewski, C. F. and M. S. Halpert, 1986: North American precipitation and temperature patterns associated with El Niño/Southern Oscillation. *Monthly Weather Review*, **114**, 2352–2362.

Ropelewski, C. F. and M. S. Halpert, 1987: Global and regional scale precipitation associated with El Niño/Southern Oscillation. *Monthly Weather Review*, **115**, 985–996.

Rutllant, J., and H. Fuenzalida, 1991: Synoptic aspects of the central Chile rainfall variability associated with the Southern Oscillation. *International Journal of Climatology*, **11**, 63–76.

Schwerdtfeger, W. (ed.), 1976: *The Climates of South America*. World Survey of Climatology, Vol. 12, Elsevier, New York, 381 pp.

Schonher, T., and S. E. Nicholson, 1989: The relationship between California rainfall and ENSO events. *Journal of Climate*, **2**, 1258–1269.

Seltzer, G. O., and Coauthors, 2002: Early warming of tropical South America at the last glacial-interglacial transition. *Science*, **296**, 1685–1686.

Shanahan, T. M., and Coauthors, 2006: Paleoclimatic variations in West Africa from a record of late Pleistocene and Holocene lake level stands of Lake Bosumtwi, Ghana. *Palaeogeography, Palaeoclimatology, Palaeoecology*, **242**, 287–302.

Shukla, J., 1981: Dynamical predictability of monthly means. *Journal of the Atmospheric Sciences*, **38**, 2547–2572.

Shuman, B., and Coauthors, 2009: Holocene lake-level trends in the Rocky Mountains, USA. *Quaternary Science Reviews*, **26**, 1861–1879.

Stahle, D. W., and M. K. Cleaveland, 1993: Southern Oscillation extremes reconstructed from tree rings of the Sierra Madre Occidental and Southern Great Plains. *Journal of Climate*, **6**, 129–140.

Stockton, C. W., and D. M. Meko, 1975: A long-term history of drought occurrence in western United States as inferred from tree rings. *Weatherwise*, **28**, 244–249.

Street, F. A., and A. T. Grove, 1976: Environmental and climatic implications of late Quaternary lake-level fluctuations in Africa. *Nature*, **261**, 385–390.

Street, F. A., and A. T. Grove, 1979: Global maps of lake-level fluctuations since 30,000 yr B.P. *Quaternary Research*, **12**, 83–118.

Swetnam, T. W., and J. L. Betancourt, 1990: Fire–Southern Oscillation relations in the southwestern United States. *Science*, **249**, 1017–1020.

Tarawneh, Z., and N. Hadadin, 2009: Reconstruction of the rainy season precipitation in central Jordan. *Hydrological Sciences Journal*, **54**, 189–198.

Taulis, E., 1934: De la distribution des pluies au Chile. In *Materoux pour l'Etude des Calamités*, Part I. Société de Géographie de Genève, pp. 3–20.

Taylor, C. M., E. F. Lambin, N. Stephenne, R. J. Hardin, and R. L. H. Essery, 2002: The influence of land use change on climate in the Sahel. *Journal of Climate*, **15**, 3615–3629.

Thomas, D. S. G., and P. A. Shaw, 1991: *The Kalahari Environment*. Cambridge University Press, Cambridge, 284 pp.

Thompson, L. G., T. Yao, E. Mosley-Thompson, M. E. Davis, K. A. Henderson, and P.-N. Lin, 2000: A high-resolution millennial record of the South Asian monsoon from Himalayan ice cores. *Science*, **289**, 1916–1919.

Toussoun, O., 1925: Mémoire sur l'histoire du Nil. *Mémoires de l'Institut d'Egypte*, **9**, 63–213.

Treydte, K. S., G. H. Schleser, G. Helle, D. C. Frank, M. Winiger, G. H. Haug, and J. Esper, 2006: The twentieth century was the wettest period in northern Pakistan over the past millennium. *Nature*, **440**, 1179–1182.

Verschuren, D., Laird, K. R., and Cumming, B. F., 2000: Rainfall and drought in equatorial east Africa during the past 1,100 years. *Nature*, 403, 410–414.

Vicuña-Mackenna, B., 1877: Ensayo-histórico sobre el clima de Chile (desde los tiempos prehistóricos hasta el grande temporal de Julio de 1877). Valparaiso.

Waisel, Y., and N. Liphschitz, 1968: Dendrochronological studies in Israel: *Juniperus phoenica* of north and central Sinai. *La-Yaaran*, **18**, 1–22.

Wang, W.-C., and K. Li, 1990: Precipitation fluctuation over semiarid region in northern China and the relationship with El Niño/Southern Oscillation. *Journal of Climate*, **3**, 769–783.

Wang, X. L., and H.-R. Cho, 1997: Spatial-temporal structures of trend and oscillatory variabilities of precipitation over northern Eurasia. *Journal of Climate*, **10**, 2285–2298.

Ward, M. N., 1998: Diagnosis and short-lead time prediction of summer rainfall in tropical North America at interannual and multidecadal timescales. *Journal of Climate*, **11**, 3167–3191.

Weaver, C. P., and R. Avissar, 2001: Atmospheric disturbances caused by human modification of the landscape. *Bulletin of the American Meteorological Society*, **82**, 269–281.

Williams, J. W., B. Shuman, and P. J. Bartlein, 2009: Rapid responses of the prairie-forest ecotone to early Holocene aridity in midcontinental North Americas. *Global and Planetary Change*, **66**, 195–207.

Winkler, M. G., and P. K. Wang, 1993: The Late-Quaternary vegetation and climate of China. In *Global Climates since the Last Glacial Maximum* (H. E. Wright Jr., J. E. Kutzbach, T. Webb III, W. F. Ruddiman, F. A. Street-Perrott, and P. J. Bartlein, eds.), University of Minnesota Press, Minneapolis, MN, pp. 221–264.

Woodhouse, C. A., and J. T. Overpeck, 1998: 2000 years of drought variability in the central United States. *Bulletin of the American Meteorological Society*, **79**, 2693–2714.

Wyrwoll, K.-H., and G. H. Miller, 2001: Initiation of the Australian summer 14,000 years ago. *Quaternary International*, **83/85**, 119–128.

Xue, Y., F. de Sales, W.-P. Li, C. R. Mechoso, C. A. Nobre, and H.-M. Juang, 2006: Role of land surface processes in South American monsoon development. *Journal of Climate*, **19**, 741–762.

Yarnal, B., and H. F. Diaz, 1986: Relationships between extremes of the Southern Oscillation and the winter climate of the Anglo-American Pacific coast. *Journal of Climate*, **6**, 197–219.

Ye, D., W. Dong, and Y. Jiang, 2003: The northward shift of climatic belts in China during the last 50 years. *Global Change Newsletter*, **53**, 7–9.

Yin, X., and S. E. Nicholson, 2002: Interpreting annual rainfall from the levels of Lake Victoria. *Journal of Hydrometeorology*, **3**, 406–416.

Zhang, J., and T. J. Crowley, 1989: Historical climate records in China and reconstruction of past changes. *Journal of Climate*, **2**, 835–849.

Index

Printed in the United States
By Bookmasters